# Fuzzy Sets, Fuzzy Logic and Their Applications

# Fuzzy Sets, Fuzzy Logic and Their Applications

Special Issue Editor

**Michael Gr. Voskoglou**

MDPI • Basel • Beijing • Wuhan • Barcelona • Belgrade • Manchester • Tokyo • Cluj • Tianjin

*Special Issue Editor*
Michael Gr. Voskoglou
Graduate Technological
Educational Institute of
Western Greece
Greece

*Editorial Office*
MDPI
St. Alban-Anlage 66
4052 Basel, Switzerland

This is a reprint of articles from the Special Issue published online in the open access journal *Mathematics* (ISSN 2227-7390) (available at: https://www.mdpi.com/journal/mathematics/special_issues/Fuzzy_Sets).

For citation purposes, cite each article independently as indicated on the article page online and as indicated below:

LastName, A.A.; LastName, B.B.; LastName, C.C. Article Title. *Journal Name* **Year**, *Article Number*, Page Range.

**ISBN 978-3-03928-520-4 (Pbk)**
**ISBN 978-3-03928-521-1 (PDF)**

# Contents

# About the Special Issue Editor

**Michael Gr. Voskoglou** (B.S., M.S., M.Phil. , Ph.D. in Mathematics) is currently an Emeritus Professor of Mathematical Sciences at the Graduate Technological Educational Institute of Western Greece. He was a Visiting Researcher at the Bulgarian Academy of Sciences (1997–2000), a Visiting Professor at the University of Warsaw (2009), of Applied Sciences of Berlin (2010), and at the National Institute of Technology of Durgapur, India (2016). Prof. Voskoglou is the author of 16 books (9 in Greek and 7 in English) and of more than 500 papers published in reputed mathematical journals and proceedings of international conferences of about 30 countries in the five continents (2018 Google Scholar Citations: 2401, H-idex 24). He is also the Editor in Chief of the International Journal of Applications of Fuzzy Sets and Artificial Intelligence (ISSN 2241-1240), reviewer of the AMS and member of the Editorial Board or referee in many mathematical journals. His research interests include algebra, fuzzy sets, Markov chains, artificial intelligence, and mathematics education.

# Preface to "Fuzzy Sets, Fuzzy Logic and Their Applications"

A few decades ago, probability theory used to be the unique tool in the hands of the experts for handling situations of uncertainty appearing in problems of science, technology, and in everyday life. Probability, which is based on the principles of traditional bivalent logic, is sufficient for tackling problems of uncertainty connected to randomness, but not those connected with imprecise or vague information. However, with the development of fuzzy set theory introduced by Zadeh in 1965 and of fuzzy logic which is based on that theory, things have changed. These new mathematical tools gave scientists the ability to model under conditions that are vague or not precisely defined, thus succeeding to mathematically solve problems whose statements are expressed in our natural language.

As a result, the spectrum of their applications has rapidly extended, covering all the physical sciences; economics and management; expert systems like financial planners, diagnostic, meteorological, information retrieval, control systems, etc.; industry; robotics; decision making; programming; medicine; biology; humanities; education and almost all the other sectors of te human activity, including human reasoning. The first major commercial application of fuzzy logic was in cement kiln control (Zadeh, 1983), followed by a navigation system for automatic cars, a fuzzy controller for automatic operation of trains, laboratory level controllers, controllers for robot vision, graphics, controllers for automated police sketchers, and many others. Fuzzy mathematics has also significantly developed at the theoretical level, providing important insights even into branches of classical mathematics, like algebra, analysis, and geometry.

The present book contains 20 papers collected from amongst 53 manuscripts submitted for the Special Issue "Fuzzy Sets, Fuzzy Logic and Their Applications" of the MDPI journal *Mathematics*. These papers appear in the book in the series in which they were accepted, and published in Volumes 7 (2019) and 8 (2020) of the journal and cover a wide range of topics and applications of fuzzy systems and fuzzy logic and their extensions and generalizations. This range includes, among others, management of uncertainty in a fuzzy environment, fuzzy assessment methods of human–machine performance, fuzzy graphs, fuzzy topological and convergence spaces, bipolar fuzzy relations, type-2 fuzzy sets, as well as intuitionistic, interval-valued, complex, picture and Pythagorean fuzzy sets, soft sets, and algebras. The applications presented are oriented to finance, fuzzy analytic hierarchy, green supply chain industries, smart health practice, and hotel selection. This wide range of topics makes the book interesting for all those working in the wider area of fuzzy sets and systems and of fuzzy logic, and for those with a mathematical background to familiarize themselves with recent advances in fuzzy mathematics, which has entered almost all sectors of human life and activity.

As the Guest Editor of this Special Issue, I am grateful to the authors of the papers for their quality contributions, to the reviewers for their valuable comments toward the improvement of the submitted works ,and to the administrative staff of MDPI for their support to complete this project. Special thanks are due to the Contact Editor of the Special Issue, Ms. Grace Du, for her excellent ollaboration and kind assistance.

**Michael Gr. Voskoglou**
*Special Issue Editor*

*Article*

# A Partial-Consensus Posterior-Aggregation FAHP Method—Supplier Selection Problem as an Example

**Yu-Cheng Wang [1,*] and Tin-Chih Toly Chen [2]**

[1] Department of Industrial Engineering and Management, Chaoyang University of Technology, Taichung 41349, Taiwan

[2] Department of Industrial Engineering and Management, National Chiao Tung University, 1001, University Road, Hsinchu 300, Taiwan; tolychen@ms37.hinet.net

* Correspondence: tony.cobra@msa.hinet.net

Received: 18 January 2019; Accepted: 11 February 2019; Published: 15 February 2019

**Abstract:** Existing fuzzy analytic hierarchy process (FAHP) methods usually aggregate the fuzzy pairwise comparison results produced by multiple decision-makers (DMs) rather than the fuzzy weights estimations. This is problematic because fuzzy pairwise comparison results are subject to uncertainty and lack consensus. To address this problem, a partial-consensus posterior-aggregation FAHP (PCPA-FAHP) approach is proposed in this study. The PCPA-FAHP approach seeks a partial consensus among most DMs instead of an overall consensus among all DMs, thereby increasing the possibility of reaching a consensus. Subsequently, the aggregation result is defuzzified using the prevalent center-of-gravity method. The PCPA-FAHP approach was applied to a supplier selection problem to validate its effectiveness. According to the experimental results, the PCPA-FAHP approach not only successfully found out the partial consensus among the DMs, but also shrunk the widths of the estimated fuzzy weights to enhance the precision of the FAHP analysis.

**Keywords:** fuzzy analytic hierarchy process; decision-making; partial consensus; posterior aggregation

## 1. Introduction

Analytic hierarchy process (AHP) has been widely applied to multi-criteria decision-making problems in various fields [1–3]. However, AHP is based on pairwise comparison results that are subjective. To solve this problem, AHP usually aggregates the pairwise comparison results by multiple decision-makers (DMs) [4–6]. In most past studies, the number of DMs ranged from 3 [7,8] to up to 30 [9]. In addition, a DM may be uncertain whether the pairwise comparison results are reflective of his/her beliefs or not. To consider such uncertainty, pairwise comparison results can be mapped to fuzzy values. The two treatments give rise to the prevalent multi-DM fuzzy AHP (FAHP) methods.

According to Forman and Peniwati [10], there are two ways to aggregate multiple DMs' judgments in FAHP. The first way, the aggregating individual judgements (AIJ) way, aggregates the pairwise comparison results. The other way, the aggregating individual priorities (AIP) way, aggregates the estimated fuzzy weights/priorities. The AIJ way is an anterior aggregation, while the AIP way is a posterior aggregation. The AIJ way considerably simplifies the required computation because the fuzzy weights are estimated just once, and therefore is more prevalent [11]. However, since pairwise comparison results are subjective and uncertain, the AIP way is problematic. In contrast, few studies have implemented AIP-type FAHP methods. Pan [12] proposed an FAHP approach in which each DM solved an individual FAHP problem using the fuzzy geometric mean (FGM) method. Then, the fuzzy weights estimated by the DMs were aggregated using the max-min operator, i.e., fuzzy intersection (FI) or the minimum T-norm. After that, the aggregation result was defuzzified using the center-of-gravity (COG) method [13]. The same operator has been widely adopted to measure the consensus among multiple DMs [14,15]. A different aggregation mechanism was proposed by [16,17] that minimized the sum of the squared deviations between each DM's judgement and the aggregation result. However, the aggregation mechanism assumed the existence of consensus, and derived the aggregation result directly. It was possible that the values of a fuzzy weight estimated by the DMs were very different, i.e., the values did not contain each other. Pan's method did not address this issue. To address this issue, several attempts have been made in the past. For example, Chen [18] proposed the concept of partial consensus that sought for a consensus among some of the DMs rather than all DMs. The method was applied to the problem of forecasting the foreign exchange rate in [19]. Recently, Chen [20] evaluated the entropy of the FI result. If the entropy was above a threshold, then the consensus was insufficient, and the DMs needed to modify their forecasts. However, these attempts were made for fuzzy collaborative forecasting or design rather than for FAHP. A fuzzy collaborative forecasting problem is usually a supervised learning problem; there are actual values. However, there are no actual values of fuzzy weights in an FAHP problem.

To address the problem of Pan's FAHP method, in this study, a partial-consensus posterior-aggregation FAHP (PCPA-FAHP) method is proposed. In the proposed PCPA-FAHP approach, each DM applies the prevalent FGM approach to estimate the fuzzy weights. Then, the partial-consensus FI (PCFI) method [18] is applied to aggregate the estimation results, so as to derive the narrowest range of the fuzzy weight. The PCFI method finds out the partial consensus among most DMs instead of the overall consensus among all DMs. The former is obviously easier than the latter. After that, the COG method is applied to defuzzify the aggregation results, generating a crisp/representative value. Compared to the existing methods, the proposed PCPA-FAHP approach has the following novel characteristics:

(1) The proposed PCPA-FAHP approach adopts a posterior aggregation. Namely, the fuzzy weights estimated by the DMs, rather than the fuzzy pairwise comparison results by them, are aggregated: A multiple-DM FAHP method can be considered as a fuzzy collaborative forecasting (FCF) approach. Most existing FCF approaches adopt a posterior aggregation, i.e., the forecasts by multiple DMs, rather than their opinions, are aggregated [21,22]. If the forecasting performance based on the aggregation result is not satisfactory, the DMs modify their opinions by referring to others' opinions [23,24]. However, existing FCF methods belong to supervised learning methods, while the PCPA-FAHP approach does not because there is no actual value of the fuzzy weight. In addition, it is not easy for the DMs to modify their pairwise comparison results by referring to others' because the relative importance levels of factors/attributes/criteria are correlated and cannot be modified in an independent way.

(2) When a consensus among all DMs cannot be achieved, the partial consensus, i.e., the consensus among most DMs, is to be sought.

The remainder of this paper is organized as follows. Section 2 introduces the PCPA-FAHP approach. A real case is used in Section 3 to illustrate the applicability of the PCPA-FAHP approach. Some existing methods are also applied to the case for a comparison in Section 4. Section 5 concludes this study. Some possible topics are also provided for further investigation.

## 2. The Proposed Methodology

The PCPA-FAHP approach is proposed in this study for comparing the relative importance levels or priorities of several factors/attributes-criteria. The PCPA-FAHP approach consists of the following steps:

Step 1. Each DM applies the FGM approach to estimate the fuzzy weights.

Step 2. Apply FI to aggregate the estimation results, so as to derive the narrowest range of each fuzzy weight.

Step 3. If the overall consensus among the DMs does not exist, go to Step 4; otherwise, go to Step 5.

Step 4. Apply PCFI to aggregate the estimation results, so as to derive the narrowest range of each fuzzy weight.

Step 5.  Apply COG to defuzzify the aggregation result, so as to generate a crisp/representative value.

The procedure of the PCPA-FAHP approach is illustrated with a flowchart in Figure 1.

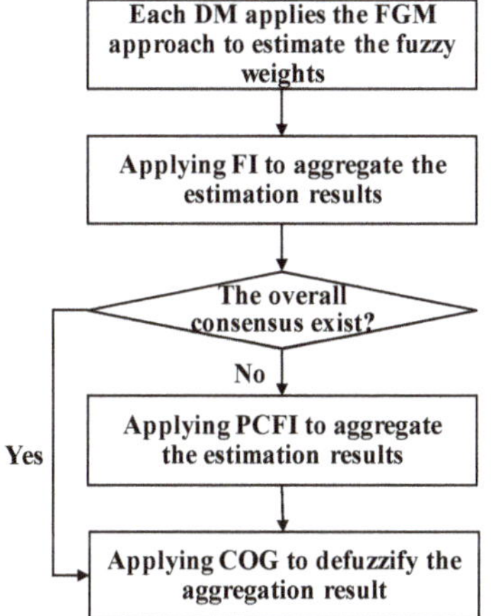


**Figure 1.** The procedure of the partial-consensus posterior-aggregation fuzzy analytic hierarchy process (PCPA-FAHP) approach.

### 2.1. Applying the FGM Approach to Estimate the Fuzzy Weights

In the FGM approach, at first each DM expresses his/her opinion on the relative importance/priority of a factor/attribute/criterion over that of another with linguistic terms such as "as equal as", "weakly more important than", "strongly more important than", "very strongly more important than", and "absolutely more important than". Without loss of generality, these linguistic terms can be mapped to triangular fuzzy numbers (TFNs) such as:

L1:  "As equal as" = (1, 1, 3);
L2:  "Weakly more important than" = (1, 3, 5);
L3:  "Strongly more important than" = (3, 5, 7);
L4:  "Very strongly more important than" = (5, 7, 9);
L5:  "Absolutely more important than" = (7, 9, 9);

which are illustrated in Figure 2. It is a theoretically challenging task to find a set of TFNs for the linguistic terms to increase the possibility of reaching a consensus.

Based on the pairwise comparison results by the $m$-th DM, a fuzzy pairwise comparison matrix is constructed as:

$$\widetilde{\mathbf{A}}_{n \times n}(m) = [\widetilde{a}_{ij}(m)]; \; i,j = 1 \sim n, \tag{1}$$

where

$$\widetilde{a}_{ij}(m) = \begin{cases} 1 & \text{if } i = j \\ \dfrac{1}{\widetilde{a}_{ji}(m)} & \text{otherwise} \end{cases} ; \; i,j = 1 \sim n, \tag{2}$$

$\widetilde{a}_{ij}(m)$ is the fuzzy pairwise comparison result embodying the relative importance of factor/attribute/criterion $i$ over factor/attribute/criterion $j$. $\widetilde{a}_{ij}(m)$ is chosen from the linguistic terms in Figure 2. $\widetilde{a}_{ij}(m)$ is a positive comparison if $\widetilde{a}_{ij}(m) \geq 1$. The fuzzy eigenvalue and eigenvector of $\widetilde{\mathbf{A}}(m)$, indicated respectively with $\widetilde{\lambda}(m)$ and $\widetilde{\mathbf{x}}(m)$, satisfy

$$\det(\widetilde{\mathbf{A}}(m)(-)\widetilde{\lambda}(m)\mathbf{I}) = 0 \tag{3}$$

and

$$(\widetilde{\mathbf{A}}(m)(-)\widetilde{\lambda}(m)\mathbf{I})(\times)\widetilde{\mathbf{x}}(m) = 0 \tag{4}$$

where $(-)$ and $(\times)$ denote fuzzy subtraction and multiplication, respectively. The fuzzy maximal eigenvalue and weight of each criterion are derived respectively as

$$\widetilde{\lambda}_{\max}(m) = \max\widetilde{\lambda}(m) \tag{5}$$

$$\widetilde{w}_i(m) = \frac{\widetilde{x}_i(m)}{\sum\limits_{j=1}^{n} \widetilde{x}_j(m)}. \tag{6}$$

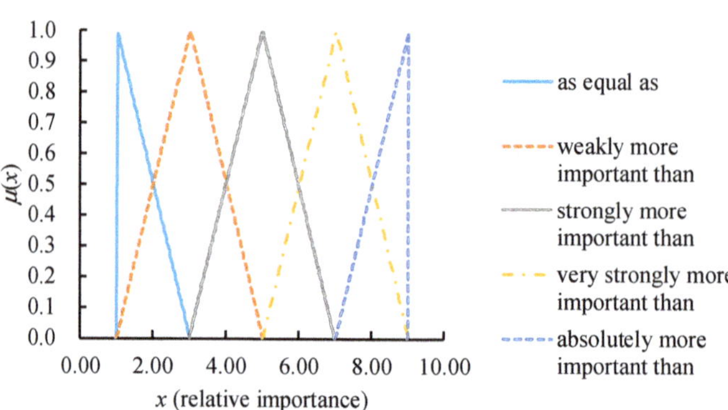


**Figure 2.** The triangular fuzzy numbers (TFNs) for the linguistic terms.

Based on $\widetilde{\lambda}_{\max}(m)$, the consistency among the fuzzy pairwise comparison results is evaluated as

$$\text{Consistency index}: \ \widetilde{C.I.}(m) = \frac{\widetilde{\lambda}_{\max}(m) - n}{n - 1} \tag{7}$$

$$\text{Consistency ratio}: \ \widetilde{C.R.}(m) = \frac{\widetilde{C.I.}(m)}{R.I.} \tag{8}$$

where $R.I.$ is the random index [25]. The fuzzy pairwise comparison results are inconsistent if $\widetilde{C.I.}(m) < 0.1$ or $\widetilde{C.R.}(m) < 0.1$ [25], which can be relaxed to $\widetilde{C.I.}(m) < 0.3$ or $\widetilde{C.R.}(m) < 0.3$ if the matrix size is large [26,27].

The FGM method estimates the values of fuzzy weights as

$$\widetilde{\mathbf{w}}(m) = [\widetilde{w}_i(m)]^t = \left[\frac{\sqrt[n]{\prod\limits_{j=1}^{n} \widetilde{a}_{ij}(m)}}{\sum\limits_{i=1}^{n} \sqrt[n]{\prod\limits_{j=1}^{n} \widetilde{a}_{ij}(m)}}\right]^t. \tag{9}$$

Based on (9), the fuzzy maximal eigenvalue can be estimated as

$$\begin{aligned}
\widetilde{\lambda}_{\max}(m) \ &= \frac{\widetilde{\mathbf{A}}(m)(\times)\widetilde{\mathbf{w}}(m)}{\widetilde{\mathbf{w}}(m)} \\
&= \frac{1}{n}\sum\limits_{i=1}^{n}\left(\frac{\sum\limits_{j=1}^{n}(\widetilde{a}_{ij}(m)(\times)\widetilde{w}_i(m))}{\widetilde{w}_i(m)}\right).
\end{aligned} \tag{10}$$

Obviously, Equations (9) and (10) are fuzzy weighted average (FWA) problems. A variant of FGM (a simplified yet more prevalent version), indicated with FGM*i*, is to ignore the dependency between the dividend and divisor of either equation.

*2.2. FI for Finding out the Overall Consensus*

As mentioned previously, a multi-DM post-aggregation FAHP problem is analogous to an unsupervised FCF problem to which a suitable consensus aggregator is critical. According to Kuncheva and Krishnapuram [28], a consensus aggregator should meet three requirements: Symmetry, selective monotonicity, and unanimity. They proposed three aggregation rules: The minimum aggregation rule, the maximum aggregation rule, and the average aggregation rule, for which the degree of consensus was measured in terms of the highest discrepancy. FI is the most prevalent consensus aggregator in FCF methods [24,29]. FI finds out the values common to those estimated by all DMs. Therefore, it can be used to find out the overall consensus among the DMs.

When the fuzzy weight estimated by each DM is approximated with a TFN, the FI result will be a polygon-shaped fuzzy number (see Figure 3) [29] that embodies the DMs' overall consensus of the weight/priority of the factor/attribute/criterion:

$$\mu_{\widetilde{w}_i}(x) = \min_m (\mu_{\widetilde{w}_i(m)}(x)). \tag{11}$$

However, it is possible that the FI result is an empty set, which means there is no overall consensus among the DMs.

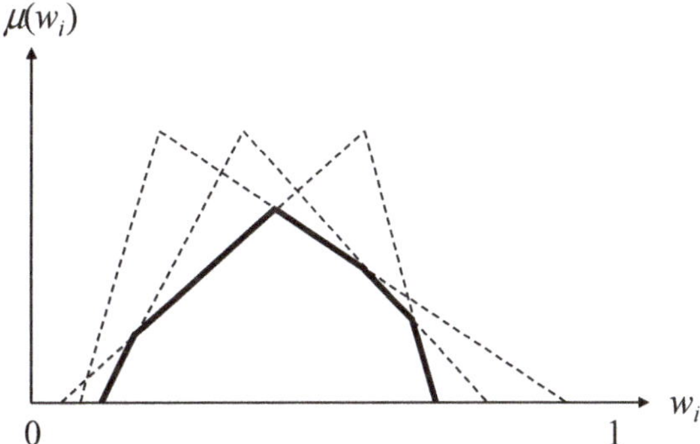


**Figure 3.** The fuzzy intersection (FI) result.

*2.3. PCFI for Finding out the Partial Consensus*

When there is no overall consensus among all DMs, the partial consensus among them, i.e., the consensus among most DMs, can be sought instead.

**Definition 1.** (PCFI) [20] *The H/M PCFI of the i-th fuzzy weight estimated by the M DMs, i.e., $\widetilde{w}_i(1) \sim \widetilde{w}_i(M)$ is indicated with $\widetilde{I}^{H/M}(\widetilde{w}_i(1), \ldots, \widetilde{w}_i(M))$ such that*

$$\mu_{\widetilde{I}^{H/M}(\widetilde{w}_i(1),\ldots,\widetilde{w}_i(M))}(x) = \max_{\text{all } g}(\min(\mu_{\widetilde{w}_1(g(1))}(x), \ldots, \mu_{\widetilde{w}_1(g(H))}(x))) \tag{12}$$

*where $g() \in Z^+; 1 \leq g() \leq M; g(p) \cap g(q) = \varnothing \; \forall \, p \neq q; H \geq 2$.*

For example, the 2/3 PCFI of $\tilde{w}_i(1) \sim \tilde{w}_i(3)$ can be obtained as

$$\mu_{\tilde{I}^{2/3}(\tilde{w}_i(1),\,\ldots,\,\tilde{w}_i(3))}(x) = \max(\min(\mu_{\tilde{w}_i(1)}(x),\,\mu_{\tilde{w}_i(2)}(x)),\,\min(\mu_{\tilde{w}_i(1)}(x),\,\mu_{\tilde{w}_i(3)}(x)),$$
$$\min(\mu_{\tilde{w}_i(2)}(x),\,\mu_{\tilde{w}_i(3)}(x))) \tag{13}$$

which is illustrated in Figure 4.

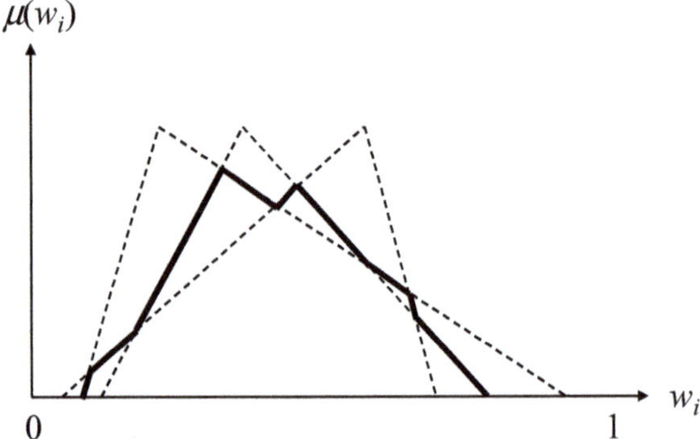


**Figure 4.** The 2/3 partial consensus fuzzy intersection (PCFI) result.

The PCFI aggregator meets four requirements: Boundary, monotonicity, commutativity, and associativity. In addition, the following properties hold for the $H/M$ PCFI result:

(1)  FI is equivalent to $M/M$ PCFI.
(2)  $H-1$ membership functions are outside the $H/M$ PCFI result. In contrast, $M-1$ membership functions are outside the FI result.
(3)  The range of $\tilde{I}^{H_1/M}(\tilde{w}_i(1),\,\ldots,\,\tilde{w}_i(M))$ is wider than that of $\tilde{I}^{H_2/M}(\tilde{w}_i(1),\,\ldots,\,\tilde{w}_i(M))$ if $H_1 < H_2$.
(4)  The range of any PCFI result is obviously wider than that of the FI result.
(5)  For the training data, every PCFI result contains the actual values [20].
(6)  For the testing data, the probability that actual values are contained is higher in a PCFI result than in the FI result [20].

The PCFI result is also a polygon-shaped fuzzy number (see Figure 4). Compared with the original TFNs, the PCFI result has a narrower range while still containing the actual value. Therefore, the precision of estimating the fuzzy weight will be improved after applying the PCFI. In addition, it is possible to find partial consensus among the DMs using the PCFI even if there is no overall consensus when the FI is applied.

*2.4. COG for Defuzzifying the Aggregation Result*

The PCFI result in Figure 4 can be represented as the union of several non-normal trapezoidal fuzzy numbers (TrFNs) (see Figure 5):

$$\tilde{w}_i = \{(x_r, \mu_{\tilde{w}_i}(x_r)) | r = 1 \sim R\} \tag{14}$$

where $(x_r, \mu_{\tilde{w}_i}(x_r))$ is the $r$-th endpoint of $\tilde{w}_i$; $x_r \leq x_{r+1}$. For any $x$ value,

$$\mu_{\tilde{w}_i}(x) = \begin{cases} 0 & if \quad x < x_1 \\ \mu_{\tilde{w}_i}(x_r) + \frac{x-x_r}{x_{r+1}-x_r}(\mu_{\tilde{w}_i}(x_{r+1}) - \mu_{\tilde{w}_i}(x_r)) & if \quad x_r \leq x < x_{r+1} \\ 0 & if \quad x_R \leq x \end{cases} \tag{15}$$

COG is applied to defuzzify $\tilde{w}_i$:

$$\begin{aligned}
\text{COG}(\tilde{w}_i) &= \frac{\int_0^1 x \cdot \mu_{\tilde{w}_i}(x)dx}{\int_0^1 \mu_{\tilde{w}_i}(x)dx} \\[2mm]
&= \frac{\sum\limits_{r=1}^{R} \int_{x_{r-1}}^{x_r} x \cdot \mu_{\tilde{w}_i}(x)dx}{\sum\limits_{r=1}^{R} \int_{x_{r-1}}^{x_r} \mu_{\tilde{w}_i}(x)dx}
\end{aligned} \tag{16}$$

to which the following theorems are helpful.

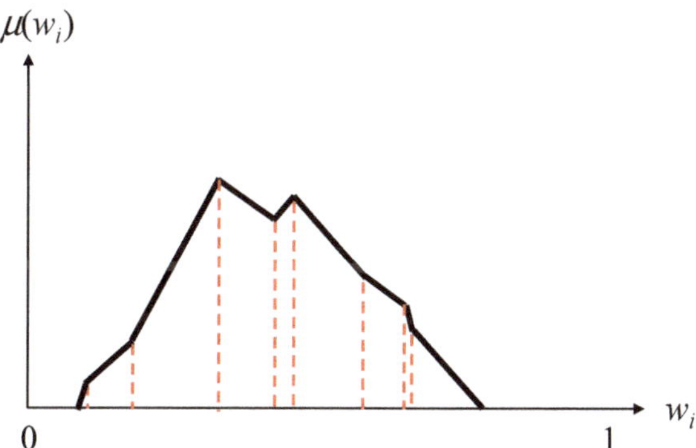


**Figure 5.** The PCFI result as a union of non-normal trapezoidal fuzzy numbers (TrFNs).

**Theorem 1.** *Let $\tilde{A}$ be a non-normal TrFN as shown in Figure 6. Then the integral of $\tilde{A}$ is*

$$\int_{x_1}^{x_2} \mu_{\tilde{A}(x)}(x)dx = \frac{\mu_2 x_2^2 + \mu_1 x_2^2 - 2\mu_2 x_1 x_2 + \mu_1 x_1^2 - 2\mu_1 x_1 x_2 + \mu_2 x_1^2}{2(x_2 - x_1)}. \tag{17}$$

**Proof.**

$$\begin{aligned}
&\int_{x_1}^{x_2} \mu_{\tilde{A}(x)}(x)dx \\[1mm]
&= \int_{x_1}^{x_2} \left( \frac{x - x_1}{x_2 - x_1}(\mu_2 - \mu_1) + \mu_1 \right)dx \\[1mm]
&= \left( \frac{\mu_2 - \mu_1}{2(x_2 - x_1)}x^2 + \frac{x_2\mu_1 - x_1\mu_2}{x_2 - x_1}x + C \right)\bigg|_{x_1}^{x_2} \\[1mm]
&= \frac{\mu_2 - \mu_1}{2(x_2 - x_1)}x_2^2 + \frac{x_2\mu_1 - x_1\mu_2}{x_2 - x_1}x_2 - \frac{\mu_2 - \mu_1}{2(x_2 - x_1)}x_1^2 - \frac{x_2\mu_1 - x_1\mu_2}{x_2 - x_1}x_1 \\[1mm]
&= \frac{\mu_2 x_2^2 + \mu_1 x_2^2 - 2\mu_2 x_1 x_2 + \mu_1 x_1^2 - 2\mu_1 x_1 x_2 + \mu_2 x_1^2}{2(x_2 - x_1)}
\end{aligned} \tag{18}$$

Theorem 1 is proved. $\square$

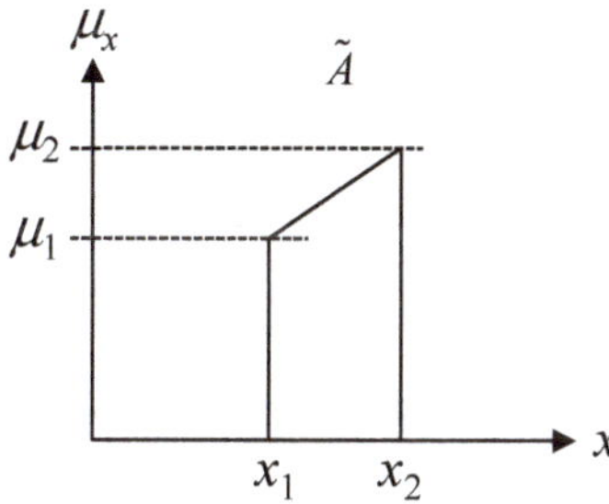


**Figure 6.** A non-normal TrFN.

**Theorem 2.** *Let $\widetilde{A}$ be a non-normal TrFN as shown in Figure 6. Then the integral of $x\widetilde{A}$ is*

$$\int_{x_1}^{x_2} x\mu_{\widetilde{A}(x)}(x)dx = \frac{2\mu_2 x_2^3 + \mu_1 x_2^3 - 3\mu_2 x_1 x_2^2 + \mu_2 x_1^3 + 2\mu_1 x_1^3 - 3\mu_1 x_1^2 x_2}{6(x_2 - x_1)}. \tag{19}$$

**Proof.**

$$\begin{aligned}
&\int_{x_1}^{x_2} x\mu_{\widetilde{A}(x)}(x)dx \\
&= \int_{x_1}^{x_2} x\left(\frac{x-x_1}{x_2-x_1}(\mu_2 - \mu_1) + \mu_1\right)dx \\
&= \left(\frac{\mu_2-\mu_1}{3(x_2-x_1)}x^3 + \frac{x_2\mu_1 - x_1\mu_2}{2(x_2-x_1)}x^2 + C\right)\Big|_{x_1}^{x_2} \\
&= \frac{\mu_2-\mu_1}{3(x_2-x_1)}x_2^3 + \frac{x_2\mu_1 - x_1\mu_2}{2(x_2-x_1)}x_2^2 - \frac{\mu_2-\mu_1}{3(x_2-x_1)}x_1^3 - \frac{x_2\mu_1 - x_1\mu_2}{2(x_2-x_1)}x_1^2. \\
&= \frac{2\mu_2 x_2^3 + \mu_1 x_2^3 - 3\mu_2 x_1 x_2^2 + \mu_2 x_1^3 + 2\mu_1 x_1^3 - 3\mu_1 x_1^2 x_2}{6(x_2-x_1)}
\end{aligned} \tag{20}$$

Theorem 2 is proved.  $\square$

### 3. Application to a Supplier Selection Problem

The supplier selection problem discussed in Lima Junior et al. [7] was used to illustrate the applicability of the proposed methodology. In the supplier selection problem, the performance of a supplier was assessed along five dimensions including quality, price, delivery, supplier profile, and supplier relationship. To aggregate the performances along the five dimensions, fuzzy weighted average (FWA) was applicable, for which the weight of each dimension needed to be specified. To this end, the PCFI-FAHP approach was applied.

Three DMs, including one industrial engineering manager, one production control manager, and one procurement department manager, were involved in the supplier selection problem. At first, each of them utilized the following linguistic terms [30] to express his/her belief about the relative importance of a dimension over another:

L1: "As equal as" = $(1, 1, 3)$;
L2: "Weakly more important than" = $(1, 3, 5)$;
L3: "Strongly more important than" = $(3, 5, 7)$;
L4: "Very strongly more important than" = $(5, 7, 9)$;
L5: "Absolutely more important than" = $(7, 9, 9)$.

Based on these inputs, the fuzzy pairwise comparison matrixes were constructed for the DMs in Table 1.

**Table 1.** The fuzzy pairwise comparison matrixes constructed for the decision-makers (DMs).

| | | | | | |
|---|---|---|---|---|---|
| | $(1,1,1)$ | $(7,9,9)$ | $(5,7,9)$ | $(7,9,9)$ | $(5,7,9)$ |
| | - | $(1,1,1)$ | - | $(1,1,3)$ | $(1,3,5)$ |
| (DM #1) | - | $(3,5,7)$ | $(1,1,1)$ | $(1,3,5)$ | $(5,7,9)$ |
| | - | - | - | $(1,1,1)$ | $(7,9,9)$ |
| | - | - | - | - | $(1,1,1)$ |
| | $(1,1,1)$ | $(1,3,5)$ | - | $(1,3,5)$ | $(5,7,9)$ |
| | - | $(1,1,1)$ | - | $(5,7,9)$ | $(3,5,7)$ |
| (DM #2) | $(1,3,5)$ | $(7,9,9)$ | $(1,1,1)$ | $(5,7,9)$ | $(3,5,7)$ |
| | - | - | - | $(1,1,1)$ | - |
| | - | - | - | $(1,3,5)$ | $(1,1,1)$ |
| | $(1,1,1)$ | $(3,5,7)$ | $(1,3,5)$ | $(1,3,5)$ | $(5,7,9)$ |
| | - | $(1,1,1)$ | - | - | - |
| (DM #3) | - | $(1,3,5)$ | $(1,1,1)$ | - | - |
| | - | $(7,9,9)$ | $(1,3,5)$ | $(1,1,1)$ | $(5,7,9)$ |
| | - | $(5,7,9)$ | $(1,1,3)$ | - | $(1,1,1)$ |

Each DM applied FGM to estimate the fuzzy maximal eigenvalue and fuzzy weights from the corresponding fuzzy pairwise comparison matrix. As a result, the estimated fuzzy maximal eigenvalues were:

$$\widetilde{\lambda}_{\max}(1) = (2.102, 5.778, 19.587),$$
$$\widetilde{\lambda}_{\max}(2) = (1.630, 5.879, 27.473),$$
$$\widetilde{\lambda}_{\max}(3) = (1.461, 5.672, 31.092).$$

The corresponding consistency indexes were

$$\widetilde{C.I.}(1) = (-0.725, 0.194, 3.647),$$
$$\widetilde{C.I.}(2) = (-0.843, 0.220, 5.618),$$
$$\widetilde{C.I.}(3) = (-0.885, 0.168, 6.523),$$

showing certain levels of consistency since $\widetilde{C.I.}(1)\sim\widetilde{C.I.}(3) \leq 0.3$. In addition, the estimated fuzzy weights are summarized in Figure 7.

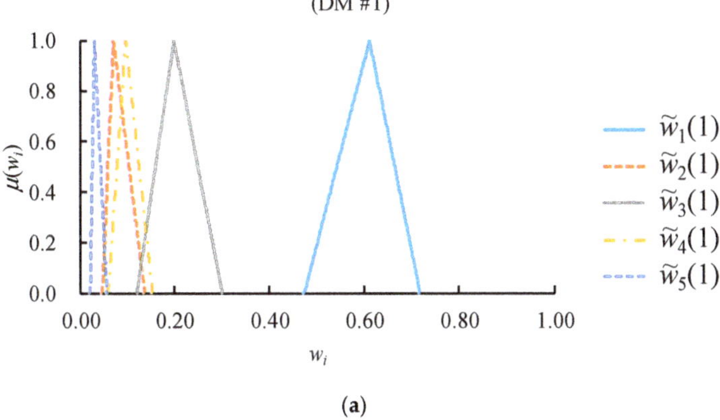


(**a**)

**Figure 7.** *Cont.*

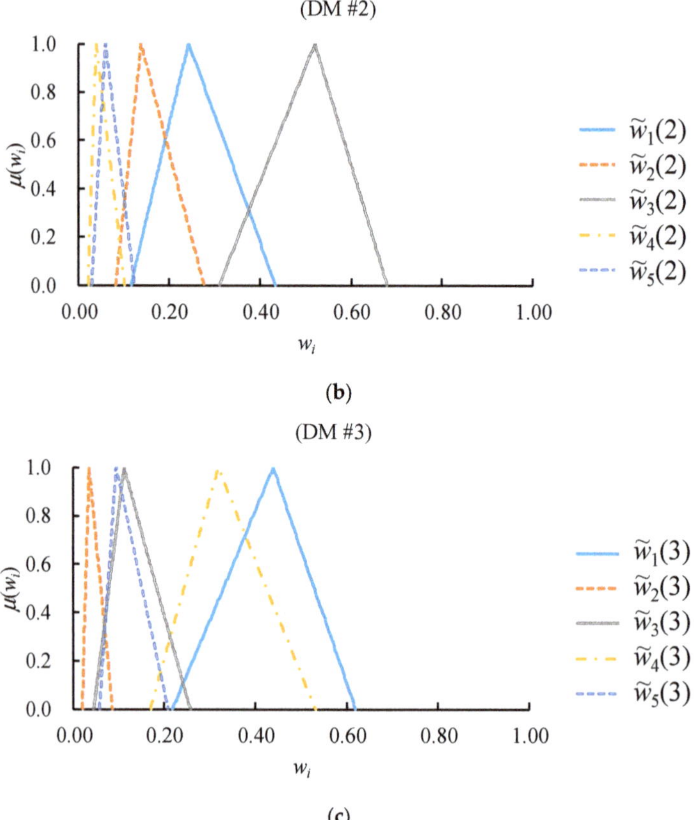



**Figure 7.** The estimated fuzzy weights: (**a**) by DM #1; (**b**) by DM #2; (**c**) by DM #3.

Subsequently, FI was applied to aggregate the fuzzy weights estimated by the DMs. However, unfortunately the overall consensus did not exist. For example, the result of aggregating the values of $\widetilde{w}_3$ estimated by the DMs is illustrated in Figure 8. The FI result was an empty set, showing a lack of (overall) consensus. In addition, overall consensus was not achieved for the values of $\widetilde{w}_1$, $\widetilde{w}_4$, or $\widetilde{w}_5$.

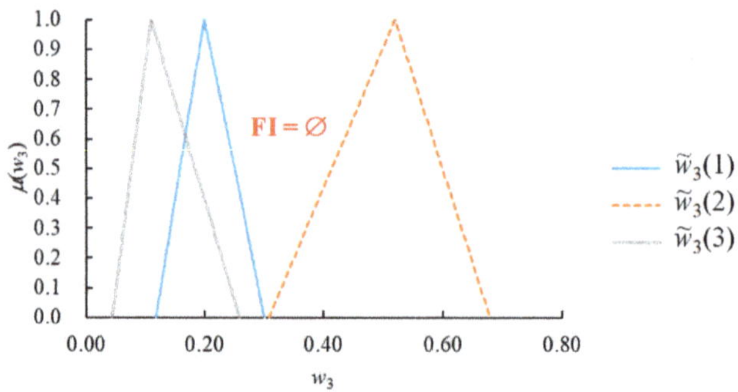



**Figure 8.** The FI result.

As a consequence, the partial consensus, rather than the overall consensus, was sought. It is noteworthy that there were only two levels of PCFI in the supplier selection problem, 2/3 PCFI and 3/3 PCFI. Between them, 3/3 PCFI was equal to FI. The result of applying 2/3 PCFI to the values of $\widetilde{w}_3$ estimated by the DMs is shown in Figure 9. There existed some partial consensus among the DMs. In addition, the DMs also achieved some partial consensus concerning the values of the other fuzzy weights.

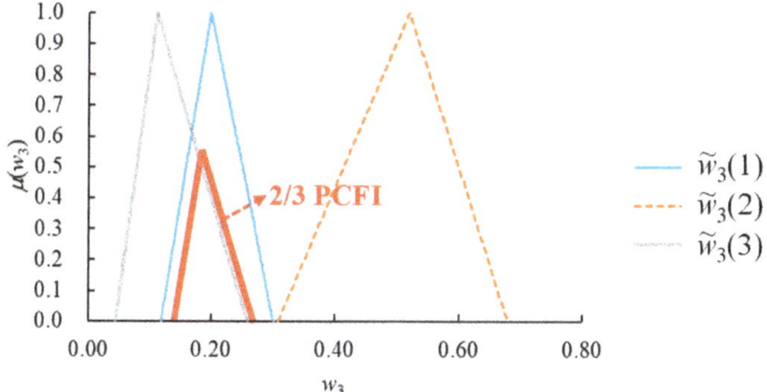


**Figure 9.** The result of applying 2/3 PCFI to the values of $\widetilde{w}_3$ estimated by the DMs.

Based on the PCFI results, the minimums, maximums, and widths of the fuzzy weights are summarized in Table 2. The minimal and average widths were 0.054 and 0.401, respectively.

**Table 2.** The minimums, maximums, and range widths of the fuzzy weights.

| Fuzzy Weight | Minimum | Maximum | Range Width |
|---|---|---|---|
| $\widetilde{w}_1$ | 0.219 | 0.620 | 0.401 |
| $\widetilde{w}_2$ | 0.044 | 0.135 | 0.091 |
| $\widetilde{w}_3$ | 0.120 | 0.258 | 0.138 |
| $\widetilde{w}_4$ | 0.056 | 0.102 | 0.054 |
| $\widetilde{w}_5$ | 0.029 | 0.124 | 0.095 |

COG was applied to defuzzify the fuzzy weights. The results are shown in Table 3.

**Table 3.** The defuzzification results.

| Fuzzy Weight | Center-Of-Gravity (COG) |
|---|---|
| $\widetilde{w}_1$ | 0.428 |
| $\widetilde{w}_2$ | 0.083 |
| $\widetilde{w}_3$ | 0.183 |
| $\widetilde{w}_4$ | 0.090 |
| $\widetilde{w}_5$ | 0.068 |

However, the sum of the defuzzified weights might not be 1 anymore, which required a re-normalization. The results are shown in Table 4.

**Table 4.** The re-normalization results.

| Weight | Re-Normalized Value |
|---|---|
| COG ($\widetilde{w}_1$) | 0.502 |
| COG ($\widetilde{w}_2$) | 0.097 |
| COG ($\widetilde{w}_3$) | 0.215 |
| COG ($\widetilde{w}_4$) | 0.106 |
| COG ($\widetilde{w}_5$) | 0.080 |

## 4. A Comparison with Some Existing Methods

To make a comparison, in this section four existing methods including FGM [31], FGM*i* [32], fuzzy extent analysis (FEA) [33], and FEA*i* [8,11,30] were also applied to the supplier selection problem:

(1)  The results obtained using FGM and FGM*i* were fuzzy, while those obtained using FEA and FEA*i* were crisp.
(2)  All the existing methods assumed there was consensus among the DMs.

In the FGM method, the fuzzy pairwise comparison results by the DMs were aggregated using FGM. Then, the fuzzy maximal eigenvalue and fuzzy weights were estimated from the aggregation result using FGM as well. The FGM*i* method was basically identical to the FGM method, except for the simplification of calculation by ignoring the dependence between the dividend and divisor. In both methods, COG was applied to defuzzify the fuzzy weights. A re-normalization was also required after defuzzification.

In the FEA method, the fuzzy pairwise comparison results by the DMs were also aggregated using FGM. Then, the fuzzy synthetic extent of each criterion was derived using fuzzy arithmetic mean. The weight of the criterion was set to its minimal degree of being the maximum. The only difference between the FEA*i* method and the FEA method is the ignorance of the dependence between the dividend and divisor. In both methods, defuzzification and re-normalization were not necessary.

The results obtained using the existing methods are summarized in Table 5. However, it was not possible to say which method was the most accurate method, since there was no actual value of a weight. Nevertheless, the weights obtained using the FGM were the closest to those obtained using the proposed methodology. In contrast, the weights obtained using the FEA*i* method were the farthest.

**Table 5.** The weights estimated using various methods.

| Weight | Fuzzy Geometric Mean (FGM) | FGM*i* | Fuzzy Extent Analysis (FEA) | FEA*i* | Partial Consensus Posterior Aggregation (PCPA)-Fuzzy Analytic Hierarchy Process (FAHP) |
|---|---|---|---|---|---|
| $w_1$ | 0.409 | 0.405 | 0.487 | 0.369 | 0.502 |
| $w_2$ | 0.107 | 0.101 | 0.064 | 0.111 | 0.097 |
| $w_3$ | 0.257 | 0.279 | 0.317 | 0.295 | 0.215 |
| $w_4$ | 0.140 | 0.135 | 0.132 | 0.162 | 0.106 |
| $w_5$ | 0.086 | 0.079 | 0.000 | 0.063 | 0.080 |

The average widths of the fuzzy weights obtained using various methods were compared and the results are shown in Figure 10. Obviously, the proposed methodology achieved the highest precision by minimizing the average width of the fuzzy weights, which was obviously due to the fuzzy collaboration mechanism.

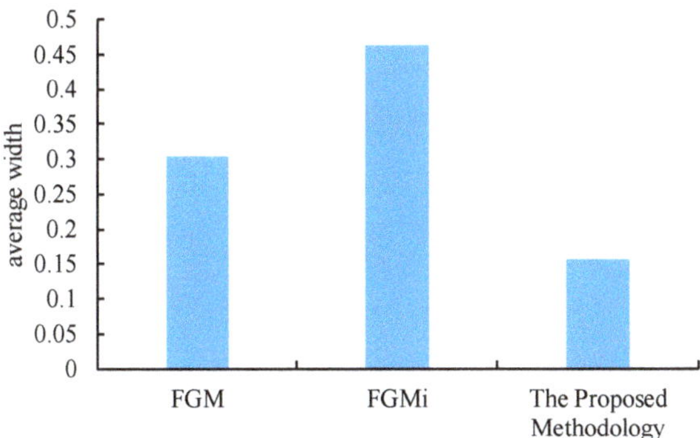

**Figure 10.** The average widths of the fuzzy weights estimated using various methods.

Compared to the existing FAHP methods, the proposed methodology was distinctive in that it excluded DM opinions that could not be included in the consensus. The proposed methodology did not attempt to compromise opposing DM opinions into neutral results. A DM whose opinion was excluded either accepted the consensus or modified his/her opinion to be included. In other words, it was important for DMs to reach consensus before aggregating their opinions. In addition, the overlap among the fuzzy weights estimated by the DMs seemed to be a very good tool for the visualization of consensus.

## 5. Conclusions

Among existing multi-DM FAHP methods, most of them aggregate the fuzzy pairwise comparison results by the DMs before deriving (or estimating) the fuzzy maximal eigenvalue and fuzzy weights. This is problematic because the fuzzy pairwise comparison results are uncertain and inconsistent. Instead, aggregating the fuzzy weights by the DMs, which are compromised results, in a posterior way seems to be more reasonable. However, a problem facing such a way is that there may be no consensus among the DMs, i.e., there was no intersection among the fuzzy weights specified by them. One way to address this problem is to seek the partial consensus among most DMs instead of the overall consensus among all DMs. From this point of view, the PCPA-FAHP approach is proposed in this study.

The PCPA-FAHP approach was applied to a supplier selection problem to validate its effectiveness. Some existing methods were also applied to the supplier selection problem to make a comparison. According to the experimental results:

(1)   Among the methods compared in the supplier selection problem, only the PCPA-FAHP approach was able to check the existence of the consensus among the DMs.
(2)   Although there was no overall consensus among all DMs in the supplier selection problem, some partial consensus did exist among most DMs.
(3)   The fuzzy collaboration mechanism employed in the PCPA-FAHP approach successfully shrunk the widths of the estimated fuzzy weights, thereby enhancing the precision of the FAHP analysis.
(4)   Among existing methods, the results obtained using FGM were the closest to those obtained using the PCPA-FAHP approach. However, the estimation precision achieved using FGM was much worse than that achieved using the PCPA-FAHP approach.

The PCPA-FAHP method needs to be applied to more real cases to further elaborate its effectiveness. In addition, more advanced collaboration mechanisms can be designed in future studies to further enhance the estimation precision. Furthermore, the uncertain pairwise comparison results can be expressed with rough numbers instead of fuzzy numbers [15,16], which constitutes a rough AHP (RAHP) problem [34]. The proposed methodology should be extended to deal with such problems.

**Author Contributions:** All authors contributed equally to the writing of this paper. All authors read and approved the final manuscript. Writing original draft, T.-C.T.C. and Y.-C.W.; Writing review & editing, T.-C.T.C. and Y.-C.W.

**Funding:** This research received no external funding.

**Acknowledgments:** We are thankful to the referees for their careful reading of the manuscript and for valuable comments and suggestions that greatly improved the presentation of this work.

**Conflicts of Interest:** The authors declare that there is no conflict of interest regarding the publication of this article.

## References

1. Van Laarhoven, P.J.M.; Pedrycz, W. A fuzzy extension of Saaty's priority theory. *Fuzzy Sets Syst.* **1983**, *11*, 229–241. [CrossRef]
2. Promentilla, M.A.B.; Furuichi, T.; Ishii, K.; Tanikawa, N. A fuzzy analytic network process for multi-criteria evaluation of contaminated site remedial countermeasures. *J. Environ. Manag.* **2008**, *88*, 479–495. [CrossRef] [PubMed]
3. Jain, V.; Sakhuja, S.; Thoduka, N.; Aggarwal, R.; Sangaiah, A.K. Supplier selection using fuzzy AHP and TOPSIS: a case study in the Indian automotive industry. *Neural Comput. Appl.* **2016**, *29*, 555–564. [CrossRef]
4. Kahraman, C.; Cebeci, U.; Ruan, D. Multi-attribute comparison of catering service companies using fuzzy AHP: The case of Turkey. *Int. J. Prod. Econ.* **2004**, *87*, 171–184. [CrossRef]
5. Ignatius, J.; Hatami-Marbini, A.; Rahman, A.; Dhamotharan, L.; Khoshnevis, P. A fuzzy decision support system for credit scoring. *Neural Comput. Appl.* **2016**, *29*, 921–937. [CrossRef]
6. Foroozesh, N.; Tavakkoli-Moghaddam, R.; Mousavi, S.M. A novel group decision model based on mean–variance–skewness concepts and interval-valued fuzzy sets for a selection problem of the sustainable warehouse location under uncertainty. *Neural Comput. Appl.* **2017**, *30*, 3277–3293. [CrossRef]
7. Junior, F.R.L.; Osiro, L.; Carpinetti, L.C.R. A comparison between Fuzzy AHP and Fuzzy TOPSIS methods to supplier selection. *Appl. Soft Comput.* **2014**, *21*, 194–209. [CrossRef]
8. Güran, A.; Uysal, M.; Ekinci, Y. An additive FAHP based sentence score function for text summarization. *ITC* **2017**, *46*, 53–69. [CrossRef]
9. Gnanavelbabu, A.; Arunagiri, P. Ranking of MUDA using AHP and Fuzzy AHP algorithm. *Mater. Today: Proc.* **2018**, *5*, 13406–13412. [CrossRef]
10. Forman, E.; Peniwati, K. Aggregating individual judgments and priorities with the analytic hierarchy process. *Eur. J. Oper. Res.* **1998**, *108*, 165–169. [CrossRef]
11. Wang, Y.-C.; Chen, T.; Yeh, Y.-L. Advanced 3D printing technologies for the aircraft industry: a fuzzy systematic approach for assessing the critical factors. *Int. J. Adv. Manuf. Technol.* **2018**, 1–11. [CrossRef]
12. Pan, N.-F. Fuzzy AHP approach for selecting the suitable bridge construction method. *Autom. Constr.* **2008**, *17*, 958–965. [CrossRef]
13. Hanss, M. *Applied Fuzzy Arithmetic*; Springer: Berlin/Heidelberg, Germany, 2005.
14. Ostrosi, E.; Haxhiaj, L.; Fukuda, S. Fuzzy modelling of consensus during design conflict resolution. *Res. Eng. Design* **2011**, *23*, 53–70. [CrossRef]
15. Chen, T. Forecasting the Unit Cost of a Product with Some Linear Fuzzy Collaborative Forecasting Models. *Algorithms* **2012**, *5*, 449–468. [CrossRef]
16. Ostrosi, E.; Bluntzer, J.-B.; Zhang, Z.; Stjepandić, J. Car style-holon recognition in computer-aided design. *J. Comput. Design Eng.* **2018**. [CrossRef]
17. Zhang, Z.; Xu, D.; Ostrosi, E.; Yu, L.; Fan, B. A systematic decision-making method for evaluating design alternatives of product service system based on variable precision rough set. *J. Intell. Manuf.* **2017**. [CrossRef]
18. Chen, T. A hybrid fuzzy and neural approach with virtual experts and partial consensus for DRAM price forecasting. *Int. J. Innov. Comput. Infor. Control* **2012**, *8*, 583–597.
19. Chen, T. Foreign exchange rate forecasting with a virtual-expert partial-consensus fuzzy-neural approach for semiconductor manufacturers in Taiwan. *IJISE* **2013**, *13*, 73–91. [CrossRef]
20. Chen, T. Estimating unit cost using agent-based fuzzy collaborative intelligence approach with entropy-consensus. *Appl. Soft Comput.* **2018**, *73*, 884–897. [CrossRef]

21. Mitra, S.; Banka, H.; Pedrycz, W. Rough–Fuzzy Collaborative Clustering. *IEEE Trans. Syst. Man Cybern. B Cybern.* **2006**, *36*, 795–805. [CrossRef]
22. Chen, T.; Romanowski, R. Forecasting the productivity of a virtual enterprise by agent-based fuzzy collaborative intelligence—With Facebook as an example. *Appl. Soft Comput.* **2014**, *24*, 511–521. [CrossRef]
23. Pedrycz, W.; Rai, P. Collaborative clustering with the use of Fuzzy C-Means and its quantification. *Fuzzy Sets Syst.* **2008**, *159*, 2399–2427. [CrossRef]
24. Chen, T.; Wang, Y.-C. An Agent-Based Fuzzy Collaborative Intelligence Approach for Precise and Accurate Semiconductor Yield Forecasting. *IEEE Trans. Fuzzy Syst.* **2014**, *22*, 201–211. [CrossRef]
25. Saaty, T.L. *The Analytic Hierarchy Process*; McGraw-Hill Education: New York, NY, USA, 1980.
26. Wedley, W.C. Consistency prediction for incomplete AHP matrices. *Math. Comput. Model.* **1993**, *17*, 151–161. [CrossRef]
27. Klaus, D. Goepel Business Performance Management Singapore. Available online: https://bpmsg.com/2013/12/ (accessed on 20 October 2018).
28. Kuncheva, L.I.; Krishnapuram, R. A fuzzy consensus aggregation operator. *Fuzzy Sets Syst.* **1996**, *79*, 347–356. [CrossRef]
29. Chen, T.; Lin, Y.-C. A fuzzy-neural system incorporating unequally important expert opinions for semiconductor yield forecasting. *Int. J. Unc. Fuzzy Knowl. Syst.* **2008**, *16*, 35–58. [CrossRef]
30. Zyoud, S.H.; Kaufmann, L.G.; Shaheen, H.; Samhan, S.; Fuchs-Hanusch, D. A framework for water loss management in developing countries under fuzzy environment: Integration of Fuzzy AHP with Fuzzy TOPSIS. *Expert Syst. Appl.* **2016**, *61*, 86–105. [CrossRef]
31. Buckley, J.J. Fuzzy hierarchical analysis. *Fuzzy Sets Syst.* **1985**, *17*, 233–247. [CrossRef]
32. Zheng, G.; Zhu, N.; Tian, Z.; Chen, Y.; Sun, B. Application of a trapezoidal fuzzy AHP method for work safety evaluation and early warning rating of hot and humid environments. *Saf. Sci.* **2012**, *50*, 228–239. [CrossRef]
33. Chang, D.-Y. Applications of the extent analysis method on fuzzy AHP. *Eur. J. Oper. Res.* **1996**, *95*, 649–655. [CrossRef]
34. Aydogan, E.K. Performance measurement model for Turkish aviation firms using the rough-AHP and TOPSIS methods under fuzzy environment. *Expert Syst. Appl.* **2011**, *38*, 3992–3998. [CrossRef]

*Article*

# $p$-Topologicalness—A Relative Topologicalness in $\top$-Convergence Spaces

**Lingqiang Li**

Department of Mathematics, Liaocheng University, Liaocheng 252059, China; lilingqiang0614@126.com;
Tel.: +86-0635-8239926

Received: 24 January 2019; Accepted: 25 February 2019; Published: 1 March 2019

**Abstract:** In this paper, $p$-topologicalness (a relative topologicalness) in $\top$-convergence spaces are studied through two equivalent approaches. One approach generalizes the Fischer's diagonal condition, the other approach extends the Gähler's neighborhood condition. Then the relationships between $p$-topologicalness in $\top$-convergence spaces and $p$-topologicalness in stratified $L$-generalized convergence spaces are established. Furthermore, the lower and upper $p$-topological modifications in $\top$-convergence spaces are also defined and discussed. In particular, it is proved that the lower (resp., upper) $p$-topological modification behaves reasonably well relative to final (resp., initial) structures.

**Keywords:** fuzzy topology; fuzzy convergence; lattice-valued convergence; $\top$-convergence space; relative topologicalness; $p$-topologcalness; diagonal condition; neighborhood condition

---

## 1. Introduction

The theory of convergence spaces [1] is natural extension of the theory of topological spaces. The topologicalness is important in the theory of convergence spaces since it mainly researches the condition of a convergence space to be a topological space. Generally, two equivalent approaches are used to characterize the topologicalness in convergence spaces. One approach is stated by the well-known Fischer's diagonal condition [2], the other approach is stated by Gähler's neighborhood condition [3]. In [4], by considering a pair of convergence spaces $(X, p)$ and $(X, q)$, Wilde and Kent investigated a kind of relative topologicalness, called $p$-topologicalness. When $p = q$, $p$-topologicalness is equivalent to topologicalness in convergence spaces. They also defined and discussed the lower and upper $p$-topological modifications in convergence spaces. Precisely, for a pair of convergence spaces $(X, p)$ and $(X, q)$, the lower (resp., upper) $p$-topological modification of $(X, q)$ is defined as the finest (resp., coarsest) $p$-topological convergence space which is coarser (resp., finer) than $(X, q)$. Similarly, a topological modification of $(X, q)$ is defined as the finest topological convergence space which is coarser than $(X, q)$.

Lattice-valued convergence spaces are common extension of convergence spaces and lattice-valued topological spaces. It should be pined out that lattice-valued convergence spaces are established on the basis of fuzzy sets. However, the lattice structure is used to replace the unit interval $[0, 1]$ as the truth table for membership degrees. In recent years, two kinds of lattice-valued convergence spaces received much attention: (1) the theory of stratified $L$-generalized convergence spaces based on $L$-filters, which is initiated by Jäger [5] and then developed by many researchers [6–25]; and (2) the theory of $\top$-convergence spaces based on $\top$-filters, which is investigated by Fang [26] in 2017. The topologicalness in stratified $L$-generalized convergence spaces was studied by Jäger [27–29] and Li [30,31], the $p$-topologicalness and $p$-topological modifications in stratified $L$-generalized convergence spaces were discussed by Li [32,33].

The topologicalness in $\mathsf{T}$-convergence spaces was researched by Fang [26] and Li [34]. In this paper, we shall consider the $p$-topologicalness and $p$-topological modifications in $\mathsf{T}$-convergence spaces.

The contents are arranged as follows. Section 2 recalls some basic notions as preliminary. Section 3 discusses the $p$-topologicalness in $\mathsf{T}$-convergence spaces by generalized Fischer's diagonal condition and generalized Gähler's neighborhood condition, respectively. Then the relationships between $p$-topologicalness in $\mathsf{T}$-convergence spaces and $p$-topologicalness in stratified $L$-generalized convergence spaces are established. Section 4 focuses on $p$-topological modifications in $\mathsf{T}$-convergence spaces. The lower and upper $p$-topological modifications in $\mathsf{T}$-convergence spaces are defined and discussed. Particularly, it is proved that the lower (resp., upper) $p$-topological modification behaves reasonably well relative to final (resp., initial) structures.

## 2. Preliminaries

Let $L$ be a complete lattice with the top element $\top$ and the bottom element $\bot$. For a commutative quantale, we mean a pair $(L, *)$ such that $*$ is a commutative semigroup operation on $L$ with the condition

$$\forall a \in L, \forall \{b_j\}_{j \in J} \subseteq L, a * \bigvee_{j \in J} b_j = \bigvee_{j \in J} (a * b_j).$$

$(L, *)$ is called integral if the top element $\top$ is the unique unit, i.e., $\forall a \in L, \top * a = a$. For any $a \in L$, each function $a * (-) : L \longrightarrow L$ has a right adjoint $a \to (-) : L \longrightarrow L$ defined as $a \to b = \bigvee \{c \in L : a * c \leq b\}$. In the following, we list the usual properties of $*$ and $\to$ [35].

(1) $a \to b = \top \Leftrightarrow a \leq b$;
(2) $a * b \leq c \Leftrightarrow b \leq a \to c$;
(3) $a * (a \to b) \leq b$;
(4) $a \to (b \to c) = (a * b) \to c$;
(5) $(\bigvee_{j \in J} a_j) \to b = \bigwedge_{j \in J} (a_j \to b)$;
(6) $a \to (\bigwedge_{j \in J} b_j) = \bigwedge_{j \in J} (a \to b_j)$.

We call $(L, *)$ to be a meet continuous lattice if the complete lattice $L$ is meet continuous [36], that is, $(L, \leq)$ satisfies the distributive law: $a \wedge (\bigvee_{i \in I} b_i) = \bigvee_{i \in I} (a \wedge b_i)$, for any $a \in L$ and any directed subsets $\{b_i | i \in I\}$ in $L$.

$L$ is said to be continuous if $(L, \leq)$ is a continuous lattice [36], that is, for any nonempty family $\{a_{j,k} | j \in J, k \in K(j)\}$ in $L$ with $\{a_{j,k} | k \in K(j)\}$ is directed for all $j \in J$, the identity

$$\textbf{(DD)} \qquad \bigwedge_{j \in J} \bigvee_{k \in K(j)} a_{j,k} = \bigvee_{h \in N} \bigwedge_{j \in J} a_{j,h(j)}$$

holds, where $N$ is the set of all choice functions on $J$ with values $h(j) \in K(j)$ for all $j \in J$. Obviously, continuity implies meet-continuity for $L$.

In this article, unless otherwise stated, we always assume that $L = (L, *)$ is a commutative, integral, and meet continuous quantale.

A function $\mu : X \to L$ is called an $L$-fuzzy set in $X$, and all $L$-fuzzy sets in $X$ is denoted as $L^X$. The operations $\vee, \wedge, *, \to$ on $L$ can be translated pointwisely onto $L^X$. Said precisely, for any $\mu, \nu \in L^X$ and any $\{\mu_t | t \in T\} \subseteq L^X$,

$$\mu \leq \nu \text{ iff } \mu(x) \leq \nu(x) \text{ for any } x \in X,$$
$$(\bigvee_{i \in I} \mu_t)(x) = \bigvee_{t \in T} \mu_t(x), (\bigwedge_{t \in T} \mu_t)(x) = \bigwedge_{t \in T} \mu_t(x),$$
$$(\mu * \nu)(x) = \mu(x) * \nu(x), (\mu \to \nu)(x) = \mu(x) \to \nu(x).$$

We don't distinguish between a constant function and its value because no confusion will occur.

Let $f : X \longrightarrow Y$ be a function. We define $f^{\rightarrow} : L^X \longrightarrow L^Y$ and $f^{\leftarrow} : L^Y \longrightarrow L^X$ [35] by $f^{\rightarrow}(\mu)(y) = \bigvee_{f(x)=y} \mu(x)$ for $\mu \in L^X$ and $y \in Y$, and $f^{\leftarrow}(\nu)(x) = \nu(f(x))$ for $\nu \in L^Y$ and $x \in X$.

Let $\mu, \nu$ be $L$-fuzzy sets in $X$. The subsethood degree [37–40] of $\mu, \nu$, denoted by $S_X(\mu, \nu)$, is defined by
$$S_X(\mu, \nu) = \bigwedge_{x \in X} (\mu(x) \rightarrow \nu(x)).$$

## 2.1. $\top$-Filters and Stratified L-Filters

A filter on a set $X$ is an upper set of $(2^X, \subseteq)$ ($2^X$ denotes the power set of $X$) wich is closed for finite meets and does not contain the empty set. The conception of filter has been generalized to the fuzzy setting in two methods; prefilters (or $\top$-filters more general) and $L$-filters. Both prefilters ($\top$-filters) and $L$-filters play important roles in the theory of fuzzy topology, see [26,27,34,35,41–44].

**Definition 1** ([35]). *A nonempty subset $\mathbb{F} \subseteq L^X$ is called a $\top$-filter on the set $X$ whenever:*

*(TF1)* $\bigvee_{x \in X} \lambda(x) = \top$ *for all* $\lambda \in \mathbb{F}$, *(TF2)* $\lambda \wedge \mu \in \mathbb{F}$ *for all* $\lambda, \mu \in \mathbb{F}$,

*(TF3) if* $\lambda \in L^X$ *such that* $\bigvee_{\mu \in \mathbb{F}} S_X(\mu, \lambda) = \top$, *then* $\lambda \in \mathbb{F}$.

*The set of all $\top$-filters on $X$ is denoted as* $\mathbb{F}_L^{\top}(X)$.

**Definition 2** ([35]). *A nonempty subset $\mathbb{B} \subseteq L^X$ is called a $\top$-filter base on the set $X$ provided:*

*(TB1)* $\bigvee_{x \in X} \lambda(x) = \top$ *for all* $\lambda \in \mathbb{B}$, *(TB2) if* $\lambda, \mu \in \mathbb{B}$, *then* $\bigvee_{\nu \in \mathbb{B}} S_X(\nu, \lambda \wedge \mu) = \top$.

Each $\top$-filter base generates a $\top$-filter $\mathbb{F}_{\mathbb{B}}$ defined by
$$\mathbb{F}_{\mathbb{B}} := \{\lambda \in L^X | \bigvee_{\mu \in \mathbb{B}} S_X(\mu, \lambda) = \top\}.$$

**Example 1** ([26,45]). *Let $f : X \longrightarrow Y$ be a function, $\mathbb{F} \in \mathbb{F}_L^{\top}(X)$ and $\mathbb{G} \in \mathbb{F}_L^{\top}(Y)$. Then*

*(1) The family $\{f^{\rightarrow}(\lambda) | \lambda \in \mathbb{F}\}$ forms a $\top$-filter base on $Y$, and the $\top$-filter $f^{\Rightarrow}(\mathbb{F})$ generated by it is called the image of $\mathbb{F}$ under $f$. It is easily seen that $\mu \in f^{\Rightarrow}(\mathbb{F}) \iff f^{\leftarrow}(\mu) \in \mathbb{F}$.*

*(2) The family $\{f^{\leftarrow}(\mu) | \mu \in \mathbb{G}\}$ forms a $\top$-filter base on $Y$ if and only if $\bigvee_{y \in f(X)} \mu(y) = \top$ holds for all $\mu \in \mathbb{G}$, and the $\top$-filter $f^{\Leftarrow}(\mathbb{G})$ (if exists) generated by it is called the inverse image of $\mathbb{G}$ under $f$. Additionally, $\mathbb{G} \subseteq f^{\Rightarrow}(f^{\Leftarrow}(\mathbb{G}))$ holds whenever $f^{\Leftarrow}(\mathbb{G})$ exists. Particularly, $f^{\Leftarrow}(\mathbb{G})$ always exists and $\mathbb{G} = f^{\Rightarrow}(f^{\Leftarrow}(\mathbb{G}))$ if $f$ is surjective.*

*(3) For any $x \in X$, the family $[x]_{\top} =: \{\lambda \in L^X | \lambda(x) = \top\}$ is a $\top$-filter on $X$, and $f^{\Rightarrow}([x]_{\top}) = [f(x)]_{\top}$.*

A stratified $L$-filter [35] on a set $X$ is a function $\mathcal{F} : L^X \longrightarrow L$ such that: $\forall \lambda, \mu \in L^X, \forall \alpha \in L$, (LF1) $\mathcal{F}(\bot) = \bot, \mathcal{F}(\top) = \top$; (LF2) $\mathcal{F}(\lambda) \wedge \mathcal{F}(\mu) = \mathcal{F}(\lambda \wedge \mu)$; (LFs) $\mathcal{F}(\alpha * \lambda) \geq \alpha * \mathcal{F}(\lambda)$.

The set of all stratified $L$-filters on $X$ is denoted as $\mathcal{F}_L^s(X)$. A stratified $L$-filter $\mathcal{F}$ is called tight if $\mathcal{F}(\alpha) = \alpha$ for each $\alpha \in L$.

**Example 2** ([35]). *Let $f : X \longrightarrow Y$ be a function, $\mathcal{F} \in \mathcal{F}_L^s(X)$ and $\mathcal{G} \in \mathcal{F}_L^s(Y)$. Then*

*(1) The function $f^{\Rightarrow}(\mathcal{F}) : L^Y \longrightarrow L$ defined by $\mu \mapsto \mathcal{F}(\mu \circ f)$ is a stratified $L$-filter on $Y$ called the image of $\mathcal{F}$ under $f$.*

*(2) For any $x \in X$, the function $[x] : L^X \longrightarrow L, [x](\lambda) = \lambda(x)$ is a stratified $L$-filter on $X$, and $f^{\Rightarrow}([x]) = [f(x)]$.*

For each $\mathbb{F} \in \mathbb{F}_L^\top(X)$, define $\Lambda(\mathbb{F}) : L^X \longrightarrow L$ as

$$\forall \lambda \in L^X, \Lambda(\mathbb{F})(\lambda) = \bigvee_{\mu \in \mathbb{F}} S_X(\mu, \lambda),$$

then $\Lambda(\mathbb{F})$ is a tightly stratified $L$-filter on $X$ [44].

Conversely, for each tightly stratified $L$-filter $\mathcal{F}$ on a set $X$, the family

$$\Gamma(\mathcal{F}) = \{\lambda \in L^X, \mathcal{F}(\lambda) = \top\}$$

is a $\top$-filter on $X$ [44]. Given $\mathbb{F} \in \mathbb{F}_L^\top(X)$, we have $\Gamma\Lambda(\mathbb{F}) = \mathbb{F}$.

**Lemma 1.** *Let* $\{\mathbb{F}_j\}_{j \in J} \subseteq \mathbb{F}_L^\top(X)$. *If $L$ is continuous then* $\Lambda(\bigcap_{j \in J} \mathbb{F}_j) = \bigwedge_{j \in J} \Lambda(\mathbb{F}_j)$.

**Proof.** For any $\lambda \in L^X$ and any $j \in J$, note that $\{S_X(\mu_j, \lambda) | \mu_j \in \mathbb{F}_j\}$ is a directed subset of $L$. Then

$$
\begin{aligned}
\bigwedge_{j \in J} \Lambda(\mathbb{F}_j)(\lambda) &= \bigwedge_{j \in J} \bigvee_{\mu_j \in \mathbb{F}_j} S_X(\mu_j, \lambda) \stackrel{\textbf{(DD)}}{=} \bigvee_{h \in N} \bigwedge_{j \in J} S_X(h(j), \lambda) \\
&= \bigvee_{h \in N} S_X(\bigvee_{j \in J} h(j), \lambda), \text{ by } \bigvee_{j \in J} h(j) \in \bigcap_{j \in J} \mathbb{F}_j \\
&\leq \bigvee_{\nu \in \bigcap_{j \in J} \mathbb{F}_j} S_X(\nu, \lambda) \leq \Lambda(\bigcap_{j \in J} \mathbb{F}_j)(\lambda).
\end{aligned}
$$

Thus $\Lambda(\bigcap_{j \in J} \mathbb{F}_j)(\lambda) = \bigwedge_{j \in J} \Lambda(\mathbb{F}_j)(\lambda)$ since $\Lambda(\bigcap_{j \in J} \mathbb{F}_j)(\lambda) \leq \bigwedge_{j \in J} \Lambda(\mathbb{F}_j)(\lambda)$ holds obviously.  $\square$

*2.2. $\top$-Convergence Spaces and Stratified L-Generalized Convergence Spaces*

**Definition 3.** *A $\top$-convergence structure* [26] *on a set $X$ is a function* $q : \mathbb{F}_L^\top(X) \longrightarrow 2^X$ *satisfying*

**(TC1)** $[x]_\top \xrightarrow{q} x$ *for every* $x \in X$; **(TC2)** *if* $\mathbb{F} \xrightarrow{q} x$ *and* $\mathbb{F} \subseteq \mathbb{G}$, *then* $\mathbb{G} \xrightarrow{q} x$, *where* $\mathbb{F} \xrightarrow{q} x$ *is shorthand for* $x \in q(\mathbb{F})$. *The pair* $(X, q)$ *is called a $\top$-convergence space.*

A function $f : X \longrightarrow X'$ between two $\top$-convergence spaces $(X, q)$, $(X', q')$ is called continuous if $f^\Rightarrow(\mathbb{F}) \xrightarrow{q'} f(x)$ whenever $\mathbb{F} \xrightarrow{q} x$.

The category whose objects are $\top$-convergence spaces and whose morphisms are continuous functions will be denoted by $\top$-**CS**. This category is topological over **SET** [26,46].

For a given source $(X \xrightarrow{f_i} (X_i, q_i))_{i \in I}$, the initial structure [47], $q$ on $X$ is defined by

$$\mathbb{F} \xrightarrow{q} x \Longleftrightarrow \forall i \in I, f_i^\Rightarrow(\mathbb{F}) \xrightarrow{q_i} f_i(x).$$

For a given sink $((X_i, q_i) \xrightarrow{f_i} X)_{i \in I}$, the final structure, $q$ on $X$ is defined as

$$\mathbb{F} \xrightarrow{q} x \Longleftrightarrow \begin{cases} \mathbb{F} \supseteq [x]_\top, & x \notin \cup_{i \in I} f_i(X_i); \\ \mathbb{F} \supseteq f_i^\Rightarrow(\mathbb{G}_i), & \exists i \in I, x_i \in X_i, \mathbb{G}_i \in \mathbb{F}_L^\top(X_i) \text{ s.t } f(x_i) = x, \mathbb{G}_i \xrightarrow{q_i} x_i. \end{cases}$$

Thus, when $X = \cup_{i \in I} f_i(X_i)$, the final structure $q$ can be simplified as

$$\mathbb{F} \xrightarrow{q} x \Longleftrightarrow \mathbb{F} \supseteq f_i^\Rightarrow(\mathbb{G}_i) \text{ for some } \mathbb{G}_i \xrightarrow{q_i} x_i \text{ with } f(x_i) = x.$$

For a nonempty set $X$, we use $\top(X)$ to denote all $\top$-convergence structures on $X$. For $p, q \in \top(X)$, we say that $q$ is finer than $p$, or $p$ is coarser than $q$, denoted by $p \leq q$ for short, if the identity $id_X : (X, q) \longrightarrow (X, p)$ is continuous, that is, $\mathbb{F} \xrightarrow{q} x \Longrightarrow \mathbb{F} \xrightarrow{p} x$. It is easily observed from [26,47] that $(\top(X), \leq)$ forms a completed lattice, and the discrete (resp., indiscrete) structure $\delta$ (resp., $\iota$) is the top (resp., bottom) element of $(\top(X), \leq)$, where $\delta$ is given by $\mathbb{F} \xrightarrow{\delta} x$ iff $\mathbb{F} \supseteq [x]_\top$; and $\iota$ is given by $\mathbb{F} \xrightarrow{\iota} x$ for all $\mathbb{F} \in \mathbb{F}_L^\top(X)$, $x \in X$.

**Definition 4.** *(Jäger [5] and Yao [25]) A stratified $L$-generalized convergence structure on a set $X$ is a function* $\lim^q : \mathcal{F}_L^s(X) \longrightarrow L^X$ *satisfying*
    **(LC1)** $\lim^q[x](x) = 1$ *for every* $x \in X$; **(LC2)** $\forall \mathcal{F}, \mathcal{G} \in \mathcal{F}_L^s(X)$, $\mathcal{F} \leq \mathcal{G} \Longrightarrow \lim^q \mathcal{F} \leq \lim^q \mathcal{G}$.
*The pair* $(X, \lim^q)$ *is called a stratified $L$-generalized convergence space.*

Let $(X, q)$ be a $\top$-convergence space. We define $\lim^q : \mathcal{F}_L^s(X) \longrightarrow L^X$ as

$$\lim^q \mathcal{F}(x) = \begin{cases} \top, & \mathcal{F} \geq \Lambda(\mathbb{F}) \text{ for some } \mathbb{F} \xrightarrow{q} x; \\ \bot, & \text{otherwise.} \end{cases}$$

Note that $[x] = \Lambda([x]_\top)$. It follows that $(X, \lim^q)$ is a stratified $L$-generalized convergence space.

**Remark 1.** *When $L = \{\bot, \top\}$, both $\top$-convergence spaces and stratified $L$-generalized convergence spaces all reduce to convergence spaces. Therefore, these two kinds of lattice-valued convergence spaces are all natural extensions of convergence spaces.*

### 3. $p$-Topologicalness in $\top$-Convergence Spaces

In this section, we shall discuss the $p$-topologicalness in $\top$-convergence spaces by generalized Fischer's diagonal condition and generalized Gähler's neighborhood condition, respectively. We also try to establish the relationships between $p$-topologicalness in $\top$-convergence spaces and $p$-topologicalness in stratified $L$-generalized convergence spaces.

*3.1. $p$-Pretopologicalness in $\top$-Convergence Spaces*

Let $(X, p)$ be a $\top$-convergence space. Then for any $x \in X$, the $\top$-filter

$$\mathbb{U}_p(x) = \cap\{\mathbb{F} \in \mathbb{F}_L^\top(X) | \mathbb{F} \xrightarrow{p} x\}$$

is called the $\top$-neighborhood with respect to $p$ at $x$. Then the family $\mathbb{U}_p := \{\mathbb{U}_p(x)\}_{x \in X}$ is called the $\top$-neighborhood system generated by $(X, p)$ [26]. It is easily seen that if $p, p' \in T(X)$ and $p \leq p'$ then $\mathbb{U}_p(x) \subseteq \mathbb{U}_{p'}(x)$ for any $x \in X$.

In the following, we shorten a pair of $\top$-convergence spaces $(X, p)$ and $(X, q)$ as $(X, p, q)$. It is easy to check that the following conditions are equivalent:
    $p$-**(TP1)**: $\forall \{\mathbb{F}_j\}_{j \in J} \subseteq \mathbb{F}_L^\top(X), \forall x \in X, \forall j \in J, \mathbb{F}_j \xrightarrow{p} x \Longrightarrow \cap_{j \in J} \mathbb{F}_j \xrightarrow{q} x$.
    $p$-**(TP2)**: $\forall \mathbb{F} \in \mathbb{F}_L^\top(X), \forall x \in X, \mathbb{F} \supseteq \mathbb{U}_p(x) \Longrightarrow \mathbb{F} \xrightarrow{q} x$.
    $p$-**(TP3)**: $\forall x \in X, \mathbb{U}_p(x) \xrightarrow{q} x$.

**Definition 5.** *Assume that $(X, p, q)$ is a pair of $\top$-convergence spaces. Then $q$ is said to be $p$-pretopological if it fulfills either of the above three conditions.*

**Remark 2.** *When $p = q$, $p$-pretopologicalness is precise the pretopologicalness in [26]. In this case, it is observed easily that the "$\Longrightarrow$" in $p$-(**TP2**) can be replaced with "$\Longleftrightarrow$". In the following, when $p = q$, we omit the prefix "$p$" in symbols $p$-(**TP1**)–$p$-(**TP3**). This simplification is also used for the subsequent $p$-topological conditions.*

**Proposition 1.** *A $\top$-convergence structure $q$ on $X$ is pretopological iff it is $p$-pretopological for any $q \in \top(X)$ with $q \leq p$.*

**Proof.** Let $(X, q)$ be pretopological and $q \leq p$. Then by $q \leq p$ we have $\mathbb{U}_p(x) \supseteq \mathbb{U}_q(x)$ for any $x \in X$. By pretopologicalness of $q$ we get that $\mathbb{U}_q(x) \xrightarrow{q} x$. It follows that $\mathbb{U}_p(x) \xrightarrow{q} x$. Thus $q$ is $p$-pretopological. The converse implication is obvious. $\square$

The following example shows there is no $p$-pretopologicalness implies pretopologicalness in general.

**Example 3.** *Let $L$ be the linearly ordered frame $(\{\bot, \alpha, \top\}, \wedge, \top)$ with $\bot < \alpha < \top$, and $X = \{x, y\}$. For each $\mathbb{F} \in \mathbb{F}_L^\top(X)$ and $z \in X$, let $\mathbb{F} \xrightarrow{p} z \Longleftrightarrow \mathbb{F} \supseteq [z]$. In [26], it is proved that $(X, p)$ is a $\top$-convergence space and for each $z \in X, \mathbb{U}_p(z) = [z]$.*

*For $x, y \in X$, it is easily seen that the subsets $\mathbb{F}_x, \mathbb{F}_y$ of $L^X$ defined by*

$$\mathbb{F}_x = \{\lambda \in L^X : \lambda(x) \geq \alpha, \lambda(y) = \top\}; \ \mathbb{F}_y(\lambda) = \{\lambda \in L^X : \lambda(y) \geq \alpha, \lambda(x) = \top\}$$

*are all $\top$-filters on $X$. For each $\mathbb{F} \in \mathbb{F}_L^\top(X)$ and each $z \in X$, let $\mathbb{F} \xrightarrow{q} z \Longleftrightarrow \mathbb{F} \supseteq [z]$ or $\mathbb{F} \supseteq \mathbb{F}_z$. Then $(X, q)$ is a $\top$-convergence space. For each $z \in X, \mathbb{U}_q(z) = [z] \cap \mathbb{F}_z = \{\top_X\}$ and so $[z] \cap \mathbb{F}_z \neq [z], \mathbb{F}_z$.*

*Obviously, $q$ satisfies $p$-(**TP3**). But $q$ is not pretopological since we have no $\mathbb{U}_q(z) \xrightarrow{q} z$.*

### 3.2. $p$-Topologicalness in $\top$-Convergence Spaces

At first, we fix the notions of diagonal $\top$-filter and neighborhood $\top$-filter to state $p$-topologicalness. Let $J, X$ be any sets and $\phi : J \longrightarrow \mathbb{F}_L^\top(X)$ be any function. Then a function $\hat{\phi} : L^X \to L^J$ is defined as

$$\forall \lambda \in L^X, \forall j \in J, \hat{\phi}(\lambda)(j) = \Lambda(\phi(j))(\lambda) = \bigvee_{\mu \in \phi(j)} S_X(\mu, \lambda).$$

For all $\mathbb{F} \in \mathbb{F}_L^\top(J)$, it is proved that a subset of $L^X$ defined by

$$k\phi\mathbb{F} := \{\lambda \in L^X | \hat{\phi}(\lambda) \in \mathbb{F}\}$$

is a $\top$-filter, called diagonal $\top$-filter of $\mathbb{F}$ under $\phi$ [26]. In addition, for any $\lambda, \mu \in L^X$, it was proved in [26] that $S_X(\lambda, \mu) \leq S_J(\hat{\phi}(\lambda), \hat{\phi}(\mu))$.

**Definition 6** ([34]). *Let $(X, p)$ be a $\top$-convergence space and $\mathbb{U}_p : X \longrightarrow \mathbb{F}_L^\top(X)$ be the $\top$-neighborhood system generated by $(X, p)$. Then for each $\mathbb{F} \in \mathbb{F}_L^\top(X)$, the $\top$-filter $\mathbb{U}_p(\mathbb{F}) := k\mathbb{U}_p\mathbb{F}$, is called neighborhood $\top$-filter of $\mathbb{F}$ w.r.t. $p$.*

Let $\mathbb{N}$ be the set of natural numbers including $0$. Let $(X, p)$ be a $\top$- convergence space and $\mathbb{F} \in \mathbb{F}_L^\top(X)$. For any $n \in \mathbb{N}$, we define $\mathbb{U}_p^0(\mathbb{F}) = \mathbb{F}$, and if $\mathbb{U}_p^n(\mathbb{F})$ has been defined, then we define the $n + 1$ th iteration of the neighborhood $\top$-filter of $\mathbb{F}$ inductive by $\mathbb{U}_p^{n+1}(\mathbb{F}) = \mathbb{U}_p(\mathbb{U}_p^n(\mathbb{F}))$.

**Proposition 2.** *Let $(X, p)$ be a $\top$- convergence space, $n \in \mathbb{N}$ and $\mathbb{F}, \mathbb{G} \in \mathbb{F}_L^\top(X)$. Then*
*(1) $\mathbb{U}_p^n(\mathbb{F}) \subseteq \mathbb{F}$,*

(2) *if* $\mathbb{F} \subseteq \mathbb{G}$, *then* $\mathbb{U}_p^n(\mathbb{F}) \subseteq \mathbb{U}_p^n(\mathbb{G})$,

(3) *if* $p' \in T(X)$ *and* $p \leq p'$, *then* $\mathbb{U}_p^n(\mathbb{F}) \subseteq \mathbb{U}_{p'}^n(\mathbb{F})$.

**Proof.** It is obvious. $\square$

**Definition 7.** *Let* $f : (X, q) \longrightarrow (Y, p)$ *be a function between* $\top$-*convergence spaces. Then* $f$ *is said to be an interior function if* $f^{\rightarrow}(\widehat{\mathbb{U}_q}(\lambda)) \leq \widehat{\mathbb{U}_p}(f^{\rightarrow}(\lambda))$ *for all* $\lambda \in L^X$.

**Proposition 3.** *Let* $f : (X, q) \longrightarrow (Y, p)$ *be a function between* $\top$- *convergence spaces and* $\mathbb{F} \in \mathbb{F}_L^{\top}(X)$.

(1) *If* $f$ *is continuous, then* $f^{\Rightarrow}(\mathbb{U}_q^n(\mathbb{F})) \supseteq \mathbb{U}_p^n(f^{\Rightarrow}(\mathbb{F}))$.

(2) *If* $f$ *is an interior function, then* $f^{\Rightarrow}(\mathbb{U}_q^n(\mathbb{F})) \subseteq \mathbb{U}_p^n(f^{\Rightarrow}(\mathbb{F}))$.

**Proof.** (1) We prove $f^{\Rightarrow}(\mathbb{U}_q^n(\mathbb{F})) \supseteq \mathbb{U}_p^n(f^{\Rightarrow}(\mathbb{F}))$ inductively.

For each $\mathbb{F} \xrightarrow{q} x$ and each $\lambda \in \mathbb{U}_p(f(x))$ we have $\lambda \in f^{\Rightarrow}(\mathbb{F})$, i.e., $f^{\leftarrow}(\lambda) \in \mathbb{F}$ and then

$$f^{\leftarrow}(\lambda) \in \cap\{\mathbb{F}|\mathbb{F} \xrightarrow{q} x\} = \mathbb{U}_q(x),$$

i.e., $\lambda \in f^{\Rightarrow}(\mathbb{U}_q(x))$. Thus $\mathbb{U}_p(f(x)) \subseteq f^{\Rightarrow}(\mathbb{U}_q(x))$.

Fixing $\lambda \in L^Y$, we get

$$\widehat{\mathbb{U}_p}(\lambda)(f(x)) = \bigvee_{\mu \in \mathbb{U}_p(f(x))} S_Y(\mu, \lambda) \leq \bigvee_{f^{\leftarrow}(\mu) \in \mathbb{U}_q(x)} S_X(f^{\leftarrow}(\mu), f^{\leftarrow}(\lambda)) \leq$$
$$\bigvee_{\nu \in \mathbb{U}_q(x)} S_X(\nu, f^{\leftarrow}(\lambda)) = \widehat{\mathbb{U}_q}(f^{\leftarrow}(\lambda))(x).$$

It follows that $f^{\leftarrow}(\widehat{\mathbb{U}_p}(\lambda)) = \widehat{\mathbb{U}_q}(f^{\leftarrow}(\lambda))$. Thus

$$\lambda \in \mathbb{U}_p(f^{\Rightarrow}(\mathbb{F})) \implies \widehat{\mathbb{U}_p}(\lambda) \in f^{\Rightarrow}(\mathbb{F}) \implies f^{\leftarrow}(\widehat{\mathbb{U}_p}(\lambda)) \in \mathbb{F} \implies \widehat{\mathbb{U}_q}(f^{\leftarrow}(\lambda)) \in \mathbb{F}$$
$$\implies f^{\leftarrow}(\lambda) \in \mathbb{U}_q(\mathbb{F}) \implies \lambda \in f^{\Rightarrow}(\mathbb{U}_q(\mathbb{F})).$$

So, $f^{\Rightarrow}(\mathbb{U}_q^n(\mathbb{F})) \supseteq \mathbb{U}_p^n(f^{\Rightarrow}(\mathbb{F}))$ when $n = 1$.

We assume that $f^{\Rightarrow}(\mathbb{U}_q^n(\mathbb{F})) \supseteq \mathbb{U}_p^n(f^{\Rightarrow}(\mathbb{F}))$ when $n = k$. Then we need to check that $f^{\Rightarrow}(\mathbb{U}_q^n(\mathbb{F})) \supseteq \mathbb{U}_p^n(f^{\Rightarrow}(\mathbb{F}))$ when $n = k + 1$. Indeed,

$$f^{\Rightarrow}(\mathbb{U}_q^{k+1}(\mathbb{F})) = f^{\Rightarrow}(\mathbb{U}_q(\mathbb{U}_q^k(\mathbb{F}))) \supseteq \mathbb{U}_p(f^{\Rightarrow}(\mathbb{U}_q^k(\mathbb{F}))) \supseteq \mathbb{U}_p(\mathbb{U}_p^k(f^{\Rightarrow}(\mathbb{F}))) = \mathbb{U}_p^{k+1}(f^{\Rightarrow}(\mathbb{F})).$$

(2) We check only the inequalities for $n = 1$.

Let $f$ be an interior function. For each $\lambda \in \mathbb{U}_q(\mathbb{F})$, i.e., $\widehat{\mathbb{U}_q}(\lambda) \in \mathbb{F}$ we have $f^{\rightarrow}(\widehat{\mathbb{U}_q}(\lambda)) \in f^{\Rightarrow}(\mathbb{F})$ and then $\widehat{\mathbb{U}_p}(f^{\rightarrow}(\lambda)) \in f^{\Rightarrow}(\mathbb{F})$ by $f$ is an interior function. That means $f^{\leftarrow}(\lambda) \in \mathbb{U}_p(f^{\Rightarrow}(\mathbb{F}))$. Thus $f^{\Rightarrow}(\mathbb{U}_q(\mathbb{F})) \subseteq \mathbb{U}_p(f^{\Rightarrow}(\mathbb{F}))$. $\square$

Now, we tend our attention to $p$-topologicalness.

We say a pair of $\top$-convergence spaces $(X, p, q)$ satisfy the Gähler $\top$-neighborhood condition if

$p$-**(TG)**: $\forall \mathbb{F} \in \mathbb{F}_L^{\top}(X), \forall x \in X, \mathbb{F} \xrightarrow{q} x \implies \mathbb{U}_p(\mathbb{F}) \xrightarrow{q} x$.

**Definition 8.** *Let* $(X, p, q)$ *be a pair of* $\top$-*convergence spaces. Then* $q$ *is called* $p$-*topological if the condition* $p$-*(TG) is satisfied.*

**Remark 3.** *When $L = \{\bot, \top\}$, the condition p-(TG) is precise the Gähler neighborhood condition in [4], which is used to define p-topological convergence spaces. Therefore, our p-topologicalness is a natural extension of crisp p-topologicalness.*

We say a pair of $\top$-convergence spaces $(X, p, q)$ satisfy the Fischer $\top$-diagonal condition if

$p$-**(TF)**: Let $J, X$ be any sets, $\psi : J \longrightarrow X$, and $\phi : J \longrightarrow \mathbb{F}_L^\top(X)$ such that $\phi(j) \xrightarrow{p} \psi(j)$, for each $j \in J$. Then for each $\mathbb{F} \in \mathbb{F}_L^\top(J)$ and each $x \in X$, $\psi^{\Rightarrow}(\mathbb{F}) \xrightarrow{q} x$ implies $k\phi\mathbb{F} \xrightarrow{q} x$.

Restricting $J = X$ and $\psi =$ id in $p$-**(TF)**, we obtain a weaker condition $p$-**(TK)**. When $p = q$, $p$-**(TF)** is precise the Fischer $\top$-diagonal condition **(TF)**, and $p$-**(TK)** is precise the Kowalsky $\top$-diagonal condition **(TK)** in [26].

**Proposition 4.** *Let $(X, p, q)$ be a pair of $\top$-convergence spaces. Then (1) p-(TF)$\Longrightarrow$p-(TP1)+p-(TK), and (2) p-(TK)$\Longrightarrow$p-(TF) if p satisfies (TP1).*

**Proof.** (1) Obviously, $p$-(TF)$\Longrightarrow p$-(TK). Now, we check $p$-(TF)$\Longrightarrow p$-(TP1). Let $\{\mathbb{F}_j\}_{j \in J} \subseteq \mathbb{F}_L^\top(X)$ and $x \in X$ satisfy $\forall j \in J, \mathbb{F}_j \xrightarrow{p} x$. Take $\psi(j) \equiv x, \phi(j) = \mathbb{F}_j$ and $\mathbb{F} = \mathbb{F}_\bot$ (i.e., $\mathbb{F}_\bot = \{\top_J\}$, the smallest $\top$-filter on $J$) in $p$-(TF), then it is easily seen that $\psi^{\Rightarrow}(\mathbb{F}_\bot) = [x]_\top$ and $k\phi\mathbb{F}_\bot = \bigcap_{j \in J} \mathbb{F}_j$. Because $\psi^{\Rightarrow}(\mathbb{F}_\bot) = [x]_\top \xrightarrow{q} x$ we have $k\phi\mathbb{F}_\bot = \bigcap_{j \in J} \mathbb{F}_j \xrightarrow{q} x$ by $p$-(TF).

(2) Let $J, X, \psi, \phi$ satisfy the condition of $p$-(TF). Then we define a function $\widetilde{\phi} : X \longrightarrow \mathbb{F}_L^\top(X)$ as $\widetilde{\phi}(x) = \bigcap\{\phi(j) : j \in J, \psi(j) = x\}$ if there exists $j \in J$ such that $\psi(j) = x$ and $\widetilde{\phi}(x) = [x]_\top$ if not so. For each $x \in X$, if $\widetilde{\phi}(x) = [x]_\top$ then $\widetilde{\phi}(x) \xrightarrow{p} x$. If $\widetilde{\phi}(x) = \bigcap\{\phi(j) : j \in J, \psi(j) = x\}$ then by $\phi(j) \xrightarrow{p} x$ and (TP1) we have $\widetilde{\phi}(x) \xrightarrow{p} x$. Let $\mathbb{F} \in \mathbb{F}_L^\top(J)$ and $\psi^{\Rightarrow}(\mathbb{F}) \xrightarrow{q} x$. Then by $p$-(TK) we obtain $k\widetilde{\phi}\psi^{\Rightarrow}(\mathbb{F}) \xrightarrow{q} x$. One can prove that $k\phi\mathbb{F} \supseteq k\widetilde{\phi}\psi^{\Rightarrow}(\mathbb{F})$. Thus $k\phi\mathbb{F} \xrightarrow{q} x$. $\square$

**Corollary 1.** *Let $(X, p, q)$ be a pair of $\top$-convergence spaces. If p satisfies (TP1) then p-(TF)$\Leftrightarrow$p-(TK)+p-(TP1). In particular, when $p = q$ we have (TF)$\Longleftrightarrow$(TK)+(TP1) [26].*

**Remark 4.** *Let $L, X$ and $\mathbb{F}_z(z \in X)$ be defined as in Example 3. Let q be defined as $\mathbb{F} \xrightarrow{q} z$ for any $\mathbb{F} \in \mathbb{F}_L^\top(X)$ and any $z \in X$, and let p be defined as $\mathbb{F} \xrightarrow{p} z \Longleftrightarrow \mathbb{F} \supset \mathbb{F}_\bot$. Then $(X, p, q)$ is a pair of $\top$-convergence spaces. Obviously, the axiom p-(TF) is satisfied. But p does not fulfill the axiom (TP1) since $\mathbb{U}_p(z) = \mathbb{F}_\bot \xrightarrow{p} z$. Thus this example shows that p-(TF) does not imply (TP1) of $(X, p)$ generally. Therefore, we guess that the additional condition (TP1) in the above corollary can not be removed.*

The following theorem shows that if we restricting the lattice-context slightly, $p$-topologicalness can be described by Fischer $\top$-diagonal condition $p$-**(TF)**.

**Theorem 1.** *Let $(X, p, q)$ be a pair of $\top$-convergence spaces. Then p-(TG)$\Longrightarrow$p-(TF), and the converse inclusion holds if L is continuous.*

**Proof.** $p$-**(TG)**$\Longrightarrow p$-**(TF)**. Let $J, X, \phi, \psi$ satisfy the condition of $p$-**(TF)**. For any $\mathbb{F} \in \mathbb{F}_L^\top(J)$, we prove below that $\mathbb{U}_p(\psi^\Rightarrow(\mathbb{F})) \subseteq k\phi\mathbb{F}$. Let $\lambda \in L^X$,

$$
\begin{aligned}
\bigvee_{\mu \in \mathbb{F}} S_X(\psi^\rightarrow(\mu), \widehat{\mathbb{U}_p}(\lambda)) &= \bigvee_{\mu \in \mathbb{F}} \bigwedge_{x \in X} ((\bigvee_{\psi(j)=x} \mu(j)) \rightarrow \bigvee_{v \in \mathbb{U}_p(x)} S_X(v, \lambda)) \\
&= \bigvee_{\mu \in \mathbb{F}} \bigwedge_{j \in J} (\mu(j) \rightarrow \bigvee_{v \in \mathbb{U}_p(\psi(j))} S_X(v, \lambda)), \text{ by } \phi(j) \xrightarrow{p} \psi(j) \\
&\leq \bigvee_{\mu \in \mathbb{F}} \bigwedge_{j \in J} (\mu(j) \rightarrow \bigvee_{v \in \phi(j)} S_X(v, \lambda)) \\
&= \bigvee_{\mu \in \mathbb{F}} \bigwedge_{j \in J} (\mu(j) \rightarrow \hat{\phi}(\lambda)(j)) = \bigvee_{\mu \in \mathbb{F}} S_J(\mu, \hat{\phi}(\lambda)).
\end{aligned}
$$

It follows that

$$
\lambda \in \mathbb{U}_p(\psi^\Rightarrow(\mathbb{F})) \implies \bigvee_{\mu \in \mathbb{F}} S_X(\psi^\rightarrow(\mu), \widehat{\mathbb{U}_p}(\lambda)) = \top \implies \bigvee_{\mu \in \mathbb{F}} S_J(\mu, \hat{\phi}(\lambda)) = \top \implies \lambda \in k\phi\mathbb{F}.
$$

Thus $\mathbb{U}_p(\psi^\Rightarrow(\mathbb{F})) \subseteq k\phi\mathbb{F}$.

If $\psi^\Rightarrow(\mathbb{F}) \xrightarrow{q} x$ then it follows by $p$-**(TG)** that $\mathbb{U}_p(\psi^\Rightarrow(\mathbb{F})) \xrightarrow{q} x$, and so $k\phi\mathbb{F} \xrightarrow{q} x$. That is, $p$-**(TF)** is satisfied.

$p$-**(TF)**$\Longrightarrow p$-**(TG)**. Note that Lemma 1 holds since $L$ is continuous. Take

$$
J = \{(\mathbb{G}, y) \in \mathbb{F}_L^\top(X) \times X | \mathbb{G} \xrightarrow{p} y\}; \psi : J \longrightarrow X, (\mathbb{G}, y) \mapsto y; \phi : J \longrightarrow \mathbb{F}_L^\top(X), (\mathbb{G}, y) \mapsto \mathbb{G}.
$$

Then $\forall j \in J, \phi(j) \xrightarrow{p} \psi(j)$. Because $[y] \xrightarrow{p} y$ we have that $\psi$ is a surjective function. Thus for each $\mathbb{F} \in \mathbb{F}_L^\top(X), \mathbb{H} = \psi^\Leftarrow(\mathbb{F}) \in \mathbb{F}_L^\top(J)$ exists and $\psi^\Rightarrow(\mathbb{H}) = \mathbb{F}$.

We prove below that $k\phi\mathbb{H} = \mathbb{U}_p(\mathbb{F})$. For any $y \in X$, denote $I_y = \{\mathbb{G} \in \mathbb{F}_L^\top(X) | \mathbb{G} \xrightarrow{p} y\}$. Then for any $\lambda \in L^X$,

$$
\begin{aligned}
\bigvee_{\mu \in \mathbb{F}} S_J(\psi^\Leftarrow(\mu), \hat{\phi}(\lambda)) &= \bigvee_{\mu \in \mathbb{F}} \bigwedge_{(\mathbb{G},y) \in J} (\psi^\Leftarrow(\mu)(\mathbb{G}, y) \rightarrow \hat{\phi}(\lambda)(\mathbb{G}, y)) \\
&= \bigvee_{\mu \in \mathbb{F}} \bigwedge_{y \in X} \bigwedge_{\mathbb{G} \in I_y} (\mu(y) \rightarrow \Lambda(\mathbb{G})(\lambda)) \\
&= \bigvee_{\mu \in \mathbb{F}} \bigwedge_{y \in X} (\mu(y) \rightarrow \bigwedge_{\mathbb{G} \in I_y} \Lambda(\mathbb{G})(\lambda)), \text{ by Lemma 1} \\
&= \bigvee_{\mu \in \mathbb{F}} \bigwedge_{y \in X} (\mu(y) \rightarrow \Lambda(\bigcap_{\mathbb{G} \in I_y} \mathbb{G})(\lambda)) \\
&= \bigvee_{\mu \in \mathbb{F}} \bigwedge_{y \in X} (\mu(y) \rightarrow \Lambda(\mathbb{U}_p(y))(\lambda)) \\
&= \bigvee_{\mu \in \mathbb{F}} \bigwedge_{y \in X} (\mu(y) \rightarrow \widehat{\mathbb{U}_p}(\lambda)(y)) = \bigvee_{\mu \in \mathbb{F}} S_X(\mu, \widehat{\mathbb{U}_p}(\lambda)).
\end{aligned}
$$

It follows that

$$
\lambda \in k\phi\mathbb{H} \iff \hat{\phi}(\lambda) \in \psi^\Leftarrow(\mathbb{F}) \iff \bigvee_{\mu \in \mathbb{F}} S_J(\psi^\Leftarrow(\mu), \hat{\phi}(\lambda)) = \top \iff \bigvee_{\mu \in \mathbb{F}} S_X(\mu, \widehat{\mathbb{U}_p}(\lambda)) = \top
$$

$$
\iff \widehat{\mathbb{U}_p}(\lambda) \in \mathbb{F} \iff \lambda \in \mathbb{U}_p(\mathbb{F}).
$$

Thus $k\phi\mathbb{H} = \mathbb{U}_p(\mathbb{F})$.

Let $\mathbb{F} = \psi^{\Rightarrow}(\mathbb{H}) \xrightarrow{q} x$. Then by $p$-**(TF)** we have $k\phi\mathbb{H} = \mathbb{U}_p(\mathbb{F}) \xrightarrow{q} x$. That is, $p$-**(TG)** holds. $\square$

The following theorem shows that for pretopological $\mathsf{T}$-convergence spaces, $p$-topologicalness can be described by Fischer $\mathsf{T}$-diagonal condition $p$-**(TF)**.

**Theorem 2.** *Let $(X, p, q)$ be a pair of $\mathsf{T}$-convergence spaces and $(X, p)$ be pretopological. Then $p$-(TF)$\Longleftrightarrow p$-(TG).*

**Proof.** Most of the proof can copy that of Theorem 1. We only check that

$$\bigvee_{\mu\in\mathbb{F}} S_J(\psi^{\leftarrow}(\mu), \hat{\phi}(\lambda)) = \bigvee_{\mu\in\mathbb{F}} S_X(\mu, \widehat{\mathbb{U}_p}(\lambda)),$$

for any $\lambda \in L^X$ in $p$-**(TF)**$\Longrightarrow p$-**(TG)**. Indeed, since $p$ is pretopological then $\mathbb{U}_p(y) \in I_y$ for any $y \in X$. Thus

$$
\begin{aligned}
\bigvee_{\mu\in\mathbb{F}} S_J(\psi^{\leftarrow}(\mu), \hat{\phi}(\lambda)) &= \bigvee_{\mu\in\mathbb{F}} \bigwedge_{(\mathbb{G},y)\in J} (\psi^{\leftarrow}(\mu)(\mathbb{G},y) \to \hat{\phi}(\lambda)(\mathbb{G},y)) \\
&= \bigvee_{\mu\in\mathbb{F}} \bigwedge_{y\in X} \bigwedge_{\mathbb{G}\in I_y} (\mu(y) \to \Lambda(\mathbb{G})(\lambda)) \\
&= \bigvee_{\mu\in\mathbb{F}} \bigwedge_{y\in X} (\mu(y) \to \bigwedge_{\mathbb{G}\in I_y} \Lambda(\mathbb{G})(\lambda)), \text{ by } \mathbb{U}_p(y) \in I_y \\
&= \bigvee_{\mu\in\mathbb{F}} \bigwedge_{y\in X} (\mu(y) \to \Lambda(\mathbb{U}_p(y))(\lambda)) \\
&= \bigvee_{\mu\in\mathbb{F}} S_X(\mu, \widehat{\mathbb{U}_p}(\lambda)). \quad \square
\end{aligned}
$$

By Corollary 1 and Theorem 2 we get the following corollary.

**Corollary 2.** *[34] Let $(X, p)$ be a $\mathsf{T}$-convergence space. Then (TF)$\Longleftrightarrow$(TG).*

**Remark 5.** *The above corollary is one of the main results in [34]. Based on this equivalence, it was proved that $\mathsf{T}$-convergence spaces with (TF) or (TG) characterize precisely the conical L-topological spaces in [44].*

The following theorem shows that $p$-topologicalness is preserved under initial constructions.

**Theorem 3.** *Let $\{(X_i, q_i, p_i)\}_{i\in I}$ be pairs of $\mathsf{T}$-convergence spaces with each $q_i$ being $p_i$-topological. If $q$ (resp., $p$) is the initial structure on $X$ relative to the source $(X \xrightarrow{f_i} (X_i, q_i))_{i\in I}$ (resp., $(X \xrightarrow{f_i} (X_i, p_i))_{i\in I}$), then $(X, q)$ is $p$-topological.*

**Proof.** Let $\mathbb{F} \xrightarrow{q} x$. Then by definition of $q$, we have $f_i^{\Rightarrow}(\mathbb{F}) \xrightarrow{q_i} f_i(x)$ for any $i \in I$. Because $q_i$ is $p_i$-topological we have $\mathbb{U}_{p_i}(f_i^{\Rightarrow}(\mathbb{F})) \xrightarrow{q_i} f_i(x)$. Then by Proposition 3 (1) we have $f_i^{\Rightarrow}(\mathbb{U}_p(\mathbb{F})) \supseteq \mathbb{U}_{p_i}(f_i^{\Rightarrow}(\mathbb{F}))$ and so $f_i^{\Rightarrow}(\mathbb{U}_p(\mathbb{F})) \xrightarrow{q_i} f_i(x)$ for all $i \in I$. That is, $\mathbb{U}_p(\mathbb{F}) \xrightarrow{q} x$. Thus $q$ is $p$-topological. $\square$

The next theorem shows that $p$-topologicalness is preserved under final constructions with some additional conditions.

**Theorem 4.** *Let $\{(X_i, q_i, p_i)\}_{i\in I}$ be pairs of $\mathsf{T}$- convergence spaces with each $q_i$ being $p_i$-topological. Let $q$ (resp., $p$) be the final structure on $X$ w.r.t. The sink $((X_i, q_i) \xrightarrow{f_i} X)_{i\in I}$ (resp., $((X_i, p_i) \xrightarrow{f_i} X)_{i\in I}$). If $X = \cup_{i\in I} f_i(X_i)$ and each $f_i : (X_i, p_i) \longrightarrow (X, p)$ is an interior function, then $(X, q)$ is $p$-topological.*

**Proof.** Let $\mathbb{F} \xrightarrow{q} x$. Then by definition of $q$, there exists $i \in I, x_i \in X_i, \mathbb{G}_i \in \mathbb{F}_L^\top(X_i)$ such that $f_i(x_i) = x, f_i^\Rightarrow(\mathbb{G}_i) \subseteq \mathbb{F}$ and $\mathbb{G}_i \xrightarrow{q_i} x_i$.

By $f_i^\Rightarrow(\mathbb{G}_i) \subseteq \mathbb{F}$ and $f_i$ is a interior function we have $f_i^\Rightarrow(\mathbb{U}_{p_i}(\mathbb{G}_i)) \subseteq \mathbb{U}_p(f_i^\Rightarrow(\mathbb{G}_i)) \subseteq \mathbb{U}_p(\mathbb{F})$.

By $\mathbb{G}_i \xrightarrow{q_i} x_i$ and $q_i$ is $p_i$-topological we have $\mathbb{U}_{p_i}(\mathbb{G}_i) \xrightarrow{q_i} x_i$, and then $f_i^\Rightarrow(\mathbb{U}_{p_i}(\mathbb{G}_i)) \xrightarrow{q} f_i(x_i) = x$.

Then it follows that $\mathbb{U}_p(\mathbb{F}) \xrightarrow{q} x$. By Theorem 1 we get that $q$ is $p$-topological. $\square$

From Theorem 3 and Theorem 4, we conclude easily the following corollary. It will tell us that $p$-topologicalness is preserved under supremum and infimum in the lattice $\top(X)$.

**Corollary 3.** *Let $\{q_i | i \in I\} \subseteq \top(X)$ and $p \in \top(X)$ such that each $(X, q_i)$ is $p$-topological. Then both $(X, \inf\{q_i\}_{i \in I})$ and $(X, \sup\{q_i\}_{i \in I})$ are all $p$-topological.*

*3.3. On the Relationship between p-Topologicalness in $\top$-Convergence Spaces and in Stratified L-Generalized Convergence Spaces*

Let $J, X$ be any set and $\Phi : J \longrightarrow \mathcal{F}_L^s(X)$ be any function. Then a function $\hat{\Phi} : L^X \to L^J$ is defined as $\forall \lambda \in L^X, \forall j \in J, \hat{\Phi}(\lambda)(j) = \Phi(j)(\lambda)$. For all $\mathcal{F} \in \mathcal{F}_L^s(J)$, it is proved that the function $K\Phi\mathcal{F} : L^X \longrightarrow L$ defined by $\forall \lambda \in L^X, K\Phi\mathcal{F}(\lambda) = \mathcal{F}(\hat{\Phi}(\lambda))$ is a stratified $L$-filter, which is called the diagonal $L$-filter of $\mathcal{F}$ under $\Phi$ [27,30].

Let $(X, \lim^p)$ be a stratified $L$-generalized convergence space. For any $\alpha \in L, x \in X$, let $\mathcal{U}_p^\alpha(x) = \bigwedge\{\mathcal{F} : \lim^p \mathcal{F}(x) \geq \alpha\}$. Take $\Phi = \mathcal{U}_p^\alpha : X \longrightarrow \mathcal{F}_L^s(X)$, then for each $\mathcal{F} \in \mathcal{F}_L^s(X)$, the stratified $L$-filter $\mathcal{U}_p^\alpha(\mathcal{F}) := k\mathcal{U}_p^\alpha\mathcal{F}$ is called $\alpha$-level neighborhood $L$-filter of $\mathcal{F}$ w.r.t. $\lim^p$ [29].

We say a pair of stratified $L$-generalized convergence spaces $(X, \lim^p, \lim^q)$ satisfy the Fischer $L$-diagonal condition if

$p$-**(LF)**: Let $J, X$ be any sets, $\Psi : J \longrightarrow X$ and $\Phi : J \longrightarrow \mathcal{F}_L^s(X)$ be functions.

$$\forall \mathcal{F} \in \mathcal{F}_L^s(J), \forall x \in X, \lim^q \Psi^\Rightarrow(\mathcal{F})(x) \wedge \bigwedge_{j \in J} \lim^p \Phi(j)(\Psi(j)) \leq \lim^q K\Phi\mathcal{F}(x).$$

We say a pair of stratified $L$-generalized convergence spaces $(X, \lim^p, \lim^q)$ satisfy the Gähler $L$-neighborhood condition if $p$-(LG): $\forall \alpha \in L, \forall \mathcal{F} \in \mathcal{F}_L^s(X), \alpha * \lim^q \mathcal{F} \leq \lim^q \mathcal{U}_p^\alpha(\mathcal{F})$.

It was proved in [32] that $p$-(LF)$\Longleftrightarrow p$-(LG).

**Definition 9** ([32])**.** *Let $(X, \lim^p, \lim^q)$ be a pair of stratified $L$-generalized convergence spaces. Then $\lim^q$ is called $p$-topological if the condition $p$-(LF) or $p$-(LG) is satisfied.*

**Lemma 2.** *Let $\phi : J \longrightarrow \mathbb{F}_L^\top(X)$ be any function and $\mathbb{F} \in \mathbb{F}_L^\top(J)$. Then*
*(1) $\Lambda(k\phi\mathbb{F}) \leq K(\Lambda \circ \phi)\Lambda(\mathbb{F})$;*
*(2) $k\phi\mathbb{F} = \Gamma(K(\Lambda \circ \phi)\Lambda(\mathbb{F}))$.*

**Proof.** (1) Let $\lambda \in L^X$. Then for any $j \in J, \hat{\phi}(\lambda)(j) = \Lambda(\phi(j))(\lambda) = (\Lambda \circ \phi)(j)(\lambda) = \widehat{\Lambda \circ \phi}(\lambda)(j)$. It follows

$$
\begin{aligned}
\Lambda(k\phi\mathbb{F})(\lambda) &= \bigvee_{\mu \in k\phi\mathbb{F}} S_X(\mu, \lambda) = \bigvee_{\hat{\phi}(\mu) \in \mathbb{F}} S_X(\mu, \lambda) \\
&\leq \bigvee_{\hat{\phi}(\mu) \in \mathbb{F}} S_J(\hat{\phi}(\mu), \hat{\phi}(\lambda)) \leq \bigvee_{\nu \in \mathbb{F}} S_J(\nu, \hat{\phi}(\lambda)) \\
&= \bigvee_{\nu \in \mathbb{F}} S_J(\nu, \widehat{\Lambda \circ \phi}(\lambda)) = \Lambda(\mathbb{F})(\widehat{\Lambda \circ \phi}(\lambda)) = K(\Lambda \circ \phi)\Lambda(\mathbb{F})(\lambda).
\end{aligned}
$$

(2) Let $\lambda \in L^X$. Then

$$\lambda \in k\phi\mathbb{F} \iff \widehat{\Lambda \circ \phi}(\lambda) \in \mathbb{F} \iff \widehat{\Lambda \circ \phi}(\lambda) \in \mathbb{F} \iff \Lambda(\mathbb{F})(\widehat{\Lambda \circ \phi}(\lambda)) = \top$$
$$\iff K(\Lambda \circ \phi)\Lambda(\mathbb{F})(\lambda) = \top \iff \lambda \in \Gamma(K(\Lambda \circ \phi)\Lambda(\mathbb{F})). \quad \square$$

**Theorem 5.** *Let $(X, p, q)$ be pair of $\top$-convergence spaces and $L$ be continuous. Then $\lim^q$ is $p$-topological iff $q$ is $p$-topological.*

**Proof.** Let $q$ be $p$-topological. We check that $(X, \lim^p, \lim^q)$ satisfies $p$-**(LG)**. Obviously, we need only prove that $\lim^q \mathcal{F}(x) = \top$ implies $\lim^q \mathcal{U}_p^\alpha(\mathcal{F})(x) = \top$ for any $\alpha \neq \bot$.

Note that for any $\alpha \neq \bot$ and any $x \in X$ we have

$$\mathcal{U}_p^\alpha(x) = \bigwedge\{\mathcal{F}|\lim^p \mathcal{F}(x) = \top\} = \bigwedge\{\mathcal{F}|\mathcal{F} \geq \Lambda(\mathbb{F}), \mathbb{F} \xrightarrow{p} x\}$$
$$= \bigwedge\{\Lambda(\mathbb{F})|\mathbb{F} \xrightarrow{p} x\} \overset{\text{Lemma}1}{=} \Lambda(\bigcap \mathbb{F}\{|\mathbb{F} \xrightarrow{p} x\}) = \Lambda(\mathbb{U}_p(x)).$$

Let $\lim^q \mathcal{F}(x) = \top$ then $\mathcal{F} \geq \Lambda(\mathbb{F})$ for some $\mathbb{F} \xrightarrow{q} x$. It follows by $p$-**(TG)** that $\mathbb{U}_p(\mathbb{F}) \xrightarrow{q} x$ and

$$\mathcal{U}_p^\alpha(\mathcal{F}) \geq \mathcal{U}_p^\alpha(\Lambda(\mathbb{F})) = K\mathcal{U}_p^\alpha\Lambda(\mathbb{F}) = K(\Lambda \circ \mathbb{U}_p)\Lambda(\mathbb{F}) \overset{\text{Lemma}2(1)}{\geq} \Lambda(k\mathbb{U}_p\mathbb{F}) = \Lambda(\mathbb{U}_p(\mathbb{F})),$$

and so $\lim^q \mathcal{U}_p^\alpha(\mathcal{F})(x) = \top$ as desired.

Conversely, let $\lim^q$ be $p$-topological. We check that $(X, p, q)$ satisfies $p$-**(TG)**.

Assume that $\mathbb{F} \xrightarrow{q} x$. It follows by $p$-**(LG)** that

$$\lim^q \mathcal{U}_p^\top(\Lambda(\mathbb{F}))(x) = \lim^q K\mathcal{U}_p^\top \Lambda(\mathbb{F}) = \lim^q K(\Lambda \circ \mathbb{U}_p)\Lambda(\mathbb{F}) = \top,$$

and then $K(\Lambda \circ \mathbb{U}_p)\Lambda(\mathbb{F}) \geq \Lambda(\mathbb{G})$ for some $\mathbb{G} \xrightarrow{q} x$. By Lemma 2(2) we have

$$k\mathbb{U}_p\mathbb{F} = \Gamma(K(\Lambda \circ \mathbb{U}_p)\Lambda(\mathbb{F})) \supseteq \Gamma\Lambda(\mathbb{G}) = \mathbb{G}.$$

So, $\mathbb{U}_p(\mathbb{F}) = k\mathbb{U}_p\mathbb{F} \xrightarrow{q} x$ as desired. $\quad \square$

## 4. Lower and Upper $p$-Topological Modifications in $\top$-Convergence Spaces

In this section, we shall discuss the $p$-topological modification in $\top$-convergence spaces.

At first, we fix a lemma for later use. The proof is obvious, so we omit it.

**Lemma 3.** (1) *If $(X, q)$ is p-topological, then $\mathbb{F} \xrightarrow{q} x$ implies $\mathbb{U}_p^n(\mathbb{F}) \xrightarrow{q} x$ for any $n \in \mathbb{N}$.*
(2) *If $(X, q)$ is p-topological, then $(X, q)$ is $p'$-topological for any $p \leq p'$.*
(3) *$(X, \iota)$ is p-topological for any $p \in \top(X)$.*

### 4.1. Lower $p$-Topological Modification

Corollary 3 shows that $p$-topologicalness is preserved under supremum in the lattice $\top(X)$. Lemma 3(3) shows that the indiscrete space $(X, \iota)$ is $p$-topological for any $p \in \top(X)$. These two results make the following definition available.

**Definition 10.** *Let $(X, p, q)$ be a pair of $\top$-convergence spaces. Then there is a finest p-topological $\top$-convergence structure $\tau_{pq}$ on $X$ which is coarser than $q$. The structure $\tau_{pq}$ is called the lower p-topological modification of $q$.*

The next theorem gives a direct characterization on lower $p$-topological modification.

**Theorem 6.** *Let $p, q \in \mathsf{T}(X)$. Then $\mathbb{F} \xrightarrow{\tau_p q} x \iff \exists n \in \mathbb{N}, \mathbb{G} \xrightarrow{q} x$ s.t. $\mathbb{F} \supseteq \mathbb{U}_p^n(\mathbb{G})$.*

**Proof.** Let $q'$ be defined as $\mathbb{F} \xrightarrow{q'} x \iff \exists n \in \mathbb{N}, \mathbb{G} \xrightarrow{q} x$ s.t. $\mathbb{F} \supseteq \mathbb{U}_p^n(\mathbb{G})$. We need only check that $\tau_p q = q'$.

It is obvious that $q' \in \mathsf{T}(X)$ and $q' \le q$. We prove that $q'$ is $p$-topological. Indeed, let $\mathbb{F} \xrightarrow{q'} x$. Then there exists $n \in \mathbb{N}, \mathbb{G} \xrightarrow{q} x$ such that $\mathbb{F} \supseteq \mathbb{U}_p^n(\mathbb{G})$. It follows that $\mathbb{U}_p(\mathbb{F}) \supseteq \mathbb{U}_p(\mathbb{U}_p^n(\mathbb{G})) = \mathbb{U}_p^{n+1}(\mathbb{G})$ and so $\mathbb{U}_p(\mathbb{F}) \xrightarrow{q'} x$, as desired. Thus $q'$ is $p$-topological.

Let $(X, r)$ be $p$-topological and $r \le q$. We prove below $r \le q'$. Indeed, let $\mathbb{F} \xrightarrow{q'} x$. Then there exists $n \in \mathbb{N}, \mathbb{G} \xrightarrow{q} x$ such that $\mathbb{F} \supseteq \mathbb{U}_p^n(\mathbb{G})$, and then $\mathbb{G} \xrightarrow{r} x$ by $q \le r$. Since $r$ is $p$-topological it follows by Lemma 3(1) we have $\mathbb{F} \supseteq \mathbb{U}_p^n(\mathbb{G}) \xrightarrow{r} x$. Thus $r \le q'$. $\square$

**Theorem 7.** *Let $f : (X, q) \longrightarrow (X', q')$ and $f : (X, p) \longrightarrow (X', p')$ be continuous function between $\mathsf{T}$-convergence spaces. Then $f : (X, \tau_p q) \longrightarrow (X', \tau_{p'} q')$ is also continuous.*

**Proof.** For any $\mathbb{F} \in \mathbb{F}_L^\top(X)$ and $x \in X$.

$$
\begin{aligned}
\mathbb{F} \xrightarrow{\tau_p q} x &\implies \exists n \in \mathbb{N}, \mathbb{G} \xrightarrow{q} x \text{ s.t. } \mathbb{F} \supseteq \mathbb{U}_p^n(\mathbb{G}) \\
&\implies \exists n \in \mathbb{N}, f^\Rightarrow(\mathbb{G}) \xrightarrow{q'} f(x) \text{ s.t. } f^\Rightarrow(\mathbb{F}) \supseteq f^\Rightarrow(\mathbb{U}_p^n(\mathbb{G})) \\
&\implies \exists n \in \mathbb{N}, f^\Rightarrow(\mathbb{G}) \xrightarrow{q'} f(x) \text{ s.t. } f^\Rightarrow(\mathbb{F}) \supseteq \mathbb{U}_{p'}^n(f^\Rightarrow(\mathbb{G})) \\
&\implies f^\Rightarrow(\mathbb{F}) \xrightarrow{\tau_{p'} q'} (f(x)),
\end{aligned}
$$

where the second implication holds for $f : (X, q) \longrightarrow (X', q')$ being continuous, and the third implication holds by $f : (X, p) \longrightarrow (X', p')$ being continuous and Proposition 3(1). $\square$

The next theorem shows that lower $p$-topological modification behaves reasonably well relative to final structures.

**Theorem 8.** *Let $\{(X_i, q_i, p_i)\}_{i \in I}$ be pairs of spaces in $\mathsf{T}$-CS and let $q$ be the final structure w.r.t. The sink $((X_i, q_i) \xrightarrow{f_i} X)_{i \in I}$ with $X = \cup_{i \in I} f_i(X_i)$. If $(X, p)$ is in $\mathsf{T}$-CS such that each $f_i : (X_i, p_i) \longrightarrow (X, p)$ is a continuous interior function, then $\tau_p q$ is the final structure w.r.t. the sink $((X_i, \tau_{p_i} q_i) \xrightarrow{f_i} X)_{i \in I}$.*

**Proof.** Let $s$ denote the final structure w.r.t. The sink $((X_i, \tau_{p_i} q_i) \xrightarrow{f_i} X)_{i \in I}$. Let $\mathbb{F} \in \mathbb{F}_L^\top(X)$ and $x \in X$. Then

$$
\begin{aligned}
\mathbb{F} \xrightarrow{s} x &\implies \exists i \in I, x_i \in X_i, f_i(x_i) = x, \mathbb{G}_i \xrightarrow{\tau_{p_i} q_i} x_i \text{ s.t. } f_i^\Rightarrow(\mathbb{G}_i) \subseteq \mathbb{F}, \text{by Theorem 6} \\
&\implies \exists i \in I, x_i \in X_i, n \in \mathbb{N}, f_i(x_i) = x, \mathbb{H}_i \xrightarrow{q_i} x_i \text{ s.t. } \mathbb{U}_{p_i}^n(\mathbb{H}_i) \subseteq \mathbb{G}_i, f_i^\Rightarrow(\mathbb{G}_i) \subseteq \mathbb{F}, \text{Proposition 3(1)} \\
&\implies \exists i \in I, x_i \in X_i, n \in \mathbb{N}, f_i^\Rightarrow(\mathbb{H}_i) \xrightarrow{q} x \text{ s.t. } \mathbb{U}_p^n(f_i^\Rightarrow(\mathbb{H}_i)) \subseteq f_i^\Rightarrow(\mathbb{U}_{p_i}^n(\mathbb{H}_i)) \subseteq f_i^\Rightarrow(\mathbb{G}_i), f_i^\Rightarrow(\mathbb{G}_i) \subseteq \mathbb{F} \\
&\implies \exists i \in I, x_i \in X_i, n \in \mathbb{N}, f_i^\Rightarrow(\mathbb{H}_i) \xrightarrow{q} x \text{ s.t. } \mathbb{U}_p^n(f_i^\Rightarrow(\mathbb{H}_i)) \subseteq \mathbb{F} \\
&\implies \mathbb{F} \xrightarrow{\tau_p q} x.
\end{aligned}
$$

Conversely,

$$\mathbb{F} \xrightarrow{\tau_p q} x \implies \exists n \in \mathbb{N}, \mathbb{G} \xrightarrow{q} x \text{ s.t. } \mathbb{U}_p^n(\mathbb{G}) \subseteq \mathbb{F}$$

$$\implies \exists i \in I, x_i \in X_i, f_i(x_i) = x, n \in \mathbb{N}, \mathbb{H}_i \xrightarrow{q_i} x_i \text{ s.t. } f_i^{\Rightarrow}(\mathbb{H}_i) \subseteq \mathbb{G}, \mathbb{U}_p^n(\mathbb{G}) \subseteq \mathbb{F}$$

$$\implies \exists i \in I, x_i \in X_i, f_i(x_i) = x, n \in \mathbb{N}, \mathbb{H}_i \xrightarrow{q_i} x_i \text{ s.t. } \mathbb{U}_p^n(f_i^{\Rightarrow}(\mathbb{H}_i)) \subseteq \mathbb{U}_p^n(\mathbb{G}), \mathbb{U}_p^n(\mathbb{G}) \subseteq \mathbb{F}$$

$$\implies \exists i \in I, x_i \in X_i, f_i(x_i) = x, n \in \mathbb{N}, \mathbb{H}_i \xrightarrow{q_i} x_i \text{ s.t. } f_i^{\Rightarrow}(\mathbb{U}_{p_i}^n(\mathbb{H}_i)) \subseteq \mathbb{U}_p^n(f_i^{\Rightarrow}(\mathbb{H}_i)) \subseteq \mathbb{F}$$

$$\implies \exists i \in I, x_i \in X_i, f_i(x_i) = x, n \in \mathbb{N}, \mathbb{U}_{p_i}^n(\mathbb{H}_i) \xrightarrow{\tau_{p_i} q_i} x_i \text{ s.t. } f_i^{\Rightarrow}(\mathbb{U}_{p_i}^n(\mathbb{H}_i)) \subseteq \mathbb{F}$$

$$\implies \mathbb{F} \xrightarrow{s} x,$$

where the fourth implication uses Proposition 3(2). $\square$

The following corollary shows that lower $p$-topological modification behaves reasonably well relative to infimum in the lattice $\mathsf{T}(X)$.

**Corollary 4.** *Let $\{q_i | i \in I\} \subseteq \mathsf{T}(X)$, $p \in \mathsf{T}(X)$ and $q = \inf\{q_i | i \in I\}$. Then $\tau_p q = \inf\{\tau_p q_i | i \in I\}$.*

At last, we give the notion of topological modification. By Corollary 3, it is observed that topologicalness is preserved under supremum in the lattice $\mathsf{T}(X)$. Since the indiscrete space is topological, the following notion is available.

**Definition 11.** *Let $(X, q)$ be a $\mathsf{T}$-convergence space. Then there exists a finest topological $\mathsf{T}$-convergence structure $\tau q$ which is coarser than $q$. The structure $\tau q$ is called the topological modification of $(X, q)$. Indeed, $\tau q = \sup\{p | p \leq q \text{ and } p \text{ is topological}\}$.*

*4.2. Upper $p$-Topological Modification*

Note that for an arbitrary $p \in \mathsf{T}(X)$, the discrete space $(X, \delta)$ is generally not $p$-topological. Thus for a given $q \in \mathsf{T}(X)$, there may not exist $p$-topological $\mathsf{T}$-convergence structure on $X$ which is finer than $q$.

**Definition 12.** *Let $(X, p, q)$ be a pair of $\mathsf{T}$-convergence spaces. If there exists a coarsest $p$-topological $\mathsf{T}$-convergence structure $\tau^p q$ on $X$ which is finer than $q$, then it is called the upper $p$-topological modification of $q$.*

From Corollary 3 we easily conclude that the existence of $\tau^p q$ depends on the existence of a $p$-topological $\mathsf{T}$-convergence structure on $X$ which is finer than $q$. Additionally, note that $\tau_p \delta$ is the finest $p$-topological $\mathsf{T}$-convergence structure on $X$. Then it follows immediately that $\tau^p q$ exists if and only if $q \leq \tau_p \delta$. Using Theorem 6, this result can be stated as below.

**Theorem 9.** *Let $(X, p, q)$ be a pair of $\mathsf{T}$-convergence spaces. Then $\tau^p q$ exists if and only if $\mathbb{U}_p^n([x]_{\mathsf{T}}) \xrightarrow{q} x$ for all $x \in X, n \in \mathbb{N}$.*

**Proof.** For each $\mathbb{F} \in \mathbb{F}_L^{\mathsf{T}}(X)$ and each $x \in X$, by Theorem 6 we have

$$\mathbb{F} \xrightarrow{\tau_p \delta} x \iff \exists n \in \mathbb{N}, \mathbb{G} \xrightarrow{\delta} x \text{ s.t. } \mathbb{U}_p^n(\mathbb{G}) \subseteq \mathbb{F}.$$

*Necessity.* Let $\tau^p q$ exist. Then $q \leq \tau_p \delta$. It follows that for all $x \in X$, $n \in \mathbb{N}$

$$[x]_\top \xrightarrow{\delta} x \Longrightarrow \mathbb{U}_p^n([x]_\top) \xrightarrow{\tau_p \delta} x \Longrightarrow \mathbb{U}_p^n([x]_\top) \xrightarrow{q} x.$$

*Sufficiency.* Let $\mathbb{U}_p^n([x]_\top) \xrightarrow{q} x$ for all $x \in X$, $n \in \mathbb{N}$. Then for all $\mathbb{F} \in \mathbb{F}_L^\top(X)$ we have

$$
\begin{aligned}
\mathbb{F} \xrightarrow{\tau_p \delta} x \quad &\Longrightarrow \quad \exists n \in \mathbb{N}, \mathbb{G} \xrightarrow{\delta} x \text{ s.t. } \mathbb{U}_p^n(\mathbb{G}) \subseteq \mathbb{F} \\
&\Longrightarrow \quad \exists n \in \mathbb{N}, [x]_\top \subseteq \mathbb{G} \text{ s.t. } \mathbb{U}_p^n(\mathbb{G}) \subseteq \mathbb{F} \\
&\stackrel{\text{Proposition2}(2)}{\Longrightarrow} \quad \exists n \in \mathbb{N}, \mathbb{U}_p^n(([x]_\top) \subseteq \mathbb{U}_p^n(\mathbb{G}) \text{ s.t. } \mathbb{U}_p^n(\mathbb{G}) \subseteq \mathbb{F} \\
&\Longrightarrow \quad \exists n \in \mathbb{N} \text{ s.t. } \mathbb{U}_p^n(([x]_\top) \subseteq \mathbb{F} \\
&\Longrightarrow \quad \mathbb{F} \xrightarrow{q} x.
\end{aligned}
$$

It follows that $q \leq \tau_p \delta$, which means that $\tau^p q$ exists. $\quad\square$

The next theorem gives a direct characterization on upper $p$-topological modification whenever it exists.

**Theorem 10.** *Let $(X, p, q)$ be a pair of $\top$-convergence spaces. If $\tau^p q$ exists, then $\mathbb{F} \xrightarrow{\tau^p q} x \iff \forall n \in \mathbb{N}, \mathbb{U}_p^n(\mathbb{F}) \xrightarrow{q} x$.*

**Proof.** Let $q'$ be defined as $\mathbb{F} \xrightarrow{q'} x \iff \forall n \in \mathbb{N}, \mathbb{U}_p^n(\mathbb{F}) \xrightarrow{q} x$.

(1) $q' \in \top(X)$.

(TC1) Let $x \in X$. Then by Theorem 9 we have $\mathbb{U}_p^n([x]_\top) \xrightarrow{q} x$ for all $n \in \mathbb{N}$, which means $[x]_\top \xrightarrow{q'} x$.
(TC2) It is obvious.

(2) $q \leq q'$. Indeed, let $\mathbb{F} \xrightarrow{q'} x$ then $\mathbb{F} = \mathbb{U}_p^0(\mathbb{F}) \xrightarrow{q} x$.

(3) $(X, q')$ is $p$-topological. Indeed, let $\mathbb{F} \xrightarrow{q'} x$. Then for any $n \in \mathbb{N}$ we have $\mathbb{U}_p^n(\mathbb{U}_p(\mathbb{F})) = \mathbb{U}_p^{n+1}(\mathbb{F}) \xrightarrow{q} x$, which means $\mathbb{U}_p(\mathbb{F}) \xrightarrow{q'} x$. Thus $(X, q')$ is $p$-topological.

(4) Let $(X, r)$ be $p$-topological and $q \leq r$. Then $q' \leq r$. Indeed, let $\mathbb{F} \xrightarrow{r} x$ then for any $n \in \mathbb{N}$, by Proposition 3(1) we have $\mathbb{U}_p^n(\mathbb{F}) \xrightarrow{r} x$ and so $\mathbb{U}_p^n(\mathbb{F}) \xrightarrow{q} x$ by $q \leq r$. That means $\mathbb{F} \xrightarrow{q'} x$.

(1)–(4) show that $q'$ is the coarsest $p$-topological $\top$-convergence structure on $X$ which is finer than $q$. Thus $\tau^p q = q'$. $\quad\square$

**Theorem 11.** *Let $f : (X, q) \longrightarrow (X', q')$ be a continuous function, and $f : (X, p) \longrightarrow (X', p')$ be an interior function between $\top$-convergence spaces. If $\tau^p q$ and $\tau^{p'} q'$ exist then $f : (X, \tau^p q) \longrightarrow (X', \tau^{p'} q')$ is also continuous.*

**Proof.** Let $\mathbb{F} \xrightarrow{\tau^p q} x$. Then $\forall n \in \mathbb{N}, \mathbb{U}_p^n(\mathbb{F}) \xrightarrow{q} x$. Because $f : (X, q) \longrightarrow (X', q')$ is a continuous function and $f : (X, p) \longrightarrow (X', p')$ is an interior function we have

$$\forall n \in \mathbb{N}, \mathbb{U}_{p'}^n(f^\Rightarrow(\mathbb{F})) \supseteq f^\Rightarrow(\mathbb{U}_p^n(\mathbb{F})) \xrightarrow{q'} f(x),$$

which means $f^\Rightarrow(\mathbb{F}) \xrightarrow{\tau^{p'} q'} f(x)$ as desired. $\quad\square$

The next theorem shows that the upper $p$-topological modification exhibits comparable behavior relative to initial structures.

**Theorem 12.** *Let $\{(X_i, q_i, p_i)\}_{i \in I}$ be pairs of spaces in $\top$-**CS** and $q$ be the initial structure w.r.t. The source $(X \xrightarrow{f_i} (X_i, q_i))_{i \in I}$. Let $(X, p)$ be in $\top$-**CS** such that each $f_i : (X, p) \longrightarrow (X_i, p_i)$ is continuous interior function. If $\tau^{p_i} q_i$ exists for all $i \in I$, then $\tau^p q$ exists and is the initial structure w.r.t. The source $(X \xrightarrow{f_i} (X_i, \tau^{p_i} q_i))_{i \in I}$.*

**Proof.** To prove $\tau^p q$ exists, it suffices, by Theorem 10, to show that $\mathbb{U}_p^n([x]_\top) \xrightarrow{q} x$ for any $x \in X, n \in \mathbb{N}$. Indeed, by the existence of $\tau^{p_i} q_i$ we have $\mathbb{U}_{p_i}^n([f_i(x)]) \xrightarrow{q_i} f_i(x)$ for any $i \in I, x \in X, n \in \mathbb{N}$. It follows by that each $f_i : (X, p) \longrightarrow (X_i, p_i)$ being a continuous interior function we get

$$f_i^{\Rightarrow}(\mathbb{U}_p^n([x]_\top)) = \mathbb{U}_{p_i}^n(f_i^{\Rightarrow}([x]_\top)) = \mathbb{U}_{p_i}^n([f_i(x)]_\top) \xrightarrow{q_i} f_i(x),$$

which means $\mathbb{U}_p^n([x]_\top) \xrightarrow{q} x$ for any $x \in X, n \in \mathbb{N}$, i.e., $\tau^p q$ exists.

Let $s$ denote the initial structure on $X$ relative the source $(X \xrightarrow{f_i} (X_i, \tau^{p_i} q_i))_{i \in I}$. Then.

$$
\begin{aligned}
\mathbb{F} \xrightarrow{s} x \quad &\Longleftrightarrow \quad \forall i \in I, f_i^{\Rightarrow}(\mathbb{F}) \xrightarrow{\tau^{p_i} q_i} f_i(x) \stackrel{\text{Theorem }10}{\Longleftrightarrow} \forall i \in I, \forall n \in \mathbb{N}, \mathbb{U}_{p_i}^n(f_i^{\Rightarrow}(\mathbb{F})) \xrightarrow{q_i} f_i(x) \\
&\stackrel{\text{Proposition }3}{\Longleftrightarrow} \quad \forall i \in I, \forall n \in \mathbb{N}, f_i^{\Rightarrow}(\mathbb{U}_p^n(\mathbb{F})) \xrightarrow{q_i} f_i(x) \\
&\Longleftrightarrow \quad \forall n \in \mathbb{N}, \mathbb{U}_p^n(\mathbb{F}) \xrightarrow{q} x \stackrel{\text{Theorem }10}{\Longleftrightarrow} \mathbb{F} \xrightarrow{\tau^p q} x. \quad \square
\end{aligned}
$$

The following corollary shows that upper $p$-topological modification exhibits comparable behavior relative to supremum in the lattice $\top(X)$.

**Corollary 5.** *Let $\{q_i | i \in I\} \subseteq \top(X)$, $p \in \top(X)$ and $q = \sup\{q_i | i \in I\}$. If $\tau^p q_i$ exists for all $i \in I$, then $\tau^p q$ exists and $\tau^p q = \sup\{\tau^p q_i | i \in I\}$.*

## 5. Conclusions

In this paper, we discussed the $p$-topologicalness in $\top$-convergence spaces by a Fischer $\top$-diagonal condition and a Gähler $\top$-neighborhood condition, respectively. We proved that the $p$-topologicalness was preserved under the initial and final structures in the category $\top$-**CS**. As a straightforward conclusion, we further obtained that $p$-topologicalness was naturally preserved under the infimum and supremum in the lattice $\top(X)$. We also established the relationship between $p$-topologicalness in $\top$-convergence spaces and $p$-topologicalness in stratified $L$-generalized convergence spaces. Furthermore, we defined and studied the lower and upper $p$-topological modifications in $\top$-convergence spaces. In particular, we proved that the lower (resp., upper) $p$-topological modification exhibited comparable behavior relative to final (resp., initial) structures.

**Funding:** This work is supported by National Natural Science Foundation of China (No. 11801248, 11501278).

**Acknowledgments:** The author thanks the reviewers and the editor for their valuable comments and suggestions.

**Conflicts of Interest:** The author declares no conflict of interest.

## References

1. Preuss, G. *Fundations of Topology*; Kluwer Academic Publishers: London, UK, 2002.
2. Kent, D.C.; Richardson, G.D. Convergence spaces and diagonal conditions. *Topol. Appl.* **1996**, *70*, 167–174. [CrossRef]
3. Gähler, W. Monadic topology—A new concept of generalized topology. In *Recent Developments of General Topology*; Mathematical Research Volume 67; Akademie Verlag: Berlin, Germany, 1992; pp. 136–149.
4. Wilde, S.A.; Kent, D.C. *p*-topological and *p*-regular: dual notions in convergence theory. *Int. J. Math. Math. Sci.* **1999**, *22*, 1–12. [CrossRef]
5. Jäger, G. A category of *L*-fuzzy convergence spaces. *Quaest. Math.* **2001**, *24*, 501–517. [CrossRef]
6. Boustique, H.; Richardson, G. A note on regularity. *Fuzzy Sets Syst.* **2011**, *162*, 64–66. [CrossRef]
7. Boustique, H.; Richardson, G. Regularity: Lattice-valued Cauchy spaces. *Fuzzy Sets Syst.* **2012**, *190*, 94–104. [CrossRef]
8. Fang, J.M. Stratified *L*-ordered convergence structures. *Fuzzy Sets Syst.* **2010**, *161*, 2130–2149. [CrossRef]
9. Flores, P.V.; Mohapatra, R.N.; Richardson, G. Lattice-valued spaces: Fuzzy convergence. *Fuzzy Sets Syst.* **2006**, *157*, 2706–2714. [CrossRef]
10. Flores, P.V.; Richardson, G. Lattice-valued convergence: Diagonal axioms. *Fuzzy Sets Syst.* **2008**, *159*, 2520–2528. [CrossRef]
11. Jäger, G. Subcategories of lattice-valued convergence spaces. *Fuzzy Sets Syst.* **2005**, *156*, 1–24. [CrossRef]
12. Jäger, G. Lattice-valued convergence spaces and regularity. *Fuzzy Sets Syst.* **2008**, *159*, 2488–2502. [CrossRef]
13. Jäger, G. Stratified *LMN*-convergence tower spaces. *Fuzzy Sets Syst.* **2016**, *282*, 62–73. [CrossRef]
14. Jin, Q.; Li, L.Q.; Meng, G.W. On the relationships between types of *L*-convergence spaces. *Iran. J. Fuzzy Syst.* **2016**, *1*, 93–103.
15. Jin, Q.; Li, L.Q.; Lv, Y.R.; Zhao, F.; Zou, J. Connectedness for lattice-valued subsets in lattice-valued convergence spaces. *Quaest. Math.* **2018**. [CrossRef]
16. Li, L.Q.; Jin, Q. On adjunctions between Lim, SL-Top, and SL-Lim. *Fuzzy Sets Syst.* **2011**, *182*, 66–78. [CrossRef]
17. Li, L.Q.; Li, Q.G. A new regularity (p-regularity) of stratified *L*-generalized convergence spaces. *J. Comput. Anal. Appl.* **2016**, *2*, 307–318.
18. Losert, B.; Boustique, H.; Richardson, G. Modifications: Lattice-valued structures. *Fuzzy Sets Syst.* **2013**, *210*, 54–62. [CrossRef]
19. Orpen, D.; Jäger, G. Lattice-valued convergence spaces: Extending the lattices context. *Fuzzy Sets Syst.* **2012**, *190*, 1–20. [CrossRef]
20. Pang, B.; Zhao, Y. Several types of enriched $(L, M)$-fuzzy convergence spaces. *Fuzzy Sets Syst.* **2017**, *321*, 55–72. [CrossRef]
21. Pang, B. Degrees of separation properties in stratified *L*-generalized convergence spaces using residual implication. *Filomat* **2017**, *31*, 6293–6305. [CrossRef]
22. Pang, B. Stratified *L*-ordered filter spaces. *Quaest. Math.* **2017**, *40*, 661–678. [CrossRef]
23. Pang, B.; Xiu, Z.Y. Stratified *L*-prefilter convergence structures in stratified *L*-topological spaces. *Soft Comput.* **2018**, *22*, 7539–7551. [CrossRef]
24. Yang, X.; Li, S. Net-theoretical convergence in $(L, M)$-fuzzy cotopological spaces. *Fuzzy Sets Syst.* **2012**, *204*, 53–65. [CrossRef]
25. Yao, W. On many-valued stratified *L*-fuzzy convergence spaces. *Fuzzy Sets Syst.* **2008**, *159*, 2503–2519. [CrossRef]
26. Fang, J.M.; Yue, Y.L. ⊤-diagonal conditions and Continuous extension theorem. *Fuzzy Sets Syst.* **2017**, *321*, 73–89. [CrossRef]
27. Jäger, G. Pretopological and topological lattice-valued convergence spaces. *Fuzzy Sets Syst.* **2007**, *158*, 424–435. [CrossRef]
28. Jäger, G. Fischer's diagonal condition for lattice-valued convergence spaces. *Quaest. Math.* **2008**, *31*, 11–25. [CrossRef]

29. Jäger, G. Gähler's neighbourhood condition for lattice-valued convergence spaces. *Fuzzy Sets Syst.* **2012**, *204*, 27–39. [CrossRef]

30. Li, L.Q.; Jin, Q. On stratified *L*-convergence spaces: Pretopological axioms and diagonal axioms. *Fuzzy Sets Syst.* **2012**, *204*, 40–52. [CrossRef]

31. Li, L.Q.; Jin, Q.; Hu, K. On Stratified *L*-Convergence Spaces: Fischer's Diagonal Axiom. *Fuzzy Sets Syst.* **2015**, *267*, 31–40. [CrossRef]

32. Li, L.Q.; Jin, Q. *p*-Topologicalness and *p*-Regularity for lattice-valued convergence spaces. *Fuzzy Sets Syst.* **2014**, *238*, 26–45. [CrossRef]

33. Li, L.Q.; Jin, Q.; Meng, G.W.; Hu, K. The lower and upper *p*-topological (*p*-regular) modifications for lattice-valued convergence spaces. *Fuzzy Sets Syst.* **2016**, *282*, 47–61. [CrossRef]

34. Li, L.Q.; Jin, Q.; Hu, K. Lattice-valued convergence associated with CNS spaces. *Fuzzy Sets Syst.* **2018**. [CrossRef]

35. Höhle, U.; Šostak, A. Axiomatic foundations of fixed-basis fuzzy topology. In *Mathematics of Fuzzy Sets: Logic, Topology and Measure Theory*; Höhle, U., Rodabaugh, S.E., Eds.; The Handbooks of Fuzzy Sets Series; Kluwer Academic Publishers: Boston, MA, USA; Dordrecht, The Netherlands; London, UK, 1999; Volume 3, pp. 123–273.

36. Gierz, G.; Hofmann, K.H.; Keimel, K.; Lawson, J.D.; Mislove, M.W.; Scott, D.S. *Continuous Lattices and Domains*; Cambridge University Press: Cambridge, UK, 2003.

37. Li, L.Q.; Jin, Q.; Hu, K.; Zhao, F.F. The axiomatic characterizations on *L*-fuzzy covering-based approximation operators. *Int. J. Gen. Syst.* **2017**, *46*, 332–353. [CrossRef]

38. Zhang, D. An enriched category approach to many valued topology. *Fuzzy Sets Syst.* **2007**, *158*, 349–366. [CrossRef]

39. Zhao, F.F.; Jin, Q.; Li, L.Q. The axiomatic characterizations on *L*-generalized fuzzy neighborhood system-based approximation operators. *Int. J. Gen. Syst.* **2018**, *47*, 155–173. [CrossRef]

40. Bělohlávek, R. *Fuzzy Relational Systems: Foundations and Principles*; Kluwer Academic Publishers: New York, NY, USA, 2002.

41. García, J.G. On stratified *L*-valued filters induced by ⊤-filters. *Fuzzy Sets Syst.* **2006**, *157*, 813–819. [CrossRef]

42. Jin, Q.; Li, L.Q. Modified Top-convergence spaces and their relationships to lattice-valued convergence spaces. *J. Intell. Fuzzy Syst.* **2018**, *35*, 2537–2546. [CrossRef]

43. Jin, Q.; Li, L.Q. Stratified lattice-valued neighborhood tower group. *Quaest. Math.* **2018**, *41*, 847–861. [CrossRef]

44. Lai, H.; Zhang, D. Fuzzy topological spaces with conical neighborhood system. *Fuzzy Sets Syst.* **2018**, *330*, 87–104. [CrossRef]

45. Reid, L.; Richardson, G. Connecting ⊤ and Lattice-Valued Convergences. *Iran. J. Fuzzy Syst.* **2018**, *15*, 151–169.

46. Adámek, J.; Herrlich, H.; Strecker, G.E. *Abstract and Concrete Categories*; Wiley: New York, NY, USA, 1990.

47. Qiu, Y.; Fang, J.M. The category of all ⊤-convergence spaces and its cartesian-closedness. *Iran. J. Fuzzy Syst.* **2017**, *14*, 121–138.

 *mathematics*

*Review*

# Methods for Assessing Human–Machine Performance under Fuzzy Conditions

**Michael Gr. Voskoglou**

Graduate Technological Educational Institute of Western Greece, 223 34 Patras, Greece; voskoglou@teiwest.gr or mvosk@hol.gr; Tel.: + 3026-1032-8631

Received: 13 January 2019; Accepted: 25 February 2019; Published: 1 March 2019

**Abstract:** The assessment of a system's performance is a very important task, enabling its designer/ user to correct its weaknesses and make it more effective. Frequently, in practice, a system's assessment is performed under fuzzy conditions, e.g., using qualitative instead of numerical grades, incomplete information about its function, etc. The present review summarizes the author's research on building assessment models for use in a fuzzy environment. Those models include the measurement of a fuzzy system's uncertainty, the application of the center of gravity defuzzification technique, the use of triangular fuzzy or grey numbers as assessment tools, and the application of the fuzzy relation equations. Examples are provided of assessing human (students and athletes) and machine (case-based reasoning systems in computers) capacities, illustrating our results. The outcomes of those examples are compared to the outcomes of the traditional methods of calculating the mean value of scores assigned to the system's components (system's mean performance) and of the grade point average index (quality performance) and useful conclusions are obtained concerning their advantages and disadvantages. The present review forms a new basis for further research on systems' assessment in a fuzzy environment.

**Keywords:** fuzzy sets (FSs); uncertainty; center of gravity (COG) defuzzification technique; triangular fuzzy numbers (TFNs); grey numbers (GNs); fuzzy relation equations (FRE); grade point average (GPA) index

---

## 1. Introduction

In the language of management, a system is understood to be a set of interacting components forming an integrated whole and working together for achieving a common target (e.g., maximum profit, low cost of production, healthcare, student learning, etc.). One can distinguish among physical, biological, social, economic, engineering, abstract knowledge, etc. systems. The evaluation of a system's performance constitutes an important topic of the system general theory [1], because it enables the correction of its weaknesses, resulting in the improvement of its effectiveness in the environment of a given situation.

A system assessment is frequently performed under fuzzy conditions, caused by incomplete information about its function and characteristics, using qualitative (linguistic) assessment grades, or various other reasons appearing in real life situations. Several efforts have been reported in the literature for the assessment under fuzzy conditions (e.g., see [2–6]). The purpose of the present work is to review its author's research on building assessment models for use in a fuzzy environment.

The rest of the article is formulated as follows. Section 2 provides a brief account of the evolution of fuzzy set theory and its generalizations and its connection to fuzzy logic. In Section 3, the use of fuzzy system uncertainty is utilized as an assessment method of its effectiveness, whereas, in Section 4, the same is done with the center of gravity defuzzification technique, the outcomes of which are compared to those of the traditional method of calculating the grade point average index. In Sections 5 and 6, assessment models using as tools triangular fuzzy numbers and grey numbers, respectively, are developed and compared to each other. The use of fuzzy relation equations for assessing mathematical modeling skills is the objective of Section 7 and the article closes with Section 8 containing the general conclusions of the review and some hints for future research.

The present review provides a new framework for research on a system's evaluation under fuzzy conditions.

## 2. Fuzzy Sets and Generalizations

The *"Laws of Thought"* [7], of Aristotle (384–322 BC, Figure 1) that dominated for centuries the human reasoning are the following:

- The principle of identity
- The law of the excluded middle
- The law of contradiction

The law of the excluded middle (i.e., yes or no) was the basis for the traditional bi-valued logic and the precision of the classical mathematics owes undoubtedly a large part of its success to it. However, there were also strong objections to this law. The Buddha Siddhartha Gautama, who lived in India a century earlier, had already argued that almost every notion contains elements from its opposite one, while Plato (427–377 BC, Figure 1) discussed the existence of a third area beyond "true" and "false", where these two opposite notions can exist together. Modern philosophers including Hegel, Marx, and Engels adopted and further cultivated the above belief.

**Figure 1.** Plato (left) and Aristotle in a Raphael's fresco (1509).

The Polish philosopher Jan Lukasiewicz (1878–1956) first proposed a systematic alternative of the bi-valued logic introducing in the early 1900s a three-valued logic by adding the term "possible" between "true" and "false" [8]. Eventually, he developed an entire notation and axiomatic system from which he hoped to derive modern mathematics. Later, he also proposed four- and five-valued logics and then finally arrived at the conclusion that, axiomatically, nothing could prevent the derivation of an infinite valued Logic. Similar ideas were also expressed by the famous Polish-American logician and mathematician Alfred Tarski (1901–1983) [9].

However, it was not until relatively recently that an infinite-valued logic was introduced, called *Fuzzy Logic (FL)* [10], because it is based on the notion of *Fuzzy Set (FS)* initiated in 1965 by Lotfi Aliasker Zadeh (1921–2017, Figure 2) [11], an electrical engineer of Iranian origin, born in Azerbaijan, USSR and Professor at the University of Berkeley, California. A *fuzzy subset A* of the set of the discourse $U$ (or, for brevity, a FS in $U$) is defined by $A = \{(x, m_A(x)): x \in U\}$, where $m_A: U \to [0, 1]$ is its *membership function*. The real number $m_A(x)$ is called the *membership degree of $x$ in $A$*. The greater is $m_A(x)$, the more $x$ satisfies the characteristic property of $A$. Many authors, for reasons of simplicity, identify it with the function $m_A$.

**Figure 2.** L.A. Zadeh (1921–2017).

A crisp subset $B$ of $U$ can be considered as a special case of a FS in $U$ with membership function $m_B(x) = \begin{cases} 1, x \in B \\ 0, x \notin B \end{cases}$ . Based on this, most notions and operations concerning the crisp sets, e.g., subset, complement, union, intersection, Cartesian product, binary and other relations, etc., can be extended to FS. For information on FS, we refer to the book [12].

Through FL, the fuzzy terminology is translated by algorithmic procedures into numerical values, operations are performed upon those values and the outcomes are returned into natural language statements in a reliable manner. As expected, this far-reaching theory aroused some objections in the scientific community. Haack [13] argued that it can be shown that FL is unnecessary in all cases. Fox [14] responded to her objections claiming that FL is useful for handling real-world situations that are inherently fuzzy, calculating the existence in such situations fuzzy data and describing the operation of the corresponding fuzzy systems. He indicated that traditional and FL theories need not be seen as competitive, but as complementary and that FL, despite the objections of classical logicians, has proved very successful in practical applications in almost all sectors of human activity (e.g., see [12] (Chapter 6), [15–19]).

It must be mentioned here that fuzzy mathematics has also been significantly developed on the theoretical level, providing important contributions even in branches of classical mathematics, such as algebra, analysis, geometry, etc. FL constitutes one of the three portals of *computational intelligence*, which is a topic of the wide field of *artificial intelligence*. The other two portals of computational intelligence are *neural networks* and the *evolutionary computing*.

Several efforts have been made to improve and generalize the FS theory. Atanassov introduced in 1986, as a complement of Zadeh's membership degree (M), the *degree of non-membership* (N) and defined the *intuitionistic FS* [20]. Smarandache introduced in 1995 the *degree of indeterminacy/neutrality* (I) and defined the *neutrosophic set* in three components (M, N, I), where M, N and I are subsets of the interval [0, 1] [21]. Alternatives to the FS theory were proposed by Deng in 1982 (*Grey Systems* [22]), Pawlak in 1991 (*Rough Sets* [23]), Molodtsov in 1999 (*Soft Sets* [24]) and others. Principles of the Grey System (GS) theory are used in Section 6.

## 3. Types of Uncertainty in Fuzzy Systems

*Uncertainty* can be defined as the shortage of precise knowledge and complete information on data that describe the state of the corresponding system. According to a fundamental principle of classical information theory, a system's uncertainty is connected to its ability to obtain new information. Therefore, the measurement of the uncertainty could be used as a method for evaluating a system's *mean performance*.

Shannon introduced in 1948 a formula for measuring the uncertainty and the information connected to it, which is based on the laws of classical probability and it is widely known as *Shannon's entropy* [25]. This term comes from the mathematical definition of information when we have equally probable cases for the evolution of the corresponding phenomenon: $-\frac{\Delta(\log P)}{\log 2}$, where $P$ is the probability of appearance of each case. This expression appears to be analogous to a well-known formula from physics: $\Delta S = \frac{\Delta Q}{T}$, where $\Delta S$ is the increase of a physical system's entropy caused by an increase of the heat $\Delta Q$, when the absolute temperature $T$ remains constant.

Let $U$ denote the universal set of the discourse. For use in a fuzzy environment, Shannon's formula (*probabilistic uncertainty*) has been adapted [26] (p.20) to the form:

$$H = -\frac{1}{\ln n} \sum_{s=1}^{n} m_s \ln m_s.$$ (1)

In Equation (1), $m_s = m(s)$ denotes the membership degree of the element $s$ of $U$ in the corresponding fuzzy set and $n$ denotes the total number of the elements of $U$. Dividing the sum by $\ln n$, one makes H take values in the interval $[0, 1]$.

The *fuzzy probability* of an element $s$ of $U$ is defined by

$$P_s = \frac{m_s}{\sum\limits_{s \in U} m_s}$$ (2)

However, according to the British economist Shackle [27] and many other researchers after him, human reasoning can be formulated more adequately by the possibility rather than by the probability theory. The *possibility*, e.g., $r_s$, of an element $s$ of $U$ is defined by

$$r_s = \frac{m_s}{\max\{m_s\}}$$ (3)

In Equation (3), $\max\{m_s\}$ denotes the maximal value of $m_s$, for all $s$ in $U$. In other words, the possibility of $s$ expresses the relative membership degree of $s$ with respect to $\max\{m_s\}$.

In terms of possibility, the uncertainty is measured by the sum of *strife (or discord)* and *non-specificity (or imprecision)*. The former is connected to the conflict created among the membership degrees, whereas the latter is connected to the conflict created among the cardinalities (sizes) of the various fuzzy subsets of $U$ [26] (p.28). Note that the *cardinality* of a fuzzy subset of $U$ is defined sa the sum $\sum\limits_{x \in U} m(x)$ of all membership degrees of the elements of $U$ with respect to it.

The following example illustrates this situation.

**Example 1.** Let U be the set of all mountains of a country and let H and L be its fuzzy subsets of the high and low mountains with membership functions $m_H$ and $m_L$, respectively. Assume further that a mountain x in U has a height of 1000 m.

Then, strife is created by the existing conflict between the membership degrees $m_H(x)$ and $m_L(x)$. In fact, if the country has high mountains in general, then $m_H(x)$ should take values near 0 and $m_l(x)$ near 1. However, the opposite could happen, if the country has low mountains in general.

On the other hand, non-specificity is connected to the question of how many elements of U should have zero membership degrees with respect to H and L.

Strife is measured [26] (p.28) by the function ST(r) on the ordered possibility distribution r: $r_1 = 1 \geq r_2 \geq \ldots \ldots \geq r_n \geq r_{n+1}$ of a group of the elements of U defined by

$$ST(r) = \frac{1}{\log 2}\left[\sum_{i=2}^{m} (r_i - r_{i+1}) \log \frac{i}{\sum\limits_{j=1}^{i} r_j}\right].$$ (4)

Under the same conditions non-specificity is measured [26] (p.28) by the function

$$N(r) = \frac{1}{\log 2}\left[\sum_{i=2}^{m} (r_i - r_{i+1}) \log i\right].$$ (5)

The sum T(r) = ST(r) + N(r) measures the fuzzy system's total possibilistic uncertainty.

**Example 2.** Table 1 depicts the performance of two equivalent student groups in a common test with respect to the linguistic grades, where A = excellent, B = very good, C = good, D = fair and F = unsatisfactory:

**Table 1.** Student results.

| Grade | $G_1$ | $G_2$ |
|-------|-------|-------|
| A | 1 | 10 |
| B | 13 | 6 |
| C | 4 | 3 |
| D | 3 | 0 |
| F | 0 | 1 |
| Total | 21 | 20 |

It is asked to evaluate the performance of the two student groups by calculating the total possibilistic and probabilistic uncertainty existing in them.

(i) *Total possibilistic uncertainty*: Defining the membership function in terms of the frequencies of the student grades, one can represent the two student groups as fuzzy sets on the set $U = \{A, B, C, D. F\}$ in the form $\{(x, \frac{n_x}{n}): x \in U \}$, where $n_x$ is the number of students who received the grade $x$ and $n$ is the total number of the students in each group. In other words, we can write

$G_1 = \{(A, \frac{1}{21}), (B, \frac{13}{21}), (C, \frac{4}{21}), (D, \frac{3}{21}), (F, 0) \}$; and
$G_2 = \{(A, \frac{10}{20}), (B, \frac{6}{20}), (C, \frac{3}{20}), (D, 0), (F, \frac{1}{20}) \}$.

The maximal membership degree in $G_1$ is equal to $\frac{13}{21}$, hence the possibilities of the elements of $U$ in $G_1$ are: $r(A) = \frac{1}{13}$, $r(B) = 1$, $r(C) = \frac{4}{13}$, $r(D) = \frac{3}{13}$, $r(F) = 0$. Therefore, the ordered possibility distribution defined on $G_1$ is

$$r: r_1 = 1 > r_2 = \frac{4}{13} > r_3 = \frac{3}{13} > r_4 = \frac{1}{13} > r_5 = 0. \tag{6}$$

Similarly, one finds that the ordered possibility distribution on $G_2$ is:

$$r: r_1 = 1 > r_2 = \frac{6}{10} > r_3 = \frac{3}{10} > r_4 = \frac{1}{10} > r_5 = 0. \tag{7}$$

Equation (4) gives in our case that

$$ST(r) = \frac{1}{\log 2}[(r_2 - r_3) \log \frac{2}{r_1 + r_2} + (r_3 - r_4) \log \frac{3}{r_1 + r_2 + r_3} + (r_4 - r_5) \log \frac{4}{r_1 + r_2 + r_3 + r_4}]$$

Replacing the values of the possibility distribution $r$ from Equation (6), one finds for $G_1$ that

$$ST(r) = \frac{1}{\log 2}[\frac{1}{13} \log(\frac{26}{17}) + \frac{2}{13} \log(\frac{39}{20}) + \frac{1}{13} \log(\frac{42}{21})] \approx 0.27.$$

In addition, Equation (5) gives for $G_1$ that

$$N(r) = \frac{1}{\log 2}[\frac{1}{13}\log 2 + \frac{2}{13}\log 3 + \frac{1}{13}\log 4] \approx 0.48$$

Therefore, the total possibilistic uncertainty for $G_1$ is $T(r) \approx 0.27 + 0.48 = 0.75$.

In the same way, replacing the values of $r$ from Equation (7), one finds that the total possibilistic uncertainty for $G_2$ is $T(r) \approx 0.33 + 0.82 = 1.15$.

Since the two student groups have been chosen to be equivalent, they have the same existing uncertainty before the test. Thus, the reduction of uncertainty is greater for Group $G_1$, which therefore demonstrates a better mean performance than Group $G_2$.

*(ii) Probabilistic uncertainty:* Substituting the membership degrees of $G_1$ into Equation (1), one finds that the probabilistic uncertainty for the experimental group is equal to:

$$H = -\frac{1}{\ln 5}\left(\frac{1}{21}\ln\frac{1}{21} + \frac{13}{21}\ln\frac{13}{21} + \frac{4}{21}\ln\frac{4}{21} + \frac{3}{21}\ln\frac{3}{21}\right) \approx 0.64.$$

Similarly, one finds that for $G_2$ the probabilistic uncertainty is approximately equal to 0.71. Therefore, Group $G_1$ demonstrates again a better performance.

**Remark 1.** From the previous example, it becomes evident that the measurement of uncertainty can be used for assessing the performance of the two groups only under the assumption that they are equivalent, which means that they have the same existing uncertainty before the test.

## 4. The Center of Gravity Defuzzification Technique

The process of solving a problem with principles and methods of FL involves the following phases:

- *Choose* the suitable universal set.
- *Fuzzify* the given data by defining suitable membership functions for the FSs involved in this procedure.
- *Elaborate* the fuzzy data with FL techniques for expressing the solution of the given problem in the form of a *unique* FS.
- *Defuzzify* the above FS by representing it with a real numerical value to "translate" the problem's solution into the natural language.

There are more than 30 defuzzification methods in use, but two of them are the most popular. In the *Maximum method*, the crisp value representing the solution's output is one of the values at which the corresponding FS has its maximum truth (e.g., [15] (Chapter 4, pp. 97–99)). On the other hand, in the *Center of Gravity (COG) technique*, the representative crisp value is obtained by calculating the coordinates of the COG of the level's area S between the graph of the membership function of the corresponding FS and the OX axis [28].

Voskoglou [29] developed in 1999 a fuzzy model for representing mathematically the process of learning a subject matter in the classroom and later, considering the class as a fuzzy system, he calculated its existing uncertainty (see Section 3) for assessing the student mean performance [30]. Subbotin et al. [31], based on the Voskoglou's model, adapted properly the COG defuzzification technique for use as an assessment method of student learning skills. Since then, Subbotin and Voskoglou applied jointly or separately the COG method, termed as *Rectangular Assessment Model (RFAM)*, in many other types of assessment problems (e.g., see [15] (Chapter 6)). Here, we present the headlines of RFAM and we compare it to the traditional assessment method of calculating the *Grade Point Average (GPA)* index.

### 4.1. The Rectangular Fuzzy Assessment Model

We choose, exactly as in Example 2, for discourse the set of linguistic grades $U = \{A, B, C, D, F\}$ and we represent a student group G as a FS in $U$ in the form $G = \{(x, \frac{n_x}{n}): x \in U\}$, where $n$ is the total number of the students of G and $n_x$ is the number of students of G whose performance is characterized by the grade $x$ in $U$.

To be able to design the graph of the membership function $y = m(x)$, we replace $U$ with a set of real intervals as: $F \to [0, 1)$, $D \to [1, 2)$, $C \to [2, 3)$, $B \to [3, 4)$, $A \to [4, 5]$. Consequently, we have that $y_1 = m(x) = m(F)$ for all $x$ in $[0, 1)$, $y_2 = m(x) = m(D)$ for all $x$ in $[1, 2)$, $y_3 = m(x) = m(C)$ for all $x$ in $[2, 3)$, $y_4 = m(x) = m(B)$ for all $x$ in $[3, 4)$ and $y_5 = m(x) = m(A)$ for all $x$ in $[4, 5)$. Since the membership values of the elements of $U$ in G have been defined in terms of the corresponding frequencies, we obviously have that: $\sum_{i=1}^{5} y_i = m(A) + m(B) + m(C) + m(D) + m(F) = 1$.

The graph of the membership function $y = m(x)$ is shown in Figure 3, where the level's area S between the graph and the OX axis is equal to the sum of the areas of the five rectangles $S_i$, i = 1, 2, 3, 4, 5, one side of which has length 1 and lies on the OX axis. That is why this method was termed as RFAM.

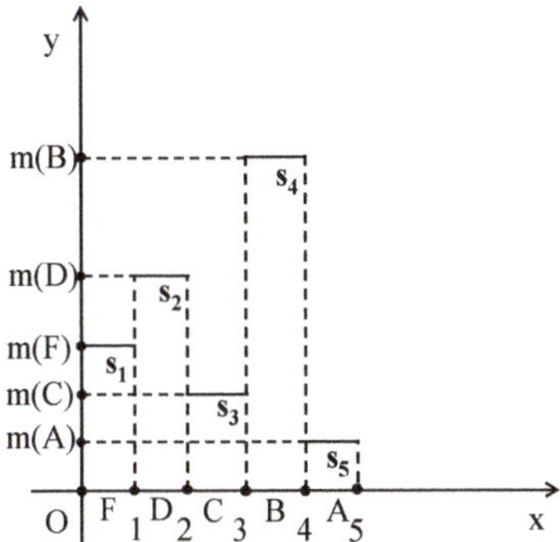

**Figure 3.** The graph of the membership function.

The COG coordinates $(x_c, y_c)$ are calculated by the well known from mechanics formulas:

$$x_c = \frac{\iint\limits_S x\,dxdy}{\iint\limits_S dxdy}, \quad y_c = \frac{\iint\limits_S y\,dxdy}{\iint\limits_S dxdy} \tag{8}$$

Here, we have that: $\iint\limits_S dxdy = \sum\limits_{i=1}^{5} y_i = 1$. In addition, $\iint\limits_S x\,dxdy = \sum\limits_{i=1}^{5} \iint\limits_{F_i} x\,dxdy = \sum\limits_{i=1}^{5} \int\limits_0^{y_i} dy \int\limits_{i-1}^{i} x\,dx$

$= \sum\limits_{i=1}^{5} y_i \int\limits_{i-1}^{i} x\,dx = \frac{1}{2}\sum\limits_{i=1}^{5}(2i-1)y_i$ and $\iint\limits_S y\,dxdy = \sum\limits_{i=1}^{5}\iint\limits_{F_i} y\,dxdy = \sum\limits_{i=1}^{5}\int\limits_0^{y_i} y\,dy \int\limits_{i-1}^{i} dx = \sum\limits_{i=1}^{n}\int\limits_0^{y_i} y\,dy = \frac{1}{2}\sum\limits_{i=1}^{n} y_i^2$.

Substituting the above values of the double integrals into Equation (8), one finds that

$$\begin{aligned} x_C &= \tfrac{1}{2}(y_1 + 3y_2 + 5y_3 + 7y_4 + 9y_5) \\ y_C &= \tfrac{1}{2}(y_1^2 + y_2^2 + y_3^2 + y_4^2 + y_5^2) \end{aligned} \tag{9}$$

However, $(y_i - y_j)^2 = y_i^2 + y_j^2 - 2y_{ij} \geq 0$, or $y_i^2 + y_j^2 \geq 2y_{ij}$ for $i, j = 1, 2, \ldots, 5$, with the equality holding if, and only if, $y_1 = y_2 = y_3 = y_4 = y_5$. Therefore, $1 = (\sum\limits_{i=1}^{5} y_i)^2 = \sum\limits_{i=1}^{n} y_i^2 + 2(y_1y_2 + y_1y_3 + \ldots + y_4y_5) \geq 5\sum\limits_{i=1}^{n} y_i^2$, or $y_1^2 + y_2^2 + y_3^2 + y_4^2 + y_5^2 \leq \tfrac{1}{5}$, with the equality holding if and only if $y_1 = y_2 = y_3 = y_4 = y_5 = \tfrac{1}{5}$. Thus, Equation (9) shows that the unique minimum $y_c = \tfrac{1}{10}$ corresponds to the COG $F_m$ $(\tfrac{5}{2}, \tfrac{1}{10})$.

The ideal case is when $y_1 = y_2 = y_3 = y_4 = 0$ and $y_5 = 1$. Then, from Equation (9), one finds that $x_c = \tfrac{9}{2}$ and $y_c = \tfrac{1}{2}$. Therefore, the COG in this case is the point $F_i$ $(\tfrac{9}{2}, \tfrac{1}{2})$.

On the other hand, the worst case is when $y_1 = 1$ and $y_2 = y_3 = y_4 = y_5 = 0$. Therefore, from Equation (9), one finds that the COG is the point $F_x$ $(\tfrac{1}{2}, \tfrac{1}{2})$.

Consequently, the area where COG $F_c$ lies is the triangle $F_xF_m F_i$ of Figure 4.

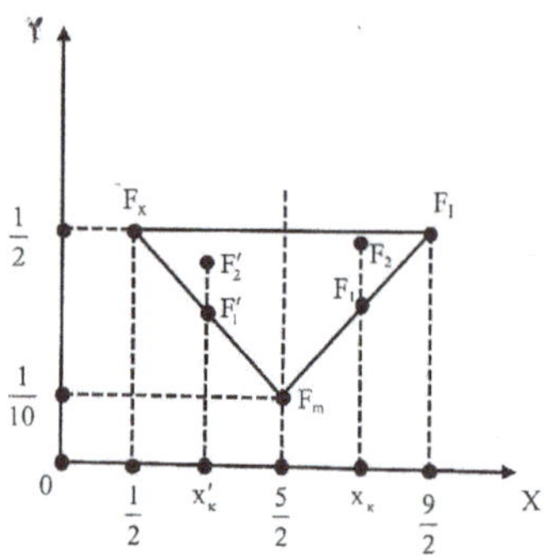

**Figure 4.** The triangle where the COG lies.

By making elementary observation on Figure 4, one obtains the following criterion for comparing the performance of two student groups:

- The group with the greater value of $x_c$ demonstrates the better performance.
- In the case of the same value of $x_c$, if $x_c \geq 2.5$, then the group with the greater value of $y_c$ performs better. On the contrary, if $x_c < 2.5$, then the group with the lower value of $y_c$ demonstrates the better performance.

Observe that a group's performance depends mainly on the value of $x_c$, while the first part of Equation (9) shows that for the calculation of $x_c$ greater coefficients are assigned to the higher scores. Therefore, the COG method focuses on the group's *quality performance*.

In addition, since the ideal group's performance corresponds to the value $x_c = \frac{9}{2}$, values greater than $\frac{9}{2} : 2 = 2.25$ could be considered as demonstrating a satisfactory performance.

### 4.2. Comparison of the RFAM with the GPA Index

Keeping the same notation as in Section 4.1, the GPA index [15] (Chapter 6, p.125) is calculated by the formula

$$\mathrm{GPA} = \frac{0n_F + 1n_D + 2n_C + 3n_B + 4n_A}{n} \tag{10}$$

In other words, since greater coefficients (weights) are assigned to the higher scores, GPA is a weighted average connected to the group's *quality performance*. In the case of the worst performance ($n_F = n$), Equation (10) gives that GPA = 0, while, in the case of the ideal performance ($n_A = n$), it gives that GPA = 4. Therefore, we have in general that $0 \leq \mathrm{GPA} \leq 4$. Consequently, values of GPA greater than 2 could be considered as indicating a satisfactory performance.

Equation (10) can be also written with respect to the frequencies in the form

$$\mathrm{GPA} = y_2 + 2y_3 + 3y_4 + 4y_5. \tag{11}$$

Therefore, $x_c = \frac{1}{2}(y_1 + 3y_2 + 5y_3 + 7y_4 + 9y_5) = \frac{1}{2}(2\mathrm{GPA} + 1)$, or

$$x_c = \mathrm{GPA} + 0.5. \tag{12}$$

As an immediate consequence of Equation (12) and the first case of the assessment criterion in Section 4.1, if the values of the GPA index are different for two groups, then the GPA index and the RFAM provide the same assessment outcomes for those groups. The following example, due to Subbotin [32], shows that, in the case of the same GPA values, the application of the GPA index cannot lead to logically-based conclusions. Therefore, in such situations, the criterion in Section 4.1 becomes useful due to its logical nature.

**Example 3.** The student grades of two classes with 60 students in each class are depicted in Table 2.

**Table 2.** Student Grades.

| Grades | Class I | Class II |
| --- | --- | --- |
| C | 10 | 0 |
| B | 0 | 20 |
| A | 50 | 40 |

The GPA index for the two classes is equal to $\frac{2\times10+4\times50}{60} = \frac{3\times20+4\times40}{60} \approx 3.67$, which means that the two classes demonstrate the same performance. However, Equation (12) gives that $x_c = 4.17$ for both classes. However, $\sum\limits_{i=1}^{5} y_i^2 = \left(\frac{1}{6}\right)^2 + \left(\frac{5}{6}\right)^2 = \frac{26}{36}$ for the first and $\sum\limits_{i=1}^{5} y_i^2 = \left(\frac{2}{6}\right)^2 + \left(\frac{4}{6}\right)^2 = \frac{20}{36}$ for the second class. Therefore, according to the second case of the assessment criterion in Section 4.1, the first class demonstrates a better performance in terms of the RFAM.

Observe now that the ratio of the students receiving B or better to the total number of students is equal to $\frac{5}{6}$ for the first class and 1 for the second class, which means that Class II demonstrates a better quality performance. However, many educators might prefer the situation in Class I having a greater number of excellent students. Conclusively, in no case is it logical to accept that the two classes demonstrate the same performance, as the equal values of the GPA index indicate.

## 5. Triangular Fuzzy Numbers

### 5.1. Preliminaries

Fuzzy Numbers (FNs), introduced by Zadeh [33], play an important role in fuzzy mathematics, analogous to the role of ordinary numbers in classical mathematics. A FN, e.g., A, is defined to be a FS on the set $R$ of real numbers, such that:

- A is *normal*, i.e., there exists x in $R$ such that $m_A(x) = 1$.
- A is *convex*, i.e., all its $a$-cuts $A^a = \{x \in R: m_A(x) \geq a\}$, with $a$ in $[0, 1]$, are closed real intervals.
- Its membership function $y = m(x)$ is a piecewise continuous function.

One can define the basic arithmetic operations on FNs in two, equivalent to each other, ways:

(i) With the help of their $a$-cuts and the representation–decomposition theorem of Ralesscou-Negoita [34] (Theorem 2.1, p.16) stating that FS A can be completely and uniquely expressed by the family of its $a$-cuts in the form $A = \sum\limits_{a\in[0,1]} aA^a$.

(ii) By applying the Zadeh's extension principle [12] (Section 1.4, p.20), which provides the means for any function f mapping the crisp set X to the crisp set Y to be generalized to map fuzzy subsets of X to fuzzy subsets of Y.

However, the above two general methods of the fuzzy arithmetic, requiring laborious calculations, are rarely used in practical applications, where the utilization of simpler forms of FNs is preferred.

A *Triangular Fuzzy Number (TFN)* (a, b, c), with a, b, and c real numbers such that a < b < c is the simplest form of a FN representing mathematically the fuzzy statement that "the value of b lies in the interval [a, c]". The membership function $y = m(x)$ of (a, b, c) is zero outside the interval [a, c], whereas its graph in [a, c] forms a triangle with the OX axis (Figure 5). Therefore, we have:

$$y = m(x) = \begin{cases} \frac{x-a}{b-a}, & x \in [a,b] \\ \frac{c-x}{c-b}, & x \in [b,c] \\ 0, & x < a \text{ or } x > c \end{cases}$$

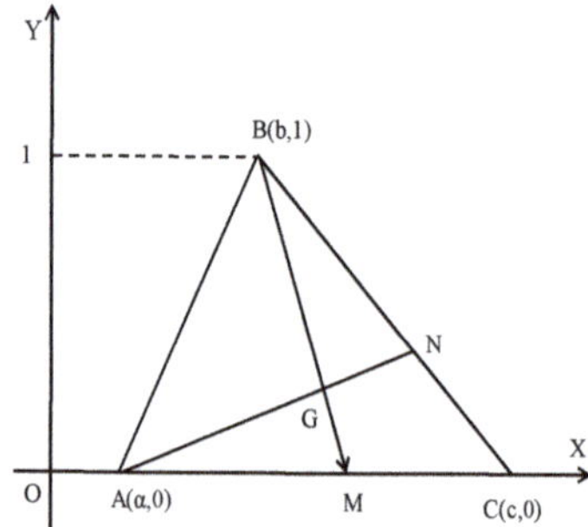


**Figure 5.** Graph and COG of the TFN (a, b, c).

The coordinates (X, Y) of the Center of Gravity (COG) of the graph of the TFN A = (a, b, c), being the intersection point G of the medians AN and BM of the triangle ABC (Figure 5), are calculated by the formulas

$$X(A) = \frac{a+b+c}{3}, \; Y(A) = \frac{1}{3}. \tag{13}$$

In fact, since M(, 0) and N($\frac{b+c}{2}, \frac{1}{2}$), the proof is easily obtained by calculating the equations of the medians AN and BM and by solving their linear system.

According to the COG technique the first part of Equation (6) can be used to defuzzify the TFN A.

The two general methods for defining arithmetic operations on FNs lead to the following simple rules for the addition and subtraction of TFNs:

Let A = (a, b, c) and B = ($a_1, b_1, c_1$) be two TFNs. Then:

- The *sum* A + B = (a+$a_1$, b+$b_1$, c+$c_1$).
- The *difference* A − B = A + (−B) = (a − $c_1$, b − $b_1$, c − $a_1$), where −B = (−$c_1$, −$b_1$, −$a_1$) is defined to be the *opposite* of B.

On the contrary, the *product* and the *quotient* of A and B are FNs, which are not TFNs in general, apart from some special cases.

The following two *scalar operations* can also be defined:

- k + A = (k + a, k + b, k + c), k $\in$ R.
- kA = (ka, kb, kc), if k > 0 and kA = (kc, kb, ka), if k < 0, k $\in$ R.

**Remark 2.** Another simple form of FNs that we have used as assessment tools [17] (Chapter 7) are the Trapezoidal Fuzzy Numbers (TpFNs). The membership function of a TpFN (a, b, c, d), with a, b, c, and d in **R** such that a < b $\leq$ c < d, is zero outside the interval [a, d], whereas its graph in [a, d] forms a trapezoid with the OX axis. The TFN (a, b, d) is a special case of the TpFN (a, b, c, d) with b = c, i.e., the TpFNs are actually generalizations of the TFNs.

For general information on FNs, we refer to the book [35].

*5.2. The Assessment Method Using TFNs*

In [36], we developed with the help of TFNs a method for assessing a system's mean performance, whose steps are the following:

- Define the *mean value* of a finite number of given TFNs $A_1, A_2, \ldots, n \geq 2$, to be the TFN A = $\frac{1}{n}$($A_1$ + $A_2$ + ... + $A_n$).
- Assign a scale of numerical scores from 1 to 100 to the linguistic grades A = excellent, B = very good, C = good, D = fair and F = unsatisfactory as follows: A (85–100), B (75–84), C (60–74), D (50–59) and F (0–49).

- Use for simplicity the same letters to represent the above grades by the TFNs A = (85, 92.5, 100), B = (75, 79.5, 84), C (60, 67, 74), D (50, 54.5, 59) and F (0, 24.5, 49), respectively, where the middle entry of each of them is equal to the mean value of its other two entries.
- Assess the individual performance of the system components by the above five qualitative grades and assign one of the TFNs A, B, C, D, F to each of those components. Then, if n is the total number of the system's components and $n_X$ denotes the number of the components corresponding to the TFN X, with X = A, B, C, D, F, the mean value M of all those TFNs is equal to the TFN

$$M\,(a,\,b,\,c) = \frac{1}{n}\,(n_A\,A + n_B\,B + n_C C + n_D D + n_F F). \tag{14}$$

Using the TFN M $(a,\,b,\,c)$ to evaluate the system's mean performance, it straightforward to check that its components a, b and c, respectively, are equal to $\frac{85n_A + 75n_B + 60n_C + 50n_D + 0n_F}{n}$ $\frac{92.5n_A + 79.5n_B + 67n_C + 54.5n_D + 24.5n_F}{n}$ and $\frac{100n_A + 84n_B + 74n_C + 59n_D + 49n_F}{n}$.
Then, Equation (13) gives that

$$X\,(M) = \frac{a + b + c}{3}\,\frac{92.5n_A + 79.5n_B + 67n_C + 54.5n_D + 24.5n_F}{n} = \frac{a + c}{2} = b. \tag{15}$$

Therefore, for the defuzzification of M $(a, b, c)$, one needs to calculate only its middle component b. The value of X (M) provides a crisp representation of the TFN M evaluating the system's mean performance.

An analogous assessment method was also developed in [37] using TpFNs instead of TFNs. Examples illustrating the above assessment method are presented in Section 6.3 to be compared with the use of grey numbers.

## 6. Grey System Theory

Because many constantly changing factors are usually involved in large and complex systems, approximate data are frequently used in many problems of everyday life, science and engineering. Nowadays, the main tool in the hands of the specialists for handling such approximate data is FL and its generalizations or alternative theories. Among those theories, the theory of *Grey System (GS)* initiated by Deng [22] in 1982 is very important.

A system that lacks information, such as structure message, operation mechanism and behavior document, is referred to as a GS. The GS theory has recently found important applications in many fields of human activity [38].

### 6.1. Grey Numbers

A *Grey Number (GN)* is a number with known range but unknown position within its boundaries. If $R$ denotes the set of real numbers, a GN A can be expressed mathematically by

$$A \in [a, b] = \{x \in R: a \leq x \leq b\}.$$

If a = b, then A is called a *white number* and if A $\in (-\infty. + \infty)$, then A is called a *black number*. A GN may enrich its uncertainty representation with respect to the interval [a, b] by a whitening function g: [a, b] $\to$ [0, 1] defining a *degree of greyness* g(x) for each x in [a, b]. The closer is g(x) to 1, the greater the probability for x to be the representative real value of the corresponding GN.

The well-known arithmetic of the real intervals [39] has been used to define the basic arithmetic operations among the GNs. More explicitly, if A $\in [a_1, a_2]$ and B $\in [b_1, b_2]$ are given GNs and k is a real number, one defines:

- *Addition* by A + B $\in [a_1 + b_1, a_2 + b_2]$.

- *Subtraction* by $A - B = A + (-B) \in [a_1 - b_2, a_2 - b_1]$, where $-B \in [-b_2, -b_1]$ is defined to be the *opposite* of B.
- *Multiplication* by $A \times B$ $[\min\{a_1b_1, a_1b_2, a_2b_1, a_2b_2\}, \max\{a_1b_1, a_1b_2, a_2b_1, a_2b_2\}]$.
- *Division* by A: $B = A \times B^{-1} \in [\min\{\frac{a_1}{b_1}, \frac{a_1}{b_2}, \frac{a_2}{b_1}, \frac{a_2}{b_2}\}, \max\{\frac{a_1}{b_1}, \frac{a_1}{b_2}, \frac{a_2}{b_1}, \frac{a_2}{b_2}\}]$. with $b_1, b_2 \neq 0$ and $B^{-1} \in [\frac{1}{b_2}, \frac{1}{b_1}]$, which is defined to be the *inverse* of B.
- *Scalar multiplication* by $kA \in [ka_1, ka_2]$, if $k \geq 0$ and by $kA \geq [ka_2, ka_1]$, if $k < 0$.

The white number whose value represents the GN $A \in [a, b]$ is denoted by W(A). The process of determining w(A) is called *whitening of A*. When the distribution of A is unknown, i.e., no whitening function has been defined for A, one usually utilizes the equally distant whitening by taking

$$W(A) = \frac{a + b}{2}. \tag{16}$$

For general information on GNs, we refer to [40].

*6.2. The Assessment Method with GNs*

When using TFNs for the assessment of a system's mean performance, we found that $X(M) = b$, which means that only the middle component b is needed for the defuzzification of the mean value M(a, b, c). This observation gives the hint to search for a "formal" assessment method that, analogous to that with TFNs, possibly reduces the required computational burden. This idea led us to utilize GNs [41] instead of TFNs for the system's assessment. The steps of our new method are the following:

- Attach the numerical scores A (100–85), B (84–75), C (74–60), D (59–50), and F (49–0) to the corresponding linguistic grades.
- Assign to each grade a GN as follows: $A \in [85, 100]$, $B \in [75, 84]$, $C \in [60, 74]$, $D \in [50, 59]$, and $F \in [0, 49]$.
- Correspond to each of the system's components one of the above GNs evaluating its performance.
- Using analogous notation with the case of TFNs, calculate the mean value

$$M* = \frac{1}{n} [n_A A + n_B B + n_C C + n_D D + n_F F] \tag{17}$$

- Since $n_A A \in [85n_A, 100n_A]$, $n_B B \in [75n_B, 84n_B]$, $n_C C \in [60n_C, 74n_C]$, $n_D D \in [50n_D, 59n_D]$, and $n_F F \in [0n_F, 49n_F]$, one obtains that $M* \in [m_1, m_2]$, where

$$m_1 = \frac{85n_A + 75n_B + 60n_C + 50n_D + 0n_F}{n}$$

$$m_2 = \frac{100n_A + 84n_B + 69n_C + 59n_D + 49n_F}{n}$$

- Since the distribution of M* is unknown, take

$$W(M*) = \frac{m_1 + m_2}{2}. \tag{18}$$

The value of W(M*) provides a crisp representation of the GN M* evaluating the system's mean performance

**Remarks 3.**

(1) From Equations (15) and (18), one obtains that $X(M) = W(M*)$. Therefore, one concludes that the assessment methods with the TFNs and the GNs are equivalent to each other, because they provide the same assessment outcomes.

(2) It is straightforward to check that, if the maximal possible numerical score corresponds to each system's component for each grade, then the mean value of those scores is equal to $c$ or $m_2$, respectively. In the same way, if the minimal possible score corresponds to each system's component for each grade, then the mean value of all scores is equal to $a$ or $m_1$, respectively. Consequently, assessment methods with TFNs and GNs give a reliable approximation of the system's mean performance.

### 6.3. Applications of the Assessment Methods with TFNs and GNs

The first application in this section concerns the assessment of *Case-Based Reasoning (CBR)* systems. CBR is the method of solving new problems based on the solution of analogous problems solved in the past (*past cases*). A case collection can be a powerful resource to use when handling a new problem. A CBR system, usually designed and functioned with the help of proper software, allows the collection of cases to develop incrementally, while its maintenance is an easy task for domain experts. The CBR approach has gained much attention and found many applications over the last 30–40 years, because, as an intelligent systems method, it enables information managers to increase efficiency and reduce cost by substantially automating processes.

The steps of the CBR process involve:

- $R_1$: *Retrieve* the most suitable past case to the new problem.
- $R_2$: *Reuse* the information of the retrieved case to solve the new problem.
- $R_3$: *Revise* the solution.
- $R_4$: *Retain* the part of the solution that could be useful for future problems.

Steps $R_1$, $R_2$ and $R_3$ are not linear, characterized by a backward–forward flow. A simple flow diagram of the CBR process is presented in Figure 6:

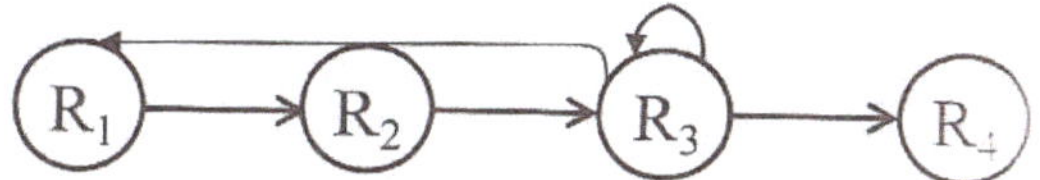


**Figure 6.** A simplified flow-chart of the CBR process.

More information and a detailed functional diagram illustrating the four steps of the CBR process are presented in [42].

**Example 4.** Consider a couple of CBR systems with a collection of 105 and 90 past cases, respectively. The designers have supplied both systems with a mechanism evaluating the degree of success of the past cases for solving new analogous problems. Table 3 depicts the performance of the past cases in the three first steps of the CBR process.

**Table 3.** Assessment of the past cases of the CBR systems.

| | First system | | | | |
|---|---|---|---|---|---|
| **Steps** | **F** | **D** | **C** | **B** | **A** |
| $R_1$ | 0 | 0 | 51 | 24 | 30 |
| $R_2$ | 18 | 18 | 48 | 21 | 0 |
| $R_3$ | 36 | 30 | 39 | 0 | 0 |
| | Second system | | | | |
| **Steps** | **F** | **D** | **C** | **B** | **A** |
| $R_1$ | 0 | 18 | 45 | 27 | 0 |
| $R_2$ | 18 | 24 | 48 | 0 | 0 |
| $R_3$ | 36 | 27 | 27 | 0 | 0 |

Here, we compare the mean performance of the two systems by applying both methods with TFNs and GNs.

*(i) Use of TFNs:* The data in Table 3 show that in Step $R_1$ we have for the first system 51 TFNs equal to C(60, 67, 74), 24 TFNs equal to B(75, 79.5, 85) and 30 TFNs equal to A(85, 92.5, 100). The calculation of the mean value of those TFNs gives that $R_1 = \frac{1}{105}(51C + 24B + 30A) = \frac{1}{105}$ [(3060, 3417, 3774) + (1800, 1908, 2016) + (2550, 2775, 3000) = $\frac{1}{105}$ (7410, 8100, 8790) ≈ (70.57, 77.14, 83.71).

Therefore, from Equation (15), one gets that $X(R_1) = 77.14$, which shows that the first system demonstrates a very good (B) performance at Step R1.

In the same way, one calculates for the first system the mean values $R_2 = \frac{1}{105}$ (18F + 18D+ 48C + 21B) ≈ (51, 60.07, 69.14) with $X(R_2) = 60.07$ and $R_3 = \frac{1}{105}$ (36F + 30D + 39C) ≈ (36.57, 48.86, 61.14) with $X(R_2) = 48.86$, thus obtaining the analogous conclusions for the system's performance at Steps $R_2$ and $R_3$ of the CBR process.

The overall system's performance can be assessed by the mean value R = $\frac{1}{3}$ ($R_1$ + $R_2$ + $R_3$) ≈ (52.71, 62.02, 71.33). Since X(R) = 62.02, the system demonstrates a good (C) mean performance.

A similar argument gives for the second system the values $R_1$ = (62.5, 68.25, 74), $R_2$ ≈ (45.33, 55.17, 65), $R_3$ = (33, 46.25, 59.5) and R ≈ (46.94, 56.56, 66.17). Therefore, analogous conclusions can be obtained for its mean performance at each step of the CBR process and for its overall mean performance, which is characterized as fair (D).

(ii) Use of GNs: Here, in Step $R_1$, we have 51 GNs equal to C ∈ [60, 74], 24 GNs equal to B ∈ [75, 84] and 30 GNs equal to A ∈ [85, 100]. The mean value of those GNs is $R_1^* = \frac{1}{105}$ (51C + 24B + 30A) ∈ [70.57, 83.71]. Therefore, $W(R_1^*) = \frac{70.57+83.71}{2} = 77.14$, etc.

**Remarks 4.**

(i)    As we show in the previous section, the use of GNs provides in general the same assessment outcomes with the use of TFNs. However, observe that, to obtain the mean value M* ∈ [$m_1$, $m_2$], one needs to calculate two components only, in contrast to the mean value M (a, b, c) where the calculation of three components is needed. Consequently, the method with GNs reduces the required computational burden.

(ii)    Another promising area for applying the above assessment methods is the fuzzy control systems (e.g., see [43], [44] (Paragraph 5.3), [45]). Traditional controllers, often implemented as PID (proportional–integral–derivative) controllers, are based on mathematical models in which the control system is described using one or more differential equations that define the system response to its inputs. They are the products of decades of development and theoretical analysis and are highly effective in general. However, in certain cases, the mathematical model of the control process may not exist, or may be too "expensive" in terms of computer processing power and memory. In such cases, a system based on empirical rules may be more effective.

Fuzzy controllers consist of *input, processing* and *output stages*. The input stage maps sensors, switches and other inputs to the appropriate membership functions (usually triangular, although trapezoidal and bell curves are also used) and truth values. The processing stage is based on a collection of logic rules in the form of IF–THEN statements. A result is generated from each rule and all these results are properly combined. For example, such a rule for a thermostat is "IF the temperature is low, THEN the heater is high". Typical fuzzy controllers have dozens of rules. Finally, the output stage converts the combined result back into a specific control output value. In many cases, fuzzy control can be used to improve existing traditional control systems by adding an extra layer of intelligence to the current control method.

**Example 5.** Six different trainers ranked with scores from 0 to 100 the performance of five athletes as follows: $A_1$ (Athlete 1): 43, 48, 49, 49, 50, and 52; $A_2$: 81, 83, 85, 88, 91, and 95; $A_3$: 76, 82, 89, 95, 95, and 98; $A_4$: 86, 86, 87, 87, 87, and 88; and $A_5$: 35, 40, 44, 52, 59, and 62.

The mean value of the $6 \times 5 = 30$ given numerical scores is approximately equal to 72.07, demonstrating a good (C) mean performance of the athletes. For reasons of comparison of the two methods, the system's mean performance is also calculated using GNs.

The given numerical scores provide 14 GNs equal to A, 4 equal to B, 1 equal to C, 4 equal to D and 7 equal to F. Their mean value is $M^* = \frac{1}{30}(14A + 4B + C + 4D + 7F) \in [60.33, 79.63]$. Therefore, Equation (18) gives that $W(M^*) = 69.98$, which shows that the five athletes demonstrate a good performance, but the exact score corresponding to their mean performance is the mean value 72.02 of the given numerical scores calculated in the traditional way.

In conclusion, the assessment methods using GNs and TFNs, although they give a satisfactory approximation of the system's mean performance, are practically useful only when the individual performance is evaluated by qualitative grades and not by numerical scores, because, in this case, the calculation of the mean value of those grades in the traditional way is not possible.

## 7. Application of Fuzzy Relation Equations to Mathematical Modeling

### 7.1. Mathematical Modeling

Until the mid-1970s, *Mathematical Modeling (MM)* was mainly a tool in the hands of scientists and engineers for solving real world problems related to their disciplines (physics, industry, constructions, economics, etc.). However, the failure to introduce "new mathematics" in school education brought the attention of researchers and educators to problem-solving processes, and in particular to the process of MM and its applications. As a result, MM appears today as a dynamic tool for teaching and learning mathematics, because it connects mathematics with our everyday life, giving the possibility to students to understand its usefulness in practice and therefore increasing their interest in it [46].

The steps of the MM process involve:

- $S_1$: *Analyze* the given problem (understanding the statement and recognizing the restrictions and requirements imposed by the corresponding real system).
- $S_2$: *Mathematize* (formulation of the problem and construction of the mathematical model).
- $S_3$: Solve the model.
- $S_4$: *Validate* (control) the model, which is usually achieved by reproducing, through the model, the behavior of the real system under the conditions existing before the solution of the model and by comparing it to the existing from the previous "history" of the real system, i.e., data. In cases of systems having no past history, an extra simulation model could be used for the validation of the mathematical model.
- $S_5$: *Interpret* the final mathematical results and implement them in the real system to give the "answer" to the real-world problem.

### 7.2. Fuzzy Relation Equations

*Fuzzy Relation Equations (FREs)* are associated with the composition of *Fuzzy Binary Relations (FBRs)*. A FBR is defined as follows:

**Definition 1.** Let X and Y be two crisp sets. Then, a FBR R(X, Y) is a FS on the Cartesian product $X \times Y$ of the form: $R(X, Y) = \{(r, mR(r): r = (x, y) \in X \times Y\}$, where $mR: X \times Y \to [0, 1]$ is the corresponding membership function. When $X = \{x_1, \ldots \ldots \ldots, x_n\}$ and $Y = \{y_1, \ldots \ldots, y_m\}$, a FBR R(X, Y) can be represented by a n X m matrix $R = [r_{ij}]$, where $r_{ij} = m_R(x_i, y_j)$, with $i = 1, \ldots, n$ and $j = 1, \ldots, m$. The matrix R is called the membership matrix of the FBR R(X, Y).

The basic ideas of fuzzy relations, which were introduced by Zadeh [47] and further investigated by other researchers, are extensively covered in the book [48].

**Definition 2.** Consider two FBR $P(X, Y)$ and $Q(Y, Z)$ with a common set Y. Then, the standard composition of these relations, denoted by $P(X, Y) \circ Q(Y, Z)$ produces a FBR $R(X, Z)$ with membership function $m_R$ defined by:

$$m_R(x_i, z_j) = \max_{y \in Y} \min \left[ m_P(x_i, y), m_Q(y, z_j) \right], \; i = 1, \ldots, n, \; j = 1, \ldots, m. \tag{19}$$

This composition is usually referred to as *max-min composition*. Compositions of FBR are conveniently performed in terms of their membership matrices. In fact, if $P = [p_{ik}]$ and $Q = [q_{kj}]$ are the membership matrices of the FBRs $P(X, Y)$ and $Q(Y, Z)$m respectively, then by Equation (19)m we get that the membership matrix of $R(X, Y) = P(X, Y) \circ Q(Y, Z)$ is the matrix $R = [r_{ij}]$, with

$$r_{ij} = \max_k \min(p_{ik}, q_{kj}). \tag{20}$$

Note that the same elements of P and Q are used in the calculation of $m_R$ as would be used in the regular multiplication of matrices, but the product and sum operations are here replaced with the min and max operations, respectively.

**Definition 3.** Consider the FBRs $P(X, Y)$, $Q(Y, Z)$ and $R(X, Z)$, defined on the sets $X = \{x_i : i \in Nn\}$, $Y = \{y_j : j \in Nm\}$, $Z = \{z_k : k \in Ns\}$, where $Nt = \{1, 2, \ldots, t\}$, for $t = n, m, k$, and let $P = [p_{ij}]$, $Q = [q_{jk}]$ and $R = [r_{ik}]$ be the membership matrices of $P(X, Y)$, $Q(Y, Z)$ and $R(X, Z)$, respectively. Assume that the above three relations constrain each other in such a way that $P \circ Q = R$, where $\circ$ denotes the max-min composition. This means that, for each i in Nn and each k in Ns,

$$r_{ik} = \max_{j \in J} \min (p_{ij}, q_{jk}). \tag{21}$$

Therefore, the matrix equation $P \circ Q = R$ encompasses nXs simultaneous equations of the form of Equation (21). When two of the components in Equation (21) are given and one is unknown, these equations are referred as FRE. The notion of FRE was first proposed by Sanchez [49] and later further investigated by other researchers (e.g., [50–52]).

*7.3. A Study of MM Skills Using FRE*

Let us consider the crisp sets $X = \{M\}$, $Y = \{A, B, C, D, F\}$ and $Z = \{S_1, S_2, S_3, S_4\}$, where M denotes the imaginary notion of the *average student* of a class; A, B, C, D, and F are the linguistic grades used for the assessment of the student performance; and $S_1, S_2, S_3$, and $S_4$ are the steps of the MM process. Without loss of generality, the validation ($S_4$) and the implementation ($S_5$) of the model, which are usually considered as two different steps, have been joined here in one step ($S_4$) to make simpler the development of our assessment method.

Let n be the total number of students of a class and let $n_i$ be the numbers of students of the class who obtained the grade i assessing their performance, $i \in Y$. Then, one can represent the average student of the class as a FS on Y in the form $M = \{(i, \frac{n_i}{n}) : i \in Y\}$.

The FS M induces a FBR $P(X, Y)$ with membership matrix $P = [\frac{n_A}{n}, \frac{n_B}{n}, \frac{n_B}{n}, \frac{n_C}{n}, \frac{n_F}{n}]$.

In an analogous way, the average student of a class can be represented as a FS on Z in the form $M' = \{(j, m(j)) : j \in Z\}$, where m: $Z \to [0, 1]$ is the corresponding membership function. In this case, the FS M' induces a FBR $R(X, Z)$ with membership matrix $R = [m(S_1) \; m(S_2) \; m(S_3) \; m(S_4)]$.

We consider also the FBR $Q(Y, Z)$ with membership matrix the $5 \times 4$ matrix $Q = [q_{ij}]$, where $q_{ij} = m_Q(i, j)$ with $i \in Y$ and $j \in Z$ and the FRE encompassed by the matrix equation $P \circ Q = R$. When the matrix Q is fixed and the row-matrix P is known, then the above equation always has a unique solution with respect to R, which enables the representation of the average student of a class as a fuzzy set on the set of the steps of the MM process. This is useful for the instructor for designing his/her future teaching plans. On the contrary, when the matrices Q and R are known, then the equation P $\circ$ Q = R may have no solution or it may have more than one solutions with respect to P, which makes the corresponding situation more complicated.

**Example 6.** The following experiment took place at the Graduate Technological Educational Institute of Western Greece, in the city of Patras, when I was teaching a group of 60 students of the School of Technological Applications (future engineers) the use of the derivative for the maximization and minimization of a function. A written test was performed after the end of the teaching process involving two mathematical modeling problems about the construction of a channel to run water and of a cylindrical tower (see the second section of [46]). The results of the test are depicted in Table 4.

**Table 4.** Student performance.

| Grade | No. of Students |
|:-----:|:---------------:|
| A | 20 |
| B | 15 |
| C | 7 |
| D | 10 |
| F | 8 |
| Total | 60 |

Therefore, the average student M of the class can be represented as a fuzzy set on Y = {A, B, C, D, F} by M = {(A, $\frac{20}{60}$), (B, $\frac{15}{60}$), (C, $\frac{7}{60}$), (D, $\frac{10}{60}$), (F, $\frac{8}{60}$)} $\approx$ {(A, 0.33), (B, 0.25), (C, 0.12), (D, 0.17), (F, 0.13)}. Thus, M induces a FBR P(X, Y), where X = {M}, with membership matrix P = [0.33 0.25 0.12 0.17 0.13].

In addition, using statistical data of the last five academic years concerning the MM skills of the students of the School of Technological Applications, we fixed the membership matrix Q of the binary fuzzy relation Q(Y, Z), where Z = {$S_1$, $S_2$, $S_3$, $S_4$}, in the form:

$$\begin{array}{c} \\ A \\ B \\ C \\ D \\ F \end{array} \begin{array}{cccc} S_1 & S_2 & S_3 & S_4 \\ \left(\begin{array}{cccc} 0.7 & 0.5 & 0.3 & 0 \\ 0.4 & 0.6 & 0.3 & 0.1 \\ 0.2 & 0.7 & 0.6 & 0.2 \\ 0.1 & 0.5 & 0.7 & 0.5 \\ 0 & 0.7 & 0.5 & 0.8 \end{array}\right) \end{array}$$

The statistical data were collected by the instructor who was inspecting the student reactions during the solution of several MM problems in the classroom.

Next, using the max-min composition of FBR, one finds that the membership matrix of R(X, Z) = P(X, Y) $\circ$ Q (Y, Z) is equal to R = P $\circ$ Q = [0.33 0.33 0.3 0.17]. Therefore, the average student of the class can be expressed as a fuzzy set on Z by M = {($S_1$, 0.33), ($S_2$, 0.33), ($S_3$, 0.3), ($S_4$, 0.17)}.

The conclusions obtained from the above expression of M are the following:

- Only $\frac{1}{3}$ of the students of the class were ready to use contents of their memory (background knowledge, etc.) to facilitate the solution of the given problems.
- All the above students were able to design the model and almost all of them were able to execute the solutions of the given problems.
- On the contrary, half of the above students could not check the correctness of the solutions found and therefore implement correctly the mathematical results to the real system.

The first conclusion was not surprising, since the majority of the students have the wrong habit to start studying the material of their courses the last month before the final exams. On the other hand, the second conclusion shows that the instructor's teaching procedure was successful, enabling the diligent students to plan and execute successfully the solutions of the given problems. Finally, the last conclusion is explained by the fact that students, when solving MM problems, frequently omit to check if their solutions are compatible to the restrictions imposed by the real system. Therefore, the instructor should emphasize during his/her lectures that the last two steps of the MM process (validation and implementation) are not a formality, but have their own importance for preventing several mistakes.

Let us now consider the case where the membership matrices $Q$ and $R$ are known and we want to determine the matrix $P$ representing the average student of the class as a fuzzy set on $Y$. This is a complicated case because we may have more than one solution or no solution at all. The following two examples illustrate this situation.

**Example 7.** Consider the membership matrices $Q$ and $R$ of the previous example and set $P = [p_1\ p_2\ p_3\ p_4\ p_5]$. Then, the matrix equation $P \circ Q = R$ encompasses the following equations:

$$\max \{\min (p_1, 0.7), \min (p_2, 0.4), \min (p_3, 0.2), \min (p_4, 0.1), (p_5, 0)\}= 0.33$$
$$\max \{\min (p_1, 0.5), \min (p_2, 0.6), \min (p_3, 0.7), \min (p_4, 0.5), \min (p_5, 0.1)\}= 0.33$$
$$\max \{\min (p_1, 0.3), \min (p_2, 0.3), \min (p_3, 0.6), \min (p_4, 0.7), (p_5, 0.5)\}= 0.3$$
$$\max \{\min (p_1, 0), \min (p_2, 0.1), \min (p_3, 0.2), \min (p_4, 0.5), \min (p_5, 0.8)\}= 0.17$$

The first of the above equations is true if, and only if, $p_1 = 0.33$ or $p_2 = 0.33$, values that satisfy the second and third equations as well. In addition, the fourth equation is true if, and only if, $p_3 = 0.17$, $p_4 = 0.17$ or $p_5 = 0.17$. Therefore, any combination of values of $p_1$, $p_2$, $p_3$, $p_4$, and $p_5$ in $[0, 1]$ such that $p_1 = 0.33$ or $p_2 = 0.33$ and $p_3 = 0.17$, $p_4 = 0.17$ or $p_5 = 0.17$ is a solution of $P \circ Q = R$.

Let $S(Q, R) = \{P: P \circ Q = R \}$ be the set of all solutions of $P \circ Q = R$. Then, one can define a partial ordering on $S(Q, R)$ by $P \leq P' \Leftrightarrow p_i \leq p'_i, \forall_i = 1, 2, 3, 4, 5.$

It is well established that, whenever $S(Q, R)$ is a non-empty set, it always contains a unique maximum solution and it may contain several minimal solutions [49]. It is further known that $S(Q, R)$ is fully characterized by the maximum and minimal solutions in the sense that all its other elements are between the maximal and each of the minimal solutions [49]. A method of determining the maximal and minimal solutions of $P \circ Q = R$ with respect to $P$ is developed in [52].

**Example 8.** Let $Q = [q_{ij}]$, $i = 1, 2, 3, 4, 5$ and $j = 1, 2, 3, 4$ be as in Example 2 and let $R = [1\ 0.33\ 0.3\ 0.17]$. Then, the first equation encompassed by the matrix equation $P \circ Q = R$ is $\max \{\min (p_1, 0.7), \min (p_2, 0.4), \min (p_3, 0.2), \min (p_4, 0.1),$ and $\min (p_5, 0)\}= 1$. In this case, it is easy to observe that the above equation has no solution with respect to $p_1$, $p_2$, $p_3$, $p_4$, and $p_5$, therefore the matrix equation $P \circ Q = R$ has no solution with respect to $P$. In general, writing $R = \{r_1\ r_2\ r_3\ r_4\}$, it becomes evident that we have no solution if $\max_j q_{ij} < r_i$.

## 8. Discussion and Conclusions

The enormous development of technology during the last years makes human life easier and more comfortable. However, the technological progress creates in parallel more and more complicated artificial systems, which are difficult to be managed by the traditional scientific methods. As a result, while 50–60 years ago probability theory used to be a unique tool in the hands of scientists for dealing with situations characterized by uncertainty and/or vagueness, today this is no loneger the rule. In fact, the introduction of FL and its generalizations (intuitionistic FS, neutrosophic sets, etc.) and of other relative theories (rough sets, soft sets, GS theory, etc.) provide new tools for dealing with such situations in real life, science and technology and enable the solution of problems with fuzzy or approximate data, which cannot be solved with traditional techniques of probability theory. The applications of FL and its relevant theories have been rapidly expanded nowadays, covering almost all sectors of human activities.

The objective of this review article is to present the author's research on developing methods for the assessment of human–machine capacities under fuzzy conditions. Those methods include:

1.  The measurement of the corresponding fuzzy system's probabilistic uncertainty (generalized Shannon's entropy) or its total possibilistic uncertainty (the sum of strife and non-specificity) for evaluating its mean performance. However, this method can be applied for comparing the performance of two different systems with respect to a common activity only under the assumption that the uncertainty in those systems is the same before the activity (equivalent systems). Moreover, the method cannot provide an exact characterization of a system's performance and it involves laborious calculations.

2.  The utilization of the COG defuzzification technique (rectangular fuzzy assessment model) for assessing a fuzzy system's quality performance. This method, initiated by Subbotin et al. [31], is useful, due to its logical nature, when comparing the performance of two systems with equal values of the traditional GPA index. In this case, the GPA index could lead to conclusions that are not close to the reality. On the contrary, for different values of the GPA index, the two methods provide the same assessment outcomes.

3.  The use of TFNs as assessment tools, a method that is easy to apply in practice and gives an exact characterization of the system's mean performance. That method is useful when qualitative grades and not numerical scores are used for the evaluation of the system's performance, which makes impossible the calculation of the mean value of those grades in a traditional way. However, a disadvantage of the method is that its understanding requires knowledge of basic principles of FS theory, which is not always easy for non-specialists.

4.  The use of GNs, instead of TFNs, as assessment tools. These two methods are equivalent to each other, providing the same assessment outcomes. However, GNs can be easily defined with the help of closed real intervals, which makes the method more accessible to non-specialists. Moreover, the use of GNs reduces significantly the required computational burden.

5.  The application of FRE for assessing MM skills. This method enables the teacher to obtain useful conclusions about student progress and was applied by the author, with the proper modifications each time, to various other assessment situations (problem-solving, learning a subject matter, human and machine reasoning, etc.) [53–55].

The above models provide an innovative framework for further research on human–machine assessment. Our future plans involve improving and extending our fuzzy and grey assessment methods and applying the principles of FL and GS theory to other fields of human activity as well. Note that such efforts have already been started by the author on solving equations, systems of equations and linear programming problems with fuzzy or grey data, connected to real life applications [56–59].

**Funding:** This research received no external funding.

**Conflicts of Interest:** The author declares no conflict of interest.

## References


1.  Balley, K.D. *Sociology and the New Systems Theory: Toward a Theoretical Synthesis*; State of New York Press: New York, NY, USA, 1994.
2.  Xu, K.; Tang, L.C.; Xie, M.; Ho, S.L.; Zhu, M.L. Fuzzy assessment of FMEA for engine systems. *Reliab. Eng. Syst. Saf.* **2002**, *75*, 17–29. [CrossRef]
3.  Liu, L.; Lee, H.-M. Fuzzy assessment method on sampling survey analysis. *Expert Syst. Appl.* **2009**, *36*, 5955–5961.
4.  Macwan, N.; Srinivas, S.P. A Linguistic Fuzzy Approach for Employee Evaluation. *Int. J. Adv. Res. Comput. Sci. Softw. Eng.* **2014**, *1*, 975–980.
5.  Liu, R.-T.; Huang, W.-C. Fuzzy Assessment on Reservoir Water Quality. *J. Mar. Sci. Technol.* **2015**, *23*, 231–239.
6.  Jevscek, M. Competencies assessment using fuzzy logic. *J. Univers. Excell.* **2016**, *5*, 187–202.
7.  Korner, S. Laws of Thought. In *Encyclopaedia of Philosophy*; Mac Millan: New York, NY, USA, 1967; Volume 4, pp. 414–417.
8.  Lejewski, C. Jan Lukasiewicz. *Encycl. Philos.* **1967**, *5*, 104–107.

9.   Tarski, A. Encyclopaedia Brittanica. 2018. Available online: www.britannica.com/biography/Alfred-Tarski (accessed on 20 December 2018).

10.  Zadeh, L.A. Outline of a new approach to the analysis of complex systems and decision processes. *IEEE Trans. Syst. Man Cybern.* **1973**, *3*, 28–44. [CrossRef]

11.  Zadeh, L.A. Fuzzy Sets. *Inf. Control* **1965**, *8*, 338–353. [CrossRef]

12.  Klir, G.J.; Folger, T.A. *Fuzzy Sets, Uncertainty and Information*; Prentice-Hall: London, UK, 1988.

13.  Haack, S. Do we need fuzzy logic? *Int. J. Man-Mach. Stud.* **1979**, *11*, 437–445. [CrossRef]

14.  Fox, J. Towards a reconciliation of fuzzy logic and standard logic. *Int. J. Man Mach. Stud.* **1981**, *15*, 213–220. [CrossRef]

15.  Tah, J.H.M.; Carr, V. A proposal for construction project risk assessment using fuzzy logic. *Constr. Manag. Econ.* **2010**, *18*, 491–500. [CrossRef]

16.  Collatta, M.; Pau, G.; Solerno, V.M.; Scata, G. A novel trust based algorithm for carpooling transportation systems. In Proceedings of the IEEE International Energy Conference and Exhibition, Florence, Italy, 9–12 September 2012.

17.  Voskoglou, M.G. *Finite Markov Chain and Fuzzy Logic Assessment Models: Emerging Research and Opportunities*; Createspace.com–Amazon: Columbia, SC, USA, 2017.

18.  Wu, C.; Liu, G.; Huang, G.; Liu, Q.; Guan, X. Ecological Vulnerability Assessment Based on Fuzzy Analytical Method and Analytic Hierarchy Process in Yellow River Delta. *Int. J. Environ. Res. Public Health* **2018**, *15*, 855. [CrossRef] [PubMed]

19.  Nilashi, M.; Cavallaro, F.; Mardani, A.; Zavadskas, E.K.; Samad, S.; Ibrahim, O. Measuring Country Sustainability Performance Using Ensembles of Neuro-Fuzzy Technique. *Sustainability* **2018**, *10*, 2707. [CrossRef]

20.  Atanassov, K.T. Intuitionistic Fuzzy Sets. *Fuzzy Sets Syst.* **1986**, *20*, 87–96. [CrossRef]

21.  Smarandache, F. *Neutrosophy/Neutrosophic Probability, Set, and Logic*; Proquest: Ann Arbor, MI, USA, 1998.

22.  Deng, J. Control Problems of Grey Systems. *Syst. Control Lett.* **1982**, *1*, 288–294.

23.  Pawlak, Z. *Rough Sets: Aspects of Reasoning about Data*; Kluer Academic Publishers: Dordrecht, The Netherlands, 1991.

24.  Molodtsov, D. Soft Set Theory–First Results. *Comput. Math. Appl.* **1999**, *37*, 19–31. [CrossRef]

25.  Shannon, C.E. A mathematical theory of communications. *Bell Syst. Tech. J.* **1948**, *27*, 379–423, 623–656. [CrossRef]

26.  Klir, G.J. Principles of Uncertainty: What are they? Why do we need them? *Fuzzy Sets Syst.* **1995**, *74*, 15–31. [CrossRef]

27.  Shackle, G.L.S. *Decision, Order and Time in Human Affairs*; Cambridge University Press: Cambridge, UK, 1961.

28.  Van Broekhoven, E.; De Baets, B. Fast and accurate centre of gravity defuzzification of fuzzy systems outputs defined on trapezoidal fuzzy partitions. *Fuzzy Sets Syst.* **2006**, *157*, 904–918. [CrossRef]

29.  Voskoglou, M.G. An Application of Fuzzy Sets to the Process of Learning. *Heuristics Didact. Exact Sci.* **1999**, *10*, 9–13.

30.  Voskoglou, M.G. Transition across Levels in the Process of Learning. *Int. J. Model. Appl.* **2009**, *1*, 37–44.

31.  Subbotin, I.Y.; Badkoobehi, H.; Bilotckii, N.N. Application of fuzzy logic to learning assessment. *Didact. Math. Probl. Investig.* **2004**, *22*, 38–41.

32.  Subbotin, I.Y. Trapezoidal Fuzzy Logic Model for Learning Assessment. *arXiv*, 2014; arXiv:1407.0823.

33.  Zadeh, L.A. The Concept of a Linguistic Variable and its Application to Approximate Reasoning, Parts 1–3. *Inf. Sci.* **1975**, *8, 9*, 43–80, 199–249, and 301–357 . [CrossRef]

34.  Sakawa, M. *Fuzzy Sets and Interactive Multiobjective Optimization*; Plenum Press: London, UK, 1993.

35.  Kaufmann, A.; Gupta, M. *Introduction to Fuzzy Arithmetic*; Van Nostrand Reinhold Company: New York, NY, USA, 1991.

36.  Voskoglou, M.G. Use of the Triangular Fuzzy Numbers for Student Assessment. *Am. J. Appl. Math. Stat.* **2015**, *3*, 146–150.

37.  Voskoglou, M.G. Assessment of Human Skills Using Trapezoidal Fuzzy Numbers. *Am. J. Appl. Math. Stat.* **2015**, *5*, 111–116.

38.  Deng, J. Introduction to Grey System Theory. *J. Grey Syst.* **1989**, *1*, 1–24.

39.  Moore, R.A.; Kearfort, R.B.; Clood, M.J. *Introduction to Interval Analysis*, 2nd ed.; SIAM: Philadelphia, PA, USA, 1995.

40. Liu, S.F.; Lin, Y. (Eds.) *Advances in Grey System Research*; Springer: Berlin/Heidelberg, Germany, 2010.
41. Voskoglou, M.G.; Theodorou, Y. Application of Grey Numbers to Assessment Processes. *Int. J. Appl. Fuzzy Sets Artif. Intell.* **2017**, *7*, 59–72.
42. Voskoglou, M.G.; Salem, A.-B.M. Analogy-Based and Case-Based Reasoning: Two Sides of the Same Coin. *Int. J. Appl. Fuzzy Sets Artif. Intell.* **2014**, *4*, 7–18.
43. Reznik, L. *Fuzzy Controllers*; Newnes: Oxford, UK, 1997.
44. Theodorou, Y.A. *Introduction to Fuzzy Logic*; Tziolas Editions: Thessaloniki, Greece, 2010. (In Greek)
45. Wikipedia. Fuzzy Control System. March 2012. Available online: http://en.wikipedia.org/wiki/Fuzzy_control_system (accessed on 20 December 2018).
46. Voskoglou, M.G. Mathematical modelling as a teaching method of mathematics. *J. Res. Innov. Teach. (Natl. Univ. CA)* **2015**, *8*, 35–50.
47. Zadeh, L.A. Similarity relations and fuzzy orderings. *Inf. Sci.* **1971**, *3*, 177–200. [CrossRef]
48. Kaufmann, A. *Introduction to the Theory of Fuzzy Subsets*; Academic Press: New York, NY, USA, 1975.
49. Sanchez, E. Resolution of Composite Fuzzy Relation Equations. *Inf. Control* **1976**, *30*, 38–43. [CrossRef]
50. Prevot, M. Algorithm for the solution of fuzzy relations. *Fuzzy Sets Syst.* **1981**, *5*, 319–322. [CrossRef]
51. Czogala, E.; Drewniak, J.; Pedryz, W. Fuzzy relation equations on a finite set. *Fuzzy Sets Syst.* **1982**, *7*, 89–101. [CrossRef]
52. Higashi, M.; Klir, G.J. Resolution of finite fuzzy relation equations. *Fuzzy Sets Syst.* **1984**, *13*, 65–82. [CrossRef]
53. Voskoglou, M.G. A Study of Student Learning Skills Using Fuzzy Relation Equations. *Egypt. Comput. Sci. J.* **2018**, *42*, 80–87.
54. Voskoglou, M.G. Application of Fuzzy Relation Equations to Assessment of Analogical Problem Solving Skills. *J. Phys. Math. Educ.* **2018**, *15*, 122–127. [CrossRef]
55. Voskoglou, M.G. Application of Fuzzy Relation Equations Student Assessment. *Am. J. Appl. Math. Stat.* **2018**, *6*, 167–171. [CrossRef]
56. Voskoglou, M.G. Solving Systems of Equations with Grey Data. *Int. J. Appl. Fuzzy Sets Artif. Intell.* **2018**, *8*, 103–111.
57. Voskoglou, M.G. Solving Linear Programming Problems with Grey Data. *Orient. J. Phys. Sci.* **2018**, *3*, 17–23.
58. Voskoglou, M.G. Fuzzy Linear Programming. *Egypt. J. Comput. Sci.* **2018**, *42*, 1–14.
59. Voskoglou, M.G. Systems of Equations with Fuzzy Coefficients. *J. Phys. Sci.* **2018**, *23*, 77–88.

*Article*

# The "Generator" of Int-Soft Filters on Residuated Lattices

**Huarong Zhang * and Minxia Luo ***

College of Science, China Jiliang University, Hangzhou 310000, Zhejiang, China
* Correspondence: hrzhang2008@cjlu.edu.cn (H.Z.); 06a0802068@cjlu.edu.cn (M.L.)

Received: 15 January 2019; Accepted: 28 February 2019; Published: 6 March 2019

**Abstract:** In this paper, we give the "generator" of int-soft filters and propose the notion of t-int-soft filters on residuated lattices. We study the properties of t-int-soft filters and obtain some commonalities (e.g., the extension property, quotient characteristics, and a triple of equivalent characteristics). We also use involution-int-soft filters as an example and show some basic properties of involution-int-soft filters. Finally, we investigate the relations among t-int-soft filters and give a simple method for judging their relations.

**Keywords:** residuated lattice; soft set; filter; t-filter; t-int-soft filter

---

## 1. Introduction

Uncertainty widely exists in many practical problems. Compared with probability theory [1], fuzzy sets [2], rough sets [3], intuitionistic fuzzy sets [4], and soft sets [5] have an unparalleled advantage in solving uncertain problems. At present, many scholars are devoted to the study of soft set theory and its applications. The theoretical work mainly focuses on some operations of soft sets [6–8], soft and fuzzy soft relations [9], soft algebraic structures [10], and distance, similarity measures, and equality of soft sets [11–13]. In applications, soft sets are extensively used in decision making problems [14] and forecasting approaches [15].

Residuated lattices [16] originated from mathematical logic without contraction. They combine the fundamental notions of multiplication, order, and residuation. In recent years, many logical algebras have been introduced as the semantic systems of logical systems, for example, Boolean algebras, MV-algebras, BL-algebras, R0-algebras, Heyting algebras, MTL-algebras, and so on. These logical algebras are all special cases of residuated lattices.

Filters correspond to sets of formulae, closed with respect to Modus ponens. So, filters play an important role in investigating the above logical algebras. At present, the filter theories of many fuzzy logical algebras have been extensively studied. On residuated lattices, the relative literature is as follows: [17–26]. In the literature, many concrete types of filters (implicative filters, fantastic filters, Boolean filters, and so on) have been introduced, their equivalent characterizations studied, and the relations among specific filters were investigated on residuated lattices. Notably, Víta [24] proposed the notion of t-filters, in order to cover the great amount of special types of filters, and obtained some basic properties of t-filters.

Recently, some scholars have applied soft sets to the filter theory of logical algebras. In [27,28], Jun et al. proposed the (strong, implicative) intersection-soft (int-soft for short) filters, and established their equivalent characterizations and the extension property of (strong) implicative int-soft filters on R0-algebras. Lin and Ma [29] proved that all int-soft filters formed a bounded distributive lattice and investigated the int-soft congruences, with respect to int-soft filters on residuated lattices. Jun et al. [30] introduced the concepts of int-soft filters, MV-int-soft filters, int-soft G-filters, and regular int-soft filters, explored their properties and characterizations, and provided the conditions for an int-soft filter to be an int-soft G-filter on residuated lattices.

T-filters provide a path toward the unifying of some types of filters. It is natural to ask whether we can give a framework to cover as many int-soft filters as possible and obtain some of their basic features. Based on this, we propose the notion of t-int-soft filters on residuated lattices, and show how particular results about many kinds of int-soft filters are unified by our framework. In addition, research about relations among kinds of int-soft filters is a hot topic. Is there a simple method to find them? In order to answer this question, we particularly investigate the quotient structure of t-int-soft filters, use it to study the relations among t-int-soft filters, and give a simple method for judging their relations. The paper is organized as follows:

In Section 2, some preliminary definitions and theorems are recalled. In Section 3, the notion of t-int-soft filters is proposed. Some characterizations of t-int-soft filters are derived. In Section 4, a specific example of a t-int-soft filter is given. In Section 5, the general principles of investigating the relations among t-int-soft filters are given.

## 2. Preliminaries

**Definition 1.** *[16] A residuated lattice is an algebra* $(L, \wedge, \vee, \otimes, \rightarrow, 0, 1)$*, such that the following conditions hold:*

R1 $(L, \wedge, \vee, 0, 1)$ *is a bounded lattice,*
R2 $(L, \otimes, 1)$ *is a commutative monoid, and*
R3 $x \otimes y \leq z$ *if and only if* $x \leq y \rightarrow z$ *for all* $x, y, z \in L$.

For $x \in L$, we define $x^* = x \rightarrow 0$ and $x^{**} = (x^*)^*$. For a natural number $n$, we define $x^0 = 1$ and $x^n = x^{n-1} \otimes x$ for $n \geq 1$.

**Definition 2.** *[17,22,31,32] Let L be a residuated lattice. Then, L is called:*

- *An involutive (or regular) residuated lattice if* $x^{**} = x$ *for* $x \in L$;
- *a Heyting algebra if* $x \otimes y = x \wedge y$ *for all* $x, y \in L$*, which is equivalent to* $x^2 = x$ *for all* $x \in L$;
- *a Rl-monoid if* $x \wedge y = x \otimes (x \rightarrow y)$ *for all* $x, y \in L$ *(axiom of divisibility);*
- *an MTL-algebra if* $(x \rightarrow y) \vee (y \rightarrow x) = 1$ *for all* $x, y \in L$ *(axiom of prelinearity);*
- *a BL-algebra if it satisfies both axioms of prelinearity and divisibility;*
- *an MV-algebra if it is a regular Rl-monoid; and*
- *a Boolean algebra if it is an idempotent MV-algebra.*

In what follows, $L$ denotes a residuated lattice, unless otherwise specified.

We give some rules of calculus on $L$, which will be needed in the rest of study.

(1)  $0^* = 1, 1^* = 0$.
(2)  $x \rightarrow y \leq (y \rightarrow z) \rightarrow (x \rightarrow z)$.
(3)  $x \otimes y \rightarrow z = x \rightarrow (y \rightarrow z)$.
(4)  $x \vee y \rightarrow y = x \rightarrow y$.
(5)  $x \leq y$ if and only if $x \rightarrow y = 1$.

**Definition 3.** *[23] A nonempty subset F of L is called a filter if*

(1)  *If* $x \in F$ *and* $x \leq y$*, then* $y \in F$*, and*
(2)  *If* $x, y \in F$*, then* $x \otimes y \in F$.

**Theorem 1.** *[23] F is a filter if and only if*

(1)  $1 \in F$*, and*
(2)  *If* $x, x \rightarrow y \in F$*, then* $y \in F$.

It is clear that $\{1\}$ and $L$ are filters. In addition, given a filter F of $L$, we can define the relation $\equiv_F$ on $L$ by $x \equiv_F y$ if and only if $x \to y \in F$ and $y \to x \in F$. We can also prove that $\equiv_F$ is a congruence relation. We use $L/F$ to denote the set of the congruence classes of $\equiv_F$ (i.e., $L/F = \{[x]_F | x \in L\}$), where $[x]_F := \{y \in L | y \equiv_F x\}$. If we define the following operations on $L/F$: $[x]_F \sqcap' [y]_F = [x \wedge y]_F$, $[x]_F \sqcup' [y]_F = [x \vee y]_F$, and $[x]_F \odot' [y]_F = [x \otimes y]_F$, $[x]_F \to' [y]_F = [x \to y]_F$, then, we have the following lemma.

**Lemma 1.** *[23,26] Let F be a filter of L. Then $(L/F, \sqcap', \sqcup', \odot', \to', [0]_F, [1]_F)$ is a residuated lattice, with respect to F.*

In what follows, we recall some basic concepts of soft sets.

Let $U$ be an initial universe set, and $E$ a set of parameters. Let $\mathscr{P}(U)$ denote the power set of $U$ and $A, B, C \cdots \subseteq E$.

**Definition 4.** *[5] A soft set $(\tilde{f}, A)$ over U is defined to be the set of ordered pairs*

$$(\tilde{f}, A) := \{(x, \tilde{f}(x)) : x \in E, \tilde{f}(x) \in \mathscr{P}(U)\},$$

*where $\tilde{f} : E \to \mathscr{P}(U)$, such that $\tilde{f}(x) = \varnothing$ if $x \notin A$.*

Now, we take a residuated lattice $L$ as the set of parameters.

**Definition 5.** *[30] A soft set $(\tilde{f}, L)$ over U is called an int-soft filter of L if it satisfies*

(1)  $\tilde{f}(x) \cap \tilde{f}(y) \subseteq \tilde{f}(x \otimes y), \forall x, y \in L$, *and*
(2)  *If $x \le y$, then $\tilde{f}(x) \subseteq \tilde{f}(y), \forall x, y \in L$.*

**Lemma 2.** *[30] Every int-soft filter $(\tilde{f}, L)$ of L satisfies*

(1)  $\tilde{f}(x) \subseteq \tilde{f}(1), \forall x \in L$, *and*
(2)  $\tilde{f}(x) \cap \tilde{f}(x \to y) \subseteq \tilde{f}(y), \forall x, y \in L$.

**Theorem 2.** *[30] A soft set $(\tilde{f}, L)$ over U is an int-soft filter of L if and only if the set*

$$\tilde{f}_\tau := \{x \in L \mid \tau \subseteq \tilde{f}(x)\}$$

*is a filter of L, for all $\tau \in \mathscr{P}(U)$, with $\tilde{f}_\tau \ne \varnothing$.*

**Lemma 3.** *A soft set $(\tilde{T}, L)$ is an int-soft filter of L, where $\tilde{T}$ has the following form*

$$\tilde{T}(x) = \begin{cases} U, & \text{if } x = 1, \\ \varnothing, & \text{if } x \ne 1. \end{cases} \tag{1}$$

**Proof.** $\forall \tau \in \mathscr{P}(U)$,

$$\tilde{T}_\tau = \begin{cases} L, & \text{if } \tau = \varnothing, \\ \{1\}, & \text{if } \tau \ne \varnothing. \end{cases} \tag{2}$$

By Theorem 2, we know that $(\tilde{T}, L)$ is an int-soft filter.  $\square$

**Lemma 4.** *Let $(\tilde{f}, L)$ be a soft set. If $(\tilde{f}, L)$ is an int-soft filter, then $\tilde{f}_{\tilde{f}(1)}$ is a filter.*

**Proof.** Let $(\tilde{f}, L)$ be an int-soft filter. Since $1 \in \tilde{f}_{\tilde{f}(1)}$, we have $\tilde{f}_{\tilde{f}(1)} \neq \varnothing$. By Theorem 2, we have $\tilde{f}_{\tilde{f}(1)}$ is a filter. $\square$

## 3. T-Filters and T-Int-Soft Filters

In this section, we use the symbol $\bar{x}$ to indicate the abbreviation of $x, y, \ldots$; that is, $\bar{x}$ is a formal listing of variables used in a given context. We use the term t to denote a term in the language of residuated lattices. Given a variety $\mathbb{B}$ of residuated lattices, we denote its subvariety satisfying the equation t = 1 by the symbol $\mathbb{B}[\text{t}]$, and we call this algebra the t-algebra.

**Definition 6.** *[24] Let t be an arbitrary term on the language of residuated lattices. A filter F of L is a t-filter if* $t(\bar{x}) \in F$ *for all* $\bar{x} \in L$.

**Example 1.** *[17,18,21,24–26] Let F be a filter of L. Then F is,*

- *an involution filter (or a regular filter) for* $t(\bar{x}) = x^{**} \to x$;
- *a Heyting filter for* $t(\bar{x}) = x \to x^2$;
- *a Rl-monoid filter (or a divisible filter) for* $t(\bar{x}) = (x \wedge y) \to (x \otimes (x \to y))$;
- *a MTL-filter for* $t(\bar{x}) = (x \to y) \vee (y \to x)$;
- *a BL-filters for* $t(\bar{x}) = ((x \to y) \to (x \to z)) \to ((x \to z) \vee (y \to z))$;
- *a MV-filter for* $t(\bar{x}) = ((x \to y) \to y) \to ((y \to x) \to x)$; and
- *a Boolean filter for* $t(\bar{x}) = x \vee x^*$.

**Remark 1.** *Many of the filters in Example 1 have different names. To avoid confusion, Buşneag and Piciu [18] proposed a new approach for classifying filters and renamed some filters.*

**Theorem 3.** *[24] Let $\mathbb{B}$ be a variety of residuated lattices and $L \in \mathbb{B}$. Then, the following statements are equivalent:*

*(1) Every filter of L is a t-filter.*
*(2) $\{1\}$ is a t-filter.*
*(3) $L \in \mathbb{B}(t)$.*

**Theorem 4.** *[24] Let $\mathbb{B}$ be a variety of residuated lattices, $L \in \mathbb{B}$, and F a filter of L. Then, F is a t-filter if and only if $L/F \in \mathbb{B}(t)$.*

**Definition 7.** *Let $(\tilde{f}, L)$ be an int-soft filter of L. $(\tilde{f}, L)$ is called a t-int-soft filter if, for all $\tau \in \mathscr{P}(U)$, $\tilde{f}_\tau$ is either empty or a t-filter.*

**Remark 2.** *According to Definition 7, we can define involution-int-soft filters, MTL-int-soft filters, and BL-int-soft filters.*

**Theorem 5.** *Let $(\tilde{f}, L)$ be an int-soft filter of L. Then, $(\tilde{f}, L)$ is a t-int-soft filter if and only if $\tilde{f}_{\tilde{f}(1)}$ is a t-filter.*

**Proof.** Let $(\tilde{f}, L)$ be a t-int-soft filter. Since $1 \in \tilde{f}_{\tilde{f}(1)}$, then $\tilde{f}_{\tilde{f}(1)} \neq \varnothing$. By Definition 7, we have that $\tilde{f}_{\tilde{f}(1)}$ is a t-filter.

Conversely, suppose $(\tilde{f}, L)$ is an int-soft filter of L. Then, by Theorem 2, $\forall \tau \in \mathscr{P}(U)$, we have if $\tilde{f}_\tau \neq \varnothing$, then $\tilde{f}_\tau$ is a filter. Let $\tilde{f}_{\tilde{f}(1)}$ be a t-filter, then $t(\bar{x}) \in \tilde{f}_{\tilde{f}(1)}$. For all $\tau \in \mathscr{P}(U)$, there exist two cases:

(i) $\tau \subseteq \tilde{f}(1)$:

Suppose $x \in \tilde{f}_{\tilde{f}(1)}$, then $\tilde{f}(x) \supseteq \tilde{f}(1) \supseteq \tau$. This shows that $x \in \tilde{f}_\tau$. Thus, $\tilde{f}_{\tilde{f}(1)} \subseteq \tilde{f}_\tau$. Hence, $t(\tilde{x}) \in \tilde{f}_\tau$. By Definition 6, we have that $\tilde{f}_\tau$ is a t-filter.

(ii)    $\tau \supset \tilde{f}(1)$:

Since $(\tilde{f}, L)$ is an int-soft filter, then $\forall x \in L$, $\tau \supset \tilde{f}(1) \supseteq \tilde{f}(x)$. Thus, $\tilde{f}_\tau = \varnothing$.

By Definition 7, we know that $(\tilde{f}, L)$ is a t-int-soft filter. $\quad\square$

**Theorem 6.** *(Extension property) Let $\tilde{f}, \tilde{g}$ be int-soft filters of $L$, $\tilde{f}(x) \subseteq \tilde{g}(x)$ for all $x \in L$, and, moreover, $\tilde{f}(1) \supseteq \tilde{g}(1)$. If $(\tilde{f}, L)$ is a t-int-soft filter, then $(\tilde{g}, L)$ is a t-int-soft filter.*

**Proof.** Let $(\tilde{f}, L)$ be a t-int-soft filter and $\tilde{f}(1) \supseteq \tilde{g}(1)$. Then, $\tilde{f}_{\tilde{f}(1)}$ is a t-filter. Thus, $t(\tilde{x}) \in \tilde{f}_{\tilde{f}(1)}$. Suppose $x \in \tilde{f}_{\tilde{f}(1)}$, then $\tilde{f}(x) \supseteq \tilde{f}(1)$. Also, $\tilde{g}(x) \supseteq \tilde{f}(x)$, and thus $\tilde{g}(x) \supseteq \tilde{f}(x) \supseteq \tilde{f}(1) \supseteq \tilde{g}(1)$. This shows that $x \in \tilde{g}_{\tilde{g}(1)}$. That is, $\tilde{f}_{\tilde{f}(1)} \subseteq \tilde{g}_{\tilde{g}(1)}$. Thus, $t(\tilde{x}) \in \tilde{g}_{\tilde{g}(1)}$. We have that $\tilde{g}_{\tilde{g}(1)}$ is a t-filter. Hence, $\tilde{g}$ is a t-int-soft filter. $\quad\square$

The next part concerns the quotient structure.

**Lemma 5.** *If $(\tilde{f}, L)$ is an int-soft filter, then $\tilde{f}(x \to z) \supseteq \tilde{f}(x \to y) \cap \tilde{f}(y \to z)$ for all $x, y, z \in L$.*

**Proof.** Assume that $(\tilde{f}, L)$ is an int-soft filter. Since $x \to y \leq (y \to z) \to (x \to z)$, it follows, from Definition 5 (2), that $\tilde{f}((y \to z) \to (x \to z)) \supseteq \tilde{f}(x \to y)$. From Lemma 2 (2), we have $\tilde{f}(x \to z) \supseteq \tilde{f}(y \to z) \cap \tilde{f}((y \to z) \to (x \to z))$. Thus, $\tilde{f}(x \to z) \supseteq \tilde{f}(x \to y) \cap \tilde{f}(y \to z)$. $\quad\square$

**Lemma 6.** *Let $(\tilde{f}, L)$ be an int-soft filter of $L$ and $x, y \in L$. For any $z \in L$, we define $\tilde{f}^x : L \to \mathscr{P}(U)$, $\tilde{f}^x(z) = \tilde{f}(x \to z) \cap \tilde{f}(z \to x)$. Then, $\tilde{f}^x = \tilde{f}^y$ if and only if $\tilde{f}(x \to y) \supseteq \tilde{f}(1)$ and $\tilde{f}(y \to x) \supseteq \tilde{f}(1)$.*

**Proof.** Suppose $\tilde{f}^x = \tilde{f}^y$, then $\tilde{f}^x(x) = \tilde{f}^y(x)$. Also, $\tilde{f}^x(x) = \tilde{f}(1)$, $\tilde{f}^y(x) = \tilde{f}(y \to x) \cap \tilde{f}(x \to y)$. Thus, $\tilde{f}(x \to y) \supseteq \tilde{f}(1)$ and $\tilde{f}(y \to x) \supseteq \tilde{f}(1)$.

Conversely, suppose $\tilde{f}(x \to y) \supseteq \tilde{f}(1)$ and $\tilde{f}(y \to x) \supseteq \tilde{f}(1)$. For any $z \in L$, $\tilde{f}^x(z) = \tilde{f}(x \to z) \cap \tilde{f}(z \to x)$, $\tilde{f}^y(z) = \tilde{f}(y \to z) \cap \tilde{f}(z \to y)$. By Lemma 5, $\tilde{f}(x \to z) \supseteq \tilde{f}(x \to y) \cap \tilde{f}(y \to z) \supseteq \tilde{f}(1) \cap \tilde{f}(y \to z) \supseteq \tilde{f}(y \to z)$. We have $\tilde{f}(z \to x) \supseteq \tilde{f}(z \to y) \cap \tilde{f}(y \to x) \supseteq \tilde{f}(z \to y) \cap \tilde{f}(1) \supseteq \tilde{f}(z \to y)$. Thus, $\tilde{f}(x \to z) \cap \tilde{f}(z \to x) \supseteq \tilde{f}(z \to y) \cap \tilde{f}(y \to z)$. Similarly, $\tilde{f}(y \to z) \cap \tilde{f}(z \to y) \supseteq \tilde{f}(x \to z) \cap \tilde{f}(z \to x)$. Therefore, $\tilde{f}(x \to z) \cap \tilde{f}(z \to x) = \tilde{f}(y \to z) \cap \tilde{f}(z \to y)$. This shows that $\tilde{f}^x(z) = \tilde{f}^y(z)$. By the arbitrariness of $z$, we have $\tilde{f}^x = \tilde{f}^y$. $\quad\square$

**Theorem 7.** *Let $(\tilde{f}, L)$ be an int-soft filter of $L$. Then, $\tilde{f}^x = \tilde{f}^y$ if and only if $x \to y \in \tilde{f}_{\tilde{f}(1)}$ and $y \to x \in \tilde{f}_{\tilde{f}(1)}$ if and only if $x \equiv_{\tilde{f}_{\tilde{f}(1)}} y$.*

**Proof.**

$$\tilde{f}^x = \tilde{f}^y \iff \tilde{f}(x \to y) \supseteq \tilde{f}(1),\ \tilde{f}(y \to x) \supseteq \tilde{f}(1)$$
$$\iff x \to y \in \tilde{f}_{\tilde{f}(1)},\ y \to x \in \tilde{f}_{\tilde{f}(1)}$$
$$\iff x \equiv_{\tilde{f}_{\tilde{f}(1)}} y.$$

$\square$

**Theorem 8.** *Let $(\tilde{f}, L)$ be an int-soft filter of $L$, $L/\tilde{f} := \{\tilde{f}^x | x \in L\}$. For any $\tilde{f}^x, \tilde{f}^y \in L/\tilde{f}$, if we define*

$\tilde{f}^x \sqcap \tilde{f}^y = \tilde{f}^{x \wedge y}, \tilde{f}^x \sqcup \tilde{f}^y = \tilde{f}^{x \vee y}, \tilde{f}^x \odot \tilde{f}^y = \tilde{f}^{x \otimes y}, \tilde{f}^x \rightharpoonup \tilde{f}^y = \tilde{f}^{x \to y}$, then $L/\tilde{f} = (L/\tilde{f}, \sqcap, \sqcup, \odot, \rightharpoonup, \tilde{f}^0, \tilde{f}^1)$ is a residuated lattice.

**Proof.** Suppose $\tilde{f}^x = \tilde{f}^s, \tilde{f}^y = \tilde{f}^t$. By Theorem 7, we have $x \equiv_{\tilde{f}_{\tilde{f}(1)}} s, y \equiv_{\tilde{f}_{\tilde{f}(1)}} t$. Since $\equiv_{\tilde{f}_{\tilde{f}(1)}}$ is a congruence relation on $L$, we have $x \vee y \equiv_{\tilde{f}_{\tilde{f}(1)}} s \vee t$, and so $\tilde{f}^{x \vee y} = \tilde{f}^{s \vee t}$. Similarly, we have $\tilde{f}^{x \wedge y} = \tilde{f}^{s \wedge t}, \tilde{f}^{x \otimes y} = \tilde{f}^{s \otimes t}, \tilde{f}^{x \to y} = \tilde{f}^{s \to t}$. This shows that the operators on $L/\tilde{f}$ are well defined.

Clearly, $L/\tilde{f}$ satisfies (R1) and (R2). We only need to prove that $(\odot, \rightharpoonup)$ is an adjoint pair. We note that the lattice order $\tilde{f}^x \preceq \tilde{f}^y$ if and only if $\tilde{f}^x \sqcup \tilde{f}^y = \tilde{f}^y$ and

$$\tilde{f}^x \preceq \tilde{f}^y \Longleftrightarrow \tilde{f}^x \sqcup \tilde{f}^y = \tilde{f}^y$$
$$\Longleftrightarrow \tilde{f}^{x \vee y} = \tilde{f}^y$$
$$\Longleftrightarrow \tilde{f}(x \vee y \to y) \supseteq \tilde{f}(1)$$
$$\Longleftrightarrow \tilde{f}(x \to y) \supseteq \tilde{f}(1).$$

Suppose $\tilde{f}^x, \tilde{f}^y, \tilde{f}^z \in L/\tilde{f}$, then

$$\tilde{f}^x \odot \tilde{f}^y \preceq \tilde{f}^z \Longleftrightarrow \tilde{f}^{x \otimes y} \preceq \tilde{f}^z$$
$$\Longleftrightarrow \tilde{f}(x \otimes y \to z) \supseteq \tilde{f}(1)$$
$$\Longleftrightarrow \tilde{f}(x \to (y \to z)) \supseteq \tilde{f}(1)$$
$$\Longleftrightarrow \tilde{f}^x \preceq \tilde{f}^{y \to z}$$
$$\Longleftrightarrow \tilde{f}^x \preceq \tilde{f}^y \rightharpoonup \tilde{f}^z.$$

Therefore, (R3) holds. This shows that $L/\tilde{f}$ is a residuated lattice. $\square$

**Theorem 9.** *Let $(\tilde{f}, L)$ be an int-soft filter. Then, the residuated lattice $L/\tilde{f} \cong L/\tilde{f}_{\tilde{f}(1)}$.*

**Proof.** Define a mapping $\varphi : L \longrightarrow L/\tilde{f}$ by $\varphi(x) = \tilde{f}^x$. Then,

$$x \in ker(\varphi) \Longleftrightarrow \varphi(x) = \tilde{f}^1$$
$$\Longleftrightarrow \tilde{f}^x = \tilde{f}^1$$
$$\Longleftrightarrow \tilde{f}(x) \supseteq \tilde{f}(1)$$
$$\Longleftrightarrow x \in \tilde{f}_{\tilde{f}(1)}.$$

Therefore, $ker(\varphi) = \tilde{f}_{\tilde{f}(1)}$. Clearly, $\varphi$ is surjective. It is easy to verify that $\varphi$ is a homomorphism. Thus, $L/\tilde{f} \cong L/\tilde{f}_{\tilde{f}(1)}$. $\square$

**Theorem 10.** *(Quotient characteristics) Let $(\tilde{f}, L)$ be **an** int-soft filter of L. Then, $(\tilde{f}, L)$ is a t-int-soft filter if and only if $L/\tilde{f} \in \mathbb{B}(t)$.*

**Proof.** $(\tilde{f}, L)$ is a t-int-soft filter if and only if $\tilde{f}_{\tilde{f}(1)}$ is a t-filter if and only if $L/\tilde{f}_{\tilde{f}(1)} \in \mathbb{B}(t)$ if and only if $L/\tilde{f} \in \mathbb{B}(t)$. $\square$

**Theorem 11.** *(Triple of equivalent characteristics) Let L be a residuated lattice. Then, the following statements are equivalent:*

*(1)   Every int-soft filter of L is a t-int-soft filter.*

(2)  $(\tilde{T}, L)$ is a t-int-soft filter.
(3)  $L \in \mathbb{B}(t)$.

**Proof.** (1)$\Longrightarrow$(2)

Since $(\tilde{T}, L)$ is an int-soft filter, the result is obvious.

(2)$\Longrightarrow$(3)

Suppose $(\tilde{T}, L)$ is a t-int-soft filter, then $\tilde{T}_{\tilde{T}(1)}$ is a t-filter. Thus, $L/\tilde{T}_{\tilde{T}(1)} \in \mathbb{B}(t)$. Also, $\tilde{T}_{\tilde{T}(1)} = \{1\}$ and $L/\{1\} \approx L$. Thus, $L \in \mathbb{B}(t)$.

(3)$\Longrightarrow$(1)

Suppose $(\tilde{f}, L)$ is an int-soft filter of $L$, then $\tilde{f}_{\tilde{f}(1)}$ is a filter. Also, $L \in \mathbb{B}(t)$, and thus $\tilde{f}_{\tilde{f}(1)}$ is a t-filter. Hence, $(\tilde{f}, L)$ is a t-int-soft filter.  $\square$

## 4. A Specific Example

In this section, we use involution-int-soft filters as an example.

**Definition 8.** *Let $(\tilde{f}, L)$ be an int-soft filter of $L$. Then, $(\tilde{f}, L)$ is called an involution-int-soft filter if, for all $\tau \in \mathscr{P}(U)$, $\tilde{f}_\tau$ is either empty or an involution filter.*

**Lemma 7.** *Let $(\tilde{f}, L)$ be an int-soft filter of $L$. Then, $(\tilde{f}, L)$ is an involution-int-soft filter if and only if $\tilde{f}(x^{**} \to x) \supseteq \tilde{f}(1)$, for all $x \in L$.*

**Proof.** Suppose $(\tilde{f}, L)$ is an involution-int-soft filter, then, by Theorem 5, we have that $\tilde{f}_{\tilde{f}(1)}$ is an involution filter. By Example 1, we have $x^{**} \to x \in \tilde{f}_{\tilde{f}(1)}$. This shows that $\tilde{f}(x^{**} \to x) \supseteq \tilde{f}(1)$.

Conversely, since $(\tilde{f}, L)$ is an int-soft filter, then $\tilde{f}_{\tilde{f}(1)}$ is a filter. Suppose $\tilde{f}(x^{**} \to x) \supseteq \tilde{f}(1)$, then $x^{**} \to x \in \tilde{f}_{\tilde{f}(1)}$. By Example 1, we know that $\tilde{f}_{\tilde{f}(1)}$ is an involution filter. Thus, $(\tilde{f}, L)$ is an involution-int-soft filter.  $\square$

**Theorem 12.** *(Extension property) Let $\tilde{f}, \tilde{g}$ be int-soft filters of $L$, $\tilde{f}(x) \subseteq \tilde{g}(x)$ for all $x \in L$, and, moreover, $\tilde{f}(1) \supseteq \tilde{g}(1)$. If $(\tilde{f}, L)$ is an involution-int-soft filter, then $(\tilde{g}, L)$ is an involution-int-soft filter.*

**Theorem 13.** *(Quotient characteristics) Let $(\tilde{f}, L)$ be an int-soft filter of $L$. Then, $(\tilde{f}, L)$ is an involution-int-soft filter if and only if $L/\tilde{f}$ is an involutive residuated lattice.*

**Theorem 14.** *(Triple of equivalent characteristics) Let $L$ be a residuated lattice. Then, the following statements are equivalent:*

(1)  *Every int-soft filter of $L$ is an involution-int-soft filter.*
(2)  *$(\tilde{T}, L)$ is an involution-int-soft filter.*
(3)  *$L$ is an involutive residuated lattice.*

## 5. The Relations among T-Int-Soft Filters on Residuated Lattices

**Lemma 8.** *[21] Let $L$ be a residuated lattice. Every Heyting algebra is a Rl-monoid.*

**Lemma 9.** *[33] Let $L$ be a residuated lattice. Then, the following statements are equivalent:*

(1)  *$L$ is a Boolean algebra.*
(2)  *$L$ is involutive and idempotent.*

**Lemma 10.** *[31,32] Let L be a residuated lattice. Then, L is an MV-algebra if and only if L is an involutive BL-algebra.*

In what follows, let $(\tilde{f}, L)$ be an int-soft filter of $L$.

**Theorem 15.** *If $(\tilde{f}, L)$ is a $t_1$-int-soft filter and $\mathbb{B}(t_1) \subseteq \mathbb{B}(t_2)$, then $(\tilde{f}, L)$ is a $t_2$-int-soft filter.*

**Proof.** $(\tilde{f}, L)$ is a $t_1$-int-soft filter $\Longrightarrow L/\tilde{f} \in \mathbb{B}(t_1) \Longrightarrow L/\tilde{f} \in \mathbb{B}(t_2) \Longrightarrow (\tilde{f}, L)$ is a $t_2$-int-soft filter. $\quad\square$

**Theorem 16.** *If $\mathbb{B}(t_1) \subseteq \mathbb{B}(t_2)$ and $\mathbb{B}(t_2) \subseteq \mathbb{B}(t_1)$, then $(\tilde{f}, L)$ is a $t_1$-int-soft filter if and only if $(\tilde{f}, L)$ is a $t_2$-int-soft filter.*

**Proof.** $(\tilde{f}, L)$ is a $t_1$-int-soft filter $\Longleftrightarrow L/\tilde{f} \in \mathbb{B}(t_1) \Longleftrightarrow L/\tilde{f} \in \mathbb{B}(t_2) \Longleftrightarrow (\tilde{f}, L)$ is a $t_2$-int-soft filter. $\quad\square$

**Remark 3.** *The above results reveal the general principles concerning the relations among t-int-soft filters. Their relations are consistent with those among the corresponding quotient algebras. Since we are familiar with the relations among these algebras, we can easily obtain the relations among t-int-soft filters. We have the following results.*

**Theorem 17.** *Let L be a residuated lattice. If $(\tilde{f}, L)$ is a Boolean-int-soft filter, then $(\tilde{f}, L)$ is a MV-int-soft filter, an involution-int-soft filter, a Heyting-int-soft filter, a Rl-monoid-int-soft filter, a MTL-int-soft filter, and a BL-int-soft filter.*

**Theorem 18.** *Let L be a residuated lattice. If $(\tilde{f}, L)$ is a MV-int-soft filter, then $(\tilde{f}, L)$ is a MTL-int-soft filter, an involution-int-soft filter, a Rl-monoid-int-soft filter, and a BL-int-soft filter.*

**Theorem 19.** *Let L be a residuated lattice. If $(\tilde{f}, L)$ is a Heyting-int-soft filter filter, then $(\tilde{f}, L)$ is a Rl-monoid-int-soft filter.*

**Theorem 20.** *Let L be a residuated lattice. Then, $(\tilde{f}, L)$ is a Boolean-int-soft filter if and only if $(\tilde{f}, L)$ is a MV-int-soft filter and a Heyting-int-soft filter.*

**Theorem 21.** *Let L be a residuated lattice. Then, $(\tilde{f}, L)$ is a MV-int-soft filter if and only if $(\tilde{f}, L)$ is an involution-int-soft filter and a Rl-monoid-int-soft filter.*

**Theorem 22.** *Let L be a residuated lattice. Then, $(\tilde{f}, L)$ is a BL-int-soft filter if and only if $(\tilde{f}, L)$ is a MTL-int-soft filter and a Rl-monoid-int-soft filter.*

## 6. Conclusions and Future Work

In this paper, we proposed the notion of t-int-soft filters on residuated lattices. In our framework, the "generator" of t-int-soft filters was given. As long as there are t-filters, we can obtain the corresponding t-int-soft filters. The characterizations of t-int-soft filters were derived. The general principles of investigating the relations among t-int-soft filters were derived. The research embodies the relationship between t-filters and t-int-soft filters. The thoughts and methods in this paper can be completely applied to special cases of residuated lattices, and the corresponding int-soft filters can be introduced and their characterizations can be obtained on these logical algebras.

Additionally, in [34–36], Chishty and Jun et al., respectively, discussed the properties of uni-soft filters on residuated lattices and MTL-algebras. In our future work, we will characterize the common features of uni-soft filters on residuated lattices.

**Author Contributions:** The author, H.Z., contributed mainly in the introduction and investigation of the properties of the t-int-soft filters. The solution process was discussed a number of times with the second author, M.L. She also provided the technical help regarding the write-up of manuscript.

**Funding:** This work is supported by the National Natural Science Foundation of China (No. 11701540, 61773019).

**Conflicts of Interest:** The authors declare no conflict of interest.

## References

1. Feller, W. An introduction to probability theory and its applications II. *Phys. Today* **1967**, *20*, 76. [CrossRef]
2. Zadeh, L.A. Fuzzy sets. *Inf. Control.* **1965**, *8*, 338–353. [CrossRef]
3. Pawlak, Z. Rough sets. *Int. J. Comput. Inf. Sci.* **1982**, *11*, 341–356. [CrossRef]
4. Atanassov, K.T. Intuitionistic fuzzy sets. *Fuzzy Sets Syst.* **1986**, *20*, 87–96. [CrossRef]
5. Molodtsov, D. Soft set theory—First results. *Comput. Math. Appl.* **1999**, *37*, 19–31. [CrossRef]
6. Ali, M.I.; Feng, F.; Liu, X. On some new operations in soft set theory. *Comput. Math. Appl.* **2009**, *57*, 1547–1553. [CrossRef]
7. Maji, P.K.; Biswas, R.; Roy, A.R. Soft set theory. *Comput. Math. Appl.* **2003**, *45*, 555–562. [CrossRef]
8. Sezgin, A.; Atagün, A.O. On operations of soft sets. *Comput. Math. Appl.* **2011**, *61*, 1457–1467. [CrossRef]
9. Som, T. On soft relation and fuzzy soft relation. *J. Fuzzy Math.* **2008**, *16*, 677–687.
10. Zhan, J.; Jun, Y.B. Soft BL-algebras based on fuzzy sets. *Comput. Math. Appl.* **2010**, *59*, 2037–2046. [CrossRef]
11. Kharal, A. Distance and similarity measures for soft sets. *New Math. Nat. Comput.* **2010**, *6*, 321–334. [CrossRef]
12. Majumdar, P.; Samanta, S.K. Similarity measure of soft sets. *New Math. Nat. Comput.* **2008**, *4*, 1–12. [CrossRef]
13. Qin, K.; Hong, Z. On soft equality. *J. Comput. Appl. Math.* **2010**, *234*, 1347–1355. [CrossRef]
14. Maji, P.K.; Roy, A.R.; Biswas, R. An application of soft sets in a decision making problem. *Comput. Math. Appl.* **2002**, *44*, 1077–1083. [CrossRef]
15. Xiao, Z.; Gong, K.; Zou, Y. A combined forecasting approach based on fuzzy soft sets. *J. Comput. Appl. Math.* **2009**, *228*, 326–333. [CrossRef]
16. Ward, M.; Dilworth, R.P. Residuated lattice. *Proc. Nat. Acad. Sci. USA* **1938**, *24*, 162–164. [CrossRef] [PubMed]
17. Buşneag, D.; Piciu, D. Semi-G-filters, Stonean filters, MTL-filters, divisible filters, BL-filters and regular filters in residuated lattices. *Iran. J. Fuzzy Syst.* **2016**, *13*, 145–160.
18. Buşneag, D.; Piciu, D. A new approach for classification of filters in residuated lattices. *Fuzzy Sets Syst.* **2015**, *260*, 121–130. [CrossRef]
19. Van Gasse, B.; Deschrijver, G.; Cornelis, C. Filters of residuated lattices and triangle algebra. *Inf. Sci.* **2010**, *180*, 3006–3020. [CrossRef]
20. Kondo, M. Characterization of some types of filters in commutative residuated lattices. *Far East J. Math. Sci.* **2010**, *2*, 193–203.
21. Ma, Z.M.; Hu, B.Q. Characterizations and new subclasses of I-filters in residuated lattices. *Fuzzy Sets Syst.* **2014**, *247*, 92–107. [CrossRef]
22. Muresan, C. Dense elements and classes of residuated lattices. *Bull. Math. Soc. Sci. Math.* **2010**, *53*, 11–24.
23. Shen, J.G.; Zhang, X.H. On filters of residuated lattice. *Chin. Q. J. Math.* **2006**, *180*, 443–447.
24. Víta, M. Why are papers about filters on residuated structures (usually) trivial? *Inf. Sci.* **2014**, *276*, 387–391. [CrossRef]
25. Zhang, X.; Zhou, H.; Mao, X. IMTL (MV)-filters and fuzzy IMTL (MV)-filters of residuated lattices. *J. Intell. Fuzzy Syst.* **2014**, *26*, 589–596.
26. Zhu, Y.; Xu, Y. On filter theory of residuated lattices. *Inf. Sci.* **2010**, *180*, 3614–3632. [CrossRef]
27. Jun, Y.B.; Ahn, S.S.; Lee, K.J. Intersection-soft filters in R0-Algebras. *Discret. Dyn. Nat. Soc.* **2013**, 950897. [CrossRef]
28. Jun, Y.B.; Ahn, S.S.; Lee, K.J. Implicative int-soft filters of R0-Algebras. *Discret. Dyn. Nat. Soc.* **2013**, *4*, 1–7. [CrossRef]
29. Lin, C.; Ma, Z. Int-Soft filters and their congruences on residuated lattices. *Op. Res. Fuzziol.* **2017**, *7*, 37–44. [CrossRef]

30. Jun, Y.B.; Ahn, S.S.; Lee, K.J. Classes of int-soft filters in residuated lattices. *Sci. World J.* **2014**, *2014*, 595160. [CrossRef] [PubMed]

31. Hájek, P. *Metamathematics of Fuzzy Logic*; Kluwer Academic Publishers: Dordrecht, The Netherlands, 1998.

32. Turunen, E. *Mathematics Behind Fuzzy Logic*; Springer: Heidelberg, Germany, 1999.

33. Zhang, H.; Li, Q. The relations among fuzzy t-filters on residuated lattices. *Sci. World J.* **2014**, *2014*, 894346. [CrossRef] [PubMed]

34. Chishty, G.M.; Al-Roqi, A. Filter theory in MTL-algebras based on Uni-soft property. *Bull. Iran. Math. Soc.* **2017**, in press.

35. Jun, J.B.; Lee, K.J.; Park, C.H.; Roh, E.H. Filters of residuated lattices based on soft set theory. *Commun. Korean Math. Soc.* **2015**, *30*, 155–168. [CrossRef]

36. Jun, J.B.; Song, S.Z. Uni-soft filters and uni-soft G-filters in residuated lattices. *J. Comput. Anal. Appl.* **2016**, *20*, 319–334.

 *mathematics*

*Article*

# On $(\alpha,\beta)$-US Sets in $BCK/BCI$-Algebras

**Chiranjibe Jana * and Madhumangal Pal**

Department of Applied Mathematics with Oceanology and Computer Programming, Vidyasagar University, Midnapore 721102, India; mmpalvu@gmail.com
* Correspondence: jana.chiranjibe7@gmail.com; Tel.: +91-9647135650

Received: 26 January 2019; Accepted: 6 March 2019; Published: 11 March 2019

**Abstract:** Molodtsov originated soft set theory, which followed a general mathematical framework for handling uncertainties, in which we encounter the data by affixing the parameterized factor during the information analysis. The aim of this paper is to establish a bridge to connect a soft set and the union operations on sets, then applying it to $BCK/BCI$-algebras. Firstly, we introduce the notion of the $(\alpha,\beta)$-Union-Soft ($(\alpha,\beta)$-US) set, with some supporting examples. Then, we discuss the soft $BCK/BCI$-algebras, which are called $(\alpha,\beta)$-US algebras, $(\alpha,\beta)$-US ideals, $(\alpha,\beta)$-US closed ideals, and $(\alpha,\beta)$-US commutative ideals. In particular, some related properties and relationships of the above algebraic structures are investigated. We also provide the condition of an $(\alpha,\beta)$-US ideal to be an $(\alpha,\beta)$-US closed ideal. Some conditions for a Union-Soft (US) ideal to be a US commutative ideal are given by means of $(\alpha,\beta)$-unions. Moreover, several characterization theorems of (closed) US ideals and US commutative ideals are given in terms of $(\alpha,\beta)$-unions. Finally, the extension property for an $(\alpha,\beta)$-US commutative ideal is established.

**Keywords:** $BCK/BCI$-algebra; $(\alpha,\beta)$-US set; $(\alpha,\beta)$-US subalgebra; $(\alpha,\beta)$-US (closed) ideal; $(\alpha,\beta)$-US commutative ideal

**MSC:** 06F35; 03G25; 06D72

---

## 1. Introduction

Most of the real-world problems in social sciences, the environment, engineering, medical sciences, economics, etc., involve data that contain uncertainties. To overcome these uncertainties, researchers are motivated to introduce some classical theories like the theories of fuzzy sets [1], rough sets [2], intuitionistic fuzzy sets [3], vague sets [4], and interval mathematics [5], i.e., by which we can have a mathematical tool to deal with uncertainties. Even though sets are very powerful to model the problems containing uncertainties, in some cases, these sets are not enough to overcome the serious types of uncertainties experienced in real-world problems. In that situation, in 1999, Molodtsov [6] posited the novel concept of soft set theory, which is a completely new approach for modeling vagueness and uncertainties. Some tremendous developments based on soft sets [7] have been recently drawn the attention of many scholars: in particular, Aktas and Çagman [8] used soft groups; Acar et al. [9] provided soft rings; and Ali et al. [10] defined new working rules on soft sets. Çagman et al. [11] studied soft-intgroups. Feng et al. [12] described soft semi-rings. Sezgin et al. [13] described soft intersection near-rings. Now, the above results have been applied to the disciplines of information sciences, decision support systems, knowledge systems, decision-making, and so on, reviewed in [14–20]. Recently, researchers have drawn attention to modeling covering-based problems on rough and soft sets with their applications to MADMproblems [21–23]. Additionally, there is a huge scope of application to develop covering-based algebraic structures in different kinds of algebras. Zhan et al. [24–26] introduced a new notion called $(M, N)$-SI-h-bi-ideals and $(M, N)$-SI-h-quasi-ideals in the environment of hemirings. At the same time, different soft algebraic structures have been developed on $BCK/BCI$-algebras, which was proposed by Imai and Iśeki [27,28]. Here, we briefly review some results of soft sets in the existing literature of $BCK/BCI$-algebras. Jana et al. [29–31], Ma and Zhan [32–34], and Senapati et al. [35,36] performed detailed investigations on $BCK/BCI$-algebras and related algebraic systems. In [37], Jun first constructed soft algebraic structure of $BCK/BCI$-algebras. Jun and Park [38] also pointed out applications of soft sets in the ideal theory of $BCK/BCI$-algebras. Jun et al. [39] also studied the soft $p$-ideal of soft $BCI$-algebras. Acar and Özürk [40] analytically studied maximal, irreducible, and prime soft ideals of $BCK/BCI$-algebras with supporting examples. First, Jun et al. [41,42] proposed a novel concept, namely union-soft sets and int-soft sets, and then implemented it to develop union-soft $BCK/BCI$-algebras and int-soft $BCK/BCI$-algebras. Sezgin [43] considered studying soft union interior ideals, quasi-ideals, and generalized bi-ideals of rings and gave their interrelationship. She also studied regular, intra-regular, regular-duo, and strongly-regular properties of rings in terms of soft-union ideals. Sezgin et al. [44] introduced a new soft classical ring theory, namely soft intersection rings, ideals, bi-ideals, interior ideals, and quasi-ideals. Furthermore, they defined their soft-union intersection product and their corresponding relationships. Jana and Pal [45] defined the concept of $(\alpha, \beta)$-soft intersection sets and then introduced this ideal to develop $(\alpha, \beta)$-soft intersectional groups structures and their various properties. Jana and Pal [46] also motivated using the same concept for the development of $(\alpha, \beta)$-soft intersectional $BCK/BCI$ algebraic structures. Again, Jana et al. [47] proposed providing $(\alpha, \beta)$-soft intersectional rings, $(\alpha, \beta)$-soft intersectional ideals, and their relationships. In this environment, different union-soft algebras and $(\alpha, \beta)$-soft intersectional algebras in different uncertain fuzzy environments under soft operations are considered by us as enough motivation to develop our proposal. There is an important issue in defining new $(\alpha, \beta)$-US sets based on soft operations and their application to develop different kinds of US algebraic structures. Therefore, based on the $(\alpha, \beta)$-US operation, how to develop $(\alpha, \beta)$-US $BCK/BCI$-algebras is a tremendous topic. To solve this problems, in this paper, we shall develop $(\alpha, \beta)$-US $BCK/BCI$-algebras introducing the concept of $(\alpha, \beta)$-US sets and their application to subalgebras, ideals, and commutative ideals in $BCK/BCI$-algebras on the basis of traditional union-soft algebras [24,25,41,42] and the results of [26,48]. We discuss the relationship between $(\alpha, \beta)$-US subalgebras, $(\alpha, \beta)$-US ideals, and $(\alpha, \beta)$-US commutative ideals in detail. We provide the condition that an $(\alpha, \beta)$-US ideal is an $(\alpha, \beta)$-US commutative ideal.

The remainder of this article is structured as follows: Section 2 proceeds with a recapitulation of all required definitions of $BCK/BCI$-algebras, basic definitions of soft sets, and subsequent discussions. In Section 3, the concepts of $(\alpha, \beta)$-US sets are proposed and illustrated by some examples. In Section 4, the notion of $(\alpha, \beta)$-US subalgebras of $BCK/BCI$-algebras is introduced and their properties discussed in detail.In Section 5, some interesting properties of $(\alpha, \beta)$-US ideals in $BCK/BCI$-algebras are introduced. Some characterization theorems of the $(\alpha, \beta)$-US commutative ideals in $BCK/BCI$-algebras are established in Section 6. Finally, in Section 7, conclusions and the scope for future research are given.

## 2. *BCK/BCI*-Algebras and Soft Sets

In this section, we introduce some elementary aspects that are necessary for this paper. For more information regarding $BCK/BCI$-algebras, the reader is referred to the monograph [49]. By a $BCI$-algebra, we mean an algebra $(X, *, 0)$ of the type $(2, 0)$ satisfying the following axioms for all $x, y, z \in X$:

($C_1$) $((x * y) * (x * z)) * (z * y) = 0$
($C_2$) $(x * (x * y)) * y = 0$
($C_3$) $x * x = 0$
($C_4$) $x * y = 0$ and $y * x = 0$ imply $x = y$.

If a *BCI*-algebra $X$ satisfies the following identity:

$(C_5)$ $0 * x = 0$,

then $X$ is called a *BCK*-algebra. Any *BCK*/*BCI*-algebra satisfies the following axioms: for all $x, y, z \in X$

$(C_6)$ $x * 0 = x$
$(C_7)$ $x \leq y \Rightarrow x * z \leq y * z, z * y \leq z * x$
$(C_8)$ $(x * y) * z = (x * z) * y$
$(C_9)$ $(x * z) * (y * z) \leq x * y$

The partial ordering is defined as $x \leq y$ if and only if $x * y = 0$. In a *BCI*-algebra $X$, the following hold:

$(C_{10})$ $(x * (x * (x * y))) = x * y$
$(C_{11})$ $(0 * (x * y)) = (0 * x) * (0 * y)$.

A non-empty subset $S$ of a *BCK*/*BCI*-algebra $X$ is called a subalgebra of $X$ if $x * y \in S$ for all $x, y \in X$. A *BCK*-algebra $X$ is said to be commutative $x \wedge y = y \wedge x$ for all $x, y \in X$, where $y \wedge x = y * (y * x)$. A subset $I$ of a *BCK*/*BCI*-algebra $X$ is called an ideal of $X$ if for all $x, y \in X$,

$(C_{12})$ $0 \in A$
$(C_{13})$ $y \in A$ and $x * y \in A \Rightarrow x \in A$.

A subset $I$ of a *BCK*-algebra $X$ is called a commutative ideal if it satisfies $(C_{12})$ and for $z \in I$
$(C_{14})$ $((x * y) * z \in I \Rightarrow (x * (y * (y * x))) \in I$.
An ideal $I$ of a *BCK*-algebra $X$ is called commutative if it satisfies the implication $x * y \in I \Rightarrow x * (y * (y * x)) \in I$.
$U$ refers to an initial universal set, and $E$ is the set of parameters. Let $\mathcal{P}(U)$ be the power set of $U$ and $A \subset E$. Molodtsov [6] introduced soft sets in the following manner:

**Definition 1** ([6]). *A pair $(\mathcal{F}, E)$ is called a soft set over $U$ if $\mathcal{F}$ is a function given by:*

$$\mathcal{F} : E \to \mathcal{P}(U).$$

*In other words, a soft set in the universe $U$ is a parameterized family of subsets of the universal set $U$. For $\varepsilon \in A$, $\mathcal{F}(\varepsilon)$ may be considered as the set of $\varepsilon$-elements of the soft $(\mathcal{F}, A)$ or as the set of $\varepsilon$-approximate elements of the soft set.*

The following example illustrates the above idea.

**Example 1.** *Let $(X, \tau)$ be a topological space, i.e., $\tau$ is a family of subsets of the set $X$ called the open sets of $X$. Then, the family of open neighborhoods $N(x)$ of point $x$, where $N(x) = \{V \in \tau | x \in V\}$, may be considered as the soft set $(N(x), \tau)$.*

**Definition 2** ([6,14]). *For a non-empty subset $A$ of $E$, a soft set $(\mathcal{F}, E)$ over $U$ satisfying the condition:*

$$\mathcal{F}(x) = \varnothing \text{ for all } x \notin A$$

is called the *A-soft set over U* and is denoted by $\mathcal{F}_A$, so an *A-soft set* $\mathcal{F}_A$ over $U$ is a function $\mathcal{F}_A : E \to \mathcal{P}(U)$ such that $\mathcal{F}_A(x) = \varnothing$ for all $x \notin A$. A soft set over $U$ can be followed by the set of ordered pairs:

$$\mathcal{F}_A = \{(x, \mathcal{F}_A(x)) : x \in E, \mathcal{F}_A(x) \in \mathcal{P}(U)\}.$$

We remark that a soft set is a parameterized family of subsets of the set $U$. A soft set $\mathcal{F}_A(x)$ may be an arbitrary, empty, and nonempty intersection. The set of all soft sets over $U$ is denoted by $S(U)$.

**Definition 3** ([14]). *Let $\mathcal{F}_A \in S(U)$. For all $x \in E$, If $\mathcal{F}_A(x) = \varnothing$, then $\mathcal{F}_A$ is said to be an empty soft set and symbolized by $\Phi_A$. If $\mathcal{F}_A(x) = U$, then $\mathcal{F}_A$ is said to be an A-universal soft set and symbolized as $\mathcal{F}_{\tilde{A}}$. If $\mathcal{F}_A(x) = U$ and $A = E$, then $\mathcal{F}_{\tilde{A}}$ is said to be a universal soft set and is denoted by $\mathcal{F}_{\tilde{E}}$.*

**Proposition 1** ([14]). *Let $\mathcal{F}_A \in S(U)$. Then,*
*(i) $\mathcal{F}_A \widetilde{\cup} \mathcal{F}_A = \mathcal{F}_A$, $\mathcal{F}_A \widetilde{\cap} \mathcal{F}_A = \mathcal{F}_A$.*
*(ii) $\mathcal{F}_A \widetilde{\cup} \Phi_A = \mathcal{F}_A$, $\mathcal{F}_A \widetilde{\cap} \Phi_A = \Phi_A$.*
*(iii) $\mathcal{F}_A \widetilde{\cup} \mathcal{F}_E = \mathcal{F}_E$, $\mathcal{F}_A \widetilde{\cap} \mathcal{F}_E = \mathcal{F}_A$.*
*(iv) $\mathcal{F}_A \widetilde{\cup} \mathcal{F}_A^c = \mathcal{F}_E$, $\mathcal{F}_{\tilde{A}}^c \widetilde{\cup} \mathcal{F}_A^c = \Phi_A$, where $\Phi_A$ is an empty set.*

**Definition 4** ([37]). *Let $E$ be a BCK/BCI-algebra and $(\mathcal{F}, A)$ be a soft set over BCK/BCI-algebra $E$. Then, $(\mathcal{F}, A)$ is called a soft BCK/BCI-algebra over $E$ if $\mathcal{F}(x)$ is a subalgebra of $E$ for all $x \in E$.*

**Definition 5** ([42]). *Let $E$ be a BCK/BCI-algebra. Let $\mathcal{F}_A \in S(U)$ for a given subalgebra $A$ of $E$. Then, $\mathcal{F}_A$ is called a US algebra of $A$ over $U$ if, for all $x, y \in A$, it satisfies the following condition:*

$$\mathcal{F}_A(x * y) \subseteq \mathcal{F}_A(x) \cup \mathcal{F}_A(y).$$

**Definition 6** ([42]). *Let $E$ be a BCK/BCI-algebra and $A$ be a subalgebra of $E$. Let $\mathcal{F}_A \in S(U)$. Then, $\mathcal{F}_A$ is called a US ideal over $U$ if, for all $x, y \in A$, it satisfies the following conditions:*
*(1) $\mathcal{F}_A(0) \subseteq \mathcal{F}_A(x)$*
*(2) $\mathcal{F}_A(x) \subseteq \mathcal{F}_A(x * y) \cup \mathcal{F}_A(y)$.*

**Definition 7** ([42]). *Let $E$ be a BCK/BCI-algebra. For a given subalgebra $A$ of $E$, let $\mathcal{F}_A \in S(U)$, then the US ideal $\mathcal{F}_A$ is said to be closed if, for all $x \in A$, it satisfies the following condition:*

$$\mathcal{F}_A(0 * x) \subseteq \mathcal{F}_A(x).$$

**Definition 8** ([42]). *Let $E$ be a BCK-algebra. For a given subalgebras $A$ of $E$, let $\mathcal{F}_A \in S(U)$. Then, $\mathcal{F}_A$ is called a US commutative ideal over $U$ if, for all $x, y, z \in A$, it satisfies the following conditions:*
*(1) $\mathcal{F}_A(0) \subseteq \mathcal{F}_A(x)$*
*(2) $\mathcal{F}_A(x * (y * (y * x))) \subseteq \mathcal{F}_A((x * y) * z) \cup \mathcal{F}_A(z)$.*

**Definition 9** ([42]). *Let $\mathcal{F}_A \in S(U)$ and $\delta \subseteq U$. Then, the $\delta$-exclusion set of $\mathcal{F}_A$, denoted by $\mathcal{F}_A^\delta$, is defined by $\mathcal{F}_A^\delta(x) = \{x \in A | \mathcal{F}_A(x) \subseteq \delta\}$.*

## 3. $(\alpha, \beta)$-US Sets

In this section, $U$ is the initial universe, $E$ is the set of parameters, and "$\rightarrowtail$" is a binary operation. Now, we let $S(U)$ be the set of all soft sets. We introduce the notion of $(\alpha, \beta)$-US sets and illustrate them by some examples. From now on, we let $\varnothing \subseteq \alpha \subset \beta \subseteq U$.

**Definition 10.** *For any non-empty subset A of E, consider the soft set $\mathcal{F}_A \in S(U)$. Then, for all $x, y \in A$, the soft set $\mathcal{F}_A$ is called an $(\alpha, \beta)$-US set over U if it satisfies the following condition:*

$$\mathcal{F}_A(x \rightarrowtail y) \cap \beta \subseteq \mathcal{F}_A(x) \cup \mathcal{F}_A(y) \cup \alpha.$$

**Example 2.** *We consider five houses in the initial universe set U, which is given by*

$$U = \{h_1, h_2, h_3, h_4, h_5\}.$$

*Let the set of parameters $E = \{\xi_1, \xi_2, \xi_3, \xi_4\}$ be the status of the set of houses, which follows for the parameters "cheap", "expensive:, "in the flooded area:, and "in the urban area", respectively, with the following binary operation:*

| $\rightarrowtail$ | $\xi_1$ | $\xi_2$ | $\xi_3$ | $\xi_4$ |
|---|---|---|---|---|
| $\xi_1$ | $\xi_1$ | $\xi_1$ | $\xi_1$ | $\xi_1$ |
| $\xi_2$ | $\xi_2$ | $\xi_1$ | $\xi_1$ | $\xi_2$ |
| $\xi_3$ | $\xi_3$ | $\xi_3$ | $\xi_1$ | $\xi_3$ |
| $\xi_4$ | $\xi_4$ | $\xi_4$ | $\xi_4$ | $\xi_1$ |

*We consider a soft set $\mathcal{F}_E$ over U, which is given as $\mathcal{F}_E(\xi_1) = \{h_3, h_5\}$, $\mathcal{F}_E(\xi_2) = \{h_3, h_4, h_5\}$, $\mathcal{F}_E(\xi_3) = \{h_2, h_3, h_4, h_5\}$, and $\mathcal{F}_E(\xi_4) = \{h_1, h_3, h_5\}$. Fix $\beta = \{h_1, h_2, h_3, h_5\}$ and $\alpha = \{h_2, h_3\}$. Then, it can be easily verified that $\mathcal{F}_E$ is an $(\alpha, \beta)$-US set over U.*

**Theorem 1.** *Let $\mathcal{F}_A, \mathcal{F}_B \in S(U)$ be soft sets such that $\mathcal{F}_A$ is a soft subset of $\mathcal{F}_B$. If $\mathcal{F}_B$ is an $(\alpha, \beta)$-US set over U, then the same holds for $\mathcal{F}_A$.*

**Proof.** Let $x, y \in A$ such that $x \rightarrowtail y \in A$. Then, $x \rightarrowtail y \in B$ since $A \subseteq B$. Thus, $\mathcal{F}_A(x \rightarrowtail y) \cap \beta \subseteq \mathcal{F}_B(x \rightarrowtail y) \cap \beta \subseteq \mathcal{F}_B(x) \cup \mathcal{F}_B(y) \cup \alpha = \mathcal{F}_A(x) \cup \mathcal{F}_A(y) \cup \alpha$. Therefore, $\mathcal{F}_A$ is an $(\alpha, \beta)$-US set over $U$. $\square$

The converse of Theorem 1 is not true in general, as can be seen in the following example.

**Example 3.** *We consider five houses in the initial universe set U, which is given by:*

$$U = \{h_1, h_2, h_3, h_4, h_5\}.$$

*Let the set of parameters $E = \{\xi_1, \xi_2, \xi_3, \xi_4\}$ be the status of the set of houses, which follows for the parameters "beautiful", "cheap", "in a good location", and "in green surroundings", respectively, with the following binary operation:*

| $\rightarrowtail$ | $\xi_1$ | $\xi_2$ | $\xi_3$ | $\xi_4$ |
|---|---|---|---|---|
| $\xi_1$ | $\xi_1$ | $\xi_2$ | $\xi_3$ | $\xi_4$ |
| $\xi_2$ | $\xi_2$ | $\xi_1$ | $\xi_3$ | $\xi_2$ |
| $\xi_3$ | $\xi_3$ | $\xi_4$ | $\xi_1$ | $\xi_2$ |
| $\xi_4$ | $\xi_4$ | $\xi_3$ | $\xi_2$ | $\xi_1$ |

*Let $A = \{\xi_1, \xi_2\} \subset E$. Consider a soft set $\mathcal{F}_E$ over U as $\mathcal{F}_A(\xi_1) = \{h_1, h_3\}$, $\mathcal{F}_A(\xi_2) = \{h_1, h_3, h_4\}$, $\mathcal{F}_A(\xi_3) = \varnothing$, $\mathcal{F}_A(\xi_4) = \varnothing$, $\beta = \{h_1, h_3, h_4, h_5\}$, and $\alpha = \{h_3, h_4\}$. Then, it can be easily verified that $\mathcal{F}_A$ is an $(\alpha, \beta)$-US set over U.*

*Consider another soft set $\mathcal{F}_B$ as $\mathcal{F}_B(\xi_1) = \{h_1, h_3\}$, $\mathcal{F}_B(\xi_2) = \{h_1, h_3, h_4\}$, $\mathcal{F}_B(\xi_3) = \{h_2, h_4\}$, and $\mathcal{F}_B(\xi_4) = \{h_4, h_5\}$. Then, $\mathcal{F}_A$ is a soft subset of $\mathcal{F}_B$. However, for $\beta = \{h_1, h_3, h_4, h_5\}$ and $\alpha = \{h_3, h_4\}$, $\mathcal{F}_B$ is not an $(\alpha, \beta)$-union soft set over U, because:*

$$\mathcal{F}_B(\xi_3 \rightarrowtail \xi_4) \cap \beta = \{h_1, h_3, h_4\} \nsubseteq \{h_2, h_3, h_4, h_5\} = \mathcal{F}_B(\xi_3) \cup \mathcal{F}_B(\xi_4) \cup \alpha.$$

### 4. $(\alpha, \beta)$-US Subalgebras in *BCK/BCI*-Algebras

In this section, we introduce the concept of the $(\alpha, \beta)$-US subalgebra of $BCK/BCI$-algebras and investigate some of its characterization. Throughout this section, $E = X$ is always a $BCK/BCI$-algebra without any specification.

**Definition 11.** *Let $E$ be a $BCK/BCI$-algebra. Let $\mathcal{F}_A \in S(U)$ for a given subalgebra $A$ of $E$. Then, $\mathcal{F}_A$ is called an $(\alpha, \beta)$-US algebra of $A$ over $U$ if, for all $x, y \in A$, it satisfies the condition:*

$$\mathcal{F}_A(x * y) \cap \beta \subseteq \mathcal{F}_A(x) \cup \mathcal{F}_A(y) \cup \alpha.$$

We consider the pre-order relation "$\subseteq''_{(\alpha,\beta)}$" on $S(U)$ as: for any $\mathcal{F}_E, \mathcal{G}_E \in S(U)$ and $\varnothing \subseteq \alpha \subset \beta \subseteq U$, we define $\mathcal{F}_E \cap \beta \subseteq \mathcal{G}_E \cup \alpha \Leftrightarrow \mathcal{F}_E(x) \cap \beta \subseteq \mathcal{G}_E(x) \cup \alpha$ for any $x \in E$. We define a relation "$=''_{(\alpha,\beta)}$" such as $\Leftrightarrow \mathcal{F}_E \cap \beta \subseteq \mathcal{G}_E \cup \alpha$ and $\mathcal{G}_E \cap \beta \subseteq \mathcal{F}_E \cup \alpha$. Using the above notion, the $(\alpha, \beta)$-US $BCK/BCI$-algebra is defined as follows:

**Definition 12.** *Let $E$ be a $BCK/BCI$-algebra. Let $\mathcal{F}_A \in S(U)$ for a given subalgebra $A$ of $E$. Then, $\mathcal{F}_A$ is called an $(\alpha, \beta)$-US algebra of $A$ over $U$ if, for all $x, y \in A$, it satisfies the condition:*

$$\mathcal{F}_A(x * y) \cap \beta \subseteq \mathcal{F}_A(x) \cup \mathcal{F}_A(y) \cup \alpha.$$

**Example 4.** *Let $X = \{0, a, b, c, d\}$ be a BCK-algebra with the following Cayley table:*

| $*$ | $0$ | $a$ | $b$ | $c$ | $d$ |
|---|---|---|---|---|---|
| $0$ | $0$ | $0$ | $0$ | $0$ | $0$ |
| $a$ | $a$ | $0$ | $0$ | $0$ | $0$ |
| $b$ | $b$ | $b$ | $0$ | $0$ | $0$ |
| $c$ | $c$ | $c$ | $c$ | $0$ | $0$ |
| $d$ | $d$ | $d$ | $c$ | $a$ | $0$ |

Let $(\mathcal{F}_A, A)$ be a soft set over $U = X$, where $E = A = X$ and $\mathcal{F}_A : A \to \mathcal{P}(U)$ is a set-valued function defined by $\mathcal{F}_A(x) = \{y \in X | y * x = 0\}$ for all $x \in A$. Then, $\mathcal{F}_A(0) = \{0\}$, $\mathcal{F}_A(a) = \{0, a\}$, $\mathcal{F}_A(b) = \{0, a, b\}$, $\mathcal{F}_A(c) = \{0, a, b, c\}$, and $\mathcal{F}_A(d) = \{0, a, b, c, d\}$. It can be easily verified that $\mathcal{F}_A$ is an $(\alpha, \beta)$-US algebra of $A$ over $U$, where $\beta = \{0, a, c, d\}$ and $\alpha = \{0, a, d\}$.

**Theorem 2.** *Let $E$ be a $BCK/BCI$-algebra, $\mathcal{F}_A \in S(U)$ be a given subalgebra $A$ of $E$, and $\beta \subseteq U$. For $\delta \in U$, $\mathcal{F}_A$ is an $(\alpha, \beta)$-US subalgebra of $A$ over $U$ if and only if each non-empty subset $B(\mathcal{F}_A : \delta)$, which is defined by:*

$$B(\mathcal{F}_A : \delta) = \{x \in A | \mathcal{F}_A(x) \subseteq \delta \cup \alpha\}$$

*where $\delta \subseteq \beta$, is a subalgebra of $A$.*

**Proof.** Let $\mathcal{F}_A$ be an $(\alpha, \beta)$-US algebra of $A$ over $U$ such that $\mathcal{F}_A(x) \subseteq \beta$ for every $x \in A$, and let $x, y \in B(\mathcal{F}_A : \delta)$. Then, $\mathcal{F}_A(x * y) \cap \beta \subseteq \mathcal{F}_A(x) \cup \mathcal{F}_A(y) \cup \alpha \subseteq \delta \cup \alpha$, which implies that $x * y \in B(\mathcal{F}_A : \delta)$. Hence, $B(\mathcal{F}_A : \delta)$ is a subalgebra of $A$.

Conversely, let each non-empty subset $B(\mathcal{F}_A : \delta)$ be a subalgebra of $A$. Then, according to our assumption on $\mathcal{F}_A$, for $x, y \in A$, there are $\delta_1, \delta_2 \subseteq \beta$ such that $\mathcal{F}_A(x) = \delta_1$ and $\mathcal{F}_A(y) = \delta_2$. Thus, $\mathcal{F}_A(x) \subseteq \delta$ and $\mathcal{F}_A(y) \subseteq \delta$ for $\delta = \delta_1 \cup \delta_2 \subseteq \beta$. Hence, $x, y \in B(\mathcal{F}_A : \delta)$. Since $B(\mathcal{F}_A : \delta)$ is a subalgebra of $A$, so $x * y \in B(\mathcal{F}_A : \delta)$. Thus, $\mathcal{F}_A(x * y) \cap \beta \subseteq \delta$ and $\mathcal{F}_A(x) \cup \mathcal{F}_A(y) \cup \alpha = \delta_1 \cup \delta_2 \cup \alpha = \delta \cup \alpha$, which implies $\mathcal{F}_A(x * y) \cap \beta \subseteq \mathcal{F}_A(x) \cup \mathcal{F}_A(y) \cup \alpha$. Hence, the proof of the theorem is completed. $\square$

**Theorem 3.** *Let $E$ be a BCK/BCI-algebra and $\mathcal{F}_A \in S(U)$ be such that $A \subseteq E$. Then, $\mathcal{F}_A$ is an $(\alpha, \beta)$-US algebra of $A$ over $U$ if, for all $x \in A$, it satisfies the condition:*

$$\mathcal{F}_A(0) \cap \beta \subseteq \mathcal{F}_A(x) \cup \alpha.$$

**Proof.** If $0 \notin A$, then $\mathcal{F}_A(0) \cap \beta = \varnothing \cap \beta \subseteq \mathcal{F}_A(x) \cup \alpha$ for all $x \in A$. If $0 \in A$, then $\mathcal{F}_A(0) \cap \beta = \mathcal{F}_A(x * x) \cap \beta \subseteq \mathcal{F}_A(x) \cup \mathcal{F}_A(x) \cup \alpha = \mathcal{F}_A(x) \cup \alpha$ for all $x \in A$. Therefore, $\mathcal{F}_A(0) \cap \beta \subseteq \mathcal{F}_A(x) \cup \alpha$ holds. $\square$

**Theorem 4.** *If a soft $\mathcal{F}_A$ over $U$ is an $(\alpha, \beta)$-US algebra of $A$, then:*

$$(\mathcal{F}_A(0) \cap \beta) \cup \alpha \subseteq (\mathcal{F}_A(x) \cap \beta) \cup \alpha, \text{ for all } x \in A.$$

**Proof.** Let $\mathcal{F}_A \in S(U)$, and by using Theorem 3, we get:

$$
\begin{aligned}
(\mathcal{F}_A(0) \cap \beta) \cup \alpha &= (\mathcal{F}_A(x * x) \cap \beta) \cup \alpha \\
&\subseteq ((\mathcal{F}_A(x) \cup \mathcal{F}_A(x) \cup \alpha) \cap \beta) \cup \alpha \\
&= ((\mathcal{F}_A(x) \cap \beta) \cup \alpha) \cup ((\mathcal{F}_A(x) \cap \beta) \cup \alpha) \\
&\subseteq (\mathcal{F}_A(x) \cap \beta) \cup \alpha.
\end{aligned}
$$

The proof of the theorem is complete. $\square$

**Theorem 5.** *Let $E$ be a BCI-algebra and $\mathcal{F}_A \in S(U)$ for a given subalgebra $A$ of $E$. Then, $\mathcal{F}_A$ is an $(\alpha, \beta)$-US algebra of $A$ over $U$ if, for all $x \in A$, it satisfies the condition:*

$$\mathcal{F}_A(x * (0 * y)) \cap \beta \subseteq \mathcal{F}_A(x) \cup \mathcal{F}_A(y) \cup \alpha.$$

**Proof.** By using Theorem 3, we have:

$$
\begin{aligned}
&\mathcal{F}_A(x * (0 * y)) \cap \beta \subseteq \mathcal{F}_A(x) \cup \mathcal{F}_A(0 * y) \cup \alpha \\
&\subseteq \mathcal{F}_A(x) \cup \mathcal{F}_A(0) \cup \mathcal{F}_A(y) \cup \alpha = \mathcal{F}_A(x) \cup \mathcal{F}_A(y) \cup \alpha.
\end{aligned}
$$

Therefore, $\mathcal{F}_A(x * (0 * y)) \cap \beta \subseteq \mathcal{F}_A(x) \cup \mathcal{F}_A(y) \cup \alpha$ holds for all $x, y \in A$. $\square$

**Proposition 2.** *Let $E$ be a BCK/BCI-algebra and $\mathcal{F}_A \in S(U)$ for a given subalgebra $A$ of $E$. Then, $\mathcal{F}_A$ is a $(\alpha, \beta)$-US algebra of $A$ over $U$ if for all $x \in A$, it satisfies the condition:*

$$\mathcal{F}_A(x * y) \cap \beta \subseteq \mathcal{F}_A(y) \cup \alpha \Leftrightarrow \mathcal{F}_A(x) \cap \beta = \mathcal{F}_A(y) \cup \alpha.$$

**Proof.** We assume that $\mathcal{F}_A(x * y) \cap \beta \subseteq \mathcal{F}_A(y) \cup \alpha$ for all $x, y \in A$. Take $y = 0$, and use $(C_6)$, which induces $\mathcal{F}_A(x) \cap \beta = \mathcal{F}_A(x * 0) \cap \beta \subseteq \mathcal{F}_A(0) \cup \alpha$. It follows from Theorem 3 that $\mathcal{F}_A(x) \cap \beta = \mathcal{F}_A(0) \cup \alpha$ for all $x \in A$.

Conversely, suppose that $\mathcal{F}_A(x) \cap \beta = \mathcal{F}_A(0) \cup \alpha$ for all $x \in A$. Then,

$$\mathcal{F}_A(x * y) \cap \beta \subseteq \mathcal{F}_A(x) \cup \mathcal{F}_A(y) \cup \alpha = \mathcal{F}_A(0) \cup \mathcal{F}_A(y) \cup \alpha = \mathcal{F}_A(y) \cup \alpha$$

for all $x, y \in A$. $\square$

For a soft set $(\mathcal{F}_A, A)$ over $E$, we consider the set:

$$X_0 = \{x \in A | \mathcal{F}_A(x) = \mathcal{F}_A(0)\}.$$

**Theorem 6.** *Let $E$ be a BCK/BCI-algebra and $A$ a subalgebra of $E$. Let $(\mathcal{F}_A, A)$ be an $(\alpha, \beta)$-US algebra over $E$. Then, the set $X_0^* = \{x \in A | (\mathcal{F}_A(x) \cap \beta) \cup \alpha = (\mathcal{F}_A(0) \cap \beta) \cup \alpha\}$ is a subalgebra of $E$.*

**Proof.** If $\mathcal{F}_A$ is an $(\alpha, \beta)$-US algebra of $A$ over $U$, then $x, y \in X_0^*$; we have $(\mathcal{F}_A(x) \cap \beta) \cup \alpha = (\mathcal{F}_A(0) \cap \beta) \cup \alpha = (\mathcal{F}_A(y) \cap \beta) \cup \alpha$. Then, from Theorem 3, we have $(\mathcal{F}_A(0) \cap \beta) \cup \alpha \subseteq (\mathcal{F}_A(x * y) \cap \beta) \cup \alpha$ for all $x, y \in A$. This also takes the following form, $(\mathcal{F}_A(x * y) \cap \beta) \cup \alpha \subseteq ((\mathcal{F}_A(x) \cup \mathcal{F}_A(y) \cup \alpha) \cap \beta) \cup \alpha = ((\mathcal{F}_A(x) \cap \beta) \cup \alpha) \cup (\mathcal{F}_A(y) \cap \beta) \cup \alpha) \subseteq (\mathcal{F}_A(0) \cap \beta) \cup \alpha$. Hence, $(\mathcal{F}_A(x * y) \cap \beta) \cup \alpha = (\mathcal{F}_A(0) \cap \beta) \cup \alpha$, and so, $x * y \in X_0^*$. Thus, $X_0^*$ is a subalgebra of $A$. $\square$

**Theorem 7.** *Let $E$ be a BCK-algebra and $\mathcal{F}_A \in S(U)$. Define a soft set $\mathcal{F}_A^*$ over $U$ by $\mathcal{F}_A^* : E \to \mathcal{P}(U)$,*

$$x \longmapsto \begin{cases} \mathcal{F}_A(x) & \text{if } x \in B(\mathcal{F}_A : \delta) \\ U & \text{otherwise.} \end{cases}$$

*If $\mathcal{F}_A$ is an $(\alpha, \beta)$-US algebra over $U$, then so is $\mathcal{F}_A^*$.*

**Proof.** If $\mathcal{F}_A$ is an $(\alpha, \beta)$-US algebra over $U$, then $B(\mathcal{F}_A : \delta)$ is a subalgebra of $A$ by Theorem 2. Let $x, y \in A$. If $x, y \in B(\mathcal{F}_A : \delta)$, then $x * y \in B(\mathcal{F}_A : \delta)$, and so,

$$\mathcal{F}_A^*(x * y) \cap \beta = \mathcal{F}_A(x * y) \cap \beta \subseteq \mathcal{F}_A(x) \cup \mathcal{F}_A(y) \cup \alpha = \mathcal{F}_A^*(x) \cup \mathcal{F}_A^*(y) \cup \alpha.$$

If $x \notin B(\mathcal{F}_A : \delta)$ or $y \notin B(\mathcal{F}_A : \delta)$, then $\mathcal{F}_A^*(x) = U$ or $\mathcal{F}_A^*(y) = U$. Thus, we have:

$$\mathcal{F}_A^*(x * y) \cap \beta \subseteq U = \mathcal{F}_A^*(x) \cup \mathcal{F}_A^*(y) \cup \alpha.$$

Therefore, $\mathcal{F}_A^*$ is an $(\alpha, \beta)$-US algebra of $A$ over $U$. $\square$

## 5. $(\alpha, \beta)$-US Ideals in *BCK/BCI*-Algebras

In this section, we define the $(\alpha, \beta)$-US ideal and $(\alpha, \beta)$-US closed ideal and characterize their properties in detail.

**Definition 13.** *Let $E$ be a BCK/BCI-algebra and $A$ be a subalgebra of $E$. Let $\mathcal{F}_A \in S(U)$, then $\mathcal{F}_A$ is called an $(\alpha, \beta)$-US ideal over $U$ if, for all $x, y \in A$, it satisfies Theorem 3 and the following condition:*
$\mathcal{F}_A(x) \cap \beta \subseteq \mathcal{F}_A(x * y) \cup \mathcal{F}_A(y) \cup \alpha.$

**Example 5.** *Let $U = Z$ (set of positive integers) be the universal set and $E = \{0, a, b, c, d\}$ be a BCK-algebra with the following Cayley table:*

| $*$ | 0 | $a$ | $b$ | $c$ | $d$ |
|---|---|---|---|---|---|
| 0 | 0 | 0 | 0 | 0 | 0 |
| $a$ | $a$ | 0 | $a$ | 0 | $a$ |
| $b$ | $b$ | $b$ | 0 | 0 | 0 |
| $c$ | $c$ | $c$ | $c$ | 0 | $c$ |
| $d$ | $d$ | $d$ | $b$ | $b$ | 0 |

*For a subalgebra $A = \{0, b, c, d\}$ of $E$, define the soft set $(\mathcal{F}_A, A)$ over $U$ as $\mathcal{F}_A(0) = \{1, 3, 4, 5, 7, 9, 11, 12\}$, $\mathcal{F}_A(b) = \{1, 2, 4, 5, 6, 7, 8, 10, 13\}$, $\mathcal{F}_A(c) = \{2, 3, 5, 6, 8, 9, 13\}$, and $\mathcal{F}_A(d) = \{1, 2, 3, 5, 8, 10, 13\}$. Then, $\mathcal{F}_A$ is an $(\alpha, \beta)$-US ideal of $A$ over $U$ where $\beta = \{1, 2, 3, 6, 7, 8, 9, 10, 11, 12, 13\}$ and $\alpha = \{1, 3, 6, 7, 9, 10, 11, 12\}$.*

**Example 6.** *Let $U = N$ (set of natural numbers) be the universal set and $E = \{0, a, b, c, d\}$ be a BCK-algebra with the following Cayley table:*

| $*$ | 0 | $a$ | $b$ | $c$ | $d$ |
|---|---|---|---|---|---|
| 0 | 0 | 0 | 0 | 0 | 0 |
| $a$ | $a$ | 0 | 0 | 0 | 0 |
| $b$ | $b$ | $b$ | 0 | 0 | 0 |
| $c$ | $c$ | $c$ | $c$ | 0 | 0 |
| $d$ | $d$ | $c$ | $c$ | $a$ | 0 |

Define the soft set $\mathcal{F}_A$ over $U$ as follows $\mathcal{F}_A(0) = N$, $\mathcal{F}_A(a) = 2N$, $\mathcal{F}_A(b) = 4N$, $\mathcal{F}_A(c) = 6N$, and $\mathcal{F}_A(d) = 8N$. Then, $\mathcal{F}_A$ is not an $(\alpha, \beta)$-US ideal over $U$, where $\beta = 12N$ and $\alpha = 24N$, because:

$$\mathcal{F}_A(0) \cap \beta = 12N \not\subseteq 8N = \mathcal{F}_A(d) \cup \alpha.$$

**Lemma 1** ([42]). *Let $E$ be a $BCK/BCI$-algebra and $A$ be a subalgebra of $E$. Let $\mathcal{F}_A \in S(U)$, if $\mathcal{F}_A$ is a US ideal over $U$, then for all $x, y \in A$:*

$$x \leq y \Rightarrow \mathcal{F}_A(x) \subseteq \mathcal{F}_A(y).$$

**Proof.** Let $x, y \in A$ be such that $x \leq y$. Then, $x * y = 0$, from which, by Definition 13 and Theorem 3, we get $\mathcal{F}_A(x) \subseteq \mathcal{F}_A(x * y) \cup \mathcal{F}_A(y) = \mathcal{F}_A(0) \cup \mathcal{F}_A(y) = \mathcal{F}_A(y)$. Hence, $\mathcal{F}_A(x) \subseteq \mathcal{F}_A(y)$. $\square$

**Lemma 2.** *Let $E$ be a $BCK/BCI$-algebra and $A$ be a subalgebra of $E$. Let $\mathcal{F}_A \in S(U)$. If $\mathcal{F}_A$ is an $(\alpha, \beta)$-US ideal over $U$, then for all $x, y \in A$:*

$$x \leq y \Rightarrow \mathcal{F}_A(x) \cap \beta \subseteq \mathcal{F}_A(y) \cup \alpha.$$

**Proof.** Let $x, y \in A$ be such that $x \leq y$. Then, $x * y = 0$, from which, by Definition 13 and Theorem 3, we get $\mathcal{F}_A(x) \cap \beta \subseteq \mathcal{F}_A(x * y) \cup \mathcal{F}_A(y) \cup \alpha = \mathcal{F}_A(0) \cup \mathcal{F}_A(y) \cup \alpha = \mathcal{F}_A(y) \cup \alpha$. Hence, $\mathcal{F}_A(x) \cap \beta \subseteq \mathcal{F}_A(y) \cup \alpha$. $\square$

**Proposition 3.** *Let $E$ be a $BCK/BCI$-algebra. For a given subalgebra $A$ of $E$, let $\mathcal{F}_A \in S(U)$. If $\mathcal{F}_A$ is an $(\alpha, \beta)$-US ideal over $U$, then for all $x, y, z \in A$, $\mathcal{F}_A$ satisfies the following conditions:*
*(1) $\mathcal{F}_A(x * y) \cap \beta \subseteq \mathcal{F}_A(x * z) \cup \mathcal{F}_A(z * y) \cup \alpha$*
*(2) $\mathcal{F}_A(x * y) = \mathcal{F}_A(0) \Rightarrow \mathcal{F}_A(x) \cap \beta \subseteq \mathcal{F}_A(y) \cup \alpha$.*

**Proof.** (1) Since $(x * y) * (x * z) \leq z * y$, then from Lemma 2, $\mathcal{F}_A((x * y) * (x * z)) \subseteq \mathcal{F}_A(z * y)$. Hence,

$$\mathcal{F}_A(x * y) \cap \beta \subseteq \mathcal{F}_A((x * y) * (x * z)) \cup \mathcal{F}_A(x * z) \cup \alpha \subseteq \mathcal{F}_A(x * z) \cup \mathcal{F}_A(z * y) \cup \alpha.$$

(2) If $\mathcal{F}_A(x * y) = \mathcal{F}_A(0)$, then for all $x, y \in A$,

$$\mathcal{F}_A(x) \cap \beta \subseteq \mathcal{F}_A(x * y) \cup \mathcal{F}_A(y) \cup \alpha = \mathcal{F}_A(0) \cup \mathcal{F}_A(y) \cup \alpha = \mathcal{F}_A(y) \cup \alpha.$$

$\square$

**Proposition 4.** *Let $E$ be a $BCK/BCI$-algebra and $A$ be a subalgebra of $E$. If $\mathcal{F}_A$ is an $(\alpha, \beta)$-US ideal over $U$, then for all $x, y, z \in A$, the following conditions are equivalent:*
*(1) $\mathcal{F}_A(x * y) \cap \beta \subseteq \mathcal{F}_A((x * y) * y) \cup \alpha$.*
*(2) $\mathcal{F}_A((x * z) * (y * z)) \cap \beta \subseteq \mathcal{F}_A((x * y) * z) \cup \alpha$.*

**Proof.** Assume that (1) holds and $x, y, z \in A$. Since $((x * (y * z)) * z) * z = ((x * z) * (y * z)) * z \leq (x * y) * z$ by (1), ($C_8$), and Lemma 2, we obtain the following equality: $\mathcal{F}_A((x * z) * (y * z)) \cap \beta = \mathcal{F}_A((x * (y * z)) * z) \cap \beta \subseteq \mathcal{F}_A(((x * (y * z)) * z) * z) \cup \alpha \subseteq \mathcal{F}_A((x * y) * z) \cup \alpha$.

Again, assume that (2) holds. If we put $y = z$ in (2), then by $(C_3)$ and $(C_6)$, we get $\mathcal{F}_A((x * z) * z) \cup \alpha \supseteq \mathcal{F}_A((x * z) * (z * z)) \cap \beta = \mathcal{F}_A((x * z) * 0) \cap \beta = \mathcal{F}_A(x * z) \cap \beta$, which implies that (1) holds. $\square$

**Theorem 8.** *Let E be a BCK/BCI-algebra and A be a given subalgebra of E. Then, every A-soft set is an* $(\alpha, \beta)$*-US ideal over U, and an A-soft set is an* $(\alpha, \beta)$*-US BCK/BCI-algebra over U.*

**Proof.** Let $\mathcal{F}_A$ be an $(\alpha, \beta)$-US ideal over $U$ and $A$ a subalgebra of $E$. From [42], we get $x * y \leq x$ for all $x, y \in A$. Then, it follows from Lemma 2 that $\mathcal{F}_A(x * y) \cap \beta \subseteq \mathcal{F}_A(x) \cup \alpha \subseteq \mathcal{F}_A(x * y) \cup \mathcal{F}_A(y) \cup \alpha \subseteq \mathcal{F}_A(x) \cup \mathcal{F}_A(y) \cup \alpha$. Hence, $\mathcal{F}_A$ is an $(\alpha, \beta)$-US BCK/BCI-algebra over $U$. $\square$

The converse of Theorem 8 is not true. This is justified by the following example.

**Example 7.** *Let* $U = N$ *be the initial universal set. Let* $E = N$ *be the set of natural numbers, and define a binary operation* $*$ *on E such that:*

$$x * y = \frac{x}{(x, y)}$$

*for all* $x, y \in X$, *where* $(x, y)$ *is the greatest common divisor of x and y. Then,* $(X; *, 1)$ *is a BCK-algebra. For a subalgebra* $A = \{1, 2, 3, 4, 5\}$ *of E, define the soft set* $(\mathcal{F}_A, A)$ *over U as follows:* $\mathcal{F}_A(1) = N$, $\mathcal{F}_A(2) = 4N$, $\mathcal{F}_A(3) = 2N$, $\mathcal{F}_A(4) = 3N$, *and* $\mathcal{F}_A(5) = 8N$. *Then,* $\mathcal{F}_A$ *is an* $(\alpha, \beta)$*-US algebra of A over U, but it is not an* $(\alpha, \beta)$*-US ideal of A over U, where* $\beta = 6N$ *and* $\alpha = 12N$, *because:*

$$\mathcal{F}_A(4) \cap \beta = 18N \nsubseteq 4N = \mathcal{F}_A(4 * 2) \cup \mathcal{F}_A(2) \cup \alpha$$

**Theorem 9.** *Let E be a BCK/BCI-algebra. Let* $\mathcal{F}_A \in S(U)$ *and A be a subalgebra of E. If* $\mathcal{F}_A$ *is an* $(\alpha, \beta)$*-US ideal over U, then for all* $x, y, z \in A$, $\mathcal{F}_A$ *satisfies the following condition:*

$$x * y \leq z \Rightarrow \mathcal{F}_A(x) \cap \beta \subseteq \mathcal{F}_A(y) \cup \mathcal{F}_A(z) \cup \alpha.$$

**Proof.** Let $x, y \in A$ be such that $x * y \leq z$, then $(x * y) * z = 0 \Rightarrow$

$$\mathcal{F}_A(x * y) \cap \beta \subseteq \mathcal{F}_A((x * y) * z) \cup \mathcal{F}_A(z) \cup \alpha = \mathcal{F}_A(0) \cup \mathcal{F}_A(z) \cup \alpha = \mathcal{F}(z) \cup \alpha.$$

Also, from which by using Definition 13 and Theorem 3, follows as $\mathcal{F}_A(x) \cap \beta \subseteq \mathcal{F}_A(x * y) \cup \mathcal{F}_A(y) \subseteq \mathcal{F}_A(y) \cup \mathcal{F}_A(z) \cup \alpha$. $\square$

The following results can be proven by induction.

**Corollary 1.** *Let E be a BCK/BCI-algebra and A be a subalgebra of E. Let* $\mathcal{F}_A \in S(U)$, *which satisfies the hypothesis of Theorem 3. Then,* $\mathcal{F}_A$ *is an* $(\alpha, \beta)$*-US ideal over U if and only if for all* $x, a_1, a_2, \ldots, a_n \in A$, *it satisfies the following condition:*

$$x * \prod_{i=1}^{n} a_i = 0 \Rightarrow \mathcal{F}_A(x) \cap \beta \subseteq \bigcup_{i=1,2,\ldots,n} \mathcal{F}_A(a_i) \cup \alpha.$$

We establish the following lemmas.

**Lemma 3.** *Let E be a BCK-algebra such that for all* $x, a, b, a_1, a_2, \ldots, a_n, b_1, b_2, \ldots, b_m \in E$, *the three conditions* $(x * a) * b = 0$, $a * \prod_{i=1}^{n} a_i = 0$ *and* $b * \prod_{j=1}^{m} b_j = 0$ *are satisfied. Then,* $(x * \prod_{i=1}^{n} a_i) * \prod_{j=1}^{m} b_j = 0$.

**Proof.** From $(x * a) * b = 0$, it follows that $x * a \leq b$.

Successively $*$-multiplying the above inequality on the right-hand side by $a_1, a_2, \ldots, a_n$ gives $\left(x * \prod_{i=1}^{n} a_i\right) * b = 0$; thus, $x * \prod_{i=1}^{n} a_i \leq b$.

Then, successively multiplying the right-hand side of the above inequality by $b_1, b_2, \ldots, b_m$ gives $\left(x * \prod_{i=1}^{n} a_i\right) * \prod_{j=1}^{m} b_j \leq b * \prod_{j=1}^{m} b_j = 0$.

Thus, $\left(x * \prod_{i=1}^{n} a_i\right) * \prod_{j=1}^{m} b_j = 0$. This completes the proof. $\square$

**Lemma 4.** *Let E be a BCK-algebra satisfying the three conditions of Lemma* 3. *If $\mathcal{F}_A$ is an $(\alpha, \beta)$-US ideal over U, then:*

$$\mathcal{F}_A(x) \cap \beta \subseteq \bigcup_{i=1,2,\ldots,n, j=1,2,\ldots,m} (\mathcal{F}_A(a_i) \cup \mathcal{F}_A(b_j)) \cup \alpha.$$

**Proof.** This follows from Lemma 3 and Corollary 1. $\square$

**Theorem 10.** *Let E be a BCK/BCI-algebra. Given a subalgebra A of E, let $\mathcal{F}_A \in S(U)$ and $\beta \subseteq U$. Then, $\mathcal{F}_A$ is an $(\alpha, \beta)$-US ideal over U if and only if the non-empty set $B(\mathcal{F}_A : \delta)$ is an ideal of A.*

**Proof.** The proof of the theorem is the same as Theorem 2. $\square$

**Definition 14.** *Let E be a BCK/BCI-algebra. For a given subalgebra A of E, let $\mathcal{F}_A \in S(U)$. An $(\alpha, \beta)$-US ideal $\mathcal{F}_A$ is said to be closed if for all $x \in A$, it satisfies the condition:*

$$\mathcal{F}_A(0 * x) \cap \beta \subseteq \mathcal{F}_A(x) \cup \alpha.$$

**Example 8.** *Let $U = Z$ (set of positive integers) be the universal set and $E = \{0, 1, 2, a, b\}$ be a BCI-algebra with the following Cayley table:*

| $*$ | 0 | 1 | 2 | $a$ | $b$ |
|---|---|---|---|---|---|
| 0 | 0 | 0 | 0 | $a$ | $a$ |
| 1 | 1 | 0 | 1 | $b$ | $a$ |
| 2 | 2 | 2 | 0 | $a$ | $a$ |
| $a$ | $a$ | $a$ | $a$ | 0 | 0 |
| $b$ | $b$ | $a$ | $b$ | 1 | 0 |

*For a subalgebra $A = \{0, 1, 2, a, b\}$ of E, define the soft set $(\mathcal{F}_A, A)$ over U as $\mathcal{F}_A(0) = \{1, 3, 4, 5, 7, 9\}$, $\mathcal{F}_A(1) = \{1, 2, 4, 5, 6, 7, 8, 10\}$, $\mathcal{F}_A(2) = \{1, 2, 4, 5, 6, 8, 12, 13\}$, $\mathcal{F}_A(a) = \{2, 3, 5, 6, 8, 9, 12, 13\}$, $\mathcal{F}_A(b) = \{1, 2, 3, 5, 7, 10, 12, 13\}$, $\beta = \{1, 2, 3, 5, 7, 9, 10, 11, 12, 13\}$, and $\alpha = \{1, 3, 5, 6, 7, 8, 9, 11\}$. Then, $\mathcal{F}_A$ is an $(\alpha, \beta)$-US closed ideal of A over U.*

**Theorem 11.** *Let E be a BCI-algebra. Then, an $(\alpha, \beta)$-US ideal over U is closed if and only if it is an $(\alpha, \beta)$-US algebra over U.*

**Proof.** Let $\mathcal{F}_A$ be an $(\alpha, \beta)$-US ideal over U. If $\mathcal{F}_A$ is closed, then $\mathcal{F}_A(0 * x) \cap \beta \subseteq \mathcal{F}_A(x) \cup \alpha$, for all $x \in A$. It follows from Definition 13 that:

$$\mathcal{F}_A(x * y) \cap \beta \subseteq \mathcal{F}_A((x * y) * x) \cup \mathcal{F}_A(x) \cup \alpha = \mathcal{F}_A(0 * y) \cup \mathcal{F}_A(x) \cup \alpha \subseteq \mathcal{F}_A(x) \cup \mathcal{F}_A(y) \cup \alpha$$

for all $x, y \in A$. Hence, $\mathcal{F}_A$ is an $(\alpha, \beta)$-US algebra of A over U.

Conversely, if $\mathcal{F}_A$ is an $(\alpha, \beta)$-US algebra over U, then,

$$\mathcal{F}_A(0 * x) \cap \beta \subseteq \mathcal{F}_A(0) \cup \mathcal{F}_A(x) \cup \alpha = \mathcal{F}_A(x) \cup \alpha.$$

for all $x \in A$. Thus, $\mathcal{F}_A$ is an $(\alpha, \beta)$-US closed ideal over $U$. $\square$

Let $E$ be a *BCI*-algebra and $B(E) = \{x \in E | 0 \leq x\}$. For any $x \in E$ and $n \in N$ ($N$ is the set of natural numbers), define $x^n$ by:

$$x^1 = x, x^{n+1} = x * (0 * x).$$

If there exists an $n \in N$ such that $x^n \in B(E)$, then we say that $x$ is finite periodic (see [50]), and its period is denoted by $|x|$ and defined by:

$$|x| = \min\{n \in N | x^n \in B(E)\}.$$

Otherwise, $x$ is infinite ordered and denoted by $|x| = infinite$.

**Theorem 12.** *Let E be a BCI-algebra in which every element is of finite period. Then, every $(\alpha, \beta)$-US ideal over U is closed.*

**Proof.** Let $\mathcal{F}_E$ be an $(\alpha, \beta)$-US ideal over $U$. Then, for any $x \in E$, suppose that $|x| = n$. Then, $x^n \in B(X)$. We get $(0 * x^{n-1}) * x = (0 * (0 * (0 * x^{n-1}))) * x = (0 * x) * (0 * (0 * x^{n-1})) = 0 * (x * (0 * x^{n-1})) = 0 * x^n = 0$, and so, $\mathcal{F}_E((0 * x^{n-1}) * x) = \mathcal{F}_E(0) \subseteq \mathcal{F}_E(x)$ by using Theorem 3. Then, from Definition 13, it follows that:

$$\mathcal{F}_E((0 * x^{n-1}) * x) \cap \beta \subseteq \mathcal{F}_E(0 * x^{n-1}) \cup \mathcal{F}_E(x) \cup \alpha = \mathcal{F}_E(0) \cup \mathcal{F}_E(x) \cup \alpha = \mathcal{F}_E(x) \cup \alpha.$$

Furthermore, it is noted that $(0 * x^{n-2}) * x = (0 * (0 * (0 * x^{n-2}))) * x = (0 * x) * (0 * (0 * x^{n-2})) = 0 * (x * (0 * x^{n-2})) = 0 * x^{n-1}$, which implies from above that:

$$\mathcal{F}_E((0 * x^{n-2}) * x) = \mathcal{F}_E(0 * x^{n-1}) \subseteq \mathcal{F}_E(x).$$

Again, by Definition 13, we have:

$$\mathcal{F}_E((0 * x^{n-2}) * x) \cap \beta \subseteq \mathcal{F}_E(0 * x^{n-2}) \cup \mathcal{F}_E(x) \cup \alpha = \mathcal{F}_E(0) \cup \mathcal{F}_E(x) \subseteq \mathcal{F}_E(x) \cup \alpha.$$

By continuation of the above process, we get $\mathcal{F}_E(0 * x) \cap \beta \subseteq \mathcal{F}_E(x) \cup \alpha$ for all $x \in E$. Hence, $\mathcal{F}_E$ is an $(\alpha, \beta)$-US closed ideal over $U$. $\square$

## 6. $(\alpha, \beta)$-US Commutative Ideals in *BCK/BCI*-Algebras

**Definition 15.** *Let E be a BCK-algebra. For a given subalgebras A of E, let $\mathcal{F}_A \in S(U)$. Then, $\mathcal{F}_A$ is called an $(\alpha, \beta)$-US commutative ideal over U if for all $x, y, z \in A$, it satisfies Theorem 3 and the following condition:*

$$\mathcal{F}_A(x * (y * (y * x))) \cap \beta \subseteq \mathcal{F}_A((x * y) * z) \cup \mathcal{F}_A(z) \cup \alpha.$$

**Example 9.** *Let $E = \{0, 1, 2, 3, 4\}$ be a BCK-algebra with the following Cayley table:*

| * | 0 | 1 | 2 | 3 | 4 |
|---|---|---|---|---|---|
| 0 | 0 | 0 | 0 | 0 | 0 |
| 1 | 1 | 0 | 1 | 1 | 0 |
| 2 | 2 | 2 | 0 | 2 | 0 |
| 3 | 3 | 3 | 3 | 0 | 0 |
| 4 | 4 | 4 | 4 | 4 | 0 |

*Let $(\mathcal{F}_A, A)$ be a soft set over $U = X$, where $A = \{1, 2, 3, 4\}$ and $\mathcal{F}_A : A \to P(X)$ is a set valued function defined by $\mathcal{F}_A(x) = \{y \in X | y * x \in \{0, 2, 3\}\}$. Then,*

$$\mathcal{F}_A(0) = \emptyset,$$

$$\mathcal{F}_A(1) = \{y \in X \mid y * 1 \in \{0,2,3\}\} = \{0,1,2,3\},$$

$$\mathcal{F}_A(2) = \{y \in X \mid y * 2 \in \{0,2,3\}\} = \{0,2,3\},$$

$$\mathcal{F}_A(3) = \{y \in X \mid y * 3 \in \{0,2,3\}\} = \{0,2,3\},$$

$$\mathcal{F}_A(4) = \{y \in X \mid y * 4 \in \{0,2,3\}\} = \{0,1,2,3,4\}.$$

Then, $\mathcal{F}_A$ is an $(\alpha, \beta)$-US commutative ideal of $A$ over $U$, where $\beta = \{0,1,3,4\}$ and $\alpha = \{0,1,3\}$.

**Theorem 13.** *Let E be a BCK-algebra. Then, any $(\alpha, \beta)$-US commutative ideal over $U$ is an $(\alpha, \beta)$-US ideal over $U$.*

**Proof.** Let $A$ be a subalgebra of $E$ and $\mathcal{F}_A$ be an $(\alpha, \beta)$-US commutative ideal over $U$. Now, we put $y = 0$ in Definition 15 and use $(C_5)$ and $(C_6)$, then we have $\mathcal{F}_A(x) \cap \beta = \mathcal{F}_A(x * (0 * (0 * x))) \cap \beta \subseteq \mathcal{F}_A((x * 0) * z) \cup \mathcal{F}_A(z) \cup \alpha = \mathcal{F}_A(x * z) \cup \mathcal{F}_A(z) \cup \alpha$ for all $x, z \in A$. Thus, $\mathcal{F}_A$ is an $(\alpha, \beta)$-US ideal over $U$. $\square$

In view of the following example, we can also establish Theorem 13.

**Example 10.** *Let $U = N$ (set of natural numbers) be the universal set and $E = \{0, a, b, c, d\}$ be a BCK-algebra with the following Cayley table:*

| $*$ | 0 | $a$ | $b$ | $c$ | $d$ |
|---|---|---|---|---|---|
| 0 | 0 | 0 | 0 | 0 | 0 |
| $a$ | $a$ | 0 | $a$ | 0 | 0 |
| $b$ | $b$ | $b$ | 0 | 0 | 0 |
| $c$ | $c$ | $c$ | $c$ | 0 | 0 |
| $d$ | $d$ | $d$ | $d$ | $c$ | 0 |

*The soft set $(\mathcal{F}_A, A)$ is defined over $U$ as follows $\mathcal{F}_A(0) = N$, $\mathcal{F}_A(a) = 3N$, $\mathcal{F}_A(b) = \mathcal{F}_A(d) = 2N$, and $\mathcal{F}_A(c) = \varnothing$. Then, $\mathcal{F}_A$ is an $(\alpha, \beta)$-US commutative ideal over $U$, as well as an $(\alpha, \beta)$-US ideal over $U$, where $\beta = 6N$ and $\alpha = 12N$.*

The following theorem provides the condition that an $(\alpha, \beta)$-US ideal over $U$ is an $(\alpha, \beta)$-US commutative ideal over $U$.

**Theorem 14.** *Let E be a BCK-algebra and $A$ be a subalgebra of $E$. Let $\mathcal{F}_A \in S(U)$, then $\mathcal{F}_A$ is an $(\alpha, \beta)$-US commutative ideal over $U$ if and only if, for all $x, y, z \in A$, $\mathcal{F}_A$ is an $(\alpha, \beta)$-US ideal over $U$ satisfying the following condition:*

$$\mathcal{F}_A(x * (y * (y * x))) \subseteq \mathcal{F}_A(x * y).$$

**Proof.** Assume that $\mathcal{F}_A$ is an $(\alpha, \beta)$-US ideal commutative ideal over $U$. Then, $\mathcal{F}_A$ is an $(\alpha, \beta)$-US soft ideal over $U$ by Theorem 13. Now, if we take $z = 0$ in Definition 15 and use $(C_5)$, then we deduce the condition given in Theorem 14.

Conversely, if $\mathcal{F}_A$ is an $(\alpha, \beta)$-US ideal over $U$ satisfying the condition of Theorem 14, then for all $x, y, z \in A$, we have $\mathcal{F}_A(x * y) \cap \beta \subseteq \mathcal{F}_A((x * y) * z) \cup \mathcal{F}_A(z) \cup \alpha$ by Definition 13. Hence, from Definition 15, we conclude that $\mathcal{F}_A$ is an $(\alpha, \beta)$-US commutative ideal over $U$. $\square$

**Corollary 2.** *Let E be a BCK-algebra and $\mathcal{F}_E \in S(U)$. Then, $\mathcal{F}_E$ is an $(\alpha, \beta)$-US commutative ideal over $U$ if and only if $\mathcal{F}_E$ is an $(\alpha, \beta)$-US ideal over $U$ satisfying the following condition for all $x, y \in A$:*

$$\mathcal{F}_E(x * (y * (y * x))) \cap \beta \subseteq \mathcal{F}_A(x * y) \cup \alpha.$$

**Theorem 15.** *Let E be a commutative BCK-algebra. Then, every $(\alpha, \beta)$-US ideal over $U$ is an $(\alpha, \beta)$-US commutative ideal over $U$.*

**Proof.** Let $\mathcal{F}_A$ be an $(\alpha, \beta)$-US ideal over $U$, where $A$ is a subalgebra of $E$. Then, for all $x, y, z \in A$, we notice that:

$$
\begin{aligned}
(((x * (y * (y * x))) * ((x * y) * z))) * z &= ((x * (y * (y * x))) * z) * ((x * y) * z) \\
&\leq ((x * (y * (y * x))) * (x * y) \\
&= (x * (x * y)) * (y * (y * x)) = 0
\end{aligned}
$$

Thus, $(x * (y * (y * x))) * ((x * y) * z) \leq z$. Then, from Theorem 9, we get $\mathcal{F}_A(x * (y * (y * x))) \cap \beta \subseteq \mathcal{F}_A((x * y) * z) \cup \mathcal{F}_A(z) \cup \alpha$. Hence, $\mathcal{F}_A$ is an $(\alpha, \beta)$-US commutative ideal over $U$. $\square$

**Theorem 16.** *Let E be a BCK-algebra and A be a subalgebra of E. Let $\mathcal{F}_A \in S(U)$. If $\mathcal{F}_A$ satisfies the following conditions:*
*(1) $x * (x * y) \leq y * (y * x)$ for all $x, y \in A$;*
*(2) $\mathcal{F}_A$ is an $(\alpha, \beta)$-US ideal over $U$;*
*then $\mathcal{F}_A$ is an $(\alpha, \beta)$-US commutative ideal over $U$.*

**Proof.** For any $x, y \in A$, we have:

$$
\mathcal{F}_A(x * (y * (y * x))) * (x * y) = (x * (x * y)) * (y * (y * x)) = 0
$$

by $(C_8)$ and (1). Therefore, $x * (y * (y * x)) \leq x * y$ for all $x, y \in A$, which indicates from Lemma 2 that $\mathcal{F}_A(x * (y * (y * x))) \cap \beta \subseteq \mathcal{F}_A(x * y) \cup \alpha$. Now, it follows from Theorem 14 that $\mathcal{F}_A$ is an $(\alpha, \beta)$-US commutative ideal of $A$ over $U$. $\square$

**Theorem 17.** *Let E be a BCK/BCI-algebra and A be a subalgebra of E. Consider $\mathcal{F}_A \in S(U)$ and $\delta \subseteq \beta \subseteq U$. Then, $\mathcal{F}_A$ is an $(\alpha, \beta)$-US commutative ideal over $U$ if and only if the non-empty set $B(\mathcal{F}_A : \delta)$ is a commutative ideal of $A$.*

**Proof.** The proof of the theorem is the same as Theorem 2. $\square$

**Theorem 18.** *Let E be a BCK-algebra and $\mathcal{F}_A \in S(U)$. Define a soft set $\mathcal{F}_A^*$ over $U$ by $\mathcal{F}_A^* : E \to \mathcal{P}(U)$,*
$$
x \longmapsto \begin{cases} \mathcal{F}_A(x) & \text{if } x \in B(\mathcal{F}_A : \delta) \\ U & \text{otherwise.} \end{cases}
$$
*If $\mathcal{F}_A$ is an $(\alpha, \beta)$-US commutative ideal over $U$, then so is $\mathcal{F}_A^*$.*

**Proof.** If $\mathcal{F}_A$ is an $(\alpha, \beta)$-US commutative ideal over $U$, then $B(\mathcal{F}_A : \delta)$ is a commutative ideal over $U$ by Theorem 17. Hence, $0 \in B(\mathcal{F}_A : \delta)$, and so, we have $\mathcal{F}_A^*(0) \cap \beta = \mathcal{F}_A(0) \cap \beta \subseteq \mathcal{F}_A(x) \cup \alpha \subseteq \mathcal{F}_A^*(x) \cup \alpha$ for all $x \in A$. Let $x, y, z \in A$. Then, $(x * y) * z \in B(\mathcal{F}_A : \delta)$ and $z \in B(\mathcal{F}_A : \delta)$; hence, $x * (y * (y * x)) \in B(\mathcal{F}_A : \delta)$, and so, we deduce the following equality:

$$
\begin{aligned}
\mathcal{F}_A^*(x * (y * (y * x))) \cap \beta &= \mathcal{F}_A(x * (y * (y * x))) \cap \beta \\
&\subseteq \mathcal{F}_A((x * y) * z) \cup \mathcal{F}_A(z) \cup \alpha \\
&= \mathcal{F}_A^*((x * y) * z) \cup \mathcal{F}_A^*(z) \cup \alpha.
\end{aligned}
$$

If $(x * y) * z \notin B(\mathcal{F}_A : \delta)$ and $z \notin B(\mathcal{F}_A : \delta)$, then $\mathcal{F}_A^*(x * (y * (y * x)))$ or $\mathcal{F}_A^*(z) = U$. Thus, we have $\mathcal{F}_A^*(x * (y * (y * x))) \cap \beta \subseteq U = \mathcal{F}_A^*((x * y) * z) \cup \mathcal{F}_A^*(z) \cup \alpha$. This shows that $\mathcal{F}_A^*$ is an $(\alpha, \beta)$-US commutative ideal of $A$ over $U$. $\square$

**Theorem 19.** *Let E be a BCK-algebra and A be a subset of E, which is a commutative ideal of E if and only if the soft subset $\mathcal{F}_A$ defined by:*

$$\mathcal{F}_A(x) = \begin{cases} \Omega & \text{if } x \in A \\ \Gamma & \text{if } x \notin A, \end{cases}$$

*where $\alpha \subseteq \Omega \subseteq \Gamma \subseteq \beta \subseteq U$, is an $(\alpha, \beta)$-US commutative ideal of A over U.*

**Proof.** Let $A$ be a commutative ideal of $E$ and if $x \in A$, then $0 \in A$. Therefore, $\mathcal{F}_A(0) = \mathcal{F}_A(x) = \Omega$, and so, $\mathcal{F}_A(0) \cap \beta = \Omega \cap \beta = \Omega$ and $\mathcal{F}_A(x) \cup \alpha = \Omega \cup \alpha = \Omega$. Thus, $\mathcal{F}_A(0) \cap \beta \subseteq \mathcal{F}_A(x) \cup \alpha$. Let for any $x, y, z \in A$ and if $(x * y) * z \in A$, $z \in A$, then $(x * (y * (y * x))) \in A$, and thus, $\mathcal{F}_A((x * y) * z) = \mathcal{F}_A(z) = \mathcal{F}_A(x * (y * (y * x))) = \Omega$. Then, $\mathcal{F}_A(x * (y * (y * x))) \cap \beta = \Omega \cap \beta = \Omega$ and $\mathcal{F}_A((x * y) * z) \cup \mathcal{F}_A(z) \cup \alpha = \Omega \cup \alpha = \Omega$, which indicates that $\mathcal{F}_A(x * (y * (y * x))) \cap \beta \subseteq \mathcal{F}_A(x * y) * z) \cup \mathcal{F}_A(z) \cup \alpha$. Now, if $x \notin A$, then $0 \in A$ or $0 \notin A$, and so, $\mathcal{F}_A(0) \cap \beta = \Omega \cap \beta = \Omega$ or $\mathcal{F}_A(0) \cap \beta = \Gamma \cap \beta = \Gamma$, but $\mathcal{F}_A(x) \cup \alpha = \Gamma \cup \alpha = \Gamma$, which implies that $\mathcal{F}_A(0) \cap \beta \subseteq \mathcal{F}_A(x) \cup \alpha$. Now, if $(x * y) * z \notin A$ or $z \notin A$, then $(x * (y * (y * x))) \in A$ or $(x * (y * (y * x))) \notin A$, and so, $\mathcal{F}_A(x * (y * (y * x))) \cap \beta = \Omega \cap \beta = \Omega$ or $\mathcal{F}_A(x * (y * (y * x))) \cap \beta = \Gamma \cap \beta = \Gamma$, but $\mathcal{F}_A((x * y) * z) \cup \mathcal{F}_A(z) \cup \alpha = \Gamma \cup \alpha = \Gamma$, which implies that $\mathcal{F}_A(x * (y * (y * x))) \cap \beta \subseteq \mathcal{F}_A((x * y) * z) \cup \mathcal{F}_A(z) \cup \alpha$. Hence, $\mathcal{F}_A$ is an $(\alpha, \beta)$-US commutative ideal of $A$ over $U$.

Conversely, assume that $\mathcal{F}_A$ is an $(\alpha, \beta)$-US commutative ideal of $A$ over $U$. If $x \in A$, then $\mathcal{F}_A(0) \cap \beta \subseteq \mathcal{F}_A(x) \cup \alpha = \Omega \cup \alpha = \Omega$. However, $\alpha \subseteq \Omega \subseteq \Gamma \subseteq \beta$; hence, $\mathcal{F}_A(0) = \Omega$, and so, $0 \in A$. Again, if $(x * y) * z \in A$ and $z \in A$, then $\mathcal{F}_A(x * (y * (y * x))) \cap \beta \subseteq \mathcal{F}_A((x * y) * z) \cup \mathcal{F}_A(z) \cup \alpha = \Omega \cup \alpha - \Omega$, and thus, $\mathcal{F}_A(x * (y * (y * x))) = \Omega$, which implies that $(x * (y * (y * x))) \in A$. Therefore, $A$ is a commutative ideal of $E$. $\square$

**Theorem 20** (Extension property). *Let E be a BCK-algebra. For two given subalgebras A and B of E, let $\mathcal{F}_A, \mathcal{F}_B \in S(U)$ such that*
*(i) $\mathcal{F}_A \subseteq \mathcal{F}_B$,*
*(ii) $\mathcal{F}_B$ is an $(\alpha, \beta)$-US ideal over U.*
*If $\mathcal{F}_A$ is an $(\alpha, \beta)$-US commutative ideal over U, then $\mathcal{F}_B$ is also an $(\alpha, \beta)$-US commutative ideal over U.*

**Proof.** Let $\delta \in U$ be such that $B(\mathcal{F}_A : \delta) \neq \emptyset$. By Condition $(ii)$ and Theorem 10, we see that $B(\mathcal{F}_A : \delta)$ is an ideal. We now consider $\mathcal{F}_A$ to be an $(\alpha, \beta)$-US commutative ideal of $A$ over $U$, then $B(\mathcal{F}_A : \delta)$ is a commutative ideal of $A$. Let $x, y \in A$ and $\delta \subseteq \beta$ be such that $x * y \in B(\mathcal{F}_A : \delta)$. Since $(x * (x * y)) * y = (x * y) * (x * y) = 0 \in B(\mathcal{F}_A : \delta)$, it follows from $(C_8)$ and $(i)$ that $(x * (y * (y * (x * (x * y))))) * (x * y) = (x * (x * y)) * (y * (y * (x * (x * y)))) \in B(\mathcal{F}_A : \delta) \subseteq B(\mathcal{F}_B : \delta)$. We see that:

$$x * (y * (y * (x * (x * y)))) \in B(\mathcal{F}_B : \delta) \tag{1}$$

as $B(\mathcal{F}_B : \delta)$ is an ideal and $x * y \in B(\mathcal{F}_B : \delta)$. Furthermore, it is noted that $x * (x * y) \leq x$, and so, we have $y * (y * (x * (x * y)))) \leq y * (y * x)$ by $(C_7)$. Thus,

$$x * (y * (y * x)) \leq x * (y * (y * (x * (x * y)))) \tag{2}$$

Hence, by using (1) and (2), we get $x * (y * (y * x)) \in B(\mathcal{F}_B : \delta)$. Therefore, $B(\mathcal{F}_B : \delta)$ is a commutative ideal, and so, $\mathcal{F}_B$ is an $(\alpha, \beta)$-US commutative ideal over $U$ by Theorem 17. $\square$

## 7. Conclusions

Soft set theory is an important mathematical notion, which easily handles uncertainties and has applications in real-life problems. In this paper, we introduce the notions of $(\alpha, \beta)$-US sets in $BCK/BCI$-algebras and $(\alpha, \beta)$-US ideals and $(\alpha, \beta)$-US commutative ideals of $BCK$-algebras. We also investigate some of their characterizations in detail. We hope that the results given in this paper will have an impact on the upcoming research in this area and other aspects of soft algebraic structures so that this leads to new horizons of interest and innovations. Our results can also be applied to other algebraic structures, such as an $(\alpha, \beta)$-US hemiring, an $(\alpha, \beta)$-US topology, $(\alpha, \beta)$-US $B$-algebras, $(\alpha, \beta)$-US $KUS$-algebras, $(\alpha, \beta)$-US Vector algebras, and $(\alpha, \beta)$-US lattices, and can be applied to other branches of pure mathematics. In addition, the recent development of fuzzy soft-covering-based [19,21,51] problems and $\beta$-covering-based rough fuzzy covering [23,52,53] problems have a huge scope of application to develop fuzzy soft covering $BCK/BCI$-algebras, fuzzy soft $\beta$-covering-based rough fuzzy $BCK/BCI$-algebras, as well as the construction of other fuzzy covering algebras.

**Author Contributions:** Conceptualization, C.J.; formal analysis, M.P.; writing, original draft preparation, C.J.; writing, review and editing, C.J. and M.P.; supervision, M.P.

**Funding:** This research received no external funding.

**Conflicts of Interest:** The authors declare no conflict of interest.

## References

1. Zadeh, L.A. Fuzzy sets. *Inform. Control* **1965**, *8*, 338–353.
2. Pawlak, Z. Rough sets. *Int. J. Inf. Comput. Sci.* **1982**, *11*, 341–356. [CrossRef]
3. Atanassov, K.T. Intuitionistic fuzzy sets. *Fuzzy Sets Syst.* **1986**, *20*, 87–96. [CrossRef]
4. Gau, W.L.; Buehrer, D.J. Vague sets. *IEEE Trans. Syst. Man Cybern.* **1993**, *23*, 610–614. [CrossRef]
5. Gorzalzany, M.B. A method of inference in approximate reasoning based on interval-valued fuzzy sets. *Fuzzy Sets Syst.* **1987**, *21*, 1–17. [CrossRef]
6. Molodtsov, D. Soft set theory-First results. *Comput. Math. Appl.* **1999**, *37*, 19–31. [CrossRef]
7. Maji, P.K.; Biswas, R.; Roy, A.R. Soft set theory. *Comput. Math. Appl.* **2003**, *45*, 555–562. [CrossRef]
8. Aktas, H.; Çagman, N. Soft sets and soft groups. *Inform. Sci.* **2007**, *177*, 2726–2735. [CrossRef]
9. Acar, U.; Koyuncu, F.; Tanay, B. Soft sets and soft rings. *Comput. Math. Appl.* **2010**, *59*, 3458–3463.
10. Ali, M.I.; Feng, F.; Liu, X.; Min, W.K.; Shabir, M. On some new operationsin in soft set theory. *Comput. Math. Appl.* **2009**, *57*, 1547–1553.
11. Çagman, N.; Citak, F.; Aktas, H. Soft-int group and its applications to group theory. *Neural Comput. Appl.* **2012**, *21*, 151–158. [CrossRef]
12. Feng, F.; Jun, Y.B.; Zhao, X. Soft semirings. *Comput. Math. Appl.* **2008**, *56*, 2621–2628.
13. Sezgin, A.; Atagun, A.O.; Çagman, N. Soft intersection near-rings with its applications. *Neural Comput. Appl.* **2012**, *21*, 221–229.
14. Çagman, N.; Enginoglu, S. Soft set theory and uni-int decision making. *Eur. J. Oper. Res.* **2010**, *207*, 848–855. [CrossRef]
15. Feng, F.; Jun, Y.B.; Liu, X.; Li, L. An adjustable approach to fuzzy soft set based decision-making. *J. Comput. Appl. Math.* **2010**, *234*, 10–20. [CrossRef]
16. Jana, C.; Pal, M. A robust single-valued neutrosophic soft aggregation operators in multi-criteria decision making. *Symmetry* **2019**, *11*, 110. [CrossRef]
17. Maji, P.K.; Roy, A.R.; Biswas, R. An application of soft sets in a decision making problem. *Comput. Math. Appl.* **2002**, *44*, 1077–1083. [CrossRef]
18. Ma, X.; Zhan, J.; Ali, M.I.; Mehmood, N. A survey of decision making methods based on two classes of hybrid soft set models. *Artif. Intell. Rev.* **2018**, *49*, 511–529. [CrossRef]
19. Zhan, J.; Sun, B.; Alcantud, J.C.R. Covering based multigranulation $(I, T)$-fuzzy rough set models and applications in multi-attribute group decision-making. *Inform. Sci.* **2019**, *476*, 290–318. [CrossRef]
20. Zou, Y.Z.; Xiao, Z. Data analysis approaches of soft sets under incomplete information. *Knowl.-Based Syst.* **2008**, *21*, 941–945. [CrossRef]
21. Zhan, J.; Wang, Q. Certain types of soft coverings based rough sets with applications. *Int. J. Mach. Learn. Cybern.* **2018**. [CrossRef]
22. Zhang, L.; Zhan, J.; Xu, Z.X. Covering-based generalized IF rough sets with applications to multi-attribute decision-making. *Inform. Sci.* **2019**, *478*, 275–302.

23. Zhang, L.; Zhan, J. Fuzzy soft $\beta$-covering based fuzzy rough sets and corresponding decision-making applications. *Int. J. Mach. Learn. Cybern.* **2018**. [CrossRef]
24. Zhan, J.; Dudek, W.A.; Neggers, J. A new soft union set: characterizations of hemirings. *Int. J. Mach. Learn. Cybern.* **2017**, *8*, 525–535.
25. Zhan, J.; Çağman, N.; Sezer, A.S. Applications of soft union sets to hemirings via *SU-h*-ideals. *J. Int. Fuzzy Syst.* **2014**, *26*, 1363–1370.
26. Zhan, J.; Yu, B. Characterizations of Hemirings via $(M, N)$-SI-h-bi-ideals and $(M, N)$-SI-h-quasi-ideals. *J. Mult.-Valued Log. Soft Comput.* **2016**, *26*, 363–388.
27. Imai, Y.; Iśeki, K. On axiom system of propositional calculi. *XIV Proc. Jpn. Acad.* **1966**, *42*, 19–22. [CrossRef]
28. Iśseki, K. An algebra related with a propositional calculus. *XIV Proc. Jpn. Acad.* **1966**, *42*, 26–29. [CrossRef]
29. Jana, C.; Senapati, T.; Pal, M. $(\in, \in \vee q)$-intuitionistic fuzzy *BCI*-subalgebras of *BCI*-algebra. *J. Int. Fuzzy Syst.* **2016**, *31*, 613–621.
30. Jana, C.; Pal, M.; Saied, A.B. $(\in, \in \vee q)$-bipolar fuzzy *BCK/BCI*-algebras. *Mol. J. Math. Sci.* **2017**, *29*, 139–160.
31. Jana, C.; Pal, M. On $(\in_\alpha, \in_\alpha \vee q_\beta)$-fuzzy soft *BCI*-algebras. *Mol. J. Math. Sci.* **2017**, *29*, 197–215.
32. Ma, X.; Zhan, J.; Jun, Y.B. Some types of $(\in, \in \vee q)$ interval valued fuzzy ideals of *BCI*-algebras. *Iran. J. Fuzzy Syst.* **2009**, *6*, 53–63.
33. Ma, X.; Zhan, J.; Jun, Y.B. Some kinds of $(\in_\gamma, \in_\gamma \vee q_\delta)$-fuzzy ideals of *BCI*-algebras. *Comput. Math. Appl.* **2011**, *61*, 1005–1015. [CrossRef]
34. Ma, X.; Zhan, J.; Jun, Y.B. New types of fuzzy ideals of *BCI*-algebras. *Neural Comput. Appl.* **2012**, *21*, S19–S27. [CrossRef]
35. Senapati, T.; Jana, C.; Pal, M.; Jun, Y.B. Cubic Intuitionistic q-Ideals of BCI-Algebras. *Symmetry* **2018**, *10*, 752. [CrossRef]
36. Senapati, T.; Bhowmik, M.; Pal, M.; Davvaz, B. Fuzzy translations of fuzzy *H*-ideals in *BCK/BCI*-algebra. *J. Indones. Math. Soc.* **2015**, *21*, 45–58. [CrossRef]
37. Jun, Y.B. Soft *BCK/BCI*-algebra. *Comput. Math. Appl.* **2008**, *56*, 1408–1413. [CrossRef]
38. Jun, Y.B.; Park, C.H. Applications of soft sets in ideal theory of *BCK/BCI*-algebra. *Inform. Sci.* **2008**, *178*, 2466–2475. [CrossRef]
39. Jun, Y.B.; Lee, K.J.; Zhan, J. Soft *p*-ideals of soft *BCI*-algebras. *Comput. Math. Appl.* **2009**, *58*, 2060–2068. [CrossRef]
40. Acar, U.; Özürk, Y. Maximal, irreducible and prime soft ideals of *BCK/BCI*-algebras. *Hacet. J. Math. Stat.* **2015**, *44*, 1–13. [CrossRef]
41. Jun, Y.B. On $(\alpha, \beta)$-fuzzy subalgebras of *BCK/BCI*-algebras. *Bull. Korean Math. Soc.* **2005**, *42*, 703–711. [CrossRef]
42. Jun, Y.B. Uion-soft sets with applications in *BCK/BCI*-algebras. *Bull. Korean Math. Soc.* **2013**, *50*, 1937–1956.
43. Sezgin, A. Soft union interior ideals, quasi-ideals and generalized Bi-ideals of rings. *Filomat* **2018**, *32*, 1–28.
44. Sezgin, A.; Çagman, N.; Atagün, A.O. A completely new view to soft intersection rings via soft uni-intproduct. *Appl. Soft Comput.* **2017**, *54*, 366–392.
45. Jana, C.; Pal, M. Applications of new soft intersection set on groups. *Ann. Fuzzy Math. Inform.* **2016**, *11*, 923–944.
46. Jana, C.; Pal, M. Applications of $(\alpha, \beta)$-soft intersectional sets on *BCK/BCI*-algebras. *Int. J. Intell. Syst. Technol. Appl.* **2017**, *16*, 269–288. [CrossRef]
47. Jana, C.; Pal, M.; Karaaslan, F.; Sezgin, A. $(\alpha, \beta)$-soft intersectional rings and ideals with their applications. *New Math. Nat. Comput.* **2018**. [CrossRef]
48. Ma, X.; Zhan, J. Applications of a new soft set to h-hemiregular hemirings via (M,N)-SI-h-ideals. *J. Int. Fuzzy Syst.* **2014**, *26*, 2515–2525.
49. Huang, Y. *BCI-Algebra*; Science Press: Beijing, China, 2006.
50. Meng, J.; Wei, S.M. Periods of elements in BCI-algebras. *Math. Jpn.* **1993**, *38*, 427–431.
51. Zhan, J.; Xu, W. Two types of coverings based multigranulation rough fuzzy sets and applications to decision making. *Artif. Intell. Rev.* **2018**. [CrossRef]
52. Jiang, H.; Zhan, J.; Chen, D. Covering based variable precision $(I, T)$-fuzzy rough sets with applications to multi-attribute decision-making. *IEEE Trans. Fuzzy Syst.* **2018**, doi:10.1109/TFUZZ.2018.2883023. [CrossRef]

53. Zhang, L.; Zhan, J. Novel classes of fuzzy soft $\beta$-coverings-based fuzzy rough sets with applications to multi-criteria fuzzy group decision making. *Soft Comput.* **2018**. [CrossRef]

 *mathematics*

*Article*

# $p$-Regularity and $p$-Regular Modification in $\top$-Convergence Spaces

**Qiu Jin [1,*], Lingqiang Li [1,2,3,*] and Guangming Lang [2,3]**

[1]  Department of Mathematics, Liaocheng University, Liaocheng 252059, China
[2]  School of Mathematics and Statistics, Changsha University of Science and Technology, Changsha 410114, China; langguangming1984@126.com
[3]  Hunan Provincial Key Laboratory of Mathematical Modeling and Analysis in Engineering, Changsha University of Science and Technology, Changsha 410114, China
*  Correspondence: jinqiu79@126.com (Q.J.); jinqiu@lcu.edu.cn (L.L.);
   Tel.: +86-150-6355-2700 (Q.J.); +86-152-0650-6635 (L.L.)

Received: 9 March 2019; Accepted: 22 April 2019; Published: 24 April 2019

**Abstract:** Fuzzy convergence spaces are extensions of convergence spaces. $\top$-convergence spaces are important fuzzy convergence spaces. In this paper, $p$-regularity (a relative regularity) in $\top$-convergence spaces is discussed by two equivalent approaches. In addition, lower and upper $p$-regular modifications in $\top$-convergence spaces are further investigated and studied. Particularly, it is shown that lower (resp., upper) $p$-regular modification and final (resp., initial) structures have good compatibility.

**Keywords:** fuzzy topology; fuzzy convergence; $\top$-convergence space; regularity

## 1. Introduction

Convergence spaces [1] are generalizations of topological spaces. Regularity is an important property in convergence spaces. In general, there are two equivalent approaches to characterize regularity. One approach is stated through a diagonal condition of filters [2,3], the other approach is represented through a closure condition of filters [4]. In [5,6], for a pair of convergence structures $p, q$ on the same underlying set, Wilde-Kent-Richardson considered a relative regularity (called $p$-regularity) both from two equivalent approaches. When $p = q$, $p$-regularity is nothing but regularity. Wilde-Kent [6] further presented a theory of lower and upper $p$-regular modifications in convergence spaces. Said precisely, for convergence structures $p, q$ on a set $X$, the lower (resp., upper) $p$-regular modification of $q$ is defined as the finest (resp., coarsest) $p$-regular convergence structure coarser (resp., finer) than $q$.

Fuzzy convergence spaces are natural extensions of convergence spaces. Quite recently, two types of fuzzy convergence spaces received wide attention: (1) stratified $L$-generalized convergence spaces (resp., stratified $L$-convergence spaces) initiated by Jäger [7] (resp., Flores [8]) and then developed by many scholars [8–30]; and (2) $\top$-convergence spaces introduced by Fang [31] and then discussed by many researchers [32–36]. Regularity in stratified $L$-generalized convergence spaces (resp., stratified $L$-convergence spaces) was studied by Jäger [37] (resp., Boustique-Richardson [38,39]), $p$-regularity and $p$-regular modifications in stratified $L$-generalized convergence spaces and that in stratified $L$-convergence spaces were discussed by Li [40,41]. Regularity in $\top$-convergence spaces by different diagonal conditions of $\top$-filters were researched by Fang [31] and Li [42], respectively. Regularity in $\top$-convergence spaces by closure condition of $\top$-filters were studied by Reid and Richardson [36]. In this paper, we shall discuss $p$-regularity and $p$-regular modifications in $\top$-convergence spaces.

The contents are arranged as follows. Section 2 recalls some notions and notations for later use. Section 3 presents $p$-regularity in $\top$-convergence spaces by a diagonal condition of $\top$-filters and a closure condition of $\top$-filters, respectively. Section 4 mainly discusses $p$-regular modifications in $\top$-convergence spaces. The lower and upper $p$-regular modifications in $\top$-convergence spaces are investigated and researched. Especially, it is shown that lower (resp., upper) $p$-regular modification and final (resp., initial) structures have good compatibility.

## 2. Preliminaries

In this paper, if not otherwise stated, $L = (L, \leq)$ is always a complete lattice with a top element $\top$ and a bottom element $\bot$, which satisfies the distributive law $\alpha \wedge (\bigvee_{i \in I} \beta_i) = \bigvee_{i \in I} (\alpha \wedge \beta_i)$. A lattice with these conditions is called a complete Heyting algebra. The operation $\rightarrow: L \times L \longrightarrow L$ given by

$$\alpha \rightarrow \beta = \bigvee \{\gamma \in L : \alpha \wedge \gamma \leq \beta\}.$$

is called the residuation with respect to $\wedge$. We collect here some basic properties of the binary operations $\wedge$ and $\rightarrow$ [43].

(1) $a \rightarrow b = \top \Leftrightarrow a \leq b$;
(2) $a \wedge b \leq c \Leftrightarrow b \leq a \rightarrow c$;
(3) $a \wedge (a \rightarrow b) \leq b$;
(4) $a \rightarrow (b \rightarrow c) = (a \wedge b) \rightarrow c$;
(5) $(\bigvee_{j \in J} a_j) \rightarrow b = \bigwedge_{j \in J} (a_j \rightarrow b)$; (6) $a \rightarrow (\bigwedge_{j \in J} b_j) = \bigwedge_{j \in J} (a \rightarrow b_j)$.

A function $\mu : X \rightarrow L$ is said to be an $L$-fuzzy set in $X$, and all $L$-fuzzy sets in $X$ are denoted as $L^X$. The operators $\vee, \wedge, \rightarrow$ on $L$ can be translated onto $L^X$ pointwisely. Precisely, for any $\mu, \nu, \mu_t (t \in T) \in L^X$,

$$\left(\bigvee_{t \in T} \mu_t\right)(x) = \bigvee_{t \in T} \mu_t(x), \quad \left(\bigwedge_{t \in T} \mu_t\right)(x) = \bigwedge_{t \in T} \mu_t(x), (\mu \rightarrow \nu)(x) = \mu(x) \rightarrow \nu(x).$$

Let $f : X \longrightarrow Y$ be a function. We define $f^{\rightarrow} : L^X \longrightarrow L^Y$ by $f^{\rightarrow}(\mu)(y) = \bigvee_{f(x)=y} \mu(x)$ for $\mu \in L^X$ and $y \in Y$, and define $f^{\leftarrow} : L^Y \longrightarrow L^X$ by $f^{\leftarrow}(\nu)(x) = \nu(f(x))$ for $\nu \in L^Y$ and $x \in X$ [43].

Let $\mu, \nu$ be $L$-fuzzy sets in $X$. The subsethood degree of $\mu, \nu$, denoted as $S_X(\mu, \nu)$, is defined by $S_X(\mu, \nu) = \bigwedge_{x \in X} (\mu(x) \rightarrow \nu(x))$ [44–46]

**Lemma 1.** *[31,42,47] Let $f : X \longrightarrow Y$ be a function and $\mu_1, \mu_2 \in L^X$, $\lambda_1, \lambda_2 \in L^Y$. Then*

*(1) $S_X(\mu_1, \mu_2) \leq S_Y(f^{\rightarrow}(\mu_1), f^{\rightarrow}(\mu_2))$,*
*(2) $S_Y(\lambda_1, \lambda_2) \leq S_X(f^{\leftarrow}(\lambda_1), f^{\leftarrow}(\lambda_2))$,*
*(3) $S_Y(f^{\rightarrow}(\mu_1), \lambda_1) = S_X(\mu_1, f^{\leftarrow}(\lambda_1))$.*

### 2.1. $\top$-Filters and $\top$-Convergence Spaces

**Definition 1.** *[43,48] A nonempty subset $\mathbb{F} \subseteq L^X$ is said to be a $\top$-filter on the set $X$ if it satisfies the following three conditions:*

*(TF1)* $\forall \lambda \in \mathbb{F}, \bigvee_{x \in X} \lambda(x) = \top$,

*(TF2)* $\forall \lambda, \mu \in \mathbb{F}, \lambda \wedge \mu \in \mathbb{F}$,

*(TF3)* *if* $\bigvee_{\mu \in \mathbb{F}} S_X(\mu, \lambda) = \top$, *then* $\lambda \in \mathbb{F}$.

*The set of all $\top$-filters on $X$ is denoted by $\mathbb{F}_L^{\top}(X)$.*

**Definition 2.** *[43] A nonempty subset $\mathbb{B} \subseteq L^X$ is referred to be a $\top$-filter base on the set $X$ if it holds that:*

*(TB1)* $\forall \lambda \in \mathbb{B}, \bigvee_{x \in X} \lambda(x) = \top$,

*(TB2)* *if* $\lambda, \mu \in \mathbb{B}$, *then* $\bigvee_{\nu \in \mathbb{B}} S_X(\nu, \lambda \wedge \mu) = \top$.

Each $\top$-filter base generates a $\top$-filter $\mathbb{F}_{\mathbb{B}}$ by

$$\mathbb{F}_{\mathbb{B}} := \{\lambda \in L^X | \bigvee_{\mu \in \mathbb{B}} S_X(\mu, \lambda) = \top\}.$$

**Example 1.** *[31,43] Let $f : X \longrightarrow Y$ be a function.*

(1) *For any $\mathbb{F} \in \mathbb{F}_L^\top(X)$, the family $\{f^\rightarrow(\lambda) | \lambda \in \mathbb{F}\}$ forms a $\top$-filter base on $Y$. The generated $\top$-filter is denoted as $f^\Rightarrow(\mathbb{F})$, called the image of $\mathbb{F}$ under $f$. It is known that $\mu \in f^\Rightarrow(\mathbb{F}) \Longleftrightarrow f^\leftarrow(\mu) \in \mathbb{F}$.*
(2) *For any $\mathbb{G} \in \mathbb{F}_L^\top(Y)$, the family $\{f^\leftarrow(\mu) | \mu \in \mathbb{G}\}$ forms a $\top$-filter base on $Y$ iff $\bigvee_{y \in f(X)} \mu(y) = \top$ holds*

   *for all $\mu \in \mathbb{G}$. The generated $\top$-filter (if exists) is denoted as $f^\Leftarrow(\mathbb{G})$, called the inverse image of $\mathbb{G}$ under $f$. It is known that $\mathbb{G} \subseteq f^\Rightarrow(f^\Leftarrow(\mathbb{G}))$ holds whenever $f^\Leftarrow(\mathbb{G})$ exists. Furthermore, $f^\Leftarrow(\mathbb{G})$ always exists and $\mathbb{G} = f^\Rightarrow(f^\Leftarrow(\mathbb{G}))$ whenever $f$ is surjective.*
(3) *For any $x \in X$, the family $[x]_\top =: \{\lambda \in L^X | \lambda(x) = \top\}$ is a $\top$-filter on $X$, and $f^\Rightarrow([x]_\top) = [f(x)]_\top$.*

**Lemma 2.** *Let $f : X \longrightarrow Y$ be a function.*

(1) *If $\mathbb{B}$ is a $\top$-filter base of $\mathbb{F} \in \mathbb{F}_L^\top(X)$, then $\{f^\rightarrow(\lambda) | \lambda \in \mathbb{B}\}$ is a $\top$-filter base of $f^\Rightarrow(\mathbb{F})$, see Example 2.9 (1) in [31].*
(2) *If $\mathbb{B}$ is a $\top$-filter base of $\mathbb{G} \in \mathbb{F}_L^\top(Y)$ and $f^\Leftarrow(\mathbb{G})$ exists, then $\{f^\leftarrow(\mu) | \mu \in \mathbb{B}\}$ is a $\top$-filter base of $f^\Leftarrow(\mathbb{G})$, see Example 2.9 (2) in [31].*
(3) *Let $\mathbb{F}, \mathbb{G} \in \mathbb{F}_L^\top(X)$ and $\mathbb{B}$ be a $\top$-filter base of $\mathbb{F}$. Then $\mathbb{B} \subseteq \mathbb{G}$ implies that $\mathbb{F} \subseteq \mathbb{G}$, see Lemma 2.5 (1) in [42].*
(4) *Let $\mathbb{F} \in \mathbb{F}_L^\top(X)$ and $\mathbb{B}$ be a $\top$-filter base of $\mathbb{F}$. Then $\bigvee_{\mu \in \mathbb{B}} S_X(\mu, \lambda) = \bigvee_{\mu \in \mathbb{F}} S_X(\mu, \lambda)$, see Lemma 3.1 in [36].*

For each $\mathbb{F} \in \mathbb{F}_L^\top(X)$, we define $\Lambda(\mathbb{F}) : L^X \longrightarrow L$ as

$$\forall \lambda \in L^X, \Lambda(\mathbb{F})(\lambda) = \bigvee_{\mu \in \mathbb{F}} S_X(\mu, \lambda),$$

then $\Lambda(\mathbb{F})$ is a tightly stratified $L$-filter on $X$ [47].

In the following, we recall some notions and notations collected in [29].

**Definition 3.** *[31] Let $X$ be a nonempty set. Then a function $q : \mathbb{F}_L^\top(X) \longrightarrow 2^X$ is said to be a $\top$-convergence structure on $X$ if it satisfies the following two conditions:*

**(TC1)** $\quad \forall x \in X, [x]_\top \xrightarrow{q} x;$
**(TC1)** $\quad$ *if $\mathbb{F} \xrightarrow{q} x$ and $\mathbb{F} \subseteq \mathbb{G}$, then $\mathbb{G} \xrightarrow{q} x$.*

*where $\mathbb{F} \xrightarrow{q} x$ is short for $x \in q(\mathbb{F})$. The pair $(X, q)$ is said to be a $\top$-convergence space.*

A function $f : X \longrightarrow X'$ between $\top$-convergence spaces $(X, q)$, $(X', q')$ is said to be continuous if $f^\Rightarrow(\mathbb{F}) \xrightarrow{q'} f(x)$ for any $\mathbb{F} \xrightarrow{q} x$.

We denote the category consisting of $\top$-convergence spaces and continuous functions as $\top$-**CS**. It has been known that $\top$-**CS** is topological over **SET** [31].

For a source $(X \xrightarrow{f_i} (X_i, q_i))_{i \in I}$, the initial structure, $q$ on $X$ is defined by
$$\mathbb{F} \xrightarrow{q} x \Longleftrightarrow \forall i \in I, f_i^\Rightarrow(\mathbb{F}) \xrightarrow{q_i} f_i(x) \ [35,49].$$

For a sink $((X_i, q_i) \xrightarrow{f_i} X)_{i \in I}$, the final structure, $q$ on $X$ is defined as

$$\mathbb{F} \xrightarrow{q} x \Longleftrightarrow \begin{cases} \mathbb{F} \supseteq [x]_\top, & x \notin \cup_{i \in I} f_i(X_i); \\ \mathbb{F} \supseteq f_i^\Rightarrow(\mathbb{G}_i), & \exists i \in I, x_i \in X_i, \mathbb{G}_i \in \mathbb{F}_L^\top(X_i) \text{ s.t } f(x_i) = x, \mathbb{G}_i \xrightarrow{q_i} x_i. \end{cases}$$

When $X = \cup_{i \in I} f_i(X_i)$, the final structure $q$ can be characterized as

$$\mathbb{F} \xrightarrow{q} x \iff \mathbb{F} \supseteq f_i^{\Rightarrow}(\mathbb{G}_i) \text{ for some } \mathbb{G}_i \xrightarrow{q_i} x_i \text{ with } f(x_i) = x.$$

Let $\top(X)$ denote the set of all $\top$-convergence structures on a set $X$. For $p, q \in \top(X)$, we say that $q$ is finer than $p$ (or $p$ is coarser than $q$), denoted as $p \leq q$ for short, if the identity $\mathrm{id}_X : (X, q) \longrightarrow (X, p)$ is continuous. It has been known that $(\top(X), \leq)$ forms a completed lattice. The discrete (resp., indiscrete) structure $\delta$ (resp., $\iota$) is the top (resp., bottom) element of $(T(X), \leq)$, where $\delta$ is defined as $\mathbb{F} \xrightarrow{\delta} x$ iff $\mathbb{F} \supseteq [x]_\top$; and $\iota$ is defined as $\mathbb{F} \xrightarrow{\iota} x$ for all $\mathbb{F} \in \mathbb{F}_L^\top(X)$, $x \in X$ [42].

**Remark 1.** *When* $L = \{\bot, \top\}$, $\top$*-convergence spaces degenerate into convergence spaces. Therefore,* $\top$*-convergence spaces are natural generalizations of convergence spaces.*

## 3. $p$-Regularity in $\top$-Convergence Spaces

In this section, we shall discuss the $p$-regularity in $\top$-convergence spaces. Two equivalent approaches are considered, one approach using diagonal $\top$-filter and the other approach using closure of $\top$-filter. Moreover, it will be proved that $p$-regularity is preserved under the initial and final structures in the category $\top$-**CS**.

At first, we recall the notions of diagonal $\top$-filter and closure of $\top$-filter to define $p$-regularity.

Let $J, X$ be any sets and $\phi : J \longrightarrow \mathbb{F}_L^\top(X)$ be any function. Then we define a function $\hat{\phi} : L^X \to L^J$ as

$$\forall \lambda \in L^X, \forall j \in J, \hat{\phi}(\lambda)(j) = \Lambda(\phi(j))(\lambda) = \bigvee_{\mu \in \phi(j)} S_X(\mu, \lambda).$$

For any $\mathbb{F} \in \mathbb{F}_L^\top(J)$, it is known that the subset of $L^X$ defined by

$$k\phi\mathbb{F} := \{\lambda \in L^X | \hat{\phi}(\lambda) \in \mathbb{F}\}$$

forms a $\top$-filter on $X$, called diagonal $\top$-filter of $\mathbb{F}$ under $\phi$ [31]. It was shown that $S_X(\lambda, \mu) \leq S_J(\hat{\phi}(\lambda), \hat{\phi}(\mu))$ for any $\lambda, \mu \in L^X$.

**Lemma 3.** *Let* $f : X \longrightarrow Y$ *and* $\phi : J \longrightarrow \mathbb{F}_L^\top(X)$ *be functions. Then for any* $\mathbb{F} \in \mathbb{F}_L^\top(J)$ *we have* $f^{\Rightarrow}(k\phi\mathbb{F}) = k(f^{\Rightarrow} \circ \phi)\mathbb{F}$.

**Proof.** $f^{\Rightarrow}(k\phi\mathbb{F}) \subseteq k(f^{\Rightarrow} \circ \phi)\mathbb{F}$. By Lemma 2 (3) we need only check that $f^{\rightarrow}(\lambda) \in k(f^{\Rightarrow} \circ \phi)\mathbb{F}$ for any $\lambda \in k\phi\mathbb{F}$. Take $\lambda \in k\phi\mathbb{F}$ then $\hat{\phi}(\lambda) \in \mathbb{F}$. Please note that $\forall j \in J$,

$$\hat{\phi}(\lambda)(j) = \bigvee_{\mu \in \phi(j)} S_X(\mu, \lambda) \leq \bigvee_{\mu \in \phi(j)} S_Y(f^{\rightarrow}(\mu), f^{\rightarrow}(\lambda)) \leq \bigvee_{\nu \in f^{\Rightarrow}(\phi(j))} S_Y(\nu, f^{\rightarrow}(\lambda)) = \widehat{f^{\Rightarrow} \circ \phi}(f^{\rightarrow}(\lambda))(j),$$

i.e., $\hat{\phi}(\lambda) \leq \widehat{f^{\Rightarrow} \circ \phi}(f^{\rightarrow}(\lambda))$, and so $\widehat{f^{\Rightarrow} \circ \phi}(f^{\rightarrow}(\lambda)) \in \mathbb{F}$, i.e., $f^{\rightarrow}(\lambda) \in k(f^{\Rightarrow} \circ \phi)\mathbb{F}$.

$k(f^{\Rightarrow} \circ \phi)\mathbb{F} \subseteq f^{\Rightarrow}(k\phi\mathbb{F})$. For any $\lambda \in k(f^{\Rightarrow} \circ \phi)\mathbb{F}$ we have $\widehat{f^{\Rightarrow} \circ \phi}(\lambda) \in \mathbb{F}$. By Lemma 2 (4), $\forall j \in J$,

$$\widehat{f^{\Rightarrow} \circ \phi}(\lambda)(j) = \bigvee_{\nu \in f^{\Rightarrow}(\phi(j))} S_Y(\nu, \lambda) = \bigvee_{\mu \in \phi(j)} S_Y(f^{\rightarrow}(\mu), \lambda) \leq \bigvee_{\mu \in \phi(j)} S_X(\mu, f^{\leftarrow}(\lambda)) = \hat{\phi}(f^{\leftarrow}(\lambda))(j),$$

i.e., $\widehat{f^{\Rightarrow} \circ \phi}(\lambda) \leq \hat{\phi}(f^{\leftarrow}(\lambda))$, and so $\hat{\phi}(f^{\leftarrow}(\lambda)) \in \mathbb{F}$, i.e., $f^{\leftarrow}(\lambda) \in k\phi\mathbb{F}$ then $\lambda \in f^{\Rightarrow}(k\phi\mathbb{F})$. $\square$

**Definition 4.** *[36] Let* $(X, p)$ *be a* $\top$*-convergence space. For each* $\lambda \in L^X$, *the L-set* $\overline{\lambda}_p \in L^X$ *defined by*

$$\forall x \in X, \overline{\lambda}_p(x) = \bigvee_{\mathbb{F} \xrightarrow{p} x} \Lambda(\mathbb{F})(\lambda) = \bigvee_{\mathbb{F} \xrightarrow{p} x} \bigvee_{\mu \in \mathbb{F}} S_X(\mu, \lambda)$$

*is called closure of $\lambda$ w.r.t $p$.*

For any $\mathbb{F} \in \mathbb{F}_L^\top(X)$, the closure of $\mathbb{F}$ regarding $p$, denoted as $cl_p(\mathbb{F})$, is defined to be the $\top$-filter generated by $\{\overline{\lambda}_p | \lambda \in \mathbb{F}\}$ as a $\top$-filter base.

**Lemma 4.** *[36] Let $(X, p)$ be a $\top$-convergence space. Then for all $\lambda, \mu \in L^X$ we get*

(1)  $\lambda \leq \overline{\lambda}_p$;
(2)  $\lambda \leq \mu$ implies $\overline{\lambda}_p \leq \overline{\mu}_p$;
(3)  $S_X(\lambda, \mu) \leq S_X(\overline{\lambda}, \overline{\mu})$.

Let $\mathbb{N}$ be the set of natural numbers containing 0 and let $(X, p)$ be a $\top$-convergence space. For any $\mathbb{F} \in \mathbb{F}_L^\top(X)$, we define $cl_p^0(\mathbb{F}) = \mathbb{F}$. Furthermore, for any $n \in \mathbb{N}$, we define the $n+1$ th iteration of the closure $\top$-filter of $\mathbb{F}$ as $cl_p^{n+1}(\mathbb{F}) = cl_p(cl_p^n(\mathbb{F}))$ if $cl_p^n(\mathbb{F})$ has been defined.

The next proposition collects the properties of closure of $\top$-filters. We omit the obvious proofs.

**Proposition 1.** *Let $(X, p)$ be a $\top$-convergence space and $\mathbb{F}, \mathbb{G} \in \mathbb{F}_L^\top(X)$. Then for any $n \in \mathbb{N}$,*

(1)  $cl_p^n(\mathbb{F}) \subseteq \mathbb{F}$,
(2)  if $\mathbb{F} \subseteq \mathbb{G}$, then $cl_p^n(\mathbb{F}) \subseteq cl_p^n(\mathbb{G})$,
(3)  if $p' \in T(X)$ and $p \leq p'$, then $cl_p^n(\mathbb{F}) \subseteq cl_{p'}^n(\mathbb{F})$.

**Definition 5.** *A function $f : (X, q) \longrightarrow (Y, p)$ between $\top$-convergence spaces is said to be a closure function if $\overline{f^\rightarrow(\lambda)}_p \leq f^\rightarrow(\overline{\lambda}_q)$ for any $\lambda \in L^X$.*

**Proposition 2.** *Suppose that $f : (X, q) \longrightarrow (Y, p)$ is a function between $\top$-convergence spaces and $\mathbb{F} \in \mathbb{F}_L^\top(X), n \in \mathbb{N}$.*

(1)  *If $f$ is a continuous function, then $f^\Rightarrow(cl_q^n(\mathbb{F})) \supseteq cl_p^n(f^\Rightarrow(\mathbb{F}))$.*
(2)  *If $f$ is a closure function, then $f^\Rightarrow(cl_q^n(\mathbb{F})) \subseteq cl_p^n(f^\Rightarrow(\mathbb{F}))$.*

**Proof.** (1) Let's prove it by mathematical induction.

Firstly, we check $\overline{f^\leftarrow(\lambda)}_q \leq f^\leftarrow(\overline{\lambda}_p)$ for any $\lambda \in L^Y$. In fact, for any $x \in X$, by continuity of $f$ we obtain

$$
\begin{aligned}
\overline{f^\leftarrow(\lambda)}_q(x) &= \bigvee_{\mathbb{G} \xrightarrow{q} x} \bigvee_{\mu \in \mathbb{G}} S_X(\mu, f^\leftarrow(\lambda)) = \bigvee_{\mathbb{G} \xrightarrow{q} x} \bigvee_{\mu \in \mathbb{G}} S_Y(f^\rightarrow(\mu), \lambda) \\
&\leq \bigvee_{f^\Rightarrow(\mathbb{G}) \xrightarrow{p} f(x)} \bigvee_{f^\rightarrow(\mu) \in f^\Rightarrow(\mathbb{G})} S_Y(f^\rightarrow(\mu), \lambda) \leq \bigvee_{\mathbb{H} \xrightarrow{p} f(x)} \bigvee_{\nu \in \mathbb{H}} S_Y(\nu, \lambda) = f^\leftarrow(\overline{\lambda}_p)(x).
\end{aligned}
$$

Secondly, we prove $f^\Rightarrow(cl_q^n(\mathbb{F})) \supseteq cl_p^n(f^\Rightarrow(\mathbb{F}))$ when $n = 1$. Let $\lambda \in f^\Rightarrow(\mathbb{F})$, i.e., $f^\leftarrow(\lambda) \in \mathbb{F}$. Then by $\overline{f^\leftarrow(\lambda)}_q \leq f^\leftarrow(\overline{\lambda}_p)$ we have $f^\leftarrow(\overline{\lambda}_p) \in cl_q(\mathbb{F})$, i.e., $\overline{\lambda}_p \in f^\Rightarrow(cl_q(\mathbb{F}))$. It follows by Lemma 2 (3) that $f^\Rightarrow(cl_q(\mathbb{F})) \supseteq cl_p(f^\Rightarrow(\mathbb{F}))$.

Thirdly, we assume that $f^\Rightarrow(cl_q^n(\mathbb{F})) \supseteq cl_p^n(f^\Rightarrow(\mathbb{F}))$ when $n = k$. Then we prove $f^\Rightarrow(cl_q^n(\mathbb{F})) \supseteq cl_p^n(f^\Rightarrow(\mathbb{F}))$ when $n = k+1$. In fact,

$$
f^\Rightarrow(cl_q^{k+1}(\mathbb{F})) = f^\Rightarrow(cl_q(cl_q^k(\mathbb{F}))) \supseteq cl_p(f^\Rightarrow(cl_q^k(\mathbb{F}))) \supseteq cl_p(cl_p^k(f^\Rightarrow(\mathbb{F}))) = cl_p^{k+1}(f^\Rightarrow(\mathbb{F})).
$$

(2) We prove only that the inequality holds for $n = 1$, and the rest of the proof is similar to (1).

For any $\lambda \in \mathbb{F}$, we have $\overline{\lambda}_q \in cl_q(\mathbb{F})$ and then $f^\rightarrow(\lambda) \in f^\Rightarrow(\mathbb{F})$. From $f$ is a closure function, we conclude that $f^\rightarrow(\overline{\lambda}_q) \geq \overline{f^\rightarrow(\lambda)}_p \in cl_p(f^\Rightarrow(\mathbb{F}))$ and so $f^\rightarrow(\overline{\lambda}_q) \in cl_p(f^\Rightarrow(\mathbb{F}))$. By Lemma 2 (1), (3) we obtain $f^\Rightarrow(cl_q(\mathbb{F})) \subseteq cl_p(f^\Rightarrow(\mathbb{F}))$.  $\square$

Now, we tend our attention to $p$-regularity and its equivalent characterization. In the following, we shorten a pair of $\top$-convergence spaces $(X, p)$ and $(X, q)$ as $(X, p, q)$.

**Definition 6.** *Let $(X, p, q)$ be a pair of $\top$-convergence spaces. Then $q$ is said to be $p$-regular if the following condition $p$-**(TC)** is fulfilled.*

$$p\text{-}\mathbf{(TC)}\colon \forall \mathbb{F} \in \mathbb{F}_L^\top(X), \forall x \in X, \mathbb{F} \xrightarrow{\ q\ } x \Longrightarrow cl_p(\mathbb{F}) \xrightarrow{\ q\ } x.$$

**Remark 2.** *When $L = \{\bot, \top\}$, a $\top$-convergence space degenerates into a convergence space, and the condition $p$-**(TC)** degenerates into the crisp $p$-regularity condition in [5]. When $p = q$, the condition $p$-**(TC)** is precisely the regular characterization in [36].*

We say a pair of $\top$-convergence spaces $(X, p, q)$ fulfill the Fischer $\top$-diagonal condition whenever

$p$-**(TR)**: Let $J, X$ be any sets, $\psi : J \longrightarrow X$, and $\phi : J \longrightarrow \mathbb{F}_L^\top(X)$ such that $\phi(j) \xrightarrow{\ p\ } \psi(j)$, for each $j \in J$. Then for each $\mathbb{F} \in \mathbb{F}_L^\top(J)$ and each $x \in X$, $k\phi\mathbb{F} \xrightarrow{\ q\ } x$ implies $\psi^{\Rightarrow}(\mathbb{F}) \xrightarrow{\ q\ } x$.

**Remark 3.** *When $L = \{\bot, \top\}$, a $\top$-convergence space degenerates into a convergence space, and the condition $p$-**(TC)** degenerates into the Fischer diagonal condition $R_{p,q}$ in [6]. When $p = q$, the condition $p$-**(TR)** is precisely the diagonal condition **(TR)** in [31].*

In the following, we shall show that $p$-regularity can be described by Fischer $\top$-diagonal condition $p$-**(TR)**.

**Lemma 5.** *Let $(X, p, q)$ be a pair of $\top$-convergence spaces and let $J, X, \phi, \psi$ be defined as in $p$-**(TR)**. Then $S_X(\overline{\mu}_p, \lambda) \leq S_J(\hat{\phi}(\mu), \psi^{\leftarrow}(\lambda))$ for all $\lambda, \mu \in L^X$.*

**Proof.** Let $\lambda, \mu \in L^X$.

$$
\begin{aligned}
S_X(\overline{\mu}_p, \lambda) &= \bigwedge_{x \in X} \left( \left[ \bigvee_{\mathbb{G} \xrightarrow{\ p\ } x} \Lambda(\mathbb{G})(\mu) \right] \to \lambda(x) \right), \text{ by } \psi(j) \in X, \phi(j) \xrightarrow{\ p\ } \psi(j) \\
&\leq \bigwedge_{j \in J} (\Lambda(\phi(j))(\mu) \to \lambda(\psi(j))) = \bigwedge_{j \in J} (\hat{\phi}(\mu)(j) \to \psi^{\leftarrow}(\lambda)(j)) \\
&= S_J(\hat{\phi}(\mu), \psi^{\leftarrow}(\lambda)). \quad \square
\end{aligned}
$$

**Theorem 1.** *(Theorem 4.8 in [36] for $p = q$) Let $(X, p, q)$ be a pair of $\top$-convergence spaces. Then $p$-**(TC)** $\Longleftrightarrow p$-**(TR)**.*

**Proof.** $p$-**(TC)** $\Longrightarrow p$-**(TR)**. Let $J, X, \phi, \psi$ be defined as in $p$-**(TR)**. Assume that $\mathbb{F} \in \mathbb{F}_L^\top(J)$ and $k\phi\mathbb{F} \xrightarrow{\ q\ } x$. Then it follows by $p$-**(TC)** that $cl_p(k\phi\mathbb{F}) \xrightarrow{\ q\ } x$.

Next we prove that $cl_p(k\phi\mathbb{F}) \subseteq \psi^{\Rightarrow}(\mathbb{F})$. Indeed, for any $\lambda \in cl_p(k\phi\mathbb{F})$, we have

$$\top = \bigvee_{\mu \in k\phi\mathbb{F}} S_X(\overline{\mu}_p, \lambda) \overset{\text{Lemma } 5}{\leq} \bigvee_{\mu \in k\phi\mathbb{F}} S_J(\hat{\phi}(\mu), \psi^{\leftarrow}(\lambda)) = \bigvee_{\hat{\phi}(\mu) \in \mathbb{F}} S_J(\hat{\phi}(\mu), \psi^{\leftarrow}(\lambda)) \leq \bigvee_{\nu \in \mathbb{F}} S_J(\nu, \psi^{\leftarrow}(\lambda)),$$

which means $\psi^{\leftarrow}(\lambda) \in \mathbb{F}$, i.e., $\lambda \in \psi^{\Rightarrow}(\mathbb{F})$.

Now we have known that $cl_p(k\phi\mathbb{F}) \xrightarrow{\ q\ } x$ and $cl_p(k\phi\mathbb{F}) \subseteq \psi^{\Rightarrow}(\mathbb{F})$. Therefore, $\psi^{\Rightarrow}(\mathbb{F}) \xrightarrow{\ q\ } x$, as desired.

$p$-**(TR)** $\Longrightarrow p$-**(TC)**. Let

$$J = \{(\mathbb{G}, y) \in \mathbb{F}_L^\top(X) \times X \mid \mathbb{G} \xrightarrow{\ p\ } y\}; \psi : J \longrightarrow X, (\mathbb{G}, y) \mapsto y; \phi : J \longrightarrow \mathbb{F}_L^\top(X), (\mathbb{G}, y) \mapsto \mathbb{G}.$$

Then $\forall j \in J, \phi(j) \xrightarrow{\ p\ } \psi(j)$. Please note that $j = (\mathbb{G}, y) \in J \Longleftrightarrow \mathbb{G} = \phi(j), y = \psi(j)$.

(1) For any $\lambda, \mu \in L^X$, $S_X(\overline{\mu}_p, \lambda) = S_J(\hat{\phi}(\mu), \psi^{\leftarrow}(\lambda))$. Indeed,

$$
\begin{aligned}
S_X(\overline{\mu}_p, \lambda) &= \bigwedge_{y \in X} ([\bigvee_{\mathbb{G} \xrightarrow{p} y} \Lambda(\mathbb{G})(\mu)] \to \lambda(y)) = \bigwedge_{y \in X} \bigwedge_{(\mathbb{G},y) \in J} (\Lambda(\mathbb{G})(\mu) \to \lambda(y)) \\
&= \bigwedge_{j=(\mathbb{G},y) \in J} (\Lambda(\phi(j))(\mu) \to \lambda(\psi(j))) = \bigwedge_{j \in J} (\hat{\phi}(\mu)(j) \to \psi^{\leftarrow}(\lambda)(j)) \\
&= S_J(\hat{\phi}(\mu), \psi^{\leftarrow}(\lambda)).
\end{aligned}
$$

(2) For each $\mathbb{F} \in \mathbb{F}_L^{\top}(X)$, the family $\{\hat{\phi}(\lambda) | \lambda \in \mathbb{F}\}$ forms a $\top$-filter base on $J$. Indeed,
(TB1): For any $\lambda \in \mathbb{F}$, by $[y]_{\top} \xrightarrow{p} y$ for any $y \in X$, we have

$$
\bigvee_{j \in J} \hat{\phi}(\lambda)(j) = \bigvee_{j \in J} \Lambda(\phi(j))(\lambda) = \bigvee_{y \in X} \bigvee_{\mathbb{G} \xrightarrow{p} y} \Lambda(\mathbb{G})(\lambda) \geq \bigvee_{y \in X} \Lambda([y]_{\top})(\lambda) = \bigvee_{y \in X} \lambda(y) = \top.
$$

(TB2): For any $\lambda, \mu \in \mathbb{F}$, note that for any $j \in J$,

$$
\begin{aligned}
\hat{\phi}(\lambda)(j) \wedge \hat{\phi}(\mu)(j) &= \bigvee_{\lambda_1 \in \phi(j)} S_X(\lambda_1, \lambda) \wedge \bigvee_{\mu_1 \in \phi(j)} S_X(\mu_1, \mu) \\
&\leq \bigvee_{\lambda_1, \mu_1 \in \phi(j)} S_X(\lambda_1 \wedge \mu_1, \lambda \wedge \mu) \\
&\leq \bigvee_{\nu \in \phi(j)} S_X(\nu, \lambda \wedge \mu) = \hat{\phi}(\lambda \wedge \mu)(j),
\end{aligned}
$$

i.e., $\hat{\phi}(\lambda) \wedge \hat{\phi}(\mu) \leq \hat{\phi}(\lambda \wedge \mu)$. It follows easily that (TB2) is satisfied. We denote the $\top$-filter generated by $\{\hat{\phi}(\lambda) | \lambda \in \mathbb{F}\}$ as $\mathbb{F}^{\phi}$.
  (3) For each $\mathbb{F} \in \mathbb{F}_L^{\top}(X)$, $k\phi\mathbb{F}^{\phi} \supseteq \mathbb{F}$. Indeed, for any $\lambda \in \mathbb{F}$, we have $\hat{\phi}(\lambda) \in \mathbb{F}^{\phi}$, i.e., $\lambda \in k\phi\mathbb{F}^{\phi}$.
  (4) For each $\mathbb{F} \in \mathbb{F}_L^{\top}(X)$, $\psi^{\Rightarrow}(\mathbb{F}^{\phi}) = cl_p(\mathbb{F})$. Indeed,

$$
\lambda \in \psi^{\Rightarrow}(\mathbb{F}^{\phi}) \iff \psi^{\leftarrow}(\lambda) \in \mathbb{F}^{\phi} \iff \bigvee_{\mu \in \mathbb{F}} S_J(\hat{\phi}(\mu), \psi^{\leftarrow}(\lambda)) = \top \overset{(1)}{\iff} \bigvee_{\mu \in \mathbb{F}} S_X(\overline{\mu}_p, \lambda) = \top \iff \lambda \in cl_p(\mathbb{F}).
$$

Assume that $\mathbb{F} \xrightarrow{q} x$, then by (3), we have $k\phi\mathbb{F}^{\phi} \supseteq \mathbb{F}$, and so $k\phi\mathbb{F}^{\phi} \xrightarrow{q} x$. From $p$-**(TR)** and (4), we get that $cl_p(\mathbb{F}) = \psi^{\Rightarrow}(\mathbb{F}^{\phi}) \xrightarrow{q} x$. Therefore, the condition $p$-**(TC)** is satisfied. $\square$

The next theorem shows that $p$-regularity is preserved under initial structures.

**Theorem 2.** *Let $\{(X_i, q_i, p_i)\}_{i \in I}$ be pairs of $\top$-convergence spaces such that each $q_i$ is $p_i$-regular. If $q$ (resp., $p$) is the initial structure on $X$ regarding the source $(X \xrightarrow{f_i} (X_i, q_i))_{i \in I}$ (resp., $(X \xrightarrow{f_i} (X_i, p_i))_{i \in I}$), then $q$ is also $p$-regular.*

**Proof.** Let $\psi : J \longrightarrow X$ and $\phi : J \longrightarrow \mathbb{F}_L^{\top}(X)$ be any function such that $\phi(j) \xrightarrow{p} \psi(j)$ for any $j \in J$. Then

$$
\forall i \in I, \forall j \in J, (f_i^{\Rightarrow} \circ \phi)(j) = f_i^{\Rightarrow}(\phi(j)) \xrightarrow{p_i} f_i(\psi(j)) = (f_i \circ \psi)(j).
$$

Let $\mathbb{F} \in \mathbb{F}_L^{\top}(J)$ satisfy $k\phi\mathbb{F} \xrightarrow{q} x$. Then by definition of $q$ and Lemma 3 we have

$$
\forall i \in I, k(f_i^{\Rightarrow} \circ \phi)\mathbb{F} = f_i^{\Rightarrow}(k\phi\mathbb{F}) \xrightarrow{q_i} f_i(x).
$$

Since $q_i$ is $p_i$-regular we have $f_i^{\Rightarrow}\psi^{\Rightarrow}(\mathbb{F}) = (f_i \circ \psi)^{\Rightarrow}(\mathbb{F}) \xrightarrow{q_i} f_i(x)$. By definition of $q$ we have $\psi^{\Rightarrow}(\mathbb{F}) \xrightarrow{q} x$. Thus $q$ is $p$-regular. $\square$

The next theorem shows that $p$-regularity is preserved under final structures with some additional assumptions.

**Theorem 3.** *Let $\{(X_i, q_i, p_i)\}_{i \in I}$ be pairs of $\top$-convergence spaces such that each $q_i$ is $p_i$-regular. Let $q$ (resp., $p$) be the final structure on $X$ relative to the sink $((X_i, q_i) \xrightarrow{f_i} X)_{i \in I}$ (resp., $((X_i, p_i) \xrightarrow{f_i} X)_{i \in I}$). If $X = \cup_{i \in I} f_i(X_i)$ and each $f_i : (X_i, p_i) \longrightarrow (X, p)$ is a closure function, then $q$ is also $p$-regular.*

**Proof.** Let $\mathbb{F} \in \mathbb{F}_L^\top(X) \xrightarrow{q} x$. Then by definition of $q$, there exists $i \in I, x_i \in X_i, \mathbb{G}_i \in \mathbb{F}_L^\top(X_i), f_i(x_i) = x$ such that $\mathbb{G}_i \xrightarrow{q_i} x_i$ and $f_i^{\Rightarrow}(\mathbb{G}_i) \subseteq \mathbb{F}$. Because $q_i$ is $p_i$-regular we get $cl_{p_i}(\mathbb{G}_i) \xrightarrow{q_i} x_i$ and then $f_i^{\Rightarrow}(cl_{p_i}(\mathbb{G}_i)) \xrightarrow{q} x$. By $f_i$ is a closure function and Proposition 2 (2) it follows that $cl_p(f_i^{\Rightarrow}(\mathbb{G}_i)) \xrightarrow{q} x$. Hence $cl_p(\mathbb{F}) \xrightarrow{q} x$ from $cl_p(f_i^{\Rightarrow}(\mathbb{G}_i)) \subseteq cl_p(\mathbb{F})$. Thus $q$ is $p$-regular. $\quad\square$

For any $\{q_i\}_{i \in I} \subseteq \top(X)$, note that the supremum (resp., infimum) of $\{q_i\}_{i \in I}$ in the lattice $\top(X)$, denoted as $\sup\{q_i | i \in I\}$ (resp., $\inf\{q_i | i \in I\}$), is precisely the initial structure (resp., final structure) regarding the source $(X \xrightarrow{id_X} (X, q_i))_{i \in I}$ (resp., the sink $((X, q_i) \xrightarrow{id_X} X)_{i \in I}$). By Theorems 2 and 3, we obtain easily the following corollary. It will show us that $p$-regularity is preserved under supremum and infimum in the lattice $\top(X)$.

**Corollary 1.** *Let $\{q_i | i \in I\} \subseteq \top(X)$ and $p \in \top(X)$ with each $(X, q_i)$ being $p$-regular. Then both $\inf\{q_i\}_{i \in I}$ and $\sup\{q_i\}_{i \in I}$ are all $p$-regular.*

## 4. Lower (Upper) $p$-Regular Modifications in $\top$-Convergence Spaces

In this section, we shall consider the $p$-regular modifications in $\top$-convergence spaces.

**Lemma 6.** *Let $p, q$ be $\top$-convergence structures on $X$.*

*(1)   If $q$ is $p$-regular, then $\mathbb{F} \xrightarrow{q} x$ implies $cl_p^n(\mathbb{F}) \xrightarrow{q} x$ for any $n \in \mathbb{N}$.*
*(2)   If $q$ is $p$-regular, then $q$ is $p'$-regular for any $p \leq p'$.*
*(3)   The indiscrete structure $\iota$ is $p$-regular for any $p \in \top(X)$.*

**Proof.** It is obvious. $\quad\square$

### 4.1. Lower p-Regular Modification

It has been known that $p$-regularity is preserved under supremum in the lattice $\top(X)$ (see Corollary 1), and the indiscrete structure $\iota$ is $p$-regular for any $p \in \top(X)$ (see Lemma 6 (3)). So, it follows easily that for a pair of $\top$-convergence spaces $(X, p, q)$, there is a finest $p$-regular $\top$-convergence structure $\gamma_p q$ on $X$ which is coarser than $q$.

**Definition 7.** *Let $(X, p, q)$ be a pair of $\top$-convergence spaces. Then the $\top$-convergence structure $\gamma_p q$ on $X$ is said to be the lower $p$-regular modification of $q$.*

The following theorem gives a characterization on lower $p$-regular modification.

**Theorem 4.** *For any $p, q \in \top(X), \mathbb{F} \xrightarrow{\gamma_p q} x \iff \exists n \in \mathbb{N}, \mathbb{G} \xrightarrow{q} x$ s.t. $\mathbb{F} \supseteq cl_p^n(\mathbb{G})$.*

**Proof.** We define $q'$ as $\mathbb{F} \xrightarrow{q'} x \iff \exists n \in \mathbb{N}, \mathbb{G} \xrightarrow{q} x$ s.t. $\mathbb{F} \supseteq cl_p^n(\mathbb{G})$, then we prove $\gamma_p q = q'$.

Obviously, $q' \in \top(X)$ and $q' \leq q$. We check that $q'$ is $p$-regular. In fact, let $\mathbb{F} \xrightarrow{q'} x$. Then there exists $n \in \mathbb{N}, \mathbb{G} \xrightarrow{q} x$ such that $\mathbb{F} \supseteq cl_p^n(\mathbb{G})$. It follows that $cl_p(\mathbb{F}) \supseteq cl_p(cl_p^n(\mathbb{G})) = cl_p^{n+1}(\mathbb{G})$, so $cl_p(\mathbb{F}) \xrightarrow{q'} x$. Now, we have proved that $q'$ is $p$-regular.

Let $r$ be $p$-regular with $r \leq q$. We prove below $r \leq q'$. In fact, let $\mathbb{F} \xrightarrow{q'} x$. Then there exists $n \in \mathbb{N}, \mathbb{G} \xrightarrow{q} x$ such that $\mathbb{F} \supseteq cl_p^n(\mathbb{G})$, so $\mathbb{G} \xrightarrow{r} x$ by $q \leq r$. Because $r$ is $p$-regular it follows by Lemma 6 (1) that $\mathbb{F} \supseteq cl_p^n(\mathbb{G}) \xrightarrow{r} x$. Therefore, $r \leq q'$. $\square$

**Theorem 5.** *If $f : (X, q) \longrightarrow (X', q')$ and $f : (X, p) \longrightarrow (X', p')$ are both continuous function between $\top$-convergence spaces then so is $f : (X, \gamma_p q) \longrightarrow (X', \gamma_{p'} q')$.*

**Proof.** For any $\mathbb{F} \in \mathbb{F}_L^\top(X)$ and $x \in X$.

$$\mathbb{F} \xrightarrow{\gamma_p q} x \implies \exists n \in \mathbb{N}, \mathbb{G} \xrightarrow{q} x \text{ s.t. } \mathbb{F} \supseteq cl_p^n(\mathbb{G})$$

$$\implies \exists n \in \mathbb{N}, f^{\Rightarrow}(\mathbb{G}) \xrightarrow{q'} f(x) \text{ s.t. } f^{\Rightarrow}(\mathbb{F}) \supseteq f^{\Rightarrow}(cl_p^n(\mathbb{G}))$$

$$\implies \exists n \in \mathbb{N}, f^{\Rightarrow}(\mathbb{G}) \xrightarrow{q'} f(x) \text{ s.t. } f^{\Rightarrow}(\mathbb{F}) \supseteq cl_{p'}^n(f^{\Rightarrow}(\mathbb{G}))$$

$$\implies f^{\Rightarrow}(\mathbb{F}) \xrightarrow{\gamma_{p'} q'} (f(x)),$$

where the second implication uses the continuity of $f : (X, q) \longrightarrow (X', q')$, and the third implication uses the continuity of $f : (X, p) \longrightarrow (X', p')$ and Proposition 2(1). $\square$

The following theorem exhibits us that lower $p$-regular modification and final structures have good compatibility.

**Theorem 6.** *Let $\{(X_i, q_i, p_i)\}_{i \in I}$ be pairs of spaces in $\top$-**CS** and let $q$ be the final structure relative to the sink $((X_i, q_i) \xrightarrow{f_i} X)_{i \in I}$ with $X = \cup_{i \in I} f_i(X_i)$. If $p \in \top(X)$ such that each $f_i : (X_i, p_i) \longrightarrow (X, p)$ is a continuous closure function, then $\gamma_p q$ is the final structure relative to the sink $((X_i, \gamma_{p_i} q_i) \xrightarrow{f_i} X)_{i \in I}$.*

**Proof.** Let $s$ denote the final structure relative to the sink $((X_i, \gamma_{p_i} q_i) \xrightarrow{f_i} X)_{i \in I}$. Then for any $\mathbb{F} \in \mathbb{F}_L^\top(X)$ and $x \in X$

$$\mathbb{F} \xrightarrow{s} x \implies \exists i \in I, x_i \in X_i, f_i(x_i) = x, \mathbb{G}_i \xrightarrow{\gamma_{p_i} q_i} x_i \text{ s.t. } f_i^{\Rightarrow}(\mathbb{G}_i) \subseteq \mathbb{F}, \text{ by Theorem 4}$$

$$\implies \exists i \in I, x_i \in X_i, n \in \mathbb{N}, f_i(x_i) = x, \mathbb{H}_i \xrightarrow{q_i} x_i \text{ s.t. } cl_{p_i}^n(\mathbb{H}_i) \subseteq \mathbb{G}_i, f_i^{\Rightarrow}(\mathbb{G}_i) \subseteq \mathbb{F}, \text{ by Proposition 2 (1)}$$

$$\implies \exists i \in I, x_i \in X_i, n \in \mathbb{N}, f_i^{\Rightarrow}(\mathbb{H}_i) \xrightarrow{q} x \text{ s.t. } cl_p^n(f_i^{\Rightarrow}(\mathbb{H}_i)) \subseteq f_i^{\Rightarrow}(cl_{p_i}^n(\mathbb{H}_i)) \subseteq f_i^{\Rightarrow}(\mathbb{G}_i), f_i^{\Rightarrow}(\mathbb{G}_i) \subseteq \mathbb{F}$$

$$\implies \exists i \in I, x_i \in X_i, n \in \mathbb{N}, f_i^{\Rightarrow}(\mathbb{H}_i) \xrightarrow{q} x \text{ s.t. } cl_p^n(f_i^{\Rightarrow}(\mathbb{H}_i)) \subseteq \mathbb{F}$$

$$\implies \mathbb{F} \xrightarrow{\gamma_p q} x.$$

Conversely,

$$\mathbb{F} \xrightarrow{\gamma_p q} x \implies \exists n \in \mathbb{N}, \mathbb{G} \xrightarrow{q} x \text{ s.t. } cl_p^n(\mathbb{G}) \subseteq \mathbb{F}$$

$$\implies \exists i \in I, x_i \in X_i, f_i(x_i) = x, n \in \mathbb{N}, \mathbb{H}_i \xrightarrow{q_i} x_i \text{ s.t. } f_i^{\Rightarrow}(\mathbb{H}_i) \subseteq \mathbb{G}, cl_p^n(\mathbb{G}) \subseteq \mathbb{F}$$

$$\implies \exists i \in I, x_i \in X_i, f_i(x_i) = x, n \in \mathbb{N}, \mathbb{H}_i \xrightarrow{q_i} x_i \text{ s.t. } cl_p^n(f_i^{\Rightarrow}(\mathbb{H}_i)) \subseteq cl_p^n(\mathbb{G}), cl_p^n(\mathbb{G}) \subseteq \mathbb{F}$$

$$\implies \exists i \in I, x_i \in X_i, f_i(x_i) = x, n \in \mathbb{N}, \mathbb{H}_i \xrightarrow{q_i} x_i \text{ s.t. } f_i^{\Rightarrow}(cl_{p_i}^n(\mathbb{H}_i)) \subseteq cl_p^n(f_i^{\Rightarrow}(\mathbb{H}_i)) \subseteq \mathbb{F}$$

$$\implies \exists i \in I, x_i \in X_i, f_i(x_i) = x, n \in \mathbb{N}, cl_{p_i}^n(\mathbb{H}_i) \xrightarrow{\gamma_{p_i} q_i} x_i \text{ s.t. } f_i^{\Rightarrow}(cl_{p_i}^n(\mathbb{H}_i)) \subseteq \mathbb{F}$$

$$\implies \mathbb{F} \xrightarrow{s} x,$$

where the fourth implication follows by Proposition 2(2). $\square$

The following corollary tells us that lower $p$-regular modification has good compatibility with infimum in the lattice $\top(X)$.

**Corollary 2.** *Assume that* $\{q_i | i \in I\} \subseteq \top(X)$, $p \in \top(X)$ *and* $q = \inf\{q_i | i \in I\}$. *Then* $\gamma_p q = \inf\{\gamma_p q_i | i \in I\}$.

*4.2. Upper p-Regular Modification*

Similar to the crisp case, the discrete structure $\delta$ is not always $p$-regular for any $p \in \top(X)$. This shows that for a given $q \in \top(X)$, there may not exist $p$-regular $\top$-convergence structure on $X$ finer than $q$.

**Definition 8.** *Let* $(X, p, q)$ *be a pair of* $\top$-convergence spaces. *If there exists a coarsest p-regular* $\top$-convergence *structure* $\gamma^p q$ *on* $X$ *finer than* $q$, *then it is said to be the upper p-regular modification of* $q$.

It has been known that the existence of $\gamma^p q$ depends on the existence of a $p$-regular $\top$-convergence structure finer than $q$ (see Corollary 1), and $\gamma_p \delta$ is the finest $p$-regular $\top$-convergence structure. So, it follows easily that $\gamma^p q$ exists if and only if $q \leq \gamma_p \delta$. By Theorem 4, we obtain the following result.

**Theorem 7.** *Let* $(X, p, q)$ *be a pair of* $\top$-convergence spaces. *Then*
$$\gamma^p q \text{ exists} \iff \forall x \in X, \forall n \in \mathbb{N}, cl_p^n([x]_\top) \xrightarrow{q} x.$$

**Proof.** For any $\mathbb{F} \in \mathbb{F}_L^\top(X)$ and any $x \in X$, from Theorem 4 we obtain

$$\mathbb{F} \xrightarrow{\gamma_p \delta} x \iff \exists n \in \mathbb{N}, \mathbb{G} \xrightarrow{\delta} x \text{ s.t. } cl_p^n(\mathbb{G}) \subseteq \mathbb{F}.$$

*Necessity.* Let $\gamma^p q$ exist. Then $q \leq \gamma_p \delta$. So, for any $x \in X, n \in \mathbb{N}$

$$[x]_\top \xrightarrow{\delta} x \implies cl_p^n([x]_\top) \xrightarrow{\gamma_p \delta} x \implies cl_p^n([x]_\top) \xrightarrow{q} x.$$

*Sufficiency.* Let $cl_p^n([x]_\top) \xrightarrow{q} x$ for any $x \in X, n \in \mathbb{N}$. Then

$$
\begin{array}{rl}
\mathbb{F} \xrightarrow{\gamma_p \delta} x & \implies \quad \exists n \in \mathbb{N}, \mathbb{G} \xrightarrow{\delta} x \text{ s.t. } cl_p^n(\mathbb{G}) \subseteq \mathbb{F} \\[2mm]
& \implies \quad \exists n \in \mathbb{N}, [x]_\top \subseteq \mathbb{G} \text{ s.t. } cl_p^n(\mathbb{G}) \subseteq \mathbb{F} \\[2mm]
& \overset{\text{Proposition 1 (2)}}{\implies} \quad \exists n \in \mathbb{N}, cl_p^n([x]_\top) \subseteq cl_p^n(\mathbb{G}) \text{ s.t. } cl_p^n(\mathbb{G}) \subseteq \mathbb{F} \\[2mm]
& \implies \quad \exists n \in \mathbb{N} \text{ s.t. } cl_p^n([x]_\top) \subseteq \mathbb{F} \\[2mm]
& \implies \quad \mathbb{F} \xrightarrow{q} x.
\end{array}
$$

It follows that $q \leq \gamma_p \delta$, so $\gamma^p q$ exists. $\square$

The following theorem gives a characterization on upper $p$-regular modification if it exists.

**Theorem 8.** *Let* $(X, p, q)$ *be a pair of* $\top$-convergence spaces and $\gamma^p q$ exists. *Then*
$$\mathbb{F} \xrightarrow{\gamma^p q} x \iff \forall n \in \mathbb{N}, cl_p^n(\mathbb{F}) \xrightarrow{q} x.$$

**Proof.** We define $q'$ as $\mathbb{F} \xrightarrow{q'} x \iff \forall n \in \mathbb{N}, cl_p^n(\mathbb{F}) \xrightarrow{q} x.$

(1)    $q' \in \top(X)$. It is obvious.

(2)    $q \leq q'$. In fact, let $\mathbb{F} \xrightarrow{q'} x$ then $\mathbb{F} = cl_p^0(\mathbb{F}) \xrightarrow{q} x$.

(3)    $q'$ is $p$-regular. In fact, let $\mathbb{F} \xrightarrow{q'} x$. Then for any $n \in \mathbb{N}$ it holds that $cl_p^n(cl_p(\mathbb{F})) = cl_p^{n+1}(\mathbb{F}) \xrightarrow{q} x$, which means $cl_p(\mathbb{F}) \xrightarrow{q'} x$. So, $q'$ is $p$-regular.

(4)    Let $r$ be $p$-regular with $q \leq r$. Then $q' \leq r$. In fact, let $\mathbb{F} \xrightarrow{r} x$ then for any $n \in \mathbb{N}$, by Proposition 6 (1) it holds that $cl_p^n(\mathbb{F}) \xrightarrow{r} x$ and so $cl_p^n(\mathbb{F}) \xrightarrow{q} x$ by $q \leq r$. That means $\mathbb{F} \xrightarrow{q'} x$.

By (1)–(4) we get that $q'$ is the coarsest $p$-regular $\top$-convergence structure finer than $q$. Therefore, $\gamma^p q = q'$. $\square$

**Theorem 9.** *Let $f : (X, q) \longrightarrow (X', q')$ be a continuous function, and $f : (X, p) \longrightarrow (X', p')$ be a closure function between $\top$-convergence spaces. If $\gamma^p q$ and $\gamma^{p'} q'$ exist then $f : (X, \gamma^p q) \longrightarrow (X', \gamma^{p'} q')$ is continuous.*

**Proof.** Let $\mathbb{F} \xrightarrow{\gamma^p q} x$. Then $\forall n \in \mathbb{N}, cl_p^n(\mathbb{F}) \xrightarrow{q} x$. Since $f : (X, q) \longrightarrow (X', q')$ is a continuous function and $f : (X, p) \longrightarrow (X', p')$ is a closure function it holds that

$$\forall n \in \mathbb{N}, cl_{p'}^n(f^\Rightarrow(\mathbb{F})) \supseteq f^\Rightarrow(cl_p^n(\mathbb{F})) \xrightarrow{q'} f(x),$$

so $f^\Rightarrow(\mathbb{F}) \xrightarrow{\gamma^{p'} q'} f(x)$, as desired. $\square$

The following theorem exhibits us that the upper $p$-regular modification has good compatibility with initial structures.

**Theorem 10.** *Let $\{(X_i, q_i, p_i)\}_{i \in I}$ be pairs of spaces in $\top$-**CS** and $q$ be the initial structure relative to the source $(X \xrightarrow{f_i} (X_i, q_i))_{i \in I}$. Let $p \in \top(X)$ such that each $f_i : (X, p) \longrightarrow (X_i, p_i)$ is continuous closure function. If $\gamma^{p_i} q_i$ exists for all $i \in I$ then so does $\gamma^p q$, and $\gamma^p q$ is precisely the initial structure relative to the source $(X \xrightarrow{f_i} (X_i, \gamma^{p_i} q_i))_{i \in I}$.*

**Proof.** At first, we show the existence of $\gamma^p q$. By Theorem 7, it suffices to check that $cl_p^n([x]_\top) \xrightarrow{q} x$ for any $x \in X, n \in \mathbb{N}$. In fact, by the existence of $\gamma^{p_i} q_i$ we have $cl_{p_i}^n([f_i(x)]) \xrightarrow{q_i} f_i(x)$ for any $i \in I, x \in X, n \in \mathbb{N}$. Then by each $f_i : (X, p) \longrightarrow (X_i, p_i)$ being a continuous closure function it holds that

$$f_i^\Rightarrow(cl_p^n([x]_\top)) = cl_{p_i}^n(f_i^\Rightarrow([x]_\top)) = cl_{p_i}^n([f_i(x)]_\top) \xrightarrow{q_i} f_i(x),$$

so $cl_p^n([x]_\top) \xrightarrow{q} x$ for any $x \in X, n \in \mathbb{N}$, i.e., $\gamma^p q$ exists.

Let $s$ denote the initial structure relative to the source $(X \xrightarrow{f_i} (X_i, \gamma^{p_i} q_i))_{i \in I}$. Then

$$\mathbb{F} \xrightarrow{s} x \iff \forall i \in I, f_i^\Rightarrow(\mathbb{F}) \xrightarrow{\gamma^{p_i} q_i} f_i(x) \overset{\text{Theorem } 8}{\iff} \forall i \in I, \forall n \in \mathbb{N}, cl_{p_i}^n(f_i^\Rightarrow(\mathbb{F})) \xrightarrow{q_i} f_i(x)$$

$$\overset{\text{Proposition } 2}{\iff} \forall i \in I, \forall n \in \mathbb{N}, f_i^\Rightarrow(cl_p^n(\mathbb{F})) \xrightarrow{q_i} f_i(x)$$

$$\iff \forall n \in \mathbb{N}, cl_p^n(\mathbb{F}) \xrightarrow{q} x \overset{\text{Theorem } 8}{\iff} \mathbb{F} \xrightarrow{\gamma^p q} x. \quad \square$$

The following corollary tells us that upper $p$-regular modification has good compatibility with supremum in the lattice $\top(X)$.

**Corollary 3.** *Assume that $\{q_i | i \in I\} \subseteq \top(X)$, $p \in \top(X)$ and $q = \sup\{q_i | i \in I\}$. If $\gamma^p q_i$ exists for all $i \in I$ then so does $\gamma^p q$ and $\gamma^p q = \sup\{\gamma^p q_i | i \in I\}$.*

## 5. Conclusions

In this paper, we studied $p$-regularity in $\top$-convergence spaces by a diagonal condition and a closure condition about $\top$-filter, respectively. We proved that $p$-regularity was preserved under the initial and final constructions in the category $\top$-**CS**. We then followed as a conclusion that $p$-regularity was preserved under the infimum and supremum in the lattice $\top(X)$. Furthermore, we defined and discussed lower (upper) $p$-regular modifications in $\top$-convergence spaces. In particular, we showed that lower (resp., upper) $p$-regular modification has good compatibility with final (resp., initial) construction.

**Author Contributions:** Both authors contributed equally in the writing of this article.

**Funding:** This work is supported by National Natural Science Foundation of China (No. 11801248, 11501278) and the Scientific Research Fund of Hunan Provincial Key Laboratory of Mathematical Modeling and Analysis in Engineering (No. 2018MMAEZD09).

**Acknowledgments:** The authors thank the reviewer and the editor for their valuable comments and suggestions.

**Conflicts of Interest:** The authors declare no conflict of interest.

## References

1.  Preuss, G. *Fundations of Topology*; Kluwer Academic Publishers: London, UK, 2002.
2.  Kent, D.C.; Richardson, G.D. Convergence spaces and diagonal conditions. *Topol. Its Appl.* **1996**, *70*, 167–174. [CrossRef]
3.  Kowalsky, H.J.; Komplettierung, L. *Math. Nachrichten* **1954**, *12*, 301–340. [CrossRef]
4.  Cook, C.H.; Fischer, H.R. Regular Convergence Spaces. *Math. Annalen* **1967**, *174*, 1–7. [CrossRef]
5.  Kent, D.C.; Richardson, G.D. $p$-regular convergence spaces. *Math. Nachrichten* **1990**, *149*, 215–222. [CrossRef]
6.  Wilde, S.A.; Kent, D.C. $p$-topological and $p$-regular: Dual notions in convergence theory. *Int. J. Math. Math. Sci.* **1999**, *22*, 1–12. [CrossRef]
7.  Jäger, G. A category of $L$-fuzzy convergence spaces. Quaestiones Mathematicae **2001**, *24*, 501–517. [CrossRef]
8.  Flores, P.V.; Mohapatra, R.N.; Richardson, G. Lattice-valued spaces: Fuzzy convergence. *Fuzzy Sets Syst.* **2006**, *157*, 2706–2714. [CrossRef]
9.  Fang, J.M. Stratified $L$-ordered convergence structures. *Fuzzy Sets Syst.* **2010**, *161*, 2130–2149. [CrossRef]
10. Flores, P.V.; Richardson, G. Lattice-valued convergence: Diagonal axioms. *Fuzzy Sets Syst.* **2008**, *159*, 2520–2528. [CrossRef]
11. Han, S.E.; Lu, L.X.; Yao, W. Quantale-valued fuzzy scott topology. *Iran. J. Fuzzy Syst.* **2018**, in press.
12. Jäger, G. Pretopological and topological lattice-valued convergence spaces. *Fuzzy Sets Syst.* **2007**, *158*, 424–435. [CrossRef]
13. Jäger, G. Fischer's diagonal condition for lattice-valued convergence spaces. *Quaest. Math.* **2008**, *31*, 11–25. [CrossRef]
14. Jäger, G. Gähler's neighbourhood condition for lattice-valued convergence spaces. *Fuzzy Sets Syst.* **2012**, *204*, 27–39. [CrossRef]
15. Jäger, G. Stratified $LMN$-convergence tower spaces. *Fuzzy Sets Syst.* **2016**, *282*, 62–73. [CrossRef]
16. Jin, Q.; Li, L.Q. Stratified lattice-valued neighborhood tower group. *Quaest. Math.* **2018**, *41*, 847-861. [CrossRef]
17. Jin, Q.; Li, L.Q.; Lv, Y.R.; Zhao, F.; Zhou, J. Connectedness for lattice-valued subsets in lattice-valued convergence spaces. *Quaest. Math.* **2019**, *42*, 135–150. [CrossRef]
18. Jin, Q.; Li, L.Q.; Meng, G.W. On the relationships between types of $L$-convergence spaces. *Iran. J. Fuzzy Syst.* **2016**, *1*, 93–103.
19. Li, L.Q.; Jin, Q. On adjunctions between Lim, SL-Top, and SL-Lim. *Fuzzy Sets Syst.* **2011**, *182*, 66–78. [CrossRef]
20. Li, L.Q.; Jin, Q. On stratified $L$-convergence spaces: Pretopological axioms and diagonal axioms. *Fuzzy Sets Syst.* **2012**, *204*, 40–52. [CrossRef]
21. Li, L.Q.; Jin, Q.; Hu, K. On Stratified $L$-Convergence Spaces: Fischer's Diagonal Axiom. *Fuzzy Sets Syst.* **2015**, *267*, 31–40. [CrossRef]

22. Losert, B.; Boustique, H.; Richardson, G. Modifications: Lattice-valued structures. *Fuzzy Sets Syst.* **2013**, *210*, 54–62. [CrossRef]

23. Orpen, D.; Jäger, G. Lattice-valued convergence spaces: Extending the lattices context. *Fuzzy Sets Syst.* **2012**, *190*, 1–20. [CrossRef]

24. Pang, B.; Zhao, Y. Several types of enriched $(L, M)$-fuzzy convergence spaces. *Fuzzy Sets Syst.* **2017**, *321*, 55–72. [CrossRef]

25. Pang, B. Degrees of Separation Properties in Stratified $L$-Generalized Convergence Spaces Using Residual Implication. *Filomat* **2017**, *31*, 6293–6305. [CrossRef]

26. Pang, B. Stratified $L$-ordered filter spaces. *Quaest. Math.* **2017**, *40*, 661–678. [CrossRef]

27. Pang, B.; Xiu, Z.Y. Stratified $L$-prefilter convergence structures in stratified $L$-topological spaces. *Soft Comput.* **2018**, *22*, 7539–7551. [CrossRef]

28. Xiu, Z.Y.; Pang, B. Base axioms and subbase axioms in $M$-fuzzifying convex spaces. *Iran. J. Fuzzy Syst.* **2018**, *15*, 75–87.

29. Yang, X.; Li, S. Net-theoretical convergence in $(L, M)$-fuzzy cotopological spaces. *Fuzzy Sets Syst.* **2012**, *204*, 53–65. [CrossRef]

30. Yao, W. On many-valued stratified $L$-fuzzy convergence spaces. *Fuzzy Sets Syst.* **2008**, *159*, 2503–2519. [CrossRef]

31. Fang, J.M.; Yue, Y.L. $\top$-diagonal conditions and Continuous extension theorem. *Fuzzy Sets Syst.* **2017**, *321*, 73–89. [CrossRef]

32. Jin, Q.; Li, L.Q. Modified Top-convergence spaces and their relationships to lattice-valued convergence spaces. *J. Intell. Fuzzy Syst.* **2018**, *35*, 2537–2546. [CrossRef]

33. Li, L.Q. $p$-Topologicalness—A Relative Topologicalness in $\top$-Convergence Spaces. *Mathematics* **2019**, *7*, 228. [CrossRef]

34. Li, L.Q.; Jin, Q.; Hu, K. Lattice-valued convergence associated with CNS spaces. *Fuzzy Sets Syst.* **2018**. [CrossRef]

35. Qiu, Y.; Fang, J.M. The category of all $\top$-convergence spaces and its cartesian-closedness. *Iran. J. Fuzzy Syst.* **2017**, *14*, 121–138.

36. Reid, L.; Richardson, G. Connecting $\top$ and Lattice-Valued Convergences. *Iran. J. Fuzzy Syst.* **2018**, *15*, 151–169.

37. Jäger, G. Lattice-valued convergence spaces and Regularity. *Fuzzy Sets Syst.* **2008**, *159*, 2488–2502. [CrossRef]

38. Boustique, H.; Richardson, G. A note on regularity. *Fuzzy Sets Syst.* **2011**, *162*, 64–66. [CrossRef]

39. Boustique, H.; Richardson, G. Regularity: Lattice-valued Cauchy spaces. *Fuzzy Sets Syst.* **2012**, *190*, 94–104. [CrossRef]

40. Li, L.Q.; Jin, Q. $p$-Topologicalness and $p$-Regularity for lattice-valued convergence spaces. *Fuzzy Sets Syst.* **2014**, *238*, 26–45. [CrossRef]

41. Li, L.Q.; Jin, Q.; Meng, G.W.; Hu, K. The lower and upper $p$-topological ($p$-regular) modifications for lattice-valued convergence spaces. *Fuzzy Sets Syst.* **2016**, *282*, 47–61. [CrossRef]

42. Li, L.Q.; Jin, Q.; Yao, B.X. Regularity of fuzzy convergence spaces. *Open Math.* **2018**, *16*, 1455–1465. [CrossRef]

43. Höhle, U.; Šostak, A. Axiomatic foundations of fixed-basis fuzzy topology. In *Mathematics of Fuzzy Sets: Logic, Topology and Measure Theory, The Handbooks of Fuzzy Sets Series*; Höhle, U., Rodabaugh, S.E., Eds.; Kluwer Academic Publishers: Boston, MA, USA; Dordrecht, The Netherlands; London, UK, 1999; Volume 3, pp. 123–273.

44. Bělohlávek, R. *Fuzzy Relational Systems: Foundations and Principles*; Kluwer Academic Publishers: New York, NY, USA, 2002.

45. Zhang, D.X. An enriched category approach to many valued Topology. *Fuzzy Sets Syst.* **2007**, *158*, 349–366. [CrossRef]

46. Zhao, F.F.; Jin, Q.; Li, L.Q. The axiomatic characterizations on $L$-generalized fuzzy neighborhood system-based approximation operators. *Int. J. Gen. Syst.* **2018**, *47*, 155–173. [CrossRef]

47. Lai, H.; Zhang, D. Fuzzy topological spaces with conical neighborhood system. *Fuzzy Sets Syst.* **2018**, *330*, 87–104. [CrossRef]

48. García, J.G. On stratified *L*-valued filters induced by T-filters. *Fuzzy Sets Syst.* **2006**, *157*, 813–819. [CrossRef]

49. Adámek, J.; Herrlich, H.; Strecker, G.E. *Abstract and Concrete Categories*; Wiley: New York, NY, USA, 1990.

 *mathematics*

Article

# Counterintuitive Test Problems for Distance-Based Similarity Measures Between Intuitionistic Fuzzy Sets

**Hui-Chin Tang * and Shen-Tai Yang ***

Department of Industrial Engineering and Management, National Kaohsiung University of Science and Technology, Kaohsiung 80778, Taiwan
* Correspondence: tang@nkust.edu.tw (H.-C.T.); 1105407103@nkust.edu.tw (S.-T.Y.)

Received: 4 April 2019; Accepted: 10 May 2019; Published: 17 May 2019

**Abstract:** This paper analyzes the counterintuitive behaviors of adopted twelve distance-based similarity measures between intuitionistic fuzzy sets. Among these distance-based similarity measures, the largest number of components of the distance in the similarity measure is four. We propose six general counterintuitive test problems to analyze their counterintuitive behaviors. The results indicate that all the distance-based similarity measures have some counterintuitive test problems. Furthermore, for the largest number of components of the distance-based similarity measure, four types of counterintuitive examples exist. Therefore, the counterintuitive behaviors are inevitable for the distance-based similarity measures between intuitionistic fuzzy sets.

**Keywords:** intuitionistic fuzzy set; similarity; counterintuitive

---

## 1. Introduction

Fuzzy sets (FSs) theory, proposed by Zadeh [1], has successfully been applied in various fields. As a generalization of FSs, intuitionistic fuzzy sets (IFSs) proposed by Atanassov [2] is characterized by a membership function and a non-membership function. IFSs have been widely applied in cluster analysis [3–5], decision-making [6–11], medical diagnosis [12,13] and pattern recognition [14–19].

A similarity measure between two IFSs represents the alignment of the two sets. There are a large number of papers discussing various similarity measures between two IFSs [20–30]. A comprehensive and accurate survey of state-of-the-art research on IFSs and similarity measures are given by Atanassov [31] and Pedrycz and Chen [9]. The degree of similarity measure is an important tool for cluster analysis [3], decision-making [6,9,11], medical diagnosis [13] and pattern recognition [14–19,22]. One of the typical definitions of a similarity measure is the one-minus distance between two vectors [17,21–30]. The first distances over intuitionistic fuzzy sets are introduced by Atanassov [32]. Szmidt and Kacprzyk [33] adopted the normalized Hamming distance and the normalized Euclidean distance to present the similarity measures for IFSs. Furthermore, Szmidt and Kacprzyk [33] used hesitancy along with the membership and non-membership values of the elements in the sets being compared. The components of Chen's [21] similarity measure and Dengfeng and Chuntian's [23] similarity measure are concerned with the difference between degrees of support. The components of the Hong and Kim's [26] similarity measure, Mitchell's [17] similarity measure, the first Liang and Shi's [30] similarity measure, Hung and Yang's [27] three similarity measures and Li, Zhongxian and Degin's [29] similarity measure are the combination of the difference between membership degrees and the difference between non-membership degrees. The components of the Li and Xu's [24] similarity measure are the combination of the difference between membership degrees, the difference between non-membership degrees and the difference between degrees of support. The second Liang and Shi's [30] similarity measure is concerned with the combination of the difference between the first quartiles of intervals of membership degree and the difference between the third quartiles of intervals of membership degree. The third Liang and Shi's [31] similarity measure considers the combination of the difference between membership degrees, the difference between non-membership degrees, the difference between degrees of support and the difference between hesitant degrees. Therefore, the components of the distance can be the combination of the difference between membership degrees, the difference between non-membership degrees, the difference between hesitant degrees, the difference between degrees of support, the difference between the first quartiles of intervals of the membership degree and the difference between the third quartiles of intervals of the membership degree. Then we apply Minkowski distance to calculate the similarity degree between two IFSs. This paper focuses on these distance-based similarity measures.

In the literature, many varied approaches for the similarity measures between two IFSs are inconsistent with our intuition [14–30]. Two typical counterintuitive examples are (1) quite distinct IFSs having high similarity to each other and (2) two very distinct IFSs appearing to have equal similarity to a third IFS. However, these two typical counterintuitive ones are simple examples in the literature. This paper proposes six general counterintuitive test problems to analyze the counterintuitive behaviors of the distance-based similarity measures. As Li et al. [28] indicates, third Liang and Shi's [30] similarity measure has " ... no counterintuitive cases". We further present four types of counterintuitive test problems for the third Liang and Shi's [30] similarity measure.

The main contribution of this paper is to propose six general counterintuitive test problems. To the best of our knowledge, the general test problems have not been yet presented to analyze the counterintuitive behaviors of the distance-based similarity measures. Moreover, the proposed general counterintuitive test problems can be used to analyze the counterintuitive behaviors of any similarity measures.

The organization of this paper is as follows. Section 2 briefly reviews the FSs and IFSs and presents the distance-based similarity measures. Section 3 presents six general counterintuitive test problems and analyzes the counterintuitive behaviors of distance-based similarity measures. Section 4 further presents the counterintuitive behaviors of the third Liang and Shi's [31] similarity measure. Finally, some concluding remarks and future research are presented.

## 2. IFSs and Similarity Measures

We firstly review the basic notations of FSs and IFSs. Let $X = \{x_1, x_2, \ldots, x_n\}$ be a non-empty universal set of real numbers $\mathcal{R}$.

**Definition 1.** *An FS A over X is defined by a membership function*

$$\mu_A : X \to [0,1] \text{ for } 1 \leq i \leq n.$$

*Then the fuzzy set A over X is the set*

$$A = \{(x_i, \mu_A(x_i)) | 1 \leq i \leq n\}. \tag{1}$$

*We denote the set of all FSs over X by FS(X).*

**Definition 2.** *An IFS A over X is defined as*

$$A = \{(x_i, \mu_A(x_i), v_A(x_i)) | 1 \leq i \leq n\}, \tag{2}$$

*where the membership function $\mu_A : X \to [0,1]$ and non-membership function $v_A : X \to [0,1]$ of the $x_i$ to the set A satisfy*

$$\mu_A(x_i) + v_A(x_i) \leq 1.$$

*The set of all IFSs over X is denoted by IFS(X). The degree of hesitancy associated with each $x_i$ is defined as*

$$\pi_A(x_i) = 1 - \mu_A(x_i) - v_A(x_i) \tag{3}$$

*measuring the lack of information or certitude.*

*We now briefly review some operations involving IFSs.*

**Definition 3.** *Let* $A = \{(x_i, \mu_A(x_i), v_A(x_i)) | 1 \leq i \leq n\}$ *and* $B = \{(x_i, \mu_B(x_i), v_B(x_i)) | 1 \leq i \leq n\}$ *be two IFSs. Then,*

1. $A \subseteq B$ *if and only if* $\mu_A(x_i) \leq \mu_B(x_i)$ *and* $v_A(x_i) \geq v_B(x_i)$ *for* $1 \leq i \leq n$.
2. $A = B$ *if and only if* $\mu_A(x_i) = \mu_B(x_i)$ *and* $v_A(x_i) = v_B(x_i)$ *for* $1 \leq i \leq n$.
3. *The complement of* $A$ *is defined as* $A_c = \{(x_i, v_A(x_i), \mu_A(x_i)) | 1 \leq i \leq n\}$.
4. *We denote the pure intuitionistic fuzzy set by* $PI = \{(x_i, 0, 0) | 1 \leq i \leq n\}$.

*For the fuzzy set* $A \in FS(X)$, $\pi_A(x_i) = 0$ *for every* $x_i \in X$, *and for the pure intuitionistic fuzzy set,* $\pi_{PI}(x_i) = 1$ *for every* $x_i \in X$.

*We now recall the definition of similarity measures between two IFSs.*

**Definition 4.** *A similarity measure* $S : IFS(X)^2 \rightarrow [0,1]$ *should satisfy the following properties:*

1. $S(A, B) \in [0, 1]$.
2. $S(A, B) = 1$ *if and only if* $A = B$.
3. $S(A, B) = S(B, A)$.
4. *If* $A \subseteq B \subseteq C$, *then* $S(A, C) \leq S(A, B)$ *and* $S(A, C) \leq S(B, C)$.

So far, in the literature, many papers have been dedicated to problems connected with the similarity measures between two IFSs and research on this area is still carried on. Let $A = \{(x_i, \mu_A(x_i), v_A(x_i)) | 1 \leq i \leq n\}$ and $B = \{(x_i, \mu_B(x_i), v_B(x_i)) | 1 \leq i \leq n\}$ be two IFSs. The existing distance-based similarity measures $S(A, B)$ between two IFSs $A$ and $B$ are defined as follows:

Chen [21]:

$$S_C(A, B) = 1 - \frac{1}{2n} \sum_{i=1}^{n} |\mu_A(x_i) - v_A(x_i) - (\mu_B(x_i) - v_B(x_i))|$$

Hong and Kim [26]:

$$S_{HK}(A, B) = 1 - \frac{1}{2n} \sum_{i=1}^{n} |\mu_A(x_i) - \mu_B(x_i)| + |v_A(x_i) - v_B(x_i)|$$

Li and Xu [24]:

$$S_{LX}(A, B) = 1 - \frac{1}{4n} \sum_{i=1}^{n} \{|\mu_A(x_i) - \mu_B(x_i)| + |v_A(x_i) - v_B(x_i)| \\ + |\mu_A(x_i) - v_A(x_i) - (\mu_B(x_i) - v_B(x_i))|\}$$

Dengfeng and Chuntian [23]:

$$S_{DC}(A, B) = 1 - \sqrt[p]{\frac{1}{2n} \sum_{i=1}^{n} |\mu_A(x_i) - v_A(x_i) - (\mu_B(x_i) - v_B(x_i))|^p}$$

Mitchell [17]:

$$S_M(A, B) = 1 - \frac{1}{2} \left\{ \sqrt[p]{\frac{1}{n} \sum_{i=1}^{n} |\mu_A(x_i) - \mu_B(x_i)|^p} + \sqrt[p]{\frac{1}{n} \sum_{i=1}^{n} |v_A(x_i) - v_B(x_i)|^p} \right\}$$

Liang and Shi [30]:

$$S_{LS1}(A,\,B) = 1 - \frac{1}{2}\sqrt[p]{\frac{1}{n}\sum_{i=1}^{n}\left(\left|\mu_A(x_i) - \mu_B(x_i)\right| + \left|v_A(x_i) - v_B(x_i)\right|\right)^p}$$

$$S_{LS2}(A,\,B) = 1 - \frac{1}{8}\left(\frac{1}{n}\sum_{i=1}^{n}\left(\left|3(\mu_A(x_i) - \mu_B(x_i)) - (v_A(x_i) - v_B(x_i))\right| + \left|(\mu_A(x_i) - \mu_B(x_i)) - 3(v_A(x_i) - v_B(x_i))\right|\right)^p\right)^{1/p}$$

$$S_{LS3}(A,\,B) = 1 - \frac{1}{2}\left(\frac{1}{3n}\sum_{i=1}^{n}\left(\left|\mu_A(x_i) - \mu_B(x_i)\right| + \left|v_A(x_i) - v_B(x_i)\right| + \left|\mu_A(x_i) - v_A(x_i) - (\mu_B(x_i) - v_B(x_i))\right| + \left|\mu_A(x_i) + v_A(x_i) - (\mu_B(x_i) + v_B(x_i))\right|\right)^p\right)^{1/p}$$

Hung and Yang [27]:

$$S_{HY1}(A,\,B) = 1 - \frac{1}{n}\sum_{i=1}^{n}\max\{\left|\mu_A(x_i) - \mu_B(x_i)\right|,\,\left|v_A(x_i) - v_B(x_i)\right|\}$$

$$S_{HY2}(A,\,B) = \frac{e^{-\frac{1}{n}\sum_{i=1}^{n}\max\{\left|\mu_A(x_i) - \mu_B(x_i)\right|,\left|v_A(x_i) - v_B(x_i)\right|\}} - e^{-1}}{1 - e^{-1}}$$

$$S_{HY3}(A,\,B) = \frac{1 - \frac{1}{n}\sum_{i=1}^{n}\max\{\left|\mu_A(x_i) - \mu_B(x_i)\right|,\,\left|v_A(x_i) - v_B(x_i)\right|\}}{1 + \frac{1}{n}\sum_{i=1}^{n}\max\{\left|\mu_A(x_i) - \mu_B(x_i)\right|,\,\left|v_A(x_i) - v_B(x_i)\right|\}}$$

Li, Zhongxian and Degin [29]:

$$S_{LZD}(A,\,B) = 1 - \sqrt{\frac{1}{2n}\sum_{i=1}^{n}\left(\mu_A(x_i) - \mu_B(x_i)\right)^2 + \left(v_A(x_i) - v_B(x_i)\right)^2}.$$

Therefore, we adopt twelve distance-based similarity measures. When $p = 1$, we have

$$S_{DC}(A,\,B) = S_C(A,\,B)$$

$$S_M(A,\,B) = 1 - \frac{1}{2n}\sum_{i=1}^{n}\left|\mu_A(x_i) - \mu_B(x_i)\right| + \left|v_A(x_i) - v_B(x_i)\right| = S_{HK}(A,\,B)$$

and

$$S_{LS1}(A,\,B) = 1 - \frac{1}{2n}\sum_{i=1}^{n}\left|\mu_A(x_i) - \mu_B(x_i)\right| + \left|v_A(x_i) - v_B(x_i)\right| = S_{HK}(A,\,B).$$

Among these distance-based similarity measures, four types of components of distance are distinguished: (I) the combination of difference between membership degrees and difference between non-membership degrees; (II) the difference between hesitant degrees; (III) the difference between degrees of support and (IV) the combination of difference between the first quartiles of intervals of membership degree and the difference between the third quartiles of intervals of membership degree. Detailed results are presented in Table 1. The similarity measures $S_{HK}(A,\,B)$, $S_M(A,\,B)$, $S_{LS1}(A,\,B)$, $S_{HY1}(A,\,B)$, $S_{HY2}(A,\,B)$, $S_{HY3}(A,\,B)$ and $S_{LZD}(A,\,B)$ are dependent on type I. The similarity measures $S_C(A,\,B)$ and $S_{DC}(A,\,B)$ are concerned with type III. $S_{LX}(A,\,B)$ is dependent on the combination of types I and III. $S_{LS2}(A,\,B)$ is concerned with type IV. $S_{LS3}(A,\,B)$ is concerned with the combination of types I, II and III. The largest number of components of the distance in the similarity measure is four for the similarity measure $S_{LS3}(A,\,B)$. Therefore, we anticipate that the intuitive behavior of $S_{LS3}(A,\,B)$ is superior to those of other similarity measures.

**Table 1.** The four types of components of distance-based similarity measures.

| Similarity | I | II | III | IV |
|---|---|---|---|---|
| $S_C(A, B)$ | | | $\checkmark$ | |
| $S_{HK}(A, B)$ | $\checkmark$ | | | |
| $S_{LX}(A, B)$ | $\checkmark$ | | $\checkmark$ | |
| $S_{DC}(A, B)$ | | | $\checkmark$ | |
| $S_M(A, B)$ | $\checkmark$ | | | |
| $S_{LS1}(A, B)$ | $\checkmark$ | | | |
| $S_{LS2}(A, B)$ | | | | $\checkmark$ |
| $S_{LS3}(A, B)$ | $\checkmark$ | $\checkmark$ | $\checkmark$ | |
| $S_{HY1}(A, B)$ | $\checkmark$ | | | |
| $S_{HY2}(A, B)$ | $\checkmark$ | | | |
| $S_{HY3}(A, B)$ | $\checkmark$ | | | |
| $S_{LZD}(A, B)$ | $\checkmark$ | | | |

## 3. General Counterintuitive Test Problems

Much literature has been written on the counterintuitive examples for the similarity measures between two IFSs. Two typical counterintuitive examples are (I) $S(A, B) = 1$ for $A \neq B$, $A, B \in IFS(X)$ and (II) $S(P_1, Q) = S(P_2, Q)$ for $P_1 \neq P_2$, $P_1, P_2, Q \in IFS(X)$. This paper proposes six general counterintuitive test problems presented in Table 2. From Definition 4, a similarity measure which fails type I is not a similarity measure.

**Table 2.** The six general counterintuitive test problems.

| # | Test Problem |
|---|---|
| T1 | $A(a, a)$, $B(b, b)$, $a \neq b$, $0 \leq a$, $b \leq 1/2$ satisfying $S(A, B) = 1$. |
| T2 | $A(a, a)$, $B(b, b)$, $C(a, b)$, $D(b, a)$, $a \neq b$, $0 \leq a$, $b \leq 1/2$ satisfying $S(A, B) = S(C, D)$. |
| T3 | $P_1(a, b)$, $P_2(\frac{a+b}{2}, \frac{a+b}{2})$, $Q(b, b)$, $a \neq b$, $0 \leq a$, $b$, $a+b \leq 1$, $b \leq 1/2$ satisfying $S(P_1, Q) = S(P_2, Q)$. |
| T4 | $P_1(1, 0)$, $P_2(a, 1-a)$, $Q(0, 0)$, $0 \leq a \leq 1$ satisfying $S(P_1, Q) = S(P_2, Q)$. |
| T5 | $P_1(b, b-\alpha)$, $P_2(b, a-\alpha)$, $Q(a, a-\alpha)$, $a \neq b$, $\alpha \leq a$, $b$, $a+b-\alpha \leq 1$, $2a-\alpha \leq 1$, $2b-\alpha \leq 1$ satisfying $S(P_1, Q) = S(P_2, Q)$. |
| T6 | $P_1\{(a+\alpha, b), (a, b+\alpha)\}$, $P_2\{(a+\alpha, b+\alpha), (a+\alpha, b+\alpha)\}$, $Q\{(a, b), (a+\alpha, b)\}$, $a \neq b$, $0 \leq a$, $b$, $\alpha$, $a+b+2\alpha \leq 1$ satisfying $S(P_1, Q) = S(P_2, Q)$. |

Test problem T1 is a type I counterintuitive one. So, a similarity measure which fails test problem T1 is not a similarity measure. Test problem T2 is to distinguish the positive difference and negative difference. Test problems T3, T4, T5 and T6 can be considered pattern recognition problems. Consider two known patterns $P_1$ and $P_2$. We want to classify an unknown pattern represented by $Q$ into one of the patterns $P_1$ and $P_2$. For test problem T3, the membership degree and non-membership degree of $P_2$ is the average of those of $P_1$ and $Q$. For $S(P_1, Q) = S(P_2, Q)$, it implies that T3 is a counterintuitive test problem. For test problem T4 with $S(P_1, Q) = S(P_2, Q)$, a crisp set $P_1$ is as similar to PI $Q$ as any FS $P_2$. So T4 is a counterintuitive test problem. For test problem T5, although

$$\left|\mu_{P_1} - \mu_Q\right| \neq \left|\mu_{P_2} - \mu_Q\right| \text{ or } \left|v_{P_1} - v_Q\right| \neq \left|v_{P_2} - v_Q\right|$$

and

$$\left|\mu_{P_1} - \mu_Q\right| + \left|v_{P_1} - v_Q\right| \neq \left|\mu_{P_2} - \mu_Q\right| + \left|v_{P_2} - v_Q\right|,$$

but

$$\left|\mu_{P_1} - \mu_Q\right| + \left|v_{P_1} - v_Q\right| + \left|\mu_{P_1} - v_{P_1} - (\mu_Q - v_Q)\right| = \left|\mu_{P_2} - \mu_Q\right| + \left|v_{P_2} - v_Q\right| + \left|\mu_{P_2} - v_{P_2} - (\mu_Q - v_Q)\right|, \tag{4}$$

then $S(P_1, Q) = S(P_2, Q)$. It follows that T5 is a counterintuitive test problem. For test problem T6, if

$$
\begin{aligned}
\sum_{i=1}^{2} & \left( \left| \mu_{P_1}(x_i) - \mu_Q(x_i) \right| + \left| v_{P_1}(x_i) - v_Q(x_i) \right| + \left| \mu_{P_1}(x_i) - v_{P_1}(x_i) - (\mu_Q(x_i) - v_Q(x_i)) \right| \right. \\
& \left. + \left| \mu_{P_1}(x_i) + v_{P_1}(x_i) - (\mu_Q(x_i) + v_Q(x_i)) \right| \right) \\
= \sum_{i=1}^{2} & \left( \left| \mu_{P_2}(x_i) - \mu_Q(x_i) \right| + \left| v_{P_2}(x_i) - v_Q(x_i) \right| + \left| \mu_{P_2}(x_i) - v_{P_2}(x_i) \right. \right. \\
& \left. \left. - (\mu_Q(x_i) - v_Q(x_i)) \right| + \left| \mu_{P_2}(x_i) + v_{P_2}(x_i) - (\mu_Q(x_i) + v_Q(x_i)) \right| \right)
\end{aligned}
\tag{5}
$$

then $S(P_1, Q) = S(P_2, Q)$. So T6 is a counterintuitive test problem.

For each distance-based similarity measure, we analyze the counterintuitive behaviors of the six general counterintuitive test problems as follows.

For similarity measure $S_C(A, B)$, since $S_C(A, B) = 1$ for test problem T1, it implies that T1 is the counterintuitive test problem for the similarity measure $S_C(A, B)$.

For similarity measure $S_{HK}(A, B)$, we have

$$
S_{HK}(A, B) = S_{HK}(C, D) = 1 - |a - b|
\tag{6}
$$

$$
S_{HK}(P_1, Q) = S_{HK}(P_2, Q) = 1 - \frac{|a - b|}{2}
\tag{7}
$$

and

$$
S_{HK}(P_1, Q) = S_{HK}(P_2, Q) = 0.5
\tag{8}
$$

for test problems T2, T3 and T4, respectively. So the counterintuitive test problems for the similarity measure $S_{HK}(A, B)$ are T2, T3 and T4.

The counterintuitive test problem for the similarity measure $S_{LX}(A, B)$ is T5, because

$$
S_{LX}(P_1, Q) = S_{LX}(P_2, Q) = 1 - \frac{|a - b|}{2}
\tag{9}
$$

A similar argument shows that the counterintuitive test problems for the similarity measures $S_{DC}(A, B)$ and $S_M(A, B)$ are test problem T1 and test problems T2, T3, T4, respectively. The counterintuitive test problems for the three Liang and Shi's similarity measures $S_{LS1}(A, B)$, $S_{LS2}(A, B)$ and $S_{LS3}(A, B)$ are test problems T2, T3, T4, test problem T5, and test problem T6, respectively. The counterintuitive test problems are T2 and T5 for the three Hung and Yang's similarity measures $S_{HY1}(A, B)$, $S_{HY2}(A, B)$ and $S_{HY3}(A, B)$, and T2 for $S_{LZD}(A, B)$. Detailed results are displayed in Table 3. $S_C(A, B)$ and $S_{DC}(A, B)$ are not similarity measures. The smallest number of counterintuitive test problems is one. These counterintuitive test problems are T2 for $S_{LZD}(A, B)$, T5 for $S_{LX}(A, B)$, $S_{LS2}(A, B)$, and T6 for $S_{LS3}(A, B)$. Therefore, the counterintuitive test problem is T1 for $S_C(A, B)$ and $S_{DC}(A, B)$, T2 for $S_{HK}(A, B)$, $S_M(A, B)$, $S_{LS1}(A, B)$, $S_{HY1}(A, B)$, $S_{HY2}(A, B)$, $S_{HY3}(A, B)$ and $S_{LZD}(A, B)$, T3 and T4 for $S_{HK}(A, B)$, $S_M(A, B)$ and $S_{LS1}(A, B)$, T5 for $S_{LX}(A, B)$, $S_{LS2}(A, B)$, $S_{HY1}(A, B)$, $S_{HY2}(A, B)$ and $S_{HY3}(A, B)$, and T6 for $S_{LS3}(A, B)$.

**Table 3.** The counterintuitive test problems of distance-based similarity measures.

| Similarity | Counterintuitive Test Problems |
|---|---|
| $S_C(A, B)$ | T1 |
| $S_{HK}(A, B)$ | T2, T3, T4 |
| $S_{LX}(A, B)$ | T5 |
| $S_{DC}(A, B)$ | T1 |
| $S_M(A, B)$ | T2, T3, T4 |
| $S_{LS1}(A, B)$ | T2, T3, T4 |
| $S_{LS2}(A, B)$ | T5 |
| $S_{LS3}(A, B)$ | T6 |
| $S_{HY1}(A, B)$ | T2, T5 |
| $S_{HY2}(A, B)$ | T2, T5 |
| $S_{HY3}(A, B)$ | T2, T5 |
| $S_{LZD}(A, B)$ | T2 |

## 4. Counterintuitive Test Problem for $S_{LS3}(A,B)$

From the analysis of counterintuitive test problems in the previous section and those of Li et al. [28], the third Liang and Shi's similarity measure $S_{LS3}(A, B)$ outperforms other similarity measures. This section further analyzes the counterintuitive behaviors of the third Liang and Shi's similarity measure $S_{LS3}(A, B)$.

Consider the four types of three IFSs $A$, $B$ and $C$ as follows.

(I) $A(a_1, a_2)$, $B(a_1, a_2 + \alpha)$ and $C(c_1, a_2 + \alpha - \beta)$ for $c_1 \geq a_1$.

(II) $A(a_1, a_2)$, $B(a_1, a_2 + \alpha)$ and $C(c_1, a_2 + \alpha - \beta)$ for $a_1 \geq c_1$.

(III) $A(a_1, a_2)$, $B(a_1, a_2 - \alpha)$ and $C(c_1, a_2 - \alpha + \beta)$ for $c_1 \geq a_1$.

(IV) $A(a_1, a_2)$, $B(a_1, a_2 - \alpha)$ and $C(c_1, a_2 - \alpha + \beta)$ for $a_1 \geq c_1$.

By fixing $A(a_1, a_2)$ and $B(a_1, a_2 \pm \alpha)$, we will present the explicit form of IFS $C(c_1, a_2 \pm \alpha \mp \beta)$ satisfying $S_{LS3}(A, B) = S_{LS3}(A, C)$. This implies that $\{A(a_1, a_2), B(a_1, a_2 \pm \alpha), C(c_1, a_2 \pm \alpha \mp \beta)\}$ is a counterintuitive example for the similarity measure $S_{LS3}(A, B)$.

We now analyze type I. From $A(a_1, a_2)$, $B(a_1, a_2 + \alpha)$, $C(c_1, a_2 + \alpha - \beta)$, and $c_1 \geq a_1$, it follows that

$$S_{LS3}(A, B) = 3\alpha \tag{10}$$

and

$$S_{LS3}(A, C) = c_1 - a_1 + |\alpha - \beta| + \left|c_1 - a_1 - (\alpha - \beta)\right| + |c_1 - a_1 + \alpha - \beta| \tag{11}$$

Four cases are considered as follows.

Case I1: $\alpha - \beta \geq 0$, $c_1 - a_1 - (\alpha - \beta) \leq 0$ and $c_1 - a_1 + \alpha - \beta \geq 0$.

We get that

$$S_{LS3}(A, C) = -a_1 + c_1 + 3\alpha - 3\beta.$$

In order to $S_{LS3}(A, B) = S_{LS3}(A, C)$, we have that $c_1 = a_1 + 3\beta$, implying

$$C(a_1 + 3\beta, a_2 + \alpha - \beta). \tag{12}$$

The following conditions should be satisfied for $A$, $B$, $C \in IFS(X)$ and $c_1 \geq a_1$,

$$a_1 - c_1 + \alpha - \beta = \alpha - 4\beta \geq 0, \ a_1 + a_2 + \alpha \leq 1, \ a_1 + a_2 + \alpha + 2\beta \leq 1, \ a_1, a_2, \alpha, \beta \geq 0.$$

Therefore, if $\alpha \geq 4\beta$, $a_1 + a_2 + \alpha + 2\beta \leq 1$, $a_1, a_2, \alpha, \beta \geq 0$, then $A(a_1, a_2)$, $B(a_1, a_2 + \alpha)$ and $C(a_1 + 3\beta, a_2 + \alpha - \beta)$ satisfying $S_{LS3}(A, B) = S_{LS3}(A, C)$.

A similar argument shows the other three cases as follows. Detailed results of $C(c_1, a_2 + \alpha - \beta)$ and conditions for $c_1 \geq a_1$ are summarized in Table 4.

Case I2: $\alpha - \beta \geq 0$, $c_1 - a_1 - (\alpha - \beta) \geq 0$ and $c_1 - a_1 + \alpha - \beta \geq 0$.

If $\beta \leq \alpha \leq 4\beta$, $a_1 + a_2 + \frac{5\alpha - 2\beta}{3} \leq 1$ and $a_1, a_2, \alpha, \beta \geq 0$, then $A(a_1, a_2)$, $B(a_1, a_2 + \alpha)$ and $C(a_1 + \frac{2\alpha + \beta}{3}, a_2 + \alpha - \beta)$ satisfying $S_{LS3}(A, B) = S_{LS3}(A, C)$.

Case I3: $\alpha - \beta \leq 0$, $c_1 - a_1 - (\alpha - \beta) \geq 0$ and $c_1 - a_1 + \alpha - \beta \geq 0$.

If $4\beta/7 \leq \alpha \leq \beta$, $a_1 + a_2 + \alpha \leq 1$ and $a_1, a_2, \alpha, \beta \geq 0$, then $A(a_1, a_2)$, $B(a_1, a_2 + \alpha)$ and $C(a_1 + \frac{4\alpha - \beta}{3}, a_2 + \alpha - \beta)$ satisfying $S_{LS3}(A, B) = S_{LS3}(A, C)$.

Case I4: $\alpha - \beta \leq 0$, $c_1 - a_1 - (\alpha - \beta) \geq 0$ and $c_1 - a_1 + \alpha - \beta \leq 0$.

If $\beta/2 \leq \alpha \leq 4\beta/7$, $a_1 + a_2 + \alpha \leq 1$ and $a_1, a_2, \alpha, \beta \geq 0$, then $A(a_1, a_2)$, $B(a_1, a_2 + \alpha)$ and $C(a_1 + 6\alpha - 3\beta, a_2 + \alpha - \beta)$ satisfying $S_{LS3}(A, B) = S_{LS3}(A, C)$.

**Table 4.** $A(a_1, a_2)$, $B(a_1, a_2 + \alpha)$ and $C$ satisfying $S_{LS3}(A, B) = S_{LS3}(A, C)$, $a_1, a_2, \alpha, \beta \geq 0$.

| | $C(c_1, a_2 + \alpha - \beta)$ | Conditions |
|---|---|---|
| I1 | $C(a_1 + 3\beta, a_2 + \alpha - \beta)$ | $c_1 \geq a_1$, $\alpha \geq 4\beta$, $a_1 + a_2 + \alpha + 2\beta \leq 1$ |
| I2 | $C(a_1 + \frac{2\alpha + \beta}{3}, a_2 + \alpha - \beta)$ | $c_1 \geq a_1$, $\beta \leq \alpha \leq 4\beta$, $a_1 + a_2 + \frac{5\alpha - 2\beta}{3} \leq 1$ |
| I3 | $C(a_1 + \frac{4\alpha - \beta}{3}, a_2 + \alpha - \beta)$ | $c_1 \geq a_1$, $4/7\beta \leq \alpha \leq \beta$, $a_1 + a_2 + \alpha \leq 1$ |
| I4 | $C(a_1 + 6\alpha - 3\beta, a_2 + \alpha - \beta)$ | $c_1 \geq a_1$, $0.5\beta \leq \alpha \leq 4/7\beta$, $a_1 + a_2 + \alpha \leq 1$ |
| II1 | $C(a_1 - 3\beta, a_2 + \alpha - \beta)$ | $a_1 \geq c_1$, $\alpha \geq 4\beta$, $a_1 + a_2 + \alpha \leq 1$ |
| II2 | $C(a_1 - \frac{2\alpha + \beta}{3}, a_2 + \alpha - \beta)$ | $a_1 \geq c_1$, $\beta \leq \alpha \leq 4\beta$, $a_1 + a_2 + \alpha \leq 1$ |
| II3 | $C(a_1 - \frac{4\alpha - \beta}{3}, a_2 + \alpha - \beta)$ | $a_1 \geq c_1$, $4/7\beta \leq \alpha \leq \beta$, $a_1 + a_2 + \alpha \leq 1$ |
| II4 | $C(a_1 - 6\alpha + 3\beta, a_2 + \alpha - \beta)$ | $a_1 \geq c_1$, $0.5\beta \leq \alpha \leq 4/7\beta$, $a_1 + a_2 + \alpha \leq 1$ |

For type II, from $A(a_1, a_2)$, $B(a_1, a_2 + \alpha)$, $C(c_1, a_2 + \alpha - \beta)$ and $a_1 \geq c_1$, it follows that Equation (10) and

$$S_{LS3}(A, C) = a_1 - c_1 + |\alpha - \beta| + |a_1 - c_1 + \alpha - \beta| + |a_1 - c_1 - \alpha + \beta| \tag{13}$$

Four cases are considered as follows.

Case II1: $\alpha - \beta \geq 0$, $a_1 - c_1 + \alpha - \beta \geq 0$ and $a_1 - c_1 - \alpha + \beta \leq 0$.

Case II2: $\alpha - \beta \geq 0$, $a_1 - c_1 + \alpha - \beta \geq 0$ and $a_1 - c_1 - \alpha + \beta \geq 0$.

Case II3: $\alpha - \beta \leq 0$, $a_1 - c_1 + \alpha - \beta \geq 0$ and $a_1 - c_1 - \alpha + \beta \geq 0$.

Case II4: $\alpha - \beta \leq 0$, $a_1 - c_1 + \alpha - \beta \leq 0$ and $a_1 - c_1 - \alpha + \beta \geq 0$.

A similar argument presents the explicit form of $C(c_1, a_2 + \alpha - \beta)$ and conditions for $a_1 \geq c_1$ shown in Table 4. Given $A(a_1, a_2)$ and $B(a_1, a_2 + \alpha)$, from $c_1 \geq a_1$ of type I to $a_1 \geq c_1$ of type II, the form of C from $C(a_1 + \gamma_i, a_2 + \alpha - \beta)$ of type I case Ii to $C(a_1 - \gamma_i, a_2 + \alpha - \beta)$ of type II case IIi, for each case, where $\gamma_i$ is the $i$th component of $\left[3\beta, \frac{2\alpha + \beta}{3}, \frac{4\alpha - \beta}{3}, 6\alpha - 3\beta\right]$.

By symmetric argument, the form of $C(c_1, a_2 - \alpha + \beta)$ and conditions for types III and IV are presented in Table 5. From Table 5, given $A(a_1, a_2)$ and $B(a_1, a_2 - \alpha)$, from $c_1 \geq a_1$ of type III to $a_1 \geq c_1$ of type IV, the form of C from $C(a_1 + \gamma_i, a_2 - \alpha + \beta)$ of type III case IIIi to $C(a_1 - \gamma_i, a_2 - \alpha + \beta)$ of type IV case IVi, for each case, where $\gamma_i$ is the $i$th component of $\left[3\beta, \frac{2\alpha + \beta}{3}, \frac{4\alpha - \beta}{3}, 6\alpha - 3\beta\right]$.

**Table 5.** $A(a_1, a_2)$, $B(a_1, a_2 - \alpha)$ and C satisfying $S_{LS3}(A, B) = S_{LS3}(A, C)$, $a_1, a_2, \alpha, \beta \geq 0$.

| | $C(c_1, a_2 - \alpha + \beta)$ | Conditions |
|---|---|---|
| III1 | $C(a_1 + 3\beta, a_2 - \alpha + \beta)$ | $c_1 \geq a_1$, $\alpha \geq 4\beta$, $a_1 + a_2 - \alpha + 4\beta \leq 1$ |
| III2 | $C(a_1 + \frac{2\alpha + \beta}{3}, a_2 - \alpha + \beta)$ | $c_1 \geq a_1$, $\beta \leq \alpha \leq 4\beta$, $a_1 + a_2 + \frac{-\alpha + 4\beta}{3} \leq 1$ |
| III3 | $C(a_1 + \frac{4\alpha - \beta}{3}, a_2 - \alpha + \beta)$ | $c_1 \geq a_1$, $4/7\beta \leq \alpha \leq \beta$, $a_1 + a_2 + \frac{\alpha + 2\beta}{3} \leq 1$ |
| III4 | $C(a_1 + 6\alpha - 3\beta, a_2 - \alpha + \beta)$ | $c_1 \geq a_1$, $0.5\beta \leq \alpha \leq 4/7\beta$, $a_1 + a_2 + 5\alpha - 2\beta \leq 1$ |
| IV1 | $C(a_1 - 3\beta, a_2 - \alpha + \beta)$ | $a_1 \geq c_1$, $\alpha \geq 4\beta$, $a_1 + a_2 - \alpha \leq 1$ |
| IV2 | $C(a_1 - \frac{2\alpha + \beta}{3}, a_2 - \alpha + \beta)$ | $a_1 \geq c_1$, $\beta \leq \alpha \leq 4\beta$, $a_1 + a_2 - \alpha \leq 1$ |
| IV3 | $C(a_1 - \frac{4\alpha - \beta}{3}, a_2 - \alpha + \beta)$ | $a_1 \geq c_1$, $4/7\beta \leq \alpha \leq \beta$, $a_1 + a_2 - \frac{7\alpha - 4\beta}{3} \leq 1$ |
| IV4 | $C(a_1 - 6\alpha + 3\beta, a_2 - \alpha + \beta)$ | $a_1 \geq c_1$, $0.5\beta \leq \alpha \leq 4/7\beta$, $a_1 + a_2 - 7\alpha + 4\beta \leq 1$ |

From Tables 4 and 5, we observe that there are 1-1 correspondences between the explicit forms of $C$ of four types. More precisely, for each case $i$, $C(a_1 + \gamma_i, a_2 + \alpha - \beta)$ of type I case Ii corresponds to $C(a_1 - \gamma_i, a_2 + \alpha - \beta)$ of type II case IIi, which, in turn, to $C(a_1 + \gamma_i, a_2 - \alpha + \beta)$ of type III case IIIi, which in turn to $C(a_1 - \gamma_i, a_2 - \alpha + \beta)$ of type IV case IVi, where $\gamma_i$ is the $i$th component of $\left[ 3\beta, \frac{2\alpha + \beta}{3}, \frac{4\alpha - \beta}{3}, 6\alpha - 3\beta \right]$.

## 5. Conclusions and Future Research

This paper considers twelve distance-based similarity measures. Among which the largest number of components of the distance in the similarity measure is four for $S_{LS3}(A, B)$. Therefore, we anticipate that the intuitive behaviors of $S_{LS3}(A, B)$ is superior to those of other similarity measures. Much literature has been written on the counterintuitive examples for the similarity measures between two IFSs. This paper proposes six general counterintuitive test problems T1–T6. The counterintuitive test problem is T1 for $S_C(A, B)$ and $S_{DC}(A, B)$, T2 for $S_{HK}(A, B)$, $S_M(A, B)$, $S_{LS1}(A, B)$, $S_{HY1}(A, B)$, $S_{HY2}(A, B)$, $S_{HY3}(A, B)$ and $S_{LZD}(A, B)$, T3 and T4 for $S_{HK}(A, B)$, $S_M(A, B)$ and $S_{LS1}(A, B)$, T5 for $S_{LX}(A, B)$, $S_{LS2}(A, B)$, $S_{HY1}(A, B)$, $S_{HY2}(A, B)$ and $S_{HY3}(A, B)$, and T6 for $S_{LS3}(A, B)$. For $S_{LS3}(A, B)$, four types of counterintuitive example $\{A(a_1, a_2), B(a_1, a_2 \pm \alpha), C(c_1, a_2 \pm \alpha \mp \beta)\}$ exist. There are 1-1 correspondences between the explicit forms of $C$ of four types.

Worthy of future research is when the counterintuitive analysis is extended to other types of similarity measures. In particular, the analysis can be extended to the similarity measures based on the transformation techniques or based on the centroid points of transformation techniques. Thus, the counterintuitive analyses of the transformation techniques based or centroid points based similarity measures is a subject of considerable ongoing research.

**Author Contributions:** H.-C.T. analyzed the method and wrote the paper. S.-T.Y. performed the experiments.

**Funding:** This research received no external funding.

**Conflicts of Interest:** The authors declare that they have no competing interests.

## References


1. Zadeh, L.A. Fuzzy sets. *Inf. Control.* **1965**, *8*, 338–356. [CrossRef]
2. Atanassov, K.T. Intuitionistic fuzzy sets. *Fuzzy Sets Syst.* **1986**, *20*, 87–96. [CrossRef]
3. Hwang, C.M.; Yang, M.S.; Hung, W.L. New similarity measures of intuitionistic fuzzy sets based on the Jaccard index with its application to clustering. *Int. J. Intell. Syst.* **2018**, *33*, 1672–1688. [CrossRef]
4. Wang, Z.; Xu, Z.; Liu, S.; Yao, Z. Direct clustering analysis based on intuitionistic fuzzy implication. *Appl. Soft Comput.* **2014**, *23*, 1–8. [CrossRef]
5. Xu, D.; Xu, Z.; Liu, S.; Zhao, H. A spectral clustering algorithm based on intuitionistic fuzzy information. *Knowl. Based Syst.* **2013**, *53*, 20–26. [CrossRef]
6. Garg, H.; Kumar, K. An advanced study on the similarity measures of intuitionistic fuzzy, sets based on the set pair analysis theory and their application in decision making. *Soft Comput.* **2018**, *22*, 4959–4970. [CrossRef]
7. He, Y.; Chen, H.; Zhou, L.; Liu, J.; Tao, Z. Intuitionistic fuzzy geometric interaction averaging operators and their application to multi-criteria decision making. *Inf. Sci.* **2014**, *269*, 142–159. [CrossRef]
8. Liu, F.; Aiwu, G.; Lukovac, V.; Vukic, M. A multicriteria model for the selection of the transport service provider: A single valued neutrosophic DEMATEL multicriteria model. *Dec. Mak. Appl. Manag. Eng.* **2018**, *1*, 121–130. [CrossRef]
9. Pedrycz, W.; Chen, S.M. *Granular Computing and Decision-Making: Interactive and Iterative Approaches*; Springer: Berlin, Germany, 2015.
10. Stanujkić, D.; Karabašević, D. An extension of the WASPAS method for decision-making problems with intuitionistic fuzzy numbers: A case of website evaluation. *Oper. Res. Eng. Sci. Theo. Appl.* **2018**, *1*, 29–39. [CrossRef]

11. Wang, J.Q.; Zhang, H.Y. Multicriteria decision-making approach based on Atanassov's intuitionistic fuzzy sets with incomplete certain information on weights. *IEEE Trans. Fuzzy Syst.* **2013**, *2*, 510–515. [CrossRef]
12. De, S.K.; Biswas, R.; Roy, A.R. An application of intuitionistic fuzzy sets in medical diagnosis. *Fuzzy Sets Syst.* **2001**, *117*, 209–213. [CrossRef]
13. Papacostas, G.A.; Hatzimichaillidis, A.G.; Kaburlasos, V.G. Distance and similarity measures between intuitionistic fuzzy sets: A comparative analysis from a pattern recognition point view. *Pattern Recognit. Lett.* **2013**, *34*, 1609–1622. [CrossRef]
14. Boran, F.E.; Akay, D. A biparametric similarity measure on intuitionistic fuzzy sets with applications to pattern recognition. *Inf. Sci.* **2014**, *255*, 45–57. [CrossRef]
15. Chen, S.M.; Chang, C.H. A novel similarity measure between Atanassov's intuitionistic fuzzy sets based on transformation techniques with applications to pattern recognition. *Inf. Sci.* **2015**, *291*, 96–114. [CrossRef]
16. Chen, S.M.; Randyanto, Y. A novel similarity measure between intuitionistic fuzzy sets and its applications. *Int. J. Pattern Recognit. Artif. Intell.* **2013**, *27*, 1350021. [CrossRef]
17. Mitchell, H.B. On the Dengfeng–Chuntian similarity measure and its application to pattern recognition. *Pattern Recognit. Lett.* **2003**, *24*, 3101–3104. [CrossRef]
18. Qian, J.; Xin, J.; Lee, S.J.; Yao, S. A new similarity/distance measure between intuitionistic fuzzy sets based on the transformed isosceles triangles and its applications to pattern recognition. *Expert Syst. Appl.* **2019**, *116*, 439–453.
19. Ye, J. Cosine similarity measures for intuitionistic fuzzy sets and their applications. *Math. Comput. Model.* **2011**, *53*, 91–97. [CrossRef]
20. Beliakov, G.; Pagola, M.; Wilkin, T. Vector valued similarity measures for Atanassov's intuitionistic fuzzy sets. *Inf. Sci.* **2014**, *280*, 352–367. [CrossRef]
21. Chen, S.M. Similarity measures between vague sets and between elements. *IEEE Trans. Syst. Man Cybern.* **1997**, *27*, 153–158. [CrossRef]
22. Chen, S.M.; Cheng, S.H.; Lan, T.C. A novel similarity measure between intuitionistic fuzzy sets based on the centroid points of transformed fuzzy numbers with applications to pattern recognition. *Inf. Sci.* **2016**, *343*, 15–40. [CrossRef]
23. Dengfeng, L.; Chuntian, C. New similarity measures of intuitionistic fuzzy sets and application to pattern recognitions. *Pattern Recognit. Lett.* **2002**, *23*, 221–226. [CrossRef]
24. Li, F.; Xu, Z. Similarity measures between vague sets. *J. Softw.* **2001**, *12*, 922–927.
25. Grzegorzewski, P. Distances between intuitionistic fuzzy sets and/or interval-valued fuzzy sets based on the Hausdorff metric. *Fuzzy Sets Syst.* **2004**, *148*, 319–328. [CrossRef]
26. Hong, D.H.; Kim, C. A note on similarity measures between vague sets and between elements. *Inf. Sci.* **1999**, *115*, 83–96. [CrossRef]
27. Hung, W.L.; Yang, M.S. Similarity measures of intuitionistic fuzzy sets based on Hausdorff distance. *Pattern Recognit. Lett.* **2004**, *26*, 1603–1611. [CrossRef]
28. Li, Y.; Olson, D.L.; Qin, Z. Similarity measures between intuitionistic fuzzy (vague) sets: A comparative analysis. *Pattern Recognit. Lett.* **2007**, *28*, 278–285. [CrossRef]
29. Li, Y.; Zhongxian, C.; Degin, Y. Similarity measures between vague sets and vague entropy. *J. Comput. Sci.* **2002**, *29*, 129–132. (In Chinese)
30. Liang, Z.; Shi, P. Similarity measures on intuitionistic fuzzy sets. *Pattern Recognit. Lett.* **2003**, *24*, 2687–2693. [CrossRef]
31. Atanassov, K. *On Intuitionistic Fuzzy Sets Theory*; Springer: Berlin, Germany, 2012.
32. Atanassov, K. *Intuitionistic Fuzzy Sets*; Springer: Heidelberg, Germany, 1999.
33. Szmidt, E.; Kacprzyk, J. Distances between intuitionistic fuzzy sets. *Fuzzy Sets Syst.* **2000**, *114*, 505–518. [CrossRef]

*Article*

# New Concepts of Picture Fuzzy Graphs with Application

**Cen Zuo [1], Anita Pal [2] and Arindam Dey [3],***

[1] Artificial Intelligence and Big Data College, Chongqing College of Electronic Engineering, Chongqing 401331, China; zuocencq@126.com
[2] Department of Mathematics, National Institute of Technology, Durgapur 713209, India; anita.buie@gmail.com
[3] Department of Computer Science and Engineering, Saroj Mohan Institute of Technology, Hooghly 712512, India
* Correspondence: arindam84nit@gmail.com; Tel.: +91-9434586091

Received: 10 May 2019; Accepted: 20 May 2019; Published: 24 May 2019

**Abstract:** The picture fuzzy set is an efficient mathematical model to deal with uncertain real life problems, in which a intuitionistic fuzzy set may fail to reveal satisfactory results. Picture fuzzy set is an extension of the classical fuzzy set and intuitionistic fuzzy set. It can work very efficiently in uncertain scenarios which involve more answers to these type: yes, no, abstain and refusal. In this paper, we introduce the idea of the picture fuzzy graph based on the picture fuzzy relation. Some types of picture fuzzy graph such as a regular picture fuzzy graph, strong picture fuzzy graph, complete picture fuzzy graph, and complement picture fuzzy graph are introduced and some properties are also described. The idea of an isomorphic picture fuzzy graph is also introduced in this paper. We also define six operations such as Cartesian product, composition, join, direct product, lexicographic and strong product on picture fuzzy graph. Finally, we describe the utility of the picture fuzzy graph and its application in a social network.

**Keywords:** fuzzy graph; picture fuzzy graph; social network; direct product; lexicographic product; strong product

**PACS:** J0101

---

## 1. Introduction

In 1965, L. A. Zadeh [1] introduced the idea of fuzzy sets which has been successfully applied to solve the various real life decision problems, which are often uncertain. Fuzzy set is a generalization of crisp set, where the elements of the set have varying degrees of membership. Since the crisp set consists of two truth values ( 0 ('false') and 1 ('true')) which is unable to work with uncertain real life problem. Fuzzy set permits its member to have the value of membership degree between 0 and 1 for better outcome, instead of considering only 0 or 1. The membership degree of a member is not same as probability, rather it indicates the belongingness degree of the element to the fuzzy set, which is a single value within the interval $[0, 1]$. However, those single values of membership degree cannot handle the uncertainties due to lack of knowledge of the problem. Atanassov [2] has proposed the intuitionistic fuzzy set to manage this kind of uncertain situation using an additional membership degree, defined as hesitation margin. Intuitionistic fuzzy set is as an extension of Zadeh's fuzzy set. The fuzzy set can consider only the membership degree of a member in a given set and sum of the membership degree and non membership is always equal to 1, while intuitionistic fuzzy set can consider independent membership degree and non membership degree, the only requirement is that sum of membership degree and non membership degree is not greater than 1. It is more flexible and efficient compared to classical fuzzy set to work with uncertainty due to the presence of hesitation margin. The intuitionistic fuzzy set is used significantly in those real life scenarios, where human perception and knowledge are required, which are generally imprecise and not fully reliable. Over the last few decades, the intuitionistic fuzzy set has been obtaining lots of attention from researchers and scientists. Intuitionistic fuzzy set is used in successfully in different field including image processing [3], decision making [4], social network [5], logic programming [6], market prediction [7], machine learning [8], medical diagnosis recognition [9], robotic systems [10], etc. The concept of neutrality degree cannot be considered in intuitionistic fuzzy set theory. However, neutrality degree needs to consider in many real life scenarios, such as democratic election station. Human beings generally give opinions having more answers of the type: yes, no, abstain and refusal. For example, in a democratic voting system, 1000 people participated in the election. The election commission issues 1000 ballot paper and one person can take only one ballot for giving his/her vote and $\alpha$ is only one candidate. The results of the election are generally divided into four groups came with the number of ballot papers namely "vote for the candidate (500)", "abstain in vote (200)", "vote against candidate (200)" and "refusal of voting (100)". The "abstain in vote" describes that ballot paper is white which contradicts both "vote for the candidate" and "vote against candidate" but it considers the vote. However, "refusal of voting" means bypassing the vote. This type of real life scenarios cannot be handled by intuitionistic fuzzy set. If we use intuitionistic fuzzy sets to describe the above voting system, the information of voting for non-candidates may be ignored. To solve this problem, Cuong and Kreinovich [11] proposed the concept of picture fuzzy set which is a modified version of fuzzy set and Intuitionistic fuzzy set. Picture fuzzy set (PFS) allows the idea degree of positive membership, degree of neutral membership and degree of negative membership of an element. Numerous papers [12–21] have been published on the picture fuzzy set. Cuong [22] described several properties of PFS and also studied a ranking method based on distance measurement between two PFSs. Phong et al. [23] have proposed several compositions of picture fuzzy relations. Then, Cuong and Hai [24] have extended some fuzzy logic operators such as conjunctions, complements, disjunctions and implications for PFSs. They proposed fuzzy inference processes for picture fuzzy systems. Peng and Dai [19] have proposed an algorithmic approach for PFS and used it in a decision making problem. Garg [20] have introduced some aggregation operations of PFS and their applications in multi-criteria decision making.

The basic idea of fuzzy graph was proposed by Rosenfeld [25], 10 years after Zadeh's significant paper on fuzzy sets. Fuzzy graphs have numerous applications in modern engineering, science and medicine especially in the areas of telecommunication, information theory, logistics, neural networks, transportation, expert systems, manufacturing, cluster analysis, social network, medical diagnosis, control theory etc. Fuzzy graph is used as an essential tool to model those problems and it gives more efficient, flexibility and compatibility to model any real world problem when compared to the classic graph. The concept of a complement fuzzy graph was introduced by Mordeson and Nair [26] and further described by Sunitha and Kumar [27]. Shannon and Atanassov [28] have proposed the definition of intuitionistic fuzzy relation and intuitionistic fuzzy graphs and described several proprieties in [28]. Parvathi et al. [29–31] described different operations on intuitionistic fuzzy graphs. Rashmanlou et al. [32] introduced several operations such as direct product, lexicographic product and strong product on intuitionistic fuzzy graphs. They have also described the join, union, composition, and Cartesian production between two intuitionistic fuzzy graphs. For more details on intuitionistic fuzzy graphs, please refer to [33–36].

Fuzzy graph has used to model many decision making problem in uncertain environment. A number of generalizations of fuzzy graphs have been introduced to deal the uncertainty of the complex real life problems. As uncertainties are well expressed using picture fuzzy set, picture fuzzy graph would be prominent research direction for modeling the uncertain optimization problems. This motivated us to introduce the idea picture fuzzy graph its related operations. In this study, we propose the definition of picture fuzzy graph based on picture fuzzy relation. We also introduce several types of picture fuzzy graph such as regular picture fuzzy graph, strong picture fuzzy graph, complete picture fuzzy graph, connected picture fuzzy graph and complement picture fuzzy graph. The idea of isomorphism of picture fuzzy graph is introduced in this paper. We describe the weak and co weak isomorphic picture fuzzy graph. We describe the degree, size and order of a picture fuzzy graph. Busy and free nodes of picture fuzzy graph are also defined in this paper. Six operations on picture fuzzy graphs, viz. Cartesian product, composition, join, direct product, lexicographic and strong product are introduced in this paper. Finally, we describe the utility of picture fuzzy graph and its application in social network.

## 2. Preliminaries

**Definition 1.** *A fuzzy set A, on the universe of discourse of X can also be described a set of ordered pairs as:*

$$A = \{x, \mu_A (x) \mid x \in X\}. \tag{1}$$

*In classical set theory, $\mu_A(x)$ has only the two membership values 0 ('false') and 1 ('true'). Instead of considering only those two truth values, fuzzy set can assign the value of $\mu_A (x)$ within the interval between 0 ('false') and 1 ('true') for better outcome.*

**Definition 2.** *Let A be a intuitionistic fuzzy set. A in X is defined by*

$$A = \{x, \mu_A (x), \nu_A (x) \mid x \in X\}, \tag{2}$$

*where $\mu_A(x) \in [0,1]$ and $\nu_A(x) \in [0,1]$ follows with the condition $0 \le \mu_A(x) + \nu_A(x) \le 1$. The $\mu_A(x)$ is used to represent the degree of membership and $\nu_A(x)$ represents the degree of non-membership of the element x in the set A. For each intuitionistic fuzzy set A in X, the intuitionistic fuzzy index is described as $\pi_A(x) = 1 - (\mu_A(x) + \nu_A(x))$.*

**Definition 3.** *Let A be a picture fuzzy set. A in X is defined by*

$$A = \{x, \mu_A (x), \eta_A (x), \nu_A (x) \mid x \in X\}, \tag{3}$$

*where $\mu_A(x)$, $\eta_A(x)$ and $\nu_A(x)$ follow the condition $0 \le \mu_A(x) + \eta_A(x) + \nu_A(x) \le 1$. The $\mu_A(x), \eta_A(x), \nu_A(x) \in [0,1]$, denote respectively the positive membership degree, neutral membership degree and negative membership degree of the element x in the set A. For each picture fuzzy set A in X, the refusal membership degree is described as $\pi_A(x) = 1 - (\mu_A(x) + \eta_A(x) + \nu_A(x))$.*

**Definition 4.** *A picture fuzzy relation R is a picture fuzzy subset of $X \times Y$ described by*

$$R = \{(x,y), \mu_R (x,y), \eta_R (x,y), \nu_R (x,y) \mid x \in X, y \in Y\}. \tag{4}$$

*Here, $\mu_R : X \times Y \to [0,1]$, $\eta_R : X \times Y \to [0,1]$ and $\nu_R : X \times Y \to [0,1]$ must satisfy the condition $0 \le \mu_R(x,y) + \eta_R(x,y) + \nu_R(x,y) \le 1$ for every $(x,y) \in X \times Y$.*

**Definition 5.** *Let A and B be two picture fuzzy sets. The union, intersection and complement are described as follows.*

1.  $A \cup B = \{x, \max (\mu_A (x), \mu_B (x)), \min (\eta_A (x), \eta_B (x)), \min (\nu_A (x), \nu_B (x)) \mid x \in X, y \in Y\}$
2.  $A \cap B = \{x, \min (\mu_A (x), \mu_B (x)), \min (\eta_A (x), \eta_B (x)), \max (\nu_A (x), \nu_B (x)) \mid x \in X, y \in Y\}$
3.  $A^c = \{\nu_A (x), \eta_A (x), \mu_A (x) \mid x \in X\}$

**Definition 6.** *Let A and B be two picture fuzzy sets on two universes $X_1$ and $X_2$. The Cartesian product of A and B is described as follows.*

$$A \times B = \{(x,y), \mu_A (x) \wedge \mu_B (y), \eta_A (x) \wedge \eta_B (y), \nu_A (x) \vee \nu_B (y) \mid x \in X_1, y \in X_2\}. \tag{5}$$

**Definition 7.** *Let $V$ be a finite nonempty set of vertices. A fuzzy graph $G$ is described by an order pair of functions $\sigma$ and $\mu$, $G = (\sigma, \mu)$. $\sigma$ represents a fuzzy subset of $V$ and $\mu$ represents a symmetric fuzzy relation on $\sigma$, i.e., $\sigma\colon V \longrightarrow [0,1]$ and $\mu\colon V \times V \longrightarrow [0,1]$ such that*

$$\mu(u,v) \leq (\sigma(u) \cap \sigma(v)), \forall u, \forall v \in V. \tag{6}$$

**Definition 8.** *A fuzzy graph $G = (\sigma, \mu)$ is called complete fuzzy graph if*

$$\mu(x,y) = (\sigma(u) \cap \sigma(v)), \forall u, \forall v \in \sigma. \tag{7}$$

**Definition 9.** *A fuzzy graph $G = (\sigma, \mu)$ is called strong fuzzy graph if*

$$\mu(x,y) = (\sigma(u) \cap \sigma(v)), \forall u, v \in \mu. \tag{8}$$

**Definition 10.** *The complement of a fuzzy graph $G = (\sigma, \mu)$ is a fuzzy graph and it is represented as $G^c = (\sigma^c, \mu^c)$, where $\sigma^c = \sigma$ and*

$$\mu^c(u,v) = \sigma(u) \wedge \sigma(v) - \mu(u,v). \tag{9}$$

**Definition 11.** *Let $V$ be a finite nonempty set of vertices. An intuitionistic fuzzy graph is described by an order pair $G = (A, B)$, where $A$ is an intuitionistic fuzzy set on $V$ and $B$ is an intuitionistic fuzzy set on $V^2$. $A$ and $B$ satisfy $\mu_B(x,y) \leq \mu_A(x) \wedge \mu_B(x)$ $v_B(x,y) \leq v_A(x) \wedge v_B(x)$, $\forall(x,y) \in V^2$ and $\mu_B(x,y) = v_B(x,y) = 0$ $\forall(x,y) \in V^2 - E$. Here, $\mu_A(x)$ and $v_A(x)$ are used to denote the degree of membership and non membership of the node $x$ and $\mu_B(x,y)$ and $v_B(x,y)$ are used to denote the degree of membership and non membership of the arcs $(x,y)$.*

**Definition 12.** *Let $G_1^* = (V_1, E_1)$ and $G_2^* = (V_2, E_2)$ be two classical (crisp) graph. The Cartesian product of two graph $G_1^*$ and $G_2^*$ is an another graph $G_3^*$ (written as $G_3^* = G_1^* \times G_2^* = (V_3, E_3)$), where $V_3 = V_1 \times V_2$ and $E_3 = \{(x, x_2), (x, y_2) \,|\, x \in V_1, x_2 y_2 \in E_2\} \cup \{(x_1, z), (y_1, z) \,|\, z \in V_2, x_1 y_1 \in E_1\}$.*

**Definition 13.** *Let $G_1^* = (V_1, E_1)$ and $G_2^* = (V_2, E_2)$ be two classical (crisp) graph. The direct product of two graph $G_1^*$ and $G_2^*$ is an another graph $G_3^*$ (written as $G_3^* = G_1^* \times G_2^* = (V_3, E_3)$), where $V_3 = V_1 \times V_2$ and $E_3 = \{(x_1, x_2), (y_1, y_2) \,|\, x_1 y_1 \in E_1, x_2 y_2 \in E_2\}$.*

**Definition 14.** *Let $G_1^* = (V_1, E_1)$ and $G_2^* = (V_2, E_2)$ be two classical (crisp) graph. The lexicographic product of two graph $G_1^*$ and $G_2^*$ is an another graph $G_3^*$ (written as $G_3^* = G_1^* \cdot G_2^* = (V_3, E_3)$), where $V_3 = V_1 \times V_2$ and $E_3 = \{(x, x_2), (x, y_2) \,|\, x \in V_1, x_2 y_2 \in E_2\} \cup \{(x_1, x_2), (y_1, y_2) \,|\, x_1 y_1 \in E_1, x_2 y_2 \in E_2\}$.*

**Definition 15.** *Let $G_1^* = (V_1, E_1)$ and $G_2^* = (V_2, E_2)$ be two classical (crisp) graph. The strong product of two graph $G_1^*$ and $G_2^*$ is an another graph $G_3^*$ (written as $G_3^* = G_1^* \cdot G_2^* = (V_3, E_3)$), where $V_3 = V_1 \times V_2$ and $E_3 = \{(x, x_2), (x, y_2) \,|\, x \in V_1, x_2 y_2 \in E_2\} \cup \{(x_1, z), (y_1, z) \,|\, z \in V_2, x_1 y_1 \in E_1 \cup (x_1, x_2), (y_1, y_2) \,|\, x_1 y_1 \in E_1, x_2 y_2 \in E_2\}\}$.*

**Definition 16.** *Let $G_1^* = (V_1, E_1)$ and $G_2^* = (V_2, E_2)$ be two classical (crisp) graph. The union of two graphs $G_1^*$ and $G_2^*$ is an another graph $G_3^*$ (written as $G_3^* = G_1^* \cup G_2^* = (V_3, E_3)$), where $V_3 = V_1 \cup V_2$ and $E_3 = E_1 \cup E_2$.*

**Definition 17.** *Let $G_1^* = (V_1, E_1)$ and $G_2^* = (V_2, E_2)$ be two classical (crisp) graph. The joining of two graphs $G_1^*$ and $G_2^*$ is an another graph $G_3^*$ (written as $G_3^* = G_1^* \cup G_2^* = (V_3, E_3)$), where $V_3 = V_1 \cup V_2$ and $E_3 = E_1 \cup E_2 \cup E'$, where $E'$ represents the set of all arcs joining the nodes $V_1$ and $V_2$.*

## 3. Picture Fuzzy Graph

In this section, we define the picture fuzzy graph and introduce different types of regular picture fuzzy graphs, strong picture fuzzy graphs, complete picture fuzzy graphs, and complement picture fuzzy graph.

**Definition 18.** *Let $G^* = (V, E)$ be a graph. A pair $G = (A, B)$ is called a picture fuzzy graph on $G^*$ where $A = (\mu_A, \eta_A, \nu_A)$ is a picture fuzzy set on $V$ and $B = (\mu_B, \eta_B, \nu_B)$ is a picture fuzzy set on $E \subseteq V \times V$ such that for each arc $uv \in E$.*

$$\mu_B(u,v) \leq \min(\mu_A(u), \mu_A(v)), \eta_B(u,v) \leq \min(\eta_A(u), \eta_A(v)), \nu_B(u,v) \geq \max(\nu_A(u), \nu_A(v)) \quad (10)$$

An example of picture fuzzy graph is shown in Figure 1.

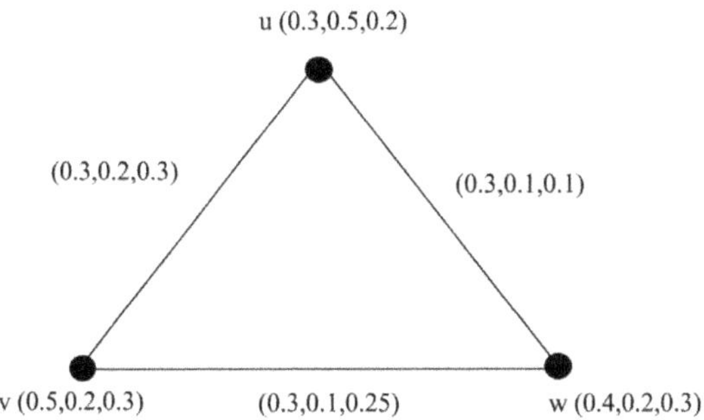


**Figure 1.** A picture fuzzy graph.

**Definition 19.** *A picture fuzzy graph $G = (A, B)$ is said to be regular picture fuzzy graph if*

$$\sum_{\substack{u \\ u \neq v}} \mu_B(u,v) = constant, \sum_{\substack{u \\ u \neq v}} \eta_B(u,v) = constant, \sum_{\substack{u \\ u \neq v}} \nu_B(u,v) = constant, \forall u, v \in E \quad (11)$$

Figure 1 is also an example of regular picture fuzzy graph.

**Definition 20.** *A picture fuzzy graph $G = (A, B)$ is defined as strong picture fuzzy graph if*

$$\mu_B(u,v) = \mu_A(u) \wedge \mu_A(v), \eta_B(u,v) = \eta_A(u) \wedge \eta_A(v), \nu_B(u,v) = \nu_A(u) \vee \nu_A(v) \quad \forall u, v \in E. \quad (12)$$

An example of a strong picture fuzzy graph is shown in Figure 2.

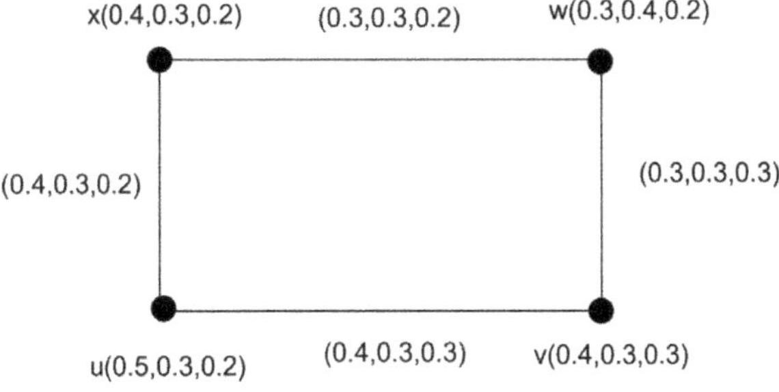


**Figure 2.** A strong picture fuzzy graph.

**Definition 21.** *A picture fuzzy graph $G = (A, B)$ is defined as complete picture fuzzy graph if*

$$\mu_B(u,v) = \mu_A(u) \wedge \mu_A(v), \eta_B(u,v) = \eta_A(u) \wedge \eta_A(v), \nu_B(u,v) = \nu_A(u) \vee \nu_A(v) \quad \forall u,v \in V. \quad (13)$$

An example of a complete picture fuzzy graph is shown in Figure 3. Note that every complete picture fuzzy graph is a strong picture fuzzy graph but not conversely.

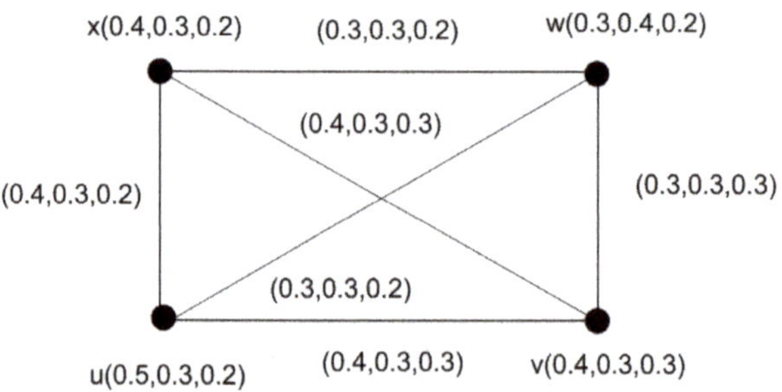


**Figure 3.** A complete picture fuzzy graph.

**Definition 22.** *A path $p$ in a picture fuzzy graph $G = (A, B)$ is a sequence of different vertices $v_0, v_1, v_2, ..., v_k$ such that*

$$(\mu_B(v_{i-1}v_i) \eta_B(v_{i-1}v_i) \nu_B(v_{i-1}v_i)) > 0, i = 1, 2, ..., k. \quad (14)$$

*Here, $k$ represents the length of path.*

**Definition 23.** *Let $G = (A, B)$ be a picture fuzzy graph. If two nodes $u$ and $v$ are connected by a path of length $k$ in $G$ such as $p: u_0, u_1, u_2, ... u_{k-1}, u_k$ then $\mu_B(u,v)$, $\eta_B(u,v)$ and $\nu_B(u,v)$ are described as follows.*

$$\mu_B^k(u,v) = \mu_B(u,u_1) \wedge \mu_B(u_1,u_2) \wedge ... \wedge \mu_B(u_{k-1},v)$$
$$\eta_B^k(u,v) = \eta_B(u,u_1) \wedge \eta_B(u_1,u_2) \wedge ... \wedge \eta_B(u_{k-1},v)$$
$$\nu_B^k(u,v) = \nu_B(u,u_1) \vee \nu_B(u_1,u_2) \vee ... \vee \nu_B(u_{k-1},v).$$

*Let $(\mu_B^\infty(u,v)), \eta_B^\infty(u,v), \nu_B^\infty(u,v)$ be the strength of connectedness between the two nodes $u$ and $v$ of a picture fuzzy graph $G$. Then $\mu_B^\infty(u,v), \eta_B^\infty(u,v),$ and $\nu_B^\infty(u,v)$ are defined as follows.*

$$\mu_B^\infty(u,v) = \sup\{\mu_B^k(u,v)|k = 1,2,...\},$$
$$\eta_B^\infty(u,v) = \sup\{\eta_B^k(u,v)|k = 1,2,...\}$$
$$\nu_B^\infty(u,v) = \inf\{\nu_B^k(u,v)|k = 1,2,...\}),$$

*where* inf *is used to find the minimum and* sup *is used to find maximum membership value. Consider a connected picture fuzzy graph as shown in the Figure 3 and the $x - u$ paths are*

$P_1:$  $x - u$ *with membership value (0.4,0.3,0.2),*
$P_2:$  $x - w - u$ *with membership value (0.3,0.3,0.2),*
$P_3:$  $x - v - u$ *with membership value (0.4,0.3,0.3),*
$P_4:$  $x - w - v - u$ *with membership value (0.3,0.3,0.3),*
$P_5:$  $x - v - w - u$ *with membership value (0.3,0.3,0.3).*

*By routine computation, we have*

$$\mu_B^\infty(x,u) = \sup\{0.4, 0.3, 0.4, 0.3, 0.3\} = 0.4,$$
$$\eta_B^\infty(x,u) = \sup\{0.3, 0.3, 0.3, 0.3, 0.3\} = 0.3,$$
$$\nu_B^\infty(x,u) = \inf\{0.2, 0.2, 0.3, 0.3, 0.3\} = 0.2.$$

*The strength of connectedness between the two nodes $x$ and $u$ of a picture fuzzy graph $G$ (Figure 3) is (0.4,0.3,0.2).*

**Definition 24.** *Let $G = (A, B)$ be a picture fuzzy graph. Then, G is said to be connected picture fuzzy graph if for every vertices $u, v \in V$, $\mu_B^\infty (u, v) > 0$ or $\eta_B^\infty (u, v) > 0$ or $\nu_B^\infty (u, v) < 1$.*

Figure 3 is an example of connected picture fuzzy graph.

**Definition 25.** *The complement of a picture fuzzy graph $G = (A, B)$ is a picture fuzzy graph $G^c = (A^c, B^c)$ if and only if follows the (15).*

$$\mu_A^c = \mu_A, \ \eta_A^c = \eta_A, \ \nu_A^c = \nu_A, \mu_B^c (u, v) = \mu_A (u) \wedge \mu_A (v) - \mu_B (u, v),$$
$$\eta_B^c (u, v) = \eta_A (u) \wedge \eta_A (v) - \eta_B (u, v), \nu_B^c (u, v) = \nu_A (u) \vee \nu_A (v) - \nu_B (u, v) \quad \forall u, v \in V. \tag{15}$$

*An example of complement picture fuzzy graph is shown in Figure 4.*

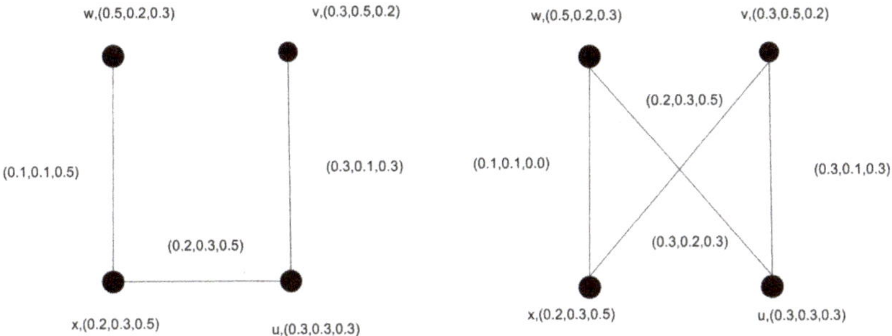


**Figure 4.** A picture fuzzy graph and its complement picture fuzzy graph.

We have

$$\mu_A^{c^c} = \mu_A^c = \mu_A, \ \eta_A^{c^c} = \eta_A^c = \eta_A, \ \nu_A^{c^c} = \nu_A^c = \nu_A \tag{16}$$

$$\mu_1^{c^c} (u, v) = \sigma_1^c (u) \wedge \sigma_1^c (v) - \mu_1^c (u, v) = \sigma_A (u) \wedge \sigma_A (v) - \mu_1^c (u, v). \tag{17}$$

Similarly, we can prove that $\mu_2^{c^c} (u, v) = \mu_2 (u, v)$ and $\mu_3^{c^c} (u, v) = \mu_2 (u, v) \ \forall u, v \in V$. Hence, $G^{c^c} = G$.

**Definition 26.** *Let $G_1 = (A_1, B_1)$ and $G_2 = (A_2, B_2)$ be two picture fuzzy graphs.*

1. *A homomorphism h from picture fuzzy graph $G_1$ to $G_2$ is a mapping function $h : V_1 \to V_2$ which always satisfy followings:*

   *(i)* $\ \forall x_1 \in V_1, x_1 y_1 \in E_1$

   *(a)* $\ \mu_{A_1} (x_1) \leq \mu_{A_2} (h (x_1)), \eta_{A_1} (x_1) \leq \eta_{A_2} (h (x_1)), \nu_{A_1} (x_1) \geq \nu_{A_2} (h (x_1)),$

   *(b)* $\ \mu_{B_1} (x_1 y_1) \leq \mu_{B_2} (h (x_1) h (x_2)), \eta_{B_1} (x_1 y_1) \leq \eta_{B_2} (h (x_1) h (x_2)),$
   $\ \ \ \ \ \nu_{B_1} (x_1 y_1) \geq \nu_{B_2} (h (x_1) h (x_2)).$

   *A isomorphism h from a picture fuzzy graph $G_1$ to $G_2$ is a bijective mapping function $h : V_1 \to V_2$ which always satisfy followings:*

2. *(i)* $\ \forall x_1 \in V_1, x_1 y_1 \in E_1$

   *(c)* $\ \mu_{A_1} (x_1) = \mu_{A_2} (h (x_1)), \eta_{A_1} (x_1) = \eta_{A_2} (h (x_1)), \nu_{A_1} (x_1) = \nu_{A_2} (h (x_1)),$

   *(d)* $\ \mu_{B_1} (x_1 y_1) = \mu_{B_2} (h (x_1) h (x_2)), \eta_{B_1} (x_1 y_1) = \eta_{B_2} (h (x_1) h (x_2)),$
   $\ \ \ \ \ \nu_{B_1} (x_1 y_1) = \nu_{B_2} (h (x_1) h (x_2)).$

*A weak isomorphism h from picture fuzzy graph $G_1$ to $G_2$ is a bijective mapping function $h : V_1 \rightarrow V_2$ which always satisfy followings:*

3.    $(i)$    $\forall x_1 \in V_1$

     $(e)$    *h is homomorphism.*

     $(f)$    $\mu_{A_1}(x_1) = \mu_{A_2}(h(x_1)), \eta_{A_1}(x_1) = \eta_{A_2}(h(x_1)), \nu_{A_1}(x_1) = \nu_{A_2}(h(x_1))$. *Thus, a weak isomorphism maintains the costs of the vertices but not necessarily the costs of the edges.*

4.    *A co weak isomorphism h from picture fuzzy graph $G_1$ to $G_2$ is a bijective mapping function $h : V_1 \rightarrow V_2$ which always satisfy followings:*

     $(i)$    $\forall x_1 y_1 \in E_1$

     $(g)$    *h is homomorphism.*

     $(h)$    $\mu_{A_1}(x_1 y_1) = \mu_{A_2}(h(x_1 y_1)), \eta_{A_1}(x_1 y_1) = \eta_{A_2}(h(x_1 y_1)), \nu_{A_1}(x_1 y_1) = \nu_{A_2}(h(x_1 y_1))$.

**Definition 27.** *A picture fuzzy graph G is said to be*

1.    *Self complementary picture fuzzy graph if and only if $G = G^c$.*
2.    *Self weak complement picture fuzzy graph if G is weak isomorphic to $G^c$.*

## 4. Busy and Free Nodes on Picture Fuzzy Graphs

In this section, we propose open degree, busy values, free node, busy node in a picture fuzzy graph.

**Definition 28.** *Let $G = (A, B)$ be picture fuzzy graph on $G^*$. The open degree of a vertex u is defined as $deg(u) = (d_\mu(u), d_\eta(u), d_\nu(u))$*

$$d_\mu(u) = \sum_{\substack{v \in V \\ u \neq v}} \mu_B(u, v), d_\eta(u) = \sum_{\substack{v \in V \\ u \neq v}} \eta_B(u, v), d_\nu(u) = \sum_{\substack{v \in V \\ u \neq v}} \nu_B(u, v), \quad \forall u, v \in E. \tag{18}$$

*The order of G is defined and denoted as*

$$O(G) = \sum_{v \in V} \mu_B(u), \sum_{v \in V} \eta_B(u), \sum_{v \in V} \nu_B(u). \tag{19}$$

*The size of picture fuzzy graph is defined as $S(G) = (S_\mu(G), S_\eta(G), S_\nu(G))$ where*

$$S_\mu(G) = \sum_{\substack{v \in V \\ u \neq v}} \mu_B(u, v), S_\eta(G) = \sum_{\substack{v \in V \\ u \neq v}} \eta_B(u, v), S_\nu(G) = \sum_{\substack{v \in V \\ u \neq v}} \nu_B(u, v). \tag{20}$$

**Definition 29.** *Let $G = (A, B)$ be a picture fuzzy graph. Then the busy value of a node v is defined by $D(v) = (D_\mu(v), D_\eta(v), D_\nu(v))$, where*

$$D_\mu(v) = \sum \mu_A(v) \wedge \mu_A(v_i), D_\eta(v) = \sum \eta_A(v) \wedge \eta_A(v_i), D_\nu(v) = \sum \nu_A(v) \vee \nu_A(v_i).$$

*Here, $v_i$ are adjacent vertices of v. The busy value of picture fuzzy graph G is the sum of all busy values of vertices of G.*

**Definition 30.** *Let $G = (A, B)$ be a picture fuzzy graph. A vertex v of G is said to be busy vertex if*

$$\mu_A(v) \leq d_\mu(u), \eta_A(u) \leq d_\eta(u), \nu_A(u) \geq d_\nu(u). \tag{21}$$

*Otherwise, it is said to be a free vertex.*

**Definition 31.** *Let $G = (A, B)$ be a picture fuzzy graph. Then an edge $u, v$ is defined as an effective edge if and only if*

$$\mu_B(u, v) = \mu_A(u) \wedge \mu_A(v), \; \eta_B(u, v) = \eta_A(u) \wedge \eta_A(v), \; \nu_B(u, v) = \nu_A(u) \vee \nu_A(v).$$

**Definition 32.** *A connected picture fuzzy graph $G$ is said to be a highly irregular picture fuzzy graph if and only if every node of $G$ is adjacent to nodes with different degrees.*

## 5. Operations on Picture Fuzzy Graph

In this section, we introduce six operations on picture fuzzy graph, viz., Cartesian product, composition, join, direct product, lexicographic and strong product.

**Definition 33.** *Let $G_1 = (A_1, B_1)$ and $G_2 = (A_2, B_2)$ are two picture fuzzy graph of $G_1^* = (V_1, E_1)$ and $G_2^* = (V_2, E_2)$ respectively. The Cartesian product $G_1 \times G_2$ of picture fuzzy graph $G_1$ and $G_2$ is defined by $(A, B)$, where $A = (\mu_A, \eta_A, \nu_A)$ and $B = (\mu_B, \eta_B, \nu_B)$ are two picture fuzzy sets on $V = (V_1 \times V_2)$, and $E = \{(x, x_2), (x, y_2) \, | \, x \in V_1, x_2 y_2 \in E_2\} \cup \{(x_1, z), (y_1, z) \, | \, z \in V_2, x_1 y_1 \in E_1\}$ respectively which satisfies the followings:*

(i) $\forall (x_1, x_2) \in V_1 \times V_2$,

  (a) $\mu_A((x_1, x_2)) = \mu_{A_1}(x_1) \wedge \mu_{A_2}(x_2)$,

  (b) $\eta_A((x_1, x_2)) = \eta_{A_1}(x_1) \wedge \eta_{A_2}(x_2)$

  (c) $\nu_A((x_1, x_2)) = \nu_{A_1}(x_1) \vee \nu_{A_2}(x_2)$,

(ii) $\forall x \in V_1$ and $\forall (x_2, y_2) \in E_2$,

  (a) $\mu_B((x, x_2)(x, y_2)) = \mu_{A_1}(x_1) \wedge \mu_{B_2}(x_2 y_2)$,

  (b) $\eta_B((x, x_2)(x, y_2)) = \eta_{A_1}(x_1) \wedge \eta_{B_2}(x_2 y_2)$,

  (c) $\nu_B((x, x_2)(x, y_2)) = \nu_{A_1}(x_1) \vee \nu_{B_2}(x_2 y_2)$.

(iii) $\forall z \in V_2$ and $\forall (x_1, x_2) \in E_1$,

  (a) $\mu_B((x_1, z)(y_1, z)) = \mu_{B_1}(x_1 y_1) \wedge \mu_{A_2}(z)$,

  (b) $\eta_B((x_1, z)(y_1, z)) = \eta_{B_1}(x_1 y_1) \wedge \eta_{A_2}(z)$,

  (c) $\nu_B((x_1, z)(y_1, z)) = \nu_{B_1}(x_1 y_1) \vee \nu_{A_2}(z)$.

**Definition 34.** *The composition $G_1 G_2$ of two picture fuzzy graph $G_1 = (A_1, B_1)$ and $G_2 = (A_2, B_2)$ defined as a pair $(A, B)$, where $A = (\mu_A, \eta_A, \nu_A)$ and $B = (\mu_B, \eta_B, \nu_B)$ are two picture fuzzy sets on $V = (V_1 \times V_2)$, and $E = \{(x, x_2), (x, y_2) \, | \, x \in V_1, x_2 y_2 \in E_2\} \cup \{(x_1, z), (y_1, z) \, | \, z \in V_2, x_1 y_1 \in E_1\} \cup \{(x_1, x_2), (y_1, y_2) \, | \, x_2 y_2 \in V_2 \neq y_2, x_1 y_1 \in E_1\}$ respectively, which satisfies the followings:*

(i) $\forall (x_1, x_2) \in V_1 \times V_2$,

  (a) $\mu_A((x_1, x_2)) = \mu_{A_1}(x_1) \wedge \mu_{A_2}(x_2)$,

  (b) $\eta_A((x_1, x_2)) = \eta_{A_1}(x_1) \wedge \eta_{A_2}(x_2)$,

  (c) $\nu_A((x_1, x_2)) = \nu_{A_1}(x_1) \vee \nu_{A_2}(x_2)$.

(ii) $\forall x \in V_1$ and $\forall (x_2, y_2) \in E_2$,

  (a) $\mu_B((x, x_2)(x, y_2)) = \mu_{A_1}(x_1) \wedge \mu_{B_2}(x_2 y_2)$,

    (b)   $\eta_B((x, x_2)(x, y_2)) = \eta_{A_1}(x_1) \wedge \eta_{B_2}(x_2 y_2),$

    (c)   $\nu_B((x, x_2)(x, y_2)) = \nu_{A_1}(x_1) \vee \nu_{B_2}(x_2 y_2).$

(iii)  $\forall z \in V_2$ *and* $\forall (x_1, x_2) \in E_1,$

    (a)   $\mu_B((x_1, z)(y_1, z)) = \mu_{B_1}(x_1 y_1) \wedge \mu_{A_2}(z),$

    (b)   $\eta_B((x_1, z)(y_1, z) = \eta_{B_1}(x_1 y_1) \wedge \eta_{A_2}(z),$

    (c)   $\nu_B((x_1, z)(y_1, z)) = \nu_{B_1}(x_1 y_1) \vee \nu_{A_2}(z).$

(iv)  $\forall x_2 y_2 \in V_2,\ x_2 \neq y_2$ *and* $\forall (x_1 y_1) \in E_1,$

    (a)   $\mu_B((x_1, x_2)(y_1, y_2)) = \mu_{A_2}(x_2) \wedge \mu_{A_2}(y_2) \wedge \mu_{B_1}(x_1 y_1),$

    (b)   $\eta_B((x_1, x_2)(y_1, y_2)) = \eta_{A_2}(x_2) \wedge \eta_{A_2}(y_2) \wedge \eta_{B_1}(x_1 y_1),$

    (c)   $\nu_B((x_1, x_2)(y_1, y_2)) = \nu_{A_2}(x_2) \wedge \nu_{A_2}(y_2) \wedge \nu_{B_1}(x_1 y_1).$

**Definition 35.** *The union $G_1 \cup G_2$ of two picture fuzzy graph $G_1 = (A_1, B_1)$ and $G_2 = (A_2, B_2)$ is defined as $(A, B)$, where $A = (\mu_A, \eta_A, \nu_A)$ is a picture fuzzy set on $V = V_1 \cup V_2$ and $B = (\mu_B, \eta_B, \nu_B)$ is an another picture fuzzy set on $E = E_1 \cup E_2$, which satisfies the following:*

(i)  (a)  $\mu_A(x) = \mu_{A_1}(x)$ *if* $x \in V_1$ *and* $x \notin V_2,$

    (b)  $\mu_A(x) = \mu_{A_2}(x)$ *if* $x \in V_2$ *and* $x \notin V_1,$

    (c)  $\mu_A(x) = \mu_{A_1}(x) \wedge \mu_{A_2}(x)$ *if* $x \in V_1 \cap V_2,$

(ii)  (a)  $\eta_A(x) = \eta_{A_1}(x)$ *if* $x \in V_1$ *and* $x \notin V_2,$

    (b)  $\eta_A(x) = \eta_{A_2}(x)$ *if* $x \in V_2$ *and* $x \notin V_1,$

    (c)  $\eta_A(x) = \eta_{A_1}(x) \wedge \eta_{A_2}(x)$ *if* $x \in V_1 \cap V_2,$

(iii)  (a)  $\nu_A(x) = \nu_{A_1}(x)$ *if* $x \in V_1$ *and* $x \notin V_2,$

    (b)  $\nu_A(x) = \nu_{A_2}(x)$ *if* $x \in V_2$ *and* $x \notin V_1,$

    (c)  $\nu_A(x) = \nu_{A_1}(x) \wedge \eta_{A_2}(x)$ *if* $x \in V_1 \cap V_2,$

(iv)  (a)  $\mu_B(xy) = \mu_{B_1}(xy)$ *if* $xy \in E_1$ *and* $xy \notin E_2,$

    (b)  $\mu_B(xy) = \mu_{B_2}(xy)$ *if* $xy \in E_2$ *and* $xy \notin E_1,$

    (c)  $\mu_B(xy) = \mu_{B_1}(xy) \wedge \mu_{B_2}(xy)$ *if* $xy \in E_1 \cap E_2,$

(v)  (a)  $\eta_B(xy) = \eta_{B_1}(xy)$ *if* $xy \in E_1$ *and* $xy \notin E_2,$

    (b)  $\eta_B(xy) = \eta_{B_2}(xy)$ *if* $xy \in E_2$ *and* $xy \notin E_1,$

    (c)  $\eta_B(xy) = \eta_{B_1}(xy) \wedge \eta_{B_2}(xy)$ *if* $xy \in E_1 \cap E_2,$

(vi)  (a)  $\nu_B(xy) = \nu_{B_1}(xy)$ *if* $xy \in E_1$ *and* $xy \notin E_2,$

    (b)  $\nu_B(xy) = \nu_{B_2}(xy)$ *if* $xy \in E_2$ *and* $xy \notin E_1,$

    (c)   $v_B(xy) = v_{B_1}(xy) \wedge v_{B_2}(xy)$ if $xy \in E_1 \cap E_2$.

**Definition 36.** *The join $G_1 + G_2$ of two picture fuzzy graph $G_1 = (A_1, B_1)$ and $G_2 = (A_2, B_2)$ is defined as $(A, B)$, where $A = (\mu_A, \eta_A, v_A)$ is a picture fuzzy set on $V = V_1 \cup V_2$ and $B = (\mu_B, \eta_B, v_B)$ is an another picture fuzzy set on $E = E_1 \cup E_2 \cup E'$ (($E'$ represents all edges joining the vertex of $V_1$ and $V_2$ ), which satisfies the following:*

(i)  (a)  $\mu_A(x) = \mu_{A_1}(x)$ if $x \in V_1$ and $x \notin V_2$

     (b)  $\mu_A(x) = \mu_{A_2}(x)$ if $x \in V_2$ and $x \notin V_1$

     (c)  $\mu_A(x) = \mu_{A_1}(x) \wedge \mu_{A_2}(x)$ if $x \in V_1 \cap V_2$

(ii)  (a)  $\eta_A(x) = \eta_{A_1}(x)$ if $x \in V_1$ and $x \notin V_2$

     (b)  $\eta_A(x) = \eta_{A_2}(x)$ if $x \in V_2$ and $x \notin V_1$

     (c)  $\eta_A(x) = \eta_{A_1}(x) \wedge \eta_{A_2}(x)$ if $x \in V_1 \cap V_2$

(iii)  (a)  $v_A(x) = v_{A_1}(x)$ if $x \in V_1$ and $x \notin V_2$

     (b)  $v_A(x) = v_{A_2}(x)$ if $x \in V_2$ and $x \notin V_1$

     (c)  $v_A(x) = v_{A_1}(x) \wedge \eta_{A_2}(x)$ if $x \in V_1 \cap V_2$

(iv)  (a)  $\mu_B(xy) = \mu_{B_1}(xy)$ if $xy \in E_1$ and $xy \notin E_2$

     (b)  $\mu_B(xy) = \mu_{B_2}(xy)$ if $xy \in E_2$ and $xy \notin E_1$

     (c)  $\mu_B(xy) = \mu_{B_1}(xy) \wedge \mu_{B_2}(xy)$ if $xy \in E_1 \cap E_2$

(v)  (a)  $\eta_B(xy) = \eta_{B_1}(xy)$ if $xy \in E_1$ and $xy \notin E_2$

     (b)  $\eta_B(xy) = \eta_{B_2}(xy)$ if $xy \in E_2$ and $xy \notin E_1$

     (c)  $\eta_B(xy) = \eta_{B_1}(xy) \wedge \eta_{B_2}(xy)$ if $xy \in E_1 \cap E_2$

(vi)  (a)  $v_B(xy) = v_{B_1}(xy)$ if $xy \in E_1$ and $xy \notin E_2$

     (b)  $v_B(xy) = v_{B_2}(xy)$ if $xy \in E_2$ and $xy \notin E_1$

     (c)  $v_B(xy) = v_{B_1}(xy) \wedge v_{B_2}(xy)$ if $xy \in E_1 \cap E_2$

(vii)  (a)  $\mu_B(xy) = \mu_{B_1}(x) \vee \mu_{B_2}(y)$ if $xy \in E'$

     (b)  $\eta_B(xy) = \eta_{B_1}(x) \vee \eta_{B_2}(y)$ if $xy \in E'$

     (c)  $v_B(xy) = v_{B_1}(x) \wedge v_{B_2}(y)$ if $xy \in E'$

**Definition 37.** *The direct product $G_1 * G_2$ of two picture fuzzy graph $G_1$ and $G_2$ is defined as a pair $(A, B)$, where $A = (\mu_A, \eta_A, v_A)$ is a picture fuzzy set on $V = V_1 \times V_2$ and $B = (\mu_B, \eta_B, v_B)$ is an another picture fuzzy set on $E = \{(x_1, x_2)(y_1, y_2) | x_1 y_1 \in E_1, x_2 y_2 \in E_2\}$, which satisfies the followings:*

(i)  $\forall (x_1, x_2) \in V_1 \times V_2$

    (a)    $\mu_A(x_1, x_2) = \mu_{A_1}(x_1) \vee \mu_{A_2}(x_2)$
    (b)    $\eta_A(x_1, x_2) = \eta_{A_1}(x_1) \vee \eta_{A_2}(x_2)$
    (c)    $\nu_A(x_1, x_2) = \nu_{A_1}(x_1) \wedge \nu_{A_2}(x_2)$

(ii)  $\forall (x_1 y_1) \in E_1, \forall (x_2 y_2) \in E_2$

    (a)    $\mu_B(x_1, x_2)(y_1, y_2) = \mu_{B_1}(x_1 y_1) \vee \mu_{B_2}(x_2 y_2)$
    (b)    $\eta_B(x_1, x_2)(y_1, y_2) = \eta_{B_1}(x_1 y_1) \vee \eta_{B_2}(x_2 y_2)$
    (c)    $\nu_B(x_1, x_2)(y_1, y_2) = \nu_{B_1}(x_1 y_1) \wedge \mu_{B_2}(x_2 y_2).$

**Definition 38.** *The lexicographic product $G_1 \cdot G_2$ of two picture fuzzy graph $G_1 = (A_1, B_1)$ and $G_2 = (A_2, B_2)$ is defined as a pair $(A, B)$, where $A = (\mu_A, \eta_A, \nu_A)$ is a picture fuzzy set on $V = V_1 \times V_2$ and $B = (\mu_B, \eta_B, \nu_B)$ is an another picture fuzzy set on $E = \{(x, x_2)(x, y_2) \,|\, x \in V_1, x_2 y_2 \in E_2\} \cup \{(x_1, x_2)(y_1, y_2) \,|\, x_1 y_1 \in E_1, x_2 y_2 \in E_2\}$ which satisfies the followings:*

(i)  $\forall (x_1, x_2)$

    (a)    $\mu_A(x_1, x_2) = \mu_{A_1}(x_1) \vee \mu_{A_2}(x_2) = \eta_{A_1}(x_1) \vee \eta_{A_2}(x_2) = \nu_{A_1}(x_1) \wedge \nu_{A_2}(x_2)$

(ii)  $\forall x \in V_1, \forall (x_2 y_2) \in E_2$

    (a)    $\mu_B(x, x_2)(x, y_2) = \mu_{A_1}(x) \vee \mu_{B_2}(x_2 y_2)$
    (b)    $\eta_B(x, x_2)(x, y_2) = \eta_{A_1}(x) \vee \eta_{B_2}(x_2 y_2)$
    (c)    $\nu_B(x, x_2)(x, y_2) = \nu_{A_1}(x) \wedge \nu_{B_2}(x_2 y_2)$

(iii)  $\forall x_1 y_1 \in E_1, \forall (x_2 y_2) \in E_2$

    (a)    $\mu_B(x_1, x_2)(y_1, y_2) = \mu_{B_1}(x_1 y_1) \vee \mu_{B_2}(x_2 y_2)$
    (b)    $\eta_B(x_1, x_2)(y_1, y_2) = \eta_{B_1}(x_1 y_1) \vee \eta_{B_2}(x_2 y_2)$
    (c)    $\nu_B(x_1, x_2)(y_1, y_2) = \nu_{B_1}(x_1 y_1) \wedge \nu_{B_2}(x_2 y_2)$

**Definition 39.** *The strong product $G_1 G_2$ of two picture fuzzy graph $G_1 = (A_1, B_1)$ and $G_2 = (A_2, B_2)$ is defined as a pair $(A, B)$, where $A = (\mu_A, \eta_A, \nu_A)$ is a picture fuzzy set on $V = V_1 \times V_2$ and $B = (\mu_B, \eta_B, \nu_B)$ is an another picture fuzzy set on $E = \{(x, x_2)(x, y_2) \,|\, x \in V_1, x_2 y_2 \in E_2\} \cup \{(x_1, z)(y_1, z) \,|\, z \in V_2, x_1 y_1 \in E_1\} \cup \{(x_1, x_2)(y_1, y_2) \,|\, x_1 y_1 \in E_1, x_2 y_2 \in E_2\}$ which satisfies the followings:*

(i)  $\forall (x_1, x_2) \in V_1 \times V_2$

    (a)    $\mu_A(x_1, x_2) = \mu_{A_1}(x_1) \vee \mu_{A_2}(x_2)$
    (b)    $\eta_A(x_1, x_2) = \eta_{A_1}(x_1) \vee \eta_{A_2}(x_2)$
    (c)    $\nu_A(x_1, x_2) = \nu_{A_1}(x_1) \wedge \nu_{A_2}(x_2)$

(ii)  $\forall x \in V_1, \forall (x_2 y_2) \in E_2$

    (a)    $\mu_B(x, x_2)(x, y_2) = \mu_{A_1}(x) \vee \mu_{B_2}(x_2 y_2)$
    (b)    $\eta_B(x, x_2)(x, y_2) = \eta_{A_1}(x) \vee \eta_{B_2}(x_2 y_2)$
    (c)    $\nu_B(x, x_2)(x, y_2) = \nu_{A_1}(x) \wedge \nu_{B_2}(x_2 y_2)$

(iii)  $\forall x_1 y_1 \in E_1, \forall (x_2 y_2) \in E_2$

    (a)    $\mu_B(x_1, x_2)(y_1, y_2) = \mu_{B_1}(x_1 y_1) \vee \mu_{B_2}(x_2 y_2)$
    (b)    $\eta_B(x_1, x_2)(y_1, y_2) = \eta_{B_1}(x_1 y_1) \vee \eta_{B_2}(x_2 y_2)$
    (c)    $\nu_B(x_1, x_2)(y_1, y_2) = \nu_{B_1}(x_1 y_1) \wedge \nu_{B_2}(x_2 y_2)$

(iv)  $\forall x_1 y_1 \in E_1, z \in V_2$

    (a)    $\mu_B(x_1, z)(y_1, z) = \mu_{B_1}(x_1 y_1) \vee \mu_{A_2}(z)$
    (b)    $\eta_B(x_1, z)(y_1, z) = \eta_{B_1}(x_1 y_1) \vee \eta_{A_2}(z)$
    (c)    $\nu_B(x_1, z)(y_1, z) = \nu_{B_1}(x_1 y_1) \wedge \nu_{A_2}(z)$

## 6. Application of Picture Fuzzy Graph in Social Network

The popularity of social networks like Facebook, Twitter, WhatsApp, ResearchGate, Instagram, and LinkedIn is increasing day by day. They are well known platforms for connecting hung number of people in everywhere in the world. We generally exchange various types of information and issues in social network. It helps us in online marketing (e-commerce and e-business), efficient social and political campaigns, future events, connecting to clients. Social networks are also used as important tools for public awareness by sending information quickly to a wide audience about any natural disaster and terrorist/criminal attack.

A social network is a collection of nodes and links. The nodes are used to represent individual, groups, country, organizations, places, enterprises, etc., and links are used to describe the relationships between nodes. We generally use classical graph to represent the social network where actors are represented by vertices and the relations/flows between vertices are represented as edges. Several manuscripts have been published on the social networks. However, social network cannot be modeled properly by classical graph. Since, in a classical graph, all vertices have equal importance. Due to this reason, all social units (individual or organization) in present social networks are considered of equal importance. But in real life, all social units do not have same the importance. Similarly, all edges (relationships) have equal strength in classical graph. In all existing social networks, it is considered that the strength of relationship between two social units are equal, but in real life it may not be true. Samanta and Pal [37] have introduced an idea to model the social network using a type 1 fuzzy graph. Many researchers [38–40] believed that these uncertainties phenomena can be modeled by a fuzzy graph. But type 1 fuzzy graph are unable to capture more complex relational states between nodes. Since the membership degree of node and edges are determined by human perception. This motivate us to introduce a new social network model using picture fuzzy graph. We define this social network as a picture fuzzy social network.

In a picture fuzzy social network, an account of a person or organization, i.e., a social unit is represented by node and if there is a relationship or flow between two social units then they are connected by one arc. In reality, each node, i.e., a social unit (person or individual) has some good, neutral and bad activities. We represent the good, neutral and bad activities using the good, neutral and bad membership values of the node and similarity the good, neutral and bad membership degree of the edge can be used to describe the strength of relationship between two nodes. For example, three persons have good knowledge of some activities such as educational activity and teaching methodology. However, they have no knowledge in some activities such administration and finance and they also have very bad knowledge in some activities such as health condition and food habit. We can easily represent this three types of node and edge membership degree using picture fuzzy set which consists each element has three membership values. This type of social network is a real life example of a picture fuzzy graph. Centrality is one of the most key ideas in social networking which finds the node effect on the social network. The centrality of a node is more central from other node. Central person is more close than other person and he can send or access more information. Freeman [41] introduced three different type of measures: degree, closeness, and betweenness of any node centrality. The degree of centrality finds the linkage of a social unit between the others social units. It basically determines the involvement of the social unit (person) in the social network. We can find this degree value any picture node as defined (28). The betweenness determines the number of paths of communication between any two social units though a unit and closeness of any node is as the inverse sum of shortest path length [42,43] to all other social nodes from a specified node. Let $I(i,j)$ represents the shortest path length from node $i$ to $j$. The diameter of a social network is defined as the longest distance between two nodes in the network. It is described as follows.

$$diameter = \max_{i,j} I(i,j). \tag{22}$$

In this study, we use picture fuzzy set to represent the arc length of a social network. The problem of finding shortest path between two social units is a fundamental and important criteria to find the closeness, betweenness and diameter of a social network. This picture social network model is more flexible and efficient than classical social network model.

The online social network can be represented by a directed or undirected weighted picture fuzzy graph. Let $\tilde{G}_{un} = (V, \tilde{E}_{un})$ be an undirected picture fuzzy graph. We can define an undirected picture fuzzy social network as a picture fuzzy relational structure $\tilde{G}_{un} = (V, \tilde{E}_{un})$, where $V = \{v_1, v_2, ..., v_n\}$ represents a non-empty set of picture vertices or actors or nodes, and $\tilde{E}_{un} = \begin{pmatrix} \tilde{e}_{11} & \cdots & \tilde{e}_{1n} \\ \vdots & \ddots & \vdots \\ \tilde{e}_{n1} & \cdots & \tilde{e}_{nn} \end{pmatrix}$ represents an undirected picture fuzzy relation on $V$. A small example of undirected picture fuzzy social network is shown in Figure 5. For undirected picture fuzzy social network where arcs are simply an absent or present undirected picture fuzzy relation with no other information attached.

Let $\tilde{G}_{dn} = (V, \tilde{E}_{dn})$ be a directed picture fuzzy graph. We can describe a directed picture fuzzy social network as a picture fuzzy relational structure $\tilde{G}_{dn} = (V, \tilde{E}_{dn})$, where $V = \{v_1, v_2, ..., v_n\}$ denotes a non-empty set of picture fuzzy nodes, and $\tilde{E}_{dn} = \begin{pmatrix} \tilde{e}_{11} & \cdots & \tilde{e}_{1n} \\ \vdots & \ddots & \vdots \\ \tilde{e}_{n1} & \cdots & \tilde{e}_{nn} \end{pmatrix}$ denotes an undirected picture fuzzy relation on $V$.

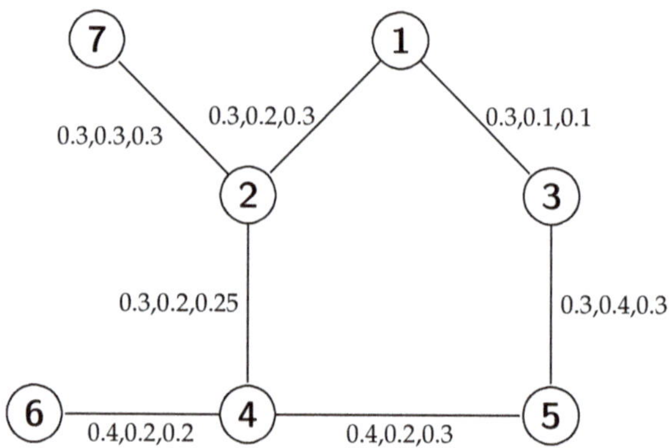


**Figure 5.** Undirected picture fuzzy social network.

The directed picture fuzzy relation is considered in directed picture fuzzy social network. Directed picture fuzzy graph is more efficient to model the social network, because where arcs are considered with directed picture fuzzy relation would contain more information. The values of $\tilde{e}_{ij}$ and $\tilde{e}_{ji}$ are equal in undirected picture fuzzy social network. However, $\tilde{e}_{ij}$ and $\tilde{e}_{ji}$ are not equal in directed picture fuzzy social network. A small example of a directed picture fuzzy social network is shown in Figure 6. Let $\tilde{G}_{dn} = (V, \tilde{E}_{dn})$ be a directed picture fuzzy social network and the picture fuzzy sets are used to describe the arc lengths of $\tilde{G}_{dn}$. The sum of the lengths of the arcs that are adjacent to a social node $v_i$ is called the picture fuzzy in degree centrality of the node $v_i$. The picture in degree centrality of node $v_i$, $\tilde{d}_I(v_i)$, is described as follows.

$$\tilde{d}_I(v_i) = \sum_{j=1, j \neq i}^{n} \tilde{e}_{ji}. \tag{23}$$

The sum of the lengths of the arcs that that are adjacent from social node $v_i$ is called the picture out degree centrality of $v_i$. The picture out degree centrality of node $v_i$, $\tilde{d}_o(v_i)$, is described as follows.

$$\tilde{d}_o(v_i) = \sum_{j=1, j \neq i}^{n} \tilde{e}_{ji}. \tag{24}$$

Here, the symbol $\sum$ is an addition operation of picture fuzzy set and $\tilde{e}_{ji}$ is a picture fuzzy set associated with the arc $(i, j)$. The sum of picture in degree centrality and picture out of degree centrality of node $v_i$ is called picture degree centrality (PDC) of $v_i$.

$$\tilde{d}(v_i) = \tilde{d}_I(v_i) \oplus \tilde{d}_o(v_i), \tag{25}$$

$\oplus$ is an addition operation of picture fuzzy set.

Let $\tilde{G}_{d7} = (V, \tilde{E}_{d7})$ be a directed picture fuzzy social network of the a research Team, where $V = \{v_1, v_2, ..., v_7\}$ represents a collection of seven researchers, $\tilde{E}_{d8}$ represents directed picture fuzzy relation between the seven researchers. This social network is shown in Figure 6. We determine the picture fuzzy in degree, picture fuzzy in degree and picture fuzzy degree centrality of the research term using (23)–(25). This three degree values are shown in Table 1. We use the raking method [44,45] of picture fuzzy set to compare the different degree values. Based on the ranking of picture fuzzy set, research (node) 4 has got highest score value of picture fuzzy in degree centrality. It means that researcher 4 has higher acceptance and good interpersonal relationship in network. The research 2 has got the highest score value of picture fuzzy out degree centrality. It describes that node 2 can nominate many other researchers.

**Table 1.** The picture fuzzy degree centrality of research team.

| Researchers | Picture Fuzzy in Degree Centrality | Picture Fuzzy out Degree Centrality | Picture Fuzzy Degree Centrality |
|---|---|---|---|
| 1 | (0,0,0) | (0.3,0.2,0.3) | (0.3,0.2,0.3) |
| 2 | (0.3,0.2,0.3) | (0.3,0.3,0.3) | (0.3,0.3,0.3) |
| 3 | (0.3,0.1,0.1) | (0.3,0.4,0.3) | (0.3,0.4,0.3) |
| 4 | (0.3,0.2,0.3) | (0,0,0) | (0.3,0.2,0.3) |
| 5 | (0.3,0.4,0.3) | (0.4,0.2,0.3) | (0.3,0.4,0.3) |
| 6 | (0,0,0) | (0.4,0.2,0.2) | (0.4,0.2,0.2) |
| 7 | (0.3,0.3,0.3) | (0,0,0) | (0.3,0.3,0.3) |

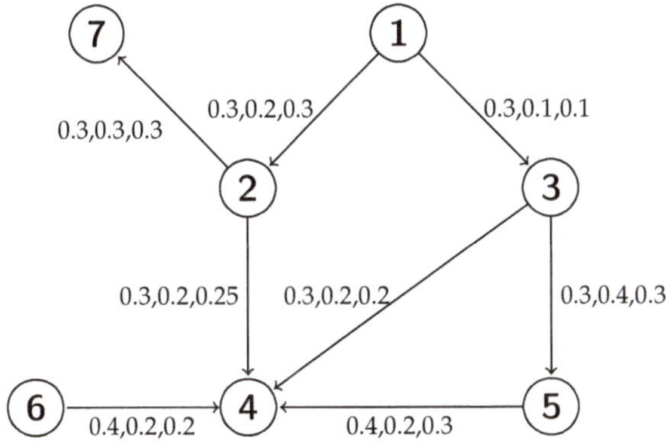


**Figure 6.** Directed picture fuzzy social network.

In this study, we have provided one simple numerical examples of picture fuzzy graph to represent one small social network. The small sized examples are very much helpful to understand the advantage of our proposed model. Social networks are based on millions of users and big data paradigm. Therefore, as future work, we need to model a large scale practical social network using the picture fuzzy graph and to compute the closeness, betweenness and diameter of the social network. Furthermore, we wish to introduce some heuristic algorithms to find those measurements of any large scale practical social network. Despite the need for future study, the proposed model discussed in this study is an important initial contributions to picture fuzzy graph and social network analysis under uncertain environment.

## 7. Conclusions

Fuzzy graph is an important tool for modeling many uncertain real world decision making problems in different field, e.g., operations research, computer science, optimization problems, number theory, algebra, geometry, etc. To deal with the complex real life problems, a number of generalizations of fuzzy graph have been studied. The picture fuzzy set is a direct extension of fuzzy set and intuitonistic fuzzy set. The main objective of this paper is to introduce the idea of picture fuzzy graph and its different operation. In this paper, first we propose the definition of a picture fuzzy graph based on the picture fuzzy relation. We also describe some different types of picture fuzzy graphs such as regular picture fuzzy graph, strong picture fuzzy graph, complete picture fuzzy graph, and complement picture fuzzy graph. We describe some different types of isomorphic picture fuzzy graph. Six different operations on picture fuzzy graph, viz. Cartesian product, composition, join, direct product, lexicographic and strong product are described in this paper. The picture fuzzy graph can increase flexibility, efficiency, precision and comparability to model the complex real life scenarios system compared to the classical fuzzy graph models. We have also described a model to represent the social network using picture fuzzy graph and we also introduce some centrality measure called picture fuzzy in degree centrality, picture fuzzy out degree centrality and picture fuzzy out degree centrality which are applicable to the picture fuzzy social network. The idea of picture fuzzy graph can be applied in several areas of database system, computer network, social network, transportation network and image processing.

**Author Contributions:** C.Z., A.P. and A.D. conceived of the presented idea. C.Z. and A.D. developed the theory and performed the computations. C.Z. and A.D. verified the analytical methods. A.P. contributed reagents/materials/analysis tools; C.Z. wrote the paper.

**Funding:** This research received no external funding.

**Acknowledgments:** The authors sincerely thank all the referee(s) for their valuable suggestions and comments, which were very much useful to improve the paper significantly.

**Conflicts of Interest:** The authors declare no conflict of interest.

## References

1. Zadeh, L.A. Fuzzy sets. *Inf. Control* **1965**, *8*, 338–353. [CrossRef]
2. Atanassov, K.T. Intuitionistic fuzzy sets. In *Intuitionistic Fuzzy Sets*; Springer: Berlin, Germany, 1999; pp. 1–137.
3. Chaira, T.; Ray, A. A new measure using intuitionistic fuzzy set theory and its application to edge detection. *Appl. Soft Comput.* **2008**, *8*, 919–927. [CrossRef]
4. Li, D.F. Multiattribute decision making models and methods using intuitionistic fuzzy sets. *J. Comput. Syst. Sci.* **2005**, *70*, 73–85. [CrossRef]
5. Wu, J.; Chiclana, F. A social network analysis trust–consensus based approach to group decision-making problems with interval-valued fuzzy reciprocal preference relations. *Knowl.-Based Syst.* **2014**, *59*, 97–107. [CrossRef]
6. Shu, M.H.; Cheng, C.H.; Chang, J.R. Using intuitionistic fuzzy sets for fault-tree analysis on printed circuit board assembly. *Microelectron. Reliab.* **2006**, *46*, 2139–2148. [CrossRef]

7. Joshi, B.P.; Kumar, S. Fuzzy time series model based on intuitionistic fuzzy sets for empirical research in stock market. *Int. J. Appl. Evol. Comput. (IJAEC)* **2012**, *3*, 71–84. [CrossRef]

8. Xu, Z.; Hu, H. Projection models for intuitionistic fuzzy multiple attribute decision making. *Int. J. Inf. Technol. Decis. Mak.* **2010**, *9*, 267–280. [CrossRef]

9. Szmidt, E.; Kacprzyk, J. A similarity measure for intuitionistic fuzzy sets and its application in supporting medical diagnostic reasoning. In Proceedings of the International Conference on Artificial Intelligence and Soft Computing, Zakopane, Poland, 7–11 June 2004; pp. 388–393.

10. Devi, K. Extension of VIKOR method in intuitionistic fuzzy environment for robot selection. *Expert Syst. Appl.* **2011**, *38*, 14163–14168. [CrossRef]

11. Cuong, B.C.; Kreinovich, V. Picture Fuzzy Sets-a new concept for computational intelligence problems. In Proceedings of the 2013 Third World Congress on Information and Communication Technologies (WICT·2013), Hanoi, Vietnam 15–18 December 2013; pp. 1–6.

12. Cuong, B.C.; Ngan, R.T.; Hai, B.D. An involutive picture fuzzy negator on picture fuzzy sets and some De Morgan triples. In Proceedings of the 2015 Seventh International Conference on Knowledge and Systems Engineering (KSE), Ho Chi Minh City, Vietnam, 8–10 October 2015; pp. 126–131.

13. Van Viet, P.; Chau, H.T.M.; Van Hai, P. Some extensions of membership graphs for picture inference systems. In Proceedings of the 2015 Seventh International Conference on Knowledge and Systems Engineering (KSE), Ho Chi Minh City, Vietnam, 8–10 October 2015; pp. 192–197.

14. Singh, P. Correlation coefficients for picture fuzzy sets. *J. Intell. Fuzzy Syst.* **2015**, *28*, 591–604.

15. Cuong, B.C.; Kreinovitch, V.; Ngan, R.T. A classification of representable t-norm operators for picture fuzzy sets. In Proceedings of the 2016 Eighth International Conference on Knowledge and Systems Engineering (KSE), Ha Noi, Vietnam, 6–8 October 2016; pp. 19–24.

16. Son, L.H. Generalized picture distance measure and applications to picture fuzzy clustering. *Appl. Soft Comput.* **2016**, *46*, 284–295. [CrossRef]

17. Son, L.H. Measuring analogousness in picture fuzzy sets: From picture distance measures to picture association measures. *Fuzzy Optim. Decis. Mak.* **2017**, *16*, 359–378. [CrossRef]

18. Van Viet, P.; Van Hai, P. Picture inference system: A new fuzzy inference system on picture fuzzy set. *Appl. Intell.* **2017**, *46*, 652–669.

19. Peng, X.; Dai, J. Algorithm for picture fuzzy multiple attribute decision-making based on new distance measure. *Int. J. Uncertain. Quantif.* **2017**, *7*, 177–187. [CrossRef]

20. Wei, G. Some cosine similarity measures for picture fuzzy sets and their applications to strategic decision making. *Informatica* **2017**, *28*, 547–564. [CrossRef]

21. Yang, Y.; Hu, J.; Liu, Y.; Chen, X. Alternative selection of end-of-life vehicle management in China: A group decision-making approach based on picture hesitant fuzzy measurements. *J. Clean. Prod.* **2019**, *206*, 631–645. [CrossRef]

22. Cuong, B.C. Picture fuzzy sets. *J. Comput. Sci. Cybern.* **2014**, *30*, 409.

23. Phong, P.H.; Hieu, D.T.; Ngan, R.; Them, P.T. Some compositions of picture fuzzy relations. In Proceedings of the 7th National Conference on Fundamental and Applied Information Technology Research (FAIR'7), Thai Nguyen, Vietnam, 19–20 June 2014; pp. 19–20.

24. Cuong, B.C.; Van Hai, P. Some fuzzy logic operators for picture fuzzy sets. In Proceedings of the 2015 Seventh International Conference on Knowledge and Systems Engineering (KSE), Ho Chi Minh City, Vietnam, 8–10 October 2015; pp. 132–137.

25. Rosenfeld, A. *Fuzzy Graphs, Fuzzy Sets and their Applications*; Zadeh, L.A.; Fu, K.S., Shimura, M., Eds.; Academic Press; New York, NY, USA, 1975.

26. Mordeson, J.N.; Nair, P.S. *Fuzzy Graphs and Fuzzy Hypergraphs*; Springer: Berlin, Germany, 2012; Voluem 46.

27. Sunitha, M.; Vijayakumar, A. Complement of a fuzzy graph. *Indian J. Pure Appl. Math.* **2002**, *33*, 1451–1464.

28. Shannon, A.; Atanassov, K. On a generalization of intuitionistic fuzzy graphs. *NIFS* **2006**, *12*, 24–29.

29. Parvathi, R.; Karunambigai, M. Intuitionistic fuzzy graphs. In *Computational Intelligence, Theory and Applications*; Springer: Berlin, Germany, 2006; pp. 139–150.

30. Parvathi, R.; Thamizhendhi, G. Domination in intuitionistic fuzzy graphs. *Notes Intuit. Fuzzy Sets* **2010**, *16*, 39–49.

31. Parvathi, R.; Karunambigai, M.; Atanassov, K.T. Operations on intuitionistic fuzzy graphs. In Proceedings of the 2009 IEEE International Conference on Fuzzy Systems, Jeju Island, Korea, 20–24 August 2009; pp. 1396–1401.

32. Rashmanlou, H.; Samanta, S.; Pal, M.; Borzooei, R.A. Intuitionistic fuzzy graphs with categorical properties. *Fuzzy Inf. Eng.* **2015**, *7*, 317–334. [CrossRef]

33. Rashmanlou, H.; Samanta, S.; Pal, M.; Borzooei, R.A. Bipolar fuzzy graphs with categorical properties. *Int. J. Comput. Intell. Syst.* **2015**, *8*, 808–818. [CrossRef]

34. Rashmanlou, H.; Borzooei, R.; Samanta, S.; Pal, M. Properties of interval valued intuitionistic (S, T)–Fuzzy graphs. *Pac. Sci. Rev. A Nat. Sci. Eng.* **2016**, *18*, 30–37. [CrossRef]

35. Sahoo, S.; Pal, M. Different types of products on intuitionistic fuzzy graphs. *Pac. Sci. Rev. A Nat. Sci. Eng.* **2015**, *17*, 87–96. [CrossRef]

36. Akram, M.; Akmal, R. Operations on intuitionistic fuzzy graph structures. *Fuzzy Inf. Eng.* **2016**, *8*, 389–410. [CrossRef]

37. Samanta, S.; Pal, M. A new approach to social networks based on fuzzy graphs. *J. Mass Commun. J.* **2014**, *5*, 078–099.

38. Kundu, S.; Pal, S.K. FGSN: Fuzzy granular social networks–model and applications. *Inf. Sci.* **2015**, *314*, 100–117. [CrossRef]

39. Kundu, S.; Pal, S.K. Fuzzy-rough community in social networks. *Pattern Recognit. Lett.* **2015**, *67*, 145–152. [CrossRef]

40. Fan, T.F.; Liau, C.J.; Lin, T.Y. Positional analysis in fuzzy social networks. In Proceedings of the 2007 IEEE International Conference on Granular Computing (GRC 2007), Fremont, CA, USA, 2–4 November 2007; p. 423.

41. Freeman, L.C. Centrality in social networks conceptual clarification. *Soc. Netw.* **1978**, *1*, 215–239. [CrossRef]

42. Dey, A.; Pal, A. Computing the shortest path with words. *Int. J. Adv. Intell. Paradig.* **2019**, *12*, 355–369. [CrossRef]

43. Dey, A.; Pradhan, R.; Pal, A.; Pal, T. A genetic algorithm for solving fuzzy shortest path problems with interval type-2 fuzzy arc lengths. *Malays. J. Comput. Sci.* **2018**, *31*, 255–270. [CrossRef]

44. Wei, G. Some similarity measures for picture fuzzy sets and their applications. *Iran. J. Fuzzy Syst.* **2018**, *15*, 77–89.

45. Wang, R.; Wang, J.; Gao, H.; Wei, G. Methods for MADM with picture fuzzy muirhead mean operators and their application for evaluating the financial investment risk. *Symmetry* **2019**, *11*, 6. [CrossRef]

 *mathematics*

*Article*

# Distance Measures between the Interval-Valued Complex Fuzzy Sets

**Songsong Dai [1], Lvqing Bi [2,*] and Bo Hu [3]**

[1] School of Electronics and Information Engineering, Taizhou University, Taizhou 318000, China; ssdai@stu.xmu.edu.cn

[2] School of Physics and Telecommunication Engineering, Guangxi Colleges and Universities Key Laboratory of Complex System Optimization and Big Data Processing, Yulin Normal University, Yulin 537000, China

[3] School of Mechanical and Electrical Engineering, Guizhou Normal University , Guiyang 550025, China; hubo@gznu.edu.cn

* Correspondence: bilvqing@stu.xmu.edu.cn

Received: 7 May 2019; Accepted: 15 June 2019; Published: 16 June 2019

**Abstract:** Complex fuzzy set (CFS) is a recent development in the field of fuzzy set (FS) theory. The significance of CFS lies in the fact that CFS assigned membership grades from a unit circle in the complex plane, i.e., in the form of a complex number whose amplitude term belongs to a $[0, 1]$ interval. The interval-valued complex fuzzy set (IVCFS) is one of the extensions of the CFS in which the amplitude term is extended from the real numbers to the interval-valued numbers. The novelty of IVCFS lies in its larger range comparative to CFS. We often use fuzzy distance measures to solve some problems in our daily life. Hence, this paper develops some series of distance measures between IVCFSs by using Hamming and Euclidean metrics. The boundaries of these distance measures for IVCFSs are obtained. Finally, we study two geometric properties include rotational invariance and reflectional invariance of these distance measures.

**Keywords:** Interval-valued complex fuzzy sets; distance measures; rotational invariance; reflectional invariance

---

## 1. Introduction

Complex fuzzy set [1] and some extensions of complex fuzzy set such as complex intuitionistic fuzzy sets [2], complex neutrosophic set [3], interval complex neutrosophic set [4], complex multi-fuzzy soft set [5], complex interval-valued intuitionistic fuzzy sets [6], complex vague soft sets [7] and interval-valued complex fuzzy sets [8,9] have been introduced and used in a variety of fields such as signal processing [10–12], decision-making [6,13–16], time series prediction [17–20] and image restoration [21].

The concept of distance measure can be defined as a numerical description of the difference between two objects. There are many distance measures introduced by numerous scholars to solve some problems in many fields to solve some problems and discover several applications. Among these distances, Euclidean distance and Hamming distance are widely used in several fields. It is well known that distance measure is also an important issue on fuzzy set and its extensions [22–29]. As defined in [1], each complex membership grade includes an amplitude term and a phase term. Hence, many researchers defined distances of CFSs by combining the difference between the phase terms and the difference between the amplitude terms. In this way, Zhang et al. [11] defined a distance measure of CFSs. Hu et al. [30] and Alkouri et al. [31] also introduced some distances for CFSs. These distance measures in [30,31] are defined by using Hamming and

Euclidean metrics. Recently, several concepts including the parallelity [32,33] and entropy measures [34] of complex fuzzy sets and distances of complex intuitionistic fuzzy sets [35], complex vague soft sets [7,36,37] and complex multi-fuzzy soft set [5] have been developed. Since Greenfield et al. [8,9] introduced the concept of an interval-valued complex fuzzy sets (IVCFSs), which is a generalization of the complex fuzzy set introduced by Ramot et al. [1], some progress has been made in various subjects related to the IVCFS. How to measure the difference between IVCFSs is an interesting subject that may be used as a useful tool for decision-making, predictions and pattern recognition in the form of IVCFSs. This paper considers the distance measures between interval-valued complex fuzzy sets.

In [38], Dick proposed the concept of rotational invariance for complex fuzzy operations, which is an intuitive and desirable feature. Then, this feature is examined for interval-valued complex fuzzy operations in [9], complex fuzzy aggregation operators in [15] and entropy measures of CFSs in [34]. Moreover, Bi et al. [15] proposed a novel feature called reflectional invariance for complex fuzzy aggregation operators. These two feature criteria serve as two criteria for choices of complex fuzzy operations and complex fuzzy aggregation operators in some real applications. After giving the distance measures for IVCFSs, we then examine the rotational invariance and reflectional invariance for distance measures of IVCFSs. In addition, the results proposed in this paper might serve as criteria for choices of distance measures of CFSs and IVCFSs.

The rest of the chapter is organized in the following way: In Section 2, we review the basic concepts relates to interval-valued complex fuzzy sets. In Section 3, we develop some distance measures of interval-valued complex fuzzy sets based on the Hamming distance and Euclidean distance. In Section 4, we study rotational invariance and reflectional invariance of these distance measures. In Section 5, we apply these distance measures to solve a decision-making problem with interval-valued complex fuzzy information. Conclusions are given in Section 6.

## 2. Preliminaries

Firstly, let us review the basic concepts relates to interval-valued complex fuzzy sets [8,9].

Let $D$ be the set of values on complex unit disk, i.e., $D = \{a \in \mathbb{C} \mid |a| \leq 1\}$. Suppose $U = \{x_1, x_2, ..., x_n\}$ is a fixed universe. A mapping $A : U \to D$ is called a complex fuzzy set on $U$.

Let $\mathbb{I}^{[0,1]}$ be the set of all closed subintervals of $[0,1]$ and $\dot{D}$ be the boundary set of $D$ i.e., $\dot{D} = \{a \in \mathbb{C} \mid |a| = 1\}$. Suppose $X, Y$ are two classical sets, and their dot product set $X \cdot Y$ is denoted as:

$$X \cdot Y = \{\alpha \cdot \beta \mid \alpha \in X \wedge \beta \in Y\}. \tag{1}$$

A mapping $A : U \to \mathbb{I}^{[0,1]} \cdot \dot{D}$ is called an interval-valued complex fuzzy set [8,9] on $U$. For any $x \in U$, the membership function $A(x)$ is

$$\mu_A(x) = \left[\underline{r_A(x)}, \overline{r_A(x)}\right] \cdot e^{j v_A(x)}, \tag{2}$$

where $j = \sqrt{-1}$, $\left[\underline{r_A(x)}, \overline{r_A(x)}\right] \in \mathbb{I}^{[0,1]}$ is the interval-valued amplitude part and $v_A(x) \in [0, 2\pi)$ is the phase part. We use the notation $IVCF(U)$ to denote the set of all IVCFSs of $U$.

## 3. Distance Measures between IVCFSs

In this section, we introduce several distances in interval-valued complex fuzzy sets case. First, we give the definition of distance between interval-valued complex fuzzy sets as follows:

**Definition 1.** *A distance of interval-valued complex fuzzy sets is a function $d : (IVCF(U) \times IVCF(U)) \to \mathbb{R}^+ \cup \{0\}$, which satisfies the following conditions: for any $A, B, C \in IVCF(U)$:*

*(i)*    $d(A, B) \geq 0$ *and* $d(A, B) = 0$ *if and only if* $A = B$,
*(ii)*   $d(A, B) = d(B, A)$,
*(iii)* $d(A, B) + d(A, C) \geq d(B, C)$.

Now, we define the Hamming, Euclidean, Normalized Hamming, and Normalized Euclidean distances in interval-valued complex fuzzy sets case as follows: for any $A, B \in IVCF(U)$,

- The Hamming distance:

$$d_H(A, B) = \frac{1}{2} \sum_{i=1}^{n} \left( \frac{1}{2} |\underline{r_A(x_i)} - \underline{r_B(x_i)}| + \frac{1}{2} |\overline{r_A(x_i)} - \overline{r_B(x_i)}| + \frac{1}{2\pi} |v_A(x_i) - v_B(x_i)| \right); \tag{3}$$

- The Euclidean distance:

$$d_E(A, B) = \sqrt{ \frac{1}{2} \sum_{i=1}^{n} \left( \frac{1}{2} |\underline{r_A(x_i)} - \underline{r_B(x_i)}|^2 + \frac{1}{2} |\overline{r_A(x_i)} - \overline{r_B(x_i)}|^2 + \frac{1}{4\pi^2} |v_A(x_i) - v_B(x_i)|^2 \right) }; \tag{4}$$

- The normalized Hamming distance:

$$d_{nH}(A, B) = \frac{1}{2n} \sum_{i=1}^{n} \left( \frac{1}{2} |\underline{r_A(x_i)} - \underline{r_B(x_i)}| + \frac{1}{2} |\overline{r_A(x_i)} - \overline{r_B(x_i)}| + \frac{1}{2\pi} |v_A(x_i) - v_B(x_i)| \right); \tag{5}$$

- The normalized Euclidean distance:

$$d_{nE}(A, B) = \sqrt{ \frac{1}{2n} \sum_{i=1}^{n} \left( \frac{1}{2} |\underline{r_A(x_i)} - \underline{r_B(x_i)}|^2 + \frac{1}{2} |\overline{r_A(x_i)} - \overline{r_B(x_i)}|^2 + \frac{1}{4\pi^2} |v_A(x_i) - v_B(x_i)|^2 \right) }. \tag{6}$$

Clearly, it is easy to notice that the above Equations (3)–(6) have the following results:

$$0 \leq d_H(A, B) \leq n, \qquad 0 \leq d_{nH}(A, B) \leq 1,$$
$$0 \leq d_E(A, B) \leq \sqrt{n}, \quad 0 \leq d_{nE}(A, B) \leq 1.$$

**Theorem 1.** *All functions defined in Equations (3)–(6) are distance measures of IVCFSs.*

**Proof.** It is easy to see that all the distance Equations (3)–(6) satisfy conditions (i) and (ii). Thus, we just go to prove the condition number (iii), i.e., the triangular inequality for $d_H(A, B), d_E(A, B), d_{nH}(A, B), d_{nE}(A, B)$.

Let $A, B, C \in IVCF(U)$; for the Hamming distance, we have

$$d_H(A,C) + d_H(C,B)$$
$$= \frac{1}{2}\sum_{i=1}^{n}\left(\frac{1}{2}|\underline{r_A(x_i)} - \underline{r_C(x_i)}| + \frac{1}{2}|\overline{r_A(x_i)} - \overline{r_C(x_i)}| + \frac{1}{2\pi}|v_A(x_i) - v_C(x_i)|\right)$$
$$+ \frac{1}{2}\sum_{i=1}^{n}\left(\frac{1}{2}|\underline{r_C(x_i)} - \underline{r_B(x_i)}| + \frac{1}{2}|\overline{r_C(x_i)} - \overline{r_B(x_i)}| + \frac{1}{2\pi}|v_C(x_i) - v_B(x_i)|\right)$$
$$= \frac{1}{2}\sum_{i=1}^{n}\left(\frac{1}{2}(|\underline{r_A(x_i)} - \underline{r_C(x_i)}| + |\underline{r_C(x_i)} - \underline{r_B(x_i)}|)\right.$$
$$+ \frac{1}{2}(|\overline{r_A(x_i)} - \overline{r_C(x_i)}| + |\overline{r_C(x_i)} - \overline{r_B(x_i)}|)$$
$$\left.+ \frac{1}{2\pi}(|v_A(x_i) - v_C(x_i)| + |v_C(x_i) - v_B(x_i)|)\right)$$
$$\geq \frac{1}{2}\sum_{i=1}^{n}\left(\frac{1}{2}(|\underline{r_A(x_i)} - \underline{r_B(x_i)}|) + \frac{1}{2}(|\overline{r_A(x_i)} - \overline{r_B(x_i)}|) + \frac{1}{2\pi}(|v_A(x_i) - v_B(x_i)|)\right)$$
$$= d_H(A,B).$$

For the Euclidean distance, we have

$$d_E(A,C) + d_E(C,B)$$
$$= \sqrt{\frac{1}{2}\sum_{i=1}^{n}\left(\frac{1}{2}|\underline{r_A(x_i)} - \underline{r_C(x_i)}|^2 + \frac{1}{2}|\overline{r_A(x_i)} - \overline{r_C(x_i)}|^2 + \frac{1}{4\pi^2}|v_A(x_i) - v_C(x_i)|^2\right)}$$
$$+ \sqrt{\frac{1}{2}\sum_{i=1}^{n}\left(\frac{1}{2}|\underline{r_C(x_i)} - \underline{r_B(x_i)}|^2 + \frac{1}{2}|\overline{r_C(x_i)} - \overline{r_B(x_i)}|^2 + \frac{1}{4\pi^2}|v_C(x_i) - v_B(x_i)|^2\right)}$$
$$\geq \left(\frac{1}{2}\sum_{i=1}^{n}\left(\frac{1}{2}|\underline{r_A(x_i)} - \underline{r_C(x_i)} + \underline{r_C(x_i)} - \underline{r_B(x_i)}|^2 + \frac{1}{2}|\overline{r_A(x_i)} - \overline{r_C(x_i)} + \overline{r_C(x_i)} - \overline{r_B(x_i)}|^2\right.\right.$$
$$\left.\left.+ \frac{1}{4\pi^2}|v_A(x_i) - v_C(x_i) + v_C(x_i) - v_B(x_i)|^2\right)\right)^{\frac{1}{2}} \quad \textit{(Using Minkowski inequality)}$$
$$= \sqrt{\frac{1}{2}\sum_{i=1}^{n}\left(\frac{1}{2}|\underline{r_A(x_i)} - \underline{r_B(x_i)}|^2 + \frac{1}{2}|\overline{r_A(x_i)} - \overline{r_B(x_i)}|^2 + \frac{1}{4\pi^2}|v_A(x_i) - v_B(x_i)|^2\right)}$$
$$= d_E(A,B),$$

analogously, for the proof of the normalized Hamming $d_{nH}(A,B)$ and the normalized Euclidean $d_{nE}(A,B)$ distances. $\square$

**Example 1.** *Let $A$ and $B$ be two IVCFSs on $X = \{x_1, x_2, x_3, x_4, x_5\}$, which are given as follows:*

$$A = \frac{[0.4, 0.5]e^{j0.3\pi}}{x_1} + \frac{[0.6, 0.9]e^{j0.5\pi}}{x_2} + \frac{[0.6, 0.9]e^{j0.1\pi}}{x_3} + \frac{[0.7, 0.9]e^{j0.2\pi}}{x_4} + \frac{[0.4, 0.7]e^{j0.5\pi}}{x_5},$$
$$B = \frac{[0.3, 0.4]e^{j1.3\pi}}{x_1} + \frac{[0.2, 0.4]e^{j0.4\pi}}{x_2} + \frac{[0.5, 0.8]e^{j0.5\pi}}{x_3} + \frac{[0.4, 0.6]e^{j0.1\pi}}{x_4} + \frac{[0.8, 0.9]e^{j0.7\pi}}{x_5}.$$

*Using Equations (3)–(6), we have the following results:*

$$
\begin{aligned}
d_H(A,B) &= \frac{1}{2}\left(\frac{1}{2}|0.4-0.3|+\frac{1}{2}|0.5-0.4|+\frac{1}{2\pi}|0.3\pi-1.3\pi|\right)\\
&+\frac{1}{2}\left(\frac{1}{2}|0.6-0.2|+\frac{1}{2}|0.9-0.4|+\frac{1}{2\pi}|0.5\pi-0.4\pi|\right)\\
&+\frac{1}{2}\left(\frac{1}{2}|0.6-0.5|+\frac{1}{2}|0.9-0.8|+\frac{1}{2\pi}|0.1\pi-0.5\pi|\right)\\
&+\frac{1}{2}\left(\frac{1}{2}|0.7-0.4|+\frac{1}{2}|0.9-0.6|+\frac{1}{2\pi}|0.2\pi-0.1\pi|\right)\\
&+\frac{1}{2}\left(\frac{1}{2}|0.4-0.8|+\frac{1}{2}|0.7-0.9|+\frac{1}{2\pi}|0.5\pi-0.7\pi|\right)\\
&= 0.3+0.25+0.15+0.175+0.2\\
&= 1.075,
\end{aligned}
$$

$$
\begin{aligned}
d_E(A,B) &= \left(\frac{1}{2}\left(\frac{1}{2}|0.4-0.3|^2+\frac{1}{2}|0.5-0.4|^2+\frac{1}{4\pi^2}|0.3\pi-1.3\pi|^2\right)\right.\\
&+\frac{1}{2}\left(\frac{1}{2}|0.6-0.2|^2+\frac{1}{2}|0.9-0.4|^2+\frac{1}{4\pi^2}|0.5\pi-0.4\pi|^2\right)\\
&+\frac{1}{2}\left(\frac{1}{2}|0.6-0.5|^2+\frac{1}{2}|0.9-0.8|^2+\frac{1}{4\pi^2}|0.1\pi-0.5\pi|^2\right)\\
&+\frac{1}{2}\left(\frac{1}{2}|0.7-0.4|^2+\frac{1}{2}|0.9-0.6|^2+\frac{1}{4\pi^2}|0.2\pi-0.1\pi|^2\right)\\
&+\left.\frac{1}{2}\left(\frac{1}{2}|0.4-0.8|^2+\frac{1}{2}|0.7-0.9|^2+\frac{1}{4\pi^2}|0.5\pi-0.7\pi|^2\right)\right)^{1/2}\\
&= (0.13+0.10375+0.025+0.0925+0.055)^{1/2}\\
&\approx 0.6374,
\end{aligned}
$$

*and*

$$
d_{nH}(A,B)=0.215,\quad d_{nE}(A,B)\approx 0.285.
$$

However, in practice, the different sets may have taken different weights i.e., $w_i \geq 0, i=1,2,...,n,$ and $\sum_{i=1}^{n} w_i = 1$, for each element $x_i \in U$. Thus, we introduce the normalized weighted Hamming and Euclidean distance measures for IVCFSs.

- The normalized weighted Hamming distance:

$$
d_{nwH}(A,B)=\frac{1}{2}\sum_{i=1}^{n} w_i\left(\frac{1}{2}|\underline{r_A(x_i)}-\underline{r_B(x_i)}|+\frac{1}{2}|\overline{r_A(x_i)}-\overline{r_B(x_i)}|+\frac{1}{2\pi}|\nu_A(x_i)-\nu_B(x_i)|\right); \tag{7}
$$

- The normalized weighted Euclidean distance:

$$
d_{nwE}(A,B)=\sqrt{\frac{1}{2}\sum_{i=1}^{n} w_i\left(\frac{1}{2}|\underline{r_A(x_i)}-\underline{r_B(x_i)}|^2+\frac{1}{2}|\overline{r_A(x_i)}-\overline{r_B(x_i)}|^2+\frac{1}{4\pi^2}|\nu_A(x_i)-\nu_B(x_i)|^2\right)}. \tag{8}
$$

**Theorem 2.** *All functions defined in Equations (7) and (8) are distance measures of IVCFSs.*

**Proof.** The proof is similar to that of Theorem 1. $\square$

Obviously, when $w_i = 1/n$ for all $i = 1, 2, ..., n$, then Equations (7) and (8) reduce to Equations (5) and (6), respectively. In addition, the above distance Equations (7) and (8) have the following results:

$$0 \leq d_{nwH}(A, B) \leq 1,$$
$$0 \leq d_{nwE}(A, B) \leq 1.$$

**Example 2.** *Let $A$ and $B$ be two IVCFSs defined in Example 1 and the weight of $x_1, x_2, ..., x_5$ is $w = (0.1, 0.1, 0.2, 0.2, 0.4)$.*

*Using Equations (7) and (8), we have the following results:*

$$
\begin{aligned}
d_{nwH}(A, B) &= \frac{0.1}{2}\left(\frac{1}{2}|0.4 - 0.3| + \frac{1}{2}|0.5 - 0.4| + \frac{1}{2\pi}|0.3\pi - 1.3\pi|\right) \\
&+ \frac{0.1}{2}\left(\frac{1}{2}|0.6 - 0.2| + \frac{1}{2}|0.9 - 0.4| + \frac{1}{2\pi}|0.5\pi - 0.4\pi|\right) \\
&+ \frac{0.2}{2}\left(\frac{1}{2}|0.6 - 0.5| + \frac{1}{2}|0.9 - 0.8| + \frac{1}{2\pi}|0.1\pi - 0.5\pi|\right) \\
&+ \frac{0.2}{2}\left(\frac{1}{2}|0.7 - 0.4| + \frac{1}{2}|0.9 - 0.6| + \frac{1}{2\pi}|0.2\pi - 0.1\pi|\right) \\
&+ \frac{0.4}{2}\left(\frac{1}{2}|0.4 - 0.8| + \frac{1}{2}|0.7 - 0.9| + \frac{1}{2\pi}|0.5\pi - 0.7\pi|\right) \\
&= 0.03 + 0.025 + 0.03 + 0.035 + 0.08 \\
&= 0.2,
\end{aligned}
$$

$$
\begin{aligned}
d_{nwE}(A, B) &= \left(\frac{0.1}{2}\left(\frac{1}{2}|0.4 - 0.3|^2 + \frac{1}{2}|0.5 - 0.4|^2 + \frac{1}{4\pi^2}|0.3\pi - 1.3\pi|^2\right)\right. \\
&+ \frac{0.1}{2}\left(\frac{1}{2}|0.6 - 0.2|^2 + \frac{1}{2}|0.9 - 0.4|^2 + \frac{1}{4\pi^2}|0.5\pi - 0.4\pi|^2\right) \\
&+ \frac{0.2}{2}\left(\frac{1}{2}|0.6 - 0.5|^2 + \frac{1}{2}|0.9 - 0.8|^2 + \frac{1}{4\pi^2}|0.1\pi - 0.5\pi|^2\right) \\
&+ \frac{0.2}{2}\left(\frac{1}{2}|0.7 - 0.4|^2 + \frac{1}{2}|0.9 - 0.6|^2 + \frac{1}{4\pi^2}|0.2\pi - 0.1\pi|^2\right) \\
&+ \left.\frac{0.4}{2}\left(\frac{1}{2}|0.4 - 0.8|^2 + \frac{1}{2}|0.7 - 0.9|^2 + \frac{1}{4\pi^2}|0.5\pi - 0.7\pi|^2\right)\right)^{1/2} \\
&= (0.013 + 0.010375 + 0.005 + 0.0185 + 0.022)^{1/2} \\
&\approx 0.2624.
\end{aligned}
$$

By the relations among IVCFS, CFS, interval-valued fuzzy sets (IVFS) and FS, we established the comparison of the proposed distance measures of IVCFSs with CFS, IVFS and FS. It is proposed that these distance measures reduce the environments of CFS, IVFS and FS. The comparison is demonstrated in Remarks 1–3.

**Remark 1.** *The distance measures proposed in Equations (3)–(6) for IVCFSs reduce to the environment of CFSs as defined in [31].*

**Remark 2.** *The distance measures proposed in Equations (3)–(6) for IVCFSs reduce to the environment of IVFSs if we considered the phase term as zero as defined in [24].*

**Remark 3.** *The distance measures proposed in Equations (7) and (8) for IVCFSs reduce to the environment of CFSs as defined in Equations (9) and (10):*

$$D_{nwH}(A, B) = \frac{1}{2} \sum_{i=1}^{n} w_i \left( |r_A(x_i) - r_B(x_i)| + \frac{1}{2\pi} |v_A(x_i) - v_B(x_i)| \right), \tag{9}$$

$$D_{nwE}(A, B) = \sqrt{\frac{1}{2} \sum_{i=1}^{n} w_i \left( |r_A(x_i) - r_B(x_i)|^2 + \frac{1}{4\pi^2} |v_A(x_i) - v_B(x_i)|^2 \right)}. \tag{10}$$

Here, we demonstrate the advantages of the proposed distance measures of IVCFSs. Firstly, the distance measures of CFSs, IVFSs, and FS could not handle the information provided in the form of IVCFSs. On the other hand, from Remarks 1 and 2, the proposed distance measures in Equations (3)–(6) can solve some problem that lies in the region of CFSs, IVFSs, and FS. Moreover, the distance measures proposed in Remark 3 for CFSs are two new distance Equations for CFSs.

## 4. Rotational Invariance and Reflectional Invariance

Let $A$ be a IVCFS on $U$ with membership function $\left[ r_A(x), \overline{r_A(x)} \right] \cdot e^{j v_A(x)}$, and then we define the following two set operations for IVCFSs:

The rotation of an IVCFS $A$ by $\theta$ radians, denoted $Rot_\theta(A)$ is defined as

$$Rot_\theta(\mu_A(x)) = \left[ r_A(x), \overline{r_A(x)} \right] \cdot e^{j\left( v_A(x) + \theta \right)}. \tag{11}$$

The reflection of an IVCFS $A$ , denoted $Ref(A)$ is defined as

$$Ref(\mu_A(x)) = \left[ r_A(x), \overline{r_A(x)} \right] \cdot e^{j(2\pi - v_A(x))}. \tag{12}$$

Moreover, we define the following:

**Definition 2.** *A distance d on IVCFS is rotationally invariant if*

$$d(Rot_\theta(A), Rot_\theta(B)) = d(A, B) \tag{13}$$

*for any $\theta$ and $A, B \in IVCF(U)$.*

**Definition 3.** *A distance d on IVCFS is reflectionally invariant if*

$$d(Ref(A), Ref(B)) = d(A, B) \tag{14}$$

*for any $A, B \in IVCF(U)$.*

Definitions 2 and 3 give two intuitive properties of distance measures of IVCFSs, named rotational invariance and reflectional invariance, respectively. They mean that the distance measure is invariant under a rotation or a reflection. More specifically, if two IVCFSs are rotated by a fixed value, rotational invariance states that the distance between two new IVCFSs will not change. If two IVCFSs are reflected, reflectional invariance states that the distance between two new IVCFSs will not change as well, which are respectively shown in Figure 1, $d(A', B')$ is equal to $d(A, B)$ after reflecting or rotating.

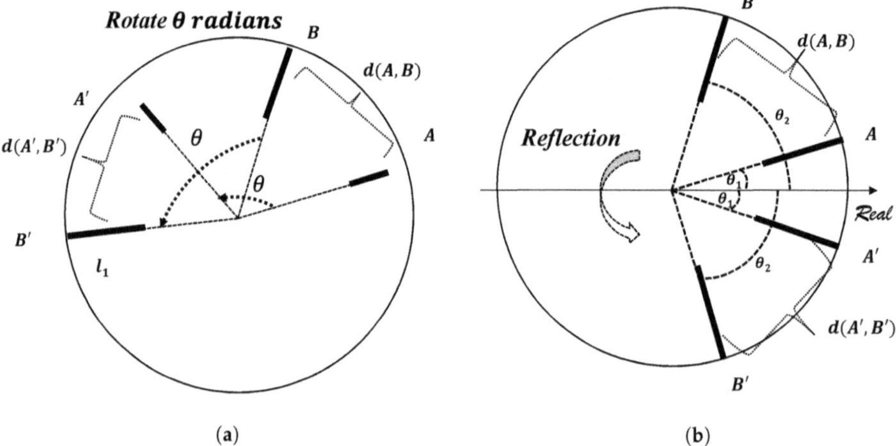


**Figure 1.** (a) reflectional invariance and (b) rotational invariance.

**Theorem 3.** *All functions defined in Equations (3)–(8) are reflectionally invariant.*

**Proof.** It is easy from that $|(2\pi - v_A(x_i)) - (2\pi - v_B(x_i))| = |v_A(x_i) - v_B(x_i)|$ for any $x_i \in U$. $\square$

**Theorem 4.** *All functions defined in Equations (3)–(8) are not rotationally invariant.*

**Proof.** Since the domain of the phase term is $v_A(x) \in [0, 2\pi)$, so the above rotation operation is addition modulo $2\pi$. Consider two IVCFSs on the set $U = \{x_1, x_2\}$; let $\mu_A(x) \equiv 1 \cdot e^{j\frac{\pi}{4}}$ and $\mu_B(x) \equiv 1 \cdot e^{j\frac{7\pi}{4}}$ for any $x \in U$. Then, $Rot_{\frac{\pi}{2}}(A) \equiv 1 \cdot e^{j\frac{3\pi}{4}}$ and $Rot_{\frac{\pi}{2}}(B) \equiv 1 \cdot e^{j\frac{\pi}{4}}$. By using Equations (3)–(8), we have

$$d_H(A, B) = \frac{3}{4}, \qquad d_H(Rot_{\frac{\pi}{2}}(A), Rot_{\frac{\pi}{2}}(B)) = \frac{1}{4},$$

$$d_E(A, B) = \frac{3}{4}, \qquad d_H(Rot_{\frac{\pi}{2}}(A), Rot_{\frac{\pi}{2}}(B)) = \frac{1}{4},$$

$$d_{nH}(A, B) = \frac{3}{8}, \qquad d_{nH}(Rot_{\frac{\pi}{2}}(A), Rot_{\frac{\pi}{2}}(B)) = \frac{1}{8},$$

$$d_{nE}(A, B) = \frac{3\sqrt{2}}{8}, \qquad d_H(Rot_{\frac{\pi}{2}}(A), Rot_{\frac{\pi}{2}}(B)) = \frac{\sqrt{2}}{8},$$

$$d_{nwH}(A, B) = \frac{3}{8}, \qquad d_{nwH}(Rot_{\frac{\pi}{2}}(A), Rot_{\frac{\pi}{2}}(B)) = \frac{1}{8},$$

$$d_{nwE}(A, B) = \frac{3\sqrt{2}}{8}, \qquad d_{nwH}(Rot_{\frac{\pi}{2}}(A), Rot_{\frac{\pi}{2}}(B)) = \frac{\sqrt{2}}{8}.$$

Thus, the distance functions defined in Equations (3)–(8) are not rotationally invariant. $\square$

Obviously, the concepts of rotational invariance and reflectional invariance can be examined for distance measures of CFSs when IVCFSs are reduced to CFSs. Then, we can study the rotational invariance and reflectional invariance of distance measures of CFSs.

**Remark 4.** *The concepts of rotational invariance and reflectional invariance proposed respectively in Definitions 2 and 3 for distance measures of IVCFSs reduce to the environment of CFSs.*

With the combination of Remark 1, we have the following results for the distance measures of CFSs.

**Corollary 1.** *Distance measures of CFSs defined in [31] are reflectionally invariant.*

**Corollary 2.** *Distance measures of CFSs defined in [31] are not rotationally invariant.*

With the combination of Remark 3, we have the following results for the distance measures of CFSs.

**Corollary 3.** *Distance functions of CFSs defined in Equations (9) and (10) are reflectionally invariant.*

**Corollary 4.** *Distance functions of CFSs defined in Equations (9) and (10) are not rotationally invariant.*

## 5. Numerical Example for Decision-Making

In this section, we consider a decision-making problem under an interval-valued complex fuzzy environment.

Assume that a customer decides to purchase a new computer for his own private use. There are four alternatives $(A_1, A_2, A_3, A_4)$ with different production dates. The customer considers five attributes, namely $C_1$: Performance, $C_2$: Appearance, $C_3$: Service, $C_4$: Experience and $C_5$: Price with the weight vectors $w = (0.2, 0.2, 0.1, 0.1, 0.4)$ for selecting computer. The corresponding rating values of attributes of alternatives are interval-valued complex fuzzy values as shown in Table 1. For instance, the rating value of the alternative $A_1$ under $C_1$ attribute is the interval-valued complex fuzzy value $[0.8, 0.9] \cdot e^{j2\pi(0.8)}$, which means that the performance of the computer $A_1$ at $C_1$ is marked by an interval value $[0.8, 0.9]$, and the customer satisfied 80% with the suitability of production date of the computer $A_1$ at $C_1$; this explanation of the phase part is referred to Garg et al.'s works on complex intuitionistic fuzzy aggregation operators [13,14]. The customer wants to select a computer among the four alternatives above. Here, we use the principle of minimum degree of difference between IVCFSs to solve the given problem.

**Table 1.** Feature values of the facial expressions.

| | $C_1$ | $C_2$ | $C_3$ | $C_4$ | $C_5$ |
|---|---|---|---|---|---|
| $A_1$ | $[0.8, 0.9] \cdot e^{j2\pi(0.8)}$ | $[0.7, 0.8] \cdot e^{j2\pi(0.8)}$ | $[0.8, 0.9] \cdot e^{j2\pi(0.9)}$ | $[0.7, 0.9] \cdot e^{j2\pi(0.9)}$ | $[0.8, 0.9] \cdot e^{j2\pi(0.6)}$ |
| $A_2$ | $[0.8, 0.9] \cdot e^{j2\pi(0.7)}$ | $[0.7, 0.8] \cdot e^{j2\pi(0.9)}$ | $[0.7, 0.8] \cdot e^{j2\pi(0.7)}$ | $[0.7, 0.8] \cdot e^{j2\pi(0.7)}$ | $[0.6, 0.7] \cdot e^{j2\pi(0.8)}$ |
| $A_3$ | $[0.6, 0.7] \cdot e^{j2\pi(0.7)}$ | $[0.8, 0.9] \cdot e^{j2\pi(0.7)}$ | $[0.7, 0.8] \cdot e^{j2\pi(0.7)}$ | $[0.6, 0.7] \cdot e^{j2\pi(0.8)}$ | $[0.7, 0.8] \cdot e^{j2\pi(0.8)}$ |
| $A_4$ | $[0.8, 0.9] \cdot e^{j2\pi(0.8)}$ | $[0.7, 0.8] \cdot e^{j2\pi(0.9)}$ | $[0.6, 0.7] \cdot e^{j2\pi(0.9)}$ | $[0.8, 0.9] \cdot e^{j2\pi(0.8)}$ | $[0.8, 0.9] \cdot e^{j2\pi(0.9)}$ |

Now, we calculate the distance between the ideal choice (each rating value is $1 \cdot e^{j2\pi(1)}$) and $A_i$ $(i = 1, 2, 3, 4)$ with the weighting vector $w$. We consider the distances as defined in Equations (3)–(8) and obtain the results as shown in Table 2.

**Table 2.** Distance results.

|       | $d_H$  | $d_E$   | $d_{nH}$ | $d_{nE}$ | $d_{nwH}$ | $d_{nwE}$ |
|-------|--------|---------|----------|----------|-----------|-----------|
| $A_1$ | 0.95   | 0.4743  | 0.19     | 0.2121   | 0.2175    | 0.2424    |
| $A_2$ | 1.225  | 0.5766  | 0.245    | 0.2579   | 0.245     | 0.2598    |
| $A_3$ | 1.325  | 0.6144  | 0.265    | 0.2748   | 0.255     | 0.2646    |
| $A_4$ | 0.875  | 0.433   | 0.175    | 0.1937   | 0.16      | 0.1761    |

From Table 2, we get an ordering of the choices for each case as shown in Table 3. Clearly, the best alternative is $A_4$, which is the one with the lowest distance to the ideal choice.

**Table 3.** Ordering of the alternatives.

|          | Ordering                                  |
|----------|-------------------------------------------|
| $d_H$    | $A_4 \succ A_1 \succ A_2 \succ A_3$       |
| $d_E$    | $A_4 \succ A_1 \succ A_2 \succ A_3$       |
| $d_{nH}$ | $A_4 \succ A_1 \succ A_2 \succ A_3$       |
| $d_{nE}$ | $A_4 \succ A_1 \succ A_2 \succ A_3$       |
| $d_{nwH}$| $A_4 \succ A_1 \succ A_2 \succ A_3$       |
| $d_{nwE}$| $A_4 \succ A_1 \succ A_2 \succ A_3$       |

## 6. Conclusions

In this paper, we have presented several distance measures in the case of the interval-valued complex fuzzy sets, which are very useful to deal with the decision information represented in interval-valued complex fuzzy values under uncertain situations. These distance measures include a family of Hamming and Euclidean distances such as the Hamming distance, the normalized Hamming distance, the normalized weighted Hamming distance, the Euclidean distance, the normalized Euclidean distance and the normalized weighted Euclidean distance. Furthermore, these distance measures are reflectionally invariant but not rotationally invariant. Finally, based on these measures, we presented an illustrative example for decision-making under interval-valued complex fuzzy information.

As future work, we can consider distance measures for interval-valued complex fuzzy sets that are both reflectionally invariant and rotationally invariant.

**Author Contributions:** Conceptualization, S.D.; Funding acquisition, L.B. and B.H.; Writing—original draft, S.D.; Writing—review and editing, L.B. and B.H.

**Funding:** This research was funded by the Opening Foundation of Guangxi Colleges and Universities Key Laboratory of Complex System Optimization and Big Data Processing OF FUNDER Grant No. 2017CSOBDP0103.

**Conflicts of Interest:** The authors declare no conflict of interest.

## References

1. Ramot, D.; Milo, R.; Friedman, M.; Kandel, A. Complex fuzzy sets. *IEEE Trans. Fuzzy Syst.* **2002**, *10*, 171–186. [CrossRef]

2. Alkouri, A.M.J.S.; Salleh, A.R. Complex intuitionistic fuzzy sets. In Proceedings of the International Conference on Fundamental and Applied Sciences (ICFAS 2012), Kuala Lumpur, Malaysia, 12–14 June, 2012; pp. 464–470.

3. Ali, M.; Smarandache, F. Complex neutrosophic set. *Neural Comput. Appl.* **2017**, *28*, 1–18. [CrossRef]

4. Ali, M.; Dat, L.Q.; Smarandache, F. Interval complex neutrosophic set: Formulation and applications in decision-making. *Int. J. Fuzzy Syst.* **2018**, *20*, 986–999. [CrossRef]

5.   Al-Qudah, Y.; Hassan, N. Complex multi-fuzzy soft set: Its entropy and similarity measure. *IEEE Access* **2018**, *6*, 65002–65017. [CrossRef]

6.   Garg, H.; Rani, D. Complex interval-valued intuitionistic fuzzy sets and their aggregation operators. *Fundam. Inform.* **2019**, *164*, 61–101. [CrossRef]

7.   Selvachandran, G.; Maji, P.K.,; Abed, I.E.; Salleh, A.R. Complex vague soft sets and its distance measures. *J. Intell. Fuzzy Syst.* **2016**, *31*, 55–68. [CrossRef]

8.   Greenfield, S.; Chiclana, F.; Dick S. Interval-valued complex fuzzy logic. In Proceedings of the 2016 IEEE International Conference on Fuzzy Systems (FUZZ-IEEE), Vancouver, BC, Canada, 24–29 July 2016; pp. 2014–2019.

9.   Greenfield, S.; Chiclana, F.; Dick S. Join and meet operations for interval-valued complex fuzzy logic. In Proceedings of the 2016 Annual Conference of the North American Fuzzy Information Processing Society (NAFIPS), El Paso, TX, USA, 31 October–4 November 2016; pp. 1–5.

10.   Ramot, D.; Friedman, M.; Langholz, G.; Kandel, A. Complex fuzzy logic. *IEEE Trans. Fuzzy Syst.* **2003**, *11*, 450–461. [CrossRef]

11.   Zhang, G.; Dillon, T.S.; Cai, K.-Y.; Ma, J.; Lu, J. Operation properties and delta-equalities of complex fuzzy sets. *Int. J. Approx. Reason.* **2009**, *50*, 1227–1249. [CrossRef]

12.   Hu, B.; Bi, L.; Dai, S. The Orthogonality between Complex Fuzzy Sets and Its Application to Signal Detection. *Symmetry* **2017**, *9*, 175. [CrossRef]

13.   Rani, D.; Garg, H. Complex intuitionistic fuzzy power aggregation operators and their applications in multicriteria decision-making. *Expert Syst.* **2018**, e12325. [CrossRef]

14.   Garg, H.; Rani, D. Some Generalized Complex Intuitionistic Fuzzy Aggregation Operators and Their Application to Multicriteria Decision-Making Process. *Arab. J. Sci. Eng.* **2019**, *44*, 2679–2698. [CrossRef]

15.   Bi, L.; Dai, S.; Hu, B. Complex fuzzy geometric aggregation operators. *Symmetry* **2018**, *10*, 251. [CrossRef]

16.   Bi, L.; Dai, S.; Hu, B.; Li, S. Complex fuzzy arithmetic aggregation operators. *J. Intell. Fuzzy Syst.* **2019**, *26*, 2765–2771. [CrossRef]

17.   Chen, Z.; Aghakhani, S.; Man, J.; Dick, S. ANCFIS: A Neuro-Fuzzy Architecture Employing Complex Fuzzy Sets. *IEEE Trans. Fuzzy Syst.* **2011**, *19*, 305–322. [CrossRef]

18.   Li, C. Complex Neuro-Fuzzy ARIMA Forecasting. A New Approach Using Complex Fuzzy Sets. *IEEE Trans. Fuzzy Syst.* **2013**, *21*, 567–584. [CrossRef]

19.   Li, C.; Chiang, T.-W.; Yeh, L.-C. A novel self-organizing complex neuro-fuzzy approach to the problem of time series forecasting. *Neurocomputing* **2013**, *99*, 467–476. [CrossRef]

20.   Ma, J.; Zhang, G.; Lu, J. A method for multiple periodic factor prediction problems using complex fuzzy sets. *IEEE Trans. Fuzzy Syst.* **2012**, *20*, 32–45.

21.   Li, C.; Wu, T.; Chan, F.-T. Self-learning complex neuro-fuzzy system with complex fuzzy sets and its application to adaptive image noise canceling. *Neurocomputing* **2012**, *94*, 121–139. [CrossRef]

22.   Liu, X. Entropy, distance measure and similarity measure of fuzzy sets and their relations. *Fuzzy Sets Syst.* **1992**, *52*, 305–318.

23.   Atanassov, K. Intuitionistic fuzzy sets. *Fuzzy Sets Syst.* **1986**, *20*, 87–96. [CrossRef]

24.   Szmidt, E.; Kacprzyk, J. Distances between intuitionistic fuzzy sets. *Fuzzy Sets Syst.* **2000**, *114*, 505–518. [CrossRef]

25.   Zhang, H.; Yu, L. New distance measures between intuitionistic fuzzy sets and interval-valued fuzzy sets. *Inf. Sci.* **2013**, *245*, 181–196. [CrossRef]

26.   Wang, W.Q.; Xin, X.L. Distance measure between intuitionistic fuzzy sets. *Pattern Recognit. Lett.* **2005**, *26*, 2063–2069. [CrossRef]

27.   Grzegorzewski, P. Distances between intuitionistic fuzzy sets and/or interval-valued fuzzy sets based on the Hausdorff metric. *Fuzzy Sets Syst.* **2004**, *148*, 319–328. [CrossRef]

28.   Xu, Z.S.; Chen, J. An overview of distance and similarity measures of intuitionistic fuzzy sets. *Int. J. Uncertain. Fuzz. Knowl.-Based Syst.* **2008**, *16*, 529–555 [CrossRef]

29.   Zeng, W.Y.; Guo, P. Normalized distance, similarity measure, inclusion measure and entropy of interval-valued fuzzy sets and their relationship. *Inf. Sci.* **2008**, *178*, 1334–1342. [CrossRef]

30. Hu, B.; Bi, L.; Dai, S.; Li, S. Distances of Complex Fuzzy Sets and Continuity of Complex Fuzzy Operations. *J. Intell. Fuzzy Syst.* **2018**, *35*, 2247–2255. [CrossRef]
31. Alkouri, A.U.M.; Salleh, A.R. Linguistic variables, hedges and several distances on complex fuzzy sets. *J. Intell. Fuzzy Syst.* **2014**, *26*, 2527–2535.
32. Bi, L.; Hu, B.; Li S.; Dai, S. The Parallelity of Complex Fuzzy sets and Parallelity Preserving Operators. *J. Intell. Fuzzy Syst.* **2018**, *34*, 4173–4180. [CrossRef]
33. Hu, B.; Bi, L.; Dai, S.; Li, S. The approximate parallelity of complex fuzzy sets. *J. Intell. Fuzzy Syst.* **2018**, *35*, 6343–6351. [CrossRef]
34. Bi, L.; Zeng, Z.; Hu, B.; Dai, S. Two Classes of Entropy Measures for Complex Fuzzy Sets. *Mathematics* **2019**, *7*, 96. [CrossRef]
35. Rani, D.; Garg, H. Distance measures between the complex intuitionistic fuzzy sets and its applications to the decision-making process. *Int. J. Uncertain. Quantif.* **2017**, *7*, 423–439. [CrossRef]
36. Selvachandran, G.; Garg, H.; Alaroud, M.H.; Salleh, A.R. Similarity measure of complex vague soft sets and its application to pattern recognition. *Int. J. Fuzzy Syst.* **2018**, *20*, 1901–1914. [CrossRef]
37. Selvachandran, G.; Garg, H.; Quek, S. Vague entropy measure for complex vague soft sets. *Entropy* **2018**, *20*, 403. [CrossRef]
38. Dick, S. Towards Complex Fuzzy Logic. *IEEE Trans. Fuzzy Syst.* **2005**, *13*, 405–414. [CrossRef]

 *mathematics*

*Article*

# An Interactive Data-Driven (Dynamic) Multiple Attribute Decision Making Model via Interval Type-2 Fuzzy Functions

**Adil Baykasoğlu** [1,*] **and İlker Gölcük** [2]

[1]   Department of Industrial Engineering, Faculty of Engineering, Dokuz Eylül University, Izmir 35397, Turkey
[2]   Department of Industrial Engineering, İzmir Bakırçay University, Izmir 35665, Turkey
*   Correspondence: adil.baykasoglu@deu.edu.tr

Received: 16 May 2019; Accepted: 27 June 2019; Published: 30 June 2019

**Abstract:** A new multiple attribute decision making (MADM) model was proposed in this paper in order to cope with the temporal performance of alternatives during different time periods. Although dynamic MADM problems are enjoying a more visible position in the literature, majority of the applications deal with combining past and present data by means of aggregation operators. There is a research gap in developing data-driven methodologies to capture the patterns and trends in the historical data. In parallel with the fact that style of decision making evolving from intuition-based to data-driven, the present study proposes a new interval type-2 fuzzy (IT2F) functions model in order to predict current performance of alternatives based on the historical decision matrices. As the availability of accurate historical data with desired quality cannot always be obtained and the data usually involves imprecision and uncertainty, predictions regarding the performance of alternatives are modeled as IT2F sets. These estimated outputs are transformed into interpretable forms by utilizing the vocabulary matching procedures. Then the interactive procedures are employed to allow decision makers to modify the predicted decision matrix based on their perceptions and subjective judgments. Finally, ranking of alternatives are performed based on past and current performance scores.

**Keywords:** dynamic multiple attribute decision making; fuzzy regression; interval type-2 fuzzy sets

---

## 1. Introduction

Managers are continuously engaged in a process of making decisions in a rapidly changing business environment. Making right decisions is crucial in order to attain organizational goals and effective use of resources. Quality of decisions relies heavily on the information processing capabilities by considering multiple and conflicting criteria. With the dramatic increase in the availability of information obtained from diverse set of resources, decision making becomes much more complicated and difficult. Multiple attribute decision making (MADM) offers a set of sophisticated techniques to help decision makers in selecting the best alternative by considering multiple, conflicting, and incommensurate criteria.

The field of MADM is rapidly expanding with the continuing proliferation of new techniques and applications. Many state of the art methods have been proposed such as multi-attribute utility theory (MAUT) [1–3], analytic hierarchy process (AHP) [4], analytic network process (ANP) [5], technique for order preference by similarity to ideal solution (TOPSIS) [6], elimination and choice translating reality (ELECTRE) [7], VlseKriterijumska Optimizacija I Kompromisno Resenje technique (VIKOR) [8], and decision-making trial and evaluation laboratory (DEMATEL) [9]. Despite many successful applications of the MADM methods currently available in the literature, the salient deficiency of these methods is their incapability to handle temporal profiles of alternatives. Unfortunately, static MADM methods cannot deal with the temporal profiles of alternatives, that is, past performance scores are not taken into consideration. In order to overcome this deficiency, data-driven, and dynamic MADM methods have been developing along with diverse applications.

In the dynamic MADM, at least two-period decision making information is considered. In addition to the alternative and criteria dimensions, time is considered as a third dimension in the dynamic MADM. With the recent advances in the information technologies, data is becoming an indispensable part of the decision making practices, which forces the pace of a paradigm shift towards data-driven decision making. As ever-more data pour through the networks of organizations, collecting, and storing performance scores of alternatives with time stamps are not cumbersome procedures anymore. As the style of decision making evolving from intuition based to data-driven, the decision makers should be supported with relevant methodologies and tools to fully capitalize the available data. However, it is evident that the field of dynamic MADM is in its infancy and the current literature is far from meeting the requirements of a fully-fledged data-driven methodology.

In one of the earliest works on the dynamic MADM, Kornbluth [10] discussed the problem of time dependence of the criteria weights, and empirical laboratory findings were used for the analysis. Decision making teams were monitored for 12 sequential decisions and the time-dependent weights were analyzed based on different scenarios. Dong et al. [11] proposed a dynamic MADM method based on relative differences between the performance scores of the subsequent time-periods. In the study, disadvantages of using absolute differences were discussed and a numerical example was provided. Lou et al. [12] proposed a dynamic MADM model to evaluate and rank country risks based on historical data. The proposed model was aimed at predicting possible credit crises in advance. The proposed model was applied to the world economy development indicators data of 32 countries. The utilités additives discriminantes (UTADIS) method was used to rank country risk scores.

Despite many new developments, the literature of the dynamic MADM field is dominated by the aggregation-operator based models. Campanella and Ribeiro [13] proposed a framework for dynamic MADM where the aggregation operators were the main computation tools, and majority of the studies in the literature employ this framework. Xu and Yager [14] proposed dynamic intuitionistic fuzzy weighted averaging and uncertain dynamic intuitionistic fuzzy weighted averaging operators. According to their model, decision matrices of the past periods are aggregated into a decision matrix, and classical MADM techniques were implemented afterwards. Park et al. [15] proposed dynamic intuitionistic fuzzy weighted geometric and uncertain dynamic intuitionistic fuzzy weighted geometric operators for dynamic MADM problems. The past decision matrices were aggregated and then the VIKOR method was used to rank the alternatives. Zhou et al. [16] hybridized dynamic triangular fuzzy weighting average operators with fuzzy VIKOR method for quality improvement pilot program selection. Dynamic feedbacks of the customers were also incorporated into the proposed model. Bali et al. [17] employed dynamic intuitionistic fuzzy weighted averaging method with TOPSIS for multi-period third-party logistics provider selection problem. Chen and Li [18] proposed a new distance measure for triangular intuitionistic fuzzy sets with an application in dynamic MADM. The weighted arithmetic averaging operator for triangular intuitionistic fuzzy numbers was used to aggregate the decision matrices of the past periods. The ranking orders were obtained by using closeness coefficients. An investment decision making was used to illustrate the proposed method. Liang et al. [19] employed evidential reasoning approach to aggregate decision matrices with incomplete information. The enterprise evaluation in a technological zone was used to illustrate the model. Bali et al. [20] proposed an integrated model based on AHP and dynamic intuitionistic fuzzy weighted averaging operator under an intuitionistic fuzzy environment. The proposed model was implemented in a personnel promotion problem.

In some of the studies, the aggregation operators were not used to aggregate the decision matrices of the past periods at the very beginning. Xu [21] proposed dynamic weighted geometric aggregation operator along with an illustrative three-period investment decision making model. Rather than aggregating the decision matrices of the past periods, aggregation was conducted based on the closeness coefficients of the different periods. Zulueta et al. [22] proposed discrete time variable index by admitting bipolar values in the aggregation. A five-period supplier selection problem was used to illustrate the proposed aggregation operator. Lin et al. [23] used grey numbers and Minkowski distance in dealing with dynamic subcontractor selection problem. The proposed method calculated the period weighted distances to the ideal and anti-ideal solutions. Similar aggregation operator-based studies in the literature, which utilize intuitionistic fuzzy numbers [24,25], 2-tuple linguistic representation [26–30], grey numbers [31,32], etc. can be found. For more information about dynamic aggregation operators, we refer to a review paper [33]. On the other hand, non-aggregation operator based studies can be summarized as follows: Saaty [34] studied time-dependent eigenvectors and approximating functional forms of relative priorities in dynamic MADM. Hashemkhani Zolfani et al. [35] emphasized the relevance and necessity of future studies in MADM problems. In the paper, scenario-based MADM papers were reviewed and analyzed. Possible changes in the experts' evaluations were expressed by using probabilities. Orji and Wei [36] integrated fuzzy logic and system dynamics simulation to sustainable supplier selection problem. Very recently, Baykasoğlu and Gölcük [37] proposed a dynamic MADM model by learning of fuzzy cognitive maps. In the study, fuzzy cognitive maps were trained by using a metaheuristic algorithm in order to capture patterns and trends in the past data. Then, the trained model was used to generate short-, medium-, and long-term future decision matrices. Finally, past, current and future decision matrices were used to rank the alternatives. The proposed model was realized in a real-life supplier selection problem.

Although a wide range of applications have been provided in the context of dynamic MADM, the literature still lacks the following considerations:

- Decision makers are expected to fill out tedious questionnaires to articulate their preferences over alternatives at each period. This is especially very time consuming and demanding when the number of criteria and alternatives are high, and the decision points are frequent, i.e., performance evaluation, risk assessment, etc.
- The models do not provide any mechanism to help decision makers making use of past decision making matrices when articulating their preferences at the current period. An interactive mechanism is needed to facilitate preference elicitation in the light of historical performance of alternatives.

Due to the availability of accurate historical data with high quality and quantity cannot always be assured and the data is usually affected by imprecision and noise, the predictions regarding the performance of alternatives should handle uncertainty properly. Interval type-2 fuzzy (IT2F) sets are very suitable tools for manipulating and reasoning with uncertain information. For that reason, the present study makes use of IT2F regression [38] to predict the current decision matrix. In order to enhance the prediction capability of the IT2F regression, a new hybrid IT2F model is proposed on the basis of highly practical method of so called "fuzzy functions" [39]. The proposed model is able to capture nonlinearities more successfully than the traditional fuzzy regression models, due to its unique and intelligent way of integrating membership grades of data points into the prediction problem.

The proposed dynamic MADM model contributes to the literature with its following features:

- A dynamic MADM model is proposed based on a new IT2F functions approach.
- An interactive procedure is provided that the current decision making matrix is predicted in forms of IT2F sets. Moreover, vocabulary matching procedure is developed so that the predicted performance scores of alternatives are recommended to the decision makers through linguistic terms such as low, medium, high, etc.
- The proposed model interacts with decision makers whose subjective judgments are combined with the notion of "let the data speak for itself". By providing decision makers with data-driven suggestions regarding the performance of alternatives, preference elicitation effort at each period is considerably reduced.

- The proposed model does not require any technical knowledge such as fuzzy sets, t-norms, t-conorms, implication functions, etc. The proposed model can be easily integrated into the legacy systems of the firms, since the crisp values are processed when providing IT2F outputs.
- A real-life personnel promotion problem is used to demonstrate the applicability of the proposed model. Rankings of employees are calculated based on past and current performance matrices with appropriate time series weights.

This paper is organized as follows: Theoretical background on the methodologies used within the scope of this paper is given in Section 2. The proposed model is provided in Section 3. The real life application of the proposed dynamic MADM model is illustrated in Section 4. Discussions are given in Section 5. Concluding remarks are given in Section 6.

## 2. Theoretical Background

### 2.1. Traditional Dynamic Multiple Attribute Decision Making

A dynamic MADM problem under study can be described as follows. Let $A = \{A_1, A_2, \ldots, A_M\}$ be a discrete set of $M$ feasible alternatives and $C = \{C_1, C_2, \ldots, C_N\}$ be a finite set of all attributes. The set of all periods is denoted by $t = \{t_1, t_2, \ldots, t_H\}$. Each alternative is evaluated in terms of $N$ attributes and $H$ periods. Each period is associated with weights that these time series weights are denoted by $\xi(t) = [\xi(t_1), \xi(t_2), \ldots, \xi(t_H)]^T$ where $\xi(t_H) \geq 0$ and $\sum_{k=1}^{H} \xi(t_k) = 1$. The weight vector of attributes are given by $[w_1(t_k), w_2(t_k), \ldots w_N(t_k)]^T$ where $w_i(t_k) \geq 0$ and $\sum_{i=1}^{N} w_i(t_k) = 1$. The decision matrix at the period $t_k$ is denoted by $A(t_k) = \left(a_{ij}(t_k)\right)_{N \times M}$ where $a_{ij}(t_k)$ is the value of alternative $A_j$ with respect to attribute $C_i$ at period $t_k$. Let $\Omega_b$ and $\Omega_c$ be the set of benefit and cost attributes, respectively. Due to the immensurability of the different attributes, decision matrix $A(t_k)$ is normalized to corresponding dimensionless decision matrix $R(t_k) = \left(r_{ij}(t_k)\right)_{N \times M}$ by using the following formulas:

$$r_{ij}(t_k) = \frac{a_{ij}(t_k)}{\max_j\{a_{ij}(t_k)\}}, \; j = 1, 2, \ldots, M; \; k = 1, 2, \ldots, H; \; i \in \Omega_b, \tag{1}$$

$$r_{ij}(t_k) = \frac{\min_j\{a_{ij}(t_k)\}}{a_{ij}(t_k)}, \; j = 1, 2, \ldots, M; \; k = 1, 2, \ldots, H; \; i \in \Omega_c, \tag{2}$$

Hence, the normalized decision matrix is obtained as:

$$R(t_k) = \begin{bmatrix} r_{11}(t_k) & r_{12}(t_k) & \cdots & r_{1M}(t_k) \\ r_{21}(t_k) & r_{22}(t_k) & \cdots & r_{2M}(t_k) \\ \vdots & \vdots & \ddots & \vdots \\ r_{N1}(t_k) & r_{N2}(t_k) & \cdots & r_{NM}(t_k) \end{bmatrix}. \tag{3}$$

The overall assessment value of the $j$th alternative is calculated by:

$$y_j = \sum_{k=1}^{H} \sum_{i=1}^{N} \xi(t_k) w_i(t_k) r_{ij}(t_k), \; j = 1, 2, \ldots, M. \tag{4}$$

Therefore, alternatives are ranked based on $y_j$ in which the best alternative is with the highest overall assessment value.

### 2.2. Possibilistic Fuzzy Regression

In this section, possibilistic fuzzy regression analysis with asymmetric fuzzy numbers is overviewed based on [38]. Although there are a variety of fuzzy regression approaches developed during the last two decades, regression models relying on possibility and necessity concepts have pivotal role in the current literature. Possibilistic models strive to minimize sum of spreads in such a way that the estimated outputs must include the given targets. It is indeed advantageous to have asymmetric fuzzy numbers in possibilistic models as the lower and upper bounds of the estimated model are not necessarily equidistant from the center, which implies superior capability to capture central tendency. A fuzzy regression model can be formalized as:

$$Y(x) = \tilde{\beta}_0 + \tilde{\beta}_1 x_1 + \cdots + \tilde{\beta}_{nv} x_{nv} = \tilde{\beta} x, \tag{5}$$

where the input vector is represented by $x = (1, x_1, \ldots, x_{nv})^t$ and $\tilde{\beta} = (\tilde{\beta}_0, \tilde{\beta}_1, \ldots, \tilde{\beta}_{nv})$ is a vector of fuzzy coefficients, and $Y(x)$ is the estimated fuzzy output. The coefficients $\tilde{\beta}_i$ are denoted as $\tilde{\beta}_i = (a_i, c_i, d_i)$ can be defined by:

$$\mu_{A_i}(x) = \begin{cases} 1 - (a_i - x)/c_i, & \text{if } a_i - c_i \le x \le a_i \\ 1 - (x - a_i)/d_i, & \text{if } a_i \le x \le a_i + d_i, \\ 0, & \text{otherwise} \end{cases} \tag{6}$$

where $a_i$ represents center, and $c_i$ and $d_i$ left- and right-spreads, respectively.

Given the input-output data as $(x_j, y_j) = (1, x_{j,1}, \ldots, x_{j,nv}; y_j)$, $j = 1, 2, \ldots, nd$, where $x_{j,nv}$ being value of the variable $nv$ of the $j$-th data-point among the total of $nd$ data-points, the estimated output $Y(x_j)$ can be calculated by using fuzzy arithmetic. Representing regression coefficients as $\tilde{\beta}_i = (a_i, c_i, d_i)$, $(i = 0, 1, \ldots, nv)$, fuzzy regression model can be expressed as:

$$Y(x_j) = (a_0, c_0, d_0) + (a_1, c_1, d_1)x_{j,1} + \cdots + (a_n, c_n, d_n)x_{j,nv}. \tag{7}$$

Equation (7) can be written in a more compact form as given in Equation (8).

$$Y(x_j) = \left(\theta_C(x_j), \theta_L(x_j), \theta_R(x_j)\right), \tag{8}$$

where the terms $\theta_C(x_j), \theta_L(x_j), \theta_R(x_j)$, are calculated as given in Equation (9).

$$\begin{aligned} \theta_C(x_j) &= \sum_{i=0}^{n} a_i x_{ji} \\ \theta_L(x_j) &= \sum_{x_{ji} \ge 0}^{n} c_i x_{ji} - \sum_{x_{ji} < 0}^{n} d_i x_{ji} \\ \theta_R(x_j) &= \sum_{x_{ji} \ge 0}^{n} d_i x_{ji} - \sum_{x_{ji} < 0}^{n} c_i x_{ji} \end{aligned} \tag{9}$$

Finally, the possibilistic fuzzy regression model can be written as:

$$\begin{aligned} J &= \sum_{j=1}^{nd} \left(y_j - a^t x_j\right)^2 + (1 - h)\sum_{j=1}^{nd} \left(c^t |x_j| + d^t |x_j|\right) + \xi\left(c^t c + d^t d\right) \\ \text{subject to} \quad & \theta_C(x_j) + (1 - h)\theta_R(x_j) \ge y_j \\ & \theta_C(x_j) - (1 - h)\theta_L(x_j) \le y_j \\ & c_i \ge 0, \, d_i \ge 0, \, i = 0, 1, \ldots, nv \end{aligned} \tag{10}$$

where $\xi$ is a small positive number. The term $\xi\left(c^t c + d^t d\right)$ is added to objective function so that the objective function becomes a quadratic function with respect to decision variables $a$, $c$, and $d$. The resulting optimization problem is a quadratic optimization problem, which involves minimizing a quadratic objective function subject to linear constrains. The possibilistic fuzzy regression analysis will be detailed in the subsequent sections.

### 2.3. Turksen's Fuzzy Functions Approach

Turksen [39] proposed a new fuzzy modeling technique as an alternative to classical fuzzy rule bases (FRB). Fuzzy rule bases (FRB) have been effectively used as a facilitator to decision makers' problem solving activities. FRBs are the one of the best currently available means to codify human knowledge. This knowledge is represented by "IF … THEN" rule structures. The "IF" part represents the antecedents and "THEN" part represents consequents. In the literature, there are different FRB system modeling strategies with unique antecedent and consequent parameter formation approaches. In these systems, membership values have also different interpretations such as "degree of fire", "degree of compatibility", "degree of belongingness", or "weight or strength of local functions". The fuzzy functions approach adds a new means of membership degrees to the list by exploiting the predictor power of membership grades [40].

The representation of each unique rule of an FRB system by means of fuzzy functions is the governing idea of the fuzzy functions approach. In the fuzzy functions approach, the membership degree of each sample vector directly affects the local fuzzy functions. One of the advantages of the fuzzy functions approach is that even non-experts can build fuzzy models as there are lower steps and parameters. It is quite practical to identify and reason with the fuzzy functions approach that some technical information regarding constructing fuzzy system models is not required such as fuzzification, t-norms and t-conorms, modus ponens, etc.

A vast array of fuzzy modeling approaches has been developed in the literature where the expert knowledge is encoded to define linguistic variables characterized by fuzzy sets. However, majority of the approaches suffers from a major drawback of being subjective and not generalizable. In order to reduce expert intervention into fuzzy system modeling, more objective methods have been developed [41–45]. In these systems, membership grades are not defined by decision makers, on the contrary, they are extracted from the dataset. There are also some approaches where some sophisticated techniques are integrated into fuzzy models so that the hybrid fuzzy system models are built. The prominent examples of these methods are neuro-fuzzy systems [46] and genetic-fuzzy systems [47].

There are still enduring challenges in the mentioned fuzzy system modeling approaches. Main disadvantages of the classical fuzzy system modeling approaches can be listed as follows:

- Membership functions pertaining to antecedent and consequent parts of the fuzzy rules should be identified.
- Aggregation of antecedents requires selection of suitable conjunction and disjunction operators (t-norms, t-conorms).
- Proper implication operators should be identified for representation of the rules, which can be a challenging issue.
- A suitable defuzzification method should be identified.

The fuzzy functions approach mainly reduces the number of fuzzy operators by taking advantage of data-driven modeling. For instance, fuzzy operators in determination of the membership functions in antecedents and consequents, fuzzification, aggregation of antecedents, implication, and in aggregation of consequents. It can be said that the fuzzy functions are more practical than their counterpart FRB models. Fuzzy function can be simply described as follows:

The training dataset is partitioned into $c$ overlapping clusters where each cluster center is represented by $v_i, i = 1, 2, \ldots, c$.

For each one of the clusters, a local fuzzy model $f_i : v_i \rightarrow \Re$ is built and one output is produced for each cluster. Here, memberships and their several transformations are added into the input space and the augmented input matrix is generated. Membership grades and their transformations are considered as new variables in the regression matrix. Practically, least square estimation is used to derive regression coefficients. Then, degree of belongingness of each given input vector is used to aggregate the local model outputs and the estimated values are produced.

General steps of the fuzzy functions approach can be given as:

Step 1: Matrix $Z$ comprises of inputs and output of the system. Inputs and output of the system are clustered by using the fuzzy c-means clustering algorithm. Fuzzy c-means clustering method can be applied by using the formulas given as:

$$v_i = \frac{\sum_{j=1}^{nd} \mu_{ij}^m z_j}{\sum_{j=1}^{nd} \mu_{ij}^m}, \ i = 1, 2, \ldots, c, \tag{11}$$

$$\mu_{ij} = \frac{1}{\sum_{h=1}^{c} \left( \frac{\|v_i - z_j\|}{\|v_h - z_j\|} \right)^{\frac{2}{m-1}}}, \ i = 1, 2, \ldots, c; \ j = 1, 2, \ldots, nd, \tag{12}$$

where $\|.\|$ represents the Euclidean distance between data point $z_j$ to cluster center $v_i$.

Step 2: In the second step, membership values of the input space are calculated. Here, the cluster centers identified in the previous step are used to calculate membership grades of the input data. Membership construction from the identified cluster centers is performed based on Equation (13).

$$\mu_{ij} = \frac{1}{\sum_{h=1}^{c} \left( \frac{\|v_i - x_j\|}{\|v_h - x_j\|} \right)^{\frac{2}{m-1}}}, \ i = 1, 2, \ldots, c; \ j = 1, 2, \ldots, nd. \tag{13}$$

Step 3: For each cluster $i$, membership values of each input data sample, $\mu_{ij}$ and original inputs are gathered together, and $i$-th local fuzzy function is obtained by predicting $Y^{(i)} = X^{(i)} \beta^{(i)} + \varepsilon^{(i)}$ based on least squares estimation. When the number of inputs is $nv$, $X^{(i)}$, and $Y^{(i)}$ matrices are as follows:

$$X^{(i)} = \begin{bmatrix} \mu_{i,1} & x_{1,1} & \cdots & x_{nv,1} \\ \mu_{i,2} & x_{1,2} & \cdots & x_{nv,2} \\ \vdots & \vdots & \ddots & \vdots \\ \mu_{i,nd} & x_{1,nd} & \cdots & x_{nv,nd} \end{bmatrix}, \ Y^{(i)} = \begin{bmatrix} y_1 \\ y_2 \\ \vdots \\ y_{nd} \end{bmatrix}. \tag{14}$$

Step 4: Output values are calculated by aggregating the results of the local fuzzy functions as follows:

$$\hat{y}_j = \frac{\sum_{i=1}^{c} \hat{y}_{ij} \mu_{ij}}{\sum_{i=1}^{c} \mu_{ij}}, \ j = 1, 2, \ldots, nd. \tag{15}$$

## 3. Developed IT2F Model

In this section, the developed IT2F model for dynamic MADM problems is given. First, the basics of the IT2F sets and necessary equations are overviewed. Afterwards, the IT2F regression model is revisited based on [38]. Then, the procedural steps of the proposed dynamic MADM model are given.

### 3.1. Interval Type-2 Fuzzy Sets

**Definition 1** ([48,49]). *A type-2 fuzzy set $\tilde{A}$ in the universe of discourse X can be represented by a type-2 membership function $\mu_{\tilde{A}}$ as:*

$$\tilde{A} = \left\{ ((x, u), \mu_{\tilde{A}}(x, u)) | \forall x \in X, \ \forall u \in J_x \subseteq [0, 1], 0 \le \mu_{\tilde{A}}(x, u) \le 1 \right\}, \tag{16}$$

where $J_x$ denotes an interval in [0,1]. Moreover, type-2 fuzzy set $\tilde{\tilde{A}}$ can also be represented as:

$$\tilde{\tilde{A}} = \int_{x\in X} \int_{u\in J_x} \mu_{\tilde{\tilde{A}}}(x,u)/(x,u), \tag{17}$$

where $J_x \subseteq [0,1]$ and $\int\int$ denotes union over all admissible $x$ and $u$.

**Definition 2** ([48,49]). *Let $\tilde{\tilde{A}}$ be a type-2 fuzzy set in the universe of discourse $X$ represented by the type-2 membership function $\mu_{\tilde{\tilde{A}}}$. If all $\mu_{\tilde{\tilde{A}}}(x,u) = 1$, then $\tilde{\tilde{A}}$ is called an IT2F set. An IT2F set $\tilde{\tilde{A}}$, which can be regarded as a special case of a type-2 fuzzy set, is represented as follows:*

$$\tilde{\tilde{A}} = \int_{x\in X} \int_{u\in J_x} 1/(x,u), \tag{18}$$

*where $J_x \subseteq [0,1]$.*

The footprint of uncertainty (FOU) is represented by the lower and upper membership functions:

$$FOU(\tilde{\tilde{A}}) = \int_{x\in X} \left[\underline{\mu}_{\tilde{A}}(x), \overline{\mu}_{\tilde{A}}(x)\right], \tag{19}$$

where $\underline{\mu}_{\tilde{A}}(x)$ and $\overline{\mu}_{\tilde{A}}(x)$ represent lower and upper membership functions, respectively.

**Definition 3** ([50]). *An IT2F set $\tilde{\tilde{A}}$ is said to be normal if $\sup\overline{\mu}_{\tilde{A}}(x) = 1$ and $\sup\underline{\mu}_{\tilde{A}}(x) = h < 1$, where $h$ represents the height of the lower membership function. An IT2F set $\tilde{\tilde{A}}$ is said to be perfectly normal if $\sup\overline{\mu}_{\tilde{A}}(x) = \sup\underline{\mu}_{\tilde{A}}(x) = 1$.*

In this study, perfectly normal IT2F sets were employed so that the basic definitions and operational laws regarding perfectly normal triangular IT2F sets were overviewed.

Considering perfectly normal triangular IT2F numbers $\tilde{\tilde{A}} = (\overline{A}, \underline{A}) = ((\overline{a}_1, \overline{a}_2, \overline{a}_3; 1)(\underline{a}_1, \underline{a}_2, \underline{a}_3; 1))$ and $\tilde{\tilde{B}} = (\overline{b}, \underline{b}) = ((\overline{b}_1, \overline{b}_2, \overline{b}_3; 1)(\underline{b}_1, \underline{b}_2, \underline{b}_3; 1))$, their operational laws are as follows [51]:

$$\tilde{\tilde{A}} \oplus \tilde{\tilde{B}} = \left( \begin{array}{c} (\overline{a}_1 + \overline{b}_1, \overline{a}_2 + \overline{b}_2, \overline{a}_3 + \overline{b}_3; 1), \\ (\underline{a}_1 + \underline{b}_1, \underline{a}_2 + \underline{b}_2, \underline{a}_3 + \underline{b}_3; 1) \end{array} \right), \tag{20}$$

$$\tilde{\tilde{A}} \ominus \tilde{\tilde{B}} = \left( \begin{array}{c} (\overline{a}_1 - \overline{b}_3, \overline{a}_2 - \overline{b}_2, \overline{a}_3 - \overline{b}_1; 1), \\ (\underline{a}_1 - \underline{b}_3, \underline{a}_2 - \underline{b}_2, \underline{a}_3 - \underline{b}_1; 1) \end{array} \right), \tag{21}$$

$$\tilde{\tilde{A}} \otimes \tilde{\tilde{B}} = \left( \begin{array}{c} (\overline{a}_1 \times \overline{b}_1, \overline{a}_2 \times \overline{b}_2, \overline{a}_3 \times \overline{b}_3; 1), \\ (\underline{a}_1 \times \underline{b}_1, \underline{a}_2 \times \underline{b}_2, \underline{a}_3 \times \underline{b}_3; 1) \end{array} \right), \tag{22}$$

$$\tilde{\tilde{A}} \oslash \tilde{\tilde{B}} = \left( \begin{array}{c} (\overline{a}_1/\overline{b}_3, \overline{a}_2/\overline{b}_2, \overline{a}_3/\overline{b}_1; 1), \\ (\underline{a}_1/\underline{b}_3, \underline{a}_2/\underline{b}_2, \underline{a}_3/\underline{b}_1; 1) \end{array} \right), \tag{23}$$

$$k \times \tilde{\tilde{A}} = \left\{ \begin{array}{ll} ((k \times \overline{a}_1, k \times \overline{a}_2, k \times \overline{a}_3; 1)(k \times \underline{a}_1, k \times \underline{a}_2, k \times \underline{a}_3; 1)), & k \geq 0 \\ ((k \times \overline{a}_3, k \times \overline{a}_2, k \times \overline{a}_1; 1)(k \times \underline{a}_3, k \times \underline{a}_2, k \times \underline{a}_1; 1)), & k \leq 0 \end{array} \right. . \tag{24}$$

**Definition 4.** *The ranking value $Rank(\tilde{\tilde{A}})$ of a IT2F set $\tilde{\tilde{A}} = \left(\overline{\tilde{A}}, \underline{\tilde{A}}\right)$ can be defined via the concept of centroid as* [52]:

$$C_{\tilde{\tilde{A}}}^{L} = \min_{\xi \in [a,b]} \frac{\int_a^{\xi} x\overline{\mu}_A(x)dx + \int_{\xi}^{b} x\underline{\mu}_A(x)dx}{\int_a^{\xi} \overline{\mu}_A(x)dx + \int_{\xi}^{b} \underline{\mu}_A(x)dx}, \tag{25}$$

$$C_{\tilde{\tilde{A}}}^{R} = \max_{\xi \in [a,b]} \frac{\int_a^{\xi} x\underline{\mu}_A(x)dx + \int_{\xi}^{b} x\overline{\mu}_A(x)dx}{\int_a^{\xi} \underline{\mu}_A(x)dx + \int_{\xi}^{b} \overline{\mu}_A(x)dx}, \tag{26}$$

*where $C_{\tilde{\tilde{A}}}^{L}$ and $C_{\tilde{\tilde{A}}}^{R}$ are the endpoints of the centroid. The ranking value of the IT2F set $\tilde{\tilde{A}}$ is calculated as:*

$$Rank(\tilde{\tilde{A}}) = \frac{C_{\tilde{\tilde{A}}}^{L} + C_{\tilde{\tilde{A}}}^{R}}{2}, \tag{27}$$

*where $Rank(\tilde{\tilde{A}})$ is the centroid-based ranking value of $\tilde{\tilde{A}}$.*

**Definition 5** ([53]). *The Jaccard similarity measure for fuzzy sets $\tilde{\tilde{A}}$ and $\tilde{\tilde{B}}$ is defined by*

$$SM\left(\tilde{\tilde{A}}, \tilde{\tilde{B}}\right) = \frac{\int_X \min\left(\overline{\mu}_A(x), \overline{\mu}_B(x)\right)dx + \int_X \min\left(\underline{\mu}_A(x), \underline{\mu}_B(x)\right)dx}{\int_X \max\left(\overline{\mu}_A(x), \overline{\mu}_B(x)\right)dx + \int_X \max\left(\underline{\mu}_A(x), \underline{\mu}_B(x)\right)dx}, \tag{28}$$

*where SM represent similarity degree of fuzzy sets with respect to Jaccard measure.*

### 3.2. IT2F Regression Model

In this section, IT2F regression model is revisited based on [38]. IT2F regression model will be utilized within the fuzzy functions approach [39] in order to increase its performance in the next section. In this section, the necessary equations are derived for IT2F regression in a step-by-step approach. IT2F regression models can be constructed based on the concepts of possibility and necessity. Here, possibility and necessity concepts are used to build an upper approximation model (UAM) and lower approximation model (LAM), respectively [54]. Building an integrated model, UAM and LAM are used to form upper and lower membership functions of the IT2F coefficients, respectively.

In mathematical terms, LAM and UAM models can be written as:

$$\begin{aligned} \text{LAM} &: \tilde{Y}_*(x_j) = \tilde{\beta}_{*0} + \tilde{\beta}_{*1}x_{j1} + \cdots + \tilde{\beta}_{*n}x_{j,nv} = \tilde{\beta}_* x_j, \; j = 1, \ldots, nd \\ \text{UAM} &: \tilde{Y}^*(x_j) = \tilde{\beta}_0^* + \tilde{\beta}_1^*x_{j1} + \cdots + \tilde{\beta}_n^*x_{j,nv} = \tilde{\beta}^* x_j, \; j = 1, \ldots, nd \end{aligned} \tag{29}$$

where coefficients $\tilde{\beta}_{*j}$ and $\tilde{\beta}_j^*$ are non-symmetric triangular fuzzy numbers. The regression coefficients $\tilde{\beta}_{*j}$ and $\tilde{\beta}_j^*$ are shown in Figure 1.

As shown in Figure 1, $\tilde{\beta}_{*i}$ and $\tilde{\beta}_i^*$ can be defined as:

$$\begin{aligned} \tilde{\beta}_{*i} &= (b_i - f_i, b_i, b_i + g_i; 1) \\ \tilde{\beta}_i^* &= (b_i - f_i - p_i, b_i, b_i + g_i + q_i; 1) \end{aligned} \tag{30}$$

where the condition $\tilde{\beta}_i^* \supseteq \tilde{\beta}_{*i}$, is satisfied for $i = 0, \ldots, nv$.

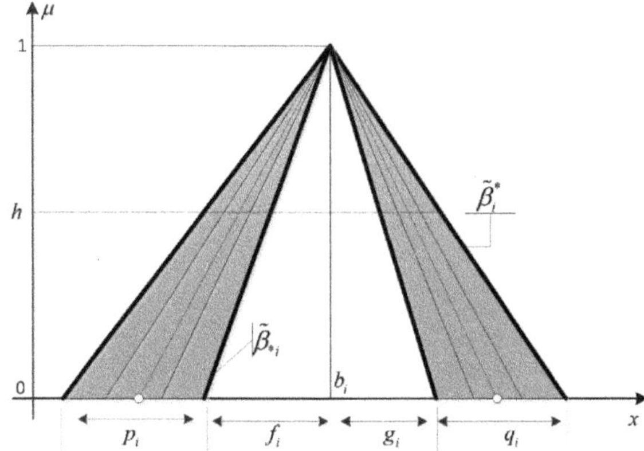

**Figure 1.** Representation of regression coefficients.

In order to increase readability of the formulations, conventional representation of the literature is adopted here, where (center, left_spread, right_spread) is used to show the LAM and UAM as given by $\tilde{\beta}_{*i} = (b_i, f_i, g_i)$ and $\tilde{\beta}^*_i = (b_i, f_i + p_i, g_i + q_i)$, respectively.

The inclusion relation between $\tilde{\beta}_{*i}$ and $\tilde{\beta}^*_i$ can be extended to $\tilde{Y}_*(x_j)$ and $\tilde{Y}^*(x_j)$, that is:

$$\tilde{Y}^*(x) \supseteq \tilde{Y}_*(x) \text{ for any } x = (1, x_1, \dots, x_{nv})^t \text{ if } \tilde{\beta}^*_i \supseteq \tilde{\beta}_{*i}. \tag{31}$$

Using the coefficients of the LAM model $\tilde{A}_{*i} = (b_i, f_i, g_i)$, $\tilde{Y}_*(x_j)$ can be expressed as:

$$\begin{aligned}
\tilde{Y}_*(x_j) &= (b_0, f_0, g_0) + (b_1, f_1, g_1)x_{j1} + \cdots + (b_{nv}, f_{nv}, g_{nv})x_{j,nv} \\
&= \left( \sum_{i=0}^{n} b_i x_{ji}, \sum_{x_{ji} \geq 0} f_i x_{ji} - \sum_{x_{ji} \leq 0} g_i x_{ji}, \sum_{x_{ji} \geq 0} g_i x_{ji} - \sum_{x_{ji} \leq 0} f_i x_{ji} \right), \\
&= \left( b^t x_j, \theta_{*L}(x_j), \theta_{*R}(x_j) \right)
\end{aligned} \tag{32}$$

where $b = (b_0, b_1, \dots, b_{nv})^t$.

Similarly, $\tilde{Y}^*(x_j)$ can be expressed as:

$$\begin{aligned}
\tilde{Y}^*(x_j) &= (b_0, f_0, g_0) + (b_1, f_1 + p_1, g_1 + q_1)x_{j1} + \cdots + (b_{nv}, f_{nv} + p_{nv}, g_{nv} + q_{nv})x_{jnv} \\
&= \left( \sum_{i=0}^{nv} b_i x_{ji}, \sum_{x_{ji} \geq 0} f_i x_{ji} + \sum_{x_{ji} \geq 0} p_i x_{ji} - \sum_{x_{ji} \leq 0} g_i x_{ji} - \sum_{x_{ji} \leq 0} q_i x_{ji}, \sum_{x_{ji} \geq 0} g_i x_{ji} + \sum_{x_{ji} \geq 0} q_i x_{ji} - \sum_{x_{ji} \leq 0} f_i x_{ji} - \sum_{x_{ji} \leq 0} p_i x_{ji} \right). \\
&= \left( b^t x_j, \theta^*_L(x_j), \theta^*_R(x_j) \right)
\end{aligned} \tag{33}$$

As the possibility and necessity concepts were employed, observed outputs were transformed into granular constructs by admitting a tolerance level for the left- and right-spreads. A user-defined tolerance_level was set, thereby possibilistic relationships between observed and estimated outputs could be defined. Generally, tolerance_level is a percentage type, i.e., assigning 20% of each output $y_j$ as the corresponding spread. In this study, left- and right-spreads of the observed outputs were called tolerance levels. The tolerance level for the $j$th data-point is denoted by $e_j$.

The possibilistic model states that the resulting output from the UAM should cover all of the observed data-points within the given tolerance- and h-level. In other words, $\left[ \tilde{Y}^*(x_j) \right]_h$ should approach to $\left[Y_j\right]_h$ from the upper side; i.e., $\left[\tilde{Y}^*(x_j)\right]_h$ should be the least interval among all feasible solutions. This brings up the following constraints:

$$
\left[\tilde{Y}^*(x_j)\right]_h \supseteq \left[Y_j\right]_h \Leftrightarrow
\left\{
\begin{array}{l}
b^t x_j + (1-h)\theta_R^*(x_j) \geq y_j + (1-h)e_j \\
b^t x_j - (1-h)\theta_L^*(x_j) \leq y_j - (1-h)e_j
\end{array}
\right\}, j = 1,\ldots,nd \quad , \tag{34}
$$
$$
f \geq 0, \, g \geq 0, \, p \geq 0, \, q \geq 0,
$$

where $x_j = \left(1, x_{j,1}, x_{j,2},\ldots, x_{j,nv}\right)$, $j = 1,\ldots,nd$ and the term $x_{j,nv}$ denotes the value of the variable $nv$ of the $j$th data point.

On the other hand, according to necessity model, the $h$-level set of the $\tilde{Y}_*(x_j)$ should be included in the h-level set of the given output $Y_j$. In other words, $\left[\tilde{Y}_*(x_j)\right]$ should approach to $\left[Y_j\right]_h$ from the lower side, i.e., $\left[\tilde{Y}_*(x_j)\right]$ should be the greatest interval among all feasible solutions. This can be written in a constraint form as:

$$
\left[\tilde{Y}_*(x_j)\right]_h \subseteq \left[Y_j\right]_h \Leftrightarrow
\left\{
\begin{array}{l}
b^t x_j + (1-h)\theta_{*R}(x_j) \leq y_j + (1-h)e_j \\
b^t x_j - (1-h)\theta_{*L}(x_j) \geq y_j - (1-h)e_j
\end{array}
\right\}, j = 1,\ldots,nd \quad . \tag{35}
$$
$$
f \geq 0, \, g \geq 0, \, i = 0,\ldots,n
$$

Integrating the possibility and necessity models by taking into account the inclusion relation $\left[\tilde{Y}_*(x_j)\right]_h \subseteq \left[\tilde{Y}^*(x_j)\right]_h$ a quadratic programming formulation of the IT2F regression model is formulated as:

$$
\min_{b,f,g,p,q} J = \sum_{j=1}^{nd}\left(y_j - b^t x_j\right)^2 + (1-h)\sum_{j=1}^{nd}\left(p^t|x_j| + q^t|x_j|\right) + \xi\left(f^t f + g^t g + p^t p + q^t q\right)
$$
$$
\text{Subject to}
\left\{
\begin{array}{l}
b^t x_j + (1-h)\theta_R^*(x_j) \geq y_j + (1-h)e_j \\
b^t x_j - (1-h)\theta_L^*(x_j) \leq y_j - (1-h)e_j \\
b^t x_j + (1-h)\theta_{*R}(x_j) \leq y_j + (1-h)e_j \\
b^t x_j - (1-h)\theta_{*L}(x_j) \geq y_j - (1-h)e_j
\end{array}
\right\}, j = 1,\ldots,nd \quad , \tag{36}
$$
$$
f \geq 0, \, g \geq 0, \, p \geq 0, \, q \geq 0
$$

where $\xi$ is a small positive number. The term $\xi\left(f^t f + g^t g + p^t p + q^t q\right)$ is inserted into the objective function so that the objective function becomes a quadratic function with respect to decision variables $b, f, g, p,$ and $q$. The obtained UAM and LAM by solving the above integrated quadratic programming model always satisfy inclusion relation $Y_*(x) \subseteq Y^*(x)$ at the h-level.

### 3.3. Dynamic MADM Model via Proposed IT2F Functions

In this section, the proposed IT2F functions approach was given in a step-by-step manner. The flowchart of the proposed model is illustrated in Figure 2.

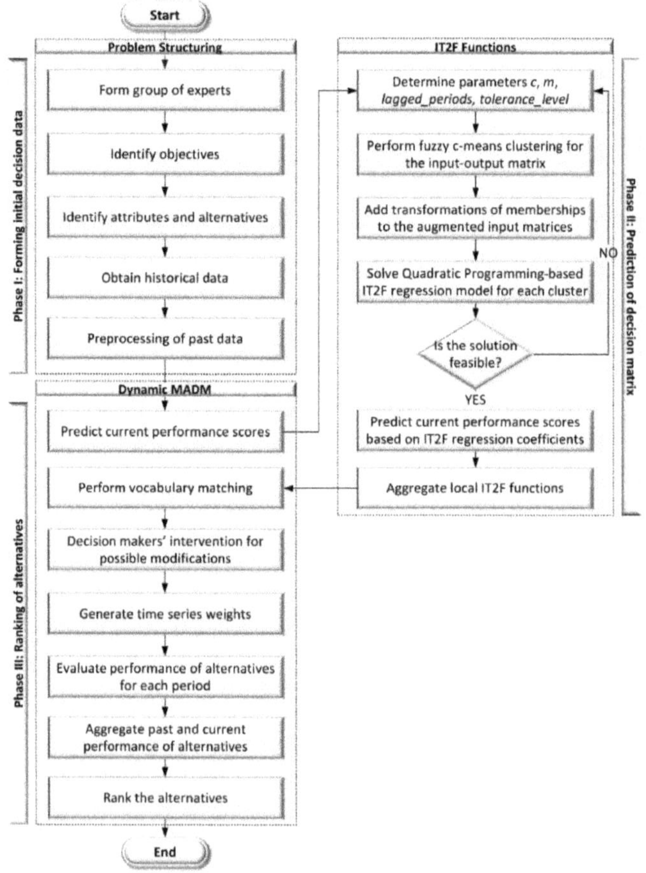

**Figure 2.** Flowchart of the proposed model.

### 3.3.1. Phase-I: Problem Structuring

Step 1: Problem-framing: In this step, a group of experts decided on the objective of the study, and the attributes and alternatives were identified. Here, expert opinions and the literature surveys helped to arrive at problem-framing.

Step 2: Obtaining historical data: Historical records were identified and the past data were fetched from the databases. Past data contained performance values of alternatives with respect to attributes at different periods as given in Equation (37).

$$A(t_1) = \begin{bmatrix} a_{11}(t_1) & a_{12}(t_1) & \cdots & a_{1M}(t_1) \\ a_{21}(t_1) & a_{22}(t_1) & \cdots & a_{2M}(t_1) \\ \vdots & \vdots & \ddots & \vdots \\ a_{N1}(t_1) & a_{N2}(t_1) & \cdots & a_{NM}(t_1) \end{bmatrix}, \ldots, A(t_H) = \begin{bmatrix} a_{11}(t_H) & a_{12}(t_H) & \cdots & a_{1M}(t_H) \\ a_{21}(t_H) & a_{22}(t_H) & \cdots & a_{2M}(t_H) \\ \vdots & \vdots & \ddots & \vdots \\ a_{N1}(t_H) & a_{N2}(t_H) & \cdots & a_{NM}(t_H) \end{bmatrix}, \quad (37)$$

where $A(t_1)$ and $A(t_H)$ are the decision matrices at the first and last period of the historical data, respectively.

This data can be unstructured so that the preprocessing is required.

Step 3: Preprocessing of historical data: In order to ensure accurate and meaningful analysis, data cleaning and preprocessing techniques are implemented in this step. Bad or missing data are eliminated by removing or replacing. Abrupt changes and local optima values are also identified. Smoothing or de-trending methods can be applied to remove noise.

Moreover, the historical decision making matrices are arranged as time series data. Here, for each alternative and attribute pair, time series data are formed. Mathematically speaking, the performance scores for a particular alternative and attribute $\left(a_{ij}(t_1), a_{ij}(t_2), \ldots, a_{ij}(t_H)\right)$ are collected from each period and the time series $y = (y_1, y_2, \ldots, y_H)^t$ is formed, where the number of points is equal to number of periods $H$.

Then, the lagged matrices are constructed where number of lagged periods is denoted by $p$. Note that $j$th data point in the input matrix $x_j = \left(x_{t-1,j}, x_{t-2,j}, \ldots, x_{t-p,j}\right)^t$ will be used to estimate $y_j$, $j = 1, 2, \ldots, nd$, where $nd$ is equal to $H - p$. The inputs and the outputs of the system are given as:

$$
X = \begin{bmatrix}
x_{t-1,1} & x_{t-2,1} & \cdots & x_{t-p,1} \\
x_{t-1,2} & x_{t-2,2} & \cdots & x_{t-p,2} \\
\vdots & \vdots & \ddots & \vdots \\
x_{t-1,nd} & x_{t-2,nd} & \cdots & x_{t-p,nd}
\end{bmatrix}, \quad
Y = \begin{bmatrix}
y_1 \\
y_2 \\
\vdots \\
y_{nd}
\end{bmatrix},
\tag{38}
$$

where $x_{t-p,nd}$ represents the value of variable $t - p$ of the data point $nd$.

Let the number of past decision matrices are four, and the lagged periods are determined as two. Suppose that decision makers are concerned with the past performance of the second alternative with respect to the first attribute. The corresponding time series data, input and output matrices are illustrated in Table 1.

**Table 1.** Illustration of the four-period example.

| Period | Past Decision Matrices | Corresponding Time Series | Input Matrix | Output Matrix |
|---|---|---|---|---|
| $t_1$ | $\begin{array}{cccc} & A_1 & A_2 & A_3 \\ C_1[ & \cdots & a_{12}(t_1)=5 & \cdots \ ] \end{array}$ | | | |
| $t_2$ | $\begin{array}{cccc} & A_1 & A_2 & A_3 \\ C_1[ & \cdots & a_{12}(t_2)=3 & \cdots \ ] \end{array}$ | $\begin{bmatrix} y_1=5 \\ y_2=3 \\ y_3=6 \\ y_4=8 \end{bmatrix}$ | $X = \begin{bmatrix} y_1 & y_2 \\ y_2 & y_3 \end{bmatrix}$ $= \begin{bmatrix} 5 & 3 \\ 3 & 6 \end{bmatrix}$ | $Y = \begin{bmatrix} y_3 \\ y_4 \end{bmatrix}$ $= \begin{bmatrix} 6 \\ 8 \end{bmatrix}$ |
| $t_3$ | $\begin{array}{cccc} & A_1 & A_2 & A_3 \\ C_1[ & \cdots & a_{12}(t_3)=6 & \cdots \ ] \end{array}$ | | | |
| $t_4$ | $\begin{array}{cccc} & A_1 & A_2 & A_3 \\ C_1[ & \cdots & a_{12}(t_4)=8 & \cdots \ ] \end{array}$ | | | |

Note that because the lagged periods are two, number of data points is $H - p = 4 - 2 = 2$.

### 3.3.2. Phase-II: Training of Fuzzy Functions Approach

Step 4: Determining parameters of the IT2F functions model: In this step, parameters of the fuzzy c-means clustering algorithm were determined. The number of clusters $c$, fuzzification coefficient $m$, lagged_periods, and tolerance_level for calculating possibility- and necessity-based constraints in IT2F regression were defined.

Step 5: Performing fuzzy c-means clustering to input-output model: In this step, inputs and outputs of the system were used to carry out fuzzy c-means clustering. Having the inputs of the system in the form of lagged variables, the next step was to form the input–output matrix $Z$. The matrix

$Z = (X, Y)$ is composed of the input matrix $X$ and output matrix $Y$. Then, elements of the $Z$ matrix $z_j$ were clustered by using the FCM algorithm. FCM was applied by using the following formulas:

$$v_i = \frac{\sum\limits_{k=1}^{nd} \mu_{ij}^m z_j}{\sum\limits_{j=1}^{nd} \mu_{ij}^m}, \quad i = 1, 2, \ldots, c, \tag{39}$$

$$\mu_{ij} = \frac{1}{\sum\limits_{h=1}^{c} \left( \frac{\|v_i - z_j\|}{\|v_h - z_j\|} \right)^{\frac{2}{m-1}}}, \quad i = 1, 2, \ldots, c; \; j = 1, 2, \ldots, nd, \tag{40}$$

where $\|.\|$ represents the Euclidian distance.

Step 6: Generating augmented input matrices: The augmented input matrix is obtained by adding memberships and their transformations into the original input matrix. Based on the cluster centers found in the Step 5, membership values of the input space are calculated as:

$$\mu_{ij} = \frac{1}{\sum\limits_{h=1}^{c} \left( \frac{\|v_i - x_j\|}{\|v_h - x_j\|} \right)^{\frac{2}{m-1}}}, \quad i = 1, 2, \ldots, c; \; j = 1, 2, \ldots, nd, \tag{41}$$

where $x$ denotes the input matrix.

Membership values of each input data sample, $\mu_{ij}$ and their transformations are augmented to the original input matrix for the $i$th cluster as:

$$\phi_i = \begin{bmatrix} 1 & \mu_{i,1} & \exp(\mu_{i,1}) & (\mu_{i,1})^p & x_{t-1,1} & \cdots & x_{t-p,1} \\ 1 & \mu_{i,2} & \exp(\mu_{i,2}) & (\mu_{i,2})^p & x_{t-1,2} & \cdots & x_{t-p,2} \\ \vdots & \vdots & \vdots & \vdots & \vdots & \vdots & \vdots \\ 1 & \mu_{i,j} & \exp(\mu_{i,j}) & (\mu_{i,j})^p & x_{t-1,j} & \cdots & x_{t-p,j} \\ \vdots & \vdots & \vdots & \vdots & \vdots & \ddots & \vdots \\ 1 & \mu_{i,nd} & \exp(\mu_{i,nd}) & (\mu_{i,nd})^p & x_{t-1,nd} & \cdots & x_{t-p,nd} \end{bmatrix}, \tag{42}$$

where $j$-th data point is represented by $\phi_{i,j} = \left( 1, \mu_{i,j}, \exp(\mu_{i,j}), \mu_{i,j}^2, x_{t-1,j}, \ldots, x_{t-p,j} \right)^t$.

The schematic representation of the proposed IT2F functions is given in Figure 3.

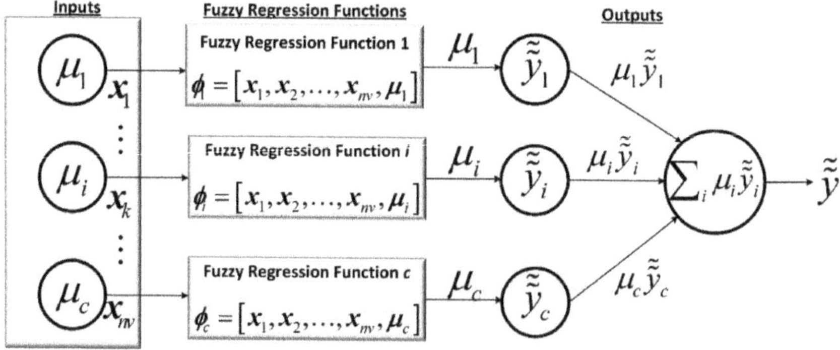


**Figure 3.** Proposed IT2F functions approach.

Step 7: Solving quadratic programming model for each cluster: Fuzzy regression coefficients are calculated for each cluster by solving a quadratic programming model:

$$\min_{b,f,g,p,q} J = \sum_{j=1}^{nd} \left(y_j - b^t\phi_{i,j}\right)^2 + (1-h)\sum_{j=1}^{nd}\left(p^t|\phi_{i,j}| + q^t|\phi_{i,j}|\right) + \xi\left(f^tf + g^tg + p^tp + q^tq\right)$$

$$\text{Subject to} \quad \left\{ \begin{array}{l} b^t\phi_{i,j} + (1-h)\theta_R^*(\phi_{i,j}) \geq y_j + (1-h)e_j \\ b^t\phi_{i,j} - (1-h)\theta_L^*(\phi_{i,j}) \leq y_j - (1-h)e_j \\ b^t\phi_{i,j} + (1-h)\theta_{*R}(\phi_{i,j}) \leq y_j + (1-h)e_j \\ b^t\phi_{i,j} - (1-h)\theta_{*L}(\phi_{i,j}) \geq y_j - (1-h)e_j \end{array} \right\} , j = 1,\ldots,nd \qquad (43)$$

$$f \geq 0, \; g \geq 0, \; p \geq 0, \; q \geq 0$$

where regression coefficients are IT2F numbers represented by $\tilde{\tilde{\beta}} = ((b, f+p, g+q), (b, f, g))$,

Step 8: Collecting predictions of local fuzzy functions: Predicted output values are calculated as:

$$\tilde{\tilde{y}}_i = \phi_i \otimes \tilde{\tilde{\beta}}_i, \qquad (44)$$

where $\tilde{\tilde{y}}_i = \left(\tilde{\tilde{y}}_{i,1}, \tilde{\tilde{y}}_{i,2}, \ldots, \tilde{\tilde{y}}_{i,j}, \ldots, \tilde{\tilde{y}}_{i,nd}\right)^t$, $\tilde{\tilde{\beta}}_i$ is the regression coefficients of the $i$th local fuzzy function and $\otimes$ denotes the fuzzy matrix multiplication.

Step 9: Aggregating local IT2F functions: Finally, outputs of the local fuzzy functions $\tilde{\tilde{y}}_i$ are weighted by the corresponding membership values and predicted IT2F output is calculated:

$$\tilde{\tilde{Y}}_j = \frac{\sum_i^c \tilde{\tilde{y}}_{i,j}\mu_{i,j}}{\sum_{i=1}^c \mu_{i,j}}, \; j = 1, 2, \ldots, nd, \qquad (45)$$

where $\tilde{\tilde{Y}}_j$ is the predicted value of the $j$th data point.

### 3.3.3. Phase-III: Ranking of Alternatives

Step 10: Performing vocabulary matching: The resulting values of the IT2F functions were inherently IT2F sets. Since experts often linguistically evaluate the objects in the decision making applications and it is difficult to analytically interpret the obtained numerical values, there was a need for transforming IT2F functions results into the linguistic terms. For that aim, similarity-based vocabulary matching was implemented in this step.

Let $V \in \{V_1, V_2, \ldots, V_U\}$ represents the vocabulary of the linguistic terms, i.e., $V_U$ denotes the linguistic term very good, $V_{U-1}$ denotes the term good, etc. The linguistic outputs of the IT2F functions approach can be given as:

$$\tilde{\tilde{A}}'(t_C) = \begin{bmatrix} \tilde{\tilde{a}}'_{11}(t_C) & \tilde{\tilde{a}}'_{12}(t_C) & \cdots & \tilde{\tilde{a}}'_{1M}(t_C) \\ \tilde{\tilde{a}}'_{21}(t_C) & \tilde{\tilde{a}}'_{22}(t_C) & \cdots & \tilde{\tilde{a}}'_{2M}(t_C) \\ \vdots & \vdots & \ddots & \vdots \\ \tilde{\tilde{a}}'_{N1}(t_C) & \tilde{\tilde{a}}'_{N2}(t_C) & \cdots & \tilde{\tilde{a}}'_{NM}(t_C) \end{bmatrix}. \qquad (46)$$

The $\tilde{\tilde{a}}'_{ij}$ values are calculated as:

$$\tilde{\tilde{a}}'_{ij} = \underset{l \in \{1,2,\ldots,U\}}{\arg\max} SM\left(\tilde{\tilde{a}}'_{ij}, V_l\right), \qquad (47)$$

where $SM$ represents the Jaccard similarity measure given earlier, and $V_l$ is the $l$th linguistic term in the vocabulary.

Step 11: Modifying solutions if necessary: In this step, the results of IT2F functions in the form of linguistic variables were presented to the decision makers. In other words, the current decision matrix was automatically generated based on the past data. The decision makers evaluated the results and made necessary modifications if needed. The illustration of this process is given in Figure 4. Here, the decision makers' perceptions had a pivotal role. For example, decision makers might decide on the fact that the performance of the first alternative with respect to the first and second attributes needs to be modified. Then the decision matrix takes the form as given in Equation (48).

$$
\tilde{A}''(t_C) = \begin{bmatrix} \tilde{\tilde{a}}''_{11}(t_C) & \tilde{\tilde{a}}'_{12}(t_C) & \cdots & \tilde{\tilde{a}}'_{1M}(t_C) \\ \tilde{\tilde{a}}''_{21}(t_C) & \tilde{\tilde{a}}'_{22}(t_C) & \cdots & \tilde{\tilde{a}}'_{2M}(t_C) \\ \vdots & \vdots & \ddots & \vdots \\ \tilde{\tilde{a}}'_{N1}(t_C) & \tilde{\tilde{a}}'_{N2}(t_C) & \cdots & \tilde{\tilde{a}}'_{NM}(t_C) \end{bmatrix},
\tag{48}
$$

where $\tilde{\tilde{a}}''_{ij}(t_C)$ represents the subjective judgments of the decision makers and $\tilde{\tilde{a}}'_{ij}(t_C)$ denotes the IT2F function result.

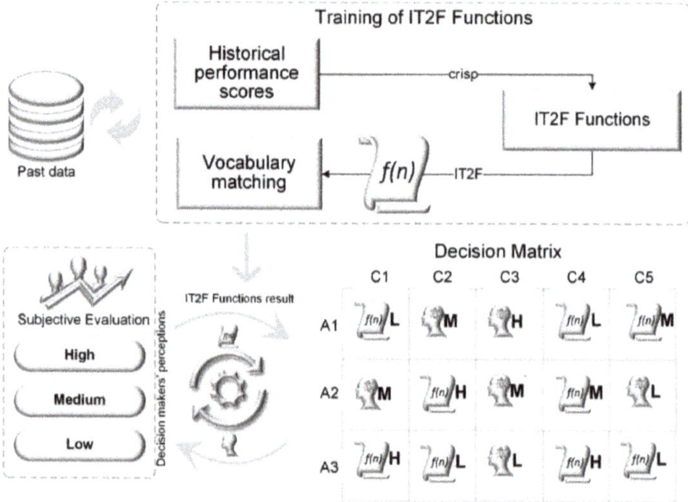


**Figure 4.** Interactive process of the proposed model.

Step 12: Generating time series weights: In this step, time series weights were generated. A basic unit-interval monotonic (BUM) function was used to generate weights. Yager [55] defined the BUM function as $Q : [0,1] \to [0,1]$, where the weights are considered as quantifiers underlying the information fusion process.

- $Q(0) = 0$,
- $Q(1) = 1$,
- $Q(x) \geq Q(y)$, if $x > y$, where $Q(x)$ is a monotonically non-decreasing function defined in the unit interval $[0,1]$.

Based on the BUM function, the time series weights are generated as:

$$
\xi(t_k) = Q\left(\frac{k}{p}\right) - Q\left(\frac{k-1}{p}\right), \quad k = 1, 2, \ldots, p,
\tag{49}
$$

where $Q(x) = \frac{e^{\alpha x} - 1}{e^{\alpha} - 1}$, $\alpha > 0$.

According to the BUM function, the more the period is closer to the current period, the higher the weight of that period, which is a desired behavior for real-world applications.

Step 13: Evaluating performance of alternatives at each period: Performance of each alternative is calculated for different periods separately. In this step, variety of MADM methods can be employed to obtain a performance indicator. For the sake of simplicity, well-known closeness coefficient measures are used to evaluate performance of alternatives at each period. As the current decision matrix comprises of IT2F evaluations, computational steps for the IT2FSs are given in this section in order to avoid repetition. Note that the required computations for the past data are the same, except the fact that numerical values are crisp. First, the decision matrices related to past periods are normalized as given earlier in Equations (1) and (2). As the current decision matrix consists of IT2F evaluations, normalization is conducted based on the Equations (50) and (51).

$$\tilde{\tilde{r}}_{ij}(t_k) = \left( \left( \frac{\bar{a}_{ij1}(t_k)}{\max_j\{\bar{a}_{ij3}(t_k)\}}, \frac{\bar{a}_{ij2}(t_k)}{\max_j\{\bar{a}_{ij3}(t_k)\}}, \frac{\bar{a}_{ij3}(t_k)}{\max_j\{\bar{a}_{ij3}(t_k)\}}; 1 \right), \left( \frac{\underline{a}_{ij1}(t_k)}{\max_j\{\bar{a}_{ij3}(t_k)\}}, \frac{\underline{a}_{ij2}(t_k)}{\max_j\{\bar{a}_{ij3}(t_k)\}}, \frac{\underline{a}_{ij3}(t_k)}{\max_j\{\bar{a}_{ij3}(t_k)\}}; 1 \right) \right), \text{if } i \in \Omega_b, \tag{50}$$

$$\tilde{\tilde{r}}_{ij}(t_k) = \left( \left( \frac{\min_j\{\underline{a}_{ij1}(t_k)\}}{\bar{a}_{ij3}(t_k)}, \frac{\min_j\{\underline{a}_{ij1}(t_k)\}}{\bar{a}_{ij2}(t_k)}, \frac{\min_j\{\underline{a}_{ij1}(t_k)\}}{\bar{a}_{ij1}(t_k)}; 1 \right), \left( \frac{\min_j\{\underline{a}_{ij1}(t_k)\}}{\underline{a}_{ij3}(t_k)}, \frac{\min_j\{\underline{a}_{ij1}(t_k)\}}{\underline{a}_{ij2}(t_k)}, \frac{\min_j\{\underline{a}_{ij1}(t_k)\}}{\underline{a}_{ij1}(t_k)}; 1 \right) \right), \text{if } i \in \Omega_c. \tag{51}$$

Then, the weighted normalized decision matrices are calculated as:

$$\tilde{\tilde{v}}_{ij}(t_k) = \tilde{\tilde{r}}_{ij}(t_k) \times w_i, \quad i = 1, 2, \ldots, N, \ j = 1, 2, \ldots, M, \ k = 1, 2, \ldots, H. \tag{52}$$

When the weighted normalized decision matrices are constructed, the next step is to calculate the positive ideal solutions (PIS) and negative ideal solutions (NIS) as:

$$\begin{aligned} \text{PIS} &= \left( v_1^+, v_2^+, \ldots, v_N^+ \right) \\ &= \left\{ \left( \max_j\{Rank(\tilde{\tilde{v}}_{ij}(t_k))\} \right) \Big| i \in \Omega_b, \left( \min_j\{Rank(\tilde{\tilde{v}}_{ij}(t_k))\} \right) \Big| i \in \Omega_c \right\}, \ k = 1, 2, \ldots, H \end{aligned} \tag{53}$$

$$\begin{aligned} \text{NIS} &= \left( v_1^-, v_2^-, \ldots, v_N^- \right) \\ &= \left\{ \left( \min_j\{Rank(\tilde{\tilde{v}}_{ij}(t_k))\} \right) \Big| i \in \Omega_b, \left( \max_j\{Rank(\tilde{\tilde{v}}_{ij}(t_k))\} \right) \Big| i \in \Omega_c \right\}, \ k = 1, 2, \ldots, H \end{aligned} \tag{54}$$

Then, separation measures are calculated by using the Euclidean distance as:

$$D_j^+(t_k) = \sqrt{\sum_{i=1}^{N} \left( Rank(\tilde{\tilde{v}}_{ij}(t_k)) - v_i^+(t_k) \right)^2}, \tag{55}$$

$$D_j^-(t_k) = \sqrt{\sum_{i=1}^{N} \left( Rank(\tilde{\tilde{v}}_{ij}(t_k)) - v_i^-(t_k) \right)^2}. \tag{56}$$

Finally, closeness coefficients are calculated as:

$$CC_j(t_k) = \frac{D_j^-(t_k)}{\left( D_j^+(t_k) + D_j^-(t_k) \right)}, \tag{57}$$

where $CC_j(t_k)$ represents the closeness coefficient of the $j$th alternative at period $t_k$.

Step 14: Aggregating past and current performance of alternatives: Finally, a dynamic weighted averaging (DWA) operator was utilized to obtain final ranking values of alternatives.

$$DWA_{\xi(t)}\big(CC_j(t_1), CC_j(t_2), \dots, CC_j(t_H)\big) = \sum_{k=1}^{H} \xi(t_k) CC_j(t_k). \tag{58}$$

Step 15: Rank the alternatives: When the past and current performance scores of alternatives were aggregated, alternatives were ranked based on their ranking values. Higher ranking value implies superiority of an alternative.

## 4. Case Study

One of the most important assets of a company is undoubtedly the human resources (HRs). Regardless of how the other resources are managed in an organization, inadequacies in the management of HRs result in poor performance of many operations. Therefore, firms have steadily recognized the importance of HRs and have been taking necessary actions to increase overall performance.

Personnel promotion is a significant task in HR management that aims to select the right person for the right job. Despite its similarity with the personnel selection problem, personnel promotion problem deals with selecting appropriate personnel for higher positions within the firms' current personnel rather than evaluating the applicants from outside the firm. Personnel promotion problem can be defined as selecting the most qualified employee among the available candidates for a vacant position by considering their performance during their employment at the firm. Hence, the personnel promotion problem is an inherently dynamic MADM problem as the temporal performance of employees are taken into consideration with respect to predetermined attributes. Unfortunately, many enterprises are not aware of the methodologies and tools to utilize historical records in their HR practices. As evaluating personnel with respect to their performance on a diverse set of criteria during their employment is cognitively demanding, firms should be supported with relevant data-driven tools.

In this study, a real-life personnel promotion problem is considered and the proposed model is implemented. The company, which was contacted within the scope of this study, was a medium sized firm involved in automobile subsidiary industry. Due to the firm policy, it was named as company A. Company A had hired three students as a part-time employee for their continuous improvement project. Upon graduation, at most two students with qualified skills will be offered to full-time job so that the supervisors have been evaluating students' performance in a monthly basis. Therefore, performance data was used to demonstrate the procedural steps of the proposed model.

### 4.1. Structering Personnel Promotion Problem

Step 1: In this step, the experts were identified based on their professional backgrounds. The heterogeneity of the experts was assured and three experts were identified. The experts were the directors of HR and production and quality control departments. After identification of the experts, participants were elucidated about the scope and details of the study. The evaluation criteria determined by the experts were employed within the scope of this study. The evaluation criteria consisted of content-specific knowledge ($C_1$), communication skills ($C_2$), job involvement ($C_3$), organizational commitment ($C_4$), and problem solving skills ($C_5$). The decision hierarchy is given in Figure 5.

Step 2: Historical performance data of the employees with respect to predefined attributes were obtained. The company had a performance evaluation system that assigned performance scores between 0–10, 0 and 10 represented the worst and the best values, respectively. In the scope of this study, performance values of the past 1.5 years (18 months) were directly used. The historical data of the performance scores of the employee 1 with respect to decision attributes is given in Table 2. Similarly, the historical records for the employee 2 and employee 3 were obtained.

Step 3: Historical data was organized in tables, and missing values were sought for. No missing values were identified within the 18-month data period. On the other hand, it was decided not to implement normalization techniques as the proposed model could handle numeric values in the range 0–10. The historical decision matrices were also transformed into time series vectors so that the IT2F functions method could be implemented.

Afterwards, the data set was divided into training and test sets. The training set consisted of the first 80% part of the historical data and the rest of the data points were allocated to the test set.

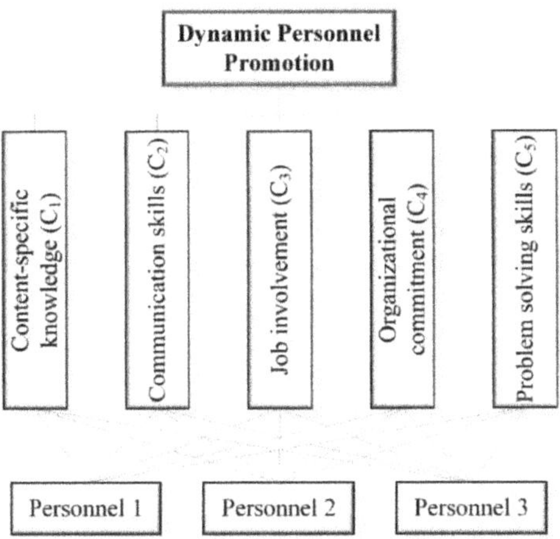


**Figure 5.** Decision hierarchy.

**Table 2.** Historical data of employee 1.

| Period | $C_1$ | $C_2$ | $C_3$ | $C_4$ | $C_5$ |
|---|---|---|---|---|---|
| 1 | 6 | 7 | 3 | 3 | 7 |
| 2 | 6 | 7 | 2 | 2 | 6 |
| 3 | 7 | 8 | 3 | 3 | 5 |
| 4 | 6 | 9 | 4 | 5 | 5 |
| 5 | 6 | 8 | 3 | 6 | 5 |
| 6 | 6 | 9 | 4 | 8 | 4 |
| 7 | 7 | 8 | 4 | 8 | 4 |
| 8 | 6 | 7 | 5 | 7 | 4 |
| 9 | 6 | 6 | 6 | 8 | 5 |
| 10 | 6 | 5 | 6 | 8 | 5 |
| 11 | 7 | 4 | 7 | 9 | 6 |
| 12 | 6 | 3 | 6 | 10 | 6 |
| 13 | 5 | 4 | 6 | 10 | 6 |
| 14 | 5 | 4 | 8 | 8 | 6 |
| 15 | 4 | 4 | 6 | 7 | 8 |
| 16 | 4 | 3 | 6 | 7 | 7 |
| 17 | 4 | 3 | 6 | 7 | 8 |
| 18 | 4 | 2 | 6 | 9 | 8 |

### 4.2. Estimating the Current Decision Matrix

Having obtained the historical data, the next step was to employ the proposed IT2F functions approach to estimate the current decision matrix based on past data.

Step 4: In this step, parameters of the model were identified. The model parameters were the number of lagged_periods, tolerance_levels, number of cluster, and the degree of fuzzification. Instead of using cluster validity indices in order to select the optimum $c$ and $m$ parameters, the model parameters were selected by using a grid search for each parameter based on Root Mean Square

Error (RMSE) performance metric. The tolerance_levels were set to 30% for being able to find feasible solutions for every combination of parameters when solving quadratic programming models.

The identified parameters are given in Table 3.

Table 3. Parameters of the model.

| Alternative | Criteria | Parameters | | |
|:---:|:---:|:---:|:---:|:---:|
| | | $c$ | $m$ | Lagged_Periods |
| | 1 | 2 | 1.6 | 5 |
| | 2 | 5 | 1.6 | 5 |
| 1 | 3 | 4 | 2.1 | 4 |
| | 4 | 5 | 2.1 | 5 |
| | 5 | 4 | 1.6 | 4 |
| | 1 | 4 | 1.6 | 5 |
| | 2 | 5 | 2.1 | 5 |
| 2 | 3 | 2 | 1.1 | 5 |
| | 4 | 5 | 2.1 | 5 |
| | 5 | 5 | 1.6 | 5 |
| | 1 | 2 | 2.1 | 5 |
| | 2 | 2 | 1.6 | 5 |
| 3 | 3 | 3 | 1.6 | 5 |
| | 4 | 5 | 2.1 | 5 |
| | 5 | 5 | 1.6 | 5 |

Step 5: In this step, the input and output matrices were combined and the FCM clustering algorithm was performed based on the parameters identified in the previous step. By utilizing FCM clustering algorithm, cluster centers were identified. These cluster centers had a key role in the Turksen's fuzzy system model in which the clusters were used as a fuzzification engine of the classical FRB systems.

Step 6: Having identified cluster centers, the next step was to find membership grades of the input data and to form the augmented input matrices by integrating membership transformations as explanatory variables. In this study, integration of membership degrees, exponentials of the memberships, and the square of the memberships were found to exhibit good performance so that these transformations were used to augment the input space.

Step 7: When the input and output matrices were formed by means of membership grades, quadratic programming model was solved in order to obtain fuzzy regression coefficients. The models were written in MATLAB 9.5.0 and quadratic programming models were solved via quadratic programming solver function cplexqp of the Cplex Optimization Studio 12.8. As the regression coefficients were defined as the distance to the center ($b$) earlier, their corresponding IT2F number representations are given in Tables 4 and 5 for the case of first alternative and first criteria.

Similarly, the same computations were performed for all of the alternative and criteria pairs. Note that number of clusters was based on the Table 3.

Step 8: When all of the IT2F regression coefficients were calculated, the output was predicted by using the obtained regression coefficients for each cluster. For the time series data of the alternative 1 and criteria 1, 2 local fuzzy functions were calculated. Since there were five clusters in the time series data of the alternative 1 and criteria 2, the total of five local fuzzy function results were obtained.

Step 9: In this step, the local fuzzy functions were aggregated and the estimated outputs were calculated. Performance of the proposed IT2F functions approach was compared with the IT2F regression model by means of RMSE and Mean Absolute Percentage Error (MAPE) metrics. Table 6 shows the performance comparison of the proposed IT2F functions approach.

**Table 4.** Cluster 1 results for the 1st alternative and 1st criteria.

| Variable | IT2F Regression Coefficients | | | | | IT2F Coefficients |
|---|---|---|---|---|---|---|
| | $b$ | $f$ | $g$ | $p$ | $q$ | |
| 1 | 10.572 | $8.65 \times 10^{-9}$ | 0.031 | $9.30 \times 10^{-11}$ | $9.13 \times 10^{-11}$ | ((10.572, 10.572, 10.603;1), (10.572, 10.572, 10.603;1)) |
| $\mu$ | −8.524 | $3.39 \times 10^{-9}$ | 0.070 | $6.45 \times 10^{-11}$ | $6.44 \times 10^{-11}$ | ((−8.524, −8.524, −8.454;1), (−8.524, −8.524, −8.454;1)) |
| $\exp(\mu)$ | −4.550 | $2.04 \times 10^{-9}$ | 0.016 | $2.79 \times 10^{-11}$ | $2.78 \times 10^{-11}$ | ((−4.55, −4.55, −4.534;1), (−4.55, −4.55, −4.534;1)) |
| $\mu^2$ | 13.466 | $7.27 \times 10^{-9}$ | 0.085 | $7.34 \times 10^{-11}$ | $7.34 \times 10^{-11}$ | ((13.466, 13.466, 13.551;1), (13.466, 13.466, 13.551;1)) |
| $x_{t-1}$ | 0.017 | 0.146043 | 0.010 | $2.67 \times 10^{-11}$ | $2.59 \times 10^{-11}$ | ((−0.129, 0.017, 0.027;1), (−0.129, 0.017, 0.027;1)) |
| $x_{t-2}$ | −0.613 | $1.24 \times 10^{-9}$ | 0.005 | 0.1 | 0.185714 | ((−0.713, −0.613, −0.423;1), (−0.613, −0.613, −0.609;1)) |
| $x_{t-3}$ | −0.211 | $1.44 \times 10^{-9}$ | 0.004 | $1.67 \times 10^{-11}$ | $1.63 \times 10^{-11}$ | ((−0.211, −0.211, −0.207;1), (−0.211, −0.211, −0.207;1)) |
| $x_{t-4}$ | 0.822 | 0.078111 | 0.126 | $1.48 \times 10^{-11}$ | $1.42 \times 10^{-11}$ | ((0.744, 0.822, 0.948;1), (0.744, 0.822, 0.948;1)) |
| $x_{t-5}$ | 0.038 | $1.44 \times 10^{-9}$ | 0.006 | $1.36 \times 10^{-11}$ | $1.34 \times 10^{-11}$ | ((0.038, 0.038, 0.044;1), (0.038, 0.038, 0.044;1)) |

**Table 5.** Cluster-2 results for the 1st alternative and 1st criteria.

| Variable | IT2F Regression Coefficients | | | | | IT2F Coefficients |
|---|---|---|---|---|---|---|
| | $b$ | $f$ | $g$ | $p$ | $q$ | |
| 1 | −10.907 | 1.28 | 1.018 | $4.25 \times 10^{-9}$ | $4.59 \times 10^{-9}$ | ((−12.184, −10.907, −9.889;1), (−12.184, −10.907, −9.889;1)) |
| $\mu$ | −18.964 | $3.03 \times 10^{-9}$ | 0.000 | $1.44 \times 10^{-8}$ | $1.44 \times 10^{-8}$ | ((−18.964, −18.964, −18.964;1), (−18.964, −18.964, −18.964;1)) |
| $\exp(\mu)$ | 15.077 | $1.79 \times 10^{-9}$ | 0.000 | $3.39 \times 10^{-9}$ | $3.61 \times 10^{-9}$ | ((15.077, 15.077, 15.077;1), (15.077, 15.077, 15.077;1)) |
| $\mu^2$ | −3.447 | $3.12 \times 10^{-9}$ | 0.000 | $7.20 \times 10^{-1}$ | 1.16 | ((−4.167, −3.447, −2.289;1), (−3.447, −3.447, −3.447;1)) |
| $x_{t-1}$ | −0.107 | $1.50 \times 10^{-9}$ | 0.000 | $8.08 \times 10^{-10}$ | $8.56 \times 10^{-10}$ | ((−0.107, −0.107, −0.107;1), (−0.107, −0.107, −0.107;1)) |
| $x_{t-2}$ | −0.725 | $9.06 \times 10^{-10}$ | 0.000 | $3.64 \times 10^{-7}$ | 0.024974 | ((−0.725, −0.725, −0.7;1), (−0.725, −0.725, −0.725;1)) |
| $x_{t-3}$ | −0.262 | $3.18 \times 10^{-9}$ | 0.000 | $6.04 \times 10^{-10}$ | $6.37 \times 10^{-10}$ | ((−0.262, −0.262, −0.262;1), (−0.262, −0.262, −0.262;1)) |
| $x_{t-4}$ | 0.735 | $3.18 \times 10^{-9}$ | 0.000 | $5.26 \times 10^{-10}$ | $5.27 \times 10^{-10}$ | ((0.735, 0.735, 0.735;1), (0.735, 0.735, 0.735;1)) |
| $x_{t-5}$ | 0.150 | $1.16 \times 10^{-5}$ | 0.001 | $5.74 \times 10^{-10}$ | $6.10 \times 10^{-10}$ | ((0.15, 0.15, 0.151;1), (0.15, 0.15, 0.151;1)) |

Despite its practicality, IT2F regression, which does not make use of FCM clustering and augmented input matrix as in the case of the proposed IT2F functions, cannot capture the trends and patterns in the historical dataset to the desired extent. Hence, both of the performance metrics were quite high. On the other hand, the proposed IT2F functions approach had successfully captured the patterns in the data thanks to its membership processing mechanisms. Figure 6 shows the estimated performance scores of the alternative 1 with respect to five criteria by means of the IT2F functions approach.

**Table 6.** Performance comparison of the proposed method and IT2F regression.

| Alternative | Criteria | Proposed IT2F Functions | | IT2F Regression | |
|---|---|---|---|---|---|
| | | **RMSE** | **MAPE** | **RMSE** | **MAPE** |
| 1 | 1 | 0.2307 | 2.3883 | 0.6974 | 11.4445 |
| | 2 | 0.1948 | 3.3149 | 0.7844 | 15.1122 |
| | 3 | 0.2519 | 4.0679 | 0.9828 | 14.0217 |
| | 4 | 0.375 | 3.4459 | 0.6952 | 7.2826 |
| | 5 | 0.3606 | 5.6463 | 0.8976 | 16.7376 |
| 2 | 1 | 0.2894 | 6.0687 | 0.4449 | 13.6456 |
| | 2 | 0.2392 | 2.4306 | 0.6793 | 7.5079 |
| | 3 | 0.3733 | 4.2811 | 0.9343 | 11.083 |
| | 4 | 0.4283 | 5.1677 | 0.5934 | 7.8132 |
| | 5 | 0.3035 | 4.0027 | 0.5657 | 7.9788 |
| 3 | 1 | 0.1457 | 1.722 | 0.7285 | 8.0463 |
| | 2 | 0.1686 | 3.9912 | 0.5833 | 17.4631 |
| | 3 | 0.2417 | 2.6302 | 0.6196 | 6.6876 |
| | 4 | 0.2142 | 2.135 | 0.3221 | 3.4379 |
| | 5 | 0.2205 | 2.7494 | 0.5567 | 7.9517 |

Note that the produced results were IT2F numbers. Another critical point is about whether the produced results satisfy the imposed constraints by possibility and necessity relationships. Figure 7 digs into the first data points of the predictions, which are illustrated in Figure 6. Note that the lower and upper tolerances covered the lower membership function, which was the property of the necessity constraints in the mathematical model. Furthermore, the lower and upper tolerances were covered by the upper membership function, which was due to the possibility constraints. It can be seen that both of the lower and upper tolerances lied within the FOU of the IT2F outputs. The estimated outputs were also inside the FOU and the centers of the IT2F outputs were quite close to the target values, which indicates the predictive power of the IT2F functions.

Based on the trained IT2F functions model, the current decision matrix was predicted. The resulting decision matrix is shown in Table 7.

As it is difficult for experts to interpret the produced results, they were transformed into linguistic terms in the next section.

**Table 7.** Resulting decision matrix of the IT2F functions.

| Criteria | Alternatives | | |
|---|---|---|---|
| | $A_1$ | $A_2$ | $A_3$ |
| $C_1$ | ((1.694, 2.99, 4.515;1), (2.064, 2.99, 3.822;1)) | ((2.163, 3.73, 5.516;1), (2.864, 3.73, 4.204;1)) | ((5.223, 7.902, 11.27;1), (5.601, 7.902, 10.396;1)) |
| $C_2$ | ((1.354, 2.255, 3.33;1), (1.891, 2.255, 2.623;1)) | ((4.992, 8.419, 12.522;1), (5.964, 8.419, 10.802;1)) | ((2.662, 3.919, 5.449;1), (3.204, 3.919, 4.596;1)) |
| $C_3$ | ((2.986, 5.358, 8.46;1), (4.13, 5.358, 6.362;1)) | ((2.194, 4.143, 5.991;1), (3.144, 4.143, 5.238;1)) | ((6.016, 9.229, 12.788;1), (6.68, 9.229, 11.589;1)) |
| $C_4$ | ((7.187, 10.353, 14.08;1), (8.084, 10.353, 12.123;1)) | ((0.469, 1.617, 3.242;1), (0.933, 1.617, 2.178;1)) | ((5.678, 8.415, 11.417;1), (6.071, 8.415, 10.508;1)) |
| $C_5$ | ((2.409, 5.252, 9.04;1), (3.372, 5.252, 7.208;1)) | ((4.427, 6.446, 9.052;1), (5.168, 6.446, 7.487;1)) | ((3.715, 6.246, 8.94;1), (4.566, 6.246, 7.472;1)) |

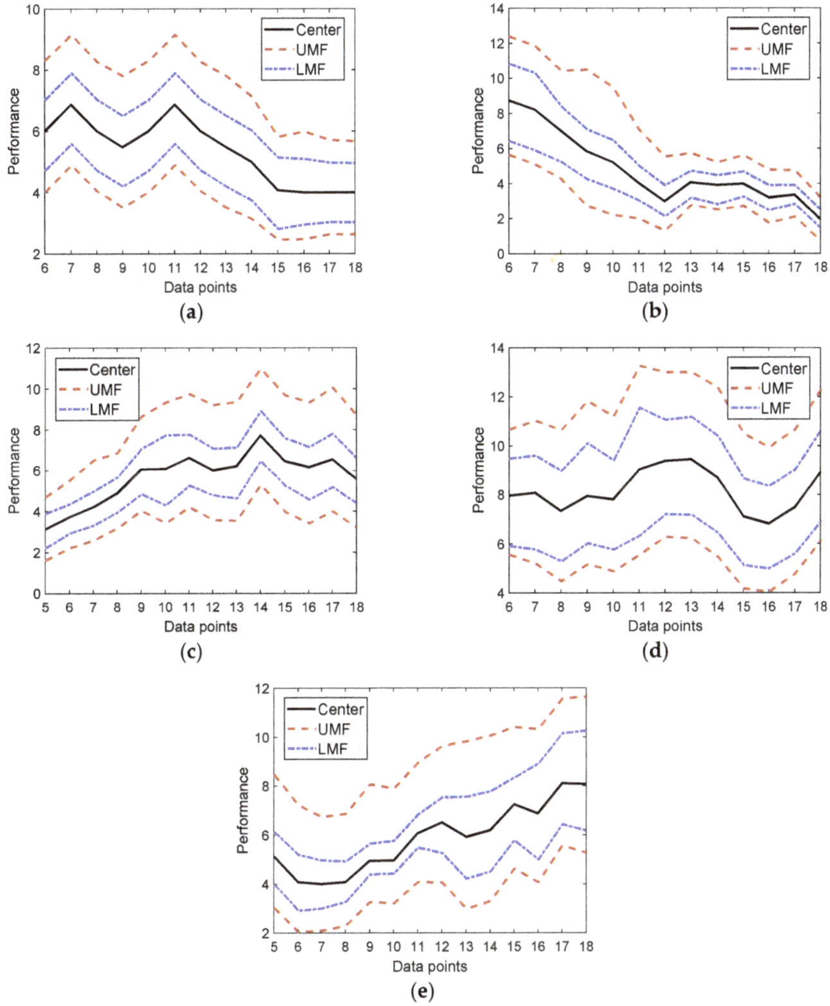

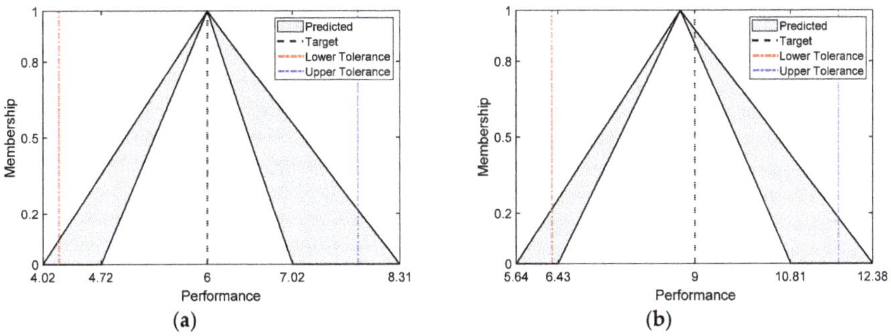

**Figure 6.** Estimations for the 1st alternative. (**a**) 1st criterion; (**b**) 2nd criterion; (**c**) 3rd criterion; (**d**) 4th criterion and (**e**) 5th criterion.

**Figure 7.** *Cont.*

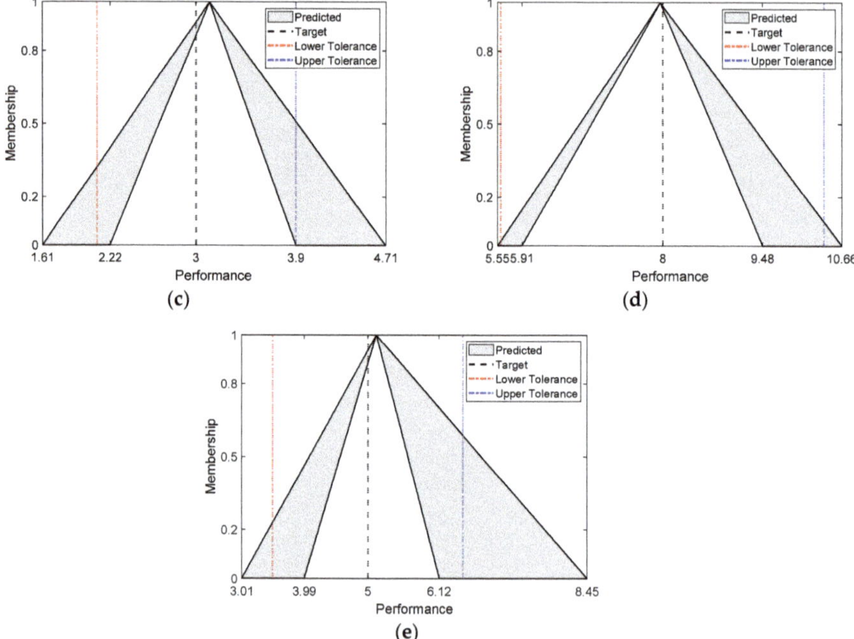

**Figure 7.** Details of the predictions for the 1st alternative. (**a**) 6th data point w.r.t. 1st criterion; (**b**) 6th data point w.r.t. 2nd criterion; (**c**) 5th data point w.r.t. 3rd criterion; (**d**) 6th data point w.r.t. 4th criterion and (**e**) 5th data point w.r.t. 5th criterion.

### 4.3. Ranking of Employees

Step 10: In this step, a similarity-based vocabulary matching was performed. First, the linguistic terms and their corresponding IT2F number were determined. Table 8 shows the linguistic terms of the vocabulary.

Based on the linguistic terms given in Table 8, similarity-based vocabulary matching was carried out. As a result, the linguistic decision matrix is given in Table 9.

Step 11: When the linguistic decision matrix was formed, the results were presented to the decision makers. Then, the decision makers were asked to change the performance values of the alternatives based on their perceptions. The decision makers might accept the decision matrix as it is or alter the whole performance values depending on their judgments. In this study, decision makers decided to change the performance values of the second and third alternatives with respect to first, fourth, and fifth criteria. The modified decision matrix is represented in Table 10. The modified linguistic terms are highlighted by gray shading.

**Table 8.** Vocabulary of linguistic terms.

| Linguistic Terms | | IT2F Number |
|---|---|---|
| **Symbol** | **Explanation** | |
| VL | Very Low | $((0, 0, 1;1), (0, 0, 0.5;1))$ |
| L | Low | $((0, 1, 3;1), (0.5, 1, 2;1))$ |
| ML | Medium Low | $((1, 3, 5;1), (2, 3, 4;1))$ |
| M | Medium | $((3, 5, 7;1), (4, 5, 6;1))$ |
| MH | Medium High | $((5, 7, 9;1), (6, 7, 8;1))$ |
| H | High | $((7, 9, 10;1), (8, 9, 9.5;1))$ |
| VH | Very High | $((9, 10, 10;1), (9.5, 10, 10;1))$ |

**Table 9.** Linguistic decision matrix.

| Criteria | Alternatives | | |
|---|---|---|---|
| | **A₁** | **A₂** | **A₃** |
| $C_1$ | ML | ML | MH |
| $C_2$ | ML | H | ML |
| $C_3$ | M | M | H |
| $C_4$ | H | L | H |
| $C_5$ | M | MH | MH |

**Table 10.** Modified decision matrix.

| Criteria | Alternatives | | |
|---|---|---|---|
| | **A₁** | **A₂** | **A₃** |
| $C_1$ | ML | MH | MH |
| $C_2$ | ML | H | ML |
| $C_3$ | M | M | H |
| $C_4$ | H | L | MH |
| $C_5$ | M | H | MH |

The decision makers accepted the linguistic terms produced by the IT2F functions model for the rest of the evaluations. By this way, required expert judgments were decreased by 80%, which dramatically increased the speed of decision making.

Step 12: In this step, time series weights were generated based on the BUM function. The only parameter required by BUM function was $\alpha$. Figure 8 illustrates the time series weights for different $\alpha$ values. In this study, $\alpha$ was selected as 0.5. Furthermore, results for the different $\alpha$ values were examined as well.

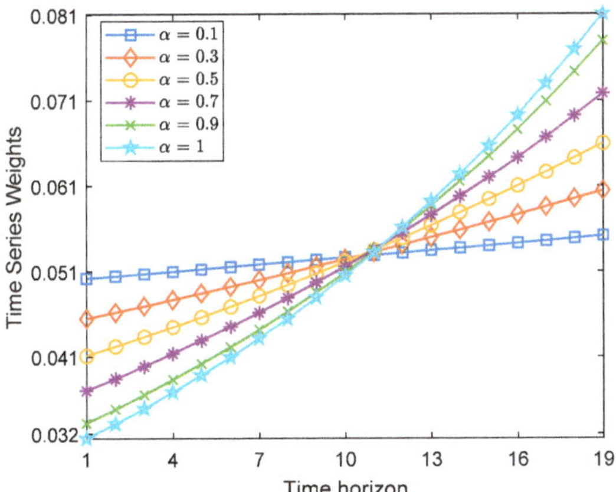


**Figure 8.** Time series weights.

Step 13: In this step, performance of each employee was evaluated at each period. Here, past and current decision matrices were evaluated one-by-one separately. As mentioned earlier, closeness coefficient is very practical to use so that for each period distance to positive and negative ideal solutions were calculated. The weights of criteria were assumed to be fixed throughout the periods and given as 0.10, 0.10, 0.15, 0.25, and 0.35, respectively. The separation measures are given in Tables 11 and 12.

**Table 11.** Distances between the performance of employees and the positive ideal solution.

| Periods | Distance to Positive Ideal Solution | | |
|---|---|---|---|
| | **Employee 1** | **Employee 2** | **Employee 3** |
| $t_1$ | 0.1458 | 0.1101 | 0.1111 |
| $t_2$ | 0.2259 | 0.1569 | 0.1084 |
| $t_3$ | 0.2155 | 0.1546 | 0.1161 |
| $t_4$ | 0.1604 | 0.1203 | 0.0850 |
| $t_5$ | 0.1730 | 0.1313 | 0.0938 |
| $t_6$ | 0.1700 | 0.0919 | 0.0833 |
| $t_7$ | 0.1700 | 0.0977 | 0.0750 |
| $t_8$ | 0.1714 | 0.0709 | 0.0643 |
| $t_9$ | 0.1256 | 0.0872 | 0.0750 |
| $t_{10}$ | 0.1303 | 0.0874 | 0.0750 |
| $t_{11}$ | 0.0960 | 0.0797 | 0.1086 |
| $t_{12}$ | 0.1146 | 0.0901 | 0.1397 |
| $t_{13}$ | 0.1204 | 0.0961 | 0.1393 |
| $t_{14}$ | 0.1004 | 0.0914 | 0.1050 |
| $t_{15}$ | 0.1226 | 0.1401 | 0.0797 |
| $t_{16}$ | 0.1264 | 0.1330 | 0.0972 |
| $t_{17}$ | 0.1422 | 0.1873 | 0.1001 |
| $t_{18}$ | 0.1404 | 0.2335 | 0.0981 |
| $t_C$ | 0.1725 | 0.1955 | 0.1143 |

**Table 12.** Distances between the performance of employees and the negative ideal solution.

| Periods | Distance to Negative Ideal Solution | | |
|---|---|---|---|
| | **Employee 1** | **Employee 2** | **Employee 3** |
| $t_1$ | 0.1409 | 0.1556 | 0.1541 |
| $t_2$ | 0.1169 | 0.1524 | 0.2462 |
| $t_3$ | 0.1161 | 0.1334 | 0.2155 |
| $t_4$ | 0.0898 | 0.0698 | 0.1613 |
| $t_5$ | 0.0995 | 0.0805 | 0.1762 |
| $t_6$ | 0.1011 | 0.1636 | 0.1836 |
| $t_7$ | 0.0983 | 0.1278 | 0.1857 |
| $t_8$ | 0.0744 | 0.1697 | 0.1807 |
| $t_9$ | 0.0684 | 0.1259 | 0.1385 |
| $t_{10}$ | 0.0563 | 0.0950 | 0.1336 |
| $t_{11}$ | 0.0811 | 0.1102 | 0.0717 |
| $t_{12}$ | 0.0997 | 0.1397 | 0.0750 |
| $t_{13}$ | 0.0817 | 0.1311 | 0.0750 |
| $t_{14}$ | 0.0680 | 0.1050 | 0.0914 |
| $t_{15}$ | 0.0927 | 0.0850 | 0.1247 |
| $t_{16}$ | 0.0810 | 0.1000 | 0.1244 |
| $t_{17}$ | 0.1218 | 0.1050 | 0.1719 |
| $t_{18}$ | 0.2143 | 0.1173 | 0.1966 |
| $t_C$ | 0.1873 | 0.1631 | 0.1751 |

Based on the distances from positive and negative ideal solutions, closeness coefficients were calculated as given in Table 13.

Step 14: In this step, closeness coefficients pertaining to the past and current performance scores of employees were aggregated by taking into account the time series weights. As there were numerous aggregation operators, the most practical one, namely the DWA operator, was used in this study. As a result of the DWA operator, rankings of employees were obtained. Moreover, for the purpose of comparison, different model components were activated and the results were examined. Table 14 summarizes the computational setting with different model components.

The proposed model integrates all of the model components. Different from the proposed model, the predicted decision matrix was not modified in model 1. In model 2, neither the linguistic decision matrix was generated nor the decision makers were allowed to change the decision matrix. Finally, model 3 made use of the IT2F functions, vocabulary matching, and modified preferences, but only the decision matrix of the current period was considered for ranking of alternatives, which corresponded to the conventional MADM approach.

Step 15: The employees were ranked and the results were interpreted in this step. As can be seen from the Table 15, employee 3 dominated the other candidates regardless of the model configurations. However, the performance of employee 1 needed more attention here as the ranking order differed by the static or dynamic MADM aspects. The performance of employee 1 has been visualized in Figure 9.

It was observed that the performance of employee 2 was better than employee 1 for all of the dynamic MADM models. However, considering only the current decision matrix in model 3, static MADM ranked employee 1 as second and employee 2 as third.

Although the performance of the employee 1 was higher than the employee 2 in the current period, considering historical performance values changed the overall rankings. Therefore, the present study provided policy makers with a broader view regarding performance of employees by considering different modeling aspects.

**Table 13.** Closeness coefficients of employees.

| Periods | Closeness Coefficients | | |
|---|---|---|---|
| | **Employee 1** | **Employee 2** | **Employee 3** |
| $t_1$ | 0.4915 | 0.5856 | 0.5811 |
| $t_2$ | 0.3411 | 0.4928 | 0.6942 |
| $t_3$ | 0.3501 | 0.4633 | 0.6499 |
| $t_4$ | 0.3588 | 0.3671 | 0.6549 |
| $t_5$ | 0.3652 | 0.3801 | 0.6527 |
| $t_6$ | 0.3729 | 0.6405 | 0.6878 |
| $t_7$ | 0.3664 | 0.5666 | 0.7123 |
| $t_8$ | 0.3028 | 0.7053 | 0.7376 |
| $t_9$ | 0.3525 | 0.5907 | 0.6487 |
| $t_{10}$ | 0.3016 | 0.5208 | 0.6404 |
| $t_{11}$ | 0.4580 | 0.5802 | 0.3978 |
| $t_{12}$ | 0.4652 | 0.6078 | 0.3493 |
| $t_{13}$ | 0.4043 | 0.5769 | 0.3500 |
| $t_{14}$ | 0.4039 | 0.5347 | 0.4653 |
| $t_{15}$ | 0.4304 | 0.3776 | 0.6099 |
| $t_{16}$ | 0.3906 | 0.4291 | 0.5614 |
| $t_{17}$ | 0.4613 | 0.3593 | 0.6321 |
| $t_{18}$ | 0.6041 | 0.3345 | 0.6671 |
| $t_C$ | 0.5206 | 0.4547 | 0.6051 |

**Table 14.** Different model configurations.

| Model Components | Models | | | |
|---|---|---|---|---|
| | **Proposed Model** | **Model 1** | **Model 2** | **Model 3** |
| IT2F Functions | Yes | Yes | Yes | Yes |
| Vocabulary Matching | Yes | Yes | No | Yes |
| Modified preferences (interactive) | Yes | No | No | Yes |
| DWA Operator | Yes | Yes | Yes | No |

**Table 15.** Overall results.

| Model | Overall Assessment | | | | | |
| --- | --- | --- | --- | --- | --- | --- |
| | Employee 1 | | Employee 2 | | Employee 3 | |
| | CC | Rank | CC | Rank | CC | Rank |
| Proposed Model | 0.4138 | 3 | 0.4984 | 2 | 0.5895 | 1 |
| Model 1 | 0.4176 | 3 | 0.4926 | 2 | 0.5967 | 1 |
| Model 2 | 0.4185 | 3 | 0.4905 | 2 | 0.5951 | 1 |
| Model 3 | 0.5206 | 2 | 0.4547 | 3 | 0.6051 | 1 |

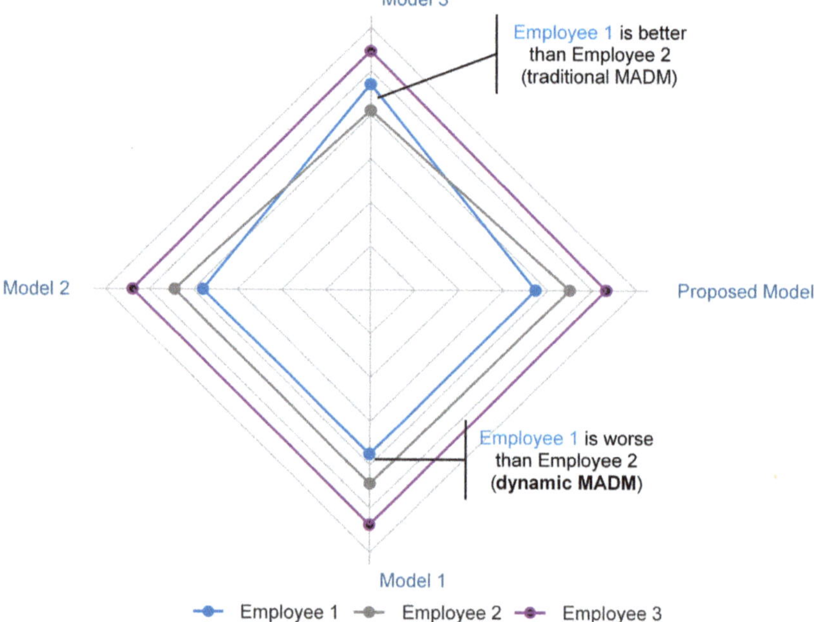


**Figure 9.** Changes in the ranking of employee 1.

## 5. Discussion

According to proposed dynamic MADM model, past performance values of employees were used to form current decision matrix. Hence, the current decision matrix was automatically generated based on the proposed IT2F functions model and the decision makers modified three performance scores out of 15. Therefore, preference elicitation procedures were dramatically improved.

The results of the case study had also shown that employee 3 had ranked first according to all of the dynamic MADM models. Although employee 1 had better performance than employee 2 in terms of current period performance, consideration of the past data made employee 1 the worst candidate for the full-time position among other employees. Although employee 2 had the worst performance without considering the past performance, with the dynamic character of the proposed model, employee 2 had become the second ranked alternative. As a result, taking into account the past and current decision matrices together provides more information to the decision makers and leads to more accurate rankings.

## 6. Conclusions

In this study, a new interactive dynamic MADM model, which combines IT2F regression and fuzzy functions approaches, was proposed. The proposed model exhibited desirable properties that helped overcome the drawbacks of the traditional dynamic MADM approaches. The proposed model was realized in a real-life personnel promotion problem. The proposed IT2F functions approach had improved the performance of IT2F regression by successfully capturing the trends and patterns in the historical records.

The advantage of the proposed model lies in its data-driven character, that is to say, the past data is used to guide preference elicitation processes. In classical MADM approaches, the decision makers are asked to fill out the entire decision matrix, which can be time consuming and cognitively demanding. On the other hand, the proposed model generates an IT2F decision matrix based on past data and allows decision makers to alter automatically generated decision matrix by using linguistic terms. Fuzziness in both the linguistic expressions and the past data were modeled by the proposed approach.

The main limitation of the presented study was that the results produced by the proposed model were highly dependent on the data obtained from the case company. Accordingly, results were difficult to be generalized. Secondly, prediction performance highly depends on available data. Nevertheless, the presented study can be improved in terms of many aspects. Hyper-parameter optimization can be used to tune the parameters of the model automatically. Different strategies can also be imposed to generate time series weights, and results can be compared. Last but not least, different MADM approaches can be used to rank the alternatives. In future studies, these considerations will be at the top of our agenda.

**Author Contributions:** The authors equally contributed to this work, thus they share first-author contribution. The authors contributed in the following manners: conceptualization, A.B. and İ.G.; methodology, A.B. and İ.G.; writing and original draft preparation, İ.G.; software, İ.G.; investigation, A.B.; supervision, A.B.; resources, A.B.; data curation, İ.G.; validation, A.B. and İ.G.

**Funding:** This research received no external funding.

**Acknowledgments:** We would like to thank the reviewers for their valuable comments.

**Conflicts of Interest:** The authors declare no conflict of interest.

## References

1. Dyer, J.S. Maut—Multiattribute Utility Theory. In *Multiple Criteria Decision Analysis: State of the Art Surveys*; Springer: New York, NY, USA, 2005; Volume 78, pp. 265–292.
2. Fishburn, P.C. *Utility Theory for Decision Making*; Wiley: New York, NY, USA, 1970.
3. Keeney, R.L. Building models of values. *Eur. J. Oper. Res.* **1988**, *37*, 149–157. [CrossRef]
4. Saaty, T.L. *The Analytic Hierarchy Process*; McGraw-Hill: New York, NY, USA, 1980.
5. Saaty, T.L. *Decision Making with Dependence and Feedback: The Analytic Network Process*; RWS Publications: Pittsburgh, PA, USA, 1996; p. 481.
6. Hwang, C.L.; Yoon, K. *Multiple Attribute Decision Making, Methods and Applications*; Springer: New York, NY, USA, 1981; Volume 186.
7. Roy, B. The Outranking Approach and the Foundations of Electre Methods. In *Readings in Multiple Criteria Decision Aid*; Bana e Costa, C., Ed.; Springer: Berlin/Heidelberg, Germany, 1990; pp. 155–183.
8. Opricovic, S.; Tzeng, G.-H. Compromise solution by MCDM methods: A comparative analysis of VIKOR and TOPSIS. *Eur. J. Oper. Res.* **2004**, *156*, 445–455. [CrossRef]
9. Fontela, E.; Gabus, A. *The DEMATEL Observer*; Battelle Geneva Research Center: Geneva, Switzerland, 1976.
10. Kornbluth, J.S.H. Dynamic multi-criteria decision making. *J. Multi-Criteria Decis. Anal.* **1992**, *1*, 81–92. [CrossRef]
11. Dong, Q.X.; Guo, Y.J.; He, Z.Y. Method of dynamic multi-criteria decision-making based on integration of absolute and relative differences. In Proceedings of the IEEE International Conference on Advanced Computer Control, Shenyang, China, 27–29 March 2010; pp. 353–356.

12. Lou, C.; Peng, Y.; Kou, G.; Ge, X. DMCDM: A dynamic multi criteria decision making model for sovereign credit default risk evaluation. In Proceedings of the 2nd International Conference on Software Engineering and Data Mining, Chengdu, China, 23–25 June 2010; pp. 489–494.

13. Campanella, G.; Ribeiro, R.A. A framework for dynamic multiple-criteria decision making. *Decis. Support Syst.* **2011**, *52*, 52–60. [CrossRef]

14. Xu, Z.; Yager, R.R. Dynamic intuitionistic fuzzy multi-attribute decision making. *Int. J. Approx. Reason.* **2008**, *48*, 246–262. [CrossRef]

15. Park, J.H.; Cho, H.J.; Kwun, Y.C. Extension of the VIKOR method to dynamic intuitionistic fuzzy multiple attribute decision making. *Comput. Math. Appl.* **2013**, *65*, 731–744. [CrossRef]

16. Zhou, F.; Wang, X.; Samvedi, A. Quality improvement pilot program selection based on dynamic hybrid MCDM approach. *Ind. Manag. Data Syst.* **2018**, *118*, 144–163. [CrossRef]

17. Bali, Ö.; Gümüş, S.; Kaya, İ. A Multi-Period Decision Making Procedure Based on Intuitionistic Fuzzy Sets for Selection Among Third-Party Logistics Providers. *J. Mult. Valued Log. Soft Comput.* **2015**, *24*, 547–569.

18. Chen, Y.; Li, B. Dynamic multi-attribute decision making model based on triangular intuitionistic fuzzy numbers. *Sci. Iran.* **2011**, *18*, 268–274. [CrossRef]

19. Liang, C.Y.; Zhang, E.Q.; Qi, X.W.; Cai, M.J. A dynamic multiple attribute decision making method under incomplete information. In Proceedings of the 6th International Conference on Natural Computation, Yantai, China, 10–12 August 2010; pp. 2720–2723.

20. Bali, O.; Dagdeviren, M.; Gumus, S. An integrated dynamic intuitionistic fuzzy MADM approach for personnel promotion problem. *Kybernetes* **2015**, *44*, 1422–1436. [CrossRef]

21. Xu, Z. A method based on the dynamic weighted geometric aggregation operator for dynamic hybrid multi-attribute group decision making. *Int. J. Uncertain. Fuzziness Knowl. Based Syst.* **2009**, *17*, 15–33. [CrossRef]

22. Zulueta, Y.; Martínez-Moreno, J.; Pérez, R.B.; Martínez, L. A discrete time variable index for supporting dynamic multi-criteria decision making. *Int. J. Uncertain. Fuzziness Knowl. Based Syst.* **2014**, *22*, 1–22. [CrossRef]

23. Lin, Y.-H.; Lee, P.-C.; Ting, H.-I. Dynamic multi-attribute decision making model with grey number evaluations. *Expert Syst. Appl.* **2008**, *35*, 1638–1644. [CrossRef]

24. Zhang, C.L. Risk assessment of supply chain finance with intuitionistic fuzzy information. *J. Intell. Fuzzy Syst.* **2016**, *31*, 1967–1975. [CrossRef]

25. Wei, G.W. Some geometric aggregation functions and their application to dynamic multiple attribute decision making in the intuitionistic fuzzy setting. *Int. J. Uncertain. Fuzziness Knowl. Based Syst.* **2009**, *17*, 179–196. [CrossRef]

26. Ai, F.Y.; Yang, J.Y. Approaches to dynamic multiple attribute decision making with 2-tuple linguistic information. *J. Intell. Fuzzy Syst.* **2014**, *27*, 2715–2723. [CrossRef]

27. Liu, H.; Jiang, L.; Martínez, L. A dynamic multi-criteria decision making model with bipolar linguistic term sets. *Expert Syst. Appl.* **2018**, *95*, 104–112. [CrossRef]

28. Liu, Y. A method for 2-tuple linguistic dynamic multiple attribute decision making with entropy weight. *J. Intell. Fuzzy Syst.* **2014**, *27*, 1803–1810. [CrossRef]

29. Zulueta-Veliz, Y.; Sanchez, P.J. Linguistic dynamic multicriteria decision making using symbolic linguistic computing models. *Granul. Comput.* **2018**. [CrossRef]

30. Xu, Z. Multi-period multi-attribute group decision-making under linguistic assessments. *Int. J. Gen. Syst.* **2009**, *38*, 823–850. [CrossRef]

31. Cui, J.; Liu, S.F.; Dang, Y.G.; Xie, N.M.; Zeng, B. A grey multi-stage dynamic multiple attribute decision making method. In Proceedings of the IEEE International Conference on Grey Systems and Intelligent Services, Nanjing, China, 15–18 September 2011; pp. 548–550.

32. Shen, J.M.; Dang, Y.G.; Zhou, W.J.; Li, X.M. Evaluation for Core Competence of Private Enterprises in Xuchang City Based on an Improved Dynamic Multiple-Attribute Decision-Making Model. *Math. Probl. Eng.* **2015**, *2015*. [CrossRef]

33. Mardani, A.; Nilashi, M.; Zavadskas, E.K.; Awang, S.R.; Zare, H.; Jamal, N.M. Decision Making Methods Based on Fuzzy Aggregation Operators: Three Decades Review from 1986 to 2017. *Int. J. Inf. Technol. Decis. Mak.* **2018**, *17*, 391–466. [CrossRef]

34. Saaty, T.L. Time dependent decision-making; dynamic priorities in the AHP/ANP: Generalizing from points to functions and from real to complex variables. *Math. Comput. Model.* **2007**, *46*, 860–891. [CrossRef]

35. Hashemkhani Zolfani, S.; Maknoon, R.; Zavadskas, E.K. An introduction to Prospective Multiple Attribute Decision Making (PMADM). *Technol. Econ. Dev. Econ.* **2016**, *22*, 309–326. [CrossRef]

36. Orji, I.J.; Wei, S. An innovative integration of fuzzy-logic and systems dynamics in sustainable supplier selection: A case on manufacturing industry. *Comput. Ind. Eng.* **2015**, *88*, 1–12. [CrossRef]

37. Baykasoğlu, A.; Gölcük, İ. A dynamic multiple attribute decision making model with learning of fuzzy cognitive maps. *Comput. Ind. Eng.* **2019**. [CrossRef]

38. Lee, H.; Tanaka, H. Fuzzy approximations with non-symmetric fuzzy parameters in fuzzy regression analysis. *J. Oper. Res. Soc. Jpn.* **1999**, *42*, 98–112. [CrossRef]

39. Turksen, I.B. Fuzzy functions with LSE. *Appl. Soft Comput. J.* **2008**, *8*, 1178–1188. [CrossRef]

40. Çelikyilmaz, A.; Turksen, I.B. Fuzzy functions with support vector machines. *Inf. Sci.* **2007**, *177*, 5163–5177. [CrossRef]

41. Kasabov, N.K.; Song, Q. DENFIS: Dynamic evolving neural-fuzzy inference system and its application for time-series prediction. *IEEE Trans. Fuzzy Syst.* **2002**, *10*, 144–154. [CrossRef]

42. Emami, M.R.; Türksen, I.B.; Goldenberg, A.A. Development of a systematic methodology of fuzzy logic modeling. *IEEE Trans. Fuzzy Syst.* **1998**, *6*, 346–361. [CrossRef]

43. Kilic, K.; Sproule, B.A.; Turksen, I.B.; Naranjo, C.A. A fuzzy system modeling algorithm for data analysis and approximate reasoning. *Robot. Auton. Syst.* **2004**, *49*, 173–180. [CrossRef]

44. Babuška, R.; Verbruggen, H.B. Constructing fuzzy models by product space clustering. In *Fuzzy Model Identification*; Springer: Berlin, Germany, 1997; pp. 53–90.

45. Zarandi, M.H.F.; Turksen, I.B.; Rezaee, B. A systematic approach to fuzzy modeling for rule generation from numerical data. In Proceedings of the IEEE Annual Meeting of the Fuzzy Information, Banff, AB, Canada, 27–30 June 2004; pp. 768–773.

46. Jang, J.-S.R. ANFIS: Adaptive-network-based fuzzy inference system. *IEEE Trans. Syst. Man Cybern.* **1993**, *23*, 665–685. [CrossRef]

47. Cordón, O.; Gomide, F.; Herrera, F.; Hoffmann, F.; Magdalena, L. Ten years of genetic fuzzy systems: Current framework and new trends. *Fuzzy Sets Syst.* **2004**, *141*, 5–31. [CrossRef]

48. Mendel, J.M.; John, R.I.; Feilong, L. Interval Type-2 Fuzzy Logic Systems Made Simple. *IEEE Trans. Fuzzy Syst.* **2006**, *14*, 808–821. [CrossRef]

49. Mendel, J.M.; Wu, D. *Perceptual Computing: Aiding People in Making Subjective Judgments*; Wiley-IEEE Press: Hoboken, NJ, USA, 2010; Volume 13.

50. Hamrawi, H.; Coupland, S. Type-2 fuzzy arithmetic using alpha-planes. In Proceedings of the IFSA-EUSFLAT Conference, Lisbon, Portugal, 20–24 July 2009; pp. 606–612.

51. Chen, S.-M.; Yang, M.-W.; Lee, L.-W.; Yang, S.-W. Fuzzy multiple attributes group decision-making based on ranking interval type-2 fuzzy sets. *Expert Syst. Appl.* **2012**, *39*, 5295–5308. [CrossRef]

52. Karnik, N.N.; Mendel, J.M. Centroid of a type-2 fuzzy set. *Inf. Sci.* **2001**, *132*, 195–220. [CrossRef]

53. Wu, D.; Mendel, J.M. A comparative study of ranking methods, similarity measures and uncertainty measures for interval type-2 fuzzy sets. *Inf. Sci.* **2009**, *179*, 1169–1192. [CrossRef]

54. Bajestani, N.S.; Kamyad, A.V.; Esfahani, E.N.; Zare, A. Prediction of retinopathy in diabetic patients using type-2 fuzzy regression model. *Eur. J. Oper. Res.* **2018**, *264*, 859–869. [CrossRef]

55. Yager, R.R. On ordered weighted averaging aggregation operators in multicriteria decisionmaking. *IEEE Trans. Syst. Man Cybern.* **1988**, *18*, 183–190. [CrossRef]

 *mathematics*

Article

# Fuzzy Counterparts of Fischer Diagonal Condition in ⊤-Convergence Spaces

**Qiu Jin, Lingqiang Li * and Jing Jiang**

School of Mathematical Sciences, Liaocheng University, Liaocheng 252059, China
* Correspondence: lilingqiang@lcu.edu.cn; Tel.: +86-152-0650-6635

Received: 4 July 2019; Accepted: 26 July 2019; Published: 31 July 2019

**Abstract:** Fischer diagonal condition plays an important role in convergence space since it precisely ensures a convergence space to be a topological space. Generally, Fischer diagonal condition can be represented equivalently both by Kowalsky compression operator and Gähler compression operator. ⊤-convergence spaces are fundamental fuzzy extensions of convergence spaces. Quite recently, by extending Gähler compression operator to fuzzy case, Fang and Yue proposed a fuzzy counterpart of Fischer diagonal condition, and proved that ⊤-convergence space with their Fischer diagonal condition just characterizes strong $L$-topology—a type of fuzzy topology. In this paper, by extending the Kowalsky compression operator, we present a fuzzy counterpart of Fischer diagonal condition, and verify that a ⊤-convergence space with our Fischer diagonal condition precisely characterizes topological generated $L$-topology—a type of fuzzy topology. Hence, although the crisp Fischer diagonal conditions based on the Kowalsky compression operator and the on Gähler compression operator are equivalent, their fuzzy counterparts are not equivalent since they describe different types of fuzzy topologies. This indicates that the fuzzy topology (convergence) is more complex and varied than the crisp topology (convergence).

**Keywords:** fuzzy topology; fuzzy convergence; ⊤-convergence; diagonal condition

---

## 1. Introduction

The notion of convergence space is investigated by generalizing the convergence in topological space [1]. For a set $X$, let $P(X)$ (respectively, $F(X)$) denote the power set (respectively, filters set) on $X$. A *convergence space* is defined by a pair $(X, c)$, where $c \subseteq F(X) \times X$ fulfills:

(C1) For every $x \in X, (\dot{x}, x) \in c, \dot{x} = \{A \in P(X) | x \in A\}$.
(C2) For all $F, G \in F(X), F \subseteq G, (F, x) \in c \Longrightarrow (G, x) \in c$.

If $(F, x) \in c$, we also denote $F \xrightarrow{c} x$, and say that F converges to $x$.

A convergence space $(X, c)$ is called pretopological if it satisfies either of the following three equivalent conditions:

(P1) For $\{F_t\}_{t \in T} \subseteq F(X)$ and $x \in X, \forall t \in T, F_t \xrightarrow{c} x \Longrightarrow \bigcap_{t \in T} F_t \xrightarrow{c} x$.

(P2) For $x \in X, U_c(x) = \bigcap\{F \in F(X) | F \xrightarrow{c} x\}$ converges to $x$. Generally, $U_c = \{U_c(x)\}_{x \in X}$ is called the neighborhood system associated with c.
(P3) For $x \in X$ and $F \in F(X), F \xrightarrow{c} x \Leftrightarrow F \supseteq U_c(x)$.

A pretopological convergence space is called topological if it fulfills the next condition:

**(U)** For any $x \in X$, if $A \in U_c(x)$ then there exists a $B \in U_c(x)$ such that $A \in U_c(y)$ for any $y \in B$.

Topological convergence spaces can also be characterized by Kowalsky diagonal condition and Fischer diagonal condition [1,2].

Let $T$ be any set and $\Phi : T \longrightarrow F(X)$ be a mapping, called a choice mapping of filters. For $F \in F(T)$, let $\Phi^{\Rightarrow}(F)$ denote the image of F under the mapping $\Phi$, i.e., the filter on $F(X)$ generated by $\{\Phi(A) | A \in F\}$ as a filter base. Then, the *Kowalsky compression operator* on $\Phi^{\Rightarrow}(F) \in F(F(X))$ is defined as

$$K\Phi F \in F(X) := \bigcup_{A \in F} \bigcap_{y \in A} \Phi(y).$$

There is an equivalent statement for $K\Phi F$ proposed by Gähler [3] (please also see the Remark in Jäger [4]):

$$A \in K\Phi F \Longleftrightarrow \{t \in T | A \in \Phi(t)\} \in F,$$

which is here called Gähler compression operator to distinguish the two descriptions.

For a convergence space $(X, c)$, the *Fischer diagonal condition* is given as below:

**(FD)** Let $T$ be any set, $\psi : T \longrightarrow X$ and $\Phi : T \longrightarrow F(X)$ with $\Phi(t) \xrightarrow{c} \psi(t)$ for every $t \in T$. For any $F \in F(X)$, $\psi^{\Rightarrow}(F) \xrightarrow{c} x \Longrightarrow K\Phi F \xrightarrow{c} x$.

Taking $T = X$ and $\psi = id_X$ in **(FD)**, we give the *Kowalsky diagonal condition* **(KD)**.

By using the Kowalsky compression operator, the condition **(U)** can be restated as:

**(U)** For any $x \in X$, $U_c(x) \subseteq KU_cU_c(x)$, where $U_c : X \longrightarrow F(X)$ is the selection mapping of filters determined by neighborhood system.

It is verified that a pretopological convergence space is topological iff it fulfills **(KD)**, and a convergence space is topological iff it fulfills **(FD)**.

It is known that a topological space $(X, \delta)$ corresponds uniquely to a topological convergence space $(X, c)$ by taking that for each $x \in X$,

$$U_c(x) = U_\delta(x) := \{A \subseteq X | \exists B \in \delta \text{ s.t. } x \in B \subseteq A\}.$$

Then, we obtain a bijection between topological convergence spaces and topological spaces. In this case, we say that topological convergence spaces characterize topological spaces, or, in other words, that it establishes the convergence theory associated with topological spaces.

In **(FD)**, change the statement

$$\psi^{\Rightarrow}(F) \xrightarrow{c} x \Longrightarrow K\Phi F \xrightarrow{c} x \text{ as } K\Phi F \xrightarrow{c} x \Longrightarrow \psi^{\Rightarrow}(F) \xrightarrow{c} x,$$

then the resulted condition is denoted as **(DFD)**, called the *dual Fischer diagonal condition*. Interestingly, the condition **(DFD)** precisely characterizes the regularity of convergence space [1,2].

To sum up, both Fischer diagonal condition **(FD)** and its dual condition **(DFD)** are based on Kowalsky (or equivalent, Gähler) compression operator, and **(FD)** describes topologicalness while **(DFD)** characterizes regularity of convergence spaces.

Fuzzy set theory, proposed by Zadeh [5], is a fundamental mathematical tool to deal with uncertain information. Fuzzy set theory has been widely used in many regards such as medical diagnosis, data mining, decision-making, machine learning and so on [6–9]. In addition, that fuzzy set combines traditional mathematics produces many new mathematical branches such as fuzzy algebra, fuzzy topology, fuzzy order, fuzzy logic, etc. In this paper, we focus on the convergence theory associated with lattice-valued topology (i.e., fuzzy topology with the membership values in a complete lattice $L$).

Lattice-valued convergence spaces are fuzzy extensions of convergence spaces. Compared with the classical convergence spaces, the lattice-value convergence spaces are more complex and varied. $L$-filters and $\top$-filter are basic tools to study lattice-valued convergence spaces. Based on $L$-filters, Jäger [10], Flores [11], Fang [12] and Li [13] defined types of $L$-convergence spaces. Based on $\top$-filters, Fang and Yue [14] introduced $\top$-convergence spaces. Nowadays, these spaces have been widely discussed and developed [4,15–42].

Many interesting lattice-valued versions of Fischer diagonal condition and dual Fischer diagonal condition were proposed to study the topologicalnesses and regularities of many kinds of lattice-valued convergence spaces. Precisely,

Under the background of $L$-filter:

(1) The Gähler compression operator was presented by Jäger [4]. Then, some types of lattice-valued Fischer diagonal conditions were proposed to discuss the topologicalnesses of different $L$-convergence spaces [13,18,28]. Particularly, it was proved in [28] that stratified $L$-convergence space with the considered Fischer diagonal condition precisely characterizes stratified $L$-topological space. Meanwhile, some lattice-valued dual Fischer diagonal conditions were introduced to study the regularities of different $L$-convergence spaces [19,27,31].

(2) The Kowalsky compression operator was presented by Flores [15]. Then, he introduced a lattice-valued Fischer diagonal condition and proved that stratified $L$-convergence space with his diagonal condition can characterize a tower of stratified $L$-topological spaces. Later, Richardson and his co-author used the dual condition of Flores's diagonal condition to describe the regularity of stratified $L$-convergence spaces [15,16].

Under the background of $\top$-filter:

(1) The Gähler compression operator was presented by Fang and Yue in $\top$-convergence space [14]. Then, they introduced a lattice-valued Fischer diagonal condition and proved that condition can characterize strong $L$-topological spaces. They also proposed a lattice-valued dual Fischer diagonal to describe a regularity of $\top$-convergence spaces. Quite recently, the author extended Fang and Yue's diagonal conditions and used them to study the relative topologicalness and the relative regularity in $\top$-convergence spaces [22,26].

(2) The Kowalsky compression operator was presented by the author in $\top$-convergence space [30]. Then, we proposed a lattice-valued version of dual Fischer diagonal condition to discuss the regularity of $\top$-convergence spaces. In preparing the paper [30], we also tried to consider a lattice-valued version of Fischer diagonal condition such that $\top$-convergence spaces with the diagonal condition can characterize a kind of lattice-valued topological spaces. However, we failed to do that.

In this paper, we resolve the above mentioned question. Precisely, we present a lattice-valued Fischer diagonal condition based on Kowalsky compression operator and prove that there is a bijection between $\top$-convergence spaces with our diagonal condition and topological generated $L$-topological spaces. Thus, our Fischer diagonal condition characterizes precisely topological generated $L$-topological spaces. Therefore, we establish the convergence theory associated with topological generated $L$-topological spaces. This is the main contribution of the paper.

This paper is organized as follows. In Section 2, we review some notions and notations as preliminary. In Section 3, we give a lattice-valued Fischer diagonal condition and use it to describe a subclass of $\mathsf{T}$-convergence spaces, called topological generated $\mathsf{T}$-convergence spaces. In Section 4, we establish the relationships between our diagonal condition and Fang and Yue's diagonal condition. Then, we further verify that there is a one-to-one correspondence between topological generated $\mathsf{T}$-convergence spaces and topological generated $L$-topological spaces. Hence, we establish the convergence theory associated with the topological generated $L$-topological spaces.

## 2. Preliminaries

In this section, we recall some notions and conclusions about $L$-fuzzy sets, $\mathsf{T}$-convergence spaces and $L$-topological spaces for later use.

By an integral, commutative quantale we mean a pair $L = (L, *)$ such that:

(i)     $(L, \leq)$ is a complete lattice, and its top (respectively, bottom) element is denoted as $\mathsf{T}$ (respectively, $\perp$).

(ii)     $*$ is a commutative semigroup operation on $L$ satisfying

$$\forall \alpha \in L, \forall \{\beta_i\}_{i \in I} \subseteq L, \alpha * \bigvee_{i \in I} \beta_i = \bigvee_{i \in I} (\alpha * \beta_i).$$

(iii)     For every $\alpha \in L, \mathsf{T} * \alpha = \alpha$.

For any $\alpha, \beta \in L$, we define

$$\alpha \to \beta = \bigvee \{\gamma \in L : \alpha * \gamma \leq \beta\}.$$

The properties of the binary operations $*$ and $\to$ can be referred to [43–45]. We list some of them used in the sequel.

(1)     $\alpha * \beta \leq \gamma \Leftrightarrow \beta \leq \alpha \to \gamma$.

(2)     $\alpha * (\alpha \to \beta) \leq \beta$.

(3)     $\alpha \to (\beta \to \gamma) = (\alpha * \beta) \to \gamma$.

(4)     $\left( \bigvee_{i \in I} \alpha_i \right) \to \beta = \bigwedge_{i \in I} (\alpha_i \to \beta)$.

(5)     $\alpha \to \left( \bigwedge_{i \in I} \beta_i \right) = \bigwedge_{i \in I} (\alpha \to \beta_i)$.

For $\alpha, \beta \in L$, we define $\alpha \ll \beta$ iff for every directed subsets $D \subseteq L, \beta \leq \vee D$ always implies the existence of $\gamma \in D$ such that $\alpha \leq \gamma$. Note that $\alpha \ll \beta$ implies $\alpha \leq \beta$. When $L = [0, 1]$, $\alpha \ll \beta \Longleftrightarrow \alpha < \beta$.

A complete lattice $(L, \leq)$ is called meet continuous if, for $\alpha \in L$ and $\{\beta_i\}_{i \in I}$ being directed in $L$, it holds that $\alpha \wedge \bigvee_{i \in I} \beta_i = \bigvee_{i \in I} (\alpha \wedge \beta_i)$. A complete lattice $(L, \leq)$ is called continuous if $\alpha = \vee \{\beta \in L | \beta \ll \alpha\}$ for each $\alpha \in L$. A continuous lattice is a natural meet continuous lattice [44].

In this paper, if not otherwise stated, we always let $L = (L, *)$ be an integral, commutative quantale with the underlying lattice $(L, \leq)$ being meet continuous.

A mapping $A : X \longrightarrow L$ is called an $L$-fuzzy set on $X$, and the family of $L$-fuzzy sets on $X$ is denoted as $L^X$. For $A \in P(X)$, let $\mathsf{T}_A$ denote its characteristic mapping. The operations $\vee, \wedge, *$ and $\to$ on $L$ can translate onto $L^X$ pointwisely. Precisely, for $A, B, A_i (i \in I) \in L^X$,

$$\left( \bigvee_{i \in I} A_i \right)(x) = \bigvee_{i \in I} A_i(x), \quad \left( \bigwedge_{i \in I} A_i \right)(x) = \bigwedge_{i \in I} A_i(x),$$

$$(A * B)(x) = A(x) * B(x), (A \rightarrow B)(x) = A(x) \rightarrow B(x).$$

Let $f : X \longrightarrow Y$ be a mapping. Then, define $f^{\rightarrow} : L^X \longrightarrow L^Y$ and $f^{\leftarrow} : L^Y \longrightarrow L^X$ as follows [45]:

$$\forall A \in L^X, \forall y \in Y, f^{\rightarrow}(A)(y) = \bigvee_{f(x)=y} A(x), \forall B \in L^Y, \forall x \in X, f^{\leftarrow}(B)(x) = B(f(x)).$$

For $A, B \in L^X$, the degree of $A$ in $B$ is defined as follows [46–48]:

$$S_X(A, B) = \bigwedge_{x \in X} (A(x) \rightarrow B(x)).$$

**Definition 1** ([45,49])**.** *A nonempty subset* $\mathbf{F} \subseteq L^X$ *is called a* $\top$*-filter on* $X$ *if it holds that:*

(1)  $A \in \mathbf{F} \Longrightarrow \bigvee_{x \in X} A(x) = \top$.

(2)  $A, B \in \mathbf{F} \Longrightarrow A \wedge B \in \mathbf{F}$.

(3)  $\bigvee_{B \in \mathbf{F}} S_X(B, A) = \top \Longrightarrow A \in \mathbf{F}$.

*The family of* $\top$*-filters on* $X$ *is denoted as* $\mathbf{F}_L^{\top}(X)$.

**Definition 2** ([45])**.** *A nonempty subset* $\mathbf{B} \subseteq L^X$ *is called a* $\top$*-filter base on* $X$ *if it satisfies:*

(1)  $A \in \mathbf{B} \Longrightarrow \bigvee_{x \in X} A(x) = \top$.

(2)  $A, B \in \mathbf{B} \Longrightarrow \bigvee_{C \in \mathbf{B}} S_X(C, A \wedge B) = \top$.

*Each* $\top$*-filter base* $\mathbf{B}$ *generates a* $\top$*-filter* $\mathbf{F_B} := \{A \in L^X | \bigvee_{B \in \mathbf{B}} S_X(B, A) = \top\}$. *In [38], it is proved that for any* $A \in L^X$, $\bigvee_{B \in \mathbf{B}} S_X(B, A) = \bigvee_{B \in \mathbf{F_B}} S_X(B, A)$.

We collect some fundamental facts about $\top$-filters in the following proposition.

**Proposition 1** ([14,45])**.**

(1)  *For every* $x \in X$, $[x]_{\top} =: \{A \in L^X | A(x) = \top\}$ *is a* $\top$*-filter.*

(2)  $\{\mathbf{F}_i\}_{i \in I} \subseteq \mathbf{F}_L^{\top}(X) \Longrightarrow \bigcap_{i \in I} \mathbf{F}_i \in \mathbf{F}_L^{\top}(X)$.

(3)  *Let* $f : X \longrightarrow Y$ *be a mapping and* $\mathbf{F} \in \mathbf{F}_L^{\top}(X)$. *Then, define* $f^{\Rightarrow}(\mathbf{F})$ *as the* $\top$*-filter on* $Y$ *generated by the* $\top$*-filter base* $\{f^{\rightarrow}(A) | A \in \mathbf{F}\}$. *Furthermore,* $B \in f^{\Rightarrow}(\mathbf{F})$ *iff* $f^{\leftarrow}(B) \in \mathbf{F}$.

Given $\mathbf{F} \in \mathbf{F}_L^{\top}(X)$, then the set $\eta(\mathbf{F}) = \{A \subseteq X | \top_A \in \mathbf{F}\}$ is a filter on $X$. Conversely, given $F \in F(X)$, then the set $\{\top_A | A \in F\}$ forms a $\top$-filter base and the associated $\top$-filter is denoted as $\omega(F)$.

**Lemma 1** (Lemma 2.6 in [30])**.** *Let* $f : X \longrightarrow Y$, $F \in F(X)$ *and* $\mathbf{F} \in \mathbf{F}_L^{\top}(X)$. *Then,*

(1)  $\eta\omega(F) = F$.

(2)  $\omega\eta(\mathbf{F}) \subseteq \mathbf{F}$.

(3)  *For every* $x \in X, \omega(\dot{x}) = [x]_{\top}$.

(4)  *For every* $x \in X, \eta([x]_{\top}) = \dot{x}$.

(5) $\quad \eta(f^{\Rightarrow}(\mathbf{F})) = f^{\Rightarrow}(\eta(\mathbf{F}))$.

**Definition 3** ([14]). *A $\top$-convergence space is a pair $(X, \mathbf{q})$, where $\mathbf{q} \subseteq \mathbf{F}_L^{\top}(X) \times X$ satisfies:*

(TC1) *For every $x \in X, ([x]_{\top}, x) \in \mathbf{q}$.*
(TC2) $(\mathbf{F}, x) \in \mathbf{q}, \mathbf{F} \subseteq \mathbf{G} \Longrightarrow (\mathbf{G}, x) \in \mathbf{q}$.

*If $(\mathbf{F}, x) \in \mathbf{q}$, we also denote $\mathbf{F} \xrightarrow{\;\mathbf{q}\;} x$ and say that $\mathbf{F}$ converges to $x$.*
*Obviously, a $\top$-convergence space reduces to a convergence space whenever $L = \{\bot, \top\}$.*

Let $T$ be any set, $\phi : T \longrightarrow \mathbf{F}_L^{\top}(X)$ and $\mathbf{F} \in \mathbf{F}_L^{\top}(T)$. In [26], the author defined an extending *Kowalsky compression operator* on $\phi^{\Rightarrow}(\mathbf{F}) \in \mathbf{F}_L^{\top}(\mathbf{F}_L^{\top}(X))$ by

$$k\phi\mathbf{F} := \bigcup_{A \in \eta(\mathbf{F})} \bigcap_{y \in A} \phi(y) \in \mathbf{F}_L^{\top}(X).$$

**Lemma 2** (Lemma 3.2 in [30]). *Let $f : X \longrightarrow Y, \phi : T \longrightarrow \mathbf{F}_L^{\top}(X)$ and $\Phi : T \longrightarrow F(X)$. Then, for every $\mathbf{F} \in \mathbf{F}_L^{\top}(T)$ and every $F \in F(T)$,*

(1) $\quad f^{\Rightarrow}(k\phi\mathbf{F}) = k(f^{\Rightarrow} \circ \phi)\mathbf{F}$.
(2) $\quad$ *put $\Phi_1 = \eta \circ \phi$, then $\eta(k\phi\mathbf{F}) = K\Phi_1\eta(\mathbf{F})$.*
(3) $\quad$ *put $\phi_1 = \omega \circ \Phi$, then $\eta(k\phi_1\omega(F)) = K\Phi F$.*
(4) $\quad$ *taking $\sigma : T \longrightarrow \mathbf{F}_L^{\top}(X)$ with $\sigma(t) \subseteq \phi(t)$ for every $t \in T$, then $k\sigma\mathbf{F} \subseteq k\phi\mathbf{F}$.*

**Definition 4** ([45,47]). *A subset $\tau \subseteq L^X$ is called an L-topology (L-Top) on $X$ if it contains $\top_X, \top_{\varnothing}$ and is closed with respect to finite meets and arbitrary joins. The pair $(X, \tau)$ is called an L-topological space. Furthermore, $\tau$ is called stratified (SL-Top) if $\alpha * A \in \tau$ for each $\alpha \in L$ and each $A \in \tau$. A stratified L-topology $\tau$ is called strong (STrL-Top) if $\alpha \to A \in \tau$ for each $\alpha \in L$ and each $A \in \tau$.*

**Definition 5** ([25,45]). *Assume that $L$ to be a continuous lattice. For a topological space $(X, \delta)$, all L-fuzzy sets $A \in L^X$ with*

$$A_\alpha = \{x \in X | \alpha \ll A(x)\} \in \delta \text{ for each } \alpha \in L$$

*form an L-topology on $X$. Such space is called a topological generated L-topological space (TGL-Top). The L-topological space generated by $(X, \delta)$ is denoted as $(X, \varrho(\delta))$. When $* = \wedge$, $\varrho(\delta)$ is precisely the L-topology generated by $\{\top_A | A \in \delta\}$ and all constant value L-fuzzy sets on $X$ as a subbase.*

In [25], Lai and Zhang introduced a kind of $L$-topological space, called conical neighborhood space (CNS). We do not give the definition of CNS here; please see Definition 5.1 in [25]. Lai and Zhang proved that the mentioned $L$-topological spaces have the following inclusive relation:

$$\text{STr}L\text{-Top} \subseteq \text{CNS} \subseteq \text{SL-Top} \subseteq L\text{-Top and TGL-Top} \subseteq \text{CNS}.$$

By suitable lattice-valued Fischer diagonal condition:

(1) $\quad$ The convergence theory associated with $SL$-Top is presented in [28].
(2) $\quad$ The convergence theory associated with $STrL$-Top is given in [14].
(3) $\quad$ The convergence theory associated with CNS is developed in [29].

In the following, we establish the convergence theory associated with TGL-Top by appropriate lattice-valued Fischer diagonal condition.

### 3. Topological Generated $\top$-Convergence Spaces vs. Fischer Diagonal Condition Based on Extending Kowalsky Compression Operator

In this section, we present a lattice-valued Fischer diagonal condition based on extending Kowalsky compression operator. We also prove that $\top$-convergence spaces with our diagonal condition precisely characterize those spaces generated by topological convergence spaces.

First, we consider the pretopological conditions for a $\top$-convergence space $(X, \mathbf{q})$. It is not difficult to verify that the following three conditions are equivalent.

**(TP1)** For every $i \in I, \mathbf{F}_i \xrightarrow{\mathbf{q}} x \Longrightarrow \bigcap_{i \in I} \mathbf{F}_i \xrightarrow{\mathbf{q}} x$.

**(TP2)** For every $x \in X, \mathbf{U_q}(x) = \bigcap \{\mathbf{F} \in \mathbf{F}_L^\top(X) | \mathbf{F} \xrightarrow{\mathbf{q}} x\}$ converges to $x$. Usually, $\mathbf{U_q} = \{\mathbf{U_q}(x)\}_{x \in X}$ is called the $\top$-neighborhood system associated with $\mathbf{q}$.

**(TP3)** For every $x \in X, \mathbf{F} \xrightarrow{\mathbf{q}} x \Leftrightarrow \mathbf{F} \supseteq \mathbf{U_q}(x)$.

**Definition 6.** *A $\top$-convergence space $(X, \mathbf{q})$ is called pretopological if it fulfills any of **(TP1)**–**(TP3)**.*

Second, we consider the topological condition for $\top$-convergence space both by extending Fischer diagonal condition and extending Kowalsky diagonal condition.

We define a lattice-valued extension of Fischer diagonal condition as below:

**(TFD)** Let $T$ be any set, $\psi : T \longrightarrow X$ and $\phi : T \longrightarrow \mathbf{F}_L^\top(X)$ satisfying $\phi(t) \xrightarrow{\mathbf{q}} \psi(t)$ for every $t \in T$. Then, for any $\mathbf{F} \in \mathbf{F}_L^\top(X), \psi^\Rightarrow(\mathbf{F}) \xrightarrow{\mathbf{q}} x \Longrightarrow k\phi\mathbf{F} \xrightarrow{\mathbf{q}} x$.

Taking $T = X$ and $\psi = id_X$ in **(TFD)**, then we get the *Lattice-valued Kowalsky diagonal condition* **(TKD)**.

Moreover, a lattice-valued version of neighborhood condition **(U)** is given as follows:

**(TU)** For any $x \in X, k\mathbf{U_q}\mathbf{U_q}(x) \supseteq \mathbf{U_q}(x)$.

The theorem below presents the relationships between lattice-valued versions of Fischer diagonal condition, Kowalsky diagonal condition and pretopological condition.

**Theorem 1.** *A $\top$-convergence space $(X, \mathbf{q})$ satisfies **(TFD)** iff it is pretopological and satisfies **(TKD)** iff it is pretopological and satisfies **(TU)**.*

**Proof. (TFD)**$\Rightarrow$**(TP1)**. Let $\{\mathbf{F}_t | t \in T\}$ be all $\top$-filters on $X$ converge to $x$. Then, define

$$\psi : T \longrightarrow X, \phi : T \longrightarrow \mathbf{F}_L^\top(X) \text{ as } \psi(t) \equiv x, \phi(t) = \mathbf{F}_t \text{ for each } t \in T.$$

Take $\mathbf{F}_\perp := \{\{\top_T\}\}$ as the least member of $\mathbf{F}_L^\top(T)$. Then, we observe easily that $\psi^\Rightarrow(\mathbf{F}_\perp) = [x]_\top$ and $k\phi\mathbf{F}_\perp = \bigcap_{t \in T} \phi(t) = \bigcap_{t \in T} \mathbf{F}_t$. By $[x]_\top \xrightarrow{\mathbf{q}} x$ and **(TFD)**, we have that $k\phi\mathbf{F}_\perp = \bigcap_{t \in T} \mathbf{F}_t \xrightarrow{\mathbf{q}} x$, i.e., **(TP1)** holds.

**(TFD)**$\Rightarrow$**(TKD)**. It is obvious.

**(TP3)**+**(TKD)**$\Rightarrow$**(TU)**. By **(TP3)**, we have that $\mathbf{U_q}(y) \xrightarrow{\mathbf{q}} y$ for any $y \in X$. Then, by $\mathbf{U_q}(x) \xrightarrow{\mathbf{q}} x$ and **(TKD)**, we get that $k\mathbf{U_q}\mathbf{U_q}(x) \xrightarrow{\mathbf{q}} x$, i.e., $k\mathbf{U_q}\mathbf{U_q}(x) \supseteq \mathbf{U_q}(x)$.

**(TP3)**+**(TU)**$\Rightarrow$**(TFD)**. Let $T$ be any set, $\psi : T \longrightarrow X$ and $\phi : T \longrightarrow \mathbf{F}_L^\top(X)$ with $\phi(t) \xrightarrow{\mathbf{q}} \psi(t)$ for every $t \in T$. It follows by **(TP3)** that $\phi(t) \supseteq \mathbf{U_q}(\psi(t))$ for any $t \in T$.

Let $\mathbf{F} \in \mathbf{F}_L^\top(X)$ and $\psi^\Rightarrow(\mathbf{F}) \xrightarrow{\mathbf{q}} x$, then it holds by **(TP3)** that $\psi^\Rightarrow(\mathbf{F}) \supseteq \mathbf{U_q}(x)$.

Next, we prove that $k\mathbf{U_q}\psi^{\Rightarrow}(\mathbf{F}) \subseteq k(\mathbf{U_q} \circ \psi)\mathbf{F}$. Indeed, let $A \in k\mathbf{U_q}\psi^{\Rightarrow}(\mathbf{F})$; then, there exists an $A \in \eta(\psi^{\Rightarrow}(\mathbf{F}))$ such that $A \in \mathbf{U_q}(y)$ for any $y \in A$. By Lemma 1(5), it holds that $A \in \psi^{\Rightarrow}(\eta(\mathbf{F}))$, i.e., $\psi^{\leftarrow}(A) \in \eta(\mathbf{F})$. Then, we get that, for every $z \in \psi^{\leftarrow}(A)$, $A \in \mathbf{U_q}(\psi(z)) = (\mathbf{U_q} \circ \psi)(z)$. Thus, $A \in k(\mathbf{U_q} \circ \psi)\mathbf{F}$.

Combinaing the above statements, it follows by Lemma 2 (4) and (**TU**) that

$$k\phi\mathbf{F} \supseteq k(\mathbf{U_q} \circ \psi)\mathbf{F} \supseteq k\mathbf{U_q}\psi^{\Rightarrow}(\mathbf{F}) \supseteq k\mathbf{U_q}\mathbf{U_q}(x) \supseteq \mathbf{U_q}(x).$$

That means $k\phi\mathbf{F} \xrightarrow{\ q\ } x$, as desired. $\square$

For a convergence space $(X, c)$, it is easy to verify that the pair $\pi(X, c) = (X, \pi(c))$ defined by

$$\forall \mathbf{F} \in \mathbf{F}_L^{\top}(X), \forall x \in X, \mathbf{F} \xrightarrow{\pi(c)} x \Leftrightarrow \eta(\mathbf{F}) \xrightarrow{\ c\ } x,$$

is a $\top$-convergence space.

**Definition 7.** *A $\top$-convergence space $(X, \mathbf{q})$ is called topological generated if $(X, \mathbf{q}) = \pi(X, c)$ for some topological convergence space $(X, c)$.*

The next theorem shows that $\top$-convergence spaces generated by convergence spaces connect diagonal condition (**FD**) and lattice-valued diagonal condition (**TFD**) well.

**Theorem 2.** *Convergence space $(X, c)$ fulfills (**FD**) iff $(X, \pi(c))$ fulfills (**TFD**).*

**Proof.** $\Longrightarrow$. Let $\psi : T \longrightarrow X$ and $\phi : T \longrightarrow \mathbf{F}_L^{\top}(X)$ with $\forall t \in T, \phi(t) \xrightarrow{\pi(c)} \psi(t)$. Putting $\Phi = \eta \circ \phi$, then, from $\phi(t) \xrightarrow{\pi(c)} \psi(t)$ we get

$$\Phi(t) = \eta(\phi(t)) \xrightarrow{\ c\ } \psi(t), \forall t \in T.$$

Taking any $\psi^{\Rightarrow}(\mathbf{F}) \xrightarrow{\pi(c)} x$, then from Lemma 1(5) we obtain

$$\psi^{\Rightarrow}(\eta(\mathbf{F})) = \eta(\psi^{\Rightarrow}(\mathbf{F})) \xrightarrow{\ c\ } x.$$

It follows from (**FD**) and Lemma 2 (2) that $\eta(k\phi\mathbf{F}) = K\Phi\eta(\mathbf{F}) \xrightarrow{\ c\ } x$, that is, $k\phi\mathbf{F} \xrightarrow{\pi(c)} x$. We verify that (**TFD**) is fulfilled.

$\Longleftarrow$. Let $\psi : T \longrightarrow X$ and $\Phi : T \longrightarrow \mathbf{F}(X)$ with $\forall t \in T, \Phi(t) \xrightarrow{\ c\ } \psi(t)$. Putting $\phi = \omega \circ \Phi$, then, from Lemma 1(1), we conclude

$$\forall t \in T, \eta \circ \phi(t) = \eta \circ \omega \circ \Phi(t) = \Phi(t) \xrightarrow{\ c\ } \psi(t), \text{ i.e., } \phi(t) \xrightarrow{\pi(c)} \psi(t).$$

Taking any $\psi^{\Rightarrow}(\mathbf{F}) \xrightarrow{\ c\ } x$, then, from Lemma 1(5), we get

$$\eta\psi^{\Rightarrow}(\omega(\mathbf{F})) = \psi^{\Rightarrow}((\eta \circ \omega)(\mathbf{F})) = \psi^{\Rightarrow}(\mathbf{F}) \xrightarrow{\ c\ } x, \text{ that is, } \psi^{\Rightarrow}(\omega(\mathbf{F})) \xrightarrow{\pi(c)} x.$$

It holds from (**TFD**) that $k\phi\omega(\mathbf{F}) \xrightarrow{\pi(c)} x$. Further, from Lemma 2 (3), we obtain $\eta(k\phi\omega(\mathbf{F})) = K\Phi\mathbf{F} \xrightarrow{\ c\ } x$. This shows that the condition (**FD**) is fulfilled. $\square$

The following theorem shows that the lattice-valued Fischer diagonal condition (**TFD**) precisely characterizes topological generated $\top$-convergence spaces.

**Theorem 3.** *A $\top$-convergence space $(X, \mathbf{q})$ is topological generated iff it satisfies* (**TFD**).

**Proof.** The necessity can be concluded from Theorem 2. We prove the sufficiency below. Assume that $(X, \mathbf{q})$ satisfies (**TFD**). Then, define $\theta(X, \mathbf{q}) = (X, \theta(\mathbf{q}))$ as

$$\forall x \in X, \forall F \in F(X), F \xrightarrow{\theta(\mathbf{q})} x \Leftrightarrow F \supseteq \eta(\mathbf{U}_{\mathbf{q}}(x)).$$

Obviously, $(X, \theta(\mathbf{q}))$ is a convergence space. Next, we verify $(X, \mathbf{q}) = \pi\theta(X, \mathbf{q})$.

(1) $\quad \mathbf{q} \subseteq \pi\theta(\mathbf{q})$. It follows by

$$F \xrightarrow{\mathbf{q}} x \overset{(\mathbf{TP3})}{\Rightarrow} F \supseteq \mathbf{U}_{\mathbf{q}}(x) \Rightarrow \eta(F) \supseteq \eta(\mathbf{U}_{\mathbf{q}}(x))$$
$$\Rightarrow \quad \eta(F) \xrightarrow{\theta(\mathbf{q})} x \Rightarrow F \xrightarrow{\pi\theta(\mathbf{q})} x.$$

(2) $\quad \pi\theta(\mathbf{q}) \subseteq \mathbf{q}$. Letting $F \xrightarrow{\pi\theta(\mathbf{q})} x$, we check below that:

(i) $\quad F \supseteq k\mathbf{U}_{\mathbf{q}}F$. Take $A \in k\mathbf{U}_{\mathbf{q}}F$, then $\exists B \in \eta(F)$ such that $A \in \mathbf{U}_{\mathbf{q}}(y) \subseteq [y]_\top$ for any $y \in B$. It follows that $A(y) = \top$ for any $y \in B$, and then $\top_B \leq A$, which means that $A \in F$.

(ii) $\quad k\mathbf{U}_{\mathbf{q}}F \supseteq k\mathbf{U}_{\mathbf{q}}\mathbf{U}_{\mathbf{q}}(x)$. By $F \xrightarrow{\pi\theta(\mathbf{q})} x$, we have $\eta(F) \xrightarrow{\theta(\mathbf{q})} x$, i.e., $\eta(F) \supseteq \eta(\mathbf{U}_{\mathbf{q}}(x))$. Then,

$$\begin{aligned}
k\mathbf{U}_{\mathbf{q}}\mathbf{U}_{\mathbf{q}}(x) &= \bigcup_{B \in \eta(\mathbf{U}_{\mathbf{q}}(x))} \bigcap_{y \in B} \mathbf{U}_{\mathbf{q}}(y) \\
&\subseteq \bigcup_{B \in \eta(F)} \bigcap_{y \in B} \mathbf{U}_{\mathbf{q}}(y) = k\mathbf{U}_{\mathbf{q}}F.
\end{aligned}$$

(iii) $\quad k\mathbf{U}_{\mathbf{q}}\mathbf{U}_{\mathbf{q}}(x) \supseteq \mathbf{U}_{\mathbf{q}}(x)$. It follow by (**TU**).

Combining (i)–(iii), we get that $F \supseteq \mathbf{U}_{\mathbf{q}}(x)$, thus it follows by (**TP3**) that $F \xrightarrow{\mathbf{q}} x$.

From Theorem 2 and that $(X, \mathbf{q}) = \pi\theta(X, \mathbf{q})$ satisfies (**TFD**), we get that $(X, \theta(\mathbf{q}))$ is topological. $\quad \square$

In the following, we give two examples of topological generated $\top$-convergence spaces.

**Example 1.** *For a set $X$, the discrete $\top$-convergence structure $\mathbf{q}_d$ is defined as $F \xrightarrow{\mathbf{q}_d} x \Longleftrightarrow F \supseteq [x]_\top$ for any $F \in F_L^\top(X)$ and any $x \in X$; and the indiscrete $\top$-convergence structure $\mathbf{q}_{ind}$ is defined by $F \xrightarrow{\mathbf{q}_{ind}} x$ for every $F \in F_L^\top(X)$ and every $x \in X$, see [14].*

(1)　$(X, \mathbf{q}_d)$ is topological generated. Indeed, for any $x \in X$, note that $\mathbf{U}_{\mathbf{q}_d}(x) = [x]_\top \xrightarrow{\mathbf{q}_d} x$ and

$$
\begin{aligned}
k\mathbf{U}_{\mathbf{q}_d}\mathbf{U}_{\mathbf{q}_d}(x) &= \bigcup_{B \in \eta(\mathbf{U}_{\mathbf{q}_d}(x))} \bigcap_{y \in B} \mathbf{U}_{\mathbf{q}_d}(y), \text{ by Lemma } 1(4) \\
&= \bigcup_{B \in \dot{x}} \bigcap_{y \in B} [y]_\top \\
&= [x]_\top \supseteq \mathbf{U}_{\mathbf{q}_d}(x).
\end{aligned}
$$

Hence, $(X, \mathbf{q}_d)$ satisfies **(TP2)** and **(TU)**. It follows by Theorems 1 and 3 that $(X, \mathbf{q}_d)$ is topological generated.

(2)　$(X, \mathbf{q}_{ind})$ is topological generated. Since any $\top$-filter converge to any point, then it follows immediately that $(X, \mathbf{q}_{ind})$ satisfies **(TFD)**. Hence, $(X, \mathbf{q}_d)$ is topological generated.

Finally, we consider the categoric properties of topological generated $\top$-convergence spaces. A mapping $f : (X, \mathbf{q}) \longrightarrow (X', \mathbf{q}')$ between $\top$-convergence spaces is called continuous if $f^\Rightarrow(\mathbf{F}) \xrightarrow{\mathbf{q}'} f(x)$ for any $\mathbf{F} \xrightarrow{\mathbf{q}} x$.

We denote the category consisting of $\top$-convergence spaces and continuous mappings as $\top$-**CON**. $\top$-**CON** is a topological category in the sense that each source $(X \xrightarrow{f_i} (X_i, \mathbf{q}_i))_{i \in I}$ has initial structure $\mathbf{q}$ on $X$ defined as follows [14,37]:

$$
\mathbf{F} \xrightarrow{\mathbf{q}} x \iff \forall i \in I, f_i^\Rightarrow(\mathbf{F}) \xrightarrow{\mathbf{q}_i} f_i(x).
$$

Let $\top(X)$ denote all $\top$-convergence structures on $X$. For $\mathbf{p}, \mathbf{q} \in \top(X)$, we say $\mathbf{q}$ is finer than $\mathbf{p}$, denoted as $\mathbf{p} \leq \mathbf{q}$, if $id_X : (X, \mathbf{q}) \longrightarrow (X, \mathbf{p})$ is continuous. It is known that $(\top(X), \leq)$ forms a completed lattice [30].

We denote **TTG-CON** as the full subcategory of **T-CON** consisting of topological generated $\top$-convergence spaces.

**Theorem 4.** **TTG-CON** *is a topological category.*

**Proof.** We verify that **TTG-CON** has initial structure. Given a source $(X \xrightarrow{f_i} (X_i, \mathbf{q}_i))_{i \in I}$ in **TTG-CON**, take $\mathbf{q}$ as the initial structure in **T-CON**, i.e.,

$$
\mathbf{F} \xrightarrow{\mathbf{q}} x \Leftrightarrow \forall i \in I, f_i^\Rightarrow(\mathbf{F}) \xrightarrow{\mathbf{q}_i} f_i(x).
$$

Next, we show $(X, \mathbf{q}) \in$ **TTG-CON**. Let $\psi : T \longrightarrow X, \phi : T \longrightarrow \mathbf{F}_L^\top(X)$ such that for every $t \in T, \phi(t) \xrightarrow{\mathbf{q}} \psi(t)$. Then, for any $i \in I, f_i^\Rightarrow(\phi(t)) \xrightarrow{\mathbf{q}_i} f_i(\psi(t))$. Putting $\phi_i = f_i^\Rightarrow \circ \phi$ and $\psi_i = f_i \circ \psi$, we obtain $\phi_i(t) \xrightarrow{\mathbf{q}_i} \psi_i(t)$.

Take any $\psi^\Rightarrow(\mathbf{F}) \xrightarrow{\mathbf{q}} x$. We have

$$
\psi_i^\Rightarrow(\mathbf{F}) = (f_i \circ \psi)^\Rightarrow(\mathbf{F}) = f_i^\Rightarrow \psi^\Rightarrow(\mathbf{F}) \xrightarrow{\mathbf{q}_i} f_i(x).
$$

Since $(X_i, \mathbf{q}_i)$ satisfies **(TFD)**, we obtain that $k\phi_i\mathbf{F} \xrightarrow{\mathbf{q}_i} x$. From Lemma 2(1), we further get for every $i \in I$,

$$
f_i^\Rightarrow(k\phi\mathbf{F}) = k(f_i^\Rightarrow \circ \phi)\mathbf{F} = k\phi_i\mathbf{F} \xrightarrow{\mathbf{q}_i} f_i(x).
$$

It follows that $k\phi\mathbf{F} \xrightarrow{\mathbf{q}} x$. Hence, $(X, \mathbf{q})$ satisfies **(TFD)**, as desired. $\square$

**Remark 1.** *Let* $\top_{TG}(X)$ *denote all topological generated* $\top$*-convergence structures on* $X$*. Then, from Theorem 4, we conclude that* $(\top_{TG}(X), \leq)$ *forms a complete lattice.*

**Theorem 5.** **TTG-CON** *is a reflective subcategory of* **T-CON**.

**Proof.** Given $(X, \mathbf{q}) \in$ **T-CON**, put

$$\mathbf{rq} = \bigvee \{\mathbf{p}' \leq \mathbf{q} | \mathbf{p}' \in \top_{TG}(X)\}.$$

Then, $\mathbf{rq} \leq \mathbf{q}$ and so $id_X : (X, \mathbf{q}) \longrightarrow (X, \mathbf{rq})$ is continuous. Moreover, from Remark 1, we get $\mathbf{rq} \in \top_{TG}(X)$.

Let $(X, \mathbf{p}) \in$ **TTG-CON** and $f : (X, \mathbf{q}) \longrightarrow (Y, \mathbf{p})$ be a continuous mapping. Take $\mathbf{s}$ as the initial structure of $f : X \longrightarrow (Y, \mathbf{p})$ in **TTG-CON**. Then, $(X, \mathbf{s}) \in$ **TTG-CON** and

$$\mathbf{s} \leq \mathbf{q} \Longrightarrow \mathbf{s} \leq \mathbf{rq} \Longrightarrow id_X : (X, \mathbf{rq}) \longrightarrow (X, \mathbf{s}) \text{ is continuous,}$$

$$f : (X, \mathbf{s}) \longrightarrow (Y, \mathbf{p}) \text{ is continuous.}$$

It follows that $f = f \circ id_X : (X, \mathbf{rq}) \longrightarrow (Y, \mathbf{p})$ is continuous. This shows that the following diagram commutes in Figure 1.

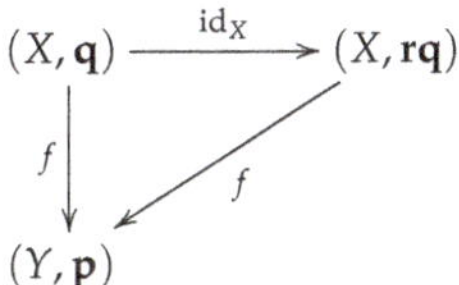

**Figure 1.** Reflective diagram.

Hence, **TTG-CON** is reflective in **T-CON**. □

## 4. Topological Generated $\top$-Convergence Spaces vs. Topological Generated $L$-Topological Spaces

In [14], Fang and Yue proposed a lattice-valued Fischer diagonal condition based on Gähler compression operator and proved that there is a bijection between $\top$-convergence spaces with their diagonal condition and strong $L$-topological spaces. In this section, we establish the relationships between our diagonal condition and Fang and Yue's diagonal condition. Then, analogizing Fang and Yue's result, we further verify that there is a one-to-one correspondence between topological generated $\top$-convergence spaces and topological generated $L$-topological spaces. Hence, we establish the convergence theory associated with the topological generated $L$-topological spaces.

Let $T$ be any set and $\phi : T \longrightarrow \mathbf{F}_L^\top(X)$. Take $\mathbf{F} \in \mathbf{F}_L^\top(T)$ and $A \in L^X$, and define $\hat{\phi}(A) \in L^T$ as

$$\forall t \in T, \hat{\phi}(A)(t) = \bigvee_{B \in \phi(t)} S_X(B, A).$$

The *Gähler compression operator* on $\phi^{\Rightarrow}(\mathbf{F}) \in \mathbf{F}_L^\top(\mathbf{F}_L^\top(X))$ is defined by

$$g\phi\mathbf{F} := \{A \in L^X | \hat{\phi}(A) \in \mathbf{F}\} \in \mathbf{F}_L^\top(X).$$

For a $\top$-convergence space $(X, \mathbf{q})$, an extension of diagonal condition **(FD)** is presented in [14]:

(**TFDW**) Let $T$ be any set, $\psi : T \longrightarrow X$ and $\phi : T \longrightarrow \mathbf{F}_L^\top(X)$ with $\phi(t) \xrightarrow{\ q\ } \psi(t)$ for every $t \in T$. For any $\mathbf{F} \in \mathbf{F}_L^\top(X)$, $\psi^\Rightarrow(\mathbf{F}) \xrightarrow{\ q\ } x \Longrightarrow g\phi\mathbf{F} \xrightarrow{\ q\ } x$.

The next theorem shows that (**TFDW**) is weaker than (**TFD**) generally, but they are equivalent when $L = \{\bot, \top\}$.

**Theorem 6.** *Let $T$ be any set, $\phi : T \longrightarrow \mathbf{F}_L^\top(X)$ and $\mathbf{F} \in \mathbf{F}_L^\top(T)$. Then,*

(1)   $k\phi\mathbf{F} \subseteq g\phi\mathbf{F}$.

(2)   $k\phi\mathbf{F} = g\phi\mathbf{F}$ *when* $L = \{\bot, \top\}$.

**Proof.** (1) Taking $A \in k\phi\mathbf{F}$, then $\exists B \in \eta(\mathbf{F})$ such that $A \in \phi(y)$ for any $y \in B$. It follows that

$$\forall y \in B, \hat{\phi}(A)(y) = \bigvee_{C \in \phi(y)} S_X(C, A) \geq S_X(A, A) = \top,$$

i.e., $\top_B \leq \hat{\phi}(A)$. By $\top_B \in \mathbf{F}$, we have $\hat{\phi}(A) \in \mathbf{F}$ and so $A \in g\phi\mathbf{F}$.

(2) We verify that $A \in g\phi\mathbf{F} \Longrightarrow A \in k\phi\mathbf{F}$. In fact, $A \in g\phi\mathbf{F}$ means $\hat{\phi}(A) \in \mathbf{F}$. Taking

$$B = \{y \in T | \hat{\phi}(A)(y) = \bigvee_{C \in \phi(y)} S_X(C, A) = \top\},$$

it follows that $A \in \phi(y)$ for any $y \in B$. Note that, when $L = \{\bot, \top\}$, $\top_B = \hat{\phi}(A) \in \mathbf{F}$, i.e., $B \in \eta(\mathbf{F})$. We get that $A \in k\phi\mathbf{F}$.   $\square$

In the following, we further verify that (**TFD**)$\Longleftrightarrow$(**TFDW**) for $\top$-convergence spaces generated by convergence spaces.

**Lemma 3.** *Let $\phi : T \longrightarrow \mathbf{F}_L^\top(X)$ and $\Phi : T \longrightarrow F(X)$.*

(1)   *Taking $\Phi_1 = \eta \circ \phi$, then for every $\mathbf{F} \in \mathbf{F}_L^\top(T)$, $\eta(g\phi\mathbf{F}) \supseteq K\Phi_1\eta(\mathbf{F})$.*

(2)   *Taking $\phi_1 = \omega \circ \Phi$, then for every $F \in F(T)$, $\eta(g\phi_1\omega(F)) = K\Phi F$.*

**Proof.** (1) It follows by

$$
\begin{aligned}
A \in K\Phi_1\eta(\mathbf{F}) \quad &\Longrightarrow \quad \exists B \in \eta(\mathbf{F}), \text{ s.t. } \forall y \in B, A \in \Phi_1(y) = (\eta \circ \phi)(y) \\
&\Longrightarrow \quad \exists \top_B \in \mathbf{F} \text{ s.t. } \forall y \in B, \top_A \in \phi(y) \\
&\Longrightarrow \quad \exists \top_B \in \mathbf{F} \text{ s.t. } \forall y \in B, \hat{\phi}(\top_A)(y) = \top \\
&\Longrightarrow \quad \exists \top_B \in \mathbf{F} \text{ s.t. } \top_B \leq \hat{\phi}(\top_A) \\
&\Longrightarrow \quad \hat{\phi}(\top_A) \in \mathbf{F} \\
&\Longrightarrow \quad \top_A \in g\phi\mathbf{F} \\
&\Longrightarrow \quad A \in \eta(g\phi\mathbf{F}).
\end{aligned}
$$

(2) Let $A \in \eta(g\phi_1\omega(\mathrm{F}))$. Then, $\top_A \in g\phi_1\omega(\mathrm{F})$ and so $\hat{\phi}_1(\top_A) \in \omega(\mathrm{F})$. Note that, for every $t \in T$,

$$
\begin{aligned}
\hat{\phi}_1(\top_A)(t) &= \bigvee_{B \in \phi_1(t)} S_X(B, \top_A) \\
&= \bigvee_{B \in \omega(\Phi(t))} S_X(B, \top_A) = \bigvee_{C \in \Phi(t)} S_X(\top_C, \top_A) \\
&= \begin{cases} \top, & \exists C \in \Phi(t), \text{ s.t. } C \subseteq A, \text{ i.e., } A \in \Phi(t); \\ \bot, & \text{otherwise.} \end{cases}
\end{aligned}
$$

Putting $C = \{t \in T | \hat{\phi}_1(\top_A)(t) = \top\}$, then $\top_C = \hat{\phi}_1(\top_A) \in \omega(\mathrm{F})$, i.e., $C \in \mathrm{F}$ and $A \in \Phi(y)$ for any $y \in C$; it follows that $A \in K\Phi\mathrm{F}$. Therefore, $\eta(g\phi_1\omega(\mathrm{F})) \subseteq K\Phi\mathrm{F}$.

Conversely, let $A \in K\Phi\mathrm{F}$. Then, there exists $B \in \mathrm{F}$ such that $A \in \Phi(y)$, i.e., $\top_A \in \omega(\Phi(y)) = \phi_1(y)$ for any $y \in B$. That means $\hat{\phi}_1(\top_A)(y) = \top$ for any $y \in B$, and so $\top_B \leq \hat{\phi}_1(\top_A)$. Then, we get $\hat{\phi}_1(\top_A) \in \omega(\mathrm{F})$, and hence $\top_A \in g\phi_1\omega(\mathrm{F})$, i.e., $A \in \eta(g\phi_1\omega(\mathrm{F}))$. Therefore, $K\Phi\mathrm{F} \subseteq \eta(g\phi_1\omega(\mathrm{F}))$. $\square$

**Theorem 7.** *Convergence space* $(X, c)$ *fulfills* **(FD)** *iff* $(X, \pi(c))$ *fulfills* **(TFDW)**.

**Proof.** $\Longrightarrow$. Let $\psi : T \longrightarrow X$, $\phi : T \longrightarrow \mathbf{F}_L^\top(X)$ with $\phi(t) \xrightarrow{\pi(c)} \psi(t)$ for every $t \in T$. Put $\Phi = \eta \circ \phi$; from $\phi(t) \xrightarrow{\pi(c)} \psi(t)$, we conclude

$$
\Phi(t) = \eta(\phi(t)) \xrightarrow{\ c\ } \psi(t).
$$

Take any $\psi^\Rightarrow(\mathbf{F}) \xrightarrow{\pi(c)} x$; from Lemma 1(5), we obtain

$$
\psi^\Rightarrow(\eta(\mathbf{F})) = \eta(\psi^\Rightarrow(\mathbf{F})) \xrightarrow{\ c\ } x.
$$

It follows by **(FD)** and Lemma 3(1) that $\eta(g\phi\mathbf{F}) \supseteq K\Phi\eta(\mathbf{F}) \xrightarrow{\ c\ } x$, i.e., $g\phi\mathbf{F} \xrightarrow{\pi(c)} x$. Thus, the condition **(TFDW)** is satisfied.

$\Longleftarrow$. Let $\psi : T \longrightarrow X$, $\Phi : T \longrightarrow F(X)$ with $\Phi(t) \xrightarrow{\ c\ } \psi(t)$ for every $t \in T$. Putting $\phi = \omega \circ \Phi$, then, from Lemma 1(1), we have

$$
\eta \circ \phi(t) = \eta \circ \omega \circ \Phi(t) = \Phi(t) \xrightarrow{\ c\ } \psi(t), \text{ i.e., } \phi(t) \xrightarrow{\pi(c)} \psi(t).
$$

Taking any $\psi^\Rightarrow(\mathrm{F}) \xrightarrow{\ c\ } x$, from Lemma 1(5), we conclude

$$
\eta\psi^\Rightarrow(\omega(\mathbf{F})) = \psi^\Rightarrow((\eta \circ \omega)(\mathbf{F})) = \psi^\Rightarrow(\mathrm{F}) \xrightarrow{\ c\ } x, \text{ i.e., } \psi^\Rightarrow(\omega(\mathbf{F})) \xrightarrow{\pi(c)} x.
$$

Thus, $g\phi\omega(\mathrm{F}) \xrightarrow{\pi(c)} x$ by **(TFDW)**. From Lemma 3(2), we further get $\eta(g\phi\omega(\mathrm{F})) = K\Phi\mathrm{F} \xrightarrow{\ c\ } x$. Hence, the condition **(FD)** is satisfied. $\square$

From Theorems 2 and 7, we get that for $\top$-convergence spaces generated by convergence spaces, conditions **(TFD)** and **(TFDW)** are equivalent.

The following example shows that there is no **(TFD)**$\Leftrightarrow$**(TFDW)** for general $\top$-convergence space.

**Example 2.** Let $X = \{x, y\}$ and $L = ([0,1], \wedge)$. Define $A_x, A_y \in L^X$ as

$$
A_x(z) = \begin{cases} 1, & z = x; \\ \frac{1}{2}, & z = y. \end{cases} \quad ; A_y(z) = \begin{cases} \frac{1}{2}, & z = x; \\ 1, & z = y. \end{cases} \quad ,
$$

and take

$$
\mathbf{F}_x = \{A \in L^X | A \geq A_x\}; \mathbf{F}_y = \{A \in L^X | A \geq A_y\}.
$$

Then, it is easily seen that $\mathbf{F}_x, \mathbf{F}_y \in \mathbf{F}_L^\top(X)$. We define

$$
\forall \mathbf{F} \in \mathbf{F}_L^\top(X), \forall z \in X, \mathbf{F} \xrightarrow{\mathbf{q}} z \Leftrightarrow \mathbf{F} \supseteq \mathbf{F}_z,
$$

then $(X, \mathbf{q})$ is a $\top$-convergence space with $\top = 1$ and $\mathbf{U_q}(z) = \mathbf{F}_z$.

(1) $(X, \mathbf{q})$ satisfies **(TFDW)**. First, it is easily seen that $(X, \mathbf{q})$ satisfies **(TP3)**.

Second, let $\psi : T \longrightarrow X$ and $\phi : T \longrightarrow \mathbf{F}_L^\top(X)$ with $\phi(t) \xrightarrow{\mathbf{q}} \psi(t)$ for every $t \in T$. From **(TP3)**, we get that $\phi(t) \supseteq \mathbf{F}_{\psi(t)}$ for any $t \in T$.

Take any $\psi^{\Rightarrow}(\mathbf{F}) \xrightarrow{\mathbf{q}} x$; then, by **(TP3)**, we have $\mathbf{F}_x \subseteq \psi^{\Rightarrow}(\mathbf{F})$, and so $A_x \in \psi^{\Rightarrow}(\mathbf{F})$, i.e., $\psi^{\leftarrow}(A_x) \in \mathbf{F}$. Hence,

$$
\begin{aligned}
\hat{\phi}(A_x)(t) &= \bigvee_{B \in \phi(t)} S_X(B, A_x) \\
&\geq \bigvee_{B \in \mathbf{F}_{\psi(t)}} S_X(B, A_x) \\
&= S_X(A_{\psi(t)}, A_x) \\
&= A_{\psi(t)}(y) \to \frac{1}{2} \\
&= \begin{cases} 1, & \psi(t) = x; \\ \frac{1}{2}, & \psi(t) = y. \end{cases} \\
&= \psi^{\leftarrow}(A_x)(t),
\end{aligned}
$$

which means that $\hat{\phi}(A_x) \geq \psi^{\leftarrow}(A_x) \in \mathbf{F}$, i.e., $A_x \in g\phi\mathbf{F}$ and so $\mathbf{F}_x \subseteq g\phi\mathbf{F}$. Thus, $g\phi\mathbf{F} \xrightarrow{\mathbf{q}} x$.

(2) $(X, \mathbf{q})$ does not satisfy **(TU)**, and thus does not satisfy **(TFD)**. Indeed, it is easily observed that $\eta(\mathbf{U_q}(z)) = \{\{X\}\}$ for any $z \in X$ and so

$$
\begin{aligned}
k\mathbf{U_q}\mathbf{U_q}(z) &= \bigcup_{A \in \eta(\mathbf{U_q}(z))} \bigcap_{z \in A} \mathbf{U_q}(z) \\
&= \bigcap_{z \in X} \mathbf{U_q}(z) = \{\{\top_X\}\}.
\end{aligned}
$$

Obviously, $\mathbf{U_q}(z) \not\subseteq k\mathbf{U_q}\mathbf{U_q}(z)$. Thus, **(TU)** is not satisfied.

In ([14], Remark 3.3), Fang and Yue proved that a strong $L$-topological space $(X, \tau)$ corresponds uniquely to a $\top$-convergence space $(X, \mathbf{q})$ with the condition **(TFDW)** by taking that for each $x \in X$,

$$
\mathbf{U_q}(x) = \mathbf{U}_\tau(x) := \{A \in L^X | \bigvee_{B \in \tau, B(x) = \top} S_X(B, A) = \top\}.
$$

Then, they constructed a bijection between strong $L$-topological spaces and $\mathsf{T}$-convergence spaces with **(TFDW)**.

Let $L$ be a continuous lattice and $(X, \delta)$ be a topological space. We prove below that, for any $x \in X$,

$$\mathbf{U}_{\pi(\delta)}(x) = \mathbf{U}_{\varrho(\delta)}(x).$$

That is, the $\mathsf{T}$-neighborhood system $\{\mathbf{U}_{\pi(\delta)}(x)\}_{x \in X}$ associated with topological generated $\mathsf{T}$-convergence space $(X, \pi(\delta))$ is equal to the $\mathsf{T}$-neighborhood system $\{\mathbf{U}_{\varrho(\delta)}(x)\}_{x \in X}$ associated with topological generated $L$-topological space $(X, \varrho(\delta))$.

**Lemma 4.** *Let $L$ be a continuous lattice. Then, for a topological space $(X, \delta)$,*

$$A \in \varrho(\delta) \iff \forall x \in X, A(x) \leq \bigvee_{x \in B \in \delta} S_X(\mathsf{T}_B, A).$$

**Proof.** Let $A \in \varrho(\delta)$ and $x \in X$. Taking any $\alpha \in L$ with $\alpha \ll A(x)$, then $x \in A_\alpha \in \delta$. It follows that

$$\alpha \leq \bigwedge_{y \in A_\alpha} A(y) = S_X(\mathsf{T}_{A_\alpha}, A) \leq \bigvee_{x \in B \in \delta} S_X(\mathsf{T}_B, A).$$

By the continuity of $L$ and the arbitrariness of $\alpha$, we obtain

$$A(x) \leq \bigvee_{x \in B \in \delta} S_X(\mathsf{T}_B, A).$$

Conversely, let $A(x) \leq \bigvee_{x \in B \in \delta} S_X(\mathsf{T}_B, A)$ for each $x \in X$. For each $\alpha \in L$, we prove below that $A_\alpha \in \delta$. Indeed, if $A_\alpha \neq \varnothing$, then taking any $x \in A_\alpha$ we have

$$\alpha \ll A(x) \leq \bigvee_{x \in B \in \delta} S_X(\mathsf{T}_B, A) = \bigvee_{x \in B \in \delta} \bigwedge_{y \in B} A(y).$$

Hence, there exists $B \in \delta$ such that $x \in B$ and $\alpha \ll A(y)$ for all $y \in B$, i.e., $B \subseteq A_\alpha$. Then, it follows that $A_\alpha \in \delta$ for any $\alpha \in L$. Therefore, $A \in \varrho(\delta)$. $\square$

**Theorem 8.** *Let $L$ be a continuous lattice. Then, for a topological space $(X, \delta)$, $\mathbf{U}_{\pi(\delta)}(x) = \mathbf{U}_{\varrho(\delta)}(x)$ for any $x \in X$.*

**Proof.** By the definition of $\pi(\delta)$, it is easily seen that $\mathbf{F} \xrightarrow{\pi(\delta)} x \iff \mathbf{F} \supseteq \omega(\mathbf{U}_\delta(x))$, and so

$$
\begin{aligned}
\mathbf{U}_{\pi(\delta)}(x) &= \omega(\mathbf{U}_\delta(x)) \\
&= \{A \in L^X \mid \bigvee_{B \in \mathbf{U}_\delta(x)} S_X(\mathsf{T}_B, A) = \top\} \\
&= \{A \in L^X \mid \bigvee_{x \in B \in \delta} S_X(\mathsf{T}_B, A) = \top\}.
\end{aligned}
$$

$$\mathbf{U}_{\varrho(\delta)}(x) = \{A \in L^X \mid \bigvee_{B \in \varrho(\delta), B(x) = \top} S_X(B, A) = \top\}.$$

Let $A \in \mathbf{U}_{\pi(\delta)}(x)$. Then,

$$
\begin{aligned}
\top &= \bigvee_{x \in B \in \delta} S_X(\top_B, A), \text{ by } \top_B \in \varrho(\delta), \top_B(x) = \top \\
&\leq \bigvee_{C \in \varrho(\delta), C(x) = \top} S_X(C, A),
\end{aligned}
$$

i.e., $A \in \mathbf{U}_{\varrho(\delta)}(x)$. Conversely, let $A \in \mathbf{U}_{\varrho(\delta)}(x)$. Note that, by Lemma 4, we get that, for each $B \in \varrho(\delta), B(x) = \top$,

$$
\top = B(x) = \bigvee_{x \in C \in \delta} S_X(\top_C, B).
$$

Hence, from $A \in \mathbf{U}_{\varrho(\delta)}(x)$ it follows that

$$
\begin{aligned}
\top &= \bigvee_{B \in \varrho(\delta), B(x) = \top} S_X(B, A) \\
&= \bigvee_{B \in \varrho(\delta), B(x) = \top} (\top * S_X(B, A)) \\
&= \bigvee_{B \in \varrho(\delta), B(x) = \top} \left( \bigvee_{x \in C \in \delta} S_X(\top_C, B) * S_X(B, A) \right) \\
&\leq \bigvee_{x \in C \in \delta} S_X(\top_C, A),
\end{aligned}
$$

i.e., $A \in \mathbf{U}_{\pi(\delta)}(x)$.  $\square$

**Remark 2.** *Similar to ([14], Remark 3.3), we get from Theorem 8 that there is a bijection between topological generated $\top$-convergence spaces and topological generated $L$-topological spaces. Thus, $\top$-convergence spaces with our Fischer diagonal condition precisely characterizes topological generated $L$-topological spaces. Therefore, we establish the convergence theory associated with topological generated $L$-topological spaces.*

## 5. Conclusions

In this paper, we present a lattice-valued Fischer diagonal condition (**TFD**) through extending Kowalsky compression operator, and verify that there is a one-to-one correspondence between $\top$-convergence spaces with (**TFD**) and topological generated $L$-topological spaces. This shows that (**TFD**) can characterize topological generated $L$-topological spaces. That is to say, we establish the convergence theory associated with topological generated $L$-topological spaces. It is well-known that Fischer diagonal condition also plays an important role in uniform convergence spaces [1]. In the further work, we shall consider the fuzzy version of Fischer diagonal condition in $\top$-uniform convergence spaces defined in [39].

**Author Contributions:** Q.J. and L.L. contributed the central idea. All authors contributed to the writing and revisions.

**Funding:** This work was supported by National Natural Science Foundation of China (Nos. 11801248 and 11501278), and the Ke Yan Foundation of Liaocheng University (318011920).

**Acknowledgments:** The authors thank the reviewers and the editor for their valuable comments and suggestions.

## References

1. Preuss, G. *Fundations of Topology*; Kluwer Academic Publishers: London, UK, 2002.
2. Kent, D.C.; Richardson, G.D. Convergence spaces and diagonal conditions. *Topol. Appl.* **1996**, *70*, 167–174. [CrossRef]
3. Gähler, W. Monadic convergence structures. In *Topological and Algebraic Structures in Fuzzy Sets*; Höhle, U., Klement, E.P., Eds.; Kluwer Academic Publishers: Boston, MA, USA; Dordrecht, The Netherlands; London, UK, 2003.
4. Jäger, G. Pretopological and topological lattice-valued convergence spaces. *Fuzzy Sets Syst.* **2007**, *158*, 424–435. [CrossRef]
5. Zadeh, L.A. Fuzzy sets. *Inf. Control* **1965**, *8*, 338–353. [CrossRef]
6. Romero, J.A.; García, P.A.G.; Marín, C.E.M.; Crespo, R.G.; Herrera-Viedma, E. Fuzzy logic models for non-programmed decision-making in personnel selection processes based on gamification. *Informatica* **2018**, *29*, 1–20.
7. Cueva-Fernandez, G.; Espada, J.P.; García-Díaz, V.; Crespo, R.G.; Garcia-Fernandez, N. Fuzzy system to adapt web voice interfaces dynamically in a vehicle sensor tracking application definition. *Soft Comput.* **2016**, *20*, 3321–3334. [CrossRef]
8. Benamina, M.; Atmani, B.; Benbelkacem, S. Diabetes diagnosis by case-based reasoning and fuzzy logic. *Int. J. Interact. Multimed. Artif. Intell.* **2018**, *5*, 72–80. [CrossRef]
9. Harish, B.S.; Kumar, S.V.A. Anomaly based Intrusion Detection using Modified Fuzzy Clustering. *Int. J. Interact. Multimed. Artif. Intell.* **2017**, *4*, 54–59. [CrossRef]
10. Jäger, G. A category of *L*-fuzzy convergence spaces. *Quaest. Math.* **2001**, *24*, 501–517. [CrossRef]
11. Flores, P.V.; Mohapatra, R.N.; Richardson, G. Lattice-valued spaces: Fuzzy convergence. *Fuzzy Sets Syst.* **2006**, *157*, 2706–2714. [CrossRef]
12. Fang, J.M. Stratified *L*-ordered convergence structures. *Fuzzy Sets Syst.* **2010**, *161*, 2130–2149. [CrossRef]
13. Li, L.Q.; Jin, Q. On stratified *L*-convergence spaces: Pretopological axioms and diagonal axioms. *Fuzzy Sets Syst.* **2012**, *204*, 40–52. [CrossRef]
14. Fang, J.M.; Yue, Y.L. $\top$-diagonal conditions and Continuous extension theorem. *Fuzzy Sets Syst.* **2017**, *321*, 73–89. [CrossRef]
15. Flores, P.V.; Richardson, G. Lattice-valued convergence: Diagonal axioms. *Fuzzy Sets Syst.* **2008**, *159*, 2520–2528. [CrossRef]
16. Boustique, H.; Richardson, G. A note on regularity. *Fuzzy Sets Syst.* **2011**, *162*, 64–66. [CrossRef]
17. Han, S.E.; Lu, L.X.; Yao, W. Quantale-valued fuzzy scott topology. *Iran. J. Fuzzy Syst.* **2019**, *16*, 175–188.
18. Jäger, G. Fischer's diagonal condition for lattice-valued convergence spaces. *Quaest. Math.* **2008**, *31*, 11–25. [CrossRef]
19. Jäger, G. Lattice-valued convergence spaces and regularity. *Fuzzy Sets Syst.* **2008**, *159*, 2488–2502. [CrossRef]
20. Jäger, G. Connectedness and local connectedness for lattice-valued convergence spaces. *Fuzzy Sets Syst.* **2016**, *300*, 134–146. [CrossRef]
21. Jin, Q.; Li, L.Q. Modified Top-convergence spaces and their relationships to lattice-valued convergence spaces. *J. Intell. Fuzzy Syst.* **2018**, *35*, 2537–2546. [CrossRef]
22. Jin, Q.; Li, L.Q.; Lang, G.M. *p*-regularity and *p*-regular modification in $\top$-convergence spaces. *Mathematics* **2019**, *7*, 370. [CrossRef]
23. Jin, Q.; Li, L.Q.; Lv, Y.R.; Zhao, F.; Zou, J. Connectedness for lattice-valued subsets in lattice-valued convergence spaces. *Quaest. Math.* **2019**, *42*, 135–150. [CrossRef]
24. Jin, Q.; Li, L.Q.; Meng, G.W. On the relationships between types of *L*-convergence spaces. *Iran. J. Fuzzy Syst.* **2016**, *1*, 93–103.
25. Lai, H.L.; Zhang, D.X. Fuzzy topological spaces with conical neighborhood system. *Fuzzy Sets Syst.* **2018**, *330*, 87–104. [CrossRef]
26. Li, L.Q. *p*-Topologicalness—A Relative Topologicalness in $\top$-Convergence Spaces. *Mathematics* **2019**, *7*, 228. [CrossRef]

27. Li, L.Q.; Jin, Q. *p*-Topologicalness and *p*-Regularity for lattice-valued convergence spaces. *Fuzzy Sets Syst.* **2014**, *238*, 26–45. [CrossRef]
28. Li, L.Q.; Jin, Q.; Hu, K. On Stratified *L*-Convergence Spaces: Fischer's Diagonal Axiom. *Fuzzy Sets Syst.* **2015**, *267*, 31–40. [CrossRef]
29. Li, L.Q.; Jin, Q.; Hu, K. Lattice-valued convergence associated with CNS spaces. *Fuzzy Sets Syst.* **2019**, *370*, 91–98. [CrossRef]
30. Li, L.Q.; Jin, Q.; Yao, B.X. Regularity of fuzzy convergence spaces. *Open Math.* **2018**, *16*, 1455–1465. [CrossRef]
31. Li, L.Q.; Jin, Q.; Meng, G.W.; Hu, K. The lower and upper *p*-topological (*p*-regular) modifications for lattice-valued convergence spaces. *Fuzzy Sets Syst.* **2016**, *282*, 47–61. [CrossRef]
32. Orpen, D.; Jäger, G. Lattice-valued convergence spaces: Extending the lattices context. *Fuzzy Sets Syst.* **2012**, *190*, 1–20. [CrossRef]
33. Pang, B. Categorical properties of *L*-Fuzzifying convergence spaces. *Filomat* **2018**. *32*, 4021–4036. [CrossRef]
34. Pang, B.; Shi, F.G. Strong inclusion orders between *L*-subsets and its applications in *L*-convex spaces. *Quaest. Math.* **2018**, *41*, 1021–1043. [CrossRef]
35. Pang, B.; Xiu, Z.Y. Stratified *L*-prefilter convergence structures in stratified *L*-topological spaces. *Soft Comput.* **2018**, *22*, 7539–7551. [CrossRef]
36. Pang, B.; Xiu, Z.X. An axiomatic approach to bases and subbases in *L*-convex spaces and their applications. *Fuzzy Sets Syst.* **2019**, *369*, 40–56. [CrossRef]
37. Qiu, Y.; Fang, J.M. The category of all $\top$-convergence spaces and its cartesian-closedness. *Iran. J. Fuzzy Syst.* **2017**, *14*, 121–138.
38. Reid, L.; Richardson, G. Connecting $\top$ and Lattice-Valued Convergences. *Iran. J. Fuzzy Syst.* **2018**, *15*, 151–169.
39. Reid, L.; Richardson, G. Lattice-valued spaces: $\top$-Completions. *Fuzzy Sets Syst.* **2019**, *369*, 1–19. [CrossRef]
40. Xiu, Z.Y.; Pang, B. Base axioms and subbase axioms in *M*-fuzzifying convex spaces. *Iran. J. Fuzzy Syst.* **2018**, *2*, 75–87.
41. Yao, W. On many-valued stratified *L*-fuzzy convergence spaces. *Fuzzy Sets Syst.* **2008**, *159*, 2503–2519. [CrossRef]
42. Yue, Y.L.; Fang, J.M. The $\top$-filter monad and its applications. *Fuzzy Sets Syst.* **2019**. [CrossRef]
43. Rosenthal, K.I. *Quantales and Their Applications*; Longman Scientific & Technical: London, UK, 1990.
44. Gierz, G.; Hofmann, K.H.; Keimel, K.; Lawson, J.D.; Mislove, M.W.; Scott, D.S. *Continuous Lattices and Domains*; Cambridge University Press: Cambridge, UK, 2003.
45. Höhle, U.; Šostak, A. Axiomatic foundations of fixed-basis fuzzy topology. In *Mathematics of Fuzzy Sets: Logic, Topology and Measure Theory*; Höhle, U., Rodabaugh, S.E., Eds.; The Handbooks of Fuzzy Sets Series; Kluwer Academic Publishers: Boston, MA, USA; Dordrecht, The Netherlands; London, UK, 1999; Volume 3, pp. 123–273.
46. Bělohlávek, R. *Fuzzy Relational Systems: Foundations and Principles*; Kluwer Academic Publishers: New York, NY, USA, 2002.
47. Zhang, D.X. An enriched category approach to many valued topology. *Fuzzy Sets Syst.* **2007**, *158*, 349–366. [CrossRef]
48. Zhao, F.F.; Jin, Q.; Li, L.Q. The axiomatic characterizations on *L*-generalized fuzzy neighborhood system-based approximation operators. *Int. J. Gen. Syst.* **2018**, *47*, 155–173. [CrossRef]
49. Gutiérrez, G.J. On stratified *L*-valued filters induced by $\top$-filters. *Fuzzy Sets Syst.* **2006**, *159*, 813–819. [CrossRef]

 *mathematics*

Article

# Degrees of *L*-Continuity for Mappings between *L*-Topological Spaces

**Zhenyu Xiu [1] and Qinghua Li [2,*]**

[1]  College of Applied Mathematics, Chengdu University of Information Technology,
    Chengdu 610000, China; xzy@cuit.edu.cn
[2]  School of Mathematics and Information Sciences, Yantai University, Yantai 264005, China
*  Correspondence:lqh@ytu.edu.cn; Tel.: +86-13880310292

Received: 15 September 2019; Accepted: 17 October 2019; Published: 24 October 2019

**Abstract:** By means of the residual implication on a frame $L$, a degree approach to $L$-continuity and $L$-closedness for mappings between $L$-cotopological spaces are defined and their properties are investigated systematically. In addition, in the situation of $L$-topological spaces, degrees of $L$-continuity and of $L$-openness for mappings are proposed and their connections are studied. Moreover, if $L$ is a frame with an order-reversing involution $'$, where $b' = b \to \perp$ for $b \in L$, then degrees of $L$-continuity for mappings between $L$-cotopological spaces and degrees of $L$-continuity for mappings between $L$-topological spaces are equivalent.

**Keywords:** $L$-cotopological space; $L$-topological space; degree of $L$-continuity; degree of $L$-closedness; degree of $L$-openness

---

## 1. Introduction

Since Chang [1] introduced fuzzy set theory to topology, fuzzy topology and its related theories have been widely investigated such as [1–18]. The degree approach that equips fuzzy topology and its related structures with some degree description is also an essential character of fuzzy set theory. This approach has been developed extensively in the theory of fuzzy topology, fuzzy convergence and fuzzy convex structure. Yue and Fang [19] introduced a degree approach to $T_1$ and $T_2$ separation properties in $(L, M)$-fuzzy topological spaces. Shi [20,21] defined the degrees of separation axioms which are compatible with $(L, M)$-fuzzy metric spaces. Li and Shi [22] introduced the degree of compactness in $L$-fuzzy topological spaces. Pang defined the compact degree of $(L, M)$-fuzzy convergence spaces [23] and degrees of $T_i$ ($i = 0, 1, 2$) separation property as well as the regular property of stratified $L$-generalized convergence spaces [24]. All of the above-mentioned research mainly equipped spatial properties with some of the degree descriptions.

Actually, special mappings between structured spaces and the structured space itself can also be endowed with some degrees. Xiu and his co-authors [25,26] defined degrees of fuzzy convex structures, fuzzy closure systems and fuzzy Alexandrov topologies and discussed their properties. Xiu and Pang [27] gave a degree approach to special mappings between $M$-fuzzifying convex spaces. In [28], Pang defined degrees of continuous mappings and open mappings between $L$-fuzzifying topological spaces to describe how a mapping between $L$-fuzzifying topological spaces becomes a continuous mapping or an open mapping in a degree sense. Liang and Shi [29] further defined the degrees of continuous mappings and open mappings between $L$-fuzzy topological spaces and investigated their relationship. Li [30] defined the degrees of special

*Mathematics* **2019**, *7*, 1013; doi:10.3390/math7111013      www.mdpi.com/journal/mathematics

mappings in the theory of $L$-convex spaces and investigated their properties. Xiu and Pang [27] developed the degree approach to $M$-fuzzifying convex spaces to define the degrees of $M$-CP mappings and $M$-CC mappings.

Following this direction, we will focus on the case of $L$-cotopological spaces and $L$-topological spaces in this paper. By means of $L$-closure operators and $L$-interior operators, we will consider degrees of $L$-continuity and $L$-closedness for mappings between $L$-cotopological spaces as well as degrees of $L$-continuity and $L$-openness for mappings between $L$-topological spaces and will investigate their properties systematically.

## 2. Preliminaries

In this paper, let $L$ be a frame, $X$ be a nonempty set and $L^X$ be the set of all $L$-subsets on $X$. The bottom element and the top element of $L$ are denoted by $\bot$ and $\top$, respectively. A residual implication can be defined by $a \rightarrow b = \bigvee\{\lambda \in L \mid a \wedge \lambda \leqslant b\}$. The operators on $L$ can be translated onto $L^X$ in a pointwise way. In this case, $L^X$ is also a complete lattice. Let $\underline{\bot}$ and $\underline{\top}$ denote the smallest element and the largest element in $L^X$, respectively

Let $f : X \longrightarrow Y$ be a mapping. Define $f_L^{\rightarrow} : L^X \longrightarrow L^Y$ and $f_L^{\leftarrow} : L^Y \longrightarrow L^X$ by $f_L^{\rightarrow}(B)(y) = \bigvee_{f(x)=y} B(x)$ for $B \in L^X$ and $y \in Y$, and $f_L^{\leftarrow}(A)(x) = A(f(x))$ for $A \in L^Y$ and $x \in X$, respectively.

Using the residual implication, the concept of fuzzy inclusion order of $L$-subsets is introduced.

**Definition 1** ([2,31]). *The fuzzy inclusion order of $L$-subsets is a mapping $\mathbb{S}_X(-,-) : L^X \times L^X \longrightarrow L$ which satisfying for each $C,\ D \in L^X$,*

$$\mathbb{S}_X(C,D) = \bigwedge_{x \in X} \left( C(x) \rightarrow D(x) \right)$$

**Lemma 1** ([2,31]). *Let $f : X \longrightarrow Y$ be a mapping. Then, for each $C, D, E \in L^X$ and $A, B \in L^Y$, the following statements hold:*

(1)  $\mathbb{S}_X(D, E) = \top$ *if and only if $D \leqslant E$.*
(2)  $D \leqslant E$ *implies $\mathbb{S}_X(D, C) \geqslant \mathbb{S}_X(E, C)$.*
(3)  $D \leqslant E$ *implies $\mathbb{S}_X(C, D) \leqslant \mathbb{S}_X(C, E)$.*
(4)  $\mathbb{S}_X(D, E) \wedge \mathbb{S}_X(E, C) \leqslant \mathbb{S}_X(D, C)$.
(5)  $\mathbb{S}_X(D, E) \leqslant \mathbb{S}_X(f_L^{\rightarrow}(D), f_L^{\rightarrow}(E))$.
(6)  $\mathbb{S}_Y(A, B) \leqslant \mathbb{S}_Y(f_L^{\leftarrow}(A), f_L^{\leftarrow}(B))$.

In the following, we will only use $\mathbb{S}$ to represent the fuzzy inclusion order of $L$-subsets on both $L^X$ and $L^Y$, not $\mathbb{S}_X$ or $\mathbb{S}_Y$. This will not lead to ambiguity in the paper.

**Definition 2** ([1,18,32,33]). *Let $\mathcal{C}$ be a subset of $L^X$. $\mathcal{C}$ is called an $L$-cotopology on $X$ if it satisfies:*
   *(LCT1) $\underline{\bot}, \underline{\top} \in \mathcal{C}$;*
   *(LCT2) if $A, B \in \mathcal{C}$, then $A \vee B \in \mathcal{C}$;*
   *(LCT3) if $\{A_j \mid j \in J\} \subseteq \mathcal{C}$, then $\bigwedge_{j \in J} A_j \in \mathcal{C}$.*
*For an $L$-cotopology $\mathcal{C}$ on $X$, the pair $(X, \mathcal{C})$ is called an $L$-cotopological space.*
   *If $\mathcal{C}$ also satisfies:*
   *(SLCT) for each $a \in L$, $\underline{a} \in \mathcal{C}$,*
*then it is called a stratified $L$-cotopology and the pair $(X, \mathcal{C})$ is called a stratified $L$-cotopological space.*
   *A mapping $f : (X, \mathcal{C}_X) \rightarrow (Y, \mathcal{C}_Y)$ is called $L$-continuous provided that, for each $B \in L^Y$, $B \in \mathcal{C}_Y$ implies $f_L^{\leftarrow}(B) \in \mathcal{C}_X$.*

*A mapping $f : (X, \mathcal{C}_X) \to (Y, \mathcal{C}_Y)$ is called L-closed provided that, for each $A \in L^Y$, $A \in \mathcal{C}_X$ implies $f_L^{\to}(B) \in \mathcal{C}_Y$.*

**Definition 3** ([18,32,33]). *An L-closure operator on X is a mapping $Cl : L^X \longrightarrow L^X$ which satisfies:*
  (LCL1) $Cl(\bot) = \bot$;
  (LCL2) $A \leqslant Cl(A)$;
  (LCL3) $Cl(Cl(A)) = Cl(A)$;
  (LCL4) $Cl(A \vee B) = Cl(A) \vee Cl(B)$.
*For an L-closure operator Cl on X, the pair $(X, Cl)$ is called an L-closure space.*

A mapping $f : (X, Cl_X) \longrightarrow (Y, Cl_Y)$ is called $L$-continuous provided that

$$\forall A \in L^X, \ f_L^{\to}(Cl_X(A)) \leqslant Cl_Y(f_L^{\to}(A)).$$

It was proved in [18,32,33] that $L$-cotopologies and $L$-closure operators are conceptually equivalent with transferring process $Cl^{\mathcal{C}}(A) = \bigwedge\{B \in L^X \mid A \leqslant B \in \mathcal{C}\}$ for each $A \in L^X$ and $\mathcal{C}^{Cl} = \{A \in L^X \mid Cl(A) = A\}$. Correspondingly, $L$-continuous mappings between $L$-cotopological spaces and $L$-continuous mappings between $L$-closure spaces are compatible. In the sequel, we treat $L$-cotopological spaces with their $L$-continuous mappings and $L$-closure spaces with their $L$-continuous mappings equivalently. We will use $Cl$ to represent $Cl^{\mathcal{C}}$ tacitly.

**Definition 4** ([1,18,32,33]). *An L-topology on X is a subset $\mathcal{T} \subseteq L^X$ which satisfies:*
  (LT1) $\bot, \top \in \mathcal{T}$;
  (LT2) if $A, B \in \mathcal{T}$, then $A \wedge B \in \mathcal{T}$;
  (LT3) if $\{A_j \mid j \in J\} \subseteq \mathcal{T}$, then $\bigvee_{j \in J} A_j \in \mathcal{T}$.
*For an L-topology $\mathcal{T}$ on X, the pair $(X, \mathcal{T})$ is called an L-topological space.*
  *A mapping $f : (X, \mathcal{T}_X) \to (Y, \mathcal{T}_Y)$ is called L-continuous provided that for each $B \in L^Y$, $B \in \mathcal{T}_Y$ implies $f_L^{\leftarrow}(B) \in \mathcal{T}_X$.*
  *A mapping $f : (X, \mathcal{T}_X) \to (Y, \mathcal{T}_Y)$ is called L-open provided that for each $A \in L^X$, $A \in \mathcal{T}_X$ implies $f_L^{\to}(B) \in \mathcal{T}_Y$.*

**Definition 5** ([18,33]). *An L-interior operator on X is a mapping $\mathcal{N} : L^X \longrightarrow L^X$ which satisfies:*
  (LN1) $\mathcal{N}(\top) = \top$;
  (LN2) $\mathcal{N}(C) \leqslant C$;
  (LN3) $\mathcal{N}(\mathcal{N}(C)) = \mathcal{N}(C)$;
  (LN4) $\mathcal{N}(C \wedge D) = \mathcal{N}(C) \wedge \mathcal{N}(D)$.
*For an L-interior operator $\mathcal{N}$ on X, the pair $(X, \mathcal{N})$ is called an L-interior space.*

A mapping $f : (X, \mathcal{N}_X) \longrightarrow (Y, \mathcal{N}_Y)$ is called $L$-continuous provided that

$$\forall A \in L^X, \ f_L^{\to}(\mathcal{N}_X(A)) \leqslant \mathcal{N}_Y(f_L^{\to}(A)).$$

It was proved in [18,33] that $L$-topologies and $L$-interior operators are conceptually equivalent with transferring process $\mathcal{N}^{\mathcal{T}}(A) = \bigvee\{B \in L^X \mid B \leqslant A, \ B \in \mathcal{T}\}$ for each $A \in L^X$ and $\mathcal{T}^{\mathcal{N}} = \{A \in L^X \mid \mathcal{N}(A) = A\}$. Correspondingly, $L$-continuous mappings between $L$-topological spaces and $L$-continuous mappings between $L$-interior spaces are compatible. In the sequel, we treat $L$-topological spaces with their $L$-continuous mappings and $L$-interior spaces with their $L$-continuous mappings equivalently. We will use $\mathcal{N}$ to represent $\mathcal{N}^{\mathcal{T}}$ tacitly.

**Definition 6** ([3,32]). *An L-filter on X is a mapping $F : L^X \longrightarrow L$ which satisfies:*
    *(F1) $F(\perp) = \perp$, $F(\top) = \top$;*
    *(F2) $F(C \wedge D) = F(C) \wedge F(D)$ for each $C, D \in L^X$.*
*Let $\mathfrak{F}_L^s(X)$ denote the family of all L-filters on X.*

**Definition 7** ([3]). *An L-fuzzy convergence on X is a mapping $\mathrm{Lim} : \mathfrak{F}_L^s(X) \longrightarrow L^X$ which satisfies:*
    *(L1) $\forall x \in X$, $\mathrm{Lim}[x](x) = \top$ ;*
    *(L2) $F_1 \leqslant F_2$ implies $\mathrm{Lim}(F_1) \leqslant \mathrm{Lim}(F_2)$.*
*For an L-fuzzy convergence Lim on X, the pair $(X, \mathrm{Lim})$ is called an L-fuzzy convergence space.*

**Theorem 1** ([3]). *For an L-topological space $(X, \mathcal{T})$, let $\mathcal{N}$ be its interior operator and $\mathcal{U}^x$ be the L-neighborhood filter defined by $\mathcal{U}^x(A) = \mathcal{N}(A)(x)$ for each $x \in X$. Then, the mapping $\mathrm{Lim} : \mathfrak{F}_L^s(X) \longrightarrow L^X$ defined by*

$$\mathrm{Lim}(F)(x) = \bigwedge_{A \in L^X} (\mathcal{U}^x(A) \to F(A))$$

*is an L-fuzzy convergence on X.*

## 3. Degrees of *L*-Continuity and *L*-Closedness for Mappings between *L*-Cotopological Spaces

In this section, we mainly define degrees of $L$-continuity of mappings and $L$-closedness of mappings to equip each mapping between $L$-cotopological spaces with some degree to be an $L$-continuous mapping and an $L$-closed mapping, respectively. Then, we will study their connections in a degree sense. Moreover, $f$ always denotes a mapping from $X$ to $Y$ and $g$ always denotes a mapping from $Y$ to $Z$ in the following sections.

**Definition 8.** *Let $(X, \mathcal{C}_X)$ and $(Y, \mathcal{C}_Y)$ be L-cotopological spaces. Then,*
    *(1) $D_{c_1}(f)$ defined by*

$$D_{c_1}(f) = \bigwedge_{B \in L^X} \mathbb{S}\left( f_L^{\rightarrow}(Cl_X(B)), Cl_Y(f_L^{\rightarrow}(B)) \right)$$

*is called the degree of L-continuity for $f$.*

    *(2) $D_b(f)$ defined by*

$$D_b(f) = \bigwedge_{A \in L^X} \mathbb{S}\left( Cl_Y(f_L^{\rightarrow}(A)), f_L^{\rightarrow}(Cl_X(A)) \right)$$

*is called the degree of L-closedness for $f$.*

**Remark 1.** *(1) If $D_{c_1}(f) = \top$, then $f_L^{\rightarrow}(Cl_X(A)) \leqslant Cl_Y(f_L^{\rightarrow}(A))$ for all $A \in L^X$, which is exactly the definition of L-continuous mappings between L-closure spaces. As we claimed that L-continuous mappings between L-cotopological spaces and L-continuous mappings between L-closure spaces are compatible, we don't distinguish them. Therefore, we defined the degree of L-continuity of mappings between L-cotopological spaces by using L-continuous mappings between their induced L-closure spaces.*
    *(2) If $D_b(f) = \top$, then $Cl_Y(f_L^{\rightarrow}(A)) \leqslant f_L^{\rightarrow}(Cl_X(A))$ for all $A \in L^X$. This is exactly the equivalent form of L-closed mappings between L-cotopological spaces by means of the corresponding L-closure operators.*

**Lemma 2.** *Let $f : X \longrightarrow Y$ and $g : Y \longrightarrow Z$ be mappings. Then, for each $A \in L^X$, $B \in L^Y$ and $C \in L^Z$, the following statements hold:*

(1)  $A \leqslant f_L^{\leftarrow}(f_L^{\rightarrow}(A))$. If $f$ is injective, then the equality holds.

(2)  $f_L^{\rightarrow}(f_L^{\leftarrow}(B)) \leqslant B$. If $f$ is surjective, then the equality holds.

(3)  $(g \circ f)_L^{\rightarrow}(A) = g_L^{\rightarrow}(f_L^{\rightarrow}(A))$.

(4)  $(g \circ f)_L^{\leftarrow}(C) = f_L^{\leftarrow}(g_L^{\leftarrow}(C))$.

**Proof.** The proofs are routine and are omitted.  $\square$

**Theorem 2.** *Let $(X, \mathcal{C}_X)$ and $(Y, \mathcal{C}_Y)$ be L-cotopological spaces. Then,*

$$D_{c_1}(f) = \bigwedge_{B \in L^Y} \mathbb{S}\Big(f_L^{\rightarrow}(Cl_X(f_L^{\leftarrow}(B))), Cl_Y(B)\Big).$$

**Proof.** It follows from the definition of $D_{c_1}(f)$ that

$$
\begin{aligned}
D_{c_1}(f) &= \bigwedge_{A \in L^X} \mathbb{S}\Big(f_L^{\rightarrow}(Cl_X(A)), Cl_Y(f_L^{\rightarrow}(A))\Big) \\
&\leqslant \bigwedge_{B \in L^Y} \mathbb{S}\Big(f_L^{\rightarrow}(Cl_X(f_L^{\leftarrow}(B))), Cl_Y(f_L^{\rightarrow}(f_L^{\leftarrow}(B)))\Big) \\
&\leqslant \bigwedge_{B \in L^Y} \mathbb{S}\Big(f_L^{\rightarrow}(Cl_X(f_L^{\leftarrow}(B))), Cl_Y(B)\Big) \\
&\leqslant \bigwedge_{A \in L^X} \mathbb{S}\Big(f_L^{\rightarrow}(Cl_X(f_L^{\leftarrow}(f_L^{\rightarrow}(A)))), Cl_Y(f_L^{\rightarrow}(A))\Big) \\
&\leqslant \bigwedge_{A \in L^X} \mathbb{S}\Big(f_L^{\rightarrow}(Cl_X(A)), Cl_Y(f_L^{\rightarrow}(A))\Big) \\
&= D_{c_1}(f).
\end{aligned}
$$

This implies

$$D_{c_1}(f) = \bigwedge_{B \in L^Y} \mathbb{S}\Big(f_L^{\rightarrow}(Cl_X(f^{\leftarrow}(B))), Cl_Y(B)\Big),$$

as desired.  $\square$

**Theorem 3.** *(1) If $id : (X, \mathcal{C}_X) \longrightarrow (X, \mathcal{C}_X)$ is the identity mapping, then $D_{c_1}(id) = \top$ and $D_b(id) = \top$.*

*(2) If $y_0 : (X, \mathcal{C}_X) \longrightarrow (Y, \mathcal{C}_Y)$ is a constant mapping between stratified L-cotopological spaces with the constant $y_0 \in Y$, then $D_{c_1}(y_0) = \top$.*

**Proof.** (1) Straightforward,

(2) It follows immediately from the definition of $D_{c_1}(y_0)$ that

$$
\begin{aligned}
D_{c_1}(y_0) &= \bigwedge_{A \in L^X} \mathbb{S}\Big(Cl_X(A), (y_0)_L^{\leftarrow}(Cl_Y((y_0)_L^{\rightarrow}(A)))\Big) \\
&= \bigwedge_{A \in L^X} \bigwedge_{x \in X} \Big(Cl_X(A)(x) \rightarrow Cl_Y((y_0)_L^{\rightarrow}(A))(y_0)\Big).
\end{aligned}
$$

Since $(X, \mathcal{C}_X)$ is stratified, we know $\bigvee_{z \in X} A(z) \in \mathcal{C}_X$. Then, for each $A \in L^X$ and $x \in X$, it follows that

$$Cl_X(A)(x) = \bigwedge\{B \in L^X \mid A \leqslant B \in \mathcal{C}_X\} \leqslant \bigvee_{z \in X} A(z).$$

Furthermore, we have

$$Cl_Y((y_0)_L^{\rightarrow}(A))(y_0) \geqslant (y_0)_L^{\rightarrow}(A)(y_0) = \bigvee_{y_0(z)=y_0} A(z) = \bigvee_{z \in X} A(z) \geqslant Cl_X(A)(x).$$

This implies that $Cl_X(A)(x) \leqslant Cl_Y((y_0)_L^{\rightarrow}(A))(y_0)$ for each $A \in L^X$ and $x \in X$. Therefore, we have $D_{c_1}(y_0) = \top$.  $\square$

Next, we give another characterizations of degrees of $L$-continuty for mappings between $L$-cotopological spaces.

**Theorem 4.** *Let $(X, C_X)$ and $(Y, C_Y)$ be $L$-cotopological spaces. Then,*

$$D_{c_1}(f) = \bigwedge_{A \in L^X} \mathbb{S}\Big(Cl_X(A), f_L^{\leftarrow}(Cl_Y(f_L^{\rightarrow}(A)))\Big).$$

**Proof.** By the definition of $\mathbb{S}$, it follows that

$$\mathbb{S}\Big(f_L^{\rightarrow}(Cl_X(A)), Cl_Y(f_L^{\rightarrow}(A))\Big)$$

$$= \bigwedge_{y \in Y} \Big(f_L^{\rightarrow}(Cl_X(A))(y) \rightarrow Cl_Y(f_L^{\rightarrow}(A))(y)\Big)$$

$$= \bigwedge_{y \in Y} \Big(\bigvee_{f(x)=y} Cl_X(A)(x) \rightarrow Cl_Y(f_L^{\rightarrow}(A))(y)\Big)$$

$$= \bigwedge_{y \in Y} \bigwedge_{f(x)=y} \Big(Cl_X(A)(x) \rightarrow Cl_Y(f_L^{\rightarrow}(A))(f(x))\Big)$$

$$= \bigwedge_{y \in Y} \bigwedge_{f(x)=y} \Big(Cl_X(A)(x) \rightarrow f_L^{\leftarrow}(Cl_Y(f_L^{\rightarrow}(A)))(x)\Big)$$

$$= \bigwedge_{x \in X} \Big(Cl_X(A)(x) \rightarrow f_L^{\leftarrow}(Cl_Y(f_L^{\rightarrow}(A)))(x)\Big)$$

$$= \mathbb{S}\Big(Cl_X(A), f_L^{\leftarrow}(Cl_Y(f_L^{\rightarrow}(A)))\Big).$$

This means

$$D_{c_1}(f) = \bigwedge_{A \in L^X} \mathbb{S}\Big(f_L^{\rightarrow}(Cl_X(A)), Cl_Y(f_L^{\rightarrow}(A))\Big)$$

$$= \bigwedge_{A \in L^X} \mathbb{S}\Big(Cl_X(A), f_L^{\leftarrow}(Cl_Y(f_L^{\rightarrow}(A)))\Big),$$

as desired.  $\square$

**Theorem 5.** *Let $(X, C_X)$ and $(Y, C_Y)$ be $L$-cotopological spaces. Then,*

$$D_{c_1}(f) = \bigwedge_{B \in L^Y} \mathbb{S}\Big(Cl_X(f_L^{\leftarrow}(B)), f_L^{\leftarrow}(Cl_Y(B))\Big).$$

**Proof.**

$$\bigwedge_{B \in L^Y} \mathbb{S}\Big(Cl_X(f_L^{\leftarrow}(B)), f_L^{\leftarrow}(Cl_Y(B))\Big)$$

$$\leq \bigwedge_{A \in L^X} \mathbb{S}\Big(Cl_X(f_L^{\leftarrow}(f_L^{\rightarrow}(A))), f_L^{\leftarrow}(Cl_Y(f_L^{\rightarrow}(A)))\Big)$$

$$\leq \bigwedge_{A \in L^X} \mathbb{S}\Big(Cl_X(A), f_L^{\leftarrow}(Cl_Y(f_L^{\rightarrow}(A)))\Big)$$

$$= D_{c_1}(f).$$

By the definition of $\mathbb{S}$, it follows that

$$\mathbb{S}\Big(Cl_X(f_L^{\leftarrow}(B)), f_L^{\leftarrow}(Cl_Y(B))\Big)$$

$$= \bigwedge_{x \in X}\Big(Cl_X(f_L^{\leftarrow}(B))(x) \rightarrow f_L^{\leftarrow}(Cl_Y(B))(x)\Big)$$

$$= \bigwedge_{x \in X}\Big(Cl_X(f_L^{\leftarrow}(B))(x) \rightarrow (Cl_Y(B))(f(x))\Big)$$

$$\geq \bigwedge_{y \in Y}\Big(Cl_X(f_L^{\leftarrow}(B))(f^{-1}(y)) \rightarrow (Cl_Y(B))(y)\Big)$$

$$\geq \bigwedge_{y \in Y}\Big(\bigvee_{f(x)=y} Cl_X(f_L^{\leftarrow}(B))(x) \rightarrow (Cl_Y(B))(y)\Big)$$

$$= \bigwedge_{y \in Y}\Big(f_L^{\rightarrow}(Cl_X(f_L^{\leftarrow}(B)))(y) \rightarrow (Cl_Y(B))(y)\Big)$$

$$= D_{c_1}(f).$$

This implies

$$D_{c_1}(f) = \bigwedge_{B \in L^Y} \mathbb{S}\Big(Cl_X(f_L^{\leftarrow}(B)), f_L^{\leftarrow}(Cl_Y(B))\Big),$$

as desired. $\square$

In $L$-cotopological spaces, compositions of $L$-continuous mappings (resp. $L$-closed mappings) are still $L$-continuous mappings (resp. $L$-closed mappings). Now, let us give a degree representation of this result.

**Theorem 6.** *For $L$-cotopological spaces $(X, \mathcal{C}_X)$, $(Y, \mathcal{C}_Y)$ and $(Z, \mathcal{C}_Z)$, the following inequalities hold:*
(1) $D_{c_1}(f) \wedge D_{c_1}(g) \leq D_{c_1}(g \circ f)$.
(2) $D_b(f) \wedge D_b(g) \leq D_b(g \circ f)$.

**Proof.** (1) By Theorem 4, we have

$$
\begin{aligned}
& D_{c_1}(f) \wedge D_{c_1}(g) \\
={}& \bigwedge_{A \in L^X} \mathbb{S}\!\left(f_L^{\rightarrow}(Cl_X(A)), Cl_Y(f_L^{\rightarrow}(A))\right) \wedge \bigwedge_{B \in L^Y} \mathbb{S}\!\left(g_L^{\rightarrow}(Cl_Y(B)), Cl_Z(g_L^{\rightarrow}(B))\right) \\
\leqslant{}& \bigwedge_{A \in L^X} \mathbb{S}\!\left(f_L^{\rightarrow}(Cl_X(A)), Cl_Y(f_L^{\rightarrow}(A))\right) \wedge \bigwedge_{B \in L^Y} \mathbb{S}\!\left(Cl_Y(B), g_L^{\leftarrow}(Cl_Z(g_L^{\rightarrow}(B)))\right) \\
\leqslant{}& \bigwedge_{A \in L^X} \left(\mathbb{S}\!\left(f_L^{\rightarrow}(Cl_X(A)), Cl_Y(f_L^{\rightarrow}(A))\right) \wedge \mathbb{S}\!\left(Cl_Y(f_L^{\rightarrow}(A)), g_L^{\leftarrow}(Cl_Z(g_L^{\rightarrow}(f_L^{\rightarrow}(A))))\right)\right) \\
\leqslant{}& \bigwedge_{A \in L^X} \mathbb{S}\!\left(f_L^{\rightarrow}(Cl_X(A)), g_L^{\leftarrow}(Cl_Z((g \circ f)_L^{\rightarrow}(A)))\right) \\
={}& \bigwedge_{A \in L^X} \mathbb{S}\!\left((g \circ f)_L^{\rightarrow}(Cl_X(A)), Cl_Z((g \circ f)_L^{\rightarrow}(A))\right) \\
={}& D_{c_1}(g \circ f).
\end{aligned}
$$

(2) Adopting the proof of (1), it can be verified directly. $\square$

Next, we investigate the connections between degrees of $L$-continuity and that of $L$-closedness.

**Theorem 7.** *For $L$-cotopological spaces $(X, \mathcal{C}_X)$, $(Y, \mathcal{C}_Y)$ and $(Z, \mathcal{C}_Z)$, if $g$ is injective, then the following inequality holds:*
$$
D_b(g \circ f) \wedge D_{c_1}(g) \leqslant D_b(f).
$$

**Proof.** If $g$ is an injective mapping, then we have $g_L^{\leftarrow}(g_L^{\rightarrow}(B)) = B$ for all $B \in L^Y$. Then, it follows that

$$
\begin{aligned}
D_b(g \circ f) \wedge D_{c_1}(g) ={}& \bigwedge_{A \in L^X} \mathbb{S}\!\left(Cl_Z((g \circ f)_L^{\rightarrow}(A)), (g \circ f)_L^{\rightarrow}(Cl_X(A))\right) \\
& \wedge \bigwedge_{B \in L^Y} \mathbb{S}\!\left(g_L^{\rightarrow}(Cl_Y(B)), Cl_Z(g_L^{\rightarrow}(B))\right) \\
\leqslant{}& \bigwedge_{A \in L^X} \mathbb{S}\!\left(Cl_Z((g \circ f)_L^{\rightarrow}(A)), (g \circ f)_L^{\rightarrow}(Cl_X(A))\right) \\
& \wedge \bigwedge_{A \in L^X} \mathbb{S}\!\left(g_L^{\rightarrow}(Cl_Y(f_L^{\rightarrow}(A))), Cl_Z(g_L^{\rightarrow}(f_L^{\rightarrow}(A)))\right) \\
\leqslant{}& \bigwedge_{A \in L^X} \mathbb{S}\!\left(Cl_Z((g \circ f)_L^{\rightarrow}(A)), (g \circ f)_L^{\rightarrow}(Cl_X(A))\right) \\
& \wedge \bigwedge_{A \in L^X} \mathbb{S}\!\left(g_L^{\rightarrow}(Cl_Y(f_L^{\rightarrow}(A))), Cl_Z((g \circ f)_L^{\rightarrow}(A))\right) \\
\leqslant{}& \bigwedge_{A \in L^X} \mathbb{S}\!\left(g_L^{\rightarrow}(Cl_Y(f_L^{\rightarrow}(A))), (g \circ f)_L^{\rightarrow}(Cl_X(A))\right) \\
={}& \bigwedge_{A \in L^X} \mathbb{S}\!\left(Cl_Y(f_L^{\rightarrow}(A)), g_L^{\leftarrow}((g \circ f)_L^{\rightarrow}(Cl_X(A)))\right) \\
={}& \bigwedge_{A \in L^X} \mathbb{S}\!\left(Cl_Y(f_L^{\rightarrow}(A)), f_L^{\rightarrow}(Cl_X(A))\right) \\
={}& D_b(f).
\end{aligned}
$$

$\square$

**Theorem 8.** *For L-cotopological spaces* $(X, \mathcal{C}_X)$, $(Y, \mathcal{C}_Y)$ *and* $(Z, \mathcal{C}_Z)$, *if* $g$ *is surjective, then the following inequality holds:*

$$D_b(g \circ f) \wedge D_{c_1}(f) \leqslant D_b(g).$$

**Proof.** If $f$ is a surjective mapping, then $f_L^{\rightarrow}(f_L^{\leftarrow}(A)) = A$ for all $A \in L^Y$. Then, it follows that

$$\bigwedge_{B \in L^Y} \mathbb{S}\left( Cl_Z(g_L^{\rightarrow}(B)), g_L^{\rightarrow}(Cl_Y(B)) \right)$$

$$= \bigwedge_{B \in L^Y} \mathbb{S}\left( Cl_Z(g_L^{\rightarrow}(f_L^{\rightarrow}(f_L^{\leftarrow}(B)))), g_L^{\rightarrow}(Cl_Y(f_L^{\rightarrow}(f_L^{\leftarrow}(B)))) \right)$$

$$\geqslant \bigwedge_{A \in L^X} \mathbb{S}\left( Cl_Z((g \circ f)_L^{\rightarrow}(A)), g_L^{\rightarrow}(Cl_Y(f_L^{\rightarrow}(A))) \right)$$

$$\geqslant \bigwedge_{B \in L^Y} \mathbb{S}\left( Cl_Z(g_L^{\rightarrow}(B)), g_L^{\rightarrow}(Cl_Y(B)) \right).$$

This shows

$$\bigwedge_{B \in L^Y} \mathbb{S}\left( Cl_Z(g_L^{\rightarrow}(B)), g_L^{\rightarrow}(Cl_Y(B)) \right) = \bigwedge_{A \in L^X} \mathbb{S}\left( Cl_Z((g \circ f)_L^{\rightarrow}(A)), g_L^{\rightarrow}(Cl_Y(f_L^{\rightarrow}(A))) \right).$$

Then, we have

$$D_b(g \circ f) \wedge D_{c_1}(f)$$

$$= \bigwedge_{A \in L^X} \mathbb{S}\left( Cl_Z((g \circ f)_L^{\rightarrow}(A)), (g \circ f)_L^{\rightarrow}(Cl_X(A)) \right)$$

$$\wedge \bigwedge_{A \in L^X} \mathbb{S}\left( f_L^{\rightarrow}(Cl_X(A)), Cl_Y(f_L^{\rightarrow}(A)) \right)$$

$$\leqslant \bigwedge_{A \in L^X} \mathbb{S}\left( Cl_Z((g \circ f)_L^{\rightarrow}(A)), (g \circ f)_L^{\rightarrow}(Cl_X(A)) \right)$$

$$\wedge \bigwedge_{A \in L^X} \mathbb{S}\left( (g \circ f)_L^{\rightarrow}(Cl_X(A)), g_L^{\rightarrow}(Cl_Y(f_L^{\rightarrow}(A))) \right)$$

$$\leqslant \bigwedge_{A \in L^X} \mathbb{S}\left( Cl_Z((g \circ f)_L^{\rightarrow}(A)), g_L^{\rightarrow}(Cl_Y(f_L^{\rightarrow}(A))) \right)$$

$$= \bigwedge_{B \in L^Y} \mathbb{S}\left( Cl_Z(g_L^{\rightarrow}(B)), g_L^{\rightarrow}(Cl_Y(B)) \right)$$

$$= D_b(g).$$

$\square$

## 4. Degrees of *L*-Continuity and *L*-Openness for Mappings between *L*-Topological Spaces

In this section, we mainly define degrees of *L*-continuity and *L*-openness to equip each mapping between *L*-topological spaces with some degree to be an *L*-continuous mapping and an *L*-open mapping, respectively. Then, we will study their connections in a degree sense.

**Definition 9.** *Let* $(X, \mathcal{T}_X)$ *and* $(X, \mathcal{T}_Y)$ *be L-topological spaces.*

*(1) $D_{c_2}(f)$ defined by*

$$D_{c_2}(f) = \bigwedge_{B \in L^Y} \mathbb{S}\left(f_L^{\leftarrow}(\mathcal{N}_Y(B)), \mathcal{N}_X(f_L^{\leftarrow}(B))\right)$$

*is called the degree of L-continuity for $f$.*

*(2) $D_k(f)$ defined by*

$$D_k(f) = \bigwedge_{A \in L^X} \mathbb{S}\left(f_L^{\rightarrow}(\mathcal{N}_X(A)), \mathcal{N}_Y(f_L^{\rightarrow}(A))\right)$$

*is called the degree of L-openness for $f$.*

**Remark 2.** *(1) If $D_{c_2}(f) = \top$, then $f_L^{\leftarrow}(\mathcal{N}_X(B)) \leqslant \mathcal{N}_Y(f_L^{\leftarrow}(B))$ for all $B \in L^Y$, which is exactly the definition of L-continuous mappings between L-interior spaces.*
*(2) If $D_k(f) = \top$, then $f_L^{\rightarrow}(\mathcal{N}_X(A)) \leqslant \mathcal{N}_Y(f_L^{\rightarrow}(A))$ for all $A \in L^X$. This is exactly the equivalent form of L-open mappings between L-topological spaces by means of the corresponding L-interior operators.*

**Theorem 9.** *Let $(X, \mathcal{T}_X)$ and $(X, \mathcal{T}_Y)$ be L-topological spaces. Then,*

$$D_{c_2}(f) = \bigwedge_{x \in X} \bigwedge_{C \in L^Y} \left(\mathcal{U}_Y^{f(x)}(C) \to f(\mathcal{U}_X^x)(C)\right).$$

**Proof.**

$$\bigwedge_{x \in X} \bigwedge_{C \in L^Y} \left(\mathcal{U}_Y^{f(x)}(C) \to f(\mathcal{U}_X^x)(C)\right)$$

$$= \bigwedge_{x \in X} \bigwedge_{C \in L^Y} \left(\mathcal{U}_Y^{f(x)}(C) \to (\mathcal{U}_X^x)(f_L^{\leftarrow}(C))\right)$$

$$= \bigwedge_{x \in X} \bigwedge_{C \in L^Y} \left(\mathcal{N}_Y(C)(f(x)) \to \mathcal{N}_X(f_L^{\leftarrow}(C))(x)\right)$$

$$= \bigwedge_{x \in X} \bigwedge_{C \in L^Y} \left(f_L^{\leftarrow}(\mathcal{N}_Y(C))(x) \to \mathcal{N}_X(f_L^{\leftarrow}(C))(x)\right)$$

$$= \bigwedge_{C \in L^Y} \bigwedge_{x \in X} \left(f_L^{\leftarrow}(\mathcal{N}_Y(C))(x) \to \mathcal{N}_X(f_L^{\leftarrow}(C))(x)\right)$$

$$= \bigwedge_{C \in L^Y} \mathbb{S}\left(f_L^{\leftarrow}(\mathcal{N}_Y(C)), \mathcal{N}_X(f_L^{\leftarrow}(C))\right)$$

$$= D_{c_2}(f).$$

$\square$

**Theorem 10.** *Let $(X, \mathcal{T}_X)$ and $(X, \mathcal{T}_Y)$ be L-topological spaces. Then,*

$$D_{c_2}(f) = \bigwedge_{F \in \mathcal{F}_L^S(X)} \mathbb{S}\left(\mathrm{Lim}_X F, f_L^{\leftarrow}(\mathrm{Lim}_Y f(F))\right).$$

**Proof.** On one hand,

$$\bigwedge_{F\in\mathfrak{F}_L^S(X)} \mathbb{S}\Big(\mathrm{Lim}_X F, f_L^{\leftarrow}(\mathrm{Lim}_Y f(F))\Big)$$

$$= \bigwedge_{F\in\mathfrak{F}_L^S(X)}\bigwedge_{x\in X}\Big(\mathrm{Lim}_X F(x)\to f_L^{\leftarrow}(\mathrm{Lim}_Y f(F))(x)\Big)$$

$$= \bigwedge_{F\in\mathfrak{F}_L^S(X)}\bigwedge_{x\in X}\Big(\mathrm{Lim}_X F(x)\to \mathrm{Lim}_Y f(F)(f(x))\Big)$$

$$= \bigwedge_{F\in\mathfrak{F}_L^S(X)}\bigwedge_{x\in X}\Big(\bigwedge_{A\in L^X}(\mathcal{U}_X^x(A)\to F(A))\to \bigwedge_{B\in L^Y}(\mathcal{U}_Y^{f(x)}(B)\to f(F)(B))\Big)$$

$$\geqslant \bigwedge_{F\in\mathfrak{F}_L^S(X)}\bigwedge_{x\in X}\Big(\bigwedge_{B\in L^Y}(\mathcal{U}_X^x(f_L^{\leftarrow}(B))\to F(f_L^{\leftarrow}(B)))\to \bigwedge_{B\in L^Y}(\mathcal{U}_Y^{f(x)}(B)\to f(F)(B))\Big)$$

$$= \bigwedge_{F\in\mathfrak{F}_L^S(X)}\bigwedge_{x\in X}\Big(\bigwedge_{B\in L^Y}(\mathcal{U}_X^x(f_L^{\leftarrow}(B))\to f(F)(B))\to \bigwedge_{B\in L^Y}(\mathcal{U}_Y^{f(x)}(B)\to f(F)(B))\Big)$$

$$\geqslant \bigwedge_{x\in X}\bigwedge_{B\in L^Y}\Big(\mathcal{U}_Y^{f(x)}(B)\to \mathcal{U}_X^x(f_L^{\leftarrow}(B))\Big)$$

$$= \bigwedge_{x\in X}\bigwedge_{B\in L^Y}\Big(\mathcal{U}_Y^{f(x)}(B)\to f(\mathcal{U}_X^x)(B)\Big)$$

$$= D_{c_2}(f).$$

On the other hand,

$$\bigwedge_{F\in\mathfrak{F}_L^S(X)} \mathbb{S}\Big(\mathrm{Lim}_X F, f_L^{\leftarrow}(\mathrm{Lim}_Y f(F))\Big)$$

$$= \bigwedge_{F\in\mathfrak{F}_L^S(X)}\bigwedge_{x\in X}\Big(\mathrm{Lim}_X F(x)\to f_L^{\leftarrow}(\mathrm{Lim}_Y f(F))(x)\Big)$$

$$= \bigwedge_{F\in\mathfrak{F}_L^S(X)}\bigwedge_{x\in X}\Big(\mathrm{Lim}_X F(x)\to \mathrm{Lim}_Y f(F)(f(x))\Big)$$

$$= \bigwedge_{F\in\mathfrak{F}_L^S(X)}\bigwedge_{x\in X}\Big(\bigwedge_{A\in L^X}(\mathcal{U}_X^x(A)\to F(A))\to \bigwedge_{B\in L^Y}(\mathcal{U}_Y^{f(x)}(B)\to f(F)(B))\Big)$$

$$\leqslant \bigwedge_{x\in X}\Big(\bigwedge_{A\in L^X}(\mathcal{U}_X^x(A)\to \mathcal{U}_X^x(A))\to \bigwedge_{B\in L^Y}(\mathcal{U}_Y^{f(x)}(B)\to f(\mathcal{U}_X^x)(B))\Big)$$

$$= \bigwedge_{x\in X}\bigwedge_{B\in L^Y}\Big(\mathcal{U}_Y^{f(x)}(B)\to f(\mathcal{U}_X^x)(B)\Big)$$

$$= D_{c_2}(f).$$

Therefore, $D_{c_2}(f) = \bigwedge_{x\in X}\bigwedge_{B\in L^Y}\Big(\mathcal{U}_Y^{f(x)}(B)\to f(\mathcal{U}_X^x)(B)\Big)$.  $\square$

In $L$-topological spaces, compositions of $L$-continuous mappings (resp. $L$-open mappings) are still $L$-continuous mappings (resp. $L$-open mappings). Now, let us give a degree representation of this result.

**Theorem 11.** *For $L$-topological spaces $(X,\mathcal{T}_X),(Y,\mathcal{T}_Y)$ and $(Z,\mathcal{T}_Z)$, the following inequalities hold:*
(1) $D_{c_2}(f)\wedge D_{c_2}(g)\leqslant D_{c_2}(g\circ f).$

(2) $D_k(f) \wedge D_k(g) \leqslant D_k(g \circ f)$.

**Proof.** (1) By Definition 9, we have

$$
\begin{aligned}
& D_{c_2}(f) \wedge D_{c_2}(g) \\
={} & \bigwedge_{A \in L^Y} \mathbb{S}\!\left(f_L^{\leftarrow}(\mathcal{N}_Y(A)), \mathcal{N}_X(f_L^{\leftarrow}(A))\right) \\
& \wedge \bigwedge_{B \in L^Z} \mathbb{S}\!\left(g_L^{\leftarrow}(\mathcal{N}_Z(B)), \mathcal{N}_Y(g_L^{\leftarrow}(B))\right) \\
\leqslant{} & \bigwedge_{C \in L^Z} \mathbb{S}\!\left(f_L^{\leftarrow}(\mathcal{N}_X(g_L^{\leftarrow}(C))), \mathcal{N}_Y(f_L^{\leftarrow}(g_L^{\leftarrow}(C)))\right) \\
& \wedge \bigwedge_{B \in L^Z} \mathbb{S}\!\left(g_L^{\leftarrow}(\mathcal{N}_Z(B)), \mathcal{N}_Y(g_L^{\leftarrow}(B))\right) \\
\leqslant{} & \bigwedge_{C \in L^Z} \mathbb{S}\!\left(f_L^{\leftarrow}(\mathcal{N}_X(g_L^{\leftarrow}(C))), \mathcal{N}_Y(f_L^{\leftarrow}(g_L^{\leftarrow}(C)))\right) \\
& \wedge \bigwedge_{B \in L^Z} \mathbb{S}\!\left(f_L^{\leftarrow}(g_L^{\leftarrow}(\mathcal{N}_Z(B))), f_L^{\leftarrow}(\mathcal{N}_Y(g_L^{\leftarrow}(B)))\right) \\
\leqslant{} & \bigwedge_{B \in L^Z} \mathbb{S}\!\left(f_L^{\leftarrow}(\mathcal{N}_X(g_L^{\leftarrow}(B))), \mathcal{N}_Y(f_L^{\leftarrow}(g_L^{\leftarrow}(B)))\right) \\
& \wedge \mathbb{S}\!\left(f_L^{\leftarrow}(g_L^{\leftarrow}(\mathcal{N}_Z(B))), f_L^{\leftarrow}(\mathcal{N}_Y(g_L^{\leftarrow}(B)))\right) \\
\leqslant{} & \bigwedge_{B \in L^Z} \mathbb{S}\!\left(f_L^{\leftarrow}(g_L^{\leftarrow}(\mathcal{N}_Z(B))), \mathcal{N}_Y(f_L^{\leftarrow}(g_L^{\leftarrow}(B)))\right) \\
={} & \bigwedge_{B \in L^Z} \mathbb{S}\!\left((g \circ f)_L^{\leftarrow}(\mathcal{N}_Z(B)), \mathcal{N}_X((g \circ f)_L^{\leftarrow}(B))\right) \\
={} & D_{c_2}(g \circ f).
\end{aligned}
$$

(2) Adopting the proof of (1), it can be verified directly. $\square$

Next, we investigate the connections between degrees of $L$-continuity and that of $L$-openness.

**Theorem 12.** *For $L$-topological spaces $(X, \mathcal{T}_X), (Y, \mathcal{T}_Y)$ and $(Z, \mathcal{T}_Z)$, if $f$ is surjective, then the following inequality holds:*
$$
D_k(g \circ f) \wedge D_{c_2}(f) \leqslant D_k(g).
$$

**Proof.** Since $f$ is surjective, we have $f_L^{\rightarrow}(f_L^{\leftarrow}(C)) = C$ for all $C \in L^Y$. Then, it follows that

$$
\begin{aligned}
& \bigwedge_{B \in L^Y} \mathbb{S}\!\left(g_L^{\rightarrow}(\mathcal{N}_Y(B)), \mathcal{N}_Z(g_L^{\rightarrow}(B))\right) \\
={} & \bigwedge_{B \in L^Y} \mathbb{S}\!\left(g_L^{\rightarrow}(\mathcal{N}_Y(f_L^{\rightarrow}(f_L^{\leftarrow}(B)))), \mathcal{N}_Z(g_L^{\rightarrow}(f_L^{\rightarrow}(f_L^{\leftarrow}(B))))\right) \\
\geqslant{} & \bigwedge_{A \in L^X} \mathbb{S}\!\left(g_L^{\rightarrow}(\mathcal{N}_Y(f_L^{\rightarrow}(A))), \mathcal{N}_Z((g \circ f)_L^{\rightarrow}(A))\right) \\
\geqslant{} & \bigwedge_{B \in L^Y} \mathbb{S}\!\left(g_L^{\rightarrow}(\mathcal{N}_Y(B)), \mathcal{N}_Z(g_L^{\rightarrow}(B))\right).
\end{aligned}
$$

This shows

$$\bigwedge_{B\in L^Y} \mathbb{S}\Big(g_L^{\rightarrow}(\mathcal{N}_Y(B)),\mathcal{N}_Z(g_L^{\rightarrow}(B))\Big) = \bigwedge_{A\in L^X} \mathbb{S}\Big(g_L^{\rightarrow}(\mathcal{N}_Y(f_L^{\rightarrow}(A))),\mathcal{N}_Z((g\circ f)_L^{\rightarrow}(A))\Big).$$

Then, we have

$$
\begin{aligned}
& D_k(g\circ f)\wedge D_{c_2}(f) \\
={} & \bigwedge_{A\in L^X} \mathbb{S}\Big((g\circ f)_L^{\rightarrow}(\mathcal{N}_X(A)),\mathcal{N}_Z((g\circ f)_L^{\rightarrow}(A))\Big) \\
& \wedge \bigwedge_{B\in L^Y} \mathbb{S}\Big(f_L^{\leftarrow}(\mathcal{N}_Y(B)),\mathcal{N}_X(f_L^{\leftarrow}(B))\Big) \\
\leqslant{} & \bigwedge_{A\in L^X} \mathbb{S}\Big((g\circ f)_L^{\rightarrow}(\mathcal{N}_X(A)),\mathcal{N}_Z((g\circ f)_L^{\rightarrow}(A))\Big) \\
& \wedge \bigwedge_{A\in L^X} \mathbb{S}\Big(g_L^{\rightarrow}(\mathcal{N}_Y(f_L^{\rightarrow}(A))),(g\circ f)_L^{\rightarrow}(\mathcal{N}_X(A))\Big) \\
\leqslant{} & \bigwedge_{A\in L^X} \mathbb{S}\Big(g_L^{\rightarrow}(\mathcal{N}_Y(f_L^{\rightarrow}(A))),\mathcal{N}_Z((g\circ f)_L^{\rightarrow}(A))\Big) \\
={} & \bigwedge_{B\in L^Y} \mathbb{S}\Big(\mathcal{N}_Z(g_L^{\rightarrow}(\mathcal{N}_Y(B))),g_L^{\rightarrow}(B))\Big) \\
={} & D_k(g).
\end{aligned}
$$

$\square$

**Theorem 13.** *For L-topological spaces* $(X,\mathcal{T}_X),(Y,\mathcal{T}_Y)$ *and* $(Z,\mathcal{T}_Z)$*, if f is injective, then the following inequality holds:*

$$D_k(g\circ f)\wedge D_{c_2}(g) \leqslant D_k(f).$$

**Proof.** Since $g$ is injective, we have $g_L^{\leftarrow}(g_L^{\rightarrow}(C)) = C$ for all $C\in L^Y$. Then, it follows that

$$
\begin{aligned}
& D_k(g\circ f)\wedge D_{c_2}(g) \\
={} & \bigwedge_{B\in L^X} \mathbb{S}\Big((g\circ f)_L^{\rightarrow}(\mathcal{N}_X(B)),\mathcal{N}_Z((g\circ f)_L^{\rightarrow}(B))\Big) \\
& \wedge \bigwedge_{C\in L^Z} \mathbb{S}\Big(g_L^{\leftarrow}(\mathcal{N}_Z(C)),\mathcal{N}_Y(g_L^{\leftarrow}(C))\Big) \\
\leqslant{} & \bigwedge_{B\in L^X} \mathbb{S}\Big(g_L^{\rightarrow}((g\circ f)_L^{\rightarrow}(\mathcal{N}_X(B))),g_L^{\rightarrow}(\mathcal{N}_Z((g\circ f)_L^{\rightarrow}(B)))\Big) \\
& \wedge \bigwedge_{B\in L^X} \mathbb{S}\Big(g_L^{\leftarrow}(\mathcal{N}_Z((g\circ f)_L^{\rightarrow}(B))),\mathcal{N}_Y(g_L^{\leftarrow}((g\circ f)_L^{\rightarrow}(B)))\Big) \\
={} & \bigwedge_{B\in L^X} \mathbb{S}\Big(f_L^{\rightarrow}(\mathcal{N}_X(B)),\mathcal{N}_Y(f_L^{\rightarrow}(B)))\Big) \\
={} & D_k(f).
\end{aligned}
$$

$\square$

**Theorem 14.** *Suppose that L is a frame with an order-reversing involution* $'$*, where* $a' = a \to \bot$ *for* $a\in L$*. For an L-topological space* $(X,\mathcal{T})$*,* $\mathcal{C} = \{B\,|\,B'\in\mathcal{T}\}$ *is an L-cotopology and* $\mathrm{Cl}(B) = (\mathcal{N}(B'))'$*.*

**Proof.** It is easy to verify that $\mathcal{C}$ is an $L$-cotopology and $Cl(A) = (\mathcal{N}(A'))'$. $\square$

**Theorem 15.** *Suppose that $L$ is a frame with an order-reversing involution $'$, where $a' = a \to \perp$ for $a \in L$. For $L$-topological spaces $(X, \mathcal{T}_X)$ and $(Y, \mathcal{T}_Y)$, $D_{c_1}(f) = D_{c_2}(f)$.*

**Proof.** By Theorems 5 and 14,

$$
\begin{aligned}
D_{c_1}(f) &= \bigwedge_{B \in L^Y} \mathbb{S}\left(Cl_X(f_L^\leftarrow(B)), f_L^\leftarrow(Cl_Y(B))\right) \\
&= \bigwedge_{B \in L^Y} \mathbb{S}\left((\mathcal{N}_X(f_L^\leftarrow(B')))', f_L^\leftarrow((\mathcal{N}_Y(B'))')\right) \\
&= \bigwedge_{B \in L^Y} \mathbb{S}\left(f_L^\leftarrow(\mathcal{N}_Y(B')), \mathcal{N}_X(f_L^\leftarrow(B'))\right) \\
&= \bigwedge_{B \in L^Y} \mathbb{S}\left(f_L^\leftarrow(\mathcal{N}_Y(B)), \mathcal{N}_X(f_L^\leftarrow(B))\right) \\
&= D_{c_2}(f).
\end{aligned}
$$

$\square$

## 5. Conclusions

In this paper, we equip each mapping between $L$-cotopological spaces with some degree to be an $L$-continuous mapping and an $L$-closed mapping, and equip each mapping between $L$-topological spaces with some degree to be an $L$-continuous mapping and an $L$-open mapping. From this aspect, we could consider the degrees of $L$-continuity, $L$-closedness and $L$-openness for a mapping even if the mapping is not a continuous mapping, a closed mapping or an open mapping. By means of these definitions, we proved that the degrees of $L$-continuity, $L$-closedness and $L$-openness for mappings naturally suggest lattice-valued logical extensions of properties related to continuous mappings, closed mappings and open mappings in classical topological spaces to fuzzy topological spaces. Moreover, if $L$ is a frame with an order-reversing involution $'$, where $b' = b \to \perp$ for $b \in L$, then degrees of $L$-continuity for mappings between $L$-cotopological spaces and degrees of $L$-continuity for mappings between $L$-topological spaces are equivalent.

As future research, we will consider the following two problems:

(1) By means of the degree method, we can also define the degrees of some topological properties. For example, we can use the convergence degree of a fuzzy ultrafilter to define the compactness degree of an $L$-topological space.

(2) Based on the degrees of $L$-continuity, $L$-openness and $L$-closedness for mappings, we can further define the degrees of $L$-homeomorphism in a degree. We only need to equip a bijective mapping with the degrees of $L$-continuity and $L$-openness.

**Author Contributions:** Z.X. contributed the central idea. Z.X. and Q.L. writed this manuscript and revised it.

**Funding:** This work is supported by the National Natural Science Foundation of China (11871097,11971448), the Project (2017M622563) funded by China Postdoctoral Science Foundation and the Project (KYTZ201631, CRF201611, 2017Z056) Supported by the Scientific Research Foundation of CUIT.

**Acknowledgments:** The authors thank the reviewers and the editor for their valuable comments and suggestions.

**Conflicts of Interest:** The authors declare no conflict of interest.

## References


1. Chang, C.L. Fuzzy topological spaces. *J. Math. Anal. Appl.* **1968**, *24*, 182–190. [CrossRef]
2. Fang, J.M. Stratified L-ordered quasiuniform limit spaces. *Fuzzy Sets Syst.* **2013**, *227*, 51–73. [CrossRef]
3. Jäger, G. A category of *L*-fuzzy convergence spaces. *Quaest. Math.* **2001**, *24*, 501–517. [CrossRef]
4. Jin, Q.; Li, L.Q.; Lv, Y.R.; Zhao, F.; Zou, J. Connectedness for lattice-valued subsets in lattice-valued convergence spaces. *Quaest. Math.* **2019**, *42*, 135–150. [CrossRef]
5. Kubiak, T. On Fuzzy Topologies. Ph.D.Thesis, Adam Mickiewicz, Poznan, Poland, 1985.
6. Lowen, E.; Lowen, R.; Wuyts, P. The categorical topological approach to fuzzy topology and fuzzy convergence. *Fuzzy Sets Syst.* **1991**, *40*, 347–373. [CrossRef]
7. Li, L.Q.; Jin, Q.; Hu, K. Lattice-valued convergence associated with CNS spaces. *Fuzzy Sets Syst.* **2019**, *370*, 91–98. [CrossRef]
8. Li, L.Q. p-Topologicalness—A Relative Topologicalness in $\mathsf{T}$-Convergence Spaces. *Mathematics* **2019**, *7*, 228. [CrossRef]
9. Pang, B. Convergence structures in *M*-fuzzifying convex spaces. *Quaest. Math.* **2019**. [CrossRef]
10. Pang, B.; Shi, F.G. Strong inclusion orders between *L*-subsets and its applications in *L*-convex spaces. *Quaest. Math.* **2018**, *41*, 1021–1043. [CrossRef]
11. Pang, B.; Shi, F.G. Fuzzy counterparts of hull operators and interval operators in the framework of *L*-convex spaces. *Fuzzy Sets Syst.* **2019**, *369*, 20–39. [CrossRef]
12. Pang, B.; Xiu, Z.Y. An axiomatic approach to bases and subbases in *L*-convex spaces and their applications. *Fuzzy Sets Syst.* **2019**, *369*, 40–56. [CrossRef]
13. Shi, F.G.; Xiu, Z.Y. $(L, M)$-fuzzy convex structures. *J. Nonlinear Sci. Appl.* **2017**, *10*, 3655–3669. [CrossRef]
14. Sostak, A.P. On a fuzzy topological structure. *Rend. Circ. Mat. Palermo (Suppl. Ser. II)* **1985**, *11*, 89–103.
15. Wang, B.; Li, Q.; Xiu, Z.Y. A categorical approach to abstract convex spaces and interval spaces. *Open Math.* **2019**, *17*, 374–384. [CrossRef]
16. Wang, L.; Pang, B. Coreflectivities of $(L, M)$-fuzzy convex structures and $(L, M)$-fuzzy cotopologies in $(L, M)$-fuzzy closure systems. *J. Intell. Fuzzy Syst.* **2019**. [CrossRef]
17. Xiu, Z.Y.; Pang, B. *M*-fuzzifying cotopological spaces and *M*-fuzzifying convex spaces as *M*-fuzzifying closure spaces. *J. Intell. Fuzzy Syst.* **2017**, *33*, 613–620. [CrossRef]
18. Zhang, D.X. An enriched category to many valued topology. *Fuzzy Sets Syst.* **2007**, *158*, 349–466. [CrossRef]
19. Yue, Y.L.; Fang, J.M. On separation axioms in *I*-fuzzy topological spaces. *Fuzzy Sets Syst.* **2006**, *157*, 780–793. [CrossRef]
20. Shi, F.G. $(L, M)$-fuzzy metric spaces. *Indian J. Math.* **2010**, *52*, 1–20.
21. Shi, F.G. Regularity and normality of $(L, M)$-fuzzy topological spaces. *Fuzzy Sets Syst.* **2011**, *182*, 37–52. [CrossRef]
22. Li, H.Y.; Shi, F.G. Degrees of fuzzy compactness in *L*-fuzzy topological spaces. *Fuzzy Sets Syst.* **2010**, *161*, 988–1001. [CrossRef]
23. Pang, B.; Shi, F.G. Degrees of compactness of $(L, M)$-fuzzy convergence spaces and its applications. *Fuzzy Sets Syst.* **2014**, *251*, 1–22. [CrossRef]
24. Pang, B. Degrees of separation properties in stratified *L*-generalized convergence spaces using residual implication. *Filomat* **2017**, *31*, 6293–6305. [CrossRef]
25. Wang, L.; Wu, X.Y.; Xiu, Z.Y. A degree approach to relationship among fuzzy convex structures, fuzzy closure systems and fuzzy Alexandrov topologies. *Open Math.* **2019**, *17*, 913–928. [CrossRef]
26. Xiu, Z.Y.; Li, Q.G. Relations among $(L, M)$-fuzzy convex structures, $(L, M)$-fuzzy closure systems and $(L, M)$-fuzzy Alexandrov topologies in a degree sense. *J. Intell. Fuzzy Syst.* **2019**, *36*, 385–396. [CrossRef]
27. Xiu, Z.Y.; Pang, B. A degree approach to special mappings between M-fuzzifying convex spaces. *J. Intell. Fuzzy Syst.* **2018**, *35*, 705–716. [CrossRef]
28. Pang, B. Degrees of continuous mappings, open mappings, and closed mappings in *L*-fuzzifying topological spaces. *J. Intell. Fuzzy Syst.* **2014**, *27*, 805–816.

29. Liang, C.Y.; Shi, F.G.; Degree of continuity for mappings of $(L, M)$-fuzzy topological spaces. *J. Intell. Fuzzy Syst.* **2014**, *27*, 2665–2677.

30. Li, Q.H.; Huang, H.L.; Xiu, Z.-Y. Degrees of special mappings in the theory of $L$-convex spaces. *J. Intell. Fuzzy Syst.* **2019**, *37*, 2265–2274. [CrossRef]

31. Bělohlávek, R. *Fuzzy Relational Systems, Foundations and Principles*; Kluwer Academic/Plenum Publishers: New York, NY, USA; Boston, MA, USA; Dordrecht, The Netherlands; London, UK; Moscow, Russia, 2002.

32. Höhle, U.; Šostak, A.P. Axiomatic foudations of fixed-basis fuzzy topology. In *Mathematics of Fuzzy Sets: Logic, Topology, and Measure Theory, Handbook Series*; Höhle, U., Rodabaugh, S.E., Eds.; Kluwer Academic Publishers: Boston, MA, USA; Dordrecht, The Netherlands; London, UK, 1999; Volume 3, pp. 123–173.

33. Liu, Y.M.; Luo, M.K. *Fuzzy Topology*; World Scientific Publication: Singapore, 1998.

 *mathematics*

Article

# Quantitative Analysis of Key Performance Indicators of Green Supply Chain in FMCG Industries Using Non-Linear Fuzzy Method

**Hamed Nozari [1], Esmaeil Najafi [2],*, Mohammad Fallah [1] and Farhad Hosseinzadeh Lotfi [3]**

[1]  Department of Industrial Engineering of Central Tehran Branch, Islamic Azad University, Tehran 1469669191, Iran; ham.nozari.eng@iauctb.ac.ir (H.N.); Mohammad.fallah43@yahoo.com (M.F.)

[2]  Department of Industrial Engineering of Science and Research Branch, Islamic Azad University, Tehran 1477893855, Iran

[3]  Department of Mathematics of Science and Research Branch, Islamic Azad University, Tehran 1477893855, Iran; hosseinzadeh_lotf@yahoo.com

*  Correspondence: e.najafi@srbiau.ac.ir

Received: 20 September 2019; Accepted: 24 October 2019; Published: 27 October 2019

**Abstract:** Nowadays, along with increasing companies' activities, one of the main environmental protective tools is green supply chain management (GSCM). Since fast-moving consumer goods (FMCG) companies are manufacturing materials that usually require special warehousing as well as different distribution systems, and since companies of food products tend to fall into this area, the safety of their manufactured materials is a vital global challenge. For this reason, organizations in addition to governments have realized the importance of the green supply chain in these industries. Therefore, the present study examines the key performance indicators (KPIs) of the green supply chain in the FMCG industry. There are several performance indicators for the green supply chain. In this study, the KPIs were extracted based on the literatures as well as the opinions of experts through which key indicators in FMCG industries were identified. Using the fuzzy decision -making trial and evaluation laboratory (DEMATEL) method, the relationships and interactions of these key indices were determined. Moreover, a fuzzy nonlinear mathematical modeling was used to investigate the significance of these indicators. It is revealed that the organizational environmental management factor has the highest priority.

**Keywords:** performance management; green supply chain; decision making; nonlinear mathematical modeling; FMCG industries; fuzzy DEMATEL

---

## 1. Introduction

Recently, due to growing consumers' demands for green products without harming the environment, the concept of green supply chain has emerged. The green supply chain not only includes environmental consciousness in business, it also guarantees the sustainable development of industries. To this end, various measures have been taken by governments and organizations to address this issue [1]. These include enforcing green laws and using eco-friendly raw materials, reducing the use of fossil and petroleum resources, recycling papers, reusing waste, etc. in organizations. The successful adoption as well as accomplishment of green ideas in supply chain business activities is such that its prospects are not easily visible. Therefore, the activities and products of the green supply chain are at different risks [2]. For this reason, accurate management and understanding of the green supply chain is of paramount importance and it will be necessary to focus on developing useful responsibilities and strategies that are effective in successfully implementing the green supply chain. Organizations help to control green supply chain risks by reducing waste, evaluating suppliers based on environmental performance, increasing production of environmentally friendly products, and other performance parameters related to green activities [3]. Optimizing the amount of energy used in the product life cycle has also become a very important issue for organizations [4]. Because of the importance of the subject as well as raising the general awareness of technologies applied in production and distribution of products in addition to the importance of goods produced in the FMCG, it seems a fundamental need to use green methods in the supply chain. It is widely recognized that the FMCG represents a key industry to provide an enormous volume of human daily demanding in which all individuals not considering the age, gender and boundaries are involved. The FMCG industry mainly consists of processed foods, prepared meals, beverages, fresh or frozen and dry foods, medicines, cosmetic products, and other non-durable products. Due to the size of this industry, its greener supply chain can lead to sustainable development and meeting the environmental standards. For this reason, green supply chain managers at leading companies in the FMCG industry strive to utilize green logistics and improve their environmental performance throughout the supply chain as a strategic weapon to gain competitive advantage [5]. Therefore, understanding the types and evaluation of KPI to improve the performance of the green supply chain can be extremely important research on this area. However, always exploring green solutions in different industries individually can be an effective way to make the industry more successful. It may be noted that there is little research that focus on KPIs for the implementation of green supply chain and no work has been done in foods and pharmaceutical parts of FMCG industries as the main sectors of these industries. Moreover, the concept of KPIs in green supply chain of the FMCG industries in Iran has not been academically investigated yet and as one the most important supply chains, it should be considered to be a major concern.

In this regard, the current study has attempted to evaluate key indicators of performance for green supply chain in FMCG industries (here are exclusively food and pharmaceutical industries). To this end, first, literature on the subject is briefly reviewed and then the key indicators affecting the green supply chain in these companies is extracted based on the literature as well as utilizing the views of experts. Afterwards, using the fuzzy decision method, the internal effects of these indicators are investigated and finally, these indicators are prioritized by using nonlinear fuzzy modeling, as a fuzzy analytic hierarchy process (AHP)-based method that was easy to use, despite the complexity and nonlinearity of the constraints. The rest of the paper is organized as follows. Section 2 presents a review of the literature in terms of FMCG industries and green supply chains. Section 3 presents the green supply chain KPIs. Section 4 illustrates the MCDM and non-linear fuzzy mathematical modeling to analyze the KPI. The obtained findings will be discussed in Section 5, and lastly, the conclusions are presented in Section 6.

## 2. Literature Review

In order to assess KPIs of green supply chain systems in foods and pharmaceutical industries of FMCG sectors, a brief review regarding FMCG industries as well as green supply chain have been presented.

### 2.1. FMCG Industries

FMCG refer to those products that are sold rapidly at a rather low cost case of low margin–high volume [6,7]. The FMCG industry trades in commodities that are classified as essential products [8]. Typically, FMCG consist of non-durable products that are not limited to general groceries, beverages and beauty, skin care, and cosmetics. However, we focus on the FMCG supply chain in foods and pharmaceutical components as an application example.

The FMCG industry comprises a big part of budget of consumers in all countries [9]. The level of involvement can further be considered as how serious a consumer is in purchasing a product and how much information they require during their decision-making process. Therefore, a functioning retail sector is crucial for daily provision of essential products at high quality and low cost [10]. Thus, since many people are frequently buying these products, the cost that households pay for

FMCG product categories is very significant, in light of the fact that products in the FMCG industry are short-lived and highly manufactured in terms of volume and variety. Therefore, there is an increasing pressure on retailers to simultaneously control cost and improve customer service [11] while reducing the environmental impact of their activities [12]. The FMCG industry primarily focuses on manufacturing, packaging goods, distributing, and some of fundamental activities including sales and marketing, financing, and purchasing [13] and it is found that the share of the FMCG industries in gross domestic product (GDP) is significant [14]. In the last few decades, the FMCG supply chain has faced more challenges due to increasing a tendency towards more demanding service level leading to higher delivery frequencies with smaller shipments sizes and consequent fragmentation of flows [11]. In addition, the FMCG environment is unpredictable and known as the most difficult part of the boom because commodities look similar without real competitive advantage and consumers tend to place a lot of values on different brands. In this industry, competition between competitors is always fierce, and the battle for market share continues [15]. Therefore, efforts for increasing the efficiency level of the FMCG supply chain have been undertaken with alternative transport solutions such as intermodal transport [16]. Colicchia et al. [9] investigated how intermodal transport in the Italian FMCG can be adopted for managing a lean and green supply chain by proposing a scenario-based estimation tool of the potential demand for intermodal transport. Singh and Acharya [17] identified supply chain flexibility dimensions from literature and field survey and they applied DEMATEL method to evaluate the flexibility dimension in Indian FMCG sector. Craggs [18] studied the impact of GSCM in the South African FMCG sector by presenting a "green supply chain maturity assessment questionnaire" as a potential answer to this need. Schoeman and Sanchez [19] investigated status quo concerning the environment and the human impact on its sustainability emphasizing on the green supply chain in FMCG industries in South Africa.

The green supply chain approach seems particularly relevant to the FMCG sector where identifying the performance indicators can potentially ensure significant benefits. Therefore, it is necessary to investigate the challenges factors and KPIs of foods and pharmaceutical companies with the consideration of green supply chain perspective.

*2.2. Green Supply Chain*

The rise of government regulations to meet environmental standards and growing consumer demand for green products has led to the emergence of the concept of GSCM, which encompasses the product life cycle from design to recycling. In recent years, GSCM as a significant approach for enterprises to be environmentally sustainable has attracted lots of attentions within both academia and industry. Among these industries, FMCG ones especially those related to foods and pharmaceutical, involves some crucial functions such as procurement, production, processing, and storage in the intermediate and final phases. In order to improve the competitiveness and profit, it was proved that supply chain management integrates internal operational activities with external customer demands [20]. Additionally, it was revealed that GSCM is the process of incorporating environmental criteria into business activities and the main point in supply chain greening is that it starts with establishing demand for greener products [21]. It was also indicated that the success of GSCM initiatives depends on proactivity and communion among supply chain members to ensure that environmental impact of the manufacturing and delivery of products and services is minimized [22]. Acevedo [23] examined contributions of interdisciplinary research to understanding interactions between environmental quality, food production, and food security. It was perceived that linking the geospatial, biotechnological, and precision agriculture technologies with the implemented models leads to achieving sustainable food production increases that maintain environmental quality. Dharni and Sharma [24] studied the status of supply chain management in food processing industry of Punjab state of India. It was revealed that logistics and supply chain management is still in its infancy in food processing sector in addition to not considering as a separate area of management. Gardas et al. [25] applied an interpretive structural modeling approach to establish the causal factors of the post-harvesting losses in the Indian context. They found that the lack of linkages between industry, government, and institution is the most significant factor. Liang et al. [26] indicated that it is crucial to evaluate the performance of GSCM in the entire supply chain. Akkerman et al. [27] reviewed quantitative operations management approaches to food distribution management, and relate this to challenges faced by the industry.

The main activities in GSCM are considered as green design, green purchasing, green manufacturing, and green transportation, which have attracted lots of attention in studies. Green design aims to design the products with enhanced biological quality that minimizes the harmful effects on the ecosystem [28,29], green purchasing presents procurement of environmentally preferable products and services [30–32], green manufacturing is introduced as a manufacturing process leading to lessening environmental impact [33–35], and green transportation refers to those ones that do not depend on diminishing natural resources [36]. Lin [37] investigated the criteria that affect GSCM practices—namely practices, performances, and external pressures—to raise awareness of environmental protection through green design, green purchasing, etc. Moreover, von Malmborg [38] argued that an environmental management system can be understood not only as a technical tool for analytical management but also as a tool for communicative action and organizational learning. Carpitella et al. [39] discussed organizational risk assessment in industrial environments by employing decision-making trial and evaluation laboratory method. Zhu and Sarkis [40] declared that GSCM tended to have win-win relationships in terms of environmental and economic performances in addition to providing an insight into the growing field of the relationship between environmental and operational practices and performance. They applied modified hierarchical regression methodology to test the various hypotheses. Chung et al. [41] examined the relationships between internal green practices, external green integration, green performance, and firm competitiveness in container shipping. Zhu et al. [42] investigated the correlation of major factors including organizational learning and management support for adopting GSCM. In order to compare organization characteristics of the two groups of respondent manufacturer, they employed Chi-Square test. Chung and Wee [43] developed an integrated inventory model with green component life cycle value design and remanufacturing considering the relevant price once implementing JIT delivery. Moreover, they implemented sensitivity analysis on the proposed time-weighted inventory deteriorating model. Yeh and Chuang [44] introduced some green criteria and developed an optimal arithmetical designing model for green partner selection including value, time, merchandise quality, and green appraisal score. Saif and Elhedhli [45] modeled the cold supply chain design problem as a mixed-integer concave minimization problem with dual objectives of minimizing the total cost—including capacity, transportation, and inventory costs—and the global warming impact through the use of a hybrid simulation-optimization approach. Bhateja et al. [46] discussed various environmental factors affecting in the manufacturing sectors and help them assess green future by considering GSCM including 17 indicators and 33 sub-indicators in which survey based on questionnaire has been utilized to collect data. Toke et al. [47] examined the measurement model of GSCM practices implementations focusing on 19 performance measure factors and applied AHP for determining relative importance and selecting appropriate approach in GSCM practice. Sharma et al. [48] suggested 13 indicators and 79 sub-indicators for implementation of GSCM in agro industry in which the quantitative phase was carried out through a survey using standard questionnaire with various agro based companies. Considering mentioned researches, it may be noted that there are still a few number of investigations that concentrate on KPIs for the implementation of GSCM and no work has been done in foods and pharmaceutical parts of FMCG industries.

## 3. Green Supply Chain KPI in FMCG's

The KPIs were explored within reviewing the literature on GSCM as well as in-depth interviews with FMCG industry experts. The expert team consisted of 18 specialists of supply chain FMCG considering foods and pharmaceutical companies in Iran and 7 university professors of supply chain management department. Then, through reviewing the literature and after discussion with experts, 12 KPIs have been identified and expressed as follows:

- Green purchasing; Since customer demands for environment-friendly or green products is increasing, green purchasing concentrates on cooperation from the suppliers to develop these eco-friendly products. Green purchasing is concerned with using eco-friendly raw materials, reuse of materials, recovery of waste, green management practices, green production processes, etc. to integrate their environmental goals. Moreover, some sub-indicators in green purchasing can be considered such as supplier's environmental management system, eco-labeling products as well as enough eco-labeling information for consumers and collaboration with suppliers and vendors for environmental targets. In addition, increasing recyclable packaging by suppliers in order to meet the desired health and environmental objectives.
- Green manufacturing; It focuses on environmental friendly procedures alongside energy efficient technologies through manufacturing stage of products. So, some certain environmental consciousness, for instance decreasing the energy consumption, employing efficient environmental management plans, etc. is considered in green manufacturing. Furthermore, some sub-indicators can be included such as system for waste minimization, impact of green manufacturing on brand image, following 3Rs—i.e., Recycle, Re-manufacture, Reuse—in addition to reduction in material cost after performing green procedures.
- Green design; It refers to design issues considering the choice of raw material, pollution prevention, design of packaging as well as redesign, conservation of resource and waste management. Additionally, life-cycle analysis is significant in view of green design concept. So, it is considered to minimize the material and energy consumption, recycle the components in addition to decreasing the usage of hazardous and toxic materials. Moreover, product designing for storage area during the transportation is should be taken into account.
- Green transportation; it refers to strategies for reducing transportation and warehousing cost as well as improving the environmental performance, considering various aspects such as sources of pollution in addition to diminishing level of fossil fuel and dependency on oil energy. Considering the effect of material, shape, and size of packaging in transportation, reducing packaging material and standardizing packaging which leading to transportation cost reduction are some of sub-indicators that can be considered in this level.
- Organizational environmental management; The engagement of greatest management supporting by other employees are really important to effectively implement the GSCM. There can be some sub-indicators in this category such as considering systems to track environmental laws and directives, documented procedure to implement curative action plan in addition to top and middle level management commitment for GSCM.
- Collaboration between customers and suppliers; A cooperation between suppliers and customers in addition to involving vendors in design phase of GSCM is an essential factor to improve the quality of the products. Furthermore, sharing of GSCM objectives with suppliers and vendors as well as having a structure of information sharing with suppliers and consumers can be considered in this indicator category in addition to taking collaboration from customer for greener production as well as green purchasing into account.
- Environmental performance; It refers to the capability of an organization to minimize various types of pollution and reduce the consumption of toxic substances. As sub-indicators, ability of GSCM to diminish emissions, effectiveness of the system to reduce water and solid pollutants, in addition to the extent to which GSCM reduces harmful material can be considered.
- Economic performance; It correlates to the economic improvement reflecting through the cost reduction in material purchasing, cost reduction connected with penalties for environmental accidents, energy consumption by GSCM as well as waste treatment and discharge. Furthermore,

it should be noticed that whether GSCM create quest for innovation and has GSCM increased income for the company?

- Operational performance; It refers to the enterprise's ability in order to efficiently produce and deliver the products to the customers, considering the number of products that timely delivered, decreasing scrap rate and inventory levels. Improvements in capacity utilization after GSCM implementation in addition to increased brand loyalty by consumers are some sub-dedicators that should be taken into account. Moreover, it should be noticed that up to what extent GSCM can enhance public relations.
- Competitive pressure; In light of the competitive marketplace, it is essential for organizations to grab the particular product's environmental impact through the implementation of the green practices. Some subsets including extent to which brand image is a major concern, effect of green schemes of competitors to implement GSCM and increase in interest of top managers by competitive green practices can be taken into account in this KPI.
- Regulatory pressure; The pressure from the government, customers and the stockholders is needed for the adoption and accomplishment of GSCM. Therefore, policies of the state government play a significant role in organizations to start up with green commencements.
- Cold storage; Regards the use of eco-friendly refrigerants in the cold storage rooms and warehouses as well as transportation vehicles in order to provide cold chain services as a crucial part in newly developing markets of foods and pharmaceutical industries. Due to the high energy waste and refrigerant gas leakages, green cold supply chains can be considered as a crucial factor in minimization of environmental problems.

The mentioned indicators, according to experts' opinions and based on the essence of the individual indicators, are decided to classify into three main branches. These main categories of indicators consist of environmental, executive, and strategic indicators, as given by Table 1.

**Table 1.** Performance indicators.

| | | |
|---|---|---|
| C11 | Organizational environmental management | **Environmental Indicators** |
| C12 | Competitive pressure | **C1** |
| C13 | Regulatory pressure | |
| C21 | Green design | |
| C22 | Green purchasing | |
| C23 | Green manufacturing | **Executive Indicators** |
| C24 | Cold storages | **C2** |
| C25 | Green transportation | |
| C31 | Operational performance | |
| C32 | Economic performance | **Strategic Indicators** |
| C33 | Environmental performance | **C3** |
| C34 | Collaboration between customers and suppliers | |

## 4. Research Methodology

The research method applied in the current study is a survey research. In terms of purpose, this study is applied because it seeks to introduce and apply fuzzy nonlinear mathematical modeling to rank the KPIs of the GSCM in FMCG industries. Solutions are also provided to understand the internal effects of these key performances. The research data were collected through library studies as well as the opinions of active experts in the field under study. The analysis in this study was carried out using questionnaires sent to specialists active in the supply chain of FMCG companies. The main categories and KPIs have been validated using the opinion of the academic professors. In this regard, according to experts' opinion, fuzzy logic is considered as one the best approaches to increase the accuracy of the questionnaires and due to using linguistic variables, the close indicators can be distinguished and therefore they can be clearly assessed. The scope of this research is the supply chain of FMCG companies (three pharmaceutical and seven food companies) in Iran. The solution methods are conceptual modeling and fuzzy decision making method in order to investigate the internal effect of indicators. In addition, a group fuzzy preference programming method is applied to understand the importance of indicators and their ranking. In order to converge the answers for the reliability of the questionnaires, we tried to monitor the dispersion of the experts' answers visually. In summary, the evaluation framework of this research consists of four stages as follows:

Step 1: Identifying and evaluating KPI in the green supply chain. In the present study, firstly, by using the literature review of the subject and related articles in the scientific databases of the world, an attempt was made to extract KPIs in the supply chain. Then, employing the views of supply chain experts (in pharmaceutical and food companies), three categories (environmental, executive, and strategic) were selected as key performance categories in the green supply chain and KPI were also identified.

Step 2: Creation of hierarchical structure of decision making. At this stage, hierarchical structure of the decision was determined using objective, criteria, and option levels (Table 1).

Step 3: Determination of internal relationships between indicators. In order to understanding how various indicators affect each other, the fuzzy DEMATEL is used.

Step 4: Calculating weight of indicators using group fuzzy preference programming method. In this study, a fuzzy nonlinear mathematical model will be used to rank the indicators based on pairwise comparisons in the AHP method. Therefore, consensus matrices of fuzzy judgments are formed based on decision makers' views. It is therefore necessary to use fuzzy numbers in explaining people's preferences and opinions.

In the following, we describe the decision making methods used in this study.

*4.1. DEMATEL Method*

DEMATEL method was first created at Science and Human Affairs Program of the Battelle Memorial Institute of Geneva. Since then, it has been widely used in many fields of studies such as evaluating core competencies, decision-making, knowledge management, operations research, and technology research [49]. The most important feature of the DEMATEL method that has been used in MCDM is the possibility of specifying the interrelationships between criteria. The DEMATEL method structure and the relevant calculation procedures can be expressed as follows:

Step 1: Constructing the direct-relation matrix. In order to evaluate the relationship between factors $i$ and $j$, the comparison scale is needed to define as an integer in the range of 0 to 4 in which 0 refers to no influence, 1 refers to low influence, 2 defines the average one, 3 describes the high influence, and 4 represents the very high influence. The integer score $x_{ij}^i$ is given by the $k$th expert where $1 \leq k \leq H$ and $H$ refers the number of experts. It shows the degree to which the criterion $i$ affects the criterion $j$. Thus, the $n \times n$ matrix $A$ can be computed by averaging individual expert's scores as written by

$$a_{ij} = \frac{1}{H} \sum_{k=1}^{H} x_{ij}^k \tag{1}$$

Step 2: Normalization of the direct-relation matrix. Based on the direct-relation matrix $A$, the normalized direct-relation matrix $D$ can be calculated as given by

$$s = \max\left( \max_{1 \leq i \leq n} \sum_{j=1}^{n} a_{ij}, \max_{1 \leq j \leq n} \sum_{i=1}^{n} a_{ij} \right) \tag{2}$$

$$D = \frac{A}{s} \tag{3}$$

It should be noted that the sum of each rows of matrix $A$ such as $I$ ($I$ is the identity matrix) indicate the direct effect of whole factor $i$ applied to other factors. Therefore, $\max\limits_{1\leq i\leq n}\sum\limits_{j=1}^{n} a_{ij}$ indicates the direct effects of total factor with the most direct effect on other factors and $\max\limits_{1\leq j\leq n}\sum\limits_{i=1}^{n} a_{ij}$ shows the most direct effect of total factors received from other factors. Matrix $D$ is obtained by dividing each elements of the matrix $A$ by the s and each element of which is between 0 to 1.

Step 3: Calculating the total-relation matrix. After obtaining the normalized direct-relation $D$, the total-relation matrix $T$ can be written as

$$T = D(I - D)^{-1} \tag{4}$$

$$T = [t_{ij}]_{n\times n} \quad i, j = 1, 2, 3, \ldots, n \tag{5}$$

Step 4: constructing a cause and effect diagram. Sum of the rows is represented by vector $r$ and sum of columns is denoted by vector $c$. The horizontal axis of vector $(r + c)$, known as 'prominence', describes the importance of the criterion and similarly, the vertical axis $(r - c)$, known as 'relation', divides criteria into a causal group and an effect group. According to the previous statements, a factor can be considered as causal if $(r - c)$ is positive, and the factor is effect when $(r - c)$ is negative. Thus, the cause and effect diagram can be achieved by mapping the dataset of $(r + c, r - c)$. Vectors $r$ and $c$ can be expressed by

$$r = \left[\sum_{j=1}^{n} t_{ij}\right]_{n\times 1} = [t_i]_{n\times 1} \tag{6}$$

$$c = \left[\sum_{j=1}^{n} t_{ij}\right]'_{1\times n} = [t_j]_{n\times 1} \tag{7}$$

Step 5: Determining the threshold value. In many studies, in order to illustrate the structural relationship between the factors, while maintaining the complexity of the system at the manageable level, it is necessary to determine the threshold value of $\alpha$ in order to filter out only the negligible effects on $T$ matrix. Threshold value $\alpha$ is determined by experts to set up the minimum value of influence level. When the correlative value in the matrix $T$ is smaller than $\alpha$, an influence relationship between two elements will be excluded from the map.

### 4.2. Fuzzy Logic

Most decisions in the real world are inaccurate due to the inaccuracy of understanding the goals, constraints, and possible actions. In light of a fuzzy environment, when a decision is made the results are highly influenced by personal judgments that can be ambiguous and inaccurate. Inaccurate sources can include non-quantifiable information, incomplete information, inaccessible information, and partial ignorance [50]. In order to find out a way to solve the problem of this inaccuracy, fuzzy set theory as a mathematical tool was proposed by Zadeh [51] in 1965 to deal with information uncertainty in decision making process. Since then, this theory has been well developed and has found many successful applications. In fuzzy logic, any number between 0 and 1 represents a part of truth, while in definite sets working with binary logic only two values of 0 and 1 are available.

Thus, fuzzy logic can express inaccurate and imprecise judgments and act mathematically with them [52]. Utilizing the conventional quantification make it difficult to express reasonably the very complicated situations, so using the linguistic variable concept is necessary in such situations. A linguistic variable is a variable whose value has the form of a phrase or sentence in natural language. Linguistic variables are also very functional in dealing with situations described in quantitative terms because these variables' values are linguistic expressions instead of numbers. In practice, linguistic values can be represented using fuzzy numbers, the most common of which are triangular fuzzy numbers (TFN). A triangular fuzzy number $\widetilde{A}$ is defined by $[(L,\ M,\ U)]$ where $L$ and $U$ are respectively top and bottom boundary of $\widetilde{A}$ as shown in Figure 1. In addition, the fuzzy linguistic scale is shown in Table 2. The fuzzy number $A$ on $R$ is a triangular fuzzy number when the membership function $\mu_{\widetilde{A}}(x) : R \rightarrow [0,1]$ is given by

$$\mu_{\widetilde{A}}(x) = \begin{cases} (x-L)/(M-L), & L \leq x \leq M \\ (U-x)/(U-M) & M \leq x \leq U \\ 0 & otherwise \end{cases} \tag{8}$$

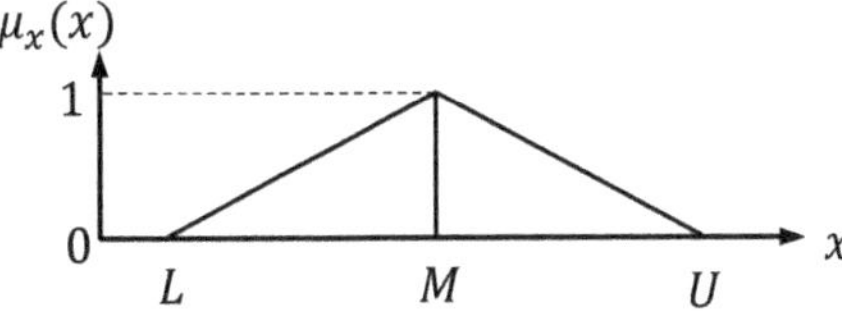

**Figure 1.** Membership function of triangular fuzzy numbers.

**Table 2.** Fuzzy linguistic scale.

| Linguistic Terms | Triangular Fuzzy Numbers |
| --- | --- |
| Very high influence (VH) | (0.75, 1.0, 1.0) |
| High influence (H) | (0.5, 0.75, 1.0) |
| Low influence (L) | (0.25, 0.5, 0.75) |
| Very low influence (VL) | (0, 0.25, 0.5) |
| No influence (No) | (0, 0, 0.25) |

### 4.3. Application of Fuzzy Logic in DEMATEL Method

One of the issues with using the DEMATEL method is to obtain the direct effect size between the two factors. The size of these concessions is always obtained by using expert surveys; but in many cases people's judgment in decision making is unclear and cannot be measured using precise numerical values; therefore, it is necessary to use fuzzy logic in dealing with issues that are ambiguous and inaccurate. To employ fuzzy logic in DEMATEL method, experts should first be asked to respond on the basis of defined linguistic variables (presented in Table 2) as pairwise comparisons to determine the effect of factors on each other. To overcome the ambiguities of human evaluations, the linguistic variable of 'influence' is used through the five expressions of very high, high, low, very low, and none, described as positive triangular fuzzy numbers $(l_{ij}, m_{ij}, r_{ij})$. After obtaining expert opinions, we calculate the fuzzy mean matrix using fuzzy averaging. Then using the existing equations to convert the fuzzy values into the non-fuzzy numbers, matrix of the final mean values is calculated.

Therefore, if we have $p$ responders, we will have fuzzy matrices as much as respondents. Now we calculate the fuzzy mean matrix, so we will have

$$\widetilde{z} = \frac{(\widetilde{z}^1 + \widetilde{z}^2 + \ldots + \widetilde{z}^p)}{p} \tag{9}$$

The $\widetilde{z}$ matrix is called the fuzzy primary relation matrix or the intermediate fuzzy matrix, which is represented as

$$\widetilde{z} = \begin{bmatrix} 0 & \widetilde{z}_{12} & \cdots & \widetilde{z}_{1n} \\ \widetilde{z}_{21} & 0 & \cdots & \widetilde{z}_{1n} \\ \vdots & \vdots & \ddots & \vdots \\ \widetilde{z}_{n2} & \widetilde{z}_{n2} & \cdots & 0 \end{bmatrix} \tag{10}$$

Now by using a deffuzication method, initial direct-relation matrix can be achieved. In order to transform TFN numbers to crisp values, we use converting fuzzy data into crisp scores (CFCS) defuzzication method [52]. In this method the final value is defined as average weight in accordance with the membership function. If $(l_{ij}, m_{ij}, r_{ij})$ indicates the effect of criterion $i$ on the criterion $j$ in the fuzzy matrix of the direct relation, then the CFCS method can be summarized in the following steps:

The first step is normalization

$$xl_{ij} = (l_{ij} - \min l_{ij}) / \Delta_{\min}^{\max} \tag{11}$$

$$xm_{ij} = (m_{ij} - \min l_{ij}) / \Delta_{\min}^{\max} \tag{12}$$

$$xr_{ij} = (r_{ij} - \min l_{ij}) / \Delta_{\min}^{\max} \tag{13}$$

where

$$\Delta_{\min}^{\max} = \max r_{ij} - \min l_{ij} \tag{14}$$

In the second step, we calculate the values on the right and left

$$xls_{ij} = xm_{ij} / (1 + xm_{ij} - xl_{ij}) \tag{15}$$

$$xrs_{ij} = xr_{ij} / (1 + xr_{ij} - xm_{ij}) \tag{16}$$

In the third step, we calculate the total normalized definite value

$$x_{ij} = [xls_{ij}(1 - xls_{ij}) + xrs_{ij}] / [1 - xls_{ij} + xrs_{ij}\} \tag{17}$$

In the last step we calculate the final definitive value

$$z_{ij} = \min l_{ij} + x_{ij}\Delta_{\min}^{\max} \tag{18}$$

*4.4. Non-Linear Fuzzy Prioritization*

In this section, a non-linear method for KPI prioritization is proposed in order to find directly crisp values of priorities from a set of comparison judgments, represented as triangular fuzzy numbers (as shown in Figure 1) [53]. We want to find a crisp priority vector so that the ratios almost satisfy the fuzzy initial judgments of $\widetilde{a}_{ij} = (l_{ij}, m_{ij}, u_{ij})$ or

$$l_{ij} \widetilde{\le} \frac{w_i}{w_j} \widetilde{\le} u_{ij} \tag{19}$$

where the symbol $\widetilde{\le}$ refers to the statement of fuzzy less or equal to.

In order to handle the above inequalities easily, they can be defined as a set of single-side fuzzy constraints as follows

$$\begin{aligned} w_i - w_j u_{ij} &\widetilde{\le} 0 \\ -w_i + w_j l_{ij} &\widetilde{\le} 0 \end{aligned} \tag{20}$$

The set of $2m$ fuzzy constraints can be written in a matrix form as

$$Rw \widetilde{\le} 0 \tag{21}$$

in which the matrix $R \in \Re^{2m \times n}$.

Then for each fuzzy judgment, a membership function can be constructed that is linear with respect to $w_i/w_j$ and can be given by

$$\mu_{ij}\left(\frac{w_i}{w_j}\right) = \begin{cases} \dfrac{\left(\frac{w_i}{w_j}-l_{ij}\right)}{m_{ij}-l_{ij}} & \frac{w_i}{w_j} \leq m_{ij} \\[2ex] \dfrac{u_{ij}-\left(\frac{w_i}{w_j}\right)}{u_{ij}-m_{ij}} & \frac{w_i}{w_j} \leq m_{ij} \end{cases} \tag{22}$$

Function (22) is linearly increasing over the interval $(-\infty, m_{ij})$ and linearly decreasing over the interval $(m_{ij}, \infty)$. It takes negative values when $w_i/w_j < l_{ij}$ or $w_i/w_j > u_{ij}$ and has a maximum value $\mu_{ij} = 1$ at $w_i/w_j = m_{ij}$. Over the range $(l_{ij}, u_{ij})$ the membership function (22) coincides with the fuzzy triangular judgment $\widetilde{a}_{ij} = (l_{ij}, m_{ij}, u_{ij})$.

Function (22) is non-linear with respect to the decision variables, but provides a fuzzy feasible area, linear in these ratios. We can define a fuzzy feasible area on $(n-1)$-dimensional simplex (23) as the intersection of all membership functions (22) and apply a max–min-approach for finding the maximizing solution.

$$Q^{n-1} = \{(w_1,\ldots,w_n)|w_i > 0, \quad w_1 + \ldots + w_n = 1\} \tag{23}$$

This leads to the following non-linear optimization problem

$$\begin{aligned} \max \quad & \lambda \\ s.t : \quad & (m_{ij} - l_{ij})\lambda w_j - w_i + l_{ij}w_j \leq 0 \\ & (u_{ij} - m_{ij})\lambda w_j + w_i - u_{ij}w_j \leq 0 \\ & \sum_{k=1}^{n} w_k = 1 \\ & w_k > 0 \quad, k = 1,2,\ldots,n; \quad i = 1,2,\ldots,n-1; j = 2,3,\ldots,n \quad, \\ & j > i \end{aligned} \tag{24}$$

The solution of the above non-linear problem (24) needs some appropriate numerical method for non-linear optimization to be employed.

To use the function (24), the pairwise comparisons matrix will be obtained by using AHP method [54] and by integrating expert opinions. Fuzzy linguistic scales will be used to obtain expert opinions. These linguistic scales for the matrix of pairwise comparisons and their fuzzy equations are shown in Table 3. The optimum positive value of the indicator in this function (24) reveals that all weight ratios are completely true to the original judgment. However, if the indicator is negative, it can be seen that fuzzy judgments were strongly incompatible.

**Table 3.** Linguistic scales for pairwise comparisons and their fuzzy equivalents.

| Triangular Fuzzy Scales | Linguistic Values for Pairwise Comparisons |
|---|---|
| Verylow | (1,2,3) |
| Low | (2,3,4) |
| Medium | (3,4,5) |
| High | (4,5,6) |
| Very high | (5,6,7) |

## 5. Research Findings

### 5.1. Internal Relationship and Severity of Impact between Performance Indicators

In this study, in order to investigate the internal effects of performance indicators, fuzzy DEMATEL method was employed. To this end, 50 questionnaires were sent to specialists working in the FMCG industry, among which 40 questionnaires were completed and came to us. To examine effective internal relationships, experts were asked to state their theories about the impact of each performance indicator on the other ones, based on linguistic options and triangular fuzzy positive numbers (as shown by Table 2) through paired comparisons between the factors obtained from the research. The fuzzy direct relation matrix was formed for the performance indicators as presented in Tables 4 and 5.

**Table 4.** Fuzzy direct relation matrix between the main categories.

|  | C1 | | | C2 | | | C3 | | |
| --- | --- | --- | --- | --- | --- | --- | --- | --- | --- |
|  | L | M | U | L | M | U | L | M | U |
| C1 | 0 | 0 | 0 | 0.8 | 0.6 | 0.4 | 0.85 | 0.75 | 0.25 |
| C2 | 0.8 | 0.75 | 0.35 | 0 | 0 | 0 | 0.85 | 0.75 | 0.55 |
| C3 | 0.5 | 0.2 | 0.15 | 0.5 | 0.25 | 0 | 0 | 0 | 0 |

**Table 5.** Fuzzy direct relationship matrix between performance indicators.

|  | C11 | | | C12 | | | C ... [1] | C33 | | | C34 | | |
| --- | --- | --- | --- | --- | --- | --- | --- | --- | --- | --- | --- | --- | --- |
|  | L | M | U | L | M | U | ... | L | M | U | L | M | U |
| C11 | 0 | 0 | 0 | 0.8 | 0.45 | 0.3 | ... | 0.65 | 0.35 | 0.1 | 0.95 | 0.45 | 0.15 |
| C12 | 0.9 | 0.95 | 0.45 | 0 | 0 | 0 | ... | 0.85 | 0.6 | 0.35 | 0.7 | 0.35 | 0.25 |
| ... | ... | ... | ... | ... | ... | ... | ... | ... | ... | ... | ... | ... | ... |
| C33 | 0.75 | 0.85 | 0.45 | 0.9 | 085 | 0.3 | ... | 0 | 0 | 0 | 0.9 | 0.85 | 0.50 |
| C34 | 0.7 | 0.65 | 0.35 | 0.9 | 0.6 | 0.45 | ... | 0.2 | 0.85 | 0.45 | 0 | 0 | 0 |

[1] The table is summarized.

In Tables 6 and 7, the general fuzzy relationship matrix for the main categories and performance indicators are demonstrated, respectively.

**Table 6.** Total fuzzy relation matrix for the main categories.

|  | C1 | | | C2 | | | C3 | | |
| --- | --- | --- | --- | --- | --- | --- | --- | --- | --- |
|  | L | M | U | L | M | U | L | M | U |
| C1 | 0.9 | 0.5 | 0.13 | 1.2 | 0.44 | 0.23 | 1.2 | 0.55 | 0.34 |
| C2 | 1.35 | 0.6 | 0.43 | 1.1 | 0.45 | 0.1 | 1.3 | 0.77 | 0.45 |
| C3 | 1.13 | 0.24 | 0.11 | 0.45 | 0.23 | 0.04 | 0.65 | 0.22 | 0.07 |

**Table 7.** Total fuzzy relation matrix for performance indicators.

|  | C11 | | | C12 | | | C ... [1] | C33 | | | C34 | | |
| --- | --- | --- | --- | --- | --- | --- | --- | --- | --- | --- | --- | --- | --- |
|  | L | M | U | L | M | U | ... | L | M | U | L | M | U |
| C11 | 0.11 | 0.1 | 0.06 | 0.23 | 0.2 | 0.07 | ... | 0.2 | 0.11 | 0.09 | 0.21 | 0.11 | 0.08 |
| C12 | 0.25 | 0.13 | 0.05 | 0.2 | 0.1 | 0.04 | ... | 0.2 | 0.13 | 0.07 | 0.22 | 0.13 | 0.07 |
| C ... | ... | ... | ... | ... | ... | ... | ... | ... | ... | ... | ... | ... | ... |
| C33 | 0.25 | 0.14 | 0.1 | 0.2 | 0.14 | 0.12 | ... | 0.15 | 0.12 | 0.04 | 0.18 | 0.2 | 0.07 |
| C34 | 0.26 | 0.16 | 0.1 | 0.24 | 0.13 | 0.1 | ... | 0.2 | 0.18 | 0.12 | 0.17 | 0.2 | 0.07 |

[1] The table is summarized.

To compile the relationship map, the sum of the elements of the columns and rows of total matrix for the main categories as well as the KPIs were calculated. These values are nominated as effective ($R$) and Influential ($D$) vectors. The results are shown in Table 8.

**Table 8.** Results of the calculations of the effect of performance indicators

| KPI | D | R | D + R | D − R |
|---|---|---|---|---|
| Environmental Indicators | 3.651 | 2.234 | 5.885 | 1.417 |
| Organizational environmental management | 0.857 | 0.611 | 1.468 | 0.246 |
| Competitive pressure | 0.872 | 0.66 | 1.532 | 0.212 |
| Regulatory pressure | 0.245 | 0.235 | 0.48 | 0.01 |
| Executive Indicators | 3.547 | 2.783 | 6.33 | 0.764 |
| Green design | 0.376 | 0.42 | 0.796 | −0.044 |
| Green purchasing | 0.387 | 0.444 | 0.831 | −0.057 |
| Green manufacturing | 0.387 | 0.413 | 0.8 | −0.026 |
| Cold storages | 0.523 | 0.397 | 0.92 | 0.126 |
| Green transportation | 0.293 | 0.28 | 0.573 | 0.013 |
| Strategic Indicators | 3.287 | 2.98 | 6.267 | 0.307 |
| Operational performance | 0.217 | 0.234 | 0.451 | −0.017 |
| Economic performance | 0.225 | 0.235 | 0.46 | −0.01 |
| Environmental performance | 0.251 | 0.254 | 0.505 | −0.003 |
| Collaboration between customers and suppliers | 0.341 | 0.333 | 0.674 | 0.008 |

Then, based on Table 8, the network relations map of the performance indicators will be obtained. This is illustrated in Figure 2. As can be seen, Figure 2a demonstrates the internal effects of main category of performance indicators. Additionally, Figure 2b–d describes the internal effects of KPIs in respectively environmental, executive and strategic categories. As can be seen in Figure 2, environmental indicators have the greatest effect on the strategic and executive indicators. At the same time, the most influential indicators among the presented categories can be clearly seen in this figure.

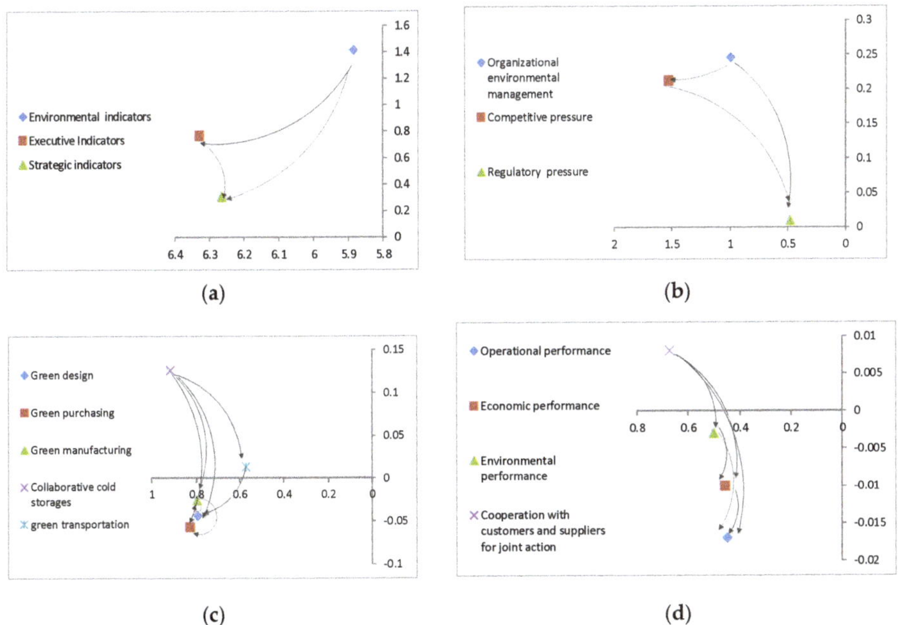


**Figure 2.** The network of influential relationships: (**a**) demonstrates the internal effects of main category of performance indicators. (**b–d**) describes the internal effects of KPIs in respectively environmental, executive, and strategic categories.

*5.2. Evaluating and Ranking Performance Indicators Using Non-Linear Fuzzy Prioritization Method*

In this study, a non-linear fuzzy prioritization method based on AHP was used to evaluate and rank the indicators obtained by using field and study methods. For this purpose, 35 specialists including 25 experts of the food industry and 10 specialists in the pharmaceutical industry, among the top supply chain managers and senior supervisors of supply chain of the FMCG industries with more than 10 years' work experience were selected and questionnaires were sent to them. Then, experts were asked to use paired comparisons using linguistic criteria (Table 3) to analyze KPIs in green supply chain and we received the answers of all experts. Therefore, evaluation and ranking of green supply chain's KPI in FMCG companies are divided into two parts.

(1) Fuzzy pairwise comparison matrix determination based on integration of expert opinions that is expressed according to linguistic criteria presented in Table 3.
(2) Solving the proposed nonlinear mathematical model (function 24) using pairwise comparisons matrices and finally gaining the weight of KPI.

The pairwise comparisons matrix for green supply chain KPI, based on the integration of expert opinion, is shown in Tables 9–12. We will use these paired comparisons for our calculations in the mathematical model.

**Table 9.** Paired comparison matrix of KPIs in the FMCG industry green supply chain, based on integration of experts.

| | Environmental | | | Executive | | | Strategic | | |
| --- | --- | --- | --- | --- | --- | --- | --- | --- | --- |
| | W1 | | | W2 | | | W3 | | |
| W1 | - | - | - | - | - | - | - | - | - |
| W2 | 2.7 | 2.9 | 6.7 | - | - | - | - | - | - |
| W3 | 2.5 | 2.6 | 3.2 | 1.1 | 1.3 | 3.1 | - | - | - |

**Table 10.** Paired comparison matrix of KPIs in the FMCG industry green supply chain, based on integration of experts in environmental section.

| | Regulatory Pressure | | | Organizational Environmental Management | | | Competitive Pressure | | |
| --- | --- | --- | --- | --- | --- | --- | --- | --- | --- |
| | W1 | | | W2 | | | W3 | | |
| W1 | - | - | - | - | - | - | - | - | - |
| W2 | 2.3 | 3.4 | 5.7 | - | - | - | - | - | - |
| W3 | 2.3 | 2.87 | 4.3 | 1 | 1.98 | 4 | - | - | - |

**Table 11.** Paired comparison matrix of KPIs in the FMCG industry green supply chain, based on integration of experts in executive section.

| | Green Purchasing | | | Green Design | | | Cold Storages | | | Green Manufacturing | | | Green Transportation | | |
| --- | --- | --- | --- | --- | --- | --- | --- | --- | --- | --- | --- | --- | --- | --- | --- |
| | W1 | | | W2 | | | W3 | | | W4 | | | W5 | | |
| W1 | - | - | - | - | - | - | - | - | - | - | - | - | - | - | - |
| W2 | 1.12 | 1.56 | 2 | - | - | - | - | - | - | - | - | - | - | - | - |
| W3 | 1.54 | 2.65 | 4 | 1.64 | 2.69 | 7.3 | - | - | - | - | - | - | - | - | - |
| W4 | 1.16 | 1.76 | 2.7 | 1 | 1.3 | 5.49 | 0.5 | 1 | 1 | - | - | - | - | - | - |
| W5 | 1.13 | 1.29 | 2.51 | 1.21 | 1.25 | 4 | 0.8 | 1 | 1.1 | 1.2 | 1.1 | 1.6 | - | - | - |

**Table 12.** Paired comparison matrix of KPIs in the FMCG industry green supply chain, based on integration of experts in strategic section.

| | Economic Performance | | | Operational Performance | | | Environmental Performance | | | Collaboration between Customers and Suppliers | | |
|---|---|---|---|---|---|---|---|---|---|---|---|---|
| | W1 | | | W2 | | | W3 | | | W4 | | |
| W1 | - | - | - | - | - | - | - | - | - | - | - | - |
| W2 | 5 | 4 | 5.5 | - | - | - | - | - | - | - | - | - |
| W3 | 3.7 | 4 | 5.5 | 5 | 2 | 5.2 | - | - | - | - | - | - |
| W4 | 2.4 | 3.65 | 5 | 2.5 | 6 | 3.2 | 1.07 | 2 | 2.5 | - | - | - |

After performing pairwise comparisons of expert opinions on different sections, we will use the data from these matrices in mathematical modeling to ranking. We put fuzzy values in the mathematical model. Since the model is nonlinear, we use GAMS software to solve the model. Thus, the weight and rank of the criteria and sub-criteria will be obtained, which is visible in Tables 13–16.

**Table 13.** Weight and ranking of KPIs in main category (taken from fuzzy nonlinear model).

| KPI | Code | Weight | Rank |
|---|---|---|---|
| Environmental | W1 | 0.4183452 | 1 |
| Executive | W2 | 0.2959360 | 2 |
| Strategic | W3 | 0.2857188 | 3 |

**Table 14.** Weight and ranking of KPIs in environmental indicators category (taken from fuzzy nonlinear model).

| KPI | Code | Weight | Rank |
|---|---|---|---|
| Regulatory pressure | W1 | 0.2583772 | 2 |
| Organizational environmental management | W2 | 0.5032456 | 1 |
| Competitive pressure | W3 | 0.2383772 | 3 |

**Table 15.** Weight and ranking of KPIs in executive indicators category (taken from fuzzy nonlinear model).

| KPI | Code | Weight | Rank |
|---|---|---|---|
| Green purchasing | W1 | 0.1243567 | 5 |
| Green design | W2 | 0.1434582 | 4 |
| Cold storages | W3 | 0.3193412 | 1 |
| Green manufacturing | W4 | 0.1900213 | 3 |
| Green transportation | W5 | 0.2228236 | 2 |

**Table 16.** Weight and ranking of KPIs in strategic indicators category (taken from fuzzy nonlinear model).

| KPI | Code | Weight | Rank |
|---|---|---|---|
| Economic performance | W1 | 0.2911781 | 2 |
| Operational performance | W2 | 0.2151470 | 3 |
| Environmental performance | W3 | 0.3213857 | 1 |
| Collaboration between customers and suppliers | W4 | 0.1722892 | 4 |

After calculating the weights of each KPI, we can normalize the weights using the information in Tables 13–16. The normalized weight for the green supply chain's KPI is shown in Table 17. The normalized weight represents the overall rank of the KPI. In addition, Figure 3 shows that the organizational environmental management and regulatory pressure are more important than the other indicators because they have a higher weight than others. Therefore, GSCM can be successfully implemented when the top level management considers green practices as their company policies in view of the systems to track environmental laws as well as documented procedure to implement corrective action plan. Regulatory pressure gets the next highest weightage which obviously shows that the government, customers, and stockholders can play a significant role in implementation of GSCM. Furthermore, the third place represents effect of competitor's green strategies to successfully implement GSCM. Cold storage achieved the fourth weightage, which shows that green cold supply chains are a major concern in minimization of environmental problems. Other KPIs occupied the resting ranks, as presented by Table 17 and Figure 3.

**Table 17.** Normalized weight and ranking of KPIs in the green supply chain (FMCG industries).

| KPI Category | Weight | KPI | Weight | Normal Weight | Rank |
|---|---|---|---|---|---|
| Environmental indicators | 0.418345 | Organizational environmental management | 0.5032456 | 0.210530381 | 1 |
| | | Competitive pressure | 0.2383772 | 0.099723957 | 3 |
| | | Regulatory pressure | 0.2583772 | 0.108090861 | 2 |
| Executive indicators | 0.295936 | Green design | 0.1434582 | 0.042454446 | 11 |
| | | Green purchasing | 0.1243567 | 0.036801624 | 12 |
| | | Green manufacturing | 0.1900213 | 0.056234143 | 9 |
| | | Cold storages | 0.3193412 | 0.094504557 | 4 |
| | | Green transportation | 0.2228236 | 0.065941525 | 7 |
| Strategic indicators | 0.285719 | Operational performance | 0.215147 | 0.061471543 | 8 |
| | | Economic performance | 0.2911781 | 0.083195057 | 6 |
| | | Environmental performance | 0.3213857 | 0.091825937 | 5 |
| | | Collaboration between customers and suppliers | 0.1722892 | 0.049226263 | 10 |

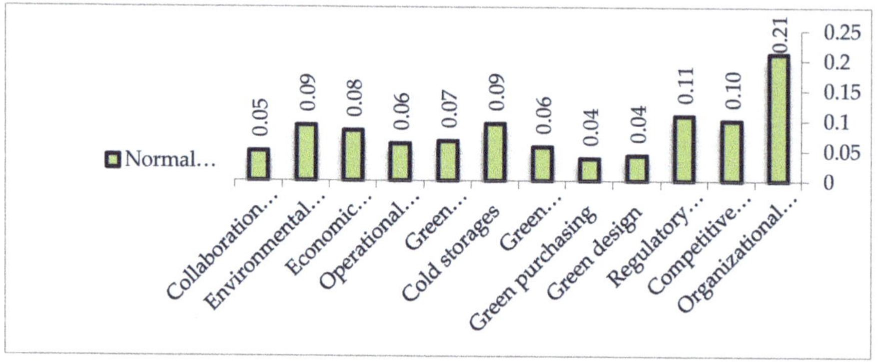


**Figure 3.** Importance of KPIs of the green supply chain in the FMCG industry.

## 6. Conclusions

Recently, many industries are moving towards being greener in their supply chain processes due to customers' needs as well as the scarce resources and policies of governments for different industries. In other world, GSCM accomplishment has become a significant issue for industries in the present competitive market that may be successfully implemented in any company through which every employee works such as a link of a chain. Therefore, reviewing the performance and explaining KPIs for the green supply chain play a great important role in many industries. Prioritizing these functions can lead to effective management of green supply chain processes in the organization. The FMCG industries are highly regarded due to the type of production of their products that are very fast consuming (the food and pharmaceutical industries fall within this range) and given the specific distribution dimensions and delivery schedules. Therefore, due to the importance of the subject, this study has attempted to identify KPIs in the FMCG green supply chain based on the literature review as well as the views of experts in the FMCG industry. In accordance with experts' opinions and based on the essence of the individual indicators, they were classified into three main categories, including 12 KIPs. Subsequently, by using fuzzy decision making approach and by distributing the questionnaire among the supply chain experts in the FMCG industry, the relationships of internal effects on the indicators were determined. Then a fuzzy nonlinear mathematical model by innovatively applying the AHP method was used to evaluate and prioritize KPIs. The results indicated that organizational environmental management is the highest priority among the KPIs in the green supply chain of FMCG industries, limited to food and pharmaceutical industries. In addition, it was found that indicators of regulatory pressures as well as competitive pressures are in the next rank and all of the first three ranks occupied by the main category of environmental indicators. Moreover, it was revealed that cold storage as an executive indicator, and afterwards the environmental performance as a strategic indicator acquired the next ranks. Therefore, it seems that FMCG industry executives and decision makers need to give more weight to these indicators to improve organizational performance in the green supply chain.

**Author Contributions:** Conceptualization, H.N. and E.N.; methodology, H.N.; software, H.N. and E.N.; validation, M.F. and F.H.; formal analysis, H.N.; investigation, H.N., E.N., M.F. and F.H.; resources, M.F. and F.H.; data curation, H.N.; writing—original draft preparation, H.N.; writing—review and editing, E.N.; supervision, E.N., M.F. and F.H.; project administration, E.N., M.F. and F.H.

**Funding:** This research received no external funding.

**Conflicts of Interest:** The authors declare no conflicts of interest.

## References

1. Mangla, S.K.; Kumar, P.; Barua, M.K. Prioritizing the responses to manage risks in green supply chain: An indian plastic manufacturing prespective. *Sustain. Prod. Consum.* **2015**, *1*, 67–86. [CrossRef]
2. Wu, H.H.; Chang, S.Y. Case study of using dematel method to identify critical factors in green supply chain management. *Appl. Math. Comput.* **2015**, *256*, 394–403. [CrossRef]
3. Cousins, P.; Lawson, B.; Petersen, K.; Fugate, B. Investigating green supply chain management practices and performance. *Int. J. Oper. Prod. Manag.* **2019**, *39*, 767–786. [CrossRef]
4. Seman, N.A.A.; Govindan, K.; Mardani, A.; Zakuan, N.; Saman, M.Z.M.; Hooker, R.E.; Ozkul, S. The mediating effect of green innovation on the relationship between green supply chain management and environmental performance. *J. Clean. Prod.* **2019**, *229*, 115–127. [CrossRef]
5. Arlbjørn, J.S. *Supply Chain Management*; Academica: London, UK, 2010.
6. Kotler, P.; Armstrong, G. *Principles of Marketing*, 13th ed.; Pearson Prentice Hall: Upper Saddle River, NJ, USA, 2011.
7. Menidjel, C.; Benhabib, A.; Bilgihan, A. Examining the moderating role of personality traits in the relationship between brand trust and brand loyalty. *J. Prod. Brand Manag.* **2017**, *6*, 631–649. [CrossRef]
8. Aljunaidi, A.; Ankrah, S. The Application of Lean Principles in the Fast Moving Consumer Goods (FMCG). *J. Oper. Supply Chain Manag.* **2014**, *7*, 1–25. [CrossRef]
9. Colicchia, C.; Creazza, A.; Dallari, F. Lean and green supply chain management through intermodal transport: Insights from the fast moving consumer goods industry. *Prod. Plan. Control* **2017**, *28*, 321–334. [CrossRef]
10. Bourlakis, M.A.; Weightman, P.W. *Food Supply Chain Management*, 1st ed.; Ames, I.A., Ed.; Blackwell Publishing: Hoboken, NJ, USA, 2004.
11. Fernie, J.; Sparks, L. *Logistics and Retail Management: Emerging Issues and New Challenges in the Retail Supply Chain*; Kogan Page: London, UK, 2014.
12. Monios, J. Integrating Intermodal Transport with Logistics: A Case Study of the UK Retail Sector. *Transp. Plan. Technol.* **2015**, *38*, 347–374. [CrossRef]
13. Green, S. Durable vs. Non-Durable Goods: What's the Difference? *2016*. Available online: https://rjofutures.rjobrien.com/market-news/2016/05/05/durable-vs-non-durable-goods-whats-the-difference (accessed on 25 June 2019).
14. Malhotra, S. A study on marketing fast moving consumer goods (FMCG). *Int. J. Innov. Res. Dev.* **2014**, *1*, 1–3.

15. Anselmsson, J.; Burt, S.; Tunca, B. An integrated retailer image and brand equity framework: Re-examining, extending, and restructuring retailer brand equity. *J. Retail. Consum. Serv.* **2017**, *38*, 194–203. [CrossRef]
16. Council, F.L. La Sostenibilità nei Trasporti e nella Logistica: Quaderno. 2016. Available online: http://www.freightleaders.org/i-quaderni/ (accessed on 25 June 2019).
17. Singh, R.K.; Acharya, P. Identification and Evaluation of Supply Chain Flexibilities in Indian FMCG Sector Using DEMATEL. *Glob. J. Flex. Syst. Manag.* **2014**, *15*, 91–100. [CrossRef]
18. Craggs, J.A. *Maturity Assessment of Green Supply Chain Management in the South African FMCG Industry*; University of Pretoria: Pretoria, South Africa, 2012.
19. Schoeman, C.; Sanchez, V.R. Green Supply Chain Overview and A South African Case Study. In Proceedings of the 28th Southern African Transport Conference, Pretoria, South Africa, 8–11 July 2009.
20. Li, X.; Wang, Q. Coordination mechanism of supply chain systems. *Eur. J. Oper. Res.* **2007**, *179*, 1e6. [CrossRef]
21. Gilber, S. *Greening Supply Chain: Enhancing Competitiveness through Green Productivity*; Asian Productivity Organization: Tapei, Taiwan, 2001.
22. Rao, P.; Holt, D. Do green supply chains lead to competitiveness and economic performance? *Int. J. Oper. Prod. Manag.* **2005**, *25*, 898. [CrossRef]
23. Acevedo, M.F. Interdisciplinary progress in food production, food security and environment. *Environ. Conserv.* **2011**, *38*, 151–171. [CrossRef]
24. Dharni, K.; Sharma, R.K. Supply chain management in food processing sector: Experience from India. *Int. J. Logist. Syst. Manag.* **2015**, *21*, 115–132. [CrossRef]
25. Gardas, B.B.; Raut, R.D.; Narkhede, B. Modeling causal factors of post-harvesting losses in vegetable and fruit supply chain: An Indian perspective. *Renew. Sustain. Energy Rev.* **2017**, *80*, 1355–1371. [CrossRef]
26. Liang, L.; Feng, F.; Cook, W.D.; Zhu, J.D. A models for supply chain efficiency evaluation. *Ann. Oper. Res.* **2006**, *145*, 35–49. [CrossRef]
27. Akkerman, R.; Farahani, P.; Grunow, M. Quality, safety and sustainability in food distribution: A review of quantitative operations management approaches and challenges. *Oper. Res. Spektrum* **2010**, *32*, 863–904. [CrossRef]
28. Bhat, V. Green Marketing Begins with Green Design. *J. Bus. Ind. Mark.* **1993**, *8*, 26–31. [CrossRef]
29. Wiley, J.A.; Benefield, J.D.; Johnson, K.H. Green Design and the Market for Commercial Office Space. *J. Real Estate Financ. Econ.* **2010**, *41*, 228–243. [CrossRef]
30. Siyavooshi, M.; Foroozanfar, A.; Sharifi, Y. Effect of Islamic values on green purchasing behavior. *J. Islam. Mark.* **2019**, *10*, 125–137. [CrossRef]
31. Chekima, B.; Azizi, K.W.S.; AisatIgau, O.; Chekima, S. Examining green consumerism motivational drivers: Does premium price and demographics matter to green purchasing? *J. Clean. Prod.* **2016**, *112*, 3436–3450. [CrossRef]
32. Foo, M.; Kanapathy, K.; Zailani, S.; Shaharudin, M.R. Green purchasing capabilities, practices and institutional pressure. *Manag. Environ. Qual.* **2019**, *30*, 1171–1189. [CrossRef]
33. Rusinko, C. Green Manufacturing: An Evaluation of Environmentally Sustainable Manufacturing Practices and Their Impact on Competitive Outcomes. *IEEE Trans. Eng. Manag.* **2007**, *54*, 445–454. [CrossRef]
34. Dornfeld, D.A. *Green Manufacturing: Fundamentals and Applications*; Springer Science & Business Media: New York, NY, USA, 2013.
35. Govindan, K.; Diabat, A.; Shankar, K.M. Analyzing the drivers of green manufacturing with fuzzy approach. *J. Clean. Prod.* **2015**, *96*, 182–193. [CrossRef]
36. Panday, A.; Bansal, B.O. Green transportation: Need, technology and challenges. *Int. J. Glob. Energy* **2015**, *37*, 304–318. [CrossRef]
37. Lin, R.J. Using fuzzy DEMATEL to evaluate the green supply chain management practices. *J. Clean. Prod.* **2013**, *40*, 32–39. [CrossRef]
38. von Malmborg, F.B. Environmental management systems, communicative action and organizational learning. *Bus. Strategy Dev.* **2002**, *11*, 312–313.
39. Carpitella, S.; Carpitella, F.; Certa, A.; Benítez, J.; Izquierdo, J. Managing Human Factors to Reduce Organisational Risk in Industry. *Math. Comput. Appl.* **2018**, *23*, 67. [CrossRef]

40. Zhu, Q.; Sarkis, J. Relationship between operational practices and performances among early adopters of Green Supply Chain Management practices in Chinese manufacturing enterprises. *J. Oper. Manag.* **2004**, *22*, 265–289. [CrossRef]

41. Chung, S.Y.; Chin, S.; JingHaider, J.J.; Peter, B. The effect of green supply chain management on green performance and firm competitiveness in the context of container shipping in Taiwan. *Transp. Res. Part E Logist. Transp. Rev.* **2013**, *55*, 55–73.

42. Zhu, Q.; Sarkis, J.; Cordeiro, J.J.; Lai, K.H. Firm-level correlates of green supply chain management practices in Chinese context. *Int. J. Manag. Sci.* **2008**, *36*, 577–591. [CrossRef]

43. Chung, C.J.; Wee, H.M. Green component life cycle values on designed and reverse manufacturing in semi-closed supply chain. *Int. J. Prod. Econ.* **2008**, *113*, 528–545. [CrossRef]

44. Yeh, W.C.; Chuang, M.C. Using multi objective genetic algorithm for partner selection in green supply chain problem. *J. Expert Syst. Appl.* **2011**, *38*, 4244–4253. [CrossRef]

45. Saif, A.; Elhedhli, S. Cold supply chain design with environmental considerations: A simulation-optimization approach. *Eur. J. Oper. Res.* **2016**, *251*, 274–287. [CrossRef]

46. Bhateja, A.K.; Babbar, R.; Singh, S.; Sachdeva, A. Study of green supply chain management in the Indian Manufacturing industry: A literature review cum an Analytical Approach for the measurement of performanc. *Int. J. Comput. Eng. Manag.* **2011**, *13*, 84–99.

47. Toke, L.K.; Gupta, R.C.; Dandekar, M. An empirical study of green supply chain management in Indian perspective. *Int. J. Appl. Sci. Eng. Res.* **2012**, *2*, 372–383.

48. Sharma, V.; Chandna, P.; Bhardwaj, A. Green supply chain management related performance indicators in Agro industry: A Review. *J. Clean. Prod.* **2017**, *141*, 1194–1208. [CrossRef]

49. Aliahmadi, A.; Sadeghi, M.E.; Nozari, H.; Jafari-Eskandari, M.; Najafi, S.E. Studying Key Factors to Creating Competitive Advantage in Science Park. In Proceedings of the Ninth International Conference on Management Science and Engineering Management, Advances in Intelligent Systems and Computing, Berlin, Germany, 21–23 July 2015; pp. 977–987.

50. Chen, S.J.; Hwang, C.L.; Hwang, F.P. *Fuzzy Multiple Attribute Decision Making Methods and Applications*; Springer: New York, NY, USA, 1992.

51. Zadeh, L. Fuzzy sets. *Inf. Control* **1956**, *8*, 338–353. [CrossRef]

52. Opricovic, S.; Tzeng, G.H. Defuzzification within a multicriteria decision model. *Int. J. Uncertain. Fuzziness Knowl. Based Syst.* **2003**, *11*, 635–652. [CrossRef]

53. Mikhailov, L. Deriving priorities from fuzzy pairwise comparison judgements. *Fuzzy Sets Syst.* **2003**, *134*, 365–385. [CrossRef]

54. Mikhailov, L. A Fuzzy Programming Method for Deriving Priorities in the Analytic Hierarchy Process. *J. Oper. Res. Soc.* **2000**, *51*, 341–349. [CrossRef]

 *mathematics*

*Article*

# Bipolar Fuzzy Relations

**Jeong-Gon Lee** [1,*] **and Kul Hur** [2]

[1]  Division of Applied Mathematics, Nanoscale Science and Technology Institute, Wonkwang University, Iksan 54538, Korea

[2]  Department of Applied Mathematics, Wonkwang University, 460, Iksan-daero, Iksan-Si, Jeonbuk 54538, Korea; kulhur@wku.ac.kr

*  Correspondence: jukolee@wku.ac.kr

Received: 22 September 2019; Accepted: 31 October 2019; Published: 3 November 2019

**Abstract:** We introduce the concepts of a bipolar fuzzy reflexive, symmetric, and transitive relation. We study bipolar fuzzy analogues of many results concerning relationships between ordinary reflexive, symmetric, and transitive relations. Next, we define the concepts of a bipolar fuzzy equivalence class and a bipolar fuzzy partition, and we prove that the set of all bipolar fuzzy equivalence classes is a bipolar fuzzy partition and that the bipolar fuzzy equivalence relation is induced by a bipolar fuzzy partition. Finally, we define an $(a, b)$-level set of a bipolar fuzzy relation and investigate some relationships between bipolar fuzzy relations and their $(a, b)$-level sets.

**Keywords:** bipolar fuzzy relation; bipolar fuzzy reflexive (resp., symmetric and transitive) relation; bipolar fuzzy equivalence relation; bipolar fuzzy partition; $(a, b)$-level set

---

## 1. Introduction

In 1965, Zadeh [1] introduced the concept of a fuzzy set as the generalization of a crisp set. In 1971, he [2] defined the notions of similarity relations and fuzzy orderings as the generalizations of crisp equivalence relations and partial orderings playing basic roles in many fields of pure and applied science. After that time, many researchers [3–11] studied fuzzy relations. Dib and Youssef [7] defined the fuzzy Cartesian product of two ordinary sets $X$ and $Y$ as the collection of all $L$-fuzzy sets of $X \times Y$, where $L = I \times I$ and $I$ denotes the unit closed interval. In particular, Lee [10] obtained many results by using the notion of fuzzy relations introduced by Dib and Youssef.

In 1994, Zhang [12] introduced the notion of a bipolar fuzzy set (refer to [13–15]). After then, Jun and Park [16], Jun et al. [17], and Lee [18] applied bipolar fuzzy sets to $BCK/BCI$-algebras. Moreover, Akram and Dudek [19] studied bipolar fuzzy graph, and Majumder [20] introduced the bipolar fuzzy $\Gamma$-semigroup. Moreover, Talebi et al. [21] investigated operations on a bipolar fuzzy graph. Azhagappan and Kamaraj [22] dealt with some properties of rw-closed sets and rw-open sets in bipolar fuzzy topological spaces. Recently, Kim et al. [23] constructed the category consisting of bipolar fuzzy set and preserving mappings between them and studied this in the sense of a topological universe. Lee et al. [24] found some properties of bases, neighborhoods, and continuities in bipolar fuzzy topological spaces. In particular, Dudziak and Pekala [25] referred to an intuitionistic fuzzy relation as a bipolar fuzzy relation and investigated some properties of equivalent bipolar fuzzy relations.

In this paper, first, we introduce a bipolar fuzzy relation from a set $X$ to $Y$ and the composition of two bipolar fuzzy relations. Furthermore, we introduce some operations between bipolar fuzzy relations and obtain some of their properties. Second, we define a bipolar fuzzy reflexive, symmetric, and transitive relation and find bipolar fuzzy analogues of many results concerning relationships between ordinary reflexive, symmetric, and transitive relations. Third, we define the concepts of a bipolar fuzzy equivalence class and a bipolar fuzzy partition, and we prove that the set of all bipolar fuzzy equivalence classes is a bipolar fuzzy partition and that the bipolar fuzzy equivalence relation is induced by a bipolar fuzzy partition. Finally, we define an $(a, b)$-level set of a bipolar fuzzy relation and investigate some relationships between bipolar fuzzy relations and their $(a, b)$-level sets.

## 2. Preliminaries

In this section, we introduce the concept of the bipolar fuzzy set, the complement of a bipolar fuzzy set, the inclusion between two bipolar fuzzy sets, and the union and the intersection of two bipolar fuzzy sets. Furthermore, we introduce the intersection and union of arbitrary bipolar fuzzy sets and list some properties.

**Definition 1.** *([13]). Let $X$ be a nonempty set. Then, a pair $A = (A^-, A^+)$ is called a bipolar-valued fuzzy set (or bipolar fuzzy set) in $X$, if $A^+ : X \to [0,1]$ and $A^- : X \to [-1,0]$ are mappings.*

*In particular, the empty fuzzy empty set (resp. the bipolar fuzzy whole set) (see [22]), denoted by $0_{bp} = (0_{bp}^-, 0_{bp}^+)$ (resp. $1_{bp} = (1_{bp}^-, 1_{bp}^+)$), is a bipolar fuzzy set in $X$ defined by: for each $x \in X$,*

$$0_{bp}^+(x) = 0 = 0_{bp}^-(x) \ (resp. \ 1_{bp}^+(x) = 1 \ and \ 1_{bp}^-(x) = -1).$$

*We will denote the set of all bipolar fuzzy sets in $X$ as $BPF(X)$.*

*For each $x \in X$, we use the positive membership degree $A^+(x)$ to denote the satisfaction degree of the element $x$ to the property corresponding to the bipolar fuzzy set $A$ and the negative membership degree $A^-(x)$ to denote the satisfaction degree of the element $x$ to some implicit counter-property corresponding to the bipolar fuzzy set $A$.*

*If $A^+(x) \neq 0$ and $A^-(x) = 0$, then it is the situation that $x$ is regarded as having only positive satisfaction for $A$. If $A^+(x) = 0$ and $A^-(x) \neq 0$, then it is the situation that $x$ does not satisfy the property of $A$, but somewhat satisfies the counter-property of $A$. It is possible for some $x \in X$ to be such that $A^+(x) \neq 0$ and $A^-(x) \neq 0$ when the membership function of the property overlaps that of its counter-property over some portion of $X$.*

*It is obvious that for each $A \in BPF(X)$ and $x \in X$, if $0 \leq A^+(x) - A^-(x) \leq 1$, then $A$ is an intuitionistic fuzzy set introduced by Atanassov [26]. In fact, $A^+(x)$ (resp. $-A^-(x)$) denotes the membership degree (resp. non-membership degree) of $x$ to $A$.*

**Definition 2.** *([13]). Let $X$ be a nonempty set, and let $A, B \in BPF(X)$.*

(i)   *We say that $A$ is a subset of $B$, denoted by $A \subset B$, if for each $x \in X$,*

$$A^+(x) \leq B^+(x) \ and \ A^-(x) \geq B^-(x).$$

(ii)  *The complement of $A$, denoted by $A^c = ((A^c)^-, (A^c)^+)$, is a bipolar fuzzy set in $X$ defined as: for each $x \in X$, $A^c(x) = (-1 - A^+(x), 1 - A^+(x))$, i.e.,*

$$(A^c)^+(x) = 1 - A^-(x), \ (A^c)^-(x) = -1 - A^-(x).$$

(iii) *The intersection of $A$ and $B$, denoted by $A \cap B$, is a bipolar fuzzy set in $X$ defined as: for each $x \in X$,*

$$(A \cap B)(x) = (A^-(x) \vee B^-(x), A^+(x) \wedge B^+(x)).$$

(iv)  *The union of $A$ and $B$, denoted by $A \cup B$, is a bipolar fuzzy set in $X$ defined as: for each $x \in X$,*

$$(A \cup B)(x) = (A^-(x) \wedge B^-(x), A^+(x) \vee B^+(x)).$$

**Definition 3.** *(see [13,22]). Let $X$ be a nonempty set, and let $A, B \in BPF(X)$. We say that $A$ is equal to $B$, denoted by $A = B$, if $A \subset B$ and $B \subset A$.*

**Result 1.** *([23], Proposition 3.5). Let $A$, $B$, $C \in BPF(X)$. Then:*

(1)  (Idempotent laws): $A \cup A = A$, $A \cap A = A$,
(2)  (Commutative laws): $A \cup B = B \cup A$, $A \cap B = B \cap A$,
(3)  (Associative laws): $A \cup (B \cup C) = (A \cup B) \cup C$, $A \cap (B \cap C) = (A \cap B) \cap C$,
(4)  (Distributive laws): $A \cup (B \cap C) = (A \cup B) \cap (A \cup C)$,
$$A \cap (B \cup C) = (A \cap B) \cup (A \cap C),$$
(5)  (Absorption laws): $A \cup (A \cap B) = A$, $A \cap (A \cup B) = A$.
(6)  (De Morgan's laws): $(A \cup B)^c = A^c \cap B^c$, $(A \cap B)^c = A^c \cup B^c$,
(7)  $(A^c)^c = A$,
(8)  $A \cap B \subset A$ and $A \cap B \subset B$,
(9)  $A \subset A \cup B$ and $B \subset A \cup B$,
(10)  *if $A \subset B$ and $B \subset C$, then $A \subset C$,*
(11)  *if $A \subset B$, then $A \cap C \subset B \cap C$ and $A \cup C \subset B \cup C$.*

**Definition 4.** *([23]). Let $X$ be a nonempty set, and let $(A_j)_{j \in J} \subset BPF(X)$.*

(i)  *The intersection of $(A_j)_{j \in J}$, denoted by $\bigcap_{j \in J} A_j$, is a bipolar fuzzy set in $X$ defined by: for each $x \in X$,*

$$\left(\bigcap_{j \in J} A_j\right)(x) = \left(\bigvee_{j \in J} A_j^-(x), \bigwedge_{j \in J} A_j^+(x)\right).$$

(ii)  *The union of $(A_j)_{j \in J}$, denoted by $\bigcup_{j \in J} A_j$, is a bipolar fuzzy set in $X$ defined by: for each $x \in X$,*

$$\left(\bigcup_{j \in J} A_j\right)(x) = \left(\bigwedge_{j \in J} A_j^-(x), \bigvee_{j \in J} A_j^+(x)\right).$$

**Result 2.** *([23], Proposition 3.8). Let $A \in BPF(X)$, and let $(A_j)_{j \in J} \subset BPF(X)$. Then:*

(1)  (Generalized distributive laws): $A \cup (\bigcap_{j \in J} A_j) = \bigcap_{j \in J}(A \cup A_j)$,
$$A \cap (\bigcup_{j \in J} A_j) = \bigcup_{j \in J}(A \cap A_j),$$
(2)  (Generalized De Morgan's laws): $(\bigcup_{j \in J} A_j)^c = \bigcap_{j \in J} A_j^c$, $(\bigcap_{j \in J} A_j)^c = \bigcup_{j \in J} A_j^c$.

From Results 1 and 2, it is obvious that $(BPF(X), \cup, \cap, {}^c, 0_{bp}, 1_{bp})$ is a complete distributive lattice satisfying De Morgan's laws.

## 3. Bipolar Fuzzy Relations

In this section, we introduce the concepts of the bipolar fuzzy relation, the composition of two bipolar fuzzy relations, and the inverse of a bipolar fuzzy relation and study some their properties.

Throughout this paper, $X, Y, Z$ denote ordinary non-empty sets, and we define the union, the intersection, and the composition between bipolar fuzzy relations by using only the min-max operator.

**Definition 5.** *$R = (R^-, R^+)$ is called a bipolar fuzzy relation (BPFR) from $X$ to $Y$, if $R^+ : X \times Y \to [0,1]$ and $R^- : X \times Y \to [-1,0]$ are mappings, i.e.,*

$$R \in BPF(X \times Y).$$

*In particular, a BPFR from from $X$ to $X$ is called a BPFR on $X$ (see [21]).*

*The empty BPFR (resp. the whole BPFR) on X, denoted by $R_0 = (R_0^-, R_0^+)$ (resp. $R_1 = (R_1^-, R_1^+)$), is defined as follows: for each $(x,y) \in X \times X$,*

$$R_0^+(x,y) = 0 = R_0^-(x,y) \text{ [resp. } R_1^+(x,y) = 1 \text{ and } R_1^-(x,y) = -1].$$

*We will denote the set of all BPFRs on X (resp. from X to Y) as $BPFR(X)$ (resp. $BPFR(X \times Y)$).*

It is obvious that if $R = (R^-, R^+) \in BPFR(X)$, then $-R^-$ and $R^+$ are fuzzy relations on X, where $(-R^{-1})(x,y) = -R^-(x,y)$ for each $(x,y) \in X \times X$.

**Definition 6.** *Let $R \in BPFR(X \times Y)$. Then:*

(i) *the inverse of R, denoted by $R^{-1} = ((R^{-1})^-, (R^{-1})^+)$, is a BPFR from Y to X defined as follows: for each $(y,x) \in Y \times X$, $R^{-1}(x,y) = R(y,x)$, i.e.,*

$$(R^{-1})^+(y,x) = R^+(x,y), \ (R^{-1})^-(y,x) = R^-(x,y).$$

(ii) *the complement of R, denoted by $R^c = ((R^c)^-, (R^c)^+)$, is a BPFR from X to Y defined as follows: for each $(x,y) \in X \times Y$,*

$$(R^c)^+(x,y) = 1 - R^+(x,y), \ (R^c)^-(x,y) = -1 - R^+(x,y).$$

**Proposition 1.** *Let $R$, $S$, $P \in BPFR(X \times Y)$. Then:*

(1) $R_0 \subset R \subset R_1$,
(2) $(R^c)^{-1} = (R^{-1})^c$,
(3) $(R^{-1})^{-1} = R$,
(4) $R \subset R \cup S$ *and* $S \subset R \cup S$,
(5) $R \cap S \subset R$ *and* $R \cap S \subset S$,
(6) *if* $R \subset S$, *then* $R^{-1} \subset S^{-1}$,
(7) *if* $R \subset P$ *and* $S \subset P$, *then* $R \cup S \subset P$,
(8) *if* $P \subset R$ *and* $P \subset S$, *then* $P \subset R \cap S$,
(9) *if* $R \subset S$, *then* $R \cup S = S$ *and* $R \cap S = R$,
(10) $(R \cup S)^{-1} = R^{-1} \cup S^{-1}$, $(R \cap S)^{-1} = R^{-1} \cap S^{-1}$,

**Proof.** (1) The proof is obvious.
(2) Let $(x,y) \in X \times Y$. Then

$$[(R^c)^{-1}]^-(x,y) = (R^c)^-(y,x) = -1 - R^-(y,x)$$
$$= -1 - (R^{-1})^-(x,y)$$
$$= [(R^{-1})^c]^-(x,y).$$

Similarly, we have $[(R^c)^{-1}]^+(x,y) = [(R^{-1})^c]^+(x,y)$. Thus, the result holds.
(3) The proof is easy by Definition 6.
(4) Let $(x,y) \in X \times Y$. Then,

$$(R \cup S)^-(x,y) = R^-(x,y) \wedge S^-(x,y) \leq R^-(x,y)$$

and

$$(R \cup S)^+(x,y) = R^-(x,y) \vee S^-(x,y) \geq R^+(x,y).$$

Thus, $R \subset R \cup S$. Similarly, we have $S \subset R \cup S$.
The remainder can be proven from Definitions 2, 3, and 6. $\quad\square$

The following is the similar results of Results 1 and 2.

**Proposition 2.** *Let* $R, S, P \in BPFR(X \times Y)$, *and let* $(R_j)_{j \in J} \subset BPFR(X \times Y)$. *Then:*

(1) (Idempotent laws): $R \cup R = R$, $R \cap R = R$,

(2) (Commutative laws): $R \cup S = S \cup R$, $R \cap S = S \cap R$,

(3) (Associative laws): $R \cup (S \cup P) = (R \cup S) \cup P$, $R \cap (S \cap P) = (R \cap S) \cap P$,

(4) (Distributive laws): $R \cup (S \cap P) = (R \cup S) \cap (R \cup P)$, $R \cap (S \cup P) = (R \cap S) \cup (R \cap P)$,

(4)$'$ (Generalized distributive laws): $R \cup (\bigcap_{j \in J} R_j) = \bigcap_{j \in J} (R \cup R_j)$, $R \cap (\bigcup_{j \in J} R_j) = \bigcup_{j \in J} (R \cap R_j)$,

(5) (Absorption laws): $R \cup (R \cap S) = R$, $R \cap (R \cup S) = R$,

(6) (De Morgan's laws): $(R \cup S)^c = R^c \cap S^c$, $(R \cap S)^c = R^c \cup S^c$,

(6)$'$ (Generalized De Morgan's laws): $(\bigcup_{j \in J} R_j)^c = \bigcap_{j \in J} R_j^c$, $(\bigcap_{j \in J} R_j)^c = \bigcup_{j \in J} R_j^c$.

(7) (Involution): $(R^c)^c = R$.

**Example 1.** *Let* $X = \{a, b, c\}$, *and let R be a BPFR on X given by the following Table* 1.

**Table 1.** $R$.

|   | a | b | c |
|---|---|---|---|
| a | $(-0.4, 0.6)$ | $(-0.7, 0.5)$ | $(-1, 0.8)$ |
| b | $(-0.3, 0.8)$ | $(-0.6, 0.3)$ | $(-0.2, 0.6)$ |
| c | $(-0.5, 0.7)$ | $(-0.8, 0.3)$ | $(-0.6, 0.3)$ |

*Then, the inverse and the complement of R are given as below Tables* 2–5.

**Table 2.** $R^{-1}$.

|   | a | b | c |
|---|---|---|---|
| a | $(-0.4, 0.6)$ | $(-0.3, 0.8)$ | $(-0.5, 0.7)$ |
| b | $(-0.7, 0.5)$ | $(-0.6, 0.3)$ | $(-0.8, 0.3)$ |
| c | $(-1, 0.8)$ | $(-0.2, 0.6)$ | $(-0.6, 0.3)$ |

**Table 3.** $R^c$.

|   | a | b | c |
|---|---|---|---|
| a | $(-0.6, 0.4)$ | $(-0.3, 0.5)$ | $(0, 0.2)$ |
| b | $(-0.7, 0.2)$ | $(-0.4, 0.7)$ | $(-0.8, 0.4)$ |
| c | $(-0.5, 0.3)$ | $(-0.2, 0.7)$ | $(-0.4, 0.7)$ |

**Table 4.** $R \cap R^c$.

|   | a | b | c |
|---|---|---|---|
| a | $(-0.4, 0.4)$ | $(-0.3, 0.5)$ | $(0, 0.2)$ |
| b | $(-0.3, 0.2)$ | $(-0.4, 0.3)$ | $(-0.2, 0.4)$ |
| c | $(-0.5, 0.3)$ | $(-0.2, 0.3)$ | $(-0.4, 0.3)$ |

**Table 5.** $R \cup R^c$.

|   | a | b | c |
|---|---|---|---|
| a | $(-0.6, 0.6)$ | $(-0.7, 0.5)$ | $(-1, 0.8)$ |
| b | $(-0.7, 0.8)$ | $(-0.6, 0.7)$ | $(-0.8, 0.6)$ |
| c | $(-0.5, 0.7)$ | $(-0.8, 0.7)$ | $(-0.6, 0.7)$ |

**Remark 1.** *For each $R \in BPFR(X)$, $R \cap R^c = R_0$ and $R \cup R^c = R_1$ do not hold, in general.*

Consider the BPFR in Example 1. Then, $R \cap R^c \neq R_0$ and $R \cup R^c \neq R_1$.

**Definition 7.** *Let $R \in BPFR(X \times Y)$, and let $S \in BPFR(Y \times Z)$. Then, the composition of $R$ and $S$, denoted by $S \circ R = ((S \circ R)^+, (S \circ R)^-)$, is a BPFR from $X$ to $Z$ defined as: for each $(x, z) \in X \times Z$,*

$$(S \circ R)^+(x, z) = (S^+ \circ R^+)(x, z) = \bigvee_{y \in Y} [R^+(x, y) \wedge S^+(y, z)],$$

$$(S \circ R)^-(x, z) = (S^- \circ R^-)(x, z) = \bigwedge_{y \in Y} [R^-(x, y) \vee S^-(y, z)].$$

We can easily see that $(-S^- \circ -R^-)(x, y) = \bigvee_{y \in Y}[-R^+(x, y) \wedge -S^-(x, y)]$.

**Proposition 3.**

(1) $P \circ (S \circ R) = (P \circ S) \circ R)$, where $R \in BPFR(X \times Y)$, $S \in BPFR(Y \times Z)$, and $P \in BPFR(Z \times W)$.
(2) $P \circ (R \cup S) = (P \circ R) \cup (P \circ S)$, $P \circ (R \cap S) = (P \circ R) \cap (P \circ S)$, where $R, S \in BPFR(X \times Y)$ and $P \in BPFR(Y \times Z)$.
(3) If $R \subset S$, then $P \circ R \subset P \circ S$, where $R, S \in BPFR(X \times Y)$ and $P \in BPFR(Y \times Z)$.
(4) $(S \circ R)^{-1} = R^{-1} \circ S^{-1}$, where $R \in BPFR(X \times Y)$ and $S \in BPFR(Y \times Z)$.

**Proof.**

(1) The proof is straightforward.
(2) The proof is straightforward.
(3) Let $R, S \in BPFR(X \times Y)$ and $P \in BPFR(Y \times Z)$. Suppose $R \subset S$, and let $(x, z) \in X \times Z$. Then

$$(P \circ R)^-(x, z) = \bigwedge_{y \in Y}[R^-(x, y) \vee P^-(y, z)]$$

$$\geq \bigwedge_{y \in Y}[S^-(x, y) \vee P^-(y, z)]$$

$$[\text{Since } R \subset S, \, R^-(x, y) \geq S^-(x, y)]$$

$$= (P \circ S)^-(x, z).$$

Similarly, we can prove that $(P \circ R)^+(x, z) \leq (P \circ S)^+(x, z)$. Furthermore, the proof of the second part is similar to the first part. Thus, the result holds.
(4) Let $R \in BPFR(X \times Y)$ and $S \in BPFR(Y \times Z)$, and let $(x, z) \in X \times Z$. Then

$$[(S \circ R)^{-1}]^-(z, x) = (S \circ R)(x, z)$$

$$= \bigwedge_{y \in Y}[R^-(x, y) \vee S^-(y, z)]$$

$$= \bigwedge_{y \in Y}[(S^{-1})^-(z, y) \vee (R^{-1})^-(y, x)]$$

$$= (R^{-1} \circ S^{-1})^-(z, x).$$

Similarly, we can see that $[(S \circ R)^{-1}]^+(z, x) = (R^{-1} \circ S^{-1})^+(z, x)$. Thus, the result holds. $\square$

**Remark 2.** *For any BPFRs $R$ and $S$, $S \circ R \neq R \circ S$, in general.*

Let $IFR(X)$ be the set of all intuitionistic fuzzy relations on a set $X$ introduced by Bustince and Burillo [27]. Then, we have the following result.

**Proposition 4.** *Let* $BPFR_b(X) = \{R \in BPFR(X) : R^+(x) - R^-(x) \le 1, \ \forall \ (x,y) \in X \times X\} \cup \{R_{0_b}, R_{1_b}\}$, *where* $R_{0_b}$ *(resp.* $R_{1_b}$*) denotes the bipolar fuzzy empty (resp. whole) relation on* $BPF_b(X)$ *defined by: for each* $x \in X$,

$$R_{0_b}(x) = (-1,0) \ (resp. \ R_{1_b} = (0,1)).$$

*We define two mappings* $f : BPFR_b(X) \to IFR(X)$ *and* $g : IFR(X) \to BPFR_b(X)$ *as follows, respectively:*

$$[f(R)](x,y) = (R^+(x,y), -R^-(x,y)), \ \forall R \in BPFR_b(X), \ \forall (x,y)x \in X \times X,$$

$$[g(S)](x,y) = (-\nu_S(x,y), \mu_S(x,y)), \ \forall S \in IFR(X), \ \forall (x,y) \in X \times X.$$

*Then* $g \circ f = 1_{BPFR_b(X)}, \ f \circ g = 1_{IFR(X)}.$

**Proof.** The proof is similar to Proposition 3.14 in [23]. $\square$

Let $IVR(X)$ be the set of all interval-valued fuzzy relations on a set $X$ (see [28]). Then, we have the following result.

**Proposition 5.** *We define two mappings* $f : IVR(X) \to IFR(X)$ *and* $g : IFR(X) \to IVR(X)$ *as follows, respectively:*

$$[f(R)](x,y) = (R^L(x,y), 1 - R^U(x,y)), \ \forall R \in IVR(X), \ \forall (x,y) \in X \times X,$$

$$[g(S)](x,y) = [\mu_S(x), 1 - \nu_S(x)], \ \forall S \in IFS(X), \ \forall (x,y) \in X \times X.$$

*Then* $g \circ f = 1_{IVR(X)}, \ f \circ g = 1_{IFR(X)}.$

**Proof.** The proof is similar to Lemma 1 in [29]. $\square$

From Propositions 4 and 5, we have the following.

**Corollary 1.** *We define two mappings* $f : BPFR_b(X) \to IVR(X)$ *and* $g : IVR(X) \to BPFR_b(X)$ *as follows, respectively:*

$$[f(R)](x,y) = [R^+(x,y), 1 + R^-(x,y)], \ \forall R \in BPFR_b(X), \ \forall (x,y) \in X \times X,$$

$$[g(S)](x,y) = (-1 + S^U(x,y), S^L(x,y)), \ \forall S \in IVR(X), \ \forall (x,y) \in X \times X.$$

*Then,* $g \circ f = 1_{BPFR_b(X)}, \ f \circ g = 1_{IVR(X)}.$

## 4. Bipolar Fuzzy Reflexive, Symmetric, and Transitive Relations

In this section, we introduce bipolar fuzzy reflexive, symmetric, and transitive relations and obtain some properties related to them.

**Definition 8.** *The bipolar fuzzy identity relation on* $X$, *denoted by* $I_X$ *(simply,* $I$*), is a BPFR on* $X$ *defined as: for each* $(x,y) \in X \times X$,

$$I_X^-(x,y) = \begin{cases} -1 & \text{if} \ \ x = y \\ 0 & \text{if} \ \ x \ne y, \end{cases} \qquad I_X^+(x,y) = \begin{cases} 1 & \text{if} \ \ x = y \\ 0 & \text{if} \ \ x \ne y. \end{cases}$$

It is clear that $I = I^{-1}$ and $I^c = (I^c)^{-1}$. Moreover, it is obvious that if $I_X$ is the bipolar fuzzy identity relation on $X$, then $-I_X^{-1}$ and $I_X^+$ are fuzzy identity relations on $X$.

**Definition 9.** $R \in BPFR(X)$ *is said to be:*

(i)  *reflexive, if for each $x \in X$, $R(x,x) = (-1,1)$, i.e.,*

$$R^-(x,x) = -1, \; R^+(x,x) = 1,$$

(ii)  *anti-reflexive, if for each $x \in X$, $R(x,x) = (0,0)$.*

From Definitions 8 and 9, it is obvious that $R$ is bipolar fuzzy reflexive if and only if $I \subset R$. The following is the immediate results of the above definition.

It is clear that $R = (R^{-1}, R^+)$ is a bipolar fuzzy reflexive (resp. anti-reflexive) relation on $X$, then $-R^-$ and $R^+$ are fuzzy reflexive (resp. anti-reflexive) relations on $X$. Thus, $R$ and $S$ are fuzzy reflexive (resp. anti-reflexive) relations on $X$ iff $(-R,S)$ or $(-S,R)$ are bipolar fuzzy reflexive (resp. anti-reflexive) relations on $X$.

**Proposition 6.** *Let $R \in BPFR(X)$.*

(1)  *$R$ is reflexive if and only if $R^{-1}$ is reflexive.*
(2)  *If $R$ is reflexive, then $R \cup S$ is reflexive, for each $S \in BPFR(X)$.*
(3)  *If $R$ is reflexive, then $R \cap S$ is reflexive if and only if $S \in BPFR(X)$ is reflexive.*

The following is the immediate result of Definitions 2, 3, 6, and 9.

**Proposition 7.** *Let $R \in BPFR(X)$.*

(1)  *$R$ is anti-reflexive if and only $R^{-1}$ is anti-reflexive.*
(2)  *If $R$ is anti-reflexive, then $R \cup S$ is anti-reflexive if and only if $S \in BPFR(X)$ is anti-reflexive.*
(3)  *If $R$ is anti-reflexive, then $R \cap S$ is anti-reflexive, for each $S \in BPFR(X)$.*

**Proposition 8.** *Let $R, \; S \in BPFR(X)$. If $R$ and $S$ are reflexive, then $S \circ R$ is reflexive.*

**Proof.** Let $x \in X$. Since $R$ and $S$ are reflexive,

$$R^-(x,x) = -1, \; R^+(x,x) = 1 \text{ and } S^-(x,x) = -1, \; S^+(x,x) = 1.$$

Thus

$$
\begin{aligned}
(S \circ R)^-(x,x) &= \textstyle\bigwedge_{y \in X}[R^-(x,y) \vee S^-(y,x)] \\
&= [\textstyle\bigwedge_{x \neq y \in X}(R^-(x,y) \vee S^-(y,x))] \wedge [R^-(x,x) \vee S^-(x,x)] \\
&= [\textstyle\bigwedge_{x \neq y \in X}(R^-(x,y) \vee S^-(y,x))] \wedge (-1 \wedge -1) \\
&= -1,
\end{aligned}
$$

$$
\begin{aligned}
(S \circ R)^+(x,x) &= \textstyle\bigvee_{y \in X}[R^+(x,y) \wedge S^+(y,x)] \\
&= [\textstyle\bigvee_{x \neq y \in X}(R^+ \wedge S^+(y,x))] \vee [R^+(x,x) \wedge S^+(x,x)] \\
&= [\textstyle\bigvee_{x \neq y \in X}(T_R(x,y) \wedge T_S(y,x))] \vee (1 \wedge 1) \\
&= 1.
\end{aligned}
$$

Therefore, $S \circ R$ is reflexive.  $\square$

**Definition 10.** *Let $R \in BPFR(X)$. Then:*

(i)  *$R$ is said to be symmetric, if for each $x,y \in X$,*

$$R(x,y) = R(y,x), \text{ i.e., } R^-(x,y) = R^-(y,x) \text{ and } R^+(x,y) = R^+(y,x),$$

(ii)  *$R$ is said to be anti-symmetric, if for each $(x,y) \in X \times X$ with $x \neq y$,*

$$R(x,y) \neq R(y,x), \text{ i.e., } R^-(x,y) \neq R^-(y,x) \text{ and } R^+(x,y) \neq R^+(y,x).$$

From Definitions 9 and 10, it is obvious that $R_0$ is a symmetric and anti-reflexive BPFR, $R_1$ and $I$ are symmetric and reflexive BPFRs, and $I^c$ is an anti-reflexive BPFR.

The following is the immediate result of Definitions 6 and 10.

**Proposition 9.** *Let $R \in BPFR(X)$. Then, $R$ is symmetric iff $R = R^{-1}$.*

**Proposition 10.** *Let $R, S \in BPFR(X)$. If $R$ and $S$ are symmetric, then $R \cup S$ and $R \cap S$ are symmetric.*

**Proof.** Let $(x, y) \in X \times X$. Then, since $R$ and $S$ and are symmetric,

$$(R \cup S)^-(x, y) = R^-(x, y) \wedge S^-(x, y) = R^-(y, x) \wedge S^-(y, x) = (R \cup S)^-(y, x)$$

and:

$$(R \cup S)^+(x, y) = R^+(x, y) \vee S^+(x, y) = R^+(y, x) \wedge S^+(y, x) = (R \cup S)^+(y, x).$$

Thus, $R \cup S$ is symmetric.

Similarly, we can prove that $R \cap S$ is symmetric. $\quad\square$

**Remark 3.** *$R$ and $S$ are symmetric, but $S \circ R$ is not symmetric, in general.*

**Example 2.** *Let $X = \{a, b, c\}$, and consider two BPFRs $R$ and $S$ on $X$ given by the following Tables 6 and 7.*

**Table 6.** *R.*

|   | a | b | c |
|---|---|---|---|
| a | $(-0.4, 0.5)$ | $(-0.3, 0.6)$ | $(-0.6, 0.3)$ |
| b | $(-0.3, 0.6)$ | $(-0.6, 0.7)$ | $(-0.2, 0.4)$ |
| c | $(-0.6, 0.3)$ | $(-0.2, 0.4)$ | $(-0.4, 0.7)$ |

**Table 7.** *S.*

|   | a | b | c |
|---|---|---|---|
| a | $(-0.6, 0.7)$ | $(-0.7, 0.5)$ | $(-1, 0.8)$ |
| b | $(-0.7, 0.5)$ | $(-0.4, 0.3)$ | $(-0.8, 0.6)$ |
| c | $(-1, 0.8)$ | $(-0.8, 0.6)$ | $(-0.6, 0.7)$ |

*Then, clearly, $R$ and $S$ are symmetric. However:*

$$(S \circ R)^+(a, b) = 0.5 \neq 0.6 = (S \circ R)^+(b, a).$$

*Thus, $S \circ R$ is not symmetric.*

The following gives the condition for its being symmetric.

**Proposition 11.** *Let $R, S \in SVNR(X)$. Let $R$ and $S$ be symmetric. Then, $S \circ R$ is symmetric if and only if $S \circ R = R \circ S$.*

**Proof.** Suppose $S \circ R$ is symmetric. Then
$$S \circ R = S^{-1} \circ R^{-1} \text{ (by Proposition 9)}$$
$$= (R \circ S)^{-1} \text{ (by Proposition 3 (4))}$$
$$= R \circ S \text{ (by the hypothesis).}$$
Conversely, suppose $S \circ R = R \circ S$. Then
$$(S \circ R)^{-1} = R^{-1} \circ S^{-1} \text{ (by Proposition 3 (4))}$$

$$= R \circ S \text{ (since } R \text{ and } S \text{ and are symmetric)}$$
$$= S \circ R \text{ (by the hypothesis).}$$

This completes the proof. $\square$

The following is the immediate result of Proposition 11.

**Corollary 2.** *If $R$ is symmetric, then $R^n$ is symmetric, for all positive integers $n$, where $R^n = R \circ R \circ \ldots$ $n$ times.*

**Definition 11.** *$R \in BPFR(X)$ is said to be transitive, if $R \circ R \subset R$, i.e., $R^2 \subset R$.*

It is clear that if $R = (R^-, R^+)$ is a bipolar fuzzy transitive relation on $X$, then $-R^-$ and $R^+$ are fuzzy transitive relations on $X$. Thus, $R$ and $S$ are fuzzy transitive relations on $X$ iff $(-R, S)$ and $(-S, R)$ are bipolar fuzzy transitive relations on $X$.

**Proposition 12.** *Let $R \in BPFR(X)$. If $R$ is transitive, then $R^{-1}$ is also.*

**Proof.** Let $(x, y) \in X \times X$. Then
$$(R^{-1})^-(x, y) = R^-(y, x)$$
$$\leq (R \circ R)^-(y, x)$$
$$= \wedge_{z \in X}[R^-(y, z) \vee R^-(z, x)]$$
$$= \wedge_{z \in X}[(R^{-1})^-(z, y) \vee (R^{-1})^-(x, z)]$$
$$= \wedge_{z \in X}[(R^{-1})^-(x, z) \vee (R^{-1})^-(z, y)]$$
$$= (R^{-1} \circ R^{-1})^-(x, y).$$

Similarly, we can prove that:

$$(R^{-1})^+(x, y) \geq (R^{-1} \circ R^{-1})^+(x, y).$$

Thus, the result holds. $\square$

**Proposition 13.** *Let $R \in BPFR(X)$. If $R$ is transitive, then so is $R^2$.*

**Proof.** Let $(x, y) \in X \times X$. Then
$$(R^2)^-(x, y) = \wedge_{z \in X}[R^-(x, z) \vee R^-(z, y)]$$
$$\leq \vee_{z \in X}[(R^2)^-(x, z) \vee (R^2)^-(z, y)]$$
$$= [(R^2)^- \circ (R^2)^-](x, y).$$
Similarly, we can see that $(R^2)^-(x, y) \geq [(R^2)^- \circ (R^2)^-](x, y)$. Thus, the result holds. $\square$

**Proposition 14.** *Let $R, S \in BPFR(X)$. If $R$ and $S$ are transitive, then $R \cap S$ is transitive.*

**Proof.** Let $(x, y) \in X \times X$. Then
$$[(R \cap S) \circ (R \cap S)]^-(x, y)$$
$$= \wedge_{z \in X}[(R \cap S)^-(y, z) \vee (R \cap S)^-(z, x)]$$
$$= \wedge_{z \in X}[(R^-(x, z) \vee S^-(x, z)) \vee (R^-(z, y) \vee S^-(z, y)]$$
$$= \wedge_{z \in X}[(R^-(x, z) \vee R^-(z, y)) \vee (S^-(x, z) \vee S^-(z, y))]$$
$$= (\wedge_{z \in X}[R^-(x, z) \wedge R^-(z, y)]) \vee (\wedge_{z \in X}[S^-(x, z) \vee S^-(z, y)])$$
$$= (R \circ R)^-(x, y) \vee (S \circ S)^-(x, y)$$
$$\geq R^-(x, y) \vee S^-(x, y) \text{ (since } R \text{ and } S \text{ are transitive)}$$
$$= (R \cap S)^-(x, y).$$
Similarly, we can prove that:

$$[(R \cap S) \circ (R \cap S)]^+(x, y) \leq R \cap S)^+(x, y).$$

Thus, $R \cap S$ is transitive. $\square$

**Remark 4.** *For two bipolar fuzzy transitive relation $R$ and $S$ in $X$, $R \cup S$ is not transitive, in general.*

**Example 3.** *Let $X = \{a, b, c\}$, and consider two BPFRs $R$ and $S$ in $X$ given by the following Tables 8 and 9.*

**Table 8.** $R$.

|   | $a$ | $b$ | $c$ |
|---|---|---|---|
| $a$ | $(-0.4, 0.5)$ | $(-0.3, 0.6)$ | $(-0.3, 0.7)$ |
| $b$ | $(-0.3, 0.6)$ | $(-0.6, 0.7)$ | $(-0.5, 0.4)$ |
| $c$ | $(-0.3, 0.7)$ | $(-0.5, 0.4)$ | $(-0.6, 0.8)$ |

**Table 9.** $S$.

|   | $a$ | $b$ | $c$ |
|---|---|---|---|
| $a$ | $(-0.3, 0.7)$ | $(-0.2, 0.5)$ | $(-0, 2, 0.6)$ |
| $b$ | $(-0.7, 0.5)$ | $(-1, 0.3)$ | $(-0.2, 0.6)$ |
| $c$ | $(-0.2, 0.6)$ | $(-0.2, 0.6)$ | $(-0.8, 0.7)$ |

*Then, we can easily see that $R$ and $S$ are transitive. Moreover, we have Table 10 as $R \cup S$.*

**Table 10.** $R \cup S$.

|   | $a$ | $b$ | $c$ |
|---|---|---|---|
| $a$ | $(-0.4, 0.7)$ | $(-0.3, 0.6)$ | $(-0.3, 0.7)$ |
| $b$ | $(-0.7, 0.6)$ | $(-1, 0.7)$ | $(-0.5, 0.6)$ |
| $c$ | $(-0.3, 0.7)$ | $(-0.5, 0.6)$ | $(-0.8, 0.8)$ |

*Thus, $[(R \cup S) \circ (R \cup S)]^-(c, a) = -0.5 \not\geq -0.3 = (R \cup S)^-(c, a)$. Therefore, $R \cup S$ is not transitive.*

## 5. Bipolar Fuzzy Equivalence Relation

In this section, we define the concept of a bipolar fuzzy equivalence class and a bipolar fuzzy partition, and we prove that the set of all bipolar fuzzy equivalence classes is a bipolar fuzzy partition and induce the bipolar fuzzy equivalence relation from a bipolar fuzzy partition.

**Definition 12.** *$R \in BPFR(X)$ is called a:*

(i) *tolerance relation on $X$, if it is reflexive and symmetric,*

(ii) *similarity (or equivalence) relation on $X$, if it is reflexive, symmetric, and transitive.*

(iii) *partial order relation on $X$, if it is reflexive, anti-symmetric, and transitive.*

*We will denote the set of all tolerance (resp., equivalence and order) relations on $X$ as $BPFT(X)$ (resp. $BPFE(X)$ and $BPFO(X)$).*

We can easily see that $R = (R^-, R^+)$ is a bipolar fuzzy tolerance (resp. similarity and partial order) relation on $X$, then $-R^-$ and $R^+$ are fuzzy tolerance (resp. similarity and partial order) relations on $X$. Furthermore, $R$ and $S$ are fuzzy tolerance (resp. similarity and partial order) relations on $X$ iff $(-R, S)$ and $(-S, R)$ are bipolar fuzzy tolerance (resp. similarity and partial order) relations on $X$.

The following is the immediate result of Propositions 6, 10, and 14.

**Proposition 15.** *Let $(R_j)_{j \in J} \subset BPFT(X)$ (resp., $BPFE(X)$ and $BPFO(X)$). Then, $\bigcap R_j \in BPFT(X)$ (resp., $BPFE(X)$ and $BPFO(X)$).*

**Proposition 16.** *Let $R \in BPFE(X)$. Then, $R = R \circ R$.*

**Proof.** From Definition 11, it is clear that $R \circ R \subset R$.

Let $(x, y) \in X \times X$. Then
$$
\begin{aligned}
(R \circ R)^-(x, y) &= \bigwedge_{z \in X}[R^-(x, z) \vee R^-(z, y)] \\
&\leq R^-(x, x) \vee R^-(x, y) \\
&= -1 \vee R^-(x, y) \text{ (since } R \text{ is reflexive)} \\
&= R^-(x, y),
\end{aligned}
$$

$$
\begin{aligned}
(R \circ R)^+(x, y) &= \bigvee_{z \in X}[R^+(x, z) \wedge R^+(z, y)] \\
&\geq R^+(x, x) \wedge R^+(x, y) \\
&= 1 \wedge R^-(x, y) \text{ (since } R \text{ is reflexive)} \\
&= R^+(x, y).
\end{aligned}
$$
Thus, $R \circ R \supset R$. Therefore, $R \circ R = R$. $\square$

**Definition 13.** *Let $A \in BPF(X)$. Then, $A$ is said to be normal, if:*

$$
\bigwedge_{x \in X} A^-(x) = -1, \quad \bigvee_{x \in X} A^+(x) = 1.
$$

**Definition 14.** *Let $R \in BPFE(X)$, and let $x \in X$. Then, the bipolar fuzzy equivalence class of $x$ by $R$, denoted by $R_x$, is a BPFS in $X$ defined as:*
$$
R_x = (R_x^-, R_x^+),
$$
*where $R_x^+ : X \to [0, 1]$ and $R_x^- : X \to [-, 0]$ are mappings defined as: for each $y \in X$,*

$$
R_x^+(y) = R^+(x, y) \text{ and } R_x^-(y) = R^-(x, y).
$$

*We will denote the set of all bipolar fuzzy equivalence classes by $R$ as $X / R$, and it will be called the bipolar fuzzy quotient set of $X$ by $R$.*

**Proposition 17.** *Let $R \in BPFE(X)$, and let $x, y \in X$. Then:*

(1) $R_x$ *is normal; in fact, $R_x \neq \mathbf{0}_{bp}$,*
(2) $R_x \cap R_y = \mathbf{0}_{bp}$ *iff $R(x, y) = (0, 0)$,*
(3) $R_x = R_y$ *iff $R(x, y) = (-1, 1)$,*
(4) $\bigcup_{x \in X} R_x = \mathbf{1}_{bp}.$

**Proof.** (1) Since $R$ is reflexive,

$$
R_x^+(x) = R^+(x, x) = 1 \text{ and } R_x^-(x) = R^-(x, x) = -1.
$$

Then, $\bigvee_{y \in X} R_x^+(y) = 1$ and $\bigwedge_{y \in X} R_x^-(y) = -1$. Therefore, $R_x$ is normal. Moreover, $R_x = (-1, 1) \neq (0, 0) = \mathbf{0}_{bp}(x)$. Hence, $R_x \neq \mathbf{0}_{bp}$.

(2) Suppose $R_x \cap R_y = \mathbf{0}_{bp}$, and let $z \in X$. Then
$$
\begin{aligned}
0 &= (R_x \cap R_y)^-(z) \\
&= R_x^-(z) \vee R_y^-(z) \\
&= R^-(x, z) \vee R^-(y, z) \text{ (by Definition 14)} \\
&= R^-(x, z) \vee R^-(z, y) \text{ (since } R \text{ is symmetric)},
\end{aligned}
$$

$$0 = (R_x \cap R_y)^+(z)$$
$$= R_x^+(z) \wedge R_y^+(z)$$
$$= R^+(x,z) \wedge R^+(y,z) \text{ (by Definition 14)}$$
$$= R^-(x,z) \wedge R^-(z,y) \text{ (since } R \text{ is symmetric).}$$

Thus

$$0 = \wedge_{z \in X}[R^-(x,z) \vee R^-(z,y)]$$
$$= (R \circ R)^-(x,y)$$
$$= R^-(x,y) \text{ (by Proposition 16),}$$

$$0 = \vee_{z \in X}[(R^+(x,z) \wedge R^+(z,y)]$$
$$= (R \circ R)^+(x,y)$$
$$= R^+(x,y) \text{ (by Proposition 16).}$$

Therefore, $R(x,y) = (0,0)$.

The sufficient condition is easily proven.

(3) Suppose $R_x = R_y$, and let $z \in X$. Then, $R(x,z) = R(y,z)$. In particular, $R(x,y) = R(y,y)$. Since $R$ is reflexive, $R(x,y) = (-1,1)$.

Conversely, suppose $R(x,y) = (-1,1)$, and let $z \in X$. Since $R$ is transitive, $R \circ R \subset R$. Then:

$$R^-(x,y) \vee R^-(y,z) \geq R^-(x,z), \ R^+(x,y) \wedge R^+(y,z) \leq R^+(x,z).$$

Since $R(x,y) = (-1,1)$, $R^+(x,y) = 1$ and $R^-(x,y) = -1$. Thus:

$$R^-(y,z) \geq R^-(x,z), \ R^+(y,z) \leq R^+(x,z).$$

Therefore, $R_y^-(z) \geq R_x^-(z)$, $R_y^+(z) \leq R_x^+(z)$. Hence, $R_y \subset R_x$. Similarly, we can see that $R_x \subset R_y$. Therefore, $R_x = R_y$.

(4) Let $y \in X$. Then

$$[\cup_{x \in X} R_x](y) = (\vee_{x \in X} R_x^+(y), \wedge_{x \in X} R_x^-(y))$$
$$= (\vee_{x \in X} R^+(x,y), \wedge_{x \in X} R-(x,y))$$
$$= (R^+(y,y), R-(y,y)) = (-1,1)$$
$$= \mathbf{1}_{bp}(y).$$

Thus, the result holds. $\square$

**Definition 15.** *Let $\Sigma = (A_j)_{j \in J} \subset BPF(X)$. Then, $\Sigma$ is called a bipolar fuzzy partition of $X$, if it satisfies the following:*

(i) *$A_j$ is normal, for each $j \in J$,*
(ii) *either $A_j = A_k$ or $A_j \neq A_k$, for any $j,k \in J$,*
(iii) *$\cup_{j \in J} A_j = \mathbf{1}_{bp}$.*

The following is the immediate result of Proposition 17 and Definition 15.

**Corollary 3.** *Let $R \in BPFE(X)$. Then, $X/R$ is a bipolar fuzzy partition of $X$.*

**Proposition 18.** *Let $\Sigma$ be a bipolar fuzzy partition of $X$. We define*
*$R(\Sigma) = (R(\Sigma)^+, R(\Sigma)^-)$ as: for each $(x,y) \in X \times X$,*

$$R(\Sigma)^+(x,y) = \bigvee_{A \in \Sigma} [A^+(x) \wedge A^+(y)], \ R(\Sigma)^-(x,y) = \bigwedge_{A \in \Sigma} [A^-(x) \vee A^-(y)],$$

*where $R(\Sigma)^+ : X \times X \to [0,1]$ and $R(\Sigma)^- : X \times X \to [-1,0]$ are mappings.*
*Then, $R(\Sigma) \in BPFE(X)$.*

**Proof.** Let $x \in X$. Then, by Definition 15 (iii),

$$R(\Sigma)^-(x,x) = \bigwedge_{A \in \Sigma} [A^-(x) \vee A^-(x)] = \bigwedge_{A \in \Sigma} A^-(x) = -1$$

and

$$R(\Sigma)^+(x,x) = \bigvee_{A \in \Sigma} [A^+(x) \wedge A^-(x)] = \bigvee_{A \in \Sigma} A^+(x) = 1.$$

Thus, $R(\Sigma)$ is reflexive.

From the definition of $R(\Sigma)$, it is clear that $R(\Sigma)$ is symmetric.

Let $(x,y) \in X \times X$. Then

$$[R(\Sigma) \circ R(\Sigma)]^-(x,y)$$
$$= \bigwedge_{z \in X}[R(\Sigma)^-(x,z) \vee R(\Sigma)^-(z,y)]$$
$$= \bigwedge_{z \in X}[\bigwedge_{A \in \Sigma}(A^-(x) \vee A^-(z)) \vee \bigwedge_{B \in \Sigma}(B^-(z) \vee B^-(y))]$$
$$= \bigwedge_{z \in X}[(\bigwedge_{A \in \Sigma} A^-(z) \vee \bigwedge_{B \in \Sigma} B^-(z)) \vee (A^-(x) \vee B^-(y))]$$
$$= \bigwedge_{z \in X}[(-1 \vee -1) \vee (A^-(x) \vee B^-(y))] \text{ (since } A \text{ and } B \text{ are normal)}$$
$$= \bigwedge_{z \in X}[A^-(x) \vee B^-(y)]$$
$$= R(\Sigma)^-(x,y).$$

Similarly, we can prove that $[R(\Sigma) \circ R(\Sigma)]^+(x,y) = R(\Sigma)^+(x,y)$. Thus, $R(\Sigma)$ is transitive. Therefore, $R(\Sigma) \in BPFE(X)$.  $\square$

**Proposition 19.** *Let $R, S \in BPFE(X)$. Then, $R \subset S$ iff $R_x \subset S_x$, for each $x \in X$.*

**Proof.** Suppose $R \subset S$, and let $x \in X$. Let $y \in X$. Then, by the hypothesis,

$$R_x^-(y) = R^-(x,y) \geq S^-(x,y) = S_x^-(y),$$

$$R_x^+(y) = R^+(x,y) \leq S^+(x,y) = S_x^+(y).$$

Thus, $R_x \subset S_x$.

The converse can be easily proven.  $\square$

**Proposition 20.** *Let $R, S \in BPFE(X)$. Then, $S \circ R \in BPFE(X)$ iff $S \circ R = R \circ S$.*

**Proof.** Suppose $S \circ R = R \circ S$. Since $R$ and $S$ are reflexive, by Proposition 8, $S \circ R$ is reflexive. Since $R$ and $S$ are symmetric, by the hypothesis and Proposition 11, $S \circ R$ is symmetric. Then, it is sufficient to show that $S \circ R$ is transitive:

$$(S \circ R) \circ (S \circ R) = S \circ (R \circ S) \circ R \text{ (by Proposition 3 (1))}$$
$$= S \circ (S \circ R) \circ R)$$
$$= (S \circ S) \circ (R \circ R)$$
$$\subset S \circ R.$$

Thus, $S \circ R$ is transitive. Therefore, $S \circ R \in BPFE(X)$.

The converse can be easily proven.  $\square$

**Proposition 21.** *Let $R, S \in BPFE(X)$. If $R \cup S = S \circ R$, then $R \cup S \in BPFE(X)$.*

**Proof.** Suppose $R \cup S = S \circ R$. Since $R$ and $S$ are reflexive, by Proposition 6 (2), $R \cup S$ is reflexive. Since $R$ and $S$ are symmetric, by the hypothesis and Proposition 10, $R \cup S$ is symmetric. Then, by the hypothesis, $S \circ R$ is symmetric. Thus, by Proposition 11, $S \circ R = R \circ S$. Therefore, by Proposition 20, $S \circ R \in BPFE(X)$. Hence, $R \cup S \in BPFE(X)$.  $\square$

## 6. Relationships between a Bipolar Fuzzy Relation and Its Level Set

Each member of $[-1,0] \times [0,1]$ will be called a bipolar point. We define the order $\leq$ and the equality $=$ between two bipolar points as follows: for any $(a,b)$, $(c,d) \in [-1,0] \times [0,1]$,

(i)  $(a,b) \leq (c,d)$ iff $a \geq c$ and $b \leq d$,
(ii) $(a,b) = (c,d)$ iff $a = c$ and $b = d$.

**Definition 16.** *Let $R \in BPFR(X \times Y)$, and let $(a,b) \in [-1,0] \times [0,1]$.*

(i)  *The strong $(a,b)$-level subset or strong $(a,b)$-cut of $R$, denoted by $[R]^*_{(a,b)}$, is an ordinary relation from $X$ to $Y$ defined as:*
$$[R]^*_{(a,b)} = \{(x,y) \in X \times Y : R^-(x,y) < a,\ R^+(x,y) > b\}.$$

(ii) *The $(a,b)$-level subset or $(a,b)$-cut of $R$, denoted by $[R]_{(a,b)}$, is an ordinary relation from $X$ to $Y$ defined as:*
$$[R]_{(a,b)} = \{(x,y) \in X \times Y : R^-(x,y) \leq a,\ R^+(x,y) \geq b\}.$$

**Example 4.** *Consider the BPFR $R$ in Example 1. Then*
$$[R]_{(-0.3,0.8)} = \{(x,y) \in X \times X : R^+(x,y) \geq 0.8,\ R^-(x,y) \leq -0.3\} = \{(a,c),(b,a)\},$$
$$[R]^*_{(-0.3,0.8)} = \{(x,y) \in X \times X : R^+(x,y) > 0.8,\ R^-(x,y) < -0.3\} = \phi,$$
$$[R]_{(-0.4,0.5)} = \{(a,a),(a,b),(a,c),(c,a)\},$$
$$[R]^*_{(-0.4,0.5)} = \{(a,c),(c,a)\}.$$

**Proposition 22.** *Let $R$, $S \in BPFR(X \times Y)$, and let $(a,b)$, $(a_1,b_1)$, $(a_2,b_2) \in [-1,0] \times [0,1]$.*

(1) *If $R \subset S$, then $[R]_{(a,b)} \subset [S]_{(a,b)}$ and $[R]^*_{((a,b)} \subset [S]^*_{(a,b)}$.*
(2) *If $(a_1,b_1) \leq (a_2,b_2)$, then $[R]_{(a_2,b_2)} \subset [R]_{(a_1,b_1)}$ and $[R]^*_{(a_2,b_2)} \subset [R]^*_{(a_1,b_1)}$.*

**Proof.** The proofs are straightforward. $\square$

**Proposition 23.** *Let $R \in BPFR(X \times Y)$. Then:*

(1) *$[R]_{(a,b)}$ is an ordinary relation from $X$ to $Y$, for each $(a,b) \in [-1,0] \times [0,1]$,*
(2) *$[R]^*_{(a,b)}$ is an ordinary relation from $X$ to $Y$, for each $(a,b) \in (-1,0] \times [0,1)$,*
(3) *$[R]_{(a,b)} = \bigcap_{(c,d)<(a,b)} [R]_{(c,d)}$, for each $(a,b) \in [-1,0) \times (0,1]$,*
(4) *$[R]^*_{(a,b)} = \bigcup_{(c,d)>(a,b)} [R]^*_{(c,d)}$, for each $(a,b) \in (-1,0] \times [0,1)$.*

**Proof.** The proofs of (1) and (2) are clear from Definition 16.

(3) From Proposition 22, it is obvious that $\{[R]_{(a,b)} : (a,b) \in [-1,0] \times [0,1]\}$ is a descending family of ordinary relations from $X$ to $Y$. Let $(a,b) \in (0,1] \times [-1,0)$ Then, clearly, $[R]_{(a,b)} \subset \bigcap_{(c,d)<(a,b)} [R]_{(c,d)}$. Assume that $(x,y) \notin [R]_{(a,b)}$. Then, $R^+(x,y) < b$ or $R^-(x,y) > a$. Thus, there is $(c,d) \in [-1,0) \times (0,1]$ such that:
$$R^+(x,y) < d < b \text{ or } R^-(x,y) > c > a.$$

Therefore, $(x,y) \notin [R]_{(c,d)}$, i.e., $(x,y) \notin \bigcap_{(c,d)<(a,b)} [R]_{(c,d)}$. Hence:
$$\bigcap_{(c,d)<(a,b)} [R]_{(c,d)} \subset [R]_{(a,b)}.$$

Therefore, the result holds.

(4) Furthermore, from Proposition 22, it is obvious that $\{[R]^*_{(a,b)} : (a,b) \in [-1,0] \times [0,1]\}$ is a descending family of ordinary relations from $X$ to $Y$. Let $(a,b) \in (-1,0] \times [0,1)$. Then, clearly,

$[R]^*_{(a,b)} \subset \bigcup_{(c,d)>(a,b)} [R]^*_{(c,d)}$. Assume that $(x,y) \notin [R]^*_{(a,b)}$. Then, $R^+(x,y) \leq b$ or $R^-(x,y) \geq a$. Thus, there exists $(c,d) \in (-1,0] \times [0,1)$ such that:

$$R^+(x,y) \leq b < c \text{ or } R^-(x,y) \geq a > d.$$

Thus, $(x,y) \notin [R]_{(c,d)}$, i.e., $(x,y) \notin \bigcup_{(c,d)>(a,b)} [R]^*_{(c,d)}$. Therefore,

$$\bigcup_{(c,d)>(a,b)} [R]^*_{(c,d)} \subset [R]^*_{(a,b)}.$$

Hence, $[R]^*_{(a,b)} = \bigcup_{(c,d)>(a,b)} [R]^*_{(c,d)}$.  $\square$

**Definition 17.** *Let $X, Y$ be non-empty sets; let $R$ be an ordinary relation from $X$ to $Y$; and let $R_B \in BPFR(X \times Y)$. Then, $R_B$ is said to be compatible with $R$, if $R = S(R_B)$, where $S(R_B) = \{(x,y) : R_B^+(x,y) > 0, R_B^-(x,y) < 0\}$.*

**Example 5.** *(1) Let $X, Y$ be non-empty sets, and let $\phi_{X \times Y}$ be the ordinary empty relation from $X$ to $Y$. Then, clearly, $S(R_0) = \phi_{X \times Y}$. Thus, $R_0$ is compatible with $\phi_{X \times Y}$.*

*(2) Let $X, Y$ be non-empty sets, and let $X \times Y$ be the whole ordinary relation from $X$ to $Y$. Then, clearly, $S(R_1) = X \times Y$. Thus, $R_1$ is compatible with $X \times Y$.*

From Definitions 9, 10, and 16, it is clear that $R \in BPFR(X)$ is reflexive (resp. symmetric), then $[R]_{(a,b)}$ and $[R]^*_{(a,b)}$ are ordinary reflexive (resp. symmetric) on $X$, for each $(a,b) \in [-1,0] \times [0,1]$.

**Proposition 24.** *Let $R \in BPFR(X)$, and let $(a,b) \in [-1,0] \times [0,1]$. If $R$ is transitive, then $[R]_{(a,b)}$ and $[R]^*_{(a,b)}$ are ordinary transitive on $X$.*

**Proof.** Suppose $R$ is transitive. Then, $R \circ R \subset R$, and let $(x,z) \in [R]_{(a,b)} \circ [R]_{(a,b)}$. Then, there exists $y \in X$ such that $(x,z), (z,y) \in [R]_{(a,b)}$. Thus:

$$R^+(x,z) \geq b, \ R^-(x,z) \leq a \text{ and } R^+(z,y) \geq b, \ R^-(z,y) \leq a.$$

Therefore, $R^+(x,z) \wedge R^+(z,y) \geq b$, $R^-(x,z) \vee R^-(z,y) \leq a$. Since $R \circ R \subset R$,

$$R^+(x,y) \geq R^+(x,z) \wedge R^+(z,y), \ R^-(x,y) \leq R^-(x,z) \vee R^-(z,y).$$

Hence, $R^+(x,y) \geq b$, $R^-(x,y) \leq a$, i.e., $(x,y) \in [R]_{(a,b)}$. Therefore, $[R]_{(a,b)}$ is ordinary transitive. The proof of the second part is similar.  $\square$

The following is the immediate results of Definitions 9, 10, and 16 and Proposition 24.

**Corollary 4.** *Let $R \in BPFE(X)$, and let $(a,b) \in [-1,0] \times [0,1]$. Then, $[R]_{(a,b)}$ and $[R]^*_{(a,b)}$ are the ordinary equivalence relation on $X$*

## 7. Conclusions

This paper dealt with the properties of bipolar fuzzy reflexive, symmetric, and transitive relations and bipolar fuzzy equivalence relations. In particular, we defined a bipolar fuzzy equivalence class of a point in a set $X$ modulo a bipolar fuzzy equivalence relation $R$ and a bipolar fuzzy partition of a set $X$. In addition, we proved that the set of all bipolar fuzzy equivalence classes is a bipolar fuzzy partition and induced the bipolar fuzzy equivalence relation by a bipolar fuzzy partition. Furthermore, we defined the $(a,b)$-level set of a BPFR and investigated some relationships between BPFRs and their $(a,b)$-level set. Then, we could see that bipolar fuzzy relations generalized fuzzy relations.

In the future, we expect that one will study bipolar fuzzy relations on a fixed BPFS $A$ and deal with a decomposition of a mapping $f : X \to Y$ by bipolar fuzzy relations. Furthermore, we think that the bipolar fuzzy relation can be applied to congruences in a semigroup, algebras, topologies, etc.

**Author Contributions:** All authors have contributed equally to this paper in all aspects. This paper was organized by the idea of K.H., J.-G.L. analyzed the related papers with this research, and they also wrote the paper. All the authors have read and approved the final manuscript.

**Funding:** This research was supported by the Basic Science Research Program through the National Research Foundation of Korea (NRF) funded by the Ministry of Education (2018R1D1A1B07049321).

**Acknowledgments:** The authors wish to thank the anonymous reviewers for their valuable comments and suggestions.



## References


1. Zadeh, L.A. Fuzzy sets. *Inf. Control* **1965**, *8*, 338–353. [CrossRef]
2. Zadeh, L.A. Similarity relations and fuzzy orderings. *Inf. Sci.* **1971**, *3*, 177–200. [CrossRef]
3. Chakraborty, M.K.; Das, M. Studies in fuzzy relations over fuzzy subsets. *Fuzzy Sets Syst.* **1983**, *9*, 79–89. [CrossRef]
4. Chakraborty, M.K.; Das, M. On fuzzy equivalence I. *Fuzzy Sets Syst.* **1983**, *11*, 185–193. [CrossRef]
5. Chakraborty, M.K.; Das, M. On fuzzy equivalence II. *Fuzzy Sets Syst.* **1983**, *11*, 299–307. [CrossRef]
6. Chakraborty, M.K.; Sarkar, S. Fuzzy Antisymmetry and Order. *Fuzzy Sets Syst.* **1987**, *21*, 169–182. [CrossRef]
7. Dib, K.A.; Youssef, N.L. Fuzzy Cartesian product, fuzzy relations and fuzzy functions. *Fuzzy Sets Syst.* **1991**, *41*, 299–315. [CrossRef]
8. Gupta, K.C.; Gupta, R.K. Fuzzy Equivalence Relation Redefined. *Fuzzy Sets Syst.* **1996**, *79*, 227–233. [CrossRef]
9. Kaufmann, A. *Introduction to the Theory of Fuzzy Subsets*; Academic Press: New York, NY, USA, 1975; Volume 1.
10. Lee, K.C. Fuzzy equivalence relations and fuzzy functions. *Int. J. Fuzzy Logic Intell. Syst.* **2009**, *9*, 20–29. [CrossRef]
11. Murali, V. Fuzzy equivalence relations. *Fuzzy Sets Syst.* **1989**, *30*, 155–163. [CrossRef]
12. Zhang, W.-R. Bipolar fuzzy sets and relations: A computational framework forcognitive modeling and multiagent decision analysis. In Proceedings of the First International Joint Conference of The North American Fuzzy Information Processing Society Biannual Conference, San Antonio, TX, USA, 18–21 December 1994; pp. 305–309.
13. Lee, K.M. Comparison of interval-valued fuzzy sets, intuitionistic fuzzy sets and bipolar-valued fuzzy sets. *J. Fuzzy Logic Intell. Syst.* **2004**, *14*, 125–129.
14. Lee, K.M. Bipolar-valued fuzzy sets and their basic operations. In Proceedings of the International Conference on Intelligent Technologies, Bangkok, Thailand, 13–15 December 2000; pp. 307–312.
15. Zhang, W.-R. (Yin)(Yang) Bipolar fuzzy sets. In Proceedings of the 1998 IEEE International Conference on Fuzzy Systems, Anchorage, AK, USA, 4–9 May 1998; pp. 835–840.
16. Jun, Y.B.; Park, C.H. Filters of $BCH$-algebras based on bipolar-valued fuzzy sets. *Int. Math. Forum* **2009**, *4*, 631–643.
17. Jun, Y.B.; Kim, H.S.; Lee, K.J. Bipolar fuzzy translations in BCK/BCI-algebras. *J. Chungcheong Math. Soc.* **2009**, *22*, 399–408.
18. Lee, K.J. Bipollar fuzzy subalgebras and bipolar fuzzy ideals of $BCK/BCI$-algebra. *Bull. Malays. Math. Soc.* **2009**, *22*, 361–373.
19. Akram, M.; Dudek, W.A. Regular bipolar fuzzy graphs. *Neural Compu. Appl.* **2012**, *21* (Suppl. 1), S197–S205. [CrossRef]
20. Majumder, S.K. Bipolar valued fuzzy sets in Γ-semigroups. *Math. Aeterna* **2012**, *2*, 203–213.
21. Talebi, A.A.; Rashmalou, H.; Jun, Y.B. Some operations on bipolar fuzzy graph. *Ann. Fuzzy Math. Inform.* **2014**, *8*, 269–289.
22. Azhagappan, M.; Kamaraj, M. Notes on bipolar valued fuzzy RW-closaed and bipolar valued fuzzy RW-open sets in bipolar valued fuzzy topological spaces. *Int. J. Math. Arch.* **2016**, *7*, 30–36.
23. Kim, J.H.; Lim, P.K.; Lee, J.G.; Hur, K. The category of bipolar fuzzy sets. *Symmetry*, to be submitted.

24. Kim, J.; Samanta, S.K.; Lim, P.K.; Lee, J.G.; Hur, K. Bipolar fuzzy topological spaces. *Ann. Fuzzy Math. Inform.* **2019**, *17*, 205–312. [CrossRef]
25. Dudziak, U.; Pekala, B. Equivalent bipolar fuzzy relation. *Fuzzy Sets Syst.* **2010**, *161*, 234–253. [CrossRef]
26. Atanassov, K.T. Intuitionistic fuzzy sets. *Fuzzy Sets Syst.* **1986**, *20*, 87–96. [CrossRef]
27. Bustince, H.; Burillo, P. Structures on intuitionistic fuzzy relations. *Fuzzy Sets Syst.* **1996**, *78*, 293–303. [CrossRef]
28. Hur, K.; Lee, J.G.; Choi, J.Y. Interval-valued fuzzy relations. *JKIIS* **2009**, *19*, 425–432. [CrossRef]
29. Atanassov, K.T.; Gargov, G. Interval-valued intuitionistic fuzzy sets. *Fuzzy Sets Syst.* **1989**, *31*, 343–349. [CrossRef]

*Review*

# Fuzzy Logic and Its Uses in Finance: A Systematic Review Exploring Its Potential to Deal with Banking Crises

Marc Sanchez-Roger [1,*], María Dolores Oliver-Alfonso [1] and Carlos Sanchís-Pedregosa [1,2]

[1] Faculty of Economics and Business, Universidad de Sevilla, Avd. Ramón y Cajal, 1, 41018 Sevilla, Spain; moliver@us.es (M.D.O.-A.); c.sanchisp@up.edu.pe (C.S.-P.)
[2] Academic Department of Business, Universidad del Pacífico, Av. Salaverry 2020, Lima 15072, Peru
[*] Correspondence: marsanrog1@alum.us.es; Tel.: +44-74-2568-0157

Received: 8 September 2019; Accepted: 23 October 2019; Published: 11 November 2019

**Abstract:** The major success of fuzzy logic in the field of remote control opened the door to its application in many other fields, including finance. However, there has not been an updated and comprehensive literature review on the uses of fuzzy logic in the financial field. For that reason, this study attempts to critically examine fuzzy logic as an effective, useful method to be applied to financial research and, particularly, to the management of banking crises. The data sources were Web of Science and Scopus, followed by an assessment of the records according to pre-established criteria and an arrangement of the information in two main axes: financial markets and corporate finance. A major finding of this analysis is that fuzzy logic has not yet been used to address banking crises or as an alternative to ensure the resolvability of banks while minimizing the impact on the real economy. Therefore, we consider this article relevant for supervisory and regulatory bodies, as well as for banks and academic researchers, since it opens the door to several new research axes on banking crisis analyses using artificial intelligence techniques.

**Keywords:** fuzzy logic; finance; banking; banking crisis management

---

## 1. Introduction

Fuzzy logic has been successfully applied in the field of finance due to its ability to address imprecise, incomplete and vague data. This methodology has also been used in the field of banking, although to a lesser extent, with particular relevance to areas such as risk management and credit scoring. However, with regard to the specific field of banking crises, the footprint of fuzzy logic has been almost non-existent. This absence could be seen as a contradiction, given the properties of fuzzy logic, which seem to fit particularly well in complex and uncertain environments. Since the global financial crisis, preventing banking crises and establishing banking resolution strategies to avoid the use of public money to rescue banks have been the main priorities within the area of finance. This priority suggests the importance to both academia and practitioners of exploring this topic through different approaches. Introducing fuzzy logic to the analysis of banking crises could represent an important step towards addressing and resolving banking crises more efficiently.

Banking crises are usually associated with significant, negative impacts on growth, output levels and asset prices, tending to lead to lower tax revenues, higher public expenditures and increased public debt [1]. In accordance with this relationship, we acknowledge the urgency to develop new tools to prevent and manage banking crises. When working with banking crises, we highlight four areas on which research has focused to date and to which the use of fuzzy logic could add value, including (i) preventing banking crises; (ii) managing banking crises and assessing their impact on the economy; (iii) defining a strong institutional and regulatory framework; and (iv) banking resolution.

As a pioneering study proposing the use of fuzzy logic in the field of banking crisis analysis, this work explores in detail the potential uses of fuzzy logic in these four areas.

Of the four points mentioned above, banking resolution is the area in which regulators and governments are currently exerting their strongest efforts. Banking resolution is understood as the restructuring of a bank led by a resolution authority, using one or more resolution tools while following several principles. These principles include safeguarding public interests, preserving the bank's critical services to the economy, ensuring financial stability and minimizing the impact on taxpayers [2]. The impact of the recent financial crisis on the global economy pushed the Basel Committee on Banking Supervision (BCBS) to review and strengthen the Basel accords. This decision led to the publication of Basel III [3], which was transposed into European law by the capital requirements regulation (CRR), as well as the capital requirements directive IV (CRD IV) and the bank recovery and resolution directive (BRRD). In particular, the BRRD addresses the resolution of troubled banking entities and proposes the bail-in tool as one of the key elements for resolving bank crises [4]. The bail-in tool is one of the key instruments of the new resolution framework, forcing banks to absorb any potential losses with capital and debt instruments and to recapitalize the bank if needed to avoid the use of public funds to rescue the entity. To ensure the credibility of the bail-in tool, all European banks are required to increase their loss absorption requirements through the minimum requirement of eligible liabilities (MREL), which is a concept similar to the well-known total loss absorption capacity (TLAC) with which global systemic important banks (G-SIBs) must comply. Although this paper is written from a European regulatory viewpoint, most of its conclusions are also useful from a broader international perspective.

Given the relevance of the bail-in tool to the new banking regulatory landscape, it is important to understand its potential spill-over effects on the real economy. As mentioned above, we propose the exploration of this topic later in this work. Only after the risk transmission mechanism between the banking sector and the real economy is well understood will it be possible to affirm that post-global financial crisis regulation is effectively improving the soundness of the European economy. However, this stability does not yet exist, and the new European resolution framework remains to be tested. The importance and complexity of this topic justify the need to search for new techniques to improve the knowledge on this subject. In accordance with this need, in an attempt to apply nonlinear advanced methods to assess the potential impact of the new resolution framework on the real economy, we have identified fuzzy logic as an adequate tool to perform such tasks. However, to the best of the authors' knowledge, there is no available, updated, comprehensive literature review of fuzzy logic in the field of finance.

The initial hypothesis of this work is, therefore, that fuzzy logic is a solid tool that can be applied in finance, banking crisis and banking resolution analyses due to its ability to manage imprecise, incomplete and vague data.

Several works explore the links between fuzzy logic and mathematics. Some studies are focused on this relationship from a historical perspective [5,6], other studies elaborate on the formal mathematic framework of the Fuzzy Sets Theory [7,8], while other studies take a step into the links between mathematics and fuzzy logic in practical applications [9].

In addition to the academic works devoted to exploring the generic links between mathematics and fuzzy logic, there are some works focused on analysing the fuzzy mathematics of finance. These works seek to show the relevance of fuzzy logic as a useful method to deal with financial research. In particular, one of the first studies suggesting a potential successful application of fuzzy logic techniques to finance proposes using fuzzy alternatives to classic financial data [10]. Other works focused on developing fuzzy mathematics in finance followed, including a generalization of the framework for fuzzy mathematics in finance [11]. From that point, as our analysis shows, the number of articles taking advantage of the foundations of fuzzy mathematics in finance increased exponentially.

In line with the above, the goal of this work is to critically examine fuzzy logic as an effective, useful method to be applied to financial research and, specifically, to analyse banking crises and resolution events. For these reasons, we conducted a bibliometric and literature review of a large group of articles indexed in Web of Science and Scopus, in which fuzzy logic has been applied to the financial field. Additionally, we discuss the potential applications of fuzzy logic in four specific areas, including the prevention of banking crises, assessments of the impact and management of banking crises, institutional settings and banking regulation, as well as banking crisis resolution.

This paper starts with a brief introduction, followed by a description of the methods. The main findings obtained are presented as bibliometric and systematic analyses results. Section 4 elaborates on the findings and provides related discussions, paying specific attention to the use of fuzzy logic in the field of banking crises. Section 5 concludes with the main findings of this study including the contributions of this work to the academia and to practitioners, and the limitations faced.

## 2. Methods

### 2.1. Fuzzy Logic

The Fuzzy Sets Theory was initially introduced in 1965 by L. A. Zadeh and can be described as a logic for dealing with uncertainly and imprecision [12]. The term "fuzzy" is appropriate to describe a mathematical environment where there are no well-defined boundaries between the variables under study [13]. The goal of fuzzy logic is to express the vagueness and imprecision of human thinking with the appropriate mathematical tools. The human way of thinking and reasoning is not binary, where everything is either yes (true) or no (false), and thus Boolean logic is not always the most efficient way to deal with real problems that human beings have to face [14]. Concepts like "danger–safe" or "hot–cold" cannot be sharply defined, and even human beings will use fuzzy language expressions like "very", "a little" or "a lot" to define temperature or dangerous situations.

The fuzzy logic theory is based on the concept of "fuzzy sets", which is a generalization of the classical set theory [14]. A crispy set can be defined by a mathematical function that only accepts binary values, meaning that it can only represent elements that fully belong to the set (represented by the value 1) and elements that do not belong to that set (represented by 0). A fuzzy set is defined by a membership function that allows every element to be represented by a different "grade of membership" specifying to which extent the element belong to the set. It is important to observe that the grades of membership are subjective and rely on the context. To illustrate this, consider a cow, which could be labelled as a "big animal" if the universe of disclosure is "farm animals", but probably will be considered a "medium-size animal" if elephants and hippopotamuses are added to the universe of disclosure.

A fuzzy set is defined from the universe of discourse, which constitutes the reference set and it cannot be fuzzy. Being $U = \{x1, x2, ..., xn\}$ the universe of disclosure, a fuzzy set F ($F \subset U$) is always defined as a set of ordered pairs, the second part of the pair being the degree of membership $\{(xi, \mu F(xi))\}$ and $\mu A$ will always take a value between 0 (non-belonging to the set) and 1 (fully belonging to the set).

Various fuzzy sets tend to be defined on the same universe of disclosure forming a partition of the universe. At this point, a linguistic expression will need to be used to label the different fuzzy sets. This is known as the linguistic variable and can be defined as a variable whose values are words instead of numbers. L.A. Zadeh defines a linguistic variable by a quintuple (X,T,U,G,M) where X is the name of the variable, T are the linguistic values of the variable, U is the universe of disclosure, G is the rule to give a name to the terms in T, and M a semantic rule which associates with each linguistic value X its meaning [15].

The mathematic foundations of fuzzy logic include basic concepts previously described, such as fuzzy sets, membership functions and the basic fuzzy operations (intersection, union and complement). There is a large number of academic papers using fuzzy logic in theoretical fields of traditional mathematics, such as topology, differential equations, probability theory, mathematics and statistics, or measure and integral theory, which shows the strong link between mathematics and fuzzy logic [8].

*2.2. Systematic Review Methodology: ProKnow-C*

Several authors have conducted literature reviews on fuzzy logic applied to different knowledge fields, such as decision making [16], social policy [17] and medical sciences [18], showing a wide range of applications in which fuzzy logic can be used. However, literature reviews of the use of fuzzy logic in finance published so far are not complete nor comprehensive enough, with some examples focusing on the application of neuro-fuzzy systems in business [19] or the uses of fuzzy logic in insurance [20]. It is noted that some interesting books on the subject exist, although they focus on particular applications rather than providing a general overview of fuzzy logic applied to finance [21–23].

There are several ways to approach a literature review, with theoretical background literature reviews being the most common, and hence the approach followed in this study. In particular, two essential forms of analysing the literature regarding a given topic are found: within-study literature analyses and between-study literature studies [24]. The former refers to analysing a specific work while the latter involves contrasting the content of several sources. Some of the benefits of a theoretical background literature review include highlighting what has been explored in a given area and what is still pending to be explored, identifying links between key concepts and listing the main analysis and methodologies that have been successfully used [25].

Rigorous literature reviews must be systematic in following a methodological approach [26]. The method used was based on the knowledge development process-constructivist (ProKnow-C) [27], a technique that describes a process analogous to a protocol. Hence, this study included an extensive search across different databases to conduct a systemic analysis to obtain information on the content of the different papers that comprise our final portfolio. A bibliometric analysis was conducted to obtain relevant data on publication trends, the most relevant authors and journals on the topic. This method has been widely applied in the literature reviews on different topics in recent years [28–30]. One of the main advantages of the methodology chosen is that it is particularly effective to deal with descriptive and exploratory tasks. Therefore, ProKnow-C is especially useful for theoretical background literature reviews such as the one developed in this analysis.

The eligibility criteria focused on identifying all relevant academic articles that somehow apply fuzzy logic to solve problems linked to the field of finance. Thus, it is important to define exactly what is meant by "finance" in this work, given that this subject is broad and therefore difficult to precisely define. Among the many definitions of finance, Drake and Fabozzi [31] defined finance as the application of economic principles to decision making that involves the allocation of money under conditions of uncertainty. The study of finance is usually divided into four main categories: corporate finance, financial markets, public finance, and personal finance. In this study, we especially focus on the fields of corporate finance and financial markets because these fields are those in which fuzzy logic has been the most frequently applied to date.

The information sources considered were the Web of Science and Scopus databases. The choice of Web of Science and Scopus was motivated by the academic acknowledgement of these databases as being two of the main academic literature collections [32,33].

The search was defined as a Boolean combination of key words grouped on two main axes, namely "fuzzy logic" and "finance". The selected key words were deliberately vague to increase the number of results in the searches.

The study selection was carefully conducted as follows. We introduced our queries of the database already defined and filtered the output by type of source, excluding all documents except academic journals and books. Then, we identified the duplicated records and excluded them from our portfolio. The records remaining were then assessed manually based on the title, abstract and full text, in that order. A full-text review at this point was also a method to reduce the risk of bias in individual studies, as it allowed better direct scanning of the articles. Another way to control risk of bias, but across studies, was the inclusion of studies from the list of references as an attempt to not exclude all grey literature without lowing the quality by considering records from the main databases. Additionally, articles with no citations in the Google Scholar database were removed. Given the small number of citations, they cannot be considered relevant from an academic viewpoint. Finally, all the references included in the final group of records were analysed with the same criteria mentioned above.

This work focused on the following data items: the number of publications per year, the most prominent journals on the topic, and the most active authors. Finally, the synthesis of the results is depicted in five main proposed areas: financial markets (forecasting, valuation financial assets, portfolio management, and trading and decision making), corporate finance (fundamental analysis, banking, insurance, investments and decision making, and others), public finance, personal finance and others.

One of the disadvantages of the methodology is that given the large number of articles on fuzzy logic applied to economics and the social sciences, settling boundaries between each and defining the eligibility criteria was a manual and complex task. As a result, some articles could have been involuntary omitted. However, our final portfolio is large enough to represent a comprehensive list of articles applying fuzzy logic in finance. It is also a large sample that helps to identify areas in which more research has been accomplished and to understand why and which methods seem to be more effective in the field of finance.

We deliberately excluded conference proceedings from our bibliographic portfolio, seeking to increase the scientific relevance of the sample under study, rather than the quantity of all the documents analysed. Both approaches are generally accepted; however, it is frequently perceived that conference proceedings are less mature than academic papers [34,35].

## 3. Results

The purpose of this section is to present the results of the analysis of the study characteristics (included in the bibliometric characteristics) and the content of the 795 final articles included, corresponding to the results of individual studies in the portfolio. We first present an analysis of the bibliometrics of the portfolio, with special attention paid to the recent trends in numbers of publications per year, as well as the most prominent journals and authors on the subject. Then, a section focused on the content of the articles follows.

Prior to presenting these results, we describe the process followed to define our final portfolio. After performing the search in both databases, filtering by type of source and excluding all documents but academic journals and books, we obtained a total of 10,941 records (5289 from Web of Science and 5652 from Scopus). Before starting the manual filtering process by the title and then by the abstract, the next step was to remove duplicates. A total of 2952 documents were found in both databases; therefore, we ended up with an initial gross bibliographic portfolio of 7985 records. The 7985 articles were then manually filtered first by title adequacy, and then followed by abstract and full-text alignment with the subject of interest. After conducting these steps, we obtained 931 papers in our final portfolio. However, all articles with no citations in the Google Scholar database were removed because, given the small number of citations, they cannot be considered relevant from an academic viewpoint. Following this approach, the number of articles decreased to 768. For the final bibliographic portfolio, all of the references included in each of the 768 papers were analysed to assess whether they could be included in the final portfolio. This increased the number of eligible papers by 27, resulting in a final portfolio of 795 articles. A flow diagram showing this study selection process at each stage is presented in Figure 1.

This section elaborates on key aspects, including the particular topics within the field of finance in which fuzzy logic has been more commonly applied to date and the combination of methods seen with greater frequency.

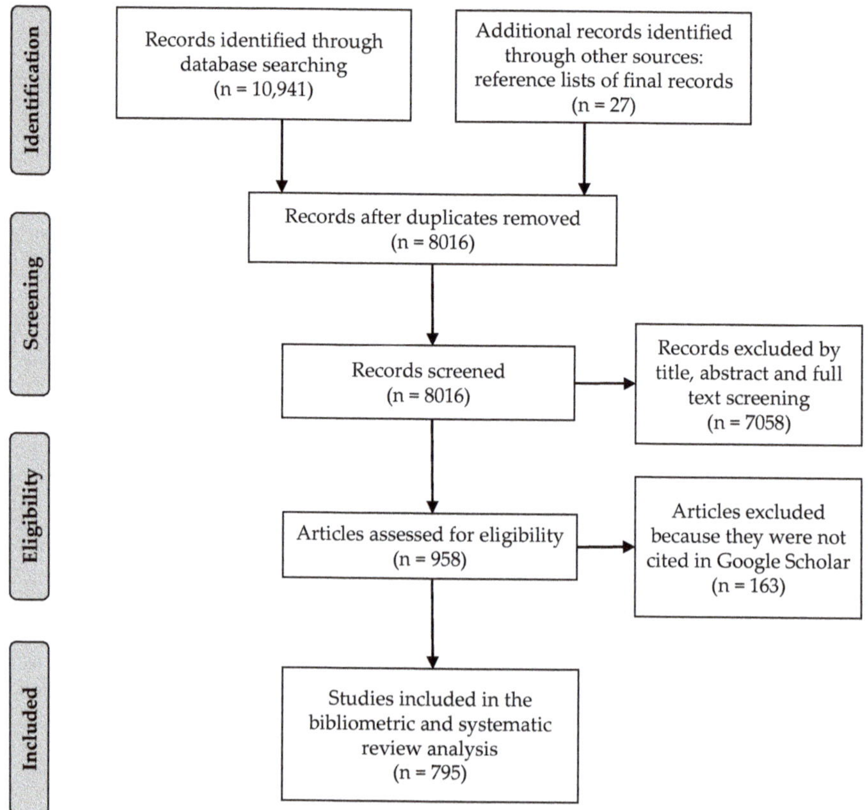



**Figure 1.** Flow diagram of the study selection process. Source: PRISMA Model [34].

### 3.1. Bibliometric Analysis

Fuzzy logic was introduced to the financial field by Buckley, who explored the mathematics of fuzzy logic in finance, applying them to the study of the time value of money [10]. Following the success of Buckley's work, Calzi [11] continued exploring the topic, widening the scope of applications to which fuzzy logic can be applied in the financial field. Several studies followed, investigating the net present value of investments [35,36]. More recently, other authors have also contributed to increasing the literature with studies related to bankruptcy forecasting, stock market prediction or portfolio management optimization, among others [37–39]. However, as shown in Figure 2, it was not until recently that the research on finance using fuzzy logic truly became sizeable. In accordance with this development, the analysis of the publications per year reveals that, from the mid-2000s, the number of articles exploring the use of fuzzy logic in finance increased significantly.

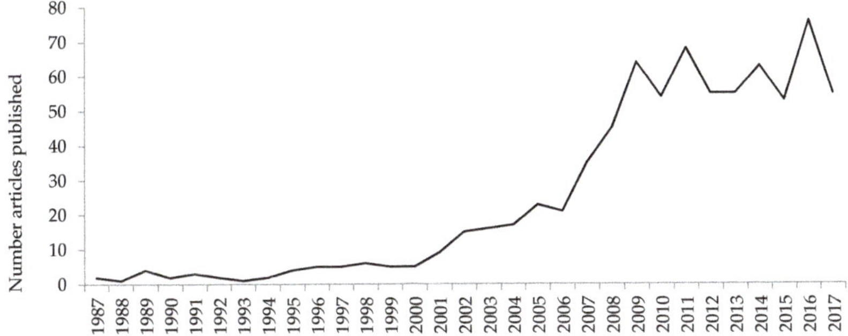

**Figure 2.** The number of articles published by year (1987–2017) exploring the use of fuzzy logic in finance has increased significantly since the mid-2000s.

In Figure 3, the academic journals that appear with greater frequency in our portfolio are shown. In addition, we included the 2017 impact factor for each article. When focusing on the most relevant journals, we found that Expert Systems with Applications is the journal with the greatest number of articles within our portfolio, with more than 90 articles and an impact factor of 3.76. Information Sciences and Fuzzy Sets and Systems followed, with 28 and 27 articles, respectively, and impact factors of 4.31 and 2.68. These findings place Expert Systems with Applications as the leading journal associated with financial research conducted using fuzzy logic methods. Overall, in our final portfolio, we found 300 different journals. However, approximately 50% of all articles are concentrated in only 25 journals, with an average impact factor of 2.7.

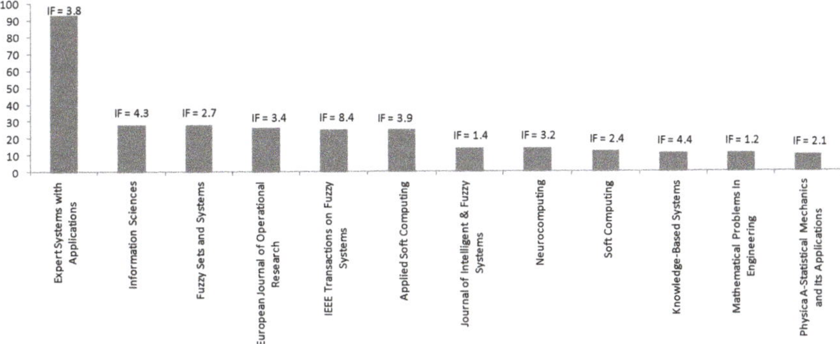

**Figure 3.** Academic journals that appear with greater frequency in our portfolio and their 2017 impact factors.

The most cited paper within our portfolio is entitled "Bankruptcy prediction in banks and firms via statistical and intelligent techniques—a review", with 849 citations. The article focuses on different approaches to address bankruptcy prediction, with fuzzy logic as one of the techniques explored. Other articles with more than 450 citations include "Effective lengths of intervals to improve forecasting in fuzzy time series", "Surveying stock market forecasting techniques—Part II: Soft-computing methods", and "The fuzzy mathematics of finance".

Table A1 summarizes the authors with the largest numbers of publications in our final portfolio. We show the aggregated number of citations per author considering only the articles included in the portfolio. Cheng, from National Yunlin University of Science and Technology (Taiwan), with 21 articles published and a total of 925 citations, is the most prominent author in the field. Other authors, with 12 articles published, are Quek and Zhang from the Nanyang Technological University (Singapore) and South China University of Technology (Guangzhou, China), respectively. Finally, we also consider the journals in which the authors have published their work, sorted first by number of articles published and second in alphabetical order.

Bibliometric analysis enables the investigation of numerous, diverse questions on the subject under study from a bibliographic viewpoint. The results presented above clearly indicate that the use of fuzzy logic in financial research has increased significantly over the last two decades, and the trend seems to indicate that the number of publications on the subject could continue rising in the near future. In addition, we obtained information regarding the most prominent journals and authors. However, no analysis of the content of the papers included in our final portfolio has yet been performed; hence, developing a systematic analysis at this point is crucial to completing the task of reviewing the literature on the topic under study. The next section focuses precisely on that topic by providing relevant insights into the most common financial topics in which fuzzy logic has been traditionally used in finance while briefly describing the main findings in each field.

### 3.2. Systematic Analysis

After a comprehensive review of the articles in which fuzzy logic has been used in the financial field and to develop the systematic analysis, we propose a classification of the articles by topic. The most frequent topic in which fuzzy logic has been applied in the field of finance is financial markets, with approximately 60% of the total articles classified in this category, followed by corporate finance, with approximately 35%. The remaining articles focused on public finance (3%), personal finance (1%) and other topics (2.52%). This fact clearly underscores the focus of the academic research on financial markets and corporate finance, and it can be explained by the greater volume of publicly available data in these fields compared with other areas, such as personal finance.

Figure 4 shows the breakdown of the main topics usually analysed when applying fuzzy logic in finance in the categories of financial markets and corporate finance. Regarding financial market research, Bahrammirzaee [40] emphasizes that the study of financial markets has been traditionally conducted following three different methodologies, namely (i) parametric statistical methods, such as discriminant analysis and regression; (ii) non-parametric statistical methods, such as nearest neighbour and decision trees; and (iii) soft-computing and artificial intelligence (AI) methods, including fuzzy logic, neural networks and genetic algorithms. The existing literature tends to agree that most of the research conducted using artificial intelligence methods, such as fuzzy logic or neural networks, generally tend to outperform parametric and non-parametric statistical methods [41,42].

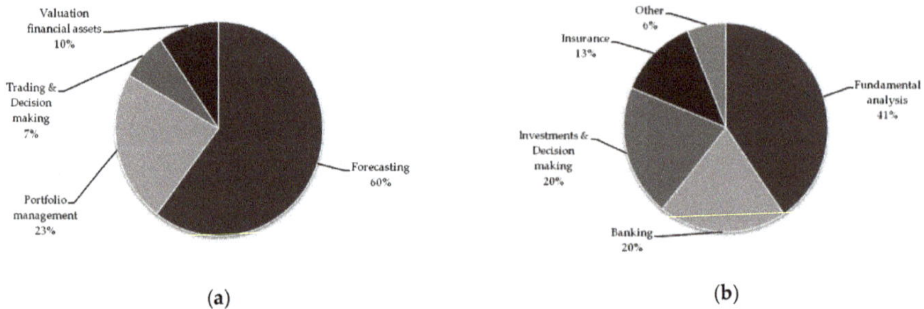


**Figure 4.** Categorization of the main topics usually analysed when applying fuzzy logic in finance: (**a**) Category of financial markets, and (**b**) category of corporate finance.

In particular, approximately 60% of papers analysing financial markets focused on forecasting. As noted by Atsalakis and Valavanis [41], the non-conventional forecasting methods, such as fuzzy logic, neural networks and genetic algorithms, outperform conventional forecasting techniques in almost all circumstances. Most of the papers reached the conclusion that optimal results are obtained when combining fuzzy logic with neural networks.

One of the areas with more practical implementations is the field of technical analysis. In this vein, one of the most cited papers proposed the use of a neuro-fuzzy system composed of an adaptive neuro-fuzzy inference system (ANFIS) to predict short-term stock market trends with encouraging results [43]. Another similar example using an ANFIS model is applied to the Istanbul Stock Exchange, and once again, the results show significant improvement in financial market predictions when the ANFIS model is used [37]. Fuzzy expert systems have also been used for financial market forecasting purposes, with Ijegwa et al. [44] proposing a pure fuzzy system to perform financial market forecasting. The authors justified the relevance of using a fuzzy system given that, in many cases, responses of technical indicators to market movements are not a definite yes or no. Hence, fuzzy reasoning is very effective in this type of environment. Van den Berg et al. proposed a probabilistic fuzzy system based on the Takagi–Sugeno method that combines the properties of probabilistic systems with the interpretability of fuzzy systems. In particular, Chang and Liu [45] developed a Takagi–Sugeno–Kang fuzzy system that predicted stock price movements across different sectors, claiming to achieve an accuracy of approximately 97.6% in the Taiwan Stock Exchange.

Portfolio management, which represents approximately 25% of all papers in financial markets research, relies heavily on Markowitz portfolio optimization theory. However, implementation of the Markowitz model requires future mean returns and correlations between assets. In real life, predicting these metrics accurately could represent a problem, making uncertainty and lack of accuracy two characteristics that all portfolio management models must address. To overcome this issue, fuzzy logic has been applied in the area of portfolio management, obtaining successful results [46–49].

Moving to the field of financial asset valuation, one of the areas in which fuzzy logic has been widely used is in options pricing. Muzzioli and De Baets [50], in a comprehensive literature review on the topic, identified that the majority of papers using fuzzy logic for option pricing have addressed the direct problem of pricing in both discrete and continuous time settings. Several articles have shown that fuzzy logic is suitable for identifying the market value of complex instruments, such as derivatives [51–53].

Trading and decision making represent approximately 7% of the articles focused on financial markets analysis. Many research papers in this area seek to find trends and identify patterns, to define complex trading strategies or to simply make, buy or sell decisions based on a set of rules. Several papers have proven the success of fuzzy logic in trading algorithms and financial decision-making processes [54–56]. In particular, Kuo et al. [57] combined genetic algorithms, fuzzy logic and neural networks to present a "buy–sell" system using both quantitative and qualitative factors. This study proved once again the advantages of combining multiple AI techniques.

The papers analysed thus far point towards the increasing use of fuzzy logic in financial market analysis in the coming years. Some of the areas in which we see the scope for further expansion include not only credit trading, particularly when managing illiquid products, but also in areas such as market analysis through human behaviour reactions (behavioural finance) and social media and trading decisions (social trading).

Once we reviewed the literature linked to financial markets, we focused on corporate finance research. However, the field of corporate finance is vast and includes a wide range of subjects. One of its predominant fields, accounting for approximately 40% of the articles, is fundamental analysis. In contrast to technical analysis, fundamental analysis relies on macroeconomic data, including interest rates, inflation rates and GDP growth, as well as on microeconomic data, which can be obtained from the annual accounts and more specifically from the profit and loss (P&L), cash flow statement and balance sheet. Therefore, fundamental analysis requires uncertain information, such as macroeconomic forecasts, and vague and imprecise data, such as management guidance and press releases, as input data. The nature of this information makes fuzzy logic one of the most suitable AI tools for managing fundamental analysis. An example can be found in the work by W. Berlin and N.F. Tseng, in which macroeconomic and company-specific data were used as linguistic variables in a fuzzy regression model to analyse the business cycle [58]. Other works include the use of fuzzy methods to address cash flow forecast analyses [59], corporate acquisition analysis [60], company evaluations [61] and credit rating analysis [62].

Another area in which fuzzy logic has been traditionally applied within corporate finance is the choice of investment opportunities, representing approximately 20% of the articles. Fuzzy logic was applied to a wide range of specific topics to evaluate the different investment options from the oil sector [63] to the real estate market [64]. These decision-making processes have been traditionally conducted in relation to multiple criteria decision aiding (MCDA) techniques. Basically, MCDA enables the inherent uncertainty linked to financial decisions to be assessed through a multidimensional process. The combination of MCDA with fuzzy logic, allowing for the treatment of variables as fuzzy, opens a wider range of possibilities to successfully use MCDA techniques [65]. Among the MCDA techniques, we highlight the frequent use of the technique for order of preference by similarity to ideal solution (TOPSIS).

Fuzzy logic has been introduced into the insurance sector by De Wit [66]. In their original work, De Wit [66] sought to identify fuzziness in underwriting. From this area, several papers studying different insurance areas in which fuzzy logic could be applied emerged. Approximately 13% of the papers in the category of corporate finance address insurance-related problems. In particular, Lemaire [67] contributed to the expansion of fuzzy logic theory applied to the insurance field by presenting some of the concepts of fuzzy logic in an insurance framework. A study published by Ostaszewski and Karwowski [68] included an in-depth analysis of the uses of fuzzy logic in actuarial sciences. Derrig and Cummins [69] explored further uses of fuzzy logic in the insurance business, with relevant contributions to the fields of property–casualty insurance forecasting and price modelling. Fuzzy logic has proven to be useful in determining insurance pricing decisions, which consider additional data on an ongoing basis [70] and the opinions of experts [71]. As observed in other knowledge areas, in its early stages, fuzzy logic was applied on a stand-alone basis in the insurance field. However, more recently, with the expansion of neural network methods, both tools have been combined, delivering in most cases even better results [72]. Shapiro also elaborated on the subject of fuzzy logic being used in the insurance sector from a broader perspective, pointing out a wide range of areas in which fuzzy theory can be successfully applied in insurance, including classification, underwriting, present value and pricing calculations, as well as asset allocation [20].

The use of fuzzy logic in public and personal finance is much more limited compared with financial markets and corporate finance. Regarding public finance, we witnessed increasing interest from scholars in the use of AI techniques applied to the public sector. Some examples of studies applying fuzzy logic to public finance address import and export forecasts [73], financing of public schools [74], forecasting countries' domestic debt using a combination of fuzzy logic and neural network techniques [75] and the analysis of the degree of effectiveness of European public policies in meeting sustainable development goals [76]. In particular, studies using AI methods in the public finance sector are particularly relevant from a regulatory and government perspective, and we should expect an increase in these types of studies in areas such as the potential impacts of quantitative easing programmes, central banks' decisions about interest rates and the impact of national regulations on the funding profiles of corporations.

Finally, focusing on the analyses of banks, the existing literature shows that the introduction of AI techniques represents a breakthrough in the analysis of several areas, such as risk management. Using big data to increase banks' knowledge of their customers, to provide loans according to certain credit scores, to measure their efficiency, to use advanced computational systems to balance their ALCO portfolios and to optimize the capital structure and liquidity and funding needs, banks have come to rely heavily on these advanced methodologies [77–82]. The growing importance of fuzzy logic in this type of analysis is demonstrated by the finding that approximately 20% of the portfolios focused on corporate finance are concerned with bank analysis.

## 4. Discussion

The analysis of the literature on fuzzy logic in the field of finance opens the door to a wide range of new and promising applications. A particular field in which the impact of fuzzy logic could be especially greater is in banking crisis and banking resolution analyses. Next, we discuss the applications of fuzzy logic in the following areas linked to banking crises and resolutions: (i) applications in banking crisis analysis and banking resolution; (ii) impact and magnitude of banking crises; (iii) institutional setting and financial regulation; and (iv) banking crises resolution.

### 4.1. Fuzzy Logic and Its Potential Applications in Banking Crisis Analyses and Banking Resolution

The existing academic literature has identified the three main factors associated with the probability and magnitude of banking crises [83]. These factors are classified as (i) pre-crisis macroeconomic conditions, since banking crises are usually preceded by credit and asset price booms, suggesting the importance of developing early warning systems to identify the build-up of imbalances at the macroeconomic level; (ii) pre-crisis characteristics of the banking sector, highlighting the importance of maintaining a sound banking sector through the entire economic cycle; and (iii) institutional settings, which should be responsible for enforcing an efficient regulatory and supervisory framework to ensure that the banking sector is solid and serves its main purposes with regard to the real economy. Of these three factors, the pre-crisis characteristics at the macroeconomic level and at the banking sector level can be grouped into one category, directly linked to the prevention of banking crises through early warning systems. The institutional setting is particularly relevant with regard to the key role that the regulatory framework and the supervisory and resolution institutions play to ensure a sound and safe banking sector; therefore, in this work, we classify it as a different topic.

In addition to the two topics mentioned above, understanding the impact of banking crises on the real economy is another key area to be addressed. Finally, since the global financial crisis, governments, regulators and banking supervisors have focused their efforts on developing a solid resolution framework to minimize the impacts of banking crises on the real economy. Therefore, the study of banking resolution is the fourth main area that we develop further in this section.

The results obtained in the previous section demonstrate that fuzzy logic could be a particularly relevant tool for addressing the four aforementioned topics. Accordingly, we next elaborate on the main topics within these four areas to which fuzzy logic could potentially add more value.

Systematic analysis reveals a reduced number of articles in which fuzzy logic has been used for banking crisis and banking resolution analysis to date. Additional analysis includes papers sorted by number of citations, author names, the journals in which the articles have been published, the year of publication, the main purpose of the articles, and the methodologies used. Most of these articles focused on early warning systems to predict banks' defaults, as shown in Table A2. However, there was also one article on systemic risk, one on contagion risk analysis and one on the analysis of bank supervisory criteria.

### 4.1.1. Preventing Banking Crises by Monitoring Pre-Crisis Macroeconomic and Banking Sector Conditions

Banking crises are considered a recurrent systemic phenomenon, often triggering deep and long recessions. They are not random events, and they tend to occur following periods of very strong credit growth (credit booms), significant increases in asset prices and other financial imbalances, such as strong correlations in risk exposures among banks [84]. Banking crises are also associated with the concept of systemic risk, which can be defined as the risk that can damage financial stability, impairing the correct functioning of the main activities of the financial system with major negative effects on the real economy. Systemic risk also builds up over a long period of time, and it is not a random event triggered suddenly. In particular, authors have emphasized that approximately one-third of banking crises are preceded by a credit boom, generally stemming from correlated risk exposures and increases in asset prices, leading to bubbles. In accordance with this observation, we suggest that early warning systems powered by fuzzy logic could be an extremely useful tool. We acknowledge the large number of variables that must be considered when attempting to prevent financial crises, ranging from macroeconomics to aggregated banking sector data or individual bank dynamics. This extremely large number of factors led us to conclude that fuzzy logic could be a useful tool for addressing this subject.

Most of the articles included in the table above elaborated on early warning systems to prevent banking crises or predict bank failure. Bankruptcy prediction is one of the areas in which fuzzy logic and other AI techniques clearly outperform classic analysis methods. The prominent work developed by Ravi-Kumar and Ravi [85] presented a list of bankruptcy prediction methods, including statistical techniques, neural networks, decision trees, evolutionary algorithms and fuzzy logic techniques. In particular, focusing on the use of fuzzy logic for bankruptcy prediction tasks, the authors highlighted that this method successfully manages imprecision and ambiguity, combining it with human expert knowledge. Interestingly, bankruptcy prediction has been traditionally studied as a classification problem. Different techniques have been applied to predict firms' defaults, including regressions, linear discriminant analysis (LDA), multiple discriminant analysis (MDA), neural networks, support vector machines and decision trees. In one of the pioneering studies on bankruptcy prediction, Altman [86] proposed a multivariate discriminant analysis technique to classify firms as solvent or bankrupt. Focusing exclusively on bank default prediction, Sinkey [87] also used MDA to detect failing banks, while Altman [88] later published a study focused on banks' insolvency prediction. The use of a fuzzy-clustering algorithm to forecast bank failure is another example [89]. Overall, the academic literature on bankruptcy prediction has tended to agree that neural networks commonly outperform other methods [90]. However, when samples are smaller, or more transparency is sought, support vector machines obtain better results [91]. Ravi-Kumar and Ravi [85] used an ensemble of classifiers to predict failures of Spanish and US banks, emphasizing that ANFIS is among the top performers. From a decision support systems viewpoint, an optimal system to predict the failure of a bank should not be a classifier on a stand-alone basis but a combination of them [92]. Therefore, a combination of two or more classifiers seems to be the most appropriate technique for predicting which banks could fail in the short or medium term. A combination of fuzzy-SVM and ANFIS could be a useful tool for determining whether a bank should be considered "failing or likely to fail" and subsequently entered into a liquidation or resolution procedure.

A particularly interesting study conducted by W.L. Tung et al. [93] elaborated on early warning systems to predict banking failures using a neuro-fuzzy system. This paper showed how the use of fuzzy logic and its combination with other AI techniques could be seen as a powerful tool for addressing banking supervision and regulatory issues. In particular, after the analysis of the literature, we propose the use of neuro-fuzzy models, particularly ANFIS, to address bankruptcy prediction.

### 4.1.2. Impact and Magnitude of Banking Crises

One of the key topics when studying banking crises is understanding their impact on the economy. Only with accurate methodologies to account for the holistic impact of such events will regulators and supervisors be able to design regulatory frameworks that could offset the existing trade-off between the benefits and costs of tight banking regulations. Currently, the academic literature is not aligned with regard to the variables that determine the impacts of banking crises [94]. As authors have pointed out, this lack of consensus could be partly explained by the proxy used as a measure of the impact on the real economy of a banking crisis. However, the large number of factors to be considered when performing such calculations is a limitation of any methodology based on traditional analysis. Following the results obtained in Section 4, we consider that fuzzy logic methods, combined with other artificial intelligence tools, such as neural networks or machine learning, could significantly increase the accuracy of such estimates.

Financial contagion is another important topic to focus on when addressing the study of banking crises. The links between banks through interbank lending and the interconnection between banks and other economic sectors must be monitored even more closely in the new resolution framework. The literature using fuzzy logic to analyse this point has been scarce; however, in a recent study, De Marco et al. [95] modelled a financial network with fuzzy numbers, which could be a good starting point for a deeper analysis of the contagion risks using fuzzy logic techniques.

The bail-in tool seems to be a promising technique with which to address banking failures. However, loss absorption requirements force banks to issue bail-in-able debt and capital instruments that are mainly acquired by non-banking entities. This process could spread financial risk across other sectors instead of achieving the goal of reducing systemic risk. Understanding the potential risk transfer from the financial sector to other sectors of the real economy and from banks to citizens is a topic that must be analysed. We propose fuzzy numbers and ANFIS as candidate methods to address contagion risk issues.

### 4.1.3. Institutional Setting and Financial Regulation

We also consider it relevant to analyse banking regulators' and supervisors' performance. In the current environment, entities such as the SRB or the ECB in Europe have significant power to address banking crises. Despite the fact that it seems necessary to have institutions in charge of the stability of banking and the financial system, the complexities of the current regulation and the power of such institutions make it necessary to monitor their tasks and decisions. Understanding whether and how banks are reaching the capital, leverage and liquidity targets imposed by regulators could be a method of assessing the banking regulators' performance. Using fuzzy MCDM and neuro-fuzzy approaches to evaluate regulators' and supervisors' performance could open a door to a new wave of studies analysing whether regulators are making the right decisions, whether the methods used are effective and whether any significant mistakes have been made thus far.

In addition, regulators should enhance banks' transparency and the accuracy of the information reported by financial entities. Fraudulent reporting is, at the time of writing, another major concern for regulators, investors and other economic agents. Proof of this concern is the interest shown by academia in studying financial reporting fraud and how to detect and prevent it [96]. Well-known examples of fraudulent reporting, such as the case of Bankia in Spain or Banco Espirito Santo in Portugal, emphasize the importance of detecting such fraudulent techniques. In the case of the Spanish entity, a potential manipulation of the financial information disclosed by the bank before the initial public offer (IPO) is currently under investigation by the Spanish High Court [97], while in the case of the Portuguese bank, the management decisions and reporting techniques are considered borderline fraud [98]. The work published in 1986 by Albrecht and Romney [99] is considered to be the first study analysing empirically the prediction of fraud in financial reports, and from this point, several articles have been published on the subject [100,101]. Fuzzy logic has proven to be useful in this field as well, with several papers published on the subject [102,103]. In particular, Lin et al. [104] proposed a neuro-fuzzy technique to detect fraudulent reports. The authors showed how neuro-fuzzy methods outperform traditional methods, such as logit models, in the detection of fraudulent reporting. The literature on this topic suggests that the application of fuzzy logic top fraud reporting will contribute to enhancing transparency in the banking sector. In particular, in accordance with other research axes, neuro-fuzzy systems seem once again to be the most efficient method for developing solid banking fraud reporting research analysis.

### 4.1.4. Banking Crisis Resolution

The triggering of a banking resolution and its effects on the real economy are very sensitive and complex processes that involve managing financial information, macroeconomic data and legal aspects. The decision to trigger the resolution of a banking entity represents a decision-making process that combines different expertise areas, in which a significant degree of uncertainty and subjectivity arises. Therefore, we suggest that the use of fuzzy logic in the banking resolution field could be a disruptive technique to model and better understand several of its implications for the real economy. However, banking resolution analysis must be extended further, with the triggering of the bank resolution representing only the first step. Setting the right loss absorption requirements, choosing the optimal resolution tool in each case and limiting the contagion from the financial sector to other sectors are some of the points that should be explored further. In accordance with these issues, due to the complexities linked to this type of analysis, in which information is uncertain, and the fact that several factors, such as politics, macroeconomic variables and regulation, participate, the impacts on the real economy of preventing a banking crisis with tools such as bail-in or deposits guarantee that schemes are difficult to analyse with traditional econometric models. We therefore suggest that the application of fuzzy logic, combined with other innovative tools, such as neural networks, could yield solid results.

To address the decision-making process of determining a financial institution "failing or likely to fail", this work considers that multicriteria decision aid (MCDA) could potentially be a key methodology for addressing such a complex decision. MCDA is a method that enables analysis of different criteria at the same time, and, according to the recent literature, it is the best method for choosing optimal solutions in probability option exercises [105]. Following this literature review, we consider that using fuzzy MCDM (and fuzzy AHP in particular) to decide whether to trigger the resolution of a bank could be a more objective and unbiased method than that currently used in Europe based on the judgement of ECB and SRB experts.

In accordance with the above discussion, it is also important to design efficient loss absorption instruments; hence, these instruments should also be a focus of study. It is also important to bear in mind that regulatory requirements, such as total loss absorption capacity (TLAC) or minimum requirement of eligible liabilities (MREL), could potentially result in higher funding costs for banks. They could also lead to a transfer of these higher funding costs to the real economy through a worsening of credit conditions. Analysing the impact of such requirements on the real economy is a complex task; however, in accordance with the existing literature on fuzzy logic, it seems that neuro-fuzzy methodologies could contribute significantly to the development of such analyses. In addition, it is essential to design instruments that are sufficiently attractive to investors to maintain a relatively low cost of funding for banks, while simultaneously, the banks' resolvability should be increased due to the loss absorption features provided by these instruments. In this regard, both fuzzy valuation methods and ANFIS seem to be good methodologies for managing this task.

## 5. Conclusions

The existing literature shows that fuzzy logic has already been applied to a wide range of areas within the field of finance. However, it is far from reaching its full potential compared with its use in other fields, such as control systems, engineering and environmental sciences, in which the number of articles using fuzzy logic is much greater. After analysing the results published in studies in which fuzzy logic has been used, one of our main conclusions is that fuzzy logic has shown to be especially efficient when addressing uncertainty and vagueness, which are two of the most common characteristics linked to financial analysis. Therefore, given the particularities of this tool in managing complex and uncertain scenarios, the first conclusion of this analysis is that we should expect a significant increase in the number of papers within the finance field using fuzzy logic in the coming years. In particular, areas such as financial market forecasting, credit market analysis, public finance and personal finance could be some of the areas benefiting the most from standardization of the use of fuzzy logic and other AI techniques in the field of business and finance.

This work also emphasizes that a combination of AI techniques, including fuzzy logic, neural networks and evolutionary programming, outperforms traditional techniques according to the results presented in Section 4. In particular, the use of fuzzy logic and neural networks as a hybrid approach tends to be the most common technique and demonstrates the best performance. We acknowledge the efficiency of this method; therefore, we expect a significant increase in the number of scholars applying such hybrid techniques in their work in the near future.

This study is, to the authors' best knowledge, pioneering in terms of the introduction of fuzzy logic to the field of banking crisis analysis and banking resolution. Consequently, this work is intended to represent only a first step in this research area, suggesting key topics in which further research would be particularly welcomed. In particular, this work proposes four research axes in which we strongly believe fuzzy logic would improve the results obtained to date when using traditional analysis methods. These topics include the (i) prevention of banking crises through fuzzy logic-powered early warning systems; (ii) management of banking crises and measurement of their impacts on the real economy; (iii) institution setting and financial regulations to address, prevent and resolve banking crises; and (iv) overcoming banking crises through efficient resolution methods.

The complexities linked to the ambitious task of proposing the use of an alternative methodology to address a well-known subject, leads to several limitations in our work. First, a limitation shared by many other literature reviews is that some articles could have been involuntary omitted; hence, our work draws conclusions from an extensive, but not exhaustive, list of papers. Another limitation is associated with the lack of papers using fuzzy logic in the field of banking crisis analysis, which means that the models that we propose have not yet been tested in a banking crisis framework. Finally, identifying the key areas within the field of banking crises to which fuzzy logic can be applied opens the door to potential overlaps between areas.

This study has relevant implications not only for researchers but also for practitioners. From an academic viewpoint, the main contributions of this paper are linked to identifying in which fields of financial research fuzzy logic has been utilized and extrapolating the results to suggest other areas where the use of fuzzy logic could bring positive advances, such as in trading and behavioural finance. It also adds value regarding the management of banking crises, since a group of articles are analysed to identify the most adequate fuzzy logic techniques to deal with specific problems linked to banking crises research. From a practitioner's standpoint, this work discussed several studies where fuzzy logic has been applied in the financial field with successful results, including financial forecasting, stock markets and public finance. It could also be useful for banking regulatory and supervisory bodies, since this work briefly explores the potential use of fuzzy logic in the field of banking regulation.

In conclusion, since banking crises are among the most devastating events from an economic viewpoint, we acknowledge the relevant efforts undertaken by the academic community to better understand the existing mechanisms to prevent and manage banking crisis. Due to the nature of the data involved in such studies, we consider fuzzy logic and its combination with neural networks particularly appropriate for the analysis of banking crises and banking resolution mechanisms. This work contributes to easing the integration of fuzzy logic into the analysis of banking crises through a comprehensive literature review and identification of the key areas for development.

**Author Contributions:** Conceptualization, M.S.-R.; methodology, M.S.-R., M.D.O.-A. and C.S.-P.; software, M.S.-R.; validation, M.D.O.-A. and C.S.-P.; formal analysis, M.S.-R.; investigation, M.S.-R., M.D.O.-A. and C.S.-P.; resources, M.D.O.-A. and C.S.-P.; data curation, M.S.-R.; writing—original draft preparation, M.S.-R.; writing—review and editing, M.S.-R., M.D.O.-A. and C.S.-P.; visualization, M.S.-R., M.D.O.-A. and C.S.-P.; supervision, M.D.O.-A. and C.S.-P.; project administration, M.S.-R., M.D.O.-A. and C.S.-P.; funding acquisition, M.D.O.-A. and C.S.-P. M.S.-R. has contributed with the conceptualization of this work, the formal analysis and the original draft. The three authors have contributed to the design of the methodology. This project has been supervised by M.D.O.-A. and C.S.-P. Funding has been obtained thanks to C.S.-P. and M.D.O.-A.

**Funding:** We acknowledge the financial support from the Regional Government of Andalusia, Spain (Research Group SEJ-555).

**Conflicts of Interest:** The authors declare no conflict of interest.

# Appendix A

**Table A1.** Main authors with the largest numbers of publications in our final portfolio.

| Author | Num. Articles | Tot. Citations | Journal 1 | Journal 2 | Journal 3 | Journal 4 | Journal 5 | Journal 6 | Journal 7 | Journal 8 | Journal 9 | Journal 10 |
|---|---|---|---|---|---|---|---|---|---|---|---|---|
| Cheng, CH | 21 | 925 | Expert Systems with Applications | International Journal of Innovative Computing Information and Control | Applied Intelligence | Applied Soft Computing | Neurocomputing | Physica A-Statistical Mechanics and Its Applications | African Journal of Business Management | Computers & Mathematics with Applications | Data & knowledge Engineering | Economic Modelling | Intelligent Automation and Soft Computing |
| Quek, C | 12 | 538 | Expert Systems with Applications | Computational Intelligence | AI Communications | Applied Intelligence | Contemporary Theory and Pragmatic Approaches In fuzzy Computing Utilization | IEEE Transactions on Evolutionary Computation | IEEE Transactions on neural networks | Neural Networks | | |
| Zhang, WG | 12 | 298 | Economic Modelling | Soft Computing | Applied Soft Computing | Automatica | European Journal of Operational Research | Fuzzy Optimization and Decision Making | Information Sciences | International Journal of Intelligent Systems | Mathematical and Computer Modelling | |
| Chen, TL | 11 | 787 | Expert Systems with Applications | Physica A-Statistical Mechanics and Its Applications | African Journal of Business Management | Computers & Mathematics with Applications | Data & knowledge Engineering | Economic Modelling | | | | |
| Wei, LY | 11 | 383 | Applied Soft Computing | Economic Modelling | International Journal of Innovative Comp. Info. and Control | African Journal of Business Management | Applied Soft Computing | Expert Systems with Applications | Neuro-computing | | | |
| Huang, XX | 10 | 871 | Journal of Computational and Applied Mathematics | Applied Mathematics and Computation | Computers & Mathematics with Applications | Expert Systems with Applications | IEEE Transactions on Fuzzy Systems | Information Sciences | International Journal of General Systems | Journal of Intelligent & Fuzzy Systems | Soft Computing | |

# Appendix B

**Table A2.** Fuzzy logic in banking resolution.

| Article | Authors | Journal | Publication Year | Purpose | Method |
|---|---|---|---|---|---|
| The use of fuzzy-clustering algorithm and self-organizing neural networks for identifying potentially failing banks: An experimental study | Alam, P; Booth, D; Lee, K; Thordarson, T | Expert Systems with Applications | 2000 | Early warning system for bank default | Fuzzy-clustering and neural networks |
| GenSo-EWS: a novel neural-fuzzy-based early warning system for predicting bank failures | Tung, WL; Quek, C; Cheng, P | Neural Networks | 2004 | Early warning system for bank default | Generic self-organizing fuzzy neural network (GenSoFNN) |
| A novel bankruptcy prediction model based on an adaptive fuzzy k-nearest neighbour method | Chen, HL; Yang, B; Wang, G; Liu, J; Xu, X; Wang, SJ; Liu, DY | Knowledge-Based Systems | 2011 | Predict bank failure | fuzzy k-nearest neighbour |
| Financial distress prediction in banks using Group Method of Data Handling neural network, counter propagation neural network and fuzzy ARTMAP | Ravisankar, P; Ravi, V | Knowledge-Based Systems | 2010 | Predict bank failure | Neural network and fuzzy ARTMAP |
| FCMAC-EWS: A bank failure early warning system based on a novel localized pattern learning and semantically associative fuzzy neural network | Ng, GS; Quek, C; Jiang, H | Expert Systems with Applications | 2008 | Early warning system for bank default | Fuzzy CMAC (cerebellar model articulation controller) model based on compositional rule of inference. (An integration of neural networks and fuzzy systems to create a hybrid structure known as a fuzzy neural network (FNN)) |
| Fuzzy Refinement Domain Adaptation for Long Term Prediction in Banking Ecosystem | Behbood, V; Lu, J; Zhang, GQ | IEEE Transactions on Industrial Informatics | 2014 | Predict bank failure | Fuzzy refinement domain adaptation algorithm |
| Multistep fuzzy Bridged Refinement Domain Adaptation Algorithm and Its Application to Bank Failure Prediction | Behbood, V; Lu, J.; Zhang, G; Pedrycz, W | IEEE Transactions on Fuzzy Systems | 2015 | Predict bank failure | Multistep fuzzy Bridged Refinement Domain Adaptation Algorithm |
| Using a Mixed Model to Explore Evaluation Criteria for Bank Supervision: A Banking Supervision Law Perspective | Tsai, SB; Chen, KY; Zhao, HR; Wei, YM; Wang, CK; Zheng, YX; Chang, LC; Wang, JT | Plos One | 2016 | Analyse bank supervision criteria | Fuzzy-DEMATEL |
| Aggregating expert knowledge for the measurement of systemic risk | Mezei, J; Sarlin, P | Decision Support Systems | 2016 | Provide a framework for measuring systemic risk by aggregating the knowledge of financial supervisors | Fuzzy cognitive maps (FCMs) and aggregation based on Choquet integrals |
| Financial institution failure prediction using adaptive neuro-fuzzy inference systems: Evidence from the east Asian economic crisis | Choensawat, W, Polsiri, P | Journal of Advanced Computational Intelligence and Intelligent Informatics | 2013 | Predict bank failure | ANFIS |
| On the measure of contagion in fuzzy financial networks | De Marco, G; Donnini, C; Gioia, F; Perla, F | Expert Systems with Applications | 2018 | Financial contagion risk | Fuzzy model using fuzzy numbers to represent the balance sheet |

## References


1. Boissay, F.; Collard, F.; Smets, F. Booms and Systemic Banking Crises. 2013.
2. Carrascosa, A.; SRB. Completing the Banking Union. 2018. Available online: http://www.europarl.europa.eu (accessed on 19 February 2019).
3. King, M.R. The Basel III Net Stable Funding Ratio and bank net interest margins. *J. Bank. Financ.* **2013**, *37*, 4144–4156. [CrossRef]
4. European Commission Regulation Proposal on Prudential Requirements for Credit Institutions and Investment Firms 2011. Available online: https://ec.europa.eu/info/publications/regulation (accessed on 2 December 2011).
5. Bělohlávek, R.; Dauben, J.W.; Klir, G.J. *Fuzzy Logic and Mathematics: A Historical Perspective*; Oxford University Press: Oxford, UK, 2017.
6. Kerre, E.E.; Mordeson, J.N. A historical overview of fuzzy mathematics. *New Math. Nat. Comput.* **2005**, *1*, 1–26. [CrossRef]
7. Zimmermann, H.J. *Fuzzy Set Theory and Its Applications*, 4th ed.; Springer Seience, Business Media: New York, NY, USA, 2001.
8. Šostaks, A. Mathematics in the context of fuzzy sets: Basic ideas, concepts, and some remarks on the history and recent trends of development. *Math. Model. Anal.* **2011**, *16*, 173–198. [CrossRef]
9. Gottwald, S. *Fuzzy Sets and Fuzzy Logic: The Foundations of Application—From a Mathematical Point of View*; Springer: Berlin, Germany, 2013.
10. Buckley, J.J. The fuzzy mathematics of finance. *Fuzzy Sets Syst.* **1985**, *21*, 257–273. [CrossRef]
11. Li Calzi, M. Towards a general setting for the fuzzy mathematics of finance. *Fuzzy Sets Syst.* **1990**, *35*, 265–280. [CrossRef]
12. Zadeh, L.A. Fuzzy sets. *Inf. Control* **1965**, *8*, 338–353. [CrossRef]
13. Venkat, N.R.; Kushal, R.N.; Sangam, S. Application of Fuzzy Logic in Financial Markets for Decision Making. *Int. J. Adv. Res. Comput. Sci.* **2017**, *8*, 382–386.
14. Werro, N. *Fuzzy Classi Cation of Online Customers*; University of Fribourg: Fribourg, Switzerland, 2008.
15. Zadeh, L.A. The concept of a linguistic variable and its application to approximate reasoning-I. *Inf. Sci.* **1975**, *8*, 199–249. [CrossRef]
16. Liu, W.; Liao, H. A Bibliometric Analysis of Fuzzy Decision Research During 1970–2015. *Int. J. Fuzzy Syst.* **2017**, *19*, 1–14. [CrossRef]
17. Lee, C.H.L.; Liu, A.; Chen, W.S. Pattern discovery of fuzzy time series for financial prediction. *IEEE Trans. Knowl. Data Eng.* **2006**, *18*, 613–625.
18. Mahfouf, M.; Abbod, M.F.; Linkens, D.A. A survey of fuzzy logic monitoring and control utilisation in medicine. *Artif. Intell. Med.* **2001**, *21*, 27–42. [CrossRef]
19. Rajab, S.; Sharma, V. A review on the applications of neuro-fuzzy systems in business. *Artif. Intell. Rev.* **2018**, *49*, 481–510. [CrossRef]
20. Shapiro, A.F. Fuzzy logic in insurance. *Insur. Math. Econ.* **2004**, *35*, 399–424. [CrossRef]
21. Von Altrock, C. *Fuzzy Logic and NeuroFuzzy Applications in Business and FInance*; Prentice-Hall, Inc.: Upper Saddle River, NJ, USA, 1996.
22. Bojadziev, G. *Fuzzy Logic for Business, Finance, and Management*; World Scientific Pub Co Inc.: Singapore, 2007.
23. Gil-Lafuente, A.M. *Fuzzy Logic in Financial Analysis*; Springer: Berlin, Germany, 2005.
24. Onwuegbuzie, A.J.; Leech, N.L.; Collins, K.M.T. Qualitative analysis techniques for the review of the literature. *Qual. Rep.* **2012**, *17*, 1–28.
25. Onwuegbuzie, A.J.; Collins, K.M.; Leech, N.L.; Dellinger, A.B.; Jiao, Q.G. A meta-framework for conducting mixed research syntheses for stress and coping researchers and beyond. In *Toward a Broader Understanding of Stress and Coping: Mixed Methods Approaches*; Information Age Publishing: Charlotte, NC, USA, 2010; pp. 169–211.
26. Okoli, C.; Schabram, K. *A Guide to Conducting a Systematic Literature Review of Information Systems Research*; Sprouts: Phoenix, AZ, USA, 2010; Volume 10.
27. Ensslin, L.; Ensslin, S.R.; Lacerda, R.T.; Tasca, J.E. ProKnow-C, Knowledge Development Process—Construtivist 2010. Available online: http://www.ucdoer.ie/index.php/Education_Theory (accessed on 20 December 2010).

28. Arruda, L.; França, S.; Quelhas, O. Mobile Computing: Opportunities for Improving Civil Constructions Productivity. *Int. Rev. Manag. Bus. Res.* **2014**, *3*, 648–655.

29. Ensslin, L.; Mussi, C.C.; Chaves, L.C.; Demetrio, S.N. Errata-"IT outsourcing management: The state of the art recognition by a constructivist process and bibliometrics". *J. Inf. Syst. Technol. Manag.* **2016**, *13*, 151. [CrossRef]

30. Thiel, G.G.; Ensslin, S.R.; Ensslin, L. Street lighting management and performance evaluation: Opportunities and challenges. *J. Local Self-Gov.* **2017**, *15*, 303–328. [CrossRef]

31. Drake, P.P.; Fabozzi, F.J. *The Basics of Finance: An Introduction to Financial Markets, Business Finance, and Portfolio Management*; John Wiley & Sons: Hoboken, NJ, USA, 2010.

32. Aghaei Chadegani, A.; Salehi, H.; Md Yunus, M.M.; Farhadi, H.; Fooladi, M.; Farhadi, M.; Ale Ebrahim, N. A comparison between two main academic literature collections: Web of science and scopus databases. *Asian Soc. Sci.* **2013**, *9*, 18–26. [CrossRef]

33. Mongeon, P.; Paul-hus, A. The journal coverage of bibliometric databases: A comparison of Scopus and Web of Science. *Scientometrics* **2014**, 1–6.

34. Moher, D.; Liberati, A.; Tetzlaff, J.; Altman, D.G. Preferred reporting items for systematic reviews and meta-analyses: The PRISMA statement. *J. Clin. Epidemiol.* **2009**, *6*, e1000097.

35. Behrens, A. Use of intervals and possibility distributions in economic analysis. *J. Oper. Res. Soc.* **1992**, *43*, 907–918.

36. Gutiérrez, I. Fuzzy numbers and net present value. *Scand. J. Manag.* **1989**, *5*, 149–159. [CrossRef]

37. Boyacioglu, M.A.; Avci, D. An adaptive network-based fuzzy inference system (ANFIS) for the prediction of stock market return: The case of the Istanbul stock exchange. *Expert Syst. Appl.* **2010**, *37*, 7908–7912. [CrossRef]

38. Tiryaki, F.; Ahlatcioglu, B. Fuzzy portfolio selection using fuzzy analytic hierarchy process. *Inf. Sci.* **2009**, *179*, 53–69. [CrossRef]

39. Verikas, A.; Kalsyte, Z.; Bacauskiene, M.; Gelzinis, A. Hybrid and ensemble-based soft computing techniques in bankruptcy prediction: A survey. *Soft Comput.* **2010**, *14*, 995–1010. [CrossRef]

40. Bahrammirzaee, A. A comparative survey of artificial intelligence applications in finance: Artificial neural networks, expert system and hybrid intelligent systems. *Neural Comput. Appl.* **2010**, *19*, 1165–1195. [CrossRef]

41. Atsalakis, G.S.; Valavanis, K.P. Surveying stock market forecasting techniques-Part I: Conventional methods. *Expert Syst. Appl.* **2009**, *36*, 5932–5941. [CrossRef]

42. Ravi Kumar, P.; Ravi, V. Bankruptcy prediction in banks and firms via statistical and intelligent techniques-A review. *Eur. J. Oper. Res.* **2006**, *180*, 1–28. [CrossRef]

43. Atsalakis, G.S.; Valavanis, K.P. Forecasting stock market short-term trends using a neuro-fuzzy based methodology. *Expert Syst. Appl.* **2009**, *36*, 10696–10707. [CrossRef]

44. Ijegwa, A.D.; Rebecca, V.O.; Olusegun, F.; Isaac, O.O. A Predictive Stock Market Technical Analysis Using Fuzzy Logic. *Comput. Inf. Sci.* **2014**, *7*, 1–17. [CrossRef]

45. Chang, P.C.; Liu, C.H. A TSK type fuzzy rule based system for stock price prediction. *Expert Syst. Appl.* **2008**, *34*, 135–144. [CrossRef]

46. Arenas Parra, M.; Bilbao Terol, A.; Rodríguez Uría, M.V. A fuzzy goal programming approach to portfolio selection. *Eur. J. Oper. Res.* **2001**, *133*, 287–297. [CrossRef]

47. Huang, X. Mean-entropy models for fuzzy portfolio selection. *IEEE Trans. Fuzzy Syst.* **2008**, *16*, 1096–1101. [CrossRef]

48. Shaverdi, M.; Ramezani, I.; Tahmasebi, R.; Rostamy, A.A.A. Combining Fuzzy AHP and Fuzzy TOPSIS with Financial Ratios to Design a Novel Performance Evaluation Model. *Int. J. Fuzzy Syst.* **2016**, *18*, 248–262. [CrossRef]

49. Nakano, M.; Takahashi, A.; Takahashi, S. Fuzzy logic-based portfolio selection with particle filtering and anomaly detection. *Knowl.-Based Syst.* **2017**, *131*, 113–124. [CrossRef]

50. Muzzioli, S.; De Baets, B. Fuzzy Approaches to Option Price Modeling. *IEEE Trans. Fuzzy Syst.* **2017**, *25*, 392–401. [CrossRef]

51. De Andrés-Sánchez, J. Pricing European Options with Triangular Fuzzy Parameters: Assessing Alternative Triangular Approximations in the Spanish Stock Option Market. *Int. J. Fuzzy Syst.* **2018**, *20*, 1624–1643. [CrossRef]

52. Li, H.; Ware, A.; Di, L.; Yuan, G.; Swishchuk, A.; Yuan, S. The application of nonlinear fuzzy parameters PDE method in pricing and hedging European options. *Fuzzy Sets Syst.* **2016**, *331*, 14–25. [CrossRef]

53. Muzzioli, S.; Torricelli, C. A multiperiod binomial model for pricing options in a vague world. *J. Econ. Dyn. Control* **2004**, *28*, 861–887. [CrossRef]

54. Dymova, L.; Sevastianov, P.; Bartosiewicz, P. A new approach to the rule-base evidential reasoning: Stock trading expert system application. *Expert Syst. Appl.* **2010**, *37*, 5564–5576. [CrossRef]

55. Huang, H.; Pasquier, M.; Quek, C. Financial market trading system with a hierarchical coevolutionary fuzzy predictive model. *IEEE Trans. Evol. Comput.* **2009**, *13*, 56–70. [CrossRef]

56. Huang, Y.; Jiang, W. Extension of TOPSIS Method and its Application in Investment. *Arab. J. Sci. Eng.* **2018**, *43*, 693–705. [CrossRef]

57. Kuo, R.J.; Chen, C.H.; Hwang, Y.C. An intelligent stock trading decision support system through integration of genetic algorithm based fuzzy neural network and artificial neural network. *Fuzzy Sets Syst.* **2001**, *118*, 21–45. [CrossRef]

58. Wu, B.; Tseng, N.-F. A new approach to fuzzy regression models with application to business cycle analysis. *Fuzzy Sets Syst.* **2002**, *130*, 33–42. [CrossRef]

59. Chiu, C.-Y.; Park, C.S. Fuzzy cash flow analysis using present worth criterion. *Eng. Econ.* **1994**, *39*, 113–138. [CrossRef]

60. McIvor, R.T.; McCloskey, A.G.; Humphreys, P.K.; Maguire, L.P. Using a fuzzy approach to support financial analysis in the corporate acquisition process. *Expert Syst. Appl.* **2004**, *27*, 533–547. [CrossRef]

61. Magni, C.A.; Malagoli, S.; Mastroleo, G. An alternative approach to firms' evaluation: Expert Systems and Fuzzy Logic. *Int. J. Inf. Technol. Decis. Mak.* **2006**, *5*, 195–225. [CrossRef]

62. Jiao, Y.; Syau, Y.R.; Lee, E.S. Modelling credit rating by fuzzy adaptive network. *Math. Comput. Model.* **2007**, *45*, 717–731. [CrossRef]

63. Amiri, M.P. Project selection for oil-fields development by using the AHP and fuzzy TOPSIS methods. *Expert Syst. Appl.* **2010**, *37*, 6218–6224. [CrossRef]

64. Mao, Y.; Wu, W. Fuzzy Real Option Evaluation of Real Estate Project Based on Risk Analysis. *Syst. Eng. Procedia* **2011**, *1*, 228–235. [CrossRef]

65. Jiménez, A.; Martín, M.C.; Mateos, A.; Pérez-Sénchez, D.; Dvorzhak, A. A fuzzy MCDA framework for safety assessment in the remediation of a uranium mill tailings site in Ukraine. Intelligent Systems and Decision Making for Risk Analysis and Crisis Response. In Proceedings of the 2013 4th International Conference on Risk Analysis and Crisis Response, RACR, Istanbul, Turkey, 27–29 August 2013; pp. 1–22.

66. De Wit, G.W. Underwriting and Uncertainty. *Insur. Math. Econ.* **1982**, *1*, 277–285. [CrossRef]

67. Lemaire, J. Fuzzy Insurance. *Astin Bull.* **1990**, *20*, 34–58. [CrossRef]

68. Ostaszewski, K.; Karwowski, W. *An Analysis of Possible Applications of Fuzzy Set Theory to the Actuarial Credibility Theory*; NASA. Johnson Space Center: Louisville, KY, USA, 1993.

69. Derrig, R.A.; Cummins, J.D. Fuzzy Trends in Property-Liability Insurance Claim Costs. *J. Risk Insur.* **1993**, *60*, 429–465.

70. Young, V.R. Insurance Rate Changing: A Fuzzy Logic Approach. *J. Risk Insur.* **1996**, *63*, 461–484. [CrossRef]

71. Casanovas, M.; Torres-Martínez, A.; Merigó, J.M. Decision making processes of non-life insurance pricing using Fuzzy Logic and OWA operators. *Econ. Comput. Econ. Cybern. Stud. Res.* **2015**. Available online: http://repositorio.uchile.cl/handle/2250/133834 (accessed on 8 June 2015).

72. Shapiro, A.F. The merging of neural networks, fuzzy logic, and genetic algorithms. *Insur. Math. Econ.* **2002**, *31*, 115–131. [CrossRef]

73. Xiao, Z.; Gong, K.; Zou, Y. A combined forecasting approach based on fuzzy soft sets. *J. Comput. Appl. Math.* **2009**, *228*, 326–333. [CrossRef]

74. Ammar, S.; Duncombe, W.; Jump, B.; Wright, R. Constructing a fuzzy-knowledge-based-system: An application for assessing the financial condition of public schools. *Expert Syst. Appl.* **2004**, *27*, 349–364. [CrossRef]

75. Keles, A.; Kolcak, M.; Keles, A. The adaptive neuro-fuzzy model for forecasting the domestic debt. *Knowl.-Based Syst.* **2008**, *21*, 951–957. [CrossRef]

76. Rivera-Lirio, J.M.; Muñoz-Torres, M.J. The effectiveness of the public support policies for the European industry financing as a contribution to sustainable development. *J. Bus. Ethics* **2010**, *94*, 489–515. [CrossRef]

77. Malhotra, R.; Malhotra, D.K. Differentiating between good credits and bad credits using neuro-fuzzy systems. *Eur. J. Oper. Res.* **2002**, *136*, 190–211. [CrossRef]

78. Wang, Y.; Wang, S.; Lai, K.K. A new fuzzy support vector machine to evaluate credit risk. *IEEE Trans. Fuzzy Syst.* **2005**, *13*, 820–831. [CrossRef]

79. Akkoç, S. An empirical comparison of conventional techniques, neural networks and the three stage hybrid Adaptive Neuro Fuzzy Inference System (ANFIS) model for credit scoring analysis: The case of Turkish credit card data. *Eur. J. Oper. Res.* **2012**, *222*, 168–178. [CrossRef]

80. Nazemi, A.; Fatemi Pour, F.; Heidenreich, K.; Fabozzi, F.J. Fuzzy decision fusion approach for loss-given-default modeling. *Eur. J. Oper. Res.* **2017**, *262*, 780–791. [CrossRef]

81. Bennouna, G.; Tkiouat, M. Fuzzy logic approach applied to credit scoring for micro finance in Morocco. *Procedia Comput. Sci.* **2018**, *127*, 274–283. [CrossRef]

82. Wanke, P.; Azad, A.K.; Emrouznejad, A. Efficiency in BRICS banking under data vagueness: A two-stage fuzzy approach. *Glob. Financ. J.* **2018**, *35*, 58–71. [CrossRef]

83. Amaglobeli, D.; End, N.; Jarmuzek, M.; Palomba, G. *From Systemic Banking Crises to Fiscal Costs: Risk Factors*; IMF Working Papers; International Monetary Fund N.W.: Washington, DC, USA; Volume WP/15/166.

84. Freixas, X.; Peydró, J.-L.; Laeven, L.; Freixas, X.; Laeven, L.; Peydró, J.-L. Systemic Risk and Macroprudential Regulation. In *Systemic Risk, Crises, and Macroprudential Regulation*; MIT Press: Cambridge, MA, USA, 2016.

85. Ravikumar, P.; Ravi, V. Bankruptcy prediction in banks by an ensemble classifier. In Proceedings of the IEEE International Conference on Industrial Technology, Mumbai, India, 15–17 December 2006; pp. 2032–2036.

86. Altman, E. Financial Ratios, Discriminant Analysis, and the Prediction of Corporate Bankruptcy. *J. Financ.* **1968**, *23*, 589–609. [CrossRef]

87. Sinkey, J.F. A Multivariate Statistical Analysis of the Characteristics of Problem Banks. *J. Financ.* **1975**, *7*, 77–91. [CrossRef]

88. Altman, E.I. Predicting performance in the savings and loan association industry. *J. Monet. Econ.* **1977**, *3*, 443–466. [CrossRef]

89. Alam, P.; Booth, D.; Lee, K.; Thordarson, T. The use of fuzzy clustering algorithm and self-organizing neural networks for identifying potentially failing banks: An experimental study. *Expert Syst. Appl.* **2000**, *18*, 185–199. [CrossRef]

90. Boyacioglu, M.A.; Kara, Y.; Baykan, Ö.K. Predicting bank financial failures using neural networks, support vector machines and multivariate statistical methods: A comparative analysis in the sample of savings deposit insurance fund (SDIF) transferred banks in Turkey. *Expert Syst. Appl.* **2009**, *36*, 3355–3366. [CrossRef]

91. Shin, K.S.; Lee, T.S.; Kim, H.J. An application of support vector machines in bankruptcy prediction model. *Expert Syst. Appl.* **2005**, *28*, 127–135. [CrossRef]

92. Olmeda, I.; Fernandez, E. Hybrid Classifiers for Financial Multicriteria Decision Making: The Case of Bankruptcy Prediction. *Comput. Econ.* **1997**, *1621*, 36–43.

93. Tung, W.L.; Quek, C.; Cheng, P. GenSo-EWS: A novel neural-fuzzy based early warning system for predicting bank failures. *Neural Netw.* **2004**, *17*, 567–587. [CrossRef] [PubMed]

94. Wilms, P.; Swank, J.; Haan, J. De Determinants of the real impact of banking crises: A review and new evidence. *N. Am. J. Econ. Financ.* **2018**, *43*, 54–70. [CrossRef]

95. De Marco, G.; Donnini, C.; Gioia, F.; Perla, F. On the measure of contagion in fuzzy financial networks. *Appl. Soft Comput. J.* **2018**, *67*, 584–595. [CrossRef]

96. Lou, Y.; Wang, M. Fraud Risk Factor Of The Fraud Triangle Assessing The Likelihood Of Fraudulent Financial Reporting. *J. Bus. Econ. Res.* **2009**, *7*, 61–78. [CrossRef]

97. Farrando, I. Bankia's IPO: Some Remarks on the Biggest Failure in the Spanish Banking System. 2018. Available online: https://ssrn.com/abstract=3176481 (accessed on 10 May 2018).

98. Sloan, T. Banco Espirito Santo and European banking regulation. 2015. Available online: http://data.europa.eu/88u/dataset/exercise-espirito-santo-financial-group-sa-esfg- (accessed on 27 July 2015).

99. Albrecht, W.S.; Romney, M.B.; Cherrington, D.J.; Payne, I.R.; Roe, A.J.; Romney, M.B. Red-flagging management fraud: A validation. *Adv. Account.* **1986**, *3*, 323–333.

100. Kalbers, L.P. Fraudulent financial reporting, corporate governance and ethics: 1987–2007. *Rev. Account. Financ.* **2009**, *15*, 65–84. [CrossRef]

101. Summers, S.L.; Sweeney, J.T. Fraudulently misstated financial statements and insider trading: An empirica analysis. *Account. Rev.* **1998**, *73*, 131–146.

102. Pathak, J.; Vidyarthi, N.; Summers, S.L. A fuzzy-based algorithm for auditors to detect elements of fraud in settled insurance claims. *Manag. Audit. J.* **2005**, *20*, 632–644. [CrossRef]
103. Ravisankar, P.; Ravi, V.; Raghava Rao, G.; Bose, I. Detection of financial statement fraud and feature selection using data mining techniques. *Decis. Support Syst.* **2011**, *50*, 491–500. [CrossRef]
104. Lin, J.W.; Hwang, M.I.; Becker, J.D. A fuzzy neural network for assessing the risk of fraudulent financial reporting. *Manag. Audit. J.* **2003**, *18*, 657–665. [CrossRef]
105. Mardani, A.; Jusoh, A.; Zavadskas, E.K. Fuzzy multiple criteria decision-making techniques and applications-Two decades review from 1994 to 2014. *Expert Syst. Appl.* **2015**, *42*, 4126–4148. [CrossRef]

 *mathematics*

Article

# On $\omega$-Limit Sets of Zadeh's Extension of Nonautonomous Discrete Systems on an Interval

**Guangwang Su [1,2], Taixiang Sun [1,2,*]**

[1] College of Information and Statistics, Guangxi University of Finance and Economics, Nanning 530003, China; s1g6w3@163.com
[2] Guangxi Key Laboratory Cultivation Base of Cross-border E-commerce Intelligent Information Processing, Nanning 530003, China
* Correspondence: stxhql@gxu.edu.cn; Tel.: +86-138-7815-8146

Received: 18 September 2019; Accepted: 13 November 2019; Published: 16 November 2019

**Abstract:** Let $I = [0,1]$ and $f_n$ be a sequence of continuous self-maps on $I$ which converge uniformly to a self-map $f$ on $I$. Denote by $\mathcal{F}(I)$ the set of fuzzy numbers on $I$, and denote by $(\mathcal{F}(I), \widehat{f})$ and $(\mathcal{F}(I), \widehat{f_n})$ the Zadeh's extensions of $(I, f)$ and $(I, f_n)$, respectively. In this paper, we study the $\omega$-limit sets of $(\mathcal{F}(I), \widehat{f_n})$ and show that, if all periodic points of $f$ are fixed points, then $\omega(A, \widehat{f_n}) \subset F(\widehat{f})$ for any $A \in \mathcal{F}(I)$, where $\omega(A, \widehat{f_n})$ is the $\omega$-limit set of $A$ under $(\mathcal{F}(I), \widehat{f_n})$ and $F(\widehat{f}) = \{A \in \mathcal{F}(I) : \widehat{f}(A) = A\}$.

**Keywords:** fuzzy number; $\omega$-limit set; periodic point; Zadeh's extension

## 1. Introduction

Research for the dynamical properties of nonautonomous discrete systems on a metric space is very interesting (see [1–11]). In [12], Kempf investigated the $\omega$-limit sets of a sequence of continuous self-maps $f_n$ on $I$ which converge uniformly to a self-map $f$ on $I$ and showed that, if $P(f) = F(f)$, then $\omega(x, f_n)$ is a closed subset of $I$ with $\omega(x, f_n) \subset F(f)$ for any $x \in I$, where $F(f)$ and $P(f)$ are the set of fixed points of $f$ and the set of periodic points of $f$, respectively, and $\omega(x, f_n)$ is the set of $\omega$-limit points of $x$ under $(X, f_n)$. Further, Cánovas [13] showed that, if $f_n$ is a sequence of continuous self-maps on $I$ which converge uniformly to a self-map $f$ on $I$ and $P(f) = F(f^{2^s})$ for some $s \in \mathbb{N}$, then $\omega(x, f_n) = \cup_{k=1}^{2^s}[p_k, q_k] \subset F(f^{2^s})$ with $f([p_k, q_k]) = [p_{k+1}, q_{k+1}]$ $(1 \leq k \leq 2^s - 1)$ and $f([p_{2^s}, q_{2^s}]) = [p_1, q_1]$ for any $x \in I$. In [14], we studied the $\omega$-limit sets of a sequence of continuous self-maps $f_n$ on a tree $T$ which converge uniformly to a self-map $f$ on $T$ and showed that, if $P(f) = F(f)$, then $\omega(x, f_n)$ is a closed connected subset of $T$ with $\omega(x, f_n) \subset F(f)$ for any $x \in T$.

It is well known [15] that the discrete dynamical system $(X, f)$ naturally induces a dynamical system $(\mathcal{F}(X), \widehat{f})$, where $\mathcal{F}(X)$ is the set of all fuzzy sets on a metric space $X$ and $\widehat{f}$ is the Zadeh's extension of continuous self-maps $f$ on $X$. It is natural to ask how the dynamical properties of $f$ is related to the dynamical properties of $\widehat{f}$. Already, there are many results for this question so far; see, e.g., References [16–21] and the related references therein, where different chaotic properties and topological entropies of Zadeh's extensions of continuous seif-maps on metric spaces were considered. Our aim in this paper is to study the $\omega$-limit sets of Zadeh's extensions of nonautonomous discrete systems on intervals.

## 2. Preliminaries

Throughout this paper, let $(X, d)$ be a metric space, write $I = [0,1]$, and denote by $\mathbb{N}$ the set of all positive integers. Let $C^0(X)$ be the set of all continuous self-maps on $X$. For a given $f \in C^0(X)$, let $f^{n+1} = f \circ f^n$ for any $n \in \mathbb{N}$. We call $F(f) = \{x \in X : f(x) = x\}$ the set of fixed points of $f$ and $P(f) = \{x \in X : f^n(x) = x$ for some $n \in \mathbb{N}\}$ the set of periodic points of $f$.

Let $f_n \in C^0(X)$ $(n \in \mathbb{N})$ and $F^0$ be the identity map of $X$, and write

$$F_n = f_n \circ f_{n-1} \circ \cdots \circ f_1 \quad \text{for any } n \in \mathbb{N}.$$

$y \in X$ is called the $\omega$-limit point of $x(\in X)$ under $(X, f_n)$ if there are $n_1 < n_2 < \cdots < n_k < \cdots$ such that

$$\lim_{k \longrightarrow \infty} F_{n_k}(x) = y.$$

Denote by $\omega(x, f_n)$ the set of $\omega$-limit points of $x$ under $(X, f_n)$. We write $f_n \Longrightarrow f$ if $f_n$ converges uniformly to $f$.

Now, let us recall some definitions for fuzzy theory which are from [15].

**Definition 1.** *Let X be a metric space. A mapping $A : X \longrightarrow [0, 1]$ is called a fuzzy set on X. For each fuzzy set A and each $\alpha \in (0, 1]$, $A_\alpha = \{t \in X : A(t) \geq \alpha\}$ is called an $\alpha$-level set of A and $A_0 = \overline{\{t \in X : A(t) > 0\}}$ is called the support of A, where $\overline{B}$ means the closure of subset B of X.*

**Definition 2.** *A fuzzy set A on I is said to be a fuzzy number if it satisfies the following conditions:*

(1) $A_1 \neq \varnothing$;
(2) *A is an upper semicontinuous function;*
(3) *For any $t_1, t_2 \in I$ and any $\lambda \in [0, 1]$, $A(\lambda t_1 + (1 - \lambda)t_2) \geq \min\{A(t_1), A(t_2)\}$;*
(4) *$A_0$ is compact.*

Let $\mathcal{F}(I)$ denote the set of fuzzy numbers on $I$. It is known that $\alpha$-level set $A_\alpha$ of $A$ determines the fuzzy number $A$ and that every $A_\alpha$ is a closed connected subset of $I$. If $A \in I$, then $A \in \mathcal{F}(I)$ with $A_\alpha = [A, A]$ for any $\alpha \in [0, 1]$.

For any $A, B \in \mathcal{F}(I)$ with $A_\alpha = [A_{l,\alpha}, A_{r,\alpha}]$ and $B_\alpha = [B_{l,\alpha}, B_{r,\alpha}]$ for any $\alpha \in (0, 1]$, we define the metric of $A$ and $B$ as follows:

$$D(A, B) = \sup_{\alpha \in (0,1]} \max\{|A_{l,\alpha} - B_{l,\alpha}|, |A_{r,\alpha} - B_{r,\alpha}|\}.$$

Obviously, we have

$$D(A, B) = \sup_{\alpha \in (0,1]} \max\{\sup_{x \in A_\alpha} d(x, B_\alpha), \sup_{y \in B_\alpha} d(y, A_\alpha)\} H(A_\alpha, B_\alpha),$$

where $d(x, J) = \inf_{y \in J} d(x, y)$ for any $x \in I$ and $J \subset I$. It is known that $(\mathcal{F}(I), D)$ is a complete metric space (refer to [15]).

Let $f \in C^0(I)$. We define the Zadeh's extension $\widehat{f} : \mathcal{F}(I) \longrightarrow \mathcal{F}(I)$ of $f$ for any $x \in I$ and $A \in \mathcal{F}(I)$ by

$$(\widehat{f}(A))(x) = \sup_{f(y)=x} A(y).$$

It follows from [15] that $f$ is continuous if and only if $\widehat{f}$ is continuous, and it follows from [22] (Lemma 2.1) that

$$[\widehat{f}(A)]_\alpha = f(A_\alpha)$$

for any $A \in \mathcal{F}(I)$ and $\alpha \in (0, 1]$. In this paper, we will show the following theorem.

**Theorem 1.** *Let $f_n$ be a sequence of continuous self-maps on I with $f_n \Longrightarrow f$. If $P(f) = F(f)$, then $\omega(A, \widehat{f_n}) \subset F(\widehat{f})$ for any $A \in \mathcal{F}(I)$.*

## 3. Proof of the Main Result

In this section, we let $f \in C^0(I)$ and $\widehat{f}$ is the Zadeh's extension of $f$. Let $\widehat{F}^0$ be the identity map of $\mathcal{F}(I)$ and for any $n \in \mathbb{N}$, we write

$$\widehat{F}_n = \widehat{f}_n \circ \widehat{f}_{n-1} \circ \cdots \circ \widehat{f}_1.$$

**Lemma 1.** *Assume that $f_n \in C^0(I)$ for any $n \in \mathbb{N}$ with $f_n \Longrightarrow f$. Then, $\widehat{f}_n \Longrightarrow \widehat{f}$ on $\mathcal{F}(I)$.*

**Proof.** Since $f_n \Longrightarrow f$ on $I$, it follows that, for any $\varepsilon > 0$, there is an $N \in \mathbb{N}$ such that, when $n \geq N$, we have

$$|f(x) - f_n(x)| < \frac{\varepsilon}{2}$$

for any $x \in I$, which implies that, for any $B \subset I$, we have $d(z, f_n(B)) < \varepsilon/2$ for any $z \in f(B)$ and $d(z, f(B)) < \varepsilon/2$ for any $z \in f_n(B)$. Thus when $n \geq N$, we have

$$D(\widehat{f}(A), \widehat{f}_n(A)) = \sup_{\alpha \in (0,1]} H(f(A_\alpha), f_n(A_\alpha)) \leq \frac{\varepsilon}{2} < \varepsilon$$

for any $A \in \mathcal{F}(I)$. Lemma 1 is proven. $\square$

**Lemma 2.** *Assume that $f_n \in C^0(I)$ for any $n \in \mathbb{N}$ with $f_n \Longrightarrow f$. If $B \in \omega(A, \widehat{f}_n)$ for some $A \in \mathcal{F}(I)$, then $\omega(x, f_n) \cap B_\alpha \neq \varnothing$ for any $\alpha \in (0,1]$ and $x \in A_\alpha$.*

**Proof.** Let $B \in \omega(A, \widehat{f}_n)$ and $\alpha \in (0,1]$. Let $n_1 < n_2 < \cdots < n_k < \cdots$ such that

$$\lim_{k \longrightarrow \infty} D(\widehat{F}n_k(A), B) = 0.$$

Then,

$$\lim_{k \longrightarrow \infty} H(Fn_k(A_\alpha), B_\alpha) = 0.$$

Let $x \in A_\alpha$. By taking a subsequence, we let $\lim_{k \longrightarrow \infty} Fn_k(x) = y \in \omega(x, f_n)$. If $y \notin B_\alpha$, then $\varepsilon = d(y, B_\alpha) > 0$. Since $\lim_{k \longrightarrow \infty} H(Fn_k(A_\alpha), B_\alpha) = 0$, there is an $N \in \mathbb{N}$ such that, when $n_k > N$, we have

$$H(Fn_k(A_\alpha), B_\alpha) < \frac{\varepsilon}{2},$$

which implies $d(Fn_k(x), B_\alpha) < \varepsilon/2$ and $\lim_{k \longrightarrow \infty} Fn_k(x) \neq y$ since $Fn_k(x) \in Fn_k(A_\alpha)$. This is a contradiction. Thus, $y \in B_\alpha$. Lemma 2 is proven. $\square$

**Proposition 1.** *Assume that $f_n \in C^0(I)$ for any $n \in \mathbb{N}$ with $f_n \Longrightarrow f$ and $P(f) = F(f)$. Then, the following statements hold:*

(1)  *If $B \in \omega(A, \widehat{f}_n)$ for some $A \in \mathcal{F}(I)$, then $\varnothing \neq f(B_\alpha) \cap B_\alpha \cap \omega(x, f_n) \subset F(f)$ for any $\alpha \in (0, 1]$ and $x \in A_\alpha$.*

(2)  *If $B \in \omega(A, \widehat{f}_n)$ for some $A \in \mathcal{F}(I)$, then $\cup_{B \in \omega(A, \widehat{f}_n)} B_\alpha \cup \omega(x, f_n)$ is a connected subset of $I$ for any $\alpha \in (0, 1]$ and $x \in A_\alpha$.*

**Proof.** It follows from Theorem 1 and Lemma 2. $\square$

**Lemma 3** (See [14] (Lemma 2)). *Assume that $f \in C^0(I)$ with $F(f) = P(f)$. Then, for any $x \in I$ and $n \in \mathbb{N}$, $f^n(x) > x$ if $f(x) > x$ and $f^n(x) < x$ if $f(x) < x$.*

Now, we show the main result of this paper.

**Proof of Theorem 1.** Let $B \in \omega(A, \widehat{f}_n)$. For any $\alpha \in (0,1]$, write $B_\alpha = [a_\alpha, b_\alpha]$ and $f(B_\alpha) = [c_\alpha, d_\alpha]$. Let $n_1 < n_2 < \cdots < n_k < \cdots$ such that

$$\lim_{k \longrightarrow \infty} D(\widehat{F}n_k(A), B) = 0. \tag{1}$$

By $\widehat{f} \in C^0(\mathcal{F}(I))$, we see that, for any $\varepsilon > 0$, there is an $\delta = \delta(\varepsilon) > 0$ such that, if $D(B,C) < \delta$ with $C \in \mathcal{F}(I)$, then

$$D(\widehat{f}(B), \widehat{f}(C)) < \frac{\varepsilon}{3}.$$

By Lemma 1, we see that there is an $N = N(\varepsilon) \in \mathbb{N}$ such that , when $n \geq N$, we have

$$D(\widehat{f}(W), \widehat{f}_n(W)) \leq \frac{\varepsilon}{3} \tag{2}$$

for any $W \in \mathcal{F}(I)$. Take $r = n_k \geq N$ such that $D(\widehat{F}_r(A), B) < \delta$. Thus,

$$D(\widehat{f}(B), \widehat{F}_{r+1}(A)) \leq D(\widehat{f}(B), \widehat{f}(\widehat{F}_r(A))) + D(\widehat{f}(\widehat{F}_r(A)), \widehat{F}_{r+1}(A)) \leq \frac{2\varepsilon}{3}. \tag{3}$$

In the following, we show that $a_\alpha = c_\alpha$ and $b_\alpha = d_\alpha$. For convenience, write $a_\alpha = a, b_\alpha = b, c_\alpha = c$, and $d_\alpha = d$.

(i) We will show $c \leq a$. Assume on the contrary that $c > a$. Then, by Proposition 1, we see $c \leq b$.

We claim that there is an $u \in (a, 1]$ such that $f(u) = a$. Indeed, if $f([a, 1]) \subset (a, 1]$, then let $\varepsilon = \min\{d(a, f([a, 1])), c - a\} > 0$. By Equation (3), we see $F_{r+1}(A_\alpha) \subset [a + \varepsilon/3, 1]$. It follows from Equation (2) that

$$H(f(F_{r+1}(A_\alpha)), F_{r+2}(A_\alpha)) \leq \frac{\varepsilon}{3}.$$

Thus, $F_{r+2}(A_\alpha) \subset [a + \varepsilon/3, 1]$. Continuing in this fashion, we have that $F_n(A_\alpha) \subset [a + \varepsilon/3, 1]$ for any $n \geq r + 1$, which contradicts Equation (1). The claim is proven.

Let $u = \min\{x \in (a, 1] : f(x) = a\}$. Then, $u > b$ since $f([a, b]) = [c, d]$ and $u > d$ (Otherwise, if $b < u \leq d$, then there exists an $u_1 \in [a, b]$ and $u_2 \in [u_1, u]$ such that $u_1 = f(u_2) < u_2 \leq d = f(u_1)$. This contradicts Lemma 3.). By Lemma 3, we see $f([a, u]) \subset [a, u)$. Write

$$
\begin{aligned}
p &= \max\{b, d, \max f([a, u])\}, \\
\varepsilon_1 &= (u - p)/2, \\
q &= \min\{c, \min f([a, p + \varepsilon_1])\}, \\
\varepsilon &= \min\{(q - a)/2, \varepsilon_1\}.
\end{aligned}
$$

By Equation (3), we see $F_{r+1}(A_\alpha) \subset [q - \varepsilon, p + \varepsilon_1]$. It follows from Equation (2) that

$$H(f(F_{r+1}(A_\alpha)), F_{r+2}(A_\alpha)) \leq \frac{\varepsilon}{3}.$$

Thus, $F_{r+2}(A_\alpha) \subset [q - \varepsilon, p + \varepsilon_1]$. Continuing in this fashion, we have that $F_n(A_\alpha) \subset [q - \varepsilon, p + \varepsilon_1]$ for any $n \geq r + 1$, which contradicts Equation (1).

(ii) In similar fashion, we can show $d \geq b$.

(iii) We will show that, if $c = a$, then $d = b$. Assume on the contrary that $d > b$. Let $u = \max\{z \in [a, b] : f(z) = d\}$ and $e = \min\{z \in [a, b] : f(z) = a\}$. Then, we have $e < u$ (Otherwise, if $e > u$, then there is an $w \in [u, e]$ satisfying $u = f(w) < w < d = f^2(w)$. This contradicts Lemma 3.) and $f(a) < u$.

We claim that there is an $v \in (u, 1]$ such that $f(v) = u$. Indeed, if $p = \min f([u, 1]) > u$, then let $\varepsilon = \min\{(d - b)/2, d(u, p)\} > 0$. By Equation (3), we see

$$D(\widehat{f}(B), \widehat{F}_{r+1}(A)) \leq \frac{2\varepsilon}{3}.$$

If $F_n(A_\alpha) \cap [a, u] \neq \varnothing$ for any $n \geq r + 1$, then we have $d - 2\varepsilon/3 \in F_n(A_\alpha)$ for any $n \geq r + 1$, which contradicts Equation (1). If $F_n(A_\alpha) \cap [a, u] = \varnothing$ for some $n \geq r + 1$, then let $m = \min\{F_n(A_\alpha) \cap [a, u] = \varnothing : n \geq r + 1\}$. Thus, $F_m(A_\alpha) \subset (u, 1]$, and it follows from Equation (2) that

$$H(f(F_m(A_\alpha)), F_{m+1}(A_\alpha)) \leq \frac{\varepsilon}{3},$$

which implies $F_{m+1}(A_\alpha) \subset [p - \varepsilon/3, 1] \subset (u, 1]$. Continuing in this fashion, we obtain that $F_n(A_\alpha) \subset [p - \varepsilon/3, 1] \subset (u, 1]$ for any $n \geq m$, which contradicts Equation (1). The claim is proven.

Let $v = \min\{x \in (u, 1] : f(x) = u\}$. Then, by Lemma 3, we see $d < v$ and $p = \max f([u, v]) < v$.

If there is an $w \in [0, a)$ satisfying $f(w) = u$, then let $w = \max\{x \in [0, a) : f(x) = u\}$. By Lemma 3, we see $q = \min f([w, a]) > w$ and $f([w, v]) = [q, p]$. Write

$$\begin{aligned}
\varepsilon_1 &= (v - p)/2, \\
z &= \min f([u, p + \varepsilon_1]) > u, \\
\varepsilon &= \min\{(d - b)/2, (q - w)/2, (z - u)/2, \varepsilon_1\}.
\end{aligned}$$

By Equation (3), we see

$$D(\widehat{f}(B), \widehat{F}_{r+1}(A)) \leq \frac{2\varepsilon}{3}.$$

This implies $F_{r+1}(A_\alpha) \subset [q - \varepsilon, p + \varepsilon_1]$. If $F_n(A_\alpha) \cap [a, u] \neq \varnothing$ for any $n \geq r + 1$, then we have $d - 2\varepsilon/3 \in F_n(A_\alpha)$ for any $n \geq r + 1$, which contradicts Equation (1). If $F_n(A_\alpha) \cap [a, u] = \varnothing$ for some $n \geq r + 1$, then let $m = \min\{F_n(A_\alpha) \cap [a, u] = \varnothing : n \geq r + 1\}$. Thus, $F_m(A_\alpha) \subset (u, p + \varepsilon_1]$, and it follows from Equation (2) that

$$H(f(F_m(A_\alpha)), F_{m+1}(A_\alpha)) \leq \frac{\varepsilon}{3},$$

which implies $F_{m+1}(A_\alpha) \subset [z - \varepsilon/3, p + \varepsilon_1] \subset (u, p + \varepsilon_1]$. Continuing in this fashion, we obtain that $F_n(A_\alpha) \subset [z - \varepsilon/3, p + \varepsilon_1] \subset (u, p + \varepsilon_1]$ for any $n \geq m$, which contradicts Equation (1).

If $\max f([0, a]) < u$, then $f([0, v]) = [0, p]$. Using the similar arguments as ones developed in the above given proof, we also obtain a conclusion which contradicts Equation (1).

(iv) We will show $c = a$. Assume on the contrary that $c < a$. Then by claim (ii), we see $b \leq d$. Using the similar arguments as ones developed in the proof of claim (iii), we can obtain $b < d$. Let $\varepsilon = \min\{(a - c)/2, (d - b)/2\}$. By Equation (3), we see $F_{r+1}(A_\alpha) \supset [a - \varepsilon, b + \varepsilon]$. It follows from Equation (2) that

$$H(f(F_{r+1}(A_\alpha)), F_{r+2}(A_\alpha)) \leq \frac{\varepsilon}{3}.$$

Thus, $F_{r+2}(A_\alpha) \supset [a - \varepsilon, b + \varepsilon]$. Continuing in this fashion, we have that $F_n(A_\alpha) \supset [a - \varepsilon, b + \varepsilon]$ for any $n \geq r + 1$, which contradicts Equation (1).

By claims (iii) and (iv), we see $f(B_\alpha) = B_\alpha$ for any $\alpha \in (0, 1]$, which implies $f(B) = B$. Theorem 1 is proven. $\square$

Using the similar arguments as ones developed in the proofs of Proposition 1.4 of [13] and Theorem 1, we may show the following result.

**Corollary 1.** *Let $f_n \in C^0(I)$ for any $n \in \mathbb{N}$ with $f_n \Longrightarrow f$. If $P(f) = F(f^{2^s})$ for some $s \in \mathbb{N}$, then $\omega(A, \widehat{f}_n) \subset F(\widehat{f}^{\,2^s})$ for any $A \in \mathcal{F}(I)$.*

The following example illustrates that there are $f_n \in C^0(I)$ for any $n \in \mathbb{N}$ such that $f_n \Longrightarrow f$ with $P(f) = F(f)$ and $\omega(A, \widehat{f_n}) = \varnothing$ for some $A \in \mathcal{F}(I)$.

**Example 1.** *Let* $f \in C^0(I)$ *with* $f(1) = 1 > 0 = f(0)$ *and* $x < f(x)$ *for any* $x \in (0,1)$ *and* $f_n \equiv f$ *for any* $n \in \mathbb{N}$. *Thus,* $f_n \Longrightarrow f$. *We define* $A \in \mathcal{F}(I)$ *for any* $x \in I$ *by*

$$A(x) = -x + 1.$$

*By calculation, we have* $A_\alpha = [0, 1 - \alpha]$ *for any* $\alpha \in (0,1]$ *and* $f^n(A_1) = \{0\}$ *for any* $n \in \mathbb{N}$. *In the following, we assume that* $\alpha \in (0,1)$ *and let* $f^n(A_\alpha) = [a_n(\alpha), b_n(\alpha)]$. *Then,* $a_n(\alpha) = 0$ *for any* $n \in \mathbb{N}$ *and* $b_n(\alpha) \leq b_{n+1}(\alpha) \leq 1$ *for any* $n \in \mathbb{N}$ *and* $b_n(\alpha) \longrightarrow 1$. *Since* $b(\alpha) = 1$ *is not left continuous at* $\alpha = 1$, *by Theorem 2.1 of [23], there is not a* $B \in \mathcal{F}(I)$ *such that* $B_\alpha = [a(\alpha), b(\alpha)] = [0,1]$ *for any* $\alpha \in (0,1]$. *Thus,* $\omega(A, \widehat{f}) = \varnothing$.

## 4. Conclusions

In this paper, we investigated the $\omega$-limit sets of Zadeh's extensions of a nonautonomous discrete system $f_n$ on an interval which converges uniformly to a map $f$ and show that, if $P(f) = F(f)$, then $\omega(A, \widehat{f_n}) \subset F(\widehat{f})$ for any $A \in \mathcal{F}(I)$.

**Author Contributions:** Conceptualization, G.S. and T.S.; methodology, G.S. and T.S.; validation, G.S. and T.S.; formal analysis, G.S.; writing—original draft preparation, T.S.; writing—review and editing, G.S. and T.S.; funding acquisition, G.S. and T.S.; the final form of this paper is approved by all authors.

**Funding:** This work is supported by NNSF of China (11761011) and NSF of Guangxi (2018GXNSFAA294010 and 2016GXNSFAA380286) and by SF of Guangxi University of Finance and Economics (2019QNB10).

**Conflicts of Interest:** The authors declare no conflict of interest.

## References

1. Balibrea, F.; Oprocha, P. Weak mixing and chaos in nonautonomous discrete systems. *Appl. Math. Lett.* **2012**, *25*, 1135–1141. [CrossRef]
2. Dvorakova, J. Chaos in nonautonomous discrete dynamical systems. *Commun. Nonlinear Sci. Numer. Simul.* **2012**, *17*, 4649–4652. [CrossRef]
3. Kawan, C.; Latushkin, Y. Some results on the entropy of non-autonomous dynamical systems. *Dynam. Syst. Inter. J.* **2016**, *31*, 251–279. [CrossRef]
4. Lan, Y.; Peris, A. Weak stability of non-autonomous discrete dynamical systems. *Topo. Appl.* **2018**, *250*, 53–60. [CrossRef]
5. Liu, L.; Chen, B. On $\omega$-limit sets and attracton of non-autonomous discrete dynamical systems. *J. Korean Math. Soc.* **2012**, *49*, 703–713. [CrossRef]
6. Liu, L.; Sun, Y. Weakly mixing sets and transitive sets for non-autonomous discrete systems. *Adv. Differ. Equ.* **2014**, *2014*, 217. [CrossRef]
7. Ma, C.; Zhu, P.; Lu, T. Some chaotic properties of non-autonomous discrete fuzzy dynamical systems. In Proceedings of the 2016 IEEE International Conference on Fuzzy Systems (FUZZ-IEEE), Vancouver, BC, Canada, 24–29 July 2016; pp. 46–49.
8. Miralles, A.; Murillo-Arcila, M.; Sanchis, M. Sensitive dependence for non-autonomous discrete dynamical systems. *J. Math. Anal. Appl.* **2018**, *463*, 268–275. [CrossRef]
9. Murillo-Arcila, M.; Peris A. Mixing properties for nonautonomous linear dynamics and invariant sets. *Appl. Math. Lett.* **2013**, *26*, 215–218. [CrossRef]
10. Rasouli, H. On the shadowing property of nonautonomous discrete systems. *Int. J. Nonlinear Anal. Appl.* **2016**, *7*, 271–277.
11. Tang, X.; Chen, G.; Lu, T. Some iterative properties of (F-1, F-2)-chaos in non-autonomous discrete systems. *Entropy* **2018**, *20*, 188. [CrossRef]

12. Kempf, R. On $\Omega$-limit sets of discrete-time dynamical systems. *J. Differ. Equ. Appl.* **2002**, *8*, 1121–1131. [CrossRef]

13. Cánovas, J.S. On $\omega$-limit sets of non-autonomous discrete systems. *J. Differ. Equ. Appl.* **2006**, *12*, 95–100. [CrossRef]

14. Sun, T. On $\omega$-limit sets of non-autonomous discrete systems on trees. *Nonlinear Anal.* **2008**, *68*, 781–784. [CrossRef]

15. Kupka, J. On fuzzifications of discrete dynamical systems. *Inform. Sci.* **2011**, *181*, 2858–2872. [CrossRef]

16. Boroński, J.P.; Kupka, J. The topology and dynamics of the hyperspaces of normal fuzzy sets and their inverse limit spaces. *Fuzzy Sets Sys.* **2017**, *321*, 90–100. [CrossRef]

17. Cánovas, J.S.; Kupka, J. Topological entropy of fuzzified dynamical systems. *Fuzzy Sets Sys.* **2011**, *165*, 67–79. [CrossRef]

18. Cánovas, J.S.; Kupka, J. On the topological entropy on the space of fuzzy numbers. *Fuzzy Sets Sys.* **2014**, *257*, 132–145. [CrossRef]

19. Cánovas, J.S.; Kupka, J. On fuzzy entropy and topological entropy of fuzzy extensions of dynamical systems. *Fuzzy Sets Sys.* **2017**, *309*, 115–130. [CrossRef]

20. Kim, C.; Chen, M.; Ju, H. Dynamics and topological entropy for Zadeh's extension of a compact system. *Fuzzy Sets Syst.* **2017**, *319*, 93–103. [CrossRef]

21. Yan, K.; Zeng, F. Conditional fuzzy entropy of fuzzy dynamical systems. *Fuzzy Sets Sys.* **2018**, *342*, 138–152. [CrossRef]

22. Papaschinopoulos, G.; Papadopoulos, B.K. On the fuzzy difference equation $x_{n+1} = A + x_n / x_{n-m}$. *Fuzzy Sets Sys.* **2002**, *129*, 73–81. [CrossRef]

23. Wu, C.; Zhang, B. Embedding problem of noncompact fuzzy number space $E^-(I)$. *Fuzzy Sets Sys.* **1999**, *105*, 165–169. [CrossRef]

*Article*

# A Fuzzy Collaborative Approach for Evaluating the Suitability of a Smart Health Practice

**Tin-Chih Toly Chen [1], Yu-Cheng Wang [2,*], Yu-Cheng Lin [3], Hsin-Chieh Wu [4] and Hai-Fen Lin [5]**

[1]   Department of Industrial Engineering and Management, National Chiao Tung University, 1001, University Road, Hsinchu 30010, Taiwan; tolychen@ms37.hinet.net

[2]   Department of Aeronautical Engineering, Chaoyang University of Technology, Taichung 41349, Taiwan

[3]   Department of Computer-Aided Industrial Design, Overseas Chinese University, Taichung 40721, Taiwan; yclin@ocu.edu.tw

[4]   Department of Industrial Engineering and Management, Chaoyang University of Technology, Taichung 41349, Taiwan; hcwul@cyut.edu.tw

[5]   Electronic Systems Research Division, National Chung-Shan Institute of Science & Technology, Taoyuan 32557, Taiwan; belinda.d853843@msa.hinet.net

*   Correspondence: tony.cobra@msa.hinet.net

Received: 12 October 2019; Accepted: 2 December 2019; Published: 3 December 2019

**Abstract:** A fuzzy collaborative approach is proposed in this study to assess the suitability of a smart health practice, which is a challenging task, as the participating decision makers may not reach a consensus. In the fuzzy collaborative approach, each decision maker first applies the alpha-cut operations method to derive the fuzzy weights of the criteria. Then, fuzzy intersection is applied to aggregate the fuzzy weights derived by all decision makers to measure the prior consensus among them. The fuzzy intersection results are then presented to the decision makers so that they can subjectively modify the pairwise comparison results to bring them closer to the fuzzy intersection results. Thereafter, the consensus among decision makers is again measured. The collaboration process will stop when no more modifications are made by any decision maker. Finally, the fuzzy weighted mean-centroid defuzzification method is applied to assess the suitability of a smart health practice. The fuzzy collaborative approach and some existing methods have been applied to assess the suitabilities of eleven smart health practices for a comparison. Among the compared practices, only the fuzzy collaborative approach could guarantee the existence of a full consensus among decision makers after the collaboration process, i.e., that the assessment results were acceptable to all decision makers.

**Keywords:** smart health; fuzzy collaborative intelligence; fuzzy analytic hierarchy process; suitability

---

## 1. Introduction

The application of smart technologies to enhance mobile health care, i.e., so-called smart health, has received a lot of attention [1,2]. As smart health practices are becoming more and more sophisticated, how to recommend suitable smart health practices to a target population becomes a challenging task. For example, Chen and Chiu [3] reviewed the literature and concluded that the most effective smart health practices adopted smart mobile services, smart phones, smart glasses/spectacles/contact lens, and smart surveillance cameras. Haymes et al. [4] applied behavior analysis to find out factors that contributed to the success of smart health practices for individuals with intellectual disabilities. Chiu and Chen [5] assessed the sustainable effectiveness of the adjustment mechanism of a ubiquitous clinic recommendation system by modelling the improvement in the successful recommendation rate as a learning process. Chen [6] put forward a hybrid methodology combining fuzzy geometric mean (FGM), alpha-cut operations (ACO), and fuzzy weighted mean (FWM) to evaluate the sustainability of a smart health practice, in which FGM, ACO, and FWM were for aggregation, prioritization, and assessment, respectively. Compared with earlier studies, the FGM-ACO-FWM method was more precise because of the application of the exact solution technique ACO. However, whether decision makers reached a consensus was not checked before aggregating their judgments, which was problematic [7]. To resolve this problem, a fuzzy collaborative approach is proposed in this study. In the fuzzy collaborative approach, decision makers' judgments will be aggregated only after they achieve a consensus.

A fuzzy collaborative approach is proposed in this study to evaluate the suitability of a smart health practice. In the proposed methodology, multiple decision makers fulfill the assessment task collaboratively. For each decision maker, ACO is applied to derive the fuzzy weights of criteria. Then, fuzzy intersection (FI) (or the minimum t-norm) is applied to aggregate the fuzzy weights derived by all decision makers. Obviously, the proposed methodology is a posterior-aggregation fuzzy analytic hierarchy process (FAHP) method, while the FGM-ACO-FWM method proposed by Chen [6] is an anterior-aggregation method. The FI results can be used to measure the prior consensus among decision makers. If the FI results are wide, i.e., many possible values are acceptable to all decision makers, then the consensus is high. Subsequently, the FI results are presented to decision makers so that they can subjectively modify the pairwise comparison results to bring them closer to the fuzzy intersection results. Thereafter, the consensus among decision makers is again measured. The collaboration process will stop when no more modifications are made by any decision maker. At last, based on the aggregation results, the FWM method [8] was applied to assess the suitability of a smart health practice. The assessment result is defuzzified using the centroid-defuzzification (CD) method for generating an absolute ranking.

The differences between our proposed methodology and some existing methods are summarized in Table 1. Our proposed methodology is a posterior-aggregation method, while the method proposed by Chen [6] is an anterior-aggregation method. In other words, our proposed methodology checks the existence of a consensus among decision makers, while the method proposed by Chen [6] does not. In addition, our proposed methodology derives the values of fuzzy weights, while the method proposed by Chen [7] only approximates the values of fuzzy weights. As a result, our proposed methodology is more precise and reliable than the method proposed by Chen [7].

**Table 1.** The differences between the proposed methodology and some existing methods. fuzzy geometric mean (FGM), alpha-cut operations (ACO), and fuzzy weighted mean (FWM), fuzzy intersection (FI).

| Method | Smart Technology | Assessment Method | Group Decision Making | Consensus | Aggregation |
|---|---|---|---|---|---|
| Chen and Chiu [3] | All | Literature review | No | - | - |
| Haymes et al. [4] | Not specified | Behavior analysis | No | - | - |
| Chiu and Chen [5] | Smart mobile services | Learning curve analysis | No | - | - |
| Chen [6] | All | FGM-ACO-FWM | Yes | Not guaranteed | Anterior-aggregation |
| Chen [7] | All | FGM-FI-FWM | Yes | Guaranteed | Posterior-aggregation |
| Our proposed methodology | All | ACO-FI-FWM | Yes | Guaranteed | Posterior-aggregation |

The remainder of this paper is organized as follows. Section 2 reviews previous works. Section 3 puts forward the fuzzy collaborative approach for assessing the suitability of a smart health practice. Section 4 provides the results of applying the fuzzy collaborative approach to assess eleven smart health practices, so as to choose the most suitable smart technology application. Three existing methods were also applied to these smart health practices for comparison. Finally, Section 5 presents concluding remarks and lists a few topics worthy of further investigation.

## 2. Previous Work

Social media, mobile devices, and sensors are valuable for communicating health-related information [9–11]. In particular, the applications of social networking apps to mobile health care are prevalent. For example, Cook et al. [12] compared the average number of Twitter posts by people with depression to that by people without depression, and found a significant difference. The survey done by Reeder and David [13] revealed that the most prevalent applications of smart watches to mobile health care included activity monitoring, heart rate monitoring, speech therapy adherence, diabetes self-management, and the detection of seizures, tremors, scratching, eating, and medication-taking behaviors. Mandal et al. [14] described the substitutable medical applications and reusable technologies (SMART) project launched by Harvard Medical School and Boston Children's Hospital jointly to increase the portability of medical applications. Based on the infrastructure, several apps have been designed. Hamidi [15] proposed a new standard for applying biometric technologies to fast identify the user of a mobile health care service.

According to the survey of Cook et al. [12], due to advances in sensor and wireless communication technologies, more than twenty-one types of sensors have been prevalent on mobile or wearable devices, much more than those in the past [16]. Eklund and Forsman [17] designed a suit of smart work clothes with embedded sensors for monitoring the heart rate and breathing of a worker, so as to provide him/her suggestions to avoid musculoskeletal disorders.

There are a number of mobile health care studies focusing on special groups, e.g., people with extremely bad vision [1], people with intellectual disabilities [4], and older adults [18]. According to the survey by Liu et al. [18], the most suitable smart health practices for older adults adopted smart homes and home-based health-monitoring technologies. Such applications were mostly used to monitor the daily activities, cognitive decline, mental health, and heart conditions of older adults with complex needs.

In the view of Eskofier et al. [19], the ability to walk was critical to the quality of life. For this reason, smart shoes with embedded sensors were applied to monitor the gait and mobility of a user, so as to support healthy living, complement medical diagnostics, and monitor therapeutic outcomes. Smart technologies can be applied to fall detection, for which machine learning and decision tress are the most prevalent data analysis techniques. In the view of Chen et al. [20], health monitoring using traditional wearable devices was not sustainable because of the uncomfortableness of long-term wearing, insufficient accuracy, etc. This problem is expected to be alleviated soon with rapid advances in wearable technologies.

## 3. Methodology

The fuzzy collaborative approach proposed in this study is composed of three major parts: ACO, FI, and FWM. A similar treatment was taken by Duman et al. [21] that combined fuzzy decision-making trial and evaluation laboratory (DEMATEL), analytic network process (ANP), and an artificial neural network (ANN). In addition, the fuzzy collaborative approach is a posterior-aggregation method. Recently, Chen [6] proposed the FGM-ACO-FWM method for a similar purpose. The differences between the two methods is highlighted by Figure 1. A recent survey of FAHP refers to [22].

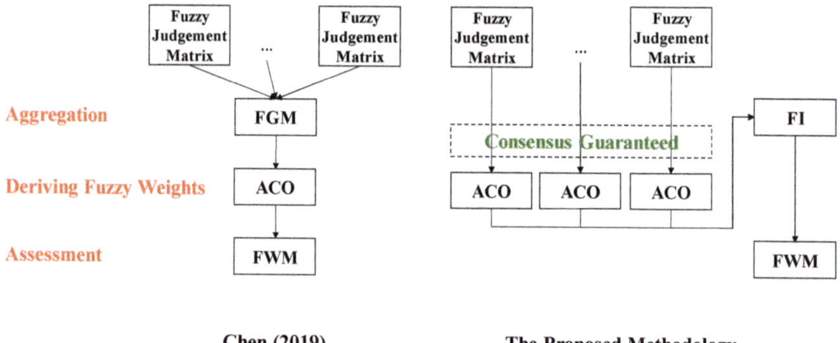



**Figure 1.** Comparison with a recent method.

### 3.1. Deriving Fuzzy Weights for Each Decision Maker Using ACO

In the proposed methodology, a team of $K$ decision makers is formed. First, each decision maker compares the relative weights of criteria for assessing the suitability of a smart health practice in pairs. The results are used to construct a fuzzy pairwise comparison matrix as

$$\widetilde{\mathbf{A}}_{n \times n}(k) = [\widetilde{a}_{ij}(k)]; \; i, \; j \; = \; 1 \; \sim \; n; \; k \; = \; 1 \; \sim \; K \tag{1}$$

where

$$\widetilde{a}_{ij}(k) = \begin{cases} 1 & \text{if} & i = j \\ \dfrac{1}{\widetilde{a}_{ji}(k)} & \text{otherwise} \end{cases} ; \; i, \; j \; = \; 1 \; \sim \; n; \; k \; = \; 1 \; \sim \; K \tag{2}$$

$\widetilde{a}_{ij}(k)$ is the relative weight of criterion $i$ over criterion $j$ judged by decision maker $k$. Equation (2) is the reciprocal requirement. $\widetilde{a}_{ij}(k)$ are mapped to triangular fuzzy numbers (TFNs) satisfying $a_{ij2}(k) = \{1, 3, 5, 7, 9\}$, $a_{ij1}(k) = \max(a_{ij2}(k) - 4, 1)$, and $a_{ij3}(k) = \min(a_{ij2}(k) + 4, 9)$ (see Figure 2). In this way, the fuzzy weights derived by decision makers are more likely to overlap before collaboration. However, that does not mean that the TFNs used in this study are better or more suitable than those adopted in [23,24]. Regardless of which set of TFNs is used, decision makers can reach a consensus after several rounds of collaboration using the proposed methodology. However, this is obviously based on the assumption that all decision makers accept that these TFNs reflect their preferences. If this assumption holds, in our view, it is more likely for each fuzzy judgment matrix to be consistent. A positive comparison satisfies $\widetilde{a}_{ij}(k) \geq 1$.

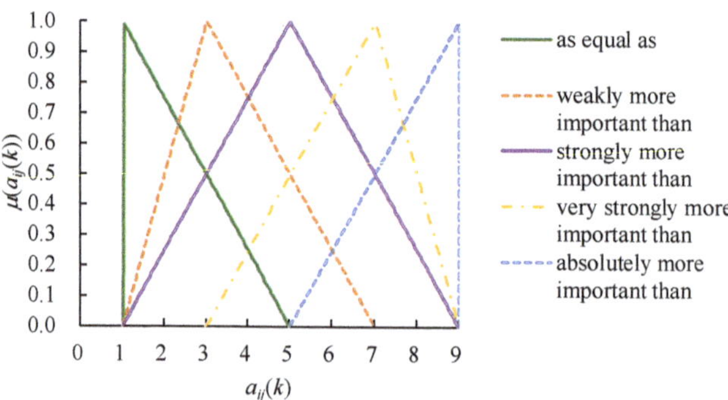



**Figure 2.** Triangular fuzzy numbers (TFNs) used in the proposed methodology.

Subsequently, a fuzzy eigen analysis is performed to derive the fuzzy eigenvalue and eigenvector of $\widetilde{\mathbf{A}}$ [25,26]:

$$\det(\widetilde{\mathbf{A}}(-)\widetilde{\lambda}\mathbf{I}) = 0 \tag{3}$$

$$(\widetilde{\mathbf{A}}(-)\widetilde{\lambda}\mathbf{I})(\times)\widetilde{\mathbf{x}} = 0 \tag{4}$$

where $(-)$ and $(\times)$ indicates fuzzy subtraction and multiplication, respectively. However, solving the two equations is a computationally intensive task. To address this, most of the past studies applied approximation techniques such as FGM [27] and fuzzy extent analysis (FEA) [28]. In contrast, an exact technique such as ACO is able to derive the values of $\widetilde{\lambda}$ and $\widetilde{\mathbf{x}}$.

The $\alpha$ cut of a fuzzy variable $y(\alpha) = [y^L(\alpha), y^R(\alpha)]$ is an interval. In ACO, the fuzzy parameters and variables in Equations (3) and (4) are replaced with their $\alpha$ cuts:

$$a_{ij}(\alpha) = [a_{ij}^L(\alpha), a_{ij}^R(\alpha)] \tag{5}$$

$$\lambda(\alpha) = [\lambda^L(\alpha), \lambda^R(\alpha)] \tag{6}$$

$$\mathbf{x}(\alpha) = [\mathbf{x}^{\mathbf{L}}(\alpha), \mathbf{x}^{\mathbf{R}}(\alpha)]. \tag{7}$$

Substituting (5–7) into (3–4) gives

$$\det(\mathbf{A}(\alpha) - \lambda(\alpha)\mathbf{I}) = 0 \tag{8}$$

$$(\mathbf{A}(\alpha) - \lambda(\alpha)\mathbf{I})\mathbf{x}(\alpha) = 0. \tag{9}$$

If the possible values of $\alpha$ are enumerated, e.g., every 0.1, then Equations (8) and (9) need to be solved $10 \cdot 2^{C_2^n} + 1$ times, from which the minimal and maximal results specify the lower and upper bounds of the $\alpha$ cut [29–31]:

$$
\begin{aligned}
w_i(\alpha) \quad &= [w_i^L(\alpha), w_i^R(\alpha)] \\
&= [\min_{*} \frac{x_i^*(\alpha)}{\sum\limits_{j=1}^{n} x_j^*(\alpha)}, \max_{*} \frac{x_i^*(\alpha)}{\sum\limits_{j=1}^{n} x_j^*(\alpha)}]
\end{aligned}
\tag{10}
$$

where $*$ can be $L$ or $R$, indicating the left or right $\alpha$ cut of the variable, respectively. The pseudo code for implementing ACO is shown in Figure 3.

```
FOR each value of α
  FOR the two α cuts of the first positive comparison
    ......
    FOR the two α cuts of the last positive comparison
      Derive the maximal eigenvalue and weight from  A(α)
      IF the derived maximal eigenvalue < the minimum of the fuzzy maximal eigenvalue THEN
        UPDATE the minimum of the fuzzy maximal eigenvalue to the derived maximal eigenvalue
      END IF
      IF the derived maximal eigenvalue > the maximum of the fuzzy maximal eigenvalue THEN
        UPDATE the maximum of the fuzzy maximal eigenvalue to the derived maximal eigenvalue
      END IF
      FOR each fuzzy weight
        IF the derived weight < the minimum of the fuzzy weight THEN
          UPDATE the minimum of the fuzzy weight to the derived weight
        END IF
        IF the derived weight > the maximum of the fuzzy weight THEN
          UPDATE the maximum of the fuzzy weight to the derived weight
        END IF
      END LOOP
    END LOOP
    ......
  END LOOP
END LOOP
```

**Figure 3.** Pseudo code for implementing ACO.

Based on $\widetilde{\lambda}$, Satty [23] suggested evaluating the consistency among fuzzy pairwise comparison results as

$$\text{Consistency index}: \widetilde{C.I.}(m) = \frac{\widetilde{\lambda}_{\max}(m) - n}{n - 1} \tag{11}$$

$$\text{Consistency ratio}: \widetilde{C.R.}(m) = \frac{\widetilde{C.I.}(m)}{R.I.} \tag{12}$$

where $R.I.$ is the randomized consistency index.

### 3.2. Aggregating the Fuzzy Weights by All Decision Makers Using FI

After the negotiation process, the FI result of the fuzzy weights derived by all decision makers is adopted to represent their consensus [32–36]. When a consensus among all decision makers does not exist, an alternative is to seek for the consensus among only some of the decision makers [37].

The membership function of the FI result is given by

$$\mu_{\widetilde{FI}(\{\widetilde{w}_i(k)\})}(x) = \min_k(\mu_{\widetilde{w}_i(k)}(x)) \tag{13}$$

as shown in Figure 4. If the TFNs for linguistic terms have narrow ranges, fuzzy weights may not overlap and $\widetilde{FI}(\{\widetilde{w}_i(k)\})$ will be an empty (null) set, as illustrated in Figure 5.

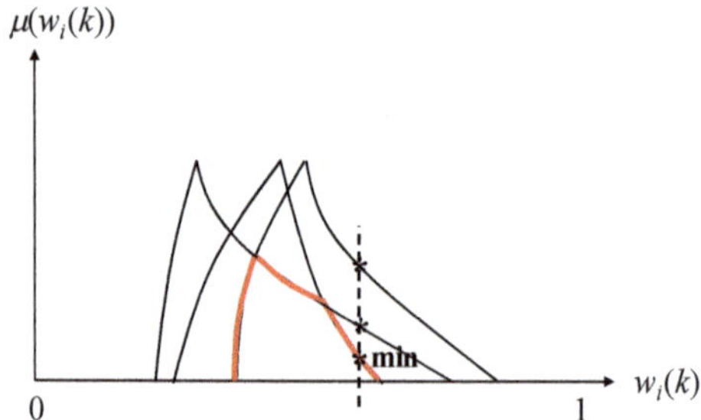


**Figure 4.** The FI result.

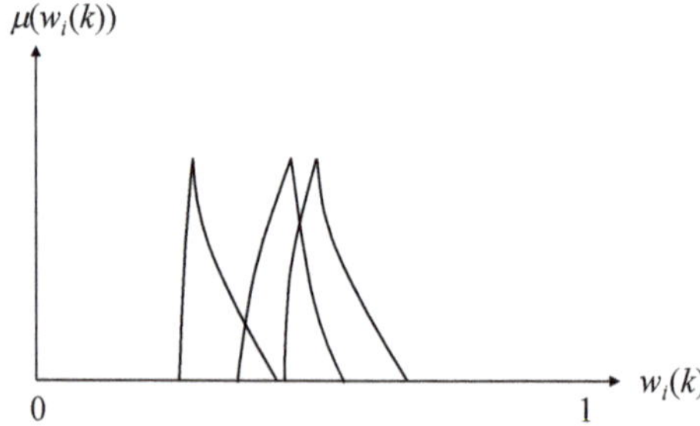


**Figure 5.** The FI result is a null set if TFNs have narrow ranges.

Alternatively, $\widetilde{FI}(\{\widetilde{w}_i(k)\})$ can be represented with its $\alpha$ cut as

$$FI^L(\{\widetilde{w}_i(k)\})(\alpha) = \max_k(w_i^L(k)(\alpha)) \tag{14}$$

$$FI^R(\{\widetilde{w}_i(k)\})(\alpha) = \min_k(w_i^R(k)(\alpha)) \tag{15}$$

as illustrated in Figure 6.

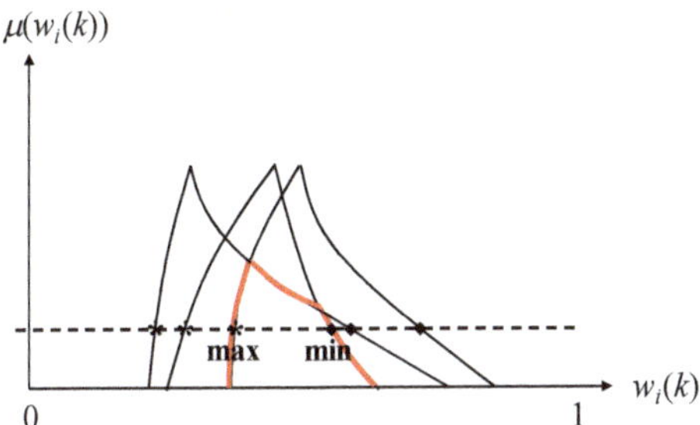


**Figure 6.** The $\alpha$ cut of the FI result.

If decision makers are of unequal importance levels, FI is not suitable. To address this issue, the fuzzy weighted intersection (FWI) can be sought for instead.

**Definition 1.** *Let $\widetilde{w}_i(1) \sim \widetilde{w}_i(K)$ be the fuzzy weights derived by K decision makers. The importance of decision maker k is $\omega(k)$. Then the fuzzy weighted intersection (FWI) of fuzzy weights, indicated with $\widetilde{FWI}(\{\widetilde{w}_i(k)|k = 1 \sim K)\}$ is expected to meet the following requirements:*

(i)  $\widetilde{FWI}(\{\widetilde{w}_i(k)\}) = \widetilde{w}_i(g)$ *if $\omega_g = 1$ for some g: If a decision maker is absolutely important, then the value of $\widetilde{w}_i$ is determined solely by the decision maker.*

(ii) $\widetilde{FWI}(\{\widetilde{w}_i(k)\}) = \widetilde{FI}(\{\widetilde{w}_i(k)\})$ *if $\omega_k = 1/K$ $\forall$ k: If all decision makers are equally important, then the value of $\widetilde{w}_i$ is determined by the consensus among the decision makers.*

(iii) $\left|\mu_{\widetilde{FWI}}(x) - \mu_{\widetilde{w}_i(g_1)}(x)\right| \geq \left|\mu_{\widetilde{FWI}}(x) - \mu_{\widetilde{w}_i(g_2)}(x)\right|$ *if $\omega_{g_1} \leq \omega_{g_2}$ $\forall$ $g_1 \neq g_2$: If decision maker $g_2$ is more important than decision maker $g_1$, then the value of $\widetilde{w}_i$ is closer to the value derived by decision maker $g_2$ than to that by decision maker $g_1$.*

In theory, there are numerous possible FWI operators. A FWI operator considers the membership function value, rather than the value, of a fuzzy weight, which is obviously distinct from the common aggregator such as FWM.

*3.3. Assessing the Suitability of a Smart Health Practice Using FWM*

Subsequently, FWM is applied to assess the suitability of a smart health practice, for which the FI result provides the required fuzzy weights:

$$\widetilde{S}_q = \frac{\sum_{i=1}^{n} (\widetilde{FI}(\{\widetilde{w}_i(k)\})(\times)\widetilde{p}_{qi})}{\sum_{i=1}^{n} \widetilde{FI}(\{\widetilde{w}_i(k)\})} \tag{16}$$

where $\widetilde{S}_q$ is the suitability of the $q$-th smart health practice; $\widetilde{p}_{qi}$ is the performance of the $q$-th smart health practice in optimizing the $i$-th criterion. A FWM problem is not easy to solve because the dividend and divisor of Equation (16) are dependent [8]. Nevertheless, for comparison, only the dividend of Equation (16) needs to be calculated, since the divisor is the same for all alternatives:

$$\widetilde{S}_q = \sum_{i=1}^{n} (\widetilde{FI}(\{\widetilde{w}_i(k)\})(\times)\widetilde{p}_{qi}). \tag{17}$$

The $\alpha$ cut of $\widetilde{S}_q$ is defined as the interval $[S_q^L, S_q^R]$ that can be derived as

$$S_q^L(\alpha) = \sum_{i=1}^{n} (FI^L(\{\widetilde{w}_i(k)\})(\alpha)p_{qi}^L(\alpha)) \tag{18}$$

$$S_q^R(\alpha) = \sum_{i=1}^{n} (FI^R(\{\widetilde{w}_i(k)\})(\alpha)p_{qi}^R(\alpha)) \tag{19}$$

according to the alpha-cut operations. The assessment result is then defuzzified using the prevalent CD method [38]. However, the alpha-cut operations method takes samples uniformly along the $y$-axis, as illustrated in Figure 7, while the CD method distributes samples evenly along the $x$-axis. To resolve this inconsistency, a possible way is to divide the range of $\widetilde{S}_q$ into $\Gamma$ equal intervals:

$$\widetilde{S}_q = \{[\frac{\Gamma - \eta + 1}{\Gamma}S_q^L(0) + \frac{\eta - 1}{\Gamma}S_q^R(0), \frac{\Gamma - \eta}{\Gamma}S_q^L(0) + \frac{\eta}{\Gamma}S_q^R(0)]|\eta = 1 \sim \Gamma\} \tag{20}$$

as illustrated in Figure 8. The center of the $\eta$-th interval is indicated with $C_q(\eta)$:

$$\begin{aligned}
C_q(\eta) &= \tfrac{1}{2}(\tfrac{\Gamma-\eta+1}{\Gamma}S_q^L(0) + \tfrac{\eta-1}{\Gamma}S_q^R(0) + \tfrac{\Gamma-\eta}{\Gamma}S_q^L(0) + \tfrac{\eta}{\Gamma}S_q^R(0)) \\
&= \tfrac{2\Gamma-2\eta+1}{2\Gamma}S_q^L(0) + \tfrac{2\eta-1}{2\Gamma}S_q^R(0)
\end{aligned} \tag{21}$$

$C_q(\eta)$ can be determined by interpolating the two closest values of $\widetilde{S}_q$:

$$\begin{aligned}
\mu_{\widetilde{S}_q}(C_q(\eta)) &= \frac{C_q(\eta) - \max\limits_{S_q^*(\alpha) \leq C_q(\eta)} S_q^*(\alpha)}{(\min\limits_{S_q^*(\alpha) \geq C_q(\eta)} S_q^*(\alpha) - \max\limits_{S_q^*(\alpha) \leq C_q(\eta)} S_q^*(\alpha))} \cdot \min\limits_{S_q^*(\alpha) \geq C_q(\eta)} \alpha \\
&\quad + \frac{\min\limits_{S_q^*(\alpha) \geq C_q(\eta)} S_q^*(\alpha) - C_q(\eta)}{(\min\limits_{S_q^*(\alpha) \geq C_q(\eta)} S_q^*(\alpha) - \max\limits_{S_q^*(\alpha) \leq C_q(\eta)} S_q^*(\alpha))} \cdot \max\limits_{S_q^*(\alpha) \leq C_q(\eta)} \alpha
\end{aligned} \tag{22}$$

where $*$ can be $R$ or $L$. Then, the centroid of $\widetilde{S}_q$ is the derived as follows:

$$COG(\widetilde{S}_q) = \frac{\sum\limits_{\eta=1}^{\Gamma} (\mu_{\widetilde{S}_q}(C_q(\eta))C_q(\eta))}{\sum\limits_{\eta=1}^{\Gamma} \mu_{\widetilde{S}_q}(C_q(\eta))}. \tag{23}$$

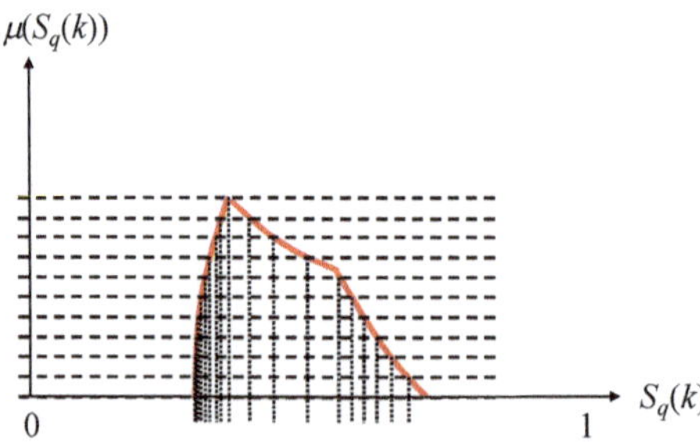


**Figure 7.** The way of taking samples in the alpha-cut operations method.

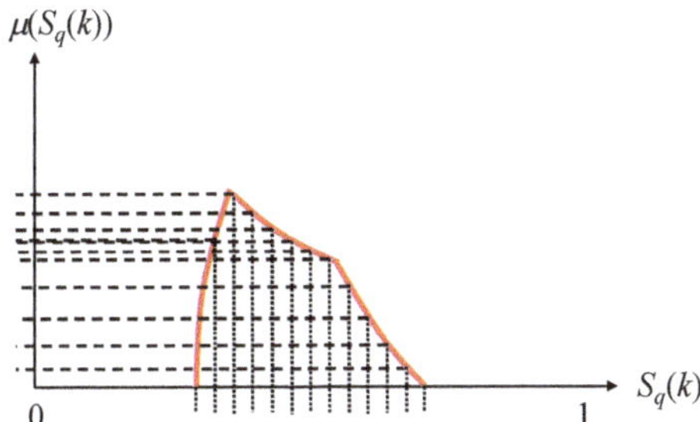

**Figure 8.** The way of taking samples in the centroid-defuzzification (CD) method.

## 4. Application

With advances in transportation, sensing, and communication technologies, smart health becomes a critical issue [39]. There have been a number of smart health practices, however, and how to choose the most suitable smart health practice is a challenging task. To fulfill this task, the proposed methodology has been applied to assess and compare the suitabilities of eleven smart health practices. Chen [6] evaluated the sustainability of a smart health practice, in which five criteria—unobtrusiveness, supporting online social networking, compliance with related medical laws, the size of the health care market, and the correct identification of a user's need and situation, were considered. Compared with sustainability, suitability is a shorter-term concept. For this reason, the following five criteria were considered in this study instead [6,40–48]:

(1)  C1: unobtrusiveness,
(2)  C2: supporting online social networking,
(3)  C3: cost effectiveness,
(4)  C4: availability of mobile health care facilities, and
(5)  C5: correct, reliable, and robust identification of a user's need and situation.

In addition, two less relevant smart health practices, smart connected vehicles and smart defense technologies, were excluded from the experiment. The proposed methodology was applied as follows.

First, a team of three decision makers, including an information management professor, a computer-aided industrial design professor, and an ambient intelligence (AmI) decision maker, was formed. Each decision maker compared the relative weights of criteria in pairs. The results are summarized in Table 2.

**Table 2.** The results of pairwise comparisons.

|  | C1 | C2 | C3 | C4 | C5 |
|---|---|---|---|---|---|
| C1 | 1 | $(1,5,9), (3,7,9),$ $(3,7,9)$ | $(1,1,5), (1,5,9),$ $(1,1,5)$ | $(3,7,9), (5,9,9),$ $(1,1,5)$ | $(1,5,9), (1,5,9),$ $(1,3,7)$ |
| C2 | - | 1 | $(1,1,5), (1,4,8),$ $(1,3,7)$ | $(3,7,9), (1,3,7),$ $(1,1,5)$ | - |
| C3 | - | - | 1 | - | $(1,5,9), (1,4,8),$ $(1,5,9)$ |
| C4 | - | - | $(1,1,5), (1,3,7),$ $(1,1,5)$ | 1 | $(3,7,9), (1,5,9),$ $(1,3,7)$ |
| C5 | - | $(1,3,7), (1,5,9),$ $(1,4,8)$ | - | - | 1 |

Each decision maker applied ACO to derive fuzzy eigenvalues and the fuzzy weights of criteria. The results are summarized in Figures 9 and 10, respectively. In our view, if the linguistic terms adopted in the proposed methodology reflected the preferences of decision makers, it was more likely for their fuzzy judgment matrixes to be consistent. To ensure this, the fuzzy consistency index $\widetilde{C.I.}(k)$ should be less than the threshold of 0.1 for each decision maker, by requesting:

$$\mu_{\widetilde{C.I.}(k)}(0.1) \geq 0 \quad \forall \, k \tag{24}$$

According to the experimental results, Condition (23) was successfully satisfied. In contrast, if the commonly adopted linguistic terms were adopted, $\widetilde{C.I.}(k)$ was always greater than 0.1. This result supported the suitability of the linguistic terms adopted in the proposed methodology.

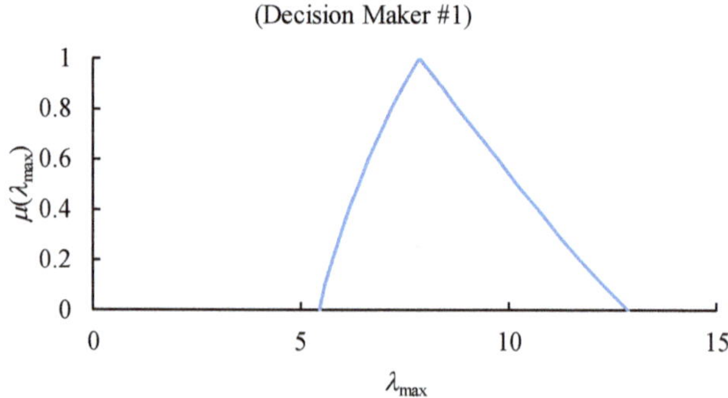


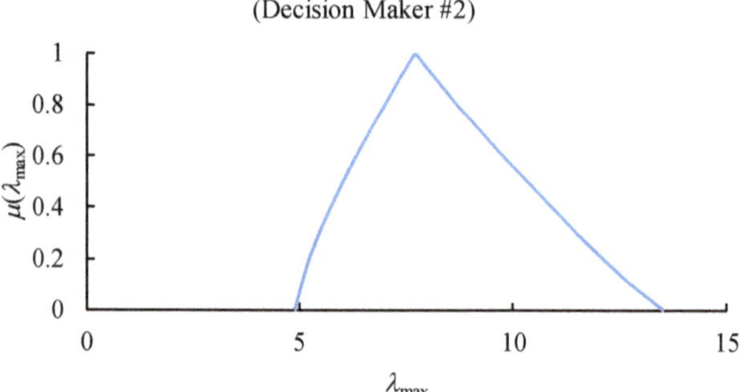


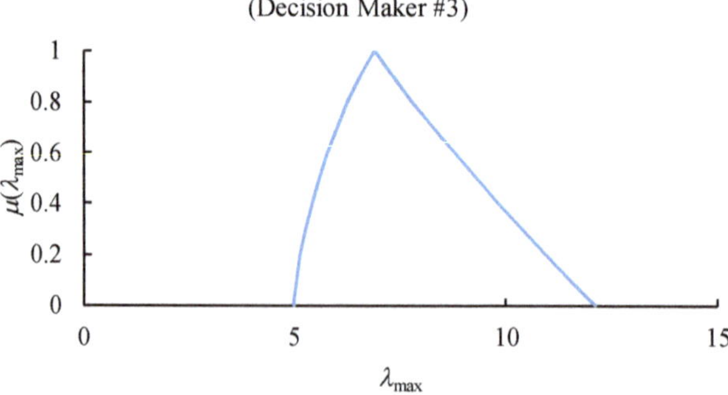


**Figure 9.** Fuzzy eigenvalues derived by decision makers.

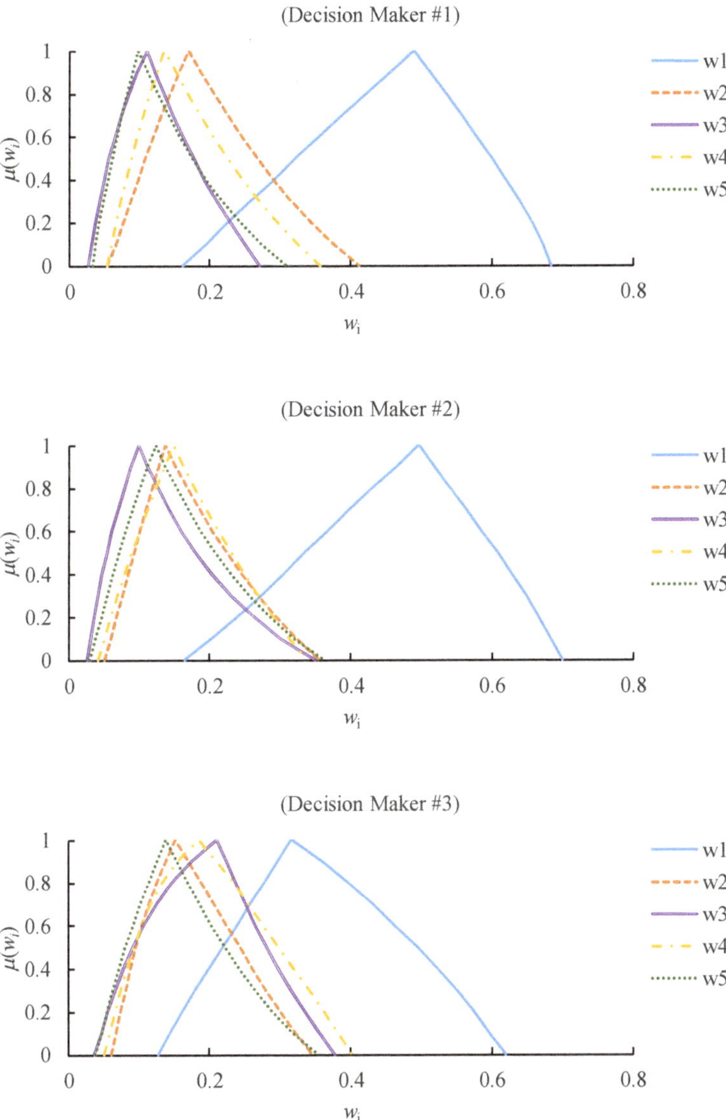


**Figure 10.** Fuzzy weights derived by decision makers.

In this experiment, decision makers were equally important. Therefore, FI was considered effective for aggregating the fuzzy weights derived by all decision makers. The results are shown in Figure 11. The results showed that a prior consensus has been achieved among decision makers regarding the values of each fuzzy weight.

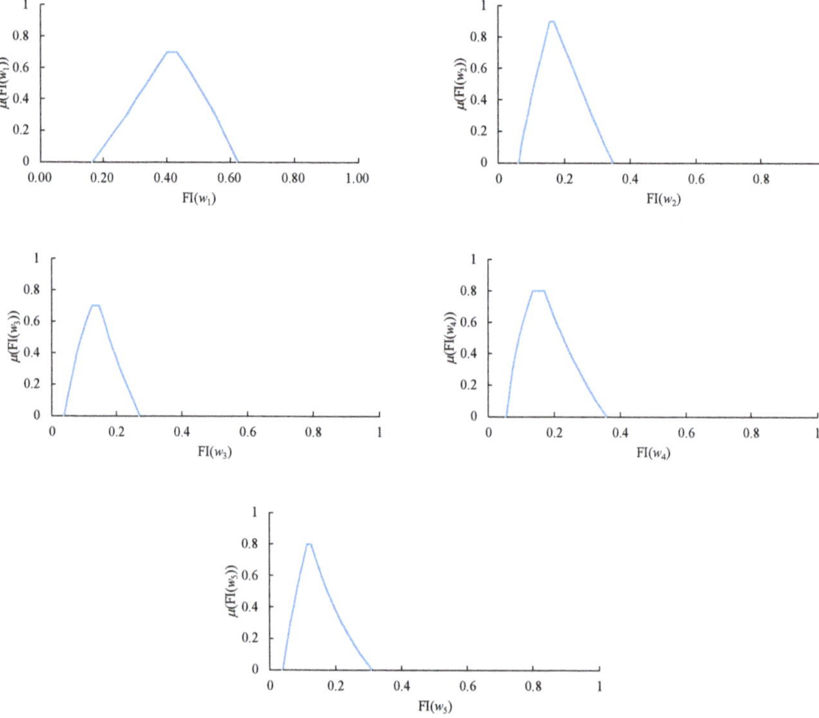

**Figure 11.** The FI results.

The FI results were presented to decision makers for them to consider when modifying their pairwise comparison results. The modified pairwise comparison results are summarized in Table 3. ACO was again applied to derive fuzzy weights for each decision maker. Then, FI was applied to measure the consensus among decision makers after collaboration. The results are summarized in Figure 12. After collaboration, the FI results became wider, showing a higher consensus since more possible values were acceptable to all decision makers. Taking $\widetilde{w}_4$ as an example, the FI results before and after collaboration are compared in Figure 13. The collaboration stopped because no decision maker made any further modification.

**Table 3.** The modified pairwise comparison results.

| | C1 | C2 | C3 | C4 | C5 |
|---|---|---|---|---|---|
| **C1** | 1 | $(1,5,9), (3,7,9),$ $(3,7,9)$ | $(1,1,5), (1,3,7),$ $(1,1,5)$ | $(3,7,9), (1,5,9),$ $(1,1,5)$ | $(1,5,9), (1,5,9),$ $(1,3,7)$ |
| **C2** | - | 1 | $(1,1,5), (1,1,5),$ $(1,3,7)$ | $(3,7,9), (1,5,9),$ $(1,1,5)$ | - |
| **C3** | - | - | 1 | - | $(1,5,9), (1,4,8),$ $(1,5,9)$ |
| **C4** | - | - | $(1,1,5), (1,3,7),$ $(1,1,5)$ | 1 | $(3,7,9), (1,4,8),$ $(1,3,7)$ |
| **C5** | - | $(1,3,7), (1,5,9),$ $(1,4,8)$ | - | - | 1 |

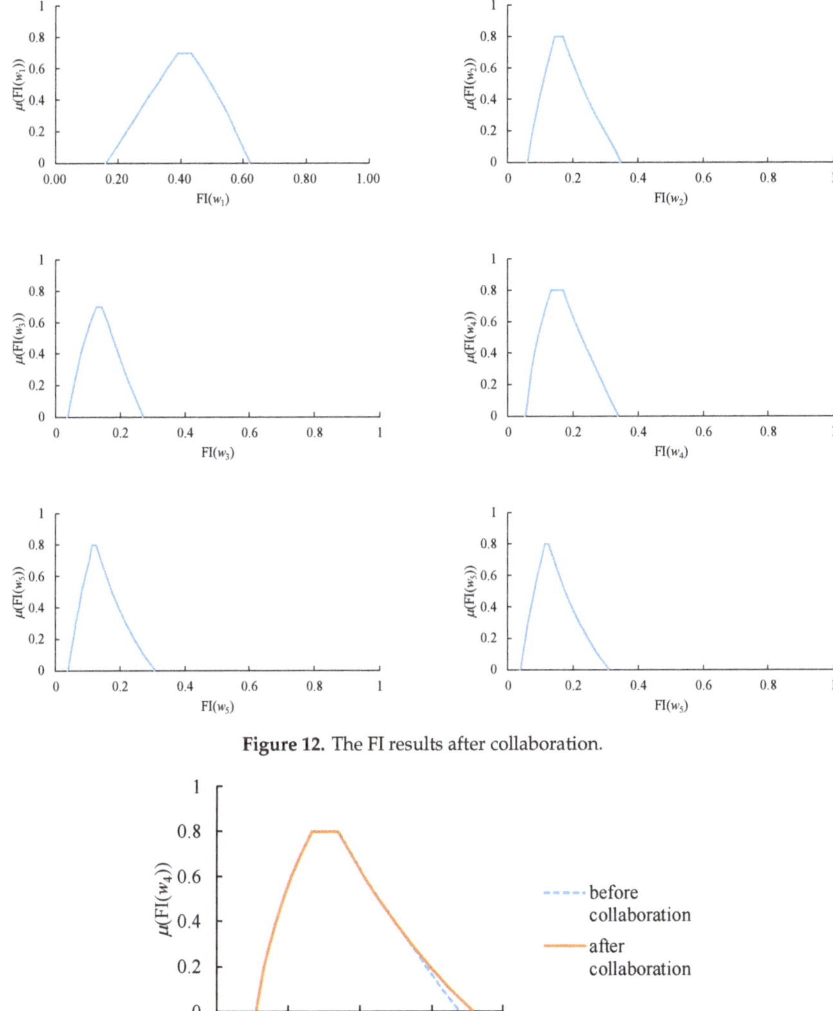

**Figure 12.** The FI results after collaboration.



**Figure 13.** The FI results of $\widetilde{w}_4$ before and after collaboration.

The suitabilities of eleven smart health practices were assessed. To this end, the performances of these smart health practices in optimizing the five criterion were evaluated by the same decision makers using the following linguistic terms [3]:

Very poor: $(0, 0, 1)$,
Poor: $(0, 1, 2)$,
Moderate: $(1.5, 2.5, 3.5)$,
Good: $(3, 4, 5)$, and
Very good: $(4, 5, 5)$.

The evaluations by decision makers were aggregated in a similar way. Specifically speaking, decision makers were asked to modify their evaluations slightly until these evaluations overlapped.

Then, the evaluations by all decision makers were averaged. The results are summarized in Table 4. It can be seen that none of the smart health practices dominated the others, causing difficulty in choosing from them. In addition, it was noteworthy that the cost effectiveness of smart phones was high, while that of smart watches was low, due to the fact that smart phones were much more prevalent than smart watches. Therefore, a decision maker did not consider buying his/her smart phone as an additional investment.

**Table 4.** Performances of eleven smart technologies along five dimensions.

| Smart Health Practice | C1 (Unobtrusiveness) | C2 (Online Social Networking) | C3 (Cost Effectiveness) | C4 (Availability of Mobile Health Care Facilities) | C5 (Correct, Reliable, and Robust Identification) |
|---|---|---|---|---|---|
| Smart body analyzers | (1.00, 2.00, 3.00) | (1.00, 2.00, 3.00) | (0.00, 1.00, 2.00) | (0.00, 1.00, 2.00) | (2.50, 3.50, 4.50) |
| Smart clothes | (0.00, 1.00, 2.00) | (0.00, 1.00, 2.00) | (0.00, 1.00, 2.00) | (0.00, 1.00, 2.00) | (2.50, 3.50, 4.50) |
| Smart glasses | (0.00, 1.00, 2.00) | (3.33, 4.33, 5.00) | (0.00, 1.00, 2.00) | (2.00, 3.00, 4.00) | (2.50, 3.50, 4.50) |
| Smart mobile services | (3.67, 4.67, 5.00) | (3.67, 4.67, 5.00) | (3.67, 4.67, 5.00) | (3.67, 4.67, 5.00) | (3.33, 4.33, 5.00) |
| Smart motion sensors | (2.00, 3.00, 4.00) | (0.00, 1.00, 2.00) | (0.00, 1.00, 2.00) | (0.00, 1.00, 2.00) | (3.67, 4.67, 5.00) |
| Smart phones | (3.67, 4.67, 5.00) | (3.67, 4.67, 5.00) | (2.50, 3.50, 4.50) | (3.67, 4.67, 5.00) | (3.67, 4.67, 5.00) |
| Smart smoke alarms | (2.00, 3.00, 4.00) | (0.00, 1.00, 2.00) | (1.00, 2.00, 3.00) | (2.00, 3.00, 4.00) | (3.67, 4.67, 5.00) |
| Smart toilets | (1.00, 2.00, 3.00) | (0.00, 1.00, 2.00) | (0.00, 1.00, 2.00) | (0.00, 1.00, 2.00) | (2.00, 3.00, 4.00) |
| Smart watches | (3.67, 4.67, 5.00) | (2.50, 3.50, 4.50) | (1.00, 2.00, 3.00) | (3.67, 4.67, 5.00) | (3.67, 4.67, 5.00) |
| Smart wheelchairs | (2.50, 3.50, 4.50) | (0.00, 1.00, 2.00) | (0.00, 1.00, 2.00) | (2.00, 3.00, 4.00) | (2.00, 3.00, 4.00) |
| Smart wigs | (2.00, 3.00, 4.00) | (0.00, 1.00, 2.00) | (0.00, 1.00, 2.00) | (0.00, 1.00, 2.00) | (2.50, 3.50, 4.50) |

FWM was applied to assess the suitability of each smart health practice, for which the fuzzy collaborative FAHP approach provided the required fuzzy weights. The results are summarized in Figure 14.

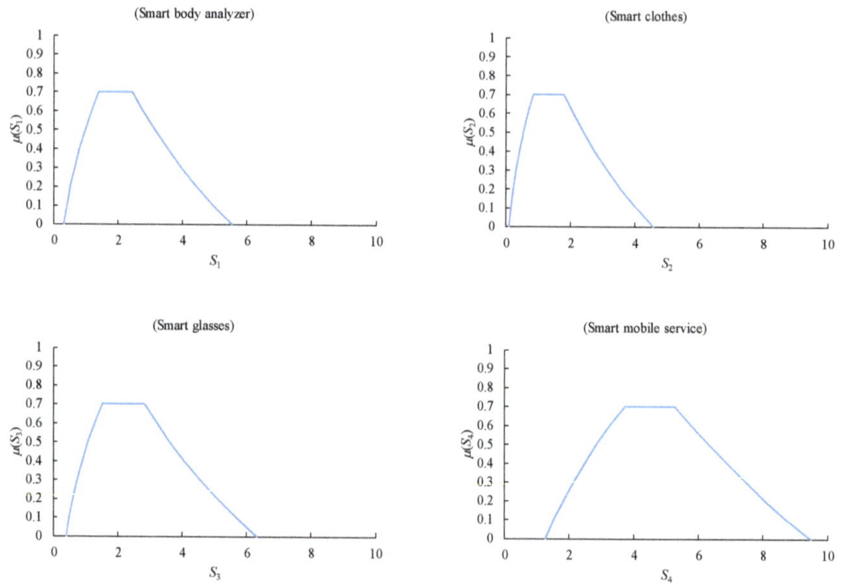


**Figure 14.** *Cont.*

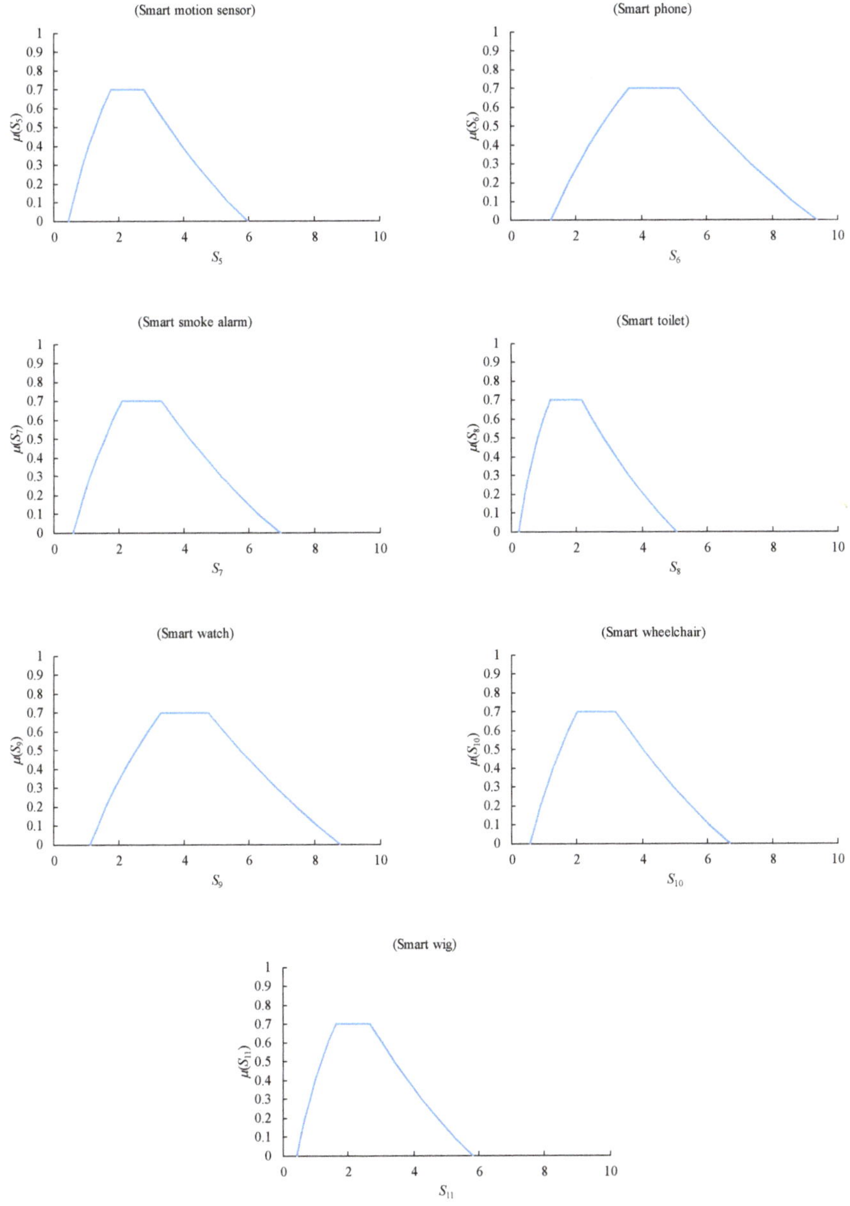



**Figure 14.** FWM results.

For generating an absolute ranking, CD was applied to defuzzify the fuzzy suitability of each smart health practice. The results are summarized in Table 5.

**Table 5.** Defuzzification results.

| Smart Health Practice | Defuzzified Suitability |
| --- | --- |
| Smart body analyzers | 2.10 |
| Smart clothes | 1.52 |
| Smart glasses | 2.40 |
| Smart mobile services | 4.64 |
| Smart motion sensors | 2.45 |
| Smart phones | 4.54 |
| Smart smoke alarms | 2.91 |
| Smart toilets | 1.86 |
| Smart watches | 4.17 |
| Smart wheelchairs | 2.79 |
| Smart wigs | 2.33 |

According to the experimental results,

(1)    Smart mobile services were the most suitable smart health practice, while smart clothes were still the least suitable smart health practice, owing to their obtrusiveness.

(2)    The suitabilities of the eleven smart health practices were ranked. The results are shown in Figure 15. The ranking results of the sustainabilities of these smart health practices, retrieved from Chen [6], are also presented in this figure for comparison. Obviously, there are some differences between the two results. For example, the suitability of smart body analyzers was low, but its suitability was high, showing the great potential of smart body analyzers in the future.

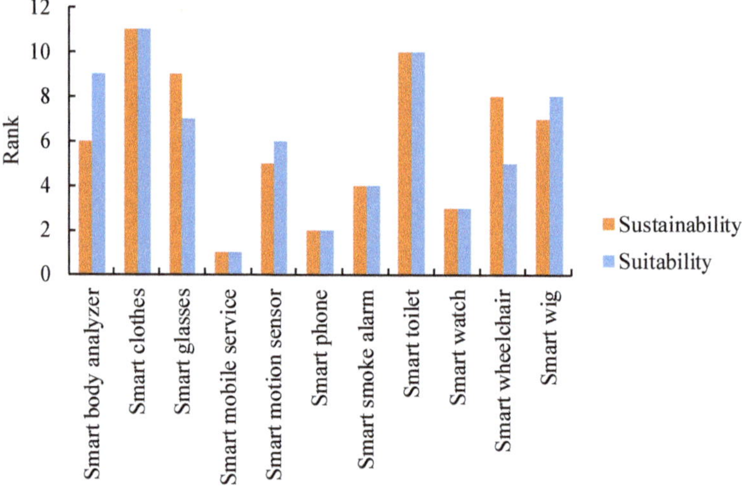


**Figure 15.** Comparing the suitability and sustainability of each smart technology application.

(3)    In the experiment, decision makers modified their fuzzy judgment results just once to achieve a higher consensus, yet this was not always the case since modifications were subjectively made. It was possible for decision makers to undergo many rounds of collaboration before achieving a higher consensus. To tackle this problem, a mechanism for facilitating the collaboration process among decision makers should be designed.

(4)    The efficiency of ACO was a problem to the application of the fuzzy collaborative approach, and needed to be enhanced somehow, e.g., by applying a genetic algorithm. In previous studies, there were two major ways of combining genetic algorithms with fuzzy analytic hierarchy analysis. The first way obtains the weights of criteria by using fuzzy analytic hierarchy analysis, which

designs the fitness function of the genetic algorithm to compare various alternatives. The second way solves a multi-objective optimization problem with a genetic algorithm to obtain multiple Pareto-optimal solutions, and then performs a fuzzy analytic hierarchy analysis to set the weights of the objective functions to further compare these Pareto-optimal solutions. The motive for applying a genetic algorithm in this study is different from those in previous studies.

(5)  Three existing methods, fuzzy ordered weighted average (FOWA), fuzzy geometric mean (FGM)-FWM, and the fuzzy extent analysis (FEA)-weighted average (WA) method proposed by Chang [29], were also applied to assess the suitability of each smart health practice for comparison. In FOWA, the moderately optimistic strategy was adopted. In FGM-FWM, fuzzy weights were approximated using FGM and expressed in terms of TFNs. In FEA-WA, since the weights estimated using FEA were crisp, WA, rather than FWM, was applied to assess the suitability of each smart health practice. Finally, the suitabilities of all smart health practices were ranked. The ranking results using various methods are compared in Figure 16. In sum, these methods came to the same conclusions about the suitabilities of smart mobile services and smart phones. In contrast, the suitabilities of other smart health practices assessed using different methods were not the same.

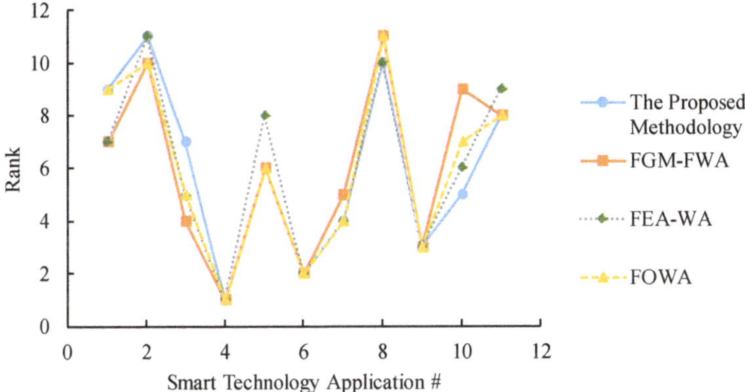


**Figure 16.** Comparing the results using various methods: fuzzy geometric mean (FGM)-fuzzy weighted average (FWA); fuzzy extent analysis (FEA)-weighted average (WA); fuzzy ordered weighted average (FOWA).

## 5. Conclusions

Smart health is the context-aware complement of mobile health within a smart city [49]. In the past work, a great deal of effort has been made to promote the smart health in a city or region [50]. However, assessing the suitability of a smart health practice is still a challenging task because the applied smart technology is still evolving. To this end, several group-based decision-making FAHP methods have been devised. However, the prerequisite for such group-based decision-making FAHP methods is the existence of a consensus among the participating decision makers, which has rarely been checked. To address this issue, this study puts forward a fuzzy collaborative approach that is the joint application of ACO, FI, and FWM. In particular, FI is applied to assess the consensus among decision makers before the collaboration process. The FI results are presented to decision makers, so that they can subjectively modify the pairwise comparison results to bring them closer to the FI results. Thereafter, the consensus among decision makers is again measured. The collaboration process continues until no further modifications are made by decision makers. Last, the FWM-CD method is applied to assess the suitability of a smart health practice.

To elaborate the effectiveness of the fuzzy collaborative approach, we assessed the suitabilities of eleven smart health practices. According to the experimental results, the following conclusions were drawn:

(1) Smart mobile services and smart clothes were evaluated as the most suitable and the least suitable smart health practices, respectively.

(2) The suitabilities of smart mobile services and smart phones evaluated using various methods were identical. In contrast, the suitabilities of other smart health practices, assessed using various methods, differed.

(3) Among the compared methods, only the fuzzy collaborative approach could guarantee the existence of a consensus among decision makers. In other words, only the results assessed using the fuzzy collaborative approach were acceptable to all decision makers.

Smart technologies are still evolving. Therefore, the suitability of a smart health practice needs to be re-assessed in the near future. In addition, other types of methods can also be applied to assess the suitability of a smart health practice.

**Author Contributions:** Data curation, T.-C.T.C. and Y.-C.W.; methodology, T.-C.T.C. and Y.-C.W.; writing—original draft preparation, T.-C.T.C. and Y.-C.W.; writing—review and editing, T.-C.T.C., Y.-C.W., Y.-C.L., H.-C.W. and H.-F.L.

**Acknowledgments:** This study was sponsored by Ambient Intelligence Association of Taiwan (AIAT) under Grant NCTU 1083RD2187.

**Conflicts of Interest:** The authors declare no conflict of interest.

## References


1. Jordan, M. What Is 'Smart' Technology? Available online: http://knowit.co.nz/2011/08/what-is-smart-technology (accessed on 12 October 2019).

2. Silva, B.M.; Rodrigues, J.J.; de la Torre Díez, I.; López-Coronado, M.; Saleem, K. Mobile-health: A review of current state in 2015. *J. Biomed. Inform.* **2015**, *56*, 265–272. [CrossRef] [PubMed]

3. Chen, T.; Chiu, M.-C. Smart technologies for assisting the life quality of persons in a mobile environment: A review. *J. Ambient. Intell. Humaniz. Comput.* **2018**, *9*, 319327. [CrossRef]

4. Haymes, L.K.; Storey, K.; Maldonado, A.; Post, M.; Montgomery, J. Using applied behavior analysis and smart technology for meeting the health needs of individuals with intellectual disabilities. *Dev. Neurorehabilit.* **2015**, *18*, 407–419. [CrossRef]

5. Chiu, M.-C.; Chen, T.-C.T. Assessing sustainable effectiveness of the adjustment mechanism of a ubiquitous clinic recommendation system. *Health Care Manag. Sci.* **2019**, 1–10. [CrossRef]

6. Chen, T.-C.T. Evaluating the sustainability of a smart technology application to mobile health care: The FGM–ACO–FWA approach. *Complex Intell. Syst.* **2019**, 1–13. [CrossRef]

7. Chen, T.-C.T. Guaranteed-consensus posterior-aggregation fuzzy analytic hierarchy process method. *Neural Comput. Appl.* **2019**, 1–12. [CrossRef]

8. Liu, F.; Mendel, J.M. Aggregation using the fuzzy weighted average as computed by the Karnik–Mendel algorithms. *IEEE Trans. Fuzzy Syst.* **2008**, *16*, 1–12.

9. Steele, R. Social media, mobile devices and sensors: Categorizing new techniques for health communication. In Proceedings of the 2011 Fifth International Conference on Sensing Technology, Palmerston North, New Zealand, 28 November–1 December 2011; pp. 187–192.

10. Hswen, Y.; Murti, V.; Vormawor, A.A.; Bhattacharjee, R.; Naslund, J.A. Virtual avatars, gaming, and social media: Designing a mobile health app to help children choose healthier food options. *J. Mob. Technol. Med.* **2013**, *2*, 8. [CrossRef] [PubMed]

11. Petersen, C.; DeMuro, P. Legal and regulatory considerations associated with use of patient-generated health data from social media and mobile health (mHealth) devices. *Appl. Clin. Inform.* **2015**, *6*, 16–26. [PubMed]

12. Cook, D.J.; Duncan, G.; Sprint, G.; Fritz, R.L. Using smart city technology to make healthcare smarter. *Proc. IEEE* **2018**, *106*, 708–722. [CrossRef]

13. Reeder, B.; David, A. Health at hand: A systematic review of smart watch uses for health and wellness. *J. Biomed. Inform.* **2016**, *63*, 269–276. [CrossRef] [PubMed]

14. Mandel, J.C.; Kreda, D.A.; Mandl, K.D.; Kohane, I.S.; Ramoni, R.B. SMART on FHIR: A standards-based, interoperable apps platform for electronic health records. *J. Am. Med Inform. Assoc.* **2016**, *23*, 899–908. [CrossRef] [PubMed]

15. Hamidi, H. An approach to develop the smart health using Internet of Things and authentication based on biometric technology. *Future Gener. Comput. Syst.* **2019**, *91*, 434–449. [CrossRef]

16. Rathod, R. Sensors Used in Smartphone. Available online: http://myphonefactor.in/2012/04/sensors-used-in-a-smartphone/ (accessed on 10 October 2019).

17. Eklund, J.; Forsman, M. Smart work clothes give better health-Through improved work technique, work organization and production technology. In *Congress of the International Ergonomics Association*; Springer: Berlin/Heidelberg, Germany, 2018; pp. 515–519.

18. Liu, L.; Stroulia, E.; Nikolaidis, I.; Miguel-Cruz, A.; Rincon, A.R. Smart homes and home health monitoring technologies for older adults: A systematic review. *Int. J. Med Inform.* **2016**, *91*, 44–59. [CrossRef] [PubMed]

19. Eskofier, B.M.; Lee, S.I.; Baron, M.; Simon, A.; Martindale, C.F.; Gaßner, H.; Klucken, J. An overview of smart shoes in the internet of health things: Gait and mobility assessment in health promotion and disease monitoring. *Appl. Sci.* **2017**, *7*, 986. [CrossRef]

20. Chen, M.; Ma, Y.; Song, J.; Lai, C.F.; Hu, B. Smart clothing: Connecting human with clouds and big data for sustainable health monitoring. *Mob. Netw. Appl.* **2016**, *21*, 825–845. [CrossRef]

21. Duman, G.M.; El-Sayed, A.; Kongar, E.; Gupta, S.M. An intelligent multiattribute group decision-making approach with preference elicitation for performance evaluation. *IEEE Trans. Eng. Manag.* **2019**, 1–17. [CrossRef]

22. Kubler, S.; Robert, J.; Derigent, W.; Voisin, A.; Le Traon, Y. A state-of the-art survey & testbed of fuzzy AHP (FAHP) applications. *Expert Syst. Appl.* **2016**, *65*, 398–422.

23. Saaty, T.L. Decision making—the analytic hierarchy and network processes (AHP/ANP). *J. Syst. Sci. Syst. Eng.* **2004**, *13*, 1–35. [CrossRef]

24. Zheng, G.; Zhu, N.; Tian, Z.; Chen, Y.; Sun, B. Application of a trapezoidal fuzzy AHP method for work safety evaluation and early warning rating of hot and humid environments. *Saf. Sci.* **2012**, *50*, 228–239. [CrossRef]

25. Michael, H. *Applied Fuzzy Arithmetic an Introduction with Engineering Applications*; Springer: Berlin/Heidelberg, Germany, 2005.

26. Golub, G.H.; Van Loan, C.F. *Matrix Computations*; Johns Hopkins: Baltimore, MD, USA, 1996; p. 694.

27. Saaty, T.L. *The Analytic Hierarchy Process*; McGraw-Hill: New York, NY, USA, 1980; p. 324.

28. Wang, Y.-C.; Chen, T.; Yeh, Y.-L. Advanced 3D printing technologies for the aircraft industry: A fuzzy systematic approach for assessing the critical factors. *Int. J. Adv. Manuf. Technol.* **2018**, 1–11. [CrossRef]

29. Chang, D.-Y. Applications of the extent analysis method on fuzzy AHP. *Eur. J. Oper. Res.* **1996**, *95*, 649–655. [CrossRef]

30. Wang, Y.C.; Chen, T. A fuzzy collaborative forecasting approach for forecasting the productivity of a factory. *Adv. Mech. Eng.* **2013**, *5*, 234571. [CrossRef]

31. Al-Refaie, A.; Aldwairi, R.; Chen, T. Optimizing performance of rigid polyurethane foam using FGP models. *J. Ambient Intell. Humaniz. Comput.* **2018**, *9*, 351–366. [CrossRef]

32. Csutora, R.; Buckley, J.J. Fuzzy hierarchical analysis: The Lambda-Max method. *Fuzzy Sets Syst.* **2001**, *120*, 181–195. [CrossRef]

33. Chen, T. An effective fuzzy collaborative forecasting approach for predicting the job cycle time in wafer fabrication. *Comput. Ind. Eng.* **2013**, *66*, 834–848. [CrossRef]

34. Yolcu, O.C.; Yolcu, U.; Egrioglu, E.; Aladag, C.H. High order fuzzy time series forecasting method based on an intersection operation. *Appl. Math. Model.* **2016**, *40*, 8750–8765. [CrossRef]

35. Chen, T. A heterogeneous fuzzy collaborative intelligence approach for forecasting the product yield. *Appl. Soft Comput.* **2017**, *57*, 210–224. [CrossRef]

36. Wang, Y.-C.; Chen, T.-C.T. A direct-solution fuzzy collaborative intelligence approach for yield forecasting in semiconductor manufacturing. *Procedia Manuf.* **2018**, *17*, 110–117. [CrossRef]

37. Wang, Y.C.; Chen, T.C.T. A partial-consensus posterior-aggregation FAHP method—supplier selection problem as an example. *Mathematics* **2019**, *7*, 179. [CrossRef]

38. Chen, T.-C.T.; Honda, K. *Fuzzy Collaborative Forecasting and Clustering: Methodology, System Architecture, and Applications*; Springer: Cham, Switzerland, 2019; p. 89.

39. Van Broekhoven, E.; De Baets, B. Fast and accurate center of gravity defuzzification of fuzzy system outputs defined on trapezoidal fuzzy partitions. *Fuzzy Sets Syst.* **2006**, *157*, 904–918. [CrossRef]

40. Ventola, C.L. Mobile devices and apps for health care professionals: Uses and benefits. *Pharm. Ther.* **2014**, *39*, 356.

41. Demiris, G.; Hensel, B.K.; Skubic, M.; Rantz, M. Senior residents' perceived need of and preferences for "smart home" sensor technologies. *Int. J. Technol. Assess. Health Care* **2008**, *24*, 120–124. [CrossRef] [PubMed]

42. Vodopivec-Jamsek, V.; de Jongh, T.; Gurol-Urganci, I.; Atun, R.; Car, J. Mobile phone messaging for preventive health care. *Cochrane Database Syst. Rev.* **2012**, *12*, CD007457. [CrossRef] [PubMed]

43. Sarasohn-Kahn, J. How Smartphones are Changing Health Care for Consumers and Providers. Available online: https://www.chcf.org/wp-content/uploads/2017/12/PDF-HowSmartphonesChangingHealthCare.pdf (accessed on 3 July 2019).

44. Bieber, G.; Kirste, T.; Urban, B. Ambient interaction by smart watches. In Proceedings of the 5th International Conference on PErvasive Technologies Related to Assistive Environments, Heraklion, Crete, Greece, 6–8 June 2012. article no. 39.

45. Free, C.; Phillips, G.; Galli, L.; Watson, L.; Felix, L.; Edwards, P.; Haines, A. The effectiveness of mobile-health technology-based health behaviour change or disease management interventions for health care consumers: A systematic review. *PLoS Med.* **2013**, *10*, e1001362. [CrossRef]

46. Porzi, L.; Messelodi, S.; Modena, C.M.; Ricci, E. A smart watch-based gesture recognition system for assisting people with visual impairments. In Proceedings of the 3rd ACM International Workshop on Interactive Multimedia on Mobile Portable Devices, Barcelona, Spain, 22 October 2013; pp. 19–24.

47. Hamel, M.B.; Cortez, N.G.; Cohen, I.G.; Kesselheim, A.S. FDA regulation of mobile health technologies. *New Engl. J. Med.* **2014**, *371*, 372.

48. Shcherbina, A.; Mattsson, C.M.; Waggott, D.; Salisbury, H.; Christle, J.W.; Hastie, T.; Ashley, E.A. Accuracy in wrist-worn, sensor-based measurements of heart rate and energy expenditure in a diverse cohort. *J. Pers. Med.* **2017**, *7*, 3. [CrossRef]

49. Solanas, A.; Patsakis, C.; Conti, M.; Vlachos, I.S.; Ramos, V.; Falcone, F.; Postolache, O.; Perez-martinez, P.A.; Pietro, R.D.; Perrea, D.N.; et al. Smart health: A context-aware health paradigm within smart cities. *IEEE Commun. Mag.* **2014**, *52*, 74–81. [CrossRef]

50. Baig, M.M.; Gholamhosseini, H. Smart health monitoring systems: An overview of design and modeling. *J. Med Syst.* **2013**, *37*, 9898. [CrossRef]

 *mathematics* 

*Article*

# Picture Fuzzy Interaction Partitioned Heronian Aggregation Operators for Hotel Selection

**Suizhi Luo and Lining Xing ***

College of Systems Engineering, National University of Defense Technology, Changsha 410073, China; szlluo@csu.edu.cn
* Correspondence: xinglining@gmail.com or xln_2002@nudt.edu.cn

Received: 27 November 2019; Accepted: 14 December 2019; Published: 18 December 2019

**Abstract:** Picture fuzzy numbers (PFNs), as the generalization of fuzzy sets, are good at fully expressing decision makers' opinions with four membership degrees. Since aggregation operators are simple but powerful tools, this study aims to explore some aggregation operators with PFNs to solve practical decision-making problems. First, new operational rules, the interaction operations of PFNs, are defined to overcome the drawbacks of existing operations. Considering that interrelationships may exist only in part of criteria, rather than all of the criteria in reality, the partitioned Heronian aggregation operator is modified with PFNs to deal with this condition. Then, desirable properties are proved and several special cases are discussed. New decision-making methods with these presented aggregation operators are suggested to process hotel selection issues. Last, their practicability and merits are certified by sensitivity analyses and comparison analyses with other existing approaches. The results indicate that our methods are feasible to address such situations where criteria interact in the same part, but are independent from each other at different parts.

**Keywords:** picture fuzzy numbers; interaction operations; partitioned Heronian; aggregation operators; hotel selection

---

## 1. Introduction

Decision making refers to selecting the optimal alternative according to some rules [1–3]. Generally, many uncertain or fuzzy factors may exist in the real decision-making process, which are not able to be described by crisp numbers [4,5]. In this case, fuzzy set theory [6], which was presented by Zadeh, can be used to express these uncertainty and vagueness. Nevertheless, because fuzzy set just has a membership degree, it cannot depict information in some circumstances, specifically where decision makers (DMs) disagree with each other. Then, Atanassov [7] put forward the concept of intuitionistic fuzzy sets (IFSs) with membership and non-membership functions, so that both people's consistency and inconsistency can be conveyed by IFSs. However, different DMs may have different attitudes for a certain decision-making issue, not just 'support' or 'opposition'. Hence, there are two inherent flaws in IFSs. (1) Except for consistent and inconsistent degrees, other possibilities, such as hesitant or refusal degrees, are not contained in the IFSs. (2) In addition, another shortcoming of IFSs is in their traditional operations: the interaction between membership degree and non-membership degree is not considered.

To overcome the first drawback, Cuong [8] first proposed picture fuzzy numbers (PFNs). PFNs can sufficiently depict DMs' diverse behaviors using four membership functions, including positive, neutral, negative, and refusal membership degrees [9]. Since then, the distance/similarity measure [10,11], cross entropy [12], and projection [13,14] of PFNs have been defined. Moreover, various decision-making methods based on them have been studied [15,16], such as the multi-attributive border approximation area comparison (MABAC) method [17] and the Vlsekriterijumska Optimizacija I Kompromisno

Resenje (VIKOR) method [13]. Clearly, PFNs show greater performances than fuzzy sets and IFSs in conveying complicated evaluation information. However, there is also an imperfection in the operations of PFNs: (3) The interactions among four different membership functions of PFNs are not taken into account fully. Thus, improper results may occur in certain cases, specifically when the neutral membership degree in a picture fuzzy number (PFN) is zero. When this happens, regardless of the value(s) of neutral membership degree(s) in other PFN(s), their aggregated neutral membership degree using the traditional operations is still zero.

To conquer disadvantage (2), He et al. [18] redefined the operational rules of IFSs through making an interaction between membership function and non-membership function. Likewise, for overcoming limitation (3), some interaction operational laws of PFNs are proposed.

When it comes to decision-making methods, information aggregation operators are basic and powerful at dealing with decision-making problems [19,20]. To date, several aggregation operators have been extended with PFNs. For example, Garg [21] defined the arithmetic weighted averaging, ordered weighted averaging, and hybrid averaging operators of PFNs; Wang et al. [22] adopted the geometric aggregation operators to aggregate picture fuzzy information; and Wei [23] integrated the Hamacher aggregation operators with PFNs to process practical problems. Nevertheless, a common defect of these operators is that (4) all of these operators presume that the parameters are standalone, and the interrelations among them are not considered at all. In reality, some inputs may be dependent on each other, but the above-mentioned operators are unable to deal with this situation. Lately, Xu [24] discussed the Muirhead mean operator in a picture fuzzy environment. The Muirhead mean operator is a useful method to capture the interrelationships among inputs. However, it may be difficult for DMs to determine a vector of parameters in the Muirhead mean operator.

In this case, the Bonferroni averaging [25,26] or Heronian averaging (HA) [27,28] operator may be a good choice to overcome limitation (4). Both of them can establish the relationships between two arguments, and only two parameter values need to be assigned by DMs. Compared with the Bonferroni averaging operator, the great advantages of the HA operator include: the correlation between inputs and itself is also taken into account, and no redundancy exists in the aggregated values. To date, the HA operator has been widely extended with various fuzzy sets, such as IFSs [29] and linguistic neutrosohpic numbers [30]. The limitations of this operator are related to two aspects. (5) Until now, no HA operator has been modified to aggregate picture fuzzy information; and (6) there is a hypothesis in the HA operator that each input is relevant with the remaining inputs, but this is not true at all times. More commonly, parts of arguments have relationships with each other, while no connections exist among several parts. Apparently, the conventional HA operator is helpless to handle this situation.

For coping with defect (6), the concepts of partitioned HA (PHA) and partitioned geometric HA (PGHA) operators are put forward by Liu et al. [31]. The function of these operators is to address such a general circumstance where the inputs are classified into several partitions, and the interrelationships are just found among the inputs in the same partition but not found among those in distinct partitions. Motivated by this idea, this study proposes some PHA operators within a picture fuzzy condition to circumvent weaknesses (5) and (6).

The main contributions of this study are outlined in the following.

First, new operational laws of PFNs are defined to capture the interactions among four membership degrees, so that limitation (3) is overcome.

Second, some PHA operators are extended with PFNs based on the new interaction operation rules. They can settle the situation where criteria need to be partitioned, as correlations can be seen only in the same part rather that in different parts. Special cases and important properties are discussed. Hence, drawbacks (4), (5), and (6) are all defeated.

Third, novel methods with the proposed aggregation operations are proposed to cope with complex decision-making issues in picture fuzzy circumstances. In the case study, PFNs are suggested to describe hotel evaluation information, and our methods are adopted to select the optimal hotel. This surmounts limitation (1) and justifies the feasibility of the proposed methods.

Fourth, full discussions with sensitivity analyses and comparison analyses are taken to confirm the strengths of our methods.

The remainder of this study is arranged as follows. In Section 2, the preliminaries of PFNs and PHA operators are introduced in brief. Section 3 defines the new interaction operation rules of PFNs, and several picture fuzzy PHA operators are based on these operations. Then, new decision-making methods with these aggregation operators are recommended in Section 4. In Section 5, a case of hotel selection is studied to show the advantages of PFNs and the practicability of our methods. Section 6 makes some discussions by analyzing the influence of parameters and comparing with other existing approaches for demonstrating the superiority of the proposed methods. Last, some necessary conclusions are provided.

## 2. Preliminaries

### 2.1. Picture Fuzzy Numbers

**Definition 1** ([8]). *The PFS (picture fuzzy set) is an object on a universe $\Omega$ with $a(\chi) = \{< \chi, p_a(\chi), m_a(\chi), n_a(\chi) > | \chi \in \Omega\}$, where $p_a(\chi) \in [0,1]$ is the positive membership degree, $m_a(\chi) \in [0,1]$ is the neutral membership degree, $n_a(\chi) \in [0,1]$ is the negative membership degree, and $v_a(\chi) = 1 - p_a(\chi) - m_a(\chi) - n_a(\chi) \in [0,1]$ is the refusal membership degree of $\chi$ in a.*

In particular, the PFS is degenerated to a PFN (picture fuzzy number) $a = (p_a, m_a, n_a)$ if $\Omega$ has only one element.

**Definition 2** ([22]). *Let $a = (p_a, m_a, n_a)$ and $b = (p_b, m_b, n_b)$ be two arbitrary PFNs (picture fuzzy numbers), then the operational laws between them are*

(1)    $a \oplus_W b = (p_a, m_a, n_a) \oplus_W (p_b, m_b, n_b) = (1 - (1 - p_a)(1 - p_b), m_a m_b, (n_a + m_a)(n_b + m_b) - m_a m_b)$;

(2)    $a \otimes_W b = (p_a, m_a, n_a) \otimes_W (p_b, m_b, n_b) = ((p_a + m_a)(p_b + m_b) - m_a m_b, m_a m_b, 1 - (1 - n_a)(1 - n_b))$;

(3)    $\delta \cdot a = (1 - (1 - p_a)^\delta, (m_a)^\delta, (n_a + m_a)^\delta - (m_a)^\delta), \delta \in (0, +\infty)$;

(4)    $a^\delta = ((p_a + m_a)^\delta - (m_a)^\delta, (m_a)^\delta, 1 - (1 - n_a)^\delta), \delta \in (0, +\infty)$.

**Example 1.** *Let $a_1 = (p_1, m_1, n_1) = (0.5, 0.3, 0.1)$ and $a_2 = (p_2, m_2, n_2) = (0.6, 0, 0.3)$ be two PFNs. According to Definition 2, the aggregation values of $a_1$ and $a_2$ are $a_1 \oplus a_2 = (0.8, 0, 0.12)$ and $a_1 \otimes a_2 = (0.48, 0, 0.37)$. That is, because $m_2 = 0$ in $a_2$, the value of $m_1$ in $a_1$ has no influence on the aggregation values of $a_1$ and $a_2$ Clearly, this is an imperfection.*

**Definition 3** ([22]). *Assume $a = (p_a, m_a, n_a)$ is a PFN, the score function and the accuracy function of a are*

$$E(a) = p_a - n_a, \tag{1}$$

$$F(a) = p_a + m_a + n_a. \tag{2}$$

**Definition 4** ([22]). *Given two PFNs $a = (p_a, m_a, n_a)$ and $b = (p_b, m_b, n_b)$, the comparison method is*

(1)    *if $E(a) < E(b)$, then $a < b$;*

(2)    *if $E(a) = E(b)$ and $F(a) > F(b)$, then $a > b$;*

(3)    *when $E(a) = E(b)$ and $F(a) = F(b)$, then $a \sim b$.*

### 2.2. Partitioned Heronian Averaging Operators

HA operator is a powerful tool to explore the relationship among inputs. The definitions of HA and geometric HA (GHA) operators are given in the following.

**Definition 5** ([29]). *Suppose $\alpha_i$ $(i = 1, 2, \ldots, y)$ is a group of non-negative real numbers and $\beta, \eta \geq 0$, then the HA aggregation operator is*

$$HA^{\beta,\eta}(\alpha_1, \alpha_2, \cdots, \alpha_y) = \left(\frac{2}{y(y+1)}\sum_{i=1}^{y}\sum_{j=1}^{y}(\alpha_i)^{\beta}(\alpha_j)^{\eta}\right)^{\frac{1}{\beta+\eta}}, \tag{3}$$

*and the GHA operator is*

$$GHA^{\beta,\eta}(\alpha_1, \alpha_2, \cdots, \alpha_y) = \frac{1}{\beta+\eta}\left(\prod_{i=1}^{y}\prod_{j=1}^{y}\beta\alpha_i + \eta\alpha_j\right)^{\frac{2}{y(y+1)}}. \tag{4}$$

To reflect a situation where argument values in the same group correlate with each other and the inputs in different groups are dissociated, Liu et al. [31] put forward the concepts of the PHA aggregation operator and PGHA operator as follows.

**Definition 6** ([31]). *Assume $\alpha_i$ $(i = 1, 2, \ldots, y)$ is a set of non-negative real numbers, and they are divided into $x$ independent groups $S_1, S_2, \ldots, S_x$, where $S_k = \{\alpha_{k1}, \alpha_{k2}, \ldots \alpha_{k|S_k|}\}$ $(k = 1, 2, \ldots, x)$, the cardinality of $S_k$ is $|S_k|$ and $\sum_{k=1}^{x}|S_k| = y$. Let $\beta, \eta \geq 0$, then the PHA operator is*

$$PHA^{\beta,\eta}(\alpha_1, \alpha_2, \cdots, \alpha_y) = \frac{1}{x}\left(\sum_{k=1}^{x}\left(\frac{2}{|S_k|(|S_k|+1)}\sum_{i=1}^{|S_k|}\sum_{j=1}^{|S_k|}(\alpha_{ki})^{\beta}(\alpha_{kj})^{\eta}\right)^{\frac{1}{\beta+\eta}}\right), \tag{5}$$

*and the PGHA operator is*

$$PGHA^{\beta,\eta}(\alpha_1, \alpha_2, \cdots, \alpha_y) = \left(\prod_{k=1}^{x}\left(\frac{1}{\beta+\eta}\left(\prod_{i=1}^{|S_k|}\prod_{j=1}^{|S_k|}\beta\alpha_{ki} + \eta\alpha_{kj}\right)^{\frac{2}{|S_k|(|S_k|+1)}}\right)\right)^{\frac{1}{x}}. \tag{6}$$

## 3. Some Picture Fuzzy Interaction Partitioned Heronian Averaging Operators

In this section, the interaction operational rules of PNFs are first defined. Thereafter, some picture fuzzy interaction PHA operators are presented based on the new rules.

### 3.1. Interaction Operational Laws of Picture Fuzzy Numbers

The In this subsection, new operational laws—the interaction operational laws of PFNs—are proposed to overcome the limitations of the existing operations mentioned in Section 2.2.

The interaction operations take the interaction among true, hesitant and false membership degrees into account, which are shown as follows.

**Definition 7.** *Given two PFNs $a_1 = (p_1, m_1, n_1)$ and $a_2 = (p_2, m_2, n_2)$, then the interaction operational laws are defined as*

(1) $a_1 \oplus a_2 = \left(1 - \prod_{i=1}^{2}(1-p_i), \prod_{i=1}^{2}(1-p_i) - \prod_{i=1}^{2}(1-p_i-m_i), \prod_{i=1}^{2}(1-p_i-m_i) - \prod_{i=1}^{2}(1-p_i-m_i-n_i)\right)$;

(2) $a_1 \otimes a_2 = \left(\prod_{i=1}^{2}(1-n_i-m_i) - \prod_{i=1}^{2}(1-n_i-m_i-p_i), \prod_{i=1}^{2}(1-n_i) - \prod_{i=1}^{2}(1-n_i-m_i), 1 - \prod_{i=1}^{2}(1-n_i)\right)$;

(3) $\delta \cdot a_1 = \left(1-(1-p_1)^{\delta}, (1-p_1)^{\delta} - (1-p_1-m_1)^{\delta}, (1-p_1-m_1)^{\delta} - (1-p_1-m_1-n_1)^{\delta}\right), \delta \in (0, +\infty)$;

(4) $(a_1)^{\delta} = \left((1-n_1-m_1)^{\delta} - (1-n_1-m_1-p_1)^{\delta}, (1-n_1)^{\delta} - (1-n_1-m_1)^{\delta}, 1-(1-n_1)^{\delta}\right), \delta \in (0, +\infty)$.

**Example 2.** *Let $a_1 = (0.5, 0.3, 0.1)$ and $a_2 = (0.6, 0, 0.3)$, which are the same as Example 1. Based on Definition 7, the interaction aggregation values of $a_1$ and $a_2$ are $a_1 \oplus a_2 = (0.8, 0.12, 0.07)$ and $a_1 \otimes a_2 = (0.41, 0.21, 0.37)$, which are more reasonable than the results in Example 1.*

### 3.2. Picture Fuzzy Interaction Partitioned Heronian Averaging Operator

In this subsection, the picture fuzzy interaction PHA (*PFIPHA*) and picture fuzzy weighted interaction PHA (*PFWIPHA*) operators are defined, and some important properties are proved.

**Definition 8.** *Assume* $a_i = (p_i, m_i, n_i)$ $(i = 1, 2, \ldots, y)$ *is a set of PFNs, and they can be partitioned into $x$ distinct sorts $S_1, S_2, \ldots, S_x$, where $S_k = \left\{ a_{k1}, a_{k2}, \ldots, a_{k|S_k|} \right\}$ $(k = 1, 2, \ldots, x)$. Then the PFIPHA operator is defined as*

$$PFIPHA^{\beta,\eta}(a_1, a_2, \ldots, a_y) = \frac{1}{x}\left( \sum_{k=1}^{x} \left( \frac{2}{|S_k|(|S_k|+1)} \sum_{i=1}^{|S_k|} \sum_{j=i}^{|S_k|} (a_{ki})^{\beta} \otimes (a_{kj})^{\eta} \right)^{\frac{1}{\beta+\eta}} \right) \tag{7}$$

*where $\beta, \eta \geq 0$, $|S_k|$ is the cardinality of $S_k$, and $\sum\limits_{k=1}^{x} |S_k| = y$.*

**Theorem 1.** *If $a_i = (p_i, m_i, n_i)$ $(i = 1, 2, \ldots, y)$ is a group of PFNs, and $\beta, \eta \geq 0$, then their aggregated result using Equation (7) is still a PFN, and the following is true:*

$$PFIPHA^{\beta,\eta}(a_1, a_2, \ldots, a_y) =$$

$$\left( 1 - \left( \prod_{k=1}^{x}\left( 1 - (1 - N_k^{\beta,\eta} + M_k^{\beta,\eta})^{\frac{1}{\beta+\eta}} + (M_k^{\beta,\eta})^{\frac{1}{\beta+\eta}} \right) \right)^{\frac{1}{x}} , \left( \prod_{k=1}^{x}\left( 1 - (1 - N_k^{\beta,\eta} + M_k^{\beta,\eta})^{\frac{1}{\beta+\eta}} + (M_k^{\beta,\eta})^{\frac{1}{\beta+\eta}} \right) \right)^{\frac{1}{x}} - \right.$$
$$\left. \left( \prod_{k=1}^{x}\left( 1 + (M_k^{\beta,\eta})^{\frac{1}{\beta+\eta}} - (1 - O_k^{\beta,\eta} + M_k^{\beta,\eta})^{\frac{1}{\beta+\eta}} \right) \right)^{\frac{1}{x}} , \left( \prod_{k=1}^{x}\left( 1 + (M_k^{\beta,\eta})^{\frac{1}{\beta+\eta}} - (1 - O_k^{\beta,\eta} + M_k^{\beta,\eta})^{\frac{1}{\beta+\eta}} \right) \right)^{\frac{1}{x}} - \left( \prod_{k=1}^{x}\left( (M_k^{\beta,\eta})^{\frac{1}{\beta+\eta}} \right) \right)^{\frac{1}{x}} \right) \tag{8}$$

*where* $M_k^{\beta,\eta} = \left( \prod\limits_{i=1,j=i}^{|S_k|} A_k^{\beta,\eta} \right)^{t_k}$, $N_k^{\beta,\eta} = \left( \prod\limits_{i=1,j=i}^{|S_k|} B_k^{\beta,\eta} \right)^{t_k}$, $O_k^{\beta,\eta} = \left( \prod\limits_{i=1,j=i}^{|S_k|} C_k^{\beta,\eta} \right)^{t_k}$, $t_k = \frac{2}{|S_k|(|S_k|+1)}$, $A_k^{\beta,\eta} = (v_{ki})^{\beta}(v_{kj})^{\eta}$, $B_k^{\beta,\eta} = 1 - (u_{ki})^{\beta}(u_{kj})^{\eta} + (v_{ki})^{\beta}(v_{kj})^{\eta}$, $C_k^{\beta,\eta} = 1 + (v_{ki})^{\beta}(v_{kj})^{\eta} - (r_{ki})^{\beta}(r_{kj})^{\eta}$, $r_{ki} = 1 - n_{ki}$, $r_{kj} = 1 - n_{kj}$, $u_{ki} = 1 - n_{ki} - m_{ki}$, $u_{kj} = 1 - n_{kj} - m_{kj}$, $v_{ki} = 1 - n_{ki} - m_{ki} - p_{ki}$ *and* $v_{kj} = 1 - n_{kj} - m_{kj} - p_{kj}$.

Note that the Proof of Theorem 1 can be seen in the Appendix A.

For the *PFIPHA* operator, the following properties should be satisfied.

**Property 1.** *(Idempotency) Suppose $a_i = (p_i, m_i, n_i)$ $(i = 1, 2, \ldots, y)$ is a collection of PFNs, and $a_i = a = (p, m, n)$ for all $i = 1, 2, \ldots, y$, then $PFIPHA^{\beta,\eta}(a_1, a_2, \ldots, a_y) = a$.*

**Proof.** Because $a_1 = a_2 = \cdots = a_y = (p, m, n)$, then for all $i = 1, 2, \ldots, y$, $r_i = 1 - n_i = r$, $u_i = 1 - n_i - m_i = u$ and $v_i = 1 - n_i - m_i - p_i = v \Rightarrow A_k^{\beta,\eta} = (v_{ki})^{\beta}(v_{kj})^{\eta} = v^{\beta+\eta}$, $B_k^{\beta,\eta} = 1 - (u_{ki})^{\beta}(u_{kj})^{\eta} + (v_{ki})^{\beta}(v_{kj})^{\eta} = 1 - u^{\beta+\eta} + v^{\beta+\eta}$ and $C_k^{\beta,\eta} = 1 + (v_{ki})^{\beta}(v_{kj})^{\eta} - (r_{ki})^{\beta}(r_{kj})^{\eta} = 1 + v^{\beta+\eta} - r^{\beta+\eta} \Rightarrow$

$$M_k^{\beta,\eta} = \left( \prod_{i=1,j=i}^{|S_k|} A_k^{\beta,\eta} \right)^{t_k} = \left( \prod_{i=1,j=i}^{|S_k|} v^{\beta+\eta} \right)^{t_k} = v^{\beta+\eta}, \quad N_k^{\beta,\eta} = \left( \prod_{i=1,j=i}^{|S_k|} B_k^{\beta,\eta} \right)^{t_k} = \left( \prod_{i=1,j=i}^{|S_k|} (1 - u^{\beta+\eta} + v^{\beta+\eta}) \right)^{t_k} =$$

$$1 - u^{\beta+\eta} + v^{\beta+\eta} \text{ and } O_k^{\beta,\eta} = \left( \prod_{i=1,j=i}^{|S_k|} C_k^{\beta,\eta} \right)^{t_k} = \left( \prod_{i=1,j=i}^{|S_k|} (1 + v^{\beta+\eta} - r^{\beta+\eta}) \right)^{t_k} = 1 + v^{\beta+\eta} - r^{\beta+\eta}, \text{ then}$$

$$PFIPHA^{\beta,\eta}(a_1, a_2, \ldots, a_y) =$$

$$\left( 1 - \left( \prod_{k=1}^{x}\left( 1 - (1 - N_k^{\beta,\eta} + M_k^{\beta,\eta})^{\frac{1}{\beta+\eta}} + (M_k^{\beta,\eta})^{\frac{1}{\beta+\eta}} \right) \right)^{\frac{1}{x}} , \left( \prod_{k=1}^{x}\left( 1 - (1 - N_k^{\beta,\eta} + M_k^{\beta,\eta})^{\frac{1}{\beta+\eta}} + (M_k^{\beta,\eta})^{\frac{1}{\beta+\eta}} \right) \right)^{\frac{1}{x}} - \right.$$
$$\left. \left( \prod_{k=1}^{x}\left( 1 + (M_k^{\beta,\eta})^{\frac{1}{\beta+\eta}} - (1 - O_k^{\beta,\eta} + M_k^{\beta,\eta})^{\frac{1}{\beta+\eta}} \right) \right)^{\frac{1}{x}} , \left( \prod_{k=1}^{x}\left( 1 + (M_k^{\beta,\eta})^{\frac{1}{\beta+\eta}} - (1 - O_k^{\beta,\eta} + M_k^{\beta,\eta})^{\frac{1}{\beta+\eta}} \right) \right)^{\frac{1}{x}} - \left( \prod_{k=1}^{x}\left( (M_k^{\beta,\eta})^{\frac{1}{\beta+\eta}} \right) \right)^{\frac{1}{x}} \right)$$
$$=$$

$$\left(1-\left(\prod_{k=1}^{x}\left(1-\left(1-(1-u^{\beta+\eta}+v^{\beta+\eta})+v^{\beta+\eta}\right)^{\frac{1}{\beta+\eta}}+(v^{\beta+\eta})^{\frac{1}{\beta+\eta}}\right)\right)\right)^{\frac{1}{x}},$$

$$\left(\prod_{k=1}^{x}\left(1-\left(1-(1-u^{\beta+\eta}+v^{\beta+\eta})+v^{\beta+\eta}\right)^{\frac{1}{\beta+\eta}}+(v^{\beta+\eta})^{\frac{1}{\beta+\eta}}\right)\right)^{\frac{1}{x}}-\left(\prod_{k=1}^{x}\left(1+(v^{\beta+\eta})^{\frac{1}{\beta+\eta}}-\left(1-(1+v^{\beta+\eta}-r^{\beta+\eta})+v^{\beta+\eta}\right)^{\frac{1}{\beta+\eta}}\right)\right)^{\frac{1}{x}},$$

$$\left(\prod_{k=1}^{x}\left(1+(v^{\beta+\eta})^{\frac{1}{\beta+\eta}}-\left(1-(1+v^{\beta+\eta}-r^{\beta+\eta})+v^{\beta+\eta}\right)^{\frac{1}{\beta+\eta}}\right)\right)^{\frac{1}{x}}-\left(\prod_{k=1}^{x}\left((v^{\beta+\eta})^{\frac{1}{\beta+\eta}}\right)\right)^{\frac{1}{x}}\right)$$

$$=$$

$$\left(1-\left(\prod_{k=1}^{x}\left(1-(u^{\beta+\eta})^{\frac{1}{\beta+\eta}}+(v^{\beta+\eta})^{\frac{1}{\beta+\eta}}\right)\right)^{\frac{1}{x}},\left(\prod_{k=1}^{x}\left(1-(u^{\beta+\eta})^{\frac{1}{\beta+\eta}}+(v^{\beta+\eta})^{\frac{1}{\beta+\eta}}\right)\right)^{\frac{1}{x}}-\left(\prod_{k=1}^{x}\left(1+(v^{\beta+\eta})^{\frac{1}{\beta+\eta}}-(r^{\beta+\eta})^{\frac{1}{\beta+\eta}}\right)\right)^{\frac{1}{x}},\right.$$

$$\left.\cdot\left(\prod_{k=1}^{x}\left(1+(v^{\beta+\eta})^{\frac{1}{\beta+\eta}}-(r^{\beta+\eta})^{\frac{1}{\beta+\eta}}\right)\right)^{\frac{1}{x}}-\left(\prod_{k=1}^{x}\left((v^{\beta+\eta})^{\frac{1}{\beta+\eta}}\right)\right)^{\frac{1}{x}}\right)$$

$$=\left(1-\left(\prod_{k=1}^{x}(1-u+v)\right)^{\frac{1}{x}},\left(\prod_{k=1}^{x}(1-u+v)\right)^{\frac{1}{x}}-\left(\prod_{k=1}^{x}(1+v-r)\right)^{\frac{1}{x}},\left(\prod_{k=1}^{x}(1+v-r)\right)^{\frac{1}{x}}-\left(\prod_{k=1}^{x}(v)\right)^{\frac{1}{x}}\right)$$

$$=(1-(1-u+v),(1-u+v)-(1+v-r),(1+v-r)-v)=(u-v,r-u,1-r)$$

$$=((1-n-m)-(1-n-m-p),(1-n)-(1-n-m),1-(1-n))=(p,m,n)=a.\ \square$$

Now, the Proof is completed.

**Property 2.** *(Commutativity) Let $a_i=(p_i,m_i,n_i)$ and $a_i^*=(p_i^*,m_i^*,n_i^*)$ $(i=1,2,\ldots,y)$ be two sets of PFNs. If $(a_1^*,a_2^*,\cdots,a_y^*)$ is an arbitrary permutation of $(a_1,a_2,\cdots,a_y)$, then $PFIPHA^{\beta,\eta}(a_1,a_2,\ldots,a_y)=PFIPHA^{\beta,\eta}(a_1^*,a_2^*,\cdots,a_y^*)$.*

**Proof.** According to Equation (8), $PFIPHA^{\beta,\eta}(a_1,a_2,\ldots,a_y)=$

$$\left(1-\left(\prod_{k=1}^{x}\left(1-(1-N_k^{\beta,\eta}+M_k^{\beta,\eta})^{\frac{1}{\beta+\eta}}+(M_k^{\beta,\eta})^{\frac{1}{\beta+\eta}}\right)\right)^{\frac{1}{x}},\left(\prod_{k=1}^{x}\left(1-(1-N_k^{\beta,\eta}+M_k^{\beta,\eta})^{\frac{1}{\beta+\eta}}+(M_k^{\beta,\eta})^{\frac{1}{\beta+\eta}}\right)\right)^{\frac{1}{x}}-\right.$$

$$\left.\left(\prod_{k=1}^{x}\left(1+(M_k^{\beta,\eta})^{\frac{1}{\beta+\eta}}-(1-O_k^{\beta,\eta}+M_k^{\beta,\eta})^{\frac{1}{\beta+\eta}}\right)\right)^{\frac{1}{x}},\left(\prod_{k=1}^{x}\left(1+(M_k^{\beta,\eta})^{\frac{1}{\beta+\eta}}-(1-O_k^{\beta,\eta}+M_k^{\beta,\eta})^{\frac{1}{\beta+\eta}}\right)\right)^{\frac{1}{x}}-\left(\prod_{k=1}^{x}\left((M_k^{\beta,\eta})^{\frac{1}{\beta+\eta}}\right)\right)^{\frac{1}{x}}\right)$$

and $PFIPHA^{\beta,\eta}(a_1^*,a_2^*,\cdots,a_y^*)=$

$$\left(1-\left(\prod_{k=1}^{x}\left(1-(1-N_k^{*\beta,\eta}+M_k^{*\beta,\eta})^{\frac{1}{\beta+\eta}}+(M_k^{*\beta,\eta})^{\frac{1}{\beta+\eta}}\right)\right)^{\frac{1}{x}},\left(\prod_{k=1}^{x}\left(1-(1-N_k^{*\beta,\eta}+M_k^{*\beta,\eta})^{\frac{1}{\beta+\eta}}+(M_k^{*\beta,\eta})^{\frac{1}{\beta+\eta}}\right)\right)^{\frac{1}{x}}-\right.$$

$$\left.\left(\prod_{k=1}^{x}\left(1+(M_k^{*\beta,\eta})^{\frac{1}{\beta+\eta}}-(1-O_k^{*\beta,\eta}+M_k^{*\beta,\eta})^{\frac{1}{\beta+\eta}}\right)\right)^{\frac{1}{x}},\left(\prod_{k=1}^{x}\left(1+(M_k^{*\beta,\eta})^{\frac{1}{\beta+\eta}}-(1-O_k^{*\beta,\eta}+M_k^{*\beta,\eta})^{\frac{1}{\beta+\eta}}\right)\right)^{\frac{1}{x}}-\left(\prod_{k=1}^{x}\left((M_k^{*\beta,\eta})^{\frac{1}{\beta+\eta}}\right)\right)^{\frac{1}{x}}\right).$$

As $(a_1^*,a_2^*,\cdots,a_y^*)$ is an arbitrary permutation of $(a_1,a_2,\cdots,a_y)$, it is clearly that $PFIPHA^{\beta,\eta}(a_1,a_2,\ldots,a_y)=PFIPHA^{\beta,\eta}(a_1^*,a_2^*,\cdots,a_y^*)$. $\square$

Several special cases of the $PFIPHA^{\beta,\eta}$ operator are discussed as follows.

**Special case 1:**

When $\eta\to0$, $PFIPHA^{\beta,0}(a_1,a_2,\ldots,a_y)=$

$$\left(1-\left(\prod_{k=1}^{x}\left(1-\left(1-\left(\prod_{i=1}^{|S_k|}\left(1-(u_{ki})^{\beta}+(v_{ki})^{\beta}\right)\right)^{t_k}+\left(\prod_{i=1}^{|S_k|}(v_{ki})^{\beta}\right)^{t_k}\right)^{\frac{1}{\beta}}+\left(\left(\prod_{i=1}^{|S_k|}(v_{ki})^{\beta}\right)^{t_k}\right)^{\frac{1}{\beta}}\right)\right)^{\frac{1}{x}},$$

$$\left(\prod_{k=1}^{x}\left(1-\left(1-\left(\prod_{i=1}^{|S_k|}\left(1-(u_{ki})^{\beta}+(v_{ki})^{\beta}\right)\right)^{t_k}+\left(\prod_{i=1}^{|S_k|}(v_{ki})^{\beta}\right)^{t_k}\right)^{\frac{1}{\beta}}+\left(\left(\prod_{i=1}^{|S_k|}(v_{ki})^{\beta}\right)^{t_k}\right)^{\frac{1}{\beta}}\right)\right)^{\frac{1}{x}}-$$

$$\left(\prod_{k=1}^{x}\left(1+\left(\left(\prod_{i=1}^{|S_k|}(v_{ki})^{\beta}\right)^{t_k}\right)^{\frac{1}{\beta}}-\left(1-\left(\prod_{i=1}^{|S_k|}\left(1+(v_{ki})^{\beta}-(r_{ki})^{\beta}\right)\right)^{t_k}+\left(\prod_{i=1}^{|S_k|}(v_{ki})^{\beta}\right)^{t_k}\right)^{\frac{1}{\beta}}\right)\right)^{\frac{1}{x}},$$

$$\left(\prod_{k=1}^{x}\left(1+\left(\left(\prod_{i=1}^{|S_k|}(v_{ki})^{\beta}\right)^{t_k}\right)^{\frac{1}{\beta}}-\left(1-\left(\prod_{i=1}^{|S_k|}\left(1+(v_{ki})^{\beta}-(r_{ki})^{\beta}\right)\right)^{t_k}+\left(\prod_{i=1}^{|S_k|}(v_{ki})^{\beta}\right)^{t_k}\right)^{\frac{1}{\beta}}\right)\right)^{\frac{1}{x}}-\left(\prod_{k=1}^{x}\left(\left(\left(\prod_{i=1}^{|S_k|}(v_{ki})^{\beta}\right)^{t_k}\right)^{\frac{1}{\beta}}\right)\right)^{\frac{1}{x}}\right).$$

**Special case 2:**

When $\beta \to 0$, then $PFIPHA^{0,\eta}(a_1, a_2, \ldots, a_y) =$

$$\left(1 - \left[\prod_{k=1}^{x}\left(1 - \left(1 - \left(\prod_{\substack{i=1,j=i}}^{|S_k|}\left(1 - (u_{kj})^{\eta} + (v_{ki})^{\eta}\right)\right)^{t_k} + \left(\prod_{\substack{i=1,j=i}}^{|S_k|}(v_{kj})^{\eta}\right)^{t_k}\right)^{\frac{1}{\eta}} + \left(\left(\prod_{\substack{i=1,j=i}}^{|S_k|}(v_{kj})^{\eta}\right)^{t_k}\right)^{\frac{1}{\eta}}\right)\right]^{\frac{1}{x}}\right),$$

$$\left(\left[\prod_{k=1}^{x}\left(1 - \left(1 - \left(\prod_{\substack{i=1,j=i}}^{|S_k|}\left(1 - (u_{kj})^{\eta} + (v_{kj})^{\eta}\right)\right)^{t_k} + \left(\prod_{\substack{i=1,j=i}}^{|S_k|}(v_{kj})^{\eta}\right)^{t_k}\right)^{\frac{1}{\eta}} + \left(\left(\prod_{\substack{i=1,j=i}}^{|S_k|}(v_{kj})^{\eta}\right)^{t_k}\right)^{\frac{1}{\eta}}\right)\right]^{\frac{1}{x}} - \right.$$

$$\left[\prod_{k=1}^{x}\left(1 + \left(\left(\prod_{\substack{i=1,j=i}}^{|S_k|}(v_{kj})^{\eta}\right)^{t_k}\right)^{\frac{1}{\eta}} - \left(1 - \left(\prod_{\substack{i=1,j=i}}^{|S_k|}\left(1 + (v_{kj})^{\eta} - (r_{kj})^{\eta}\right)\right)^{t_k} + \left(\prod_{\substack{i=1,j=i}}^{|S_k|}(v_{kj})^{\eta}\right)^{t_k}\right)^{\frac{1}{\eta}}\right)\right]^{\frac{1}{x}},$$

$$\left.\left[\prod_{k=1}^{x}\left(1 + \left(\left(\prod_{\substack{i=1,j=i}}^{|S_k|}(v_{kj})^{\eta}\right)^{t_k}\right)^{\frac{1}{\eta}} - \left(1 - \left(\prod_{\substack{i=1,j=i}}^{|S_k|}\left(1 + (v_{kj})^{\eta} - (r_{kj})^{\eta}\right)\right)^{t_k} + \left(\prod_{\substack{i=1,j=i}}^{|S_k|}(v_{kj})^{\eta}\right)^{t_k}\right)^{\frac{1}{\eta}}\right)\right]^{\frac{1}{x}} - \prod_{k=1}^{x}\left(\left(\left(\prod_{\substack{i=1,j=i}}^{|S_k|}(v_{kj})^{\eta}\right)^{t_k}\right)^{\frac{1}{\eta}}\right)^{\frac{1}{x}}\right)$$

**Special case 3:**

When $\beta = 1$ and $\eta \to 0$, then $PFIPHA^{1,0}(a_1, a_2, \ldots, a_y) =$

$$\left(1 - \left(\prod_{k=1}^{x}\left(\prod_{i=1}^{|S_k|}(1 - u_{ki} + v_{ki})\right)^{t_k}\right)^{\frac{1}{x}}, \left(\prod_{k=1}^{x}\left(\prod_{i=1}^{|S_k|}(1 - u_{ki} + v_{ki})\right)^{t_k}\right)^{\frac{1}{x}}\right) - \left(\prod_{k=1}^{x}\left(\prod_{i=1}^{|S_k|}(1 + v_{ki} - r_{ki})\right)^{t_k}\right)^{\frac{1}{x}}, \left(\prod_{k=1}^{x}\left(\prod_{i=1}^{|S_k|}(1 + v_{ki} - r_{ki})\right)^{t_k}\right)^{\frac{1}{x}} - $$

$$\left(\prod_{k=1}^{x}\left(\prod_{i=1}^{|S_k|}v_{ki}\right)^{t_k}\right)^{\frac{1}{x}}.$$

**Special case 4:**

When $\beta = \eta = 1$, then $PFIPHA^{1,0}(a_1, a_2, \ldots, a_y) =$

$$\left(1 - \left[\prod_{k=1}^{x}\left(1 - \left(1 - \left(\prod_{\substack{i=1,j=i}}^{|S_k|}\left(1 - u_{ki}u_{kj} + v_{ki}v_{kj}\right)\right)^{t_k} + \left(\prod_{\substack{i=1,j=i}}^{|S_k|}v_{ki}v_{kj}\right)^{t_k}\right)^{\frac{1}{2}} + \left(\left(\prod_{\substack{i=1,j=i}}^{|S_k|}v_{ki}v_{kj}\right)^{t_k}\right)^{\frac{1}{2}}\right)\right]^{\frac{1}{x}}\right),$$

$$\left(\left[\prod_{k=1}^{x}\left(1 - \left(1 - \left(\prod_{\substack{i=1,j=i}}^{|S_k|}\left(1 - u_{ki}u_{kj} + v_{ki}v_{kj}\right)\right)^{t_k} + \left(\prod_{\substack{i=1,j=i}}^{|S_k|}v_{ki}v_{kj}\right)^{t_k}\right)^{\frac{1}{2}} + \left(\left(\prod_{\substack{i=1,j=i}}^{|S_k|}v_{ki}v_{kj}\right)^{t_k}\right)^{\frac{1}{2}}\right)\right]^{\frac{1}{x}} - \right.$$

$$\left[\prod_{k=1}^{x}\left(1 + \left(\left(\prod_{\substack{i=1,j=i}}^{|S_k|}v_{ki}v_{kj}\right)^{t_k}\right)^{\frac{1}{2}} - \left(1 - \left(\prod_{\substack{i=1,j=i}}^{|S_k|}\left(1 + v_{ki}v_{kj} - r_{ki}r_{kj}\right)\right)^{t_k} + \left(\prod_{\substack{i=1,j=i}}^{|S_k|}v_{ki}v_{kj}\right)^{t_k}\right)^{\frac{1}{2}}\right)\right]^{\frac{1}{x}},$$

$$\left.\left[\prod_{k=1}^{x}\left(1 + \left(\left(\prod_{\substack{i=1,j=i}}^{|S_k|}v_{ki}v_{kj}\right)^{t_k}\right)^{\frac{1}{2}} - \left(1 - \left(\prod_{\substack{i=1,j=i}}^{|S_k|}\left(1 + v_{ki}v_{kj} - r_{ki}r_{kj}\right)\right)^{t_k} + \left(\prod_{\substack{i=1,j=i}}^{|S_k|}v_{ki}v_{kj}\right)^{t_k}\right)^{\frac{1}{2}}\right)\right]^{\frac{1}{x}} - \prod_{k=1}^{x}\left(\left(\left(\prod_{\substack{i=1,j=i}}^{|S_k|}v_{ki}v_{kj}\right)^{t_k}\right)^{\frac{1}{2}}\right)^{\frac{1}{x}}\right).$$

**Definition 9.** *If $a_i = (p_i, m_i, n_i)$ $(i = 1, 2, \ldots, y)$ is a set of PFNs, and they can be partitioned into $x$ distinct sorts $S_1, S_2, \ldots, S_x$, where $S_k = \left\{a_{k1}, a_{k2}, \ldots, a_{k|S_k|}\right\}$ $(k = 1, 2, \ldots, x)$; $(w_1, w_2, \cdots, w_y)$ is the weight vector of $(a_1, a_2, \cdots, a_y)$, where $w_i \in [0,1]$ and $\sum\limits_{i=1}^{y} w_i = 1$, then the PFWIPHA operator is defined as*

$$PFWIPHA^{\beta,\eta}(a_1, a_2, \ldots, a_y) = \frac{1}{x}\left(\sum_{k=1}^{x}\left(\frac{2}{|S_k|(|S_k|+1)}\sum_{i=1}^{|S_k|}\sum_{j=i}^{|S_k|}(w_i a_{ki})^{\beta} \otimes (w_j a_{kj})^{\eta}\right)^{\frac{1}{\beta+\eta}}\right) \tag{9}$$

*where $\beta, \eta \geq 0$, $|S_k|$ is the cardinality of $S_k$, and $\sum\limits_{k=1}^{x} |S_k| = y$.*

**Theorem 2.** *Suppose $a_i = (p_i, m_i, n_i)$ $(i = 1, 2, \ldots, y)$ is a group of PFNs, $(w_1, w_2, \cdots, w_y)$ is the weight vector of $(a_1, a_2, \cdots, a_y)$, $w_i \in [0, 1]$, $\sum_{i=1}^{y} w_i = 1$ and $\beta, \eta \geq 0$, then their aggregated result using Equation (9) is still a PFN, and the following is true:*

$$
PFWIPHA^{\beta,\eta}(a_1, a_2, \ldots, a_y) =
$$

$$
\left( 1 - \left( \prod_{k=1}^{x} \left( 1 - (1 + J_k^{\beta,\eta} - L_k^{\beta,\eta})^{\frac{1}{\beta+\eta}} + (J_k^{\beta,\eta})^{\frac{1}{\beta+\eta}} \right) \right)^{\frac{1}{x}}, \left( \prod_{k=1}^{x} \left( 1 - (1 + J_k^{\beta,\eta} - L_k^{\beta,\eta})^{\frac{1}{\beta+\eta}} + (J_k^{\beta,\eta})^{\frac{1}{\beta+\eta}} \right) \right)^{\frac{1}{x}} - \right.
$$

$$
\left. \left( \prod_{k=1}^{x} \left( 1 + (J_k^{\beta,\eta})^{\frac{1}{\beta+\eta}} - (1 - Q_k^{\beta,\eta} + J_k^{\beta,\eta})^{\frac{1}{\beta+\eta}} \right) \right)^{\frac{1}{x}}, \left( \prod_{k=1}^{x} \left( 1 + (J_k^{\beta,\eta})^{\frac{1}{\beta+\eta}} - (1 - Q_k^{\beta,\eta} + J_k^{\beta,\eta})^{\frac{1}{\beta+\eta}} \right) \right)^{\frac{1}{x}} - \left( \prod_{k=1}^{x} \left( (J_k^{\beta,\eta})^{\frac{1}{\beta+\eta}} \right) \right)^{\frac{1}{x}} \right)
$$

$$(10)$$

*where* $J_k^{\beta,\eta} = \left( \prod_{i=1, j=i}^{|S_k|} G_k^{\beta,\eta} \right)^{t_k}$, $L_k^{\beta,\eta} = \left( \prod_{i=1, j=i}^{|S_k|} \left( 1 - H_k^{\beta,\eta} + G_k^{\beta,\eta} \right) \right)^{t_k}$, $Q_k^{\beta,\eta} = \left( \prod_{i=1, j=i}^{|S_k|} \left( 1 + G_k^{\beta,\eta} - I_k^{\beta,\eta} \right) \right)^{t_k}$,

$t_k = \frac{2}{|S_k|(|S_k|+1)}$, $G_k^{\beta,\eta} = \left( (v_{ki})^{w_i} \right)^{\beta} \left( (v_{kj})^{w_j} \right)^{\eta}$, $H_k^{\beta,\eta} = \left( 1 + (v_{ki})^{w_i} - (e_{ki})^{w_i} \right)^{\beta} \left( 1 + (v_{kj})^{w_j} - (e_{kj})^{w_j} \right)^{\eta}$,

$I_k^{\beta,\eta} = \left( 1 - (f_{ki})^{w_i} + (v_{ki})^{w_i} \right)^{\beta} \left( 1 - (f_{kj})^{w_j} + (v_{kj})^{w_j} \right)^{\eta}$, $e_{ki} = 1 - p_{ki}$, $e_{kj} = 1 - p_{kj}$, $f_{ki} = 1 - p_{ki} - m_{ki}$,

$f_{kj} = 1 - p_{kj} - m_{kj}$, $v_{ki} = 1 - p_{ki} - m_{ki} - n_{ki}$ *and* $v_{kj} = 1 - p_{kj} - m_{kj} - n_{kj}$.

Note that the Proof of Theorem 2 can be seen in the Appendix B.

### 3.3. Picture Fuzzy Interaction Partitioned Geometric Heronian Averaging Operator

In this subsection, the picture fuzzy interaction partitioned geometric HA (*PFIPGHA*) and the picture fuzzy weighted interaction partitioned geometric HA (*PFWIPGHA*) operators are discussed. Different from arithmetic aggregation operators, the geometric aggregation operators emphasize the equilibrium of all inputs and the harmonization (rather than the complementarity) among their individual values [32]. Therefore, the following discussion of *PFIPGHA* and *PFWIPGHA* operators are also necessary.

**Definition 10.** *Assume $a_i = (p_i, m_i, n_i)$ $(i = 1, 2, \ldots, y)$ is a set of PFNs, and they can be partitioned into $x$ distinct sorts $S_1, S_2, \ldots, S_x$, where $S_k = \{a_{k1}, a_{k2}, \ldots, a_{k|S_K|}\}$ $(k = 1, 2, \ldots, x)$. Then the PFIPGHA operator is defined as*

$$
PFIPGHA^{\beta,\eta}(a_1, a_2, \ldots, a_y) = \left( \prod_{k=1}^{x} \left( \frac{1}{\beta + \eta} \left( \prod_{i=1, j=i}^{|S_k|} (\beta a_{ki} \oplus \eta a_{kj}) \right)^{\frac{2}{|S_k|(|S_k|+1)}} \right) \right)^{\frac{1}{x}}
$$

$$(11)$$

*where $\beta, \eta \geq 0$, $|S_k|$ is the cardinality of $S_k$ and $\sum_{k=1}^{x} |S_k| = y$.*

**Theorem 3.** *If $a_i = (p_i, m_i, n_i)$ $(i = 1, 2, \ldots, y)$ is a group of PFNs, and $\beta, \eta \geq 0$, then their aggregated result using Equation (11) is still a PFN, and the following is true:*

$$
PFIPGHA^{\beta,\eta}(a_1, a_2, \ldots, a_y) =
$$

$$
\left( \left( \prod_{k=1}^{x} \left( 1 + \left( (R_k^{\beta,\eta})^{t_k} \right)^{\frac{1}{\beta+\eta}} - \left( 1 - (U_k^{\beta,\eta})^{t_k} + (R_k^{\beta,\eta})^{t_k} \right)^{\frac{1}{\beta+\eta}} \right) \right)^{\frac{1}{x}} - \left( \prod_{k=1}^{x} \left( \left( (R_k^{\beta,\eta})^{t_k} \right)^{\frac{1}{\beta+\eta}} \right) \right)^{\frac{1}{x}}, \left( \prod_{k=1}^{x} \left( 1 - \left( 1 + (R_k^{\beta,\eta})^{t_k} - (V_k^{\beta,\eta})^{t_k} \right)^{\frac{1}{\beta+\eta}} + \left( (R_k^{\beta,\eta})^{t_k} \right)^{\frac{1}{\beta+\eta}} \right) \right)^{\frac{1}{x}} - \right.
$$

$$
\left. \left( \prod_{k=1}^{x} \left( 1 + \left( (R_k^{\beta,\eta})^{t_k} \right)^{\frac{1}{\beta+\eta}} - \left( 1 - (U_k^{\beta,\eta})^{t_k} + (R_k^{\beta,\eta})^{t_k} \right)^{\frac{1}{\beta+\eta}} \right) \right)^{\frac{1}{x}}, 1 - \left( \prod_{k=1}^{x} \left( 1 - \left( 1 + (R_k^{\beta,\eta})^{t_k} - (V_k^{\beta,\eta})^{t_k} \right)^{\frac{1}{\beta+\eta}} + \left( (R_k^{\beta,\eta})^{t_k} \right)^{\frac{1}{\beta+\eta}} \right) \right)^{\frac{1}{x}} \right)
$$

$$(12)$$

here
$$R_k^{\beta,\eta} = \prod_{i=1,j=i}^{|S_k|}\left((v_{ki})^\beta (v_{kj})^\eta\right), \quad U_k^{\beta,\eta} = \prod_{i=1,j=i}^{|S_k|}\left(1+(v_{ki})^\beta(v_{kj})^\eta - (e_{ki})^\beta(e_{kj})^\eta\right), \quad V_k^{\beta,\eta} =$$
$$\prod_{i=1,j=i}^{|S_k|}\left(1-(f_{ki})^\beta(f_{kj})^\eta + (v_{ki})^\beta(v_{kj})^\eta\right), \quad t_k = \frac{2}{|S_k|(|S_k|+1)}, \quad e_{ki} = 1-p_{ki}, \quad e_{kj} = 1-p_{kj}, \quad f_{ki} = 1-p_{ki}-m_{ki},$$
$$f_{kj} = 1-p_{kj}-m_{kj}, \quad v_{ki} = 1-p_{ki}-m_{ki}-n_{ki} \text{ and } v_{kj} = 1-p_{kj}-m_{kj}-n_{kj}.$$

Note that the Proof of Theorem 1 can be seen in the Appendix C.

For the *PFIPGHA* operator, the following properties should be satisfied.

**Property 3.** *(Idempotency) Suppose $a_i = (p_i, m_i, n_i)$ $(i = 1, 2, \ldots, y)$ is a collection of PFNs, and $a_i = a = (p, m, n)$ for all $i = 1, 2, \ldots, y$, then $PFIPGHA^{\beta,\eta}(a_1, a_2, \ldots, a_y) = a$.*

**Property 4.** *(Commutativity) Let $a_i = (p_i, m_i, n_i)$ and $a_i^* = (p_i^*, m_i^*, n_i^*)$ $(i = 1, 2, \ldots, y)$ be two sets of PFNs. If $(a_1^*, a_2^*, \cdots, a_y^*)$ is an arbitrary permutation of $(a_1, a_2, \cdots, a_y)$, then $PFIPGHA^{\beta,\eta}(a_1, a_2, \ldots, a_y) = PFIPGHA^{\beta,\eta}(a_1^*, a_2^*, \cdots, a_y^*)$.*

Note that the proofs of Property 3 and 4 are similar to those of Property 1 and 2, respectively. Thus, they are omitted to save space in this study.

Several special cases of the $PFIPGHA^{\beta,\eta}$ operator are discussed as follows.

**Special case 1:**

When $\eta \to 0$, $PFIPGHA^{\beta,0}(a_1, a_2, \ldots, a_y) =$

$$\left(\left(\left(\prod_{k=1}^{x}\left(1+\left(\left(\prod_{i=1}^{|S_k|}\left((g_{ki})^\beta\right)\right)^{t_k}\right)^{\frac{1}{\beta}} - \left(1-\left(\prod_{i=1}^{|S_k|}\left(1+(g_{ki})^\beta - (e_{ki})^\beta\right)\right)^{t_k} + \left(\prod_{i=1}^{|S_k|}\left((g_{ki})^\beta\right)\right)^{t_k}\right)^{\frac{1}{\beta}}\right)\right)^{\frac{1}{x}} - \prod_{k=1}^{x}\left(\left(\left(\prod_{i=1}^{|S_k|}\left((g_{ki})^\beta\right)\right)^{t_k}\right)^{\frac{1}{\beta}}\right)\right)^{\frac{1}{x}},$$

$$\left(\prod_{k=1}^{x}\left(1-\left(1+\left(\prod_{i=1}^{|S_k|}\left((g_{ki})^\beta\right)\right)^{t_k} - \left(\prod_{i=1}^{|S_k|}\left(1-(f_{ki})^\beta + (g_{ki})^\beta\right)\right)^{t_k}\right)^{\frac{1}{\beta}} + \left(\left(\prod_{i=1}^{|S_k|}\left((g_{ki})^\beta\right)\right)^{t_k}\right)^{\frac{1}{\beta}}\right)\right)^{\frac{1}{x}} -$$

$$\left(\prod_{k=1}^{x}\left(1+\left(\left(\prod_{i=1}^{|S_k|}\left((g_{ki})^\beta\right)\right)^{t_k}\right)^{\frac{1}{\beta}} - \left(1-\left(\prod_{i=1}^{|S_k|}\left(1+(g_{ki})^\beta - (e_{ki})^\beta\right)\right)^{t_k} + \left(\prod_{i=1}^{|S_k|}\left((g_{ki})^\beta\right)\right)^{t_k}\right)^{\frac{1}{\beta}}\right)\right)^{\frac{1}{x}},$$

$$1-\left(\prod_{k=1}^{x}\left(1-\left(1+\left(\prod_{i=1}^{|S_k|}\left((g_{ki})^\beta\right)\right)^{t_k} - \left(\prod_{i=1}^{|S_k|}\left(1-(f_{ki})^\beta + (g_{ki})^\beta\right)\right)^{t_k}\right)^{\frac{1}{\beta}} + \left(\left(\prod_{i=1}^{|S_k|}\left((g_{ki})^\beta\right)\right)^{t_k}\right)^{\frac{1}{\beta}}\right)\right)^{\frac{1}{x}}\right).$$

**Special case 2:**

When $\beta \to 0$, then $PFIPGHA^{0,\eta}(a_1, a_2, \ldots, a_y) =$

$$\left(\left(\left(\prod_{k=1}^{x}\left(1+\left(\left(\prod_{i=1,j=i}^{|S_k|}\left((g_{kj})^\eta\right)\right)^{t_k}\right)^{\frac{1}{\eta}} - \left(1-\left(\prod_{i=1,j=i}^{|S_k|}\left(1+(g_{kj})^\eta - (e_{kj})^\eta\right)\right)^{t_k} + \left(\prod_{i=1,j=i}^{|S_k|}\left((g_{kj})^\beta\right)\right)^{t_k}\right)^{\frac{1}{\eta}}\right)\right)^{\frac{1}{x}} - \prod_{k=1}^{x}\left(\left(\left(\prod_{i=1,j=i}^{|S_k|}\left((g_{kj})^\eta\right)\right)^{t_k}\right)^{\frac{1}{\eta}}\right)\right)^{\frac{1}{x}},$$

$$\left(\prod_{k=1}^{x}\left(1-\left(1+\left(\prod_{i=1,j=i}^{|S_k|}\left((g_{kj})^\eta\right)\right)^{t_k} - \left(\prod_{i=1,j=i}^{|S_k|}\left(1-(f_{kj})^\eta + (g_{kj})^\eta\right)\right)^{t_k}\right)^{\frac{1}{\eta}} + \left(\left(\prod_{i=1,j=i}^{|S_k|}\left((g_{kj})^\eta\right)\right)^{t_k}\right)^{\frac{1}{\eta}}\right)\right)^{\frac{1}{x}} -$$

$$\left(\prod_{k=1}^{x}\left(1+\left(\left(\prod_{i=1,j=i}^{|S_k|}\left((g_{kj})^\eta\right)\right)^{t_k}\right)^{\frac{1}{\eta}} - \left(1-\left(\prod_{i=1,j=i}^{|S_k|}\left(1+(g_{kj})^\eta - (e_{kj})^\eta\right)\right)^{t_k} + \left(\prod_{i=1,j=i}^{|S_k|}\left((g_{kj})^\eta\right)\right)^{t_k}\right)^{\frac{1}{\eta}}\right)\right)^{\frac{1}{x}},$$

$$1-\left(\prod_{k=1}^{x}\left(1-\left(1+\left(\prod_{i=1,j=i}^{|S_k|}\left((g_{kj})^\eta\right)\right)^{t_k} - \left(\prod_{i=1,j=i}^{|S_k|}\left(1-(f_{kj})^\eta + (g_{kj})^\eta\right)\right)^{t_k}\right)^{\frac{1}{\eta}} + \left(\left(\prod_{i=1,j=i}^{|S_k|}\left((g_{kj})^\eta\right)\right)^{t_k}\right)^{\frac{1}{\eta}}\right)\right)^{\frac{1}{x}}\right).$$

**Special case 3:**

When $\beta = 1$ and $\eta \to 0$, then $PFIPGHA^{1,0}(a_1, a_2, \ldots, a_y) =$

$$\left(\left(\left(\prod_{k=1}^{x}\left(\prod_{i=1}^{|S_k|}(1 + g_{ki} - e_{ki})\right)^{t_k}\right)^{\frac{1}{x}} - \left(\prod_{k=1}^{x}\left(\prod_{i=1}^{|S_k|} g_{ki}\right)^{t_k}\right)^{\frac{1}{x}}\right),\; \left(\prod_{k=1}^{x}\left(\prod_{i=1}^{|S_k|}(1 - f_{ki} + g_{ki})\right)^{t_k}\right)^{\frac{1}{x}} -$$

$$\left(\prod_{k=1}^{x}\left(\prod_{i=1}^{|S_k|}(1 + g_{ki} - e_{ki})\right)^{t_k}\right)^{\frac{1}{x}},\; 1 - \left(\prod_{k=1}^{x}\left(\prod_{i=1}^{|S_k|}(1 - f_{ki} + g_{ki})\right)^{t_k}\right)^{\frac{1}{x}}\right).$$

**Special case 4:**

When $\beta = \eta = 1$, then $PFIPHA^{1,0}(a_1, a_2, \ldots, a_y) =$

$$\left(\left(\left(\prod_{k=1}^{x}\left(1 + \left(\left(\prod_{i=1,j=i}^{|S_k|}((g_{ki})(g_{kj}))\right)^{t_k}\right)^{\frac{1}{2}}\right) - \left(1 - \left(\prod_{i=1,j=i}^{|S_k|}(1 + (g_{ki})(g_{kj}) - (e_{ki})(e_{kj}))\right)^{t_k} + \left(\prod_{i=1,j=i}^{|S_k|}((g_{ki})(g_{kj}))\right)^{t_k}\right)^{\frac{1}{2}}\right)\right)^{\frac{1}{x}} - \left(\prod_{k=1}^{x}\left(\left(\left(\prod_{i=1,j=i}^{|S_k|}((g_{ki})(g_{kj}))\right)^{t_k}\right)^{\frac{1}{2}}\right)\right)^{\frac{1}{x}}\right),$$

$$\left(\prod_{k=1}^{x}\left(1 - \left(1 + \left(\prod_{i=1,j=i}^{|S_k|}((g_{ki})(g_{kj}))\right)^{t_k} - \left(\prod_{i=1,j=i}^{|S_k|}(1 - (f_{ki})(f_{kj}) + (g_{ki})(g_{kj}))\right)^{t_k}\right)^{\frac{1}{2}} + \left(\left(\prod_{i=1,j=i}^{|S_k|}((g_{ki})(g_{kj}))\right)^{t_k}\right)^{\frac{1}{2}}\right)\right)^{\frac{1}{x}} -$$

$$\left(\prod_{k=1}^{x}\left(1 + \left(\left(\prod_{i=1,j=i}^{|S_k|}((g_{ki})(g_{kj}))\right)^{t_k}\right)^{\frac{1}{2}}\right) - \left(1 - \left(\prod_{i=1,j=i}^{|S_k|}(1 + (g_{ki})(g_{kj}) - (e_{ki})(e_{kj}))\right)^{t_k} + \left(\prod_{i=1,j=i}^{|S_k|}((g_{ki})(g_{kj}))\right)^{t_k}\right)^{\frac{1}{2}}\right)^{\frac{1}{x}},$$

$$1 - \left(\prod_{k=1}^{x}\left(1 - \left(1 + \left(\prod_{i=1,j=i}^{|S_k|}((g_{ki})(g_{kj}))\right)^{t_k} - \left(\prod_{i=1,j=i}^{|S_k|}(1 - (f_{ki})(f_{kj}) + (g_{ki})(g_{kj}))\right)^{t_k}\right)^{\frac{1}{2}} + \left(\left(\prod_{i=1,j=i}^{|S_k|}((g_{ki})(g_{kj}))\right)^{t_k}\right)^{\frac{1}{2}}\right)\right)^{\frac{1}{x}}\right).$$

**Definition 11.** *If $a_i = (p_i, m_i, n_i)$ $(i = 1, 2, \ldots, y)$ is a set of PFNs, and they can be partitioned into $x$ distinct sorts $S_1, S_2, \ldots, S_x$, where $S_k = \{a_{k1}, a_{k2}, \ldots, a_{k|S_k|}\}$ $(k = 1, 2, \ldots, x)$; $(w_1, w_2, \cdots, w_y)$ is the weight vector of $(a_1, a_2, \cdots, a_y)$, where $w_i \in [0,1]$ and $\sum_{i=1}^{y} w_i = 1$, then the PFWIPGHA operator is defined as*

$$PFWIPGHA^{\beta,\eta}(a_1, a_2, \ldots, a_y) = \left(\prod_{k=1}^{x}\left(\frac{1}{\beta + \eta}\left(\prod_{i=1,j=i}^{|S_k|}\beta(a_{ki})^{w_i} \oplus \eta(a_{kj})^{w_j}\right)^{\frac{2}{|S_k|(|S_k|+1)}}\right)\right)^{\frac{1}{x}}, \tag{13}$$

*here $\beta, \eta \geq 0$, $|S_k|$ is the cardinality of $S_k$ and $\sum_{k=1}^{x}|S_k| = y$.*

**Theorem 4.** *Suppose $a_i = (p_i, m_i, n_i)$ $(i = 1, 2, \ldots, y)$ is a group of PFNs, $(w_1, w_2, \cdots, w_y)$ is the weight vector of $(a_1, a_2, \cdots, a_y)$, $w_i \in [0,1]$, $\sum_{i=1}^{y} w_i = 1$ and $\beta, \eta \geq 0$, then their aggregated result using Equation (13) is still a PFN, and the following is true:*

$$PFWIPGHA^{\beta,\eta}(a_1, a_2, \ldots, a_y) =$$

$$\left(\left(\prod_{k=1}^{x}\left(1 - (O_k^{\beta,\eta})^{\frac{1}{\beta+\eta}} - (1 - P_k^{\beta,\eta} + O_k^{\beta,\eta})^{\frac{1}{\beta+\eta}}\right)\right)^{\frac{1}{x}} - \left(\prod_{k=1}^{x}\left((O_k^{\beta,\eta})^{\frac{1}{\beta+\eta}}\right)\right)^{\frac{1}{x}},\; \left(\prod_{k=1}^{x}\left(1 - (1 + O_k^{\beta,\eta} - Y_k^{\beta,\eta})^{\frac{1}{\beta+\eta}} + (O_k^{\beta,\eta})^{\frac{1}{\beta+\eta}}\right)\right)^{\frac{1}{x}}\right.$$

$$\left. - \left(\prod_{k=1}^{x}\left(1 - (O_k^{\beta,\eta})^{\frac{1}{\beta+\eta}} - (1 - P_k^{\beta,\eta} + O_k^{\beta,\eta})^{\frac{1}{\beta+\eta}}\right)\right)^{\frac{1}{x}},\; 1 - \left(\prod_{k=1}^{x}\left(1 - (1 + O_k^{\beta,\eta} - Y_k^{\beta,\eta})^{\frac{1}{\beta+\eta}} + (O_k^{\beta,\eta})^{\frac{1}{\beta+\eta}}\right)\right)^{\frac{1}{x}}\right) \tag{14}$$

*where $O_k^{\beta,\eta} = \left(\prod_{i=1,j=i}^{|S_k|} G_k^{\beta,\eta}\right)^{t_k}$, $P_k^{\beta,\eta} = \left(\prod_{i=1,j=i}^{|S_k|}(1 + G_k^{\beta,\eta} - T_k^{\beta,\eta})\right)^{t_k}$, $Y_k^{\beta,\eta} = \left(\prod_{i=1,j=i}^{|S_k|}(1 - Z_k^{\beta,\eta} + G_k^{\beta,\eta})\right)^{t_k}$, $t_k = \frac{2}{|S_k|(|S_k|+1)}$, $G_k^{\beta,\eta} = ((v_{ki})^{w_i})^{\beta}((v_{kj})^{w_j})^{\eta}$, $T_k^{\beta,\eta} = \left(1 - (u_{ki})^{w_i} + (v_{ki})^{w_i}\right)^{\beta}\left(1 - (u_{kj})^{w_j} + (v_{kj})^{w_j}\right)^{\eta}$, $Z_k^{\beta,\eta} = \left(1 + (v_{ki})^{w_i} - (r_{ki})^{w_i}\right)^{\beta}\left(1 + (v_{kj})^{w_j} - (r_{kj})^{w_j}\right)^{\eta}$, $r_{ki} = 1 - n_{ki}$, $r_{kj} = 1 - n_{kj}$, $u_{ki} = 1 - n_{ki} - m_{ki}$, $u_{kj} = 1 - n_{kj} - m_{kj}$, $v_{ki} = 1 - n_{ki} - m_{ki} - p_{ki}$ and $v_{kj} = 1 - n_{kj} - m_{kj} - p_{kj}$.*

Note that the Proof of Theorem 4 can be seen in the Appendix D.

## 4. Decision Making Methods with the Proposed Aggregation Operators

In this section, decision making methods with the proposed aggregation operators are explored to address complex decision making issues in a picture fuzzy environment.

### 4.1. Problem Description

With respect to a group decision making problem with picture fuzzy information, assume that there are $h$ alternatives, denoted as $\{\pi_1, \pi_2, \cdots, \pi_h\}$, and $l$ criteria, denoted as $\{\phi_1, \phi_2, \cdots, \phi_l\}$. The weight vector of these criteria is $\{w_1, w_2, \cdots, w_l\}$, where $w_1 + w_2 + \cdots + w_l = 1$ and $0 \leq w_1, w_2, \cdots, w_l \leq 1$. Then, the decision makers are asked to rank all the alternatives or select the best alternative. Suppose the evaluation matrix provided by decision makers is $A = (a_{ij})_{h \times l}$, where $a_{ij} = (p_{ij}, m_{ij}, n_{ij})$, which is a PFN, represents the evaluation value of alternative $\pi_i$ $(i = 1, 2, \cdots, h)$ under criterion $\phi_j$ $(j = 1, 2, \cdots, l)$.

### 4.2. Decision Making Procedures

In this subsection, decision making methods based on the proposed aggregation operators are suggested to solve the decision making problem described in Section 4.1. The specific decision-making procedures are stated as follows.

Step 1: Normalize the initial decision making matrix.

Because benefit and cost criteria may be contained in an initial evaluation matrix at the same time, they are usually transferred into the same type in the first step. The normalization equation of PFNs is

$$b_{ij} = \begin{cases} (p_{ij}, m_{ij}, n_{ij}) & \text{for } benefit\ criteria\ \phi_j \\ (n_{ij}, m_{ij}, p_{ij}) & \text{for } \cos t\ criteria\ \phi_j \end{cases}. \tag{15}$$

Thus, the normalized evaluation matrix can be denoted as $B = (b_{ij})_{h \times l}$.

Step 2: Calculate the overall preference degree of each alternative.

Based on the *PFWIPHA* or *PFWIPGHA* operator defined in Section 3, the evaluation values in each row of evaluation matrix $B$ are aggregated, and then the overall preference degree of each alternative is calculated as

$$a_i = PFWIPHA^{\beta,\eta}(a_{i1}, a_{i2}, \ldots, a_{il})\ (i = 1, 2, \cdots, h), \tag{16}$$

or

$$a_i = PFWIPGHA^{\beta,\eta}(a_{i1}, a_{i2}, \ldots, a_{il})\ (i = 1, 2, \cdots, h). \tag{17}$$

Step 3: Compute the score function or accuracy function.

Based on Equation (1), the score function $E(a_i)$ can be calculated. If two score function values are equal, then the accuracy function $F(a_i)$ should be computed using Equation (2).

Step 4: Obtain the ranking order.

According to the comparison method defined in Definition 4, the ranking order of all the alternatives is obtained, and the best alternative is selected as $\pi^*$.

## 5. Case Study

In this section, a hotel selection case is studied to justify the practicability of the proposed method.

Recently, five college students plan to book a hotel in advance for their trip next week. They browse hotel evaluation information in TripAdvisor.com. As a very popular tourism website, TripAdvisor.com has many true comments about hotels, restaurants and tourist attractions. Accordingly, they choose four satisfactory hotels based on their price, comfortability, service, location, and convenience. In the following, the presented methods are suggested to select the optimal hotel. The program codes of the presented method run under the MATLAB R2016b software. The operation platform is a laptop with a Windows 10 operating system, an Intel(R) Core(TM) i5-8250U CPU with 1.80 GHz, and 12G random-access memory (RAM). The advantage of MATLAB lies in numerical calculation. It can efficiently solve complex problems, dynamically simulate the system, and display the numerical results with powerful graphics functions.

First, they need to fill out a questionnaire of selected hotels (See Appendix E) for giving their respective opinions of these four alternatives (denoted as $\pi_1, \pi_2, \pi_3, \pi_4$) under six evaluation criteria. They are $\phi_1$ (price), $\phi_2$ (comfortability), $\phi_3$ (service), $\phi_4$ (location) and $\phi_5$ (convenience). Then, their answers can be counted and represented by PFNs. For example, if two students think the price of hotel $\pi_1$ is high, one student holds that the price is medium, and two students believe that the price is low, then their evaluations can be described by a PFN $a_{11} = (0.4, 0.2, 0.4)$. Thus, when they give their all evaluations, an original evaluation matrix $A$ with PFNs can be obtained, as shown in Table 1.

**Table 1.** Initial decision making matrix $A$.

| $A$ | $\phi_1$ | $\phi_2$ | $\phi_3$ | $\phi_4$ | $\phi_5$ |
|---|---|---|---|---|---|
| $\pi_1$ | $(0.4, 0.2, 0.4)$ | $(0.6, 0.2, 0.2)$ | $(0.4, 0.2, 0.2)$ | $(0.6, 0, 0.2)$ | $(0.6, 0, 0.4)$ |
| $\pi_2$ | $(0.2, 0.4, 0.4)$ | $(0.8, 0.2, 0)$ | $(0.4, 0.4, 0.2)$ | $(0.6, 0.4, 0)$ | $(0.4, 0.2, 0.2)$ |
| $\pi_3$ | $(0.4, 0.4, 0.2)$ | $(0.6, 0, 0.4)$ | $(0.8, 0, 0.2)$ | $(0.6, 0.2, 0.2)$ | $(0.4, 0.2, 0.2)$ |
| $\pi_4$ | $(0.2, 0.4, 0.2)$ | $(0.4, 0.4, 0.2)$ | $(0.6, 0.4, 0)$ | $(0.4, 0.2, 0.4)$ | $(0.8, 0.2, 0)$ |

**Case 1:** Select the best hotel with the method based on the *PFWIPHA* operator.

Step 1: Normalize the initial decision making matrix.

Since $\phi_2$, $\phi_3$, $\phi_4$ and $\phi_5$ are benefit criteria, while $\phi_1$ belongs to cost criterion, they should be normalized based on Equation (15). The normalized evaluation matrix $B$ is shown in Table 2.

**Table 2.** Normalized evaluation matrix $B$.

| $B$ | $\phi_1$ | $\phi_2$ | $\phi_3$ | $\phi_4$ | $\phi_5$ |
|---|---|---|---|---|---|
| $\pi_1$ | $(0.4, 0.2, 0.4)$ | $(0.6, 0.2, 0.2)$ | $(0.4, 0.2, 0.2)$ | $(0.6, 0, 0.2)$ | $(0.6, 0, 0.4)$ |
| $\pi_2$ | $(0.4, 0.4, 0.2)$ | $(0.8, 0.2, 0)$ | $(0.4, 0.4, 0.2)$ | $(0.6, 0.4, 0)$ | $(0.4, 0.2, 0.2)$ |
| $\pi_3$ | $(0.2, 0.4, 0.4)$ | $(0.6, 0, 0.4)$ | $(0.8, 0, 0.2)$ | $(0.6, 0.2, 0.2)$ | $(0.4, 0.2, 0.2)$ |
| $\pi_4$ | $(0.2, 0.4, 0.2)$ | $(0.4, 0.4, 0.2)$ | $(0.6, 0.4, 0)$ | $(0.4, 0.2, 0.4)$ | $(0.8, 0.2, 0)$ |

Step 2: Calculate the overall preference degree of each alternative.

Suppose $\beta = \eta = 1$, and all criteria have the same importance, namely, $w_1 = w_2 = w_3 = w_4 = w_5 = 0.2$. In general, a higher price means a large probability of good comfortability and service. Likewise, the convenience of transport may have great relations with the location of a hotel. Thus, according to this correlation pattern, these criteria are partitioned into two parts: $S_1 = \{\phi_1, \phi_2, \phi_3\}$ and $S_2 = \{\phi_4, \phi_5\}$. Then, based on the *PFWIPHA* operator in Equation (16), the overall preference degree of each hotel is computed as: $a_2 = (0.2492, 0.7506, 0)$, $a_3 = (0.2526, 0.0859, 0.6615)$ and $a_4 = (0.2142, 0.7858, 0)$.

Step 3: Compute the score function.

Using Equation (1), the score function values are calculated as: $E(a_1) = -0.2809$, $E(a_2) = 0.2494$, $E(a_3) = -0.4088$ and $E(a_4) = 0.2142$.

Step 4: Obtain the ranking order.

since $E(a_2) > E(a_4) > E(a_1) > E(a_3)$, the final ranking order is $\pi_2 \succ \pi_4 \succ \pi_1 \succ \pi_3$, and the optimal hotel is $\pi_2$.

**Case 2:** Select the best hotel with the method based on the *PFWIPGHA* operator.

Step 1: Normalize the initial decision making matrix.

The normalized evaluation matrix is the same as matrix $B$, which is shown in Table 2.

Step 2: Calculate the overall preference degree of each alternative.

Suppose $\beta = \eta = 1$, and all criteria have the same importance, namely, $w_1 = w_2 = w_3 = w_4 = w_5 = 0.2$. In accordance with the correlation pattern, these criteria are partitioned into two parts: $S_1 = \{\phi_1, \phi_2, \phi_3\}$ and $S_2 = \{\phi_4, \phi_5\}$. Then using the *PFWIPGHA* operator in Equation (17), the overall preference degree of each hotel is computed as: $a_1 = (0.7502, 0.0511, 0.1987)$, $a_2 = (0.7884, 0.1197, 0.0919)$, $a_3 = (0.7675, 0.1094, 0.1231)$ and $a_4 = (0.7805, 0.1251, 0.0943)$.

Step 3: Compute the score function and accuracy function.

Using Equation (1), the score function values are calculated as: $E(a_1) = 0.5514$, $E(a_2) = 0.6965$, $E(a_3) = 0.6443$ and $E(a_4) = 0.6862$.

Step 4: Obtain the ranking order.

Since $E(a_2) > E(a_4) > E(a_3) > E(a_1)$, the final ranking order is $\pi_2 > \pi_4 > \pi_3 > \pi_1$, and the optimal hotel is $\pi_2$.

It can be seen that the best alternative is always $\pi_2$ no matter which aggregation operator (*PFWIPHA* or *PFWIPGHA*) is used.

## 6. Discussions

In this section, the influence of parameters in the proposed method is investigated by sensitivity analyses and the advantages of our method are demonstrated through comparison analyses

### 6.1. Sensitivity Analyses

In this subsection, the impacts of parameters $\beta$ and $\eta$ on the final ranking orders under *PFWIPHA* and *PFWIPGHA* operators are discussed, respectively.

First, the score function values of each alternative under *PFWIPHA* and *PFWIPGHA* operators are calculated by assigning different $\beta$ and $\eta$ values, as shown in Figures 1 and 2.

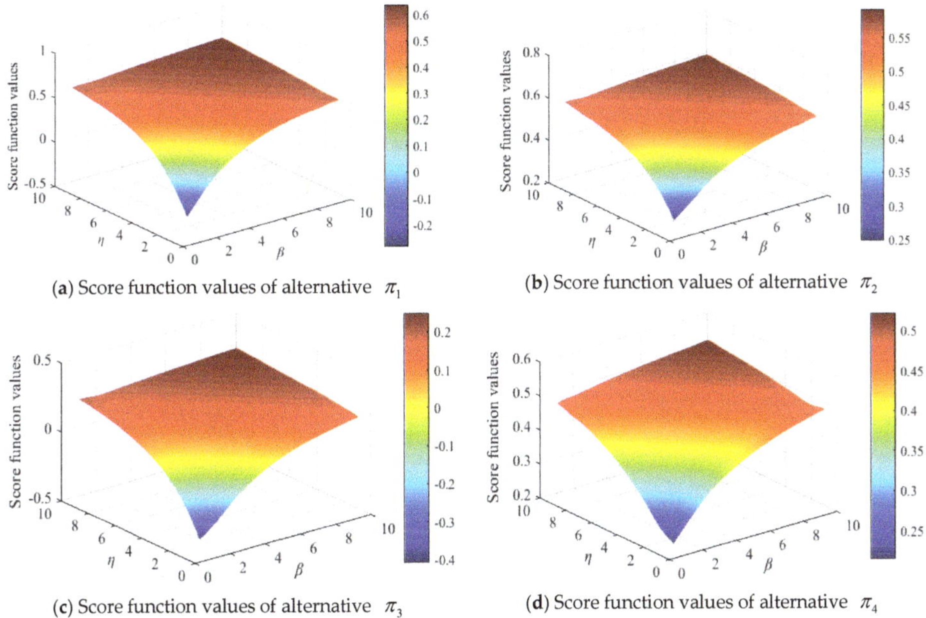

(**a**) Score function values of alternative $\pi_1$

(**b**) Score function values of alternative $\pi_2$

(**c**) Score function values of alternative $\pi_3$

(**d**) Score function values of alternative $\pi_4$

**Figure 1.** Score function values of four alternatives with $\beta, \eta \in (1,10)$ under *PFWIPHA* operators.

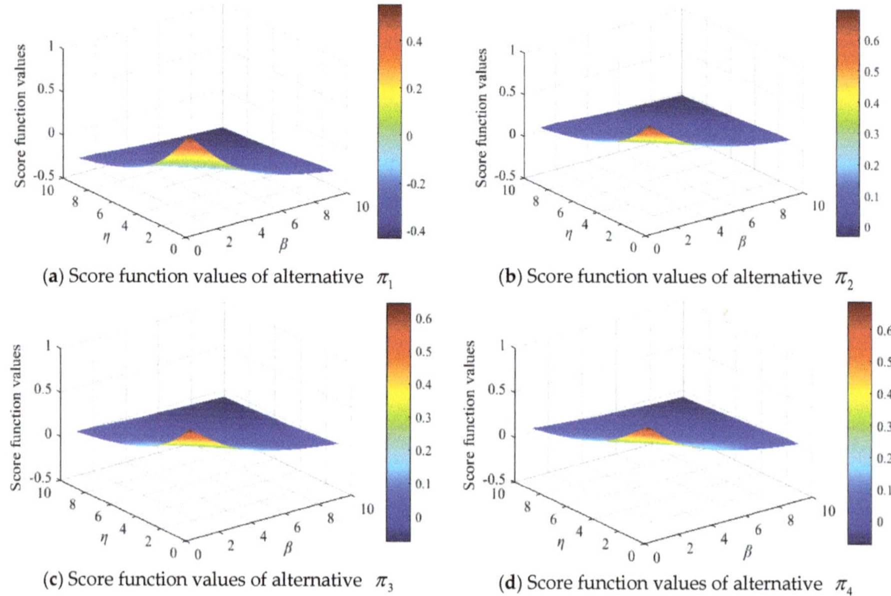

(**a**) Score function values of alternative $\pi_1$

(**b**) Score function values of alternative $\pi_2$

(**c**) Score function values of alternative $\pi_3$

(**d**) Score function values of alternative $\pi_4$

**Figure 2.** Score function values of alternative $\pi_1$ with $\beta,\eta \in (1,10)$ under *PFWIPGHA* operators.

From Figure 1, it can be seen that the ranking orders may change when dissimilar $\beta$ or $\eta$ values are allocated under the *PFWIPHA* operator. Even though, the best alternative is $\pi_2$ in many cases, and the worst alternative is always $\pi_3$. And the preference relations between alternative $\pi_1$ and $\pi_3$ (that is, $\pi_1 > \pi_3$), and that among $\pi_2$, $\pi_3$ and $\pi_4$ (that is, $\pi_2 > \pi_4 > \pi_3$) are always true.

From Figure 2, similar conclusions can be reached under the *PFWIPGHA* operator. For example, dissimilar rankings are derived with different values of $\beta$ and $\eta$, but the optimal alternative is still $\pi_2$ in most circumstances whereas the worst alternative is always $\pi_1$. Likewise, the preference relations between alternative $\pi_1$ and $\pi_4$ (that is, $\pi_4 > \pi_1$), and that among $\pi_1$, $\pi_2$ and $\pi_3$ (that is, $\pi_2 > \pi_3 > \pi_1$) always exist.

All of these indicate that both *PFWIPHA* and *PFWIPGHA* operators have a certain degree of stability, and do not lose flexibility at the same time. Moreover, an interesting phenomenon is that with the growth of $\beta$ or $\eta$ value, the score function of the aggregated values increase under the *PFWIPHA* operator, while they decrease under the *PFWIPGHA* operator. Thus, for an optimistic DM, he/she can choose a larger $\beta$ or $\eta$ value under the *PFWIPHA* operator, or a smaller $\beta$ or $\eta$ value under the *PFWIPGHA* operator. In contrast, for a pessimistic DM, he/she can choose a smaller $\beta$ or $\eta$ value under the *PFWIPHA* operator, or a larger $\beta$ or $\eta$ value under the *PFWIPGHA* operator.

### 6.2. Comparison Analyses

In this subsection, different approaches based on the existing aggregation operators of PFNs are compared in terms of ranking results and other characteristics to demonstrate the advantages of our method.

(1) Comparisons in terms of ranking results

First, the ranking results using the existing approaches with dissimilar aggregation operators (the picture fuzzy weighted averaging (*PFWA*), picture fuzzy weighted geometric averaging (*PFWGA*), *PFWIPHA* and *PFWIPGHA* operators) are listed in Table 3.

**Table 3.** Ranking orders with dissimilar aggregation operators.

| Aggregation Operators | Score Function Values | Ranking Orders | Best Alternatives |
|---|---|---|---|
| *PFWA* operator [33] | $E(a_1) = 0.1519, E(a_2) = 0.3488,$ $E(a_3) = 0.1067, E(a_4) = 0.3886.$ | $\pi_4 > \pi_2 > \pi_1 > \pi_3$ | $\phi_2$ |
| *PFWGA* operator [22] | $E(a_1) = 0.3486, E(a_2) = 0.3226,$ $E(a_3) = 0.4169, E(a_4) = 0.3023.$ | $\pi_3 > \pi_1 > \pi_2 > \pi_4$ | $\pi_3$ |
| *PFWIPHA* operator | $E(a_1) = -0.2809, E(a_2) = 0.2494,$ $E(a_3) = -0.4088, E(a_4) = 0.2142.$ | $\pi_2 > \pi_4 > \pi_1 > \pi_3$ | $\pi_2$ |
| *PFWIPGHA* operator | $E(a_1) = 0.5514, E(a_2) = 0.6965,$ $E(a_3) = 0.6443, E(a_4) = 0.6862.$ | $\pi_2 > \pi_4 > \pi_3 > \pi_1$ | $\pi_2$ |

It is clear that the four ranking orders in Table 3 are dissimilar with each other. For seeking out the optimal ranking order among them, the technique proposed by Jahan et al. [34] is employed. As exhibited in Table 4, the numbers of times a host is assigned to diverse ranks is counted. For example, the hotel $\pi_1$ has once a ranking of 2, twice a ranking of 3 and once a ranking of 4.

**Table 4.** Numbers of times a hotel assigned to diverse ranks.

| Hotels | Ranks | | | |
|---|---|---|---|---|
| | 1 | 2 | 3 | 4 |
| $\pi_1$ | | 1 | 2 | 1 |
| $\pi_2$ | 2 | 1 | 1 | |
| $\pi_3$ | 1 | | 1 | 2 |
| $\pi_4$ | 1 | 2 | | 1 |

Then, based on Table 4, the smoothing of hotels assignment over ranks is computed (See Table 5).

**Table 5.** Smoothing of hotels assignment over ranks $\Lambda_{iZ}$.

| Hotels | Ranks | | | |
|---|---|---|---|---|
| | 1 | 2 | 3 | 4 |
| $\pi_1$ | 0 | 1 | 3 | 4 |
| $\pi_2$ | 2 | 3 | 4 | 4 |
| $\pi_3$ | 1 | 1 | 2 | 4 |
| $\pi_4$ | 1 | 3 | 3 | 4 |

According to Table 5, the following maximizing objective function is performed:

$$Max\ \Theta = \sum_{i=1}^{4} \sum_{z=1}^{4} \left( \Lambda_{iz} \cdot \frac{4^2}{z} \cdot \nabla_{iz} \right)$$

$$s.t. \begin{cases} \sum_{i=1}^{4} \nabla_{iz} = 1, z = 1, 2, 3, 4 \\ \sum_{z=1}^{4} \nabla_{iz} = 1, i = 1, 2, 3, 4 \\ \quad \nabla_{iz} = 0\ or\ 1,\ \forall i, z \end{cases} \tag{18}$$

Through solving the programming model (18), the optimal ranking order $\pi_2 > \pi_4 > \pi_1 > \pi_3$ is derived.

For clarity, the optimal ranking order and four ranking orders in Table 3 are simultaneously depicted in the same figure (See Figure 3). It is clear that the optimal ranking is the same with the ranking order through adopting *PFWIPHA* operator. The difference between the optimal ranking and the ranking result by utilizing *PFWIPGHA* operator is small. Conversely, a larger discrepancy can be seen among the optimal ranking order, the ranking order with *PFWA* operator, and the ranking order with *PFWGA* operator. All of these indicate that the presented method is more suitable than the existing methods in disposing decision making problems with interactions and partitions of inputs.

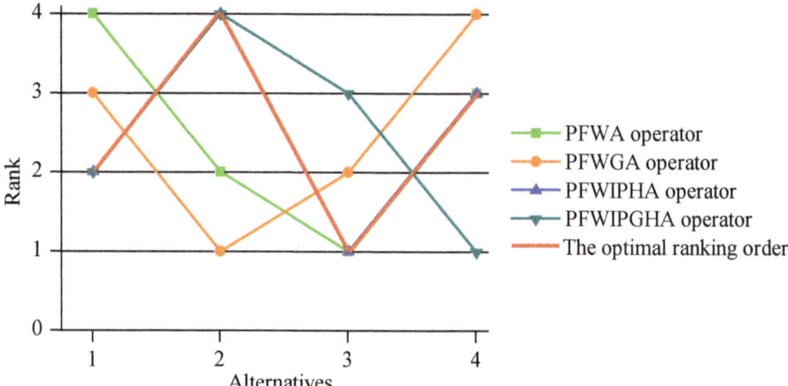


**Figure 3.** Optimal ranking order among different orders.

(2)   Further comparisons in terms of other features

In view of other characteristics (such as whether consider the relationships among membership functions, the correlations among criteria, and the partition the inputs), further comparisons are made in Table 6.

**Table 6.** Further comparisons in terms of other features.

| Aggregation Operators | Operational Rules | Interactions among Membership Functions | Interrelationships among Aggregating Arguments | Partition of the Inputs |
|---|---|---|---|---|
| *PFWA* operator [33] | Tradition | Unconsidered | Unconsidered | Unconsidered |
| *PFWGA* operator [22] | Tradition | Unconsidered | Unconsidered | Unconsidered |
| *PFWIPHA* operator | Interaction | Considered | Considered | Considered |
| *PFWIPGHA* operator | Interaction | Considered | Considered | Considered |

From Table 6, it can be seen that both the *PFWA* [21] and *PFWGA* [8] operators are based on the traditional operational laws of PFNs, where the interactions among membership functions are not considered. They cannot guarantee the accuracy of aggregated values when 'zero' occurs in the neutral membership degrees. In contrast, the present method (the *PFWIPHA* or *PFWIPGHA* operator) is based on new interaction operational rules of PFNs, which considers the interactions among membership functions. Furthermore, interrelationships among criteria are considered in our method. It is true that traditional HA and GHA operators take the interrelationships among aggregating arguments into account as well, but they hold the supposition that each argument is in relation to the rest of the arguments. Compared with them, a great advantage of our method is that the partition of the input arguments is allowed, so that the interactions of inputs are only considered in the same part rather than different parts.

(3)   Strengths analyses and summary

Based on the above comparison analyses, the strengths of our method are summarized as follows.

First, the *PFWIPHA* and *PFWIPGHA* operators suggested in this study are based on new interaction operational rules of PFNs. That is, they consider the interactions among dissimilar membership functions, and can avoid irrational outcomes even when the neutral membership degree is allocated as zero.

Second, the proposed method borrows from the ideas of PHA and *PGHA* operators, so that it can do well with the situation where arguments are divided into several portions. Thus, when the interaction operations are combined, our method can weigh the interrelationships of relevant inputs in the same panel, and prevent any consideration of the interrelationships of uncorrelated inputs in different panels.

Third, as our method under the *PFWIPHA* or *PFWIPGHA* operator is modified by two parameters, it is more flexible than the *PFWA* and *PFWGA* operators. On the other hand, when only one part exists in the inputs, the *PFWIPGA* or *PFWIPGHA* operator is regarded as the *IHA* or *GHA* operator with PFNs, respectively. In this sense, our method is more universal and powerful.

## 7. Conclusions

HA operators are effective aggregation operators to dispose interrelationships among criteria, while PHA operators aim to do with such correlations only in the same partition. As a result, several aggregation operators, such as *PFIPHA*, *PFWIPHA*, *PFIPGHA* and *PFWIPGHA* operators were presented by combining the PHA operators with picture fuzzy information. Many significant properties, such as the properties of idempotency and commutativity, were proved. Four special cases of *PFIPHA* and *PFIPGHA* with different parameter values were discussed, respectively. Moreover, the interaction operational rules of PFNs were first proposed to overcome the shortcoming of the existing operations. Then, the new decision-making methods based on these operators were used to solve the hotel selection problem by a case study. Sensitivity analyses and comparison analyses revealed that our methods are more flexible and general when processing complicated evaluation issues within a picture fuzzy environment.

However, one weakness of our methods is that the criteria weights are assigned previously and directly. This may be not very reasonable in some cases. Thus, it is worth exploring the weights determination methods in the future. On the other hand, the extended PHA operators can be well used to solve other real problems under picture fuzzy conditions. It is something that we are heading toward.

**Author Contributions:** Conceptualization, S.L. and L.X.; Funding acquisition, L.X.; Methodology, S.L.; Validation, L.X.; Writing—original draft, S.L.; Writing—review and editing, L.X. All authors have read and agreed to the published version of the manuscript.

**Funding:** This work is supported by the National Natural Science Foundation of China (No. 61773120, 61525304, 71690233 and 61472089).

**Conflicts of Interest:** The authors declare no conflict of interest.

## Appendix A. The Proof of Theorem 1

**Proof.** Based on the interaction operational laws defined in Definition 7, $(a_{ki})^\beta = \left((1-n_{ki}-m_{ki})^\beta - (1-n_{ki}-m_{ki}-p_{ki})^\beta, (1-n_{ki})^\beta - (1-n_{ki}-m_{ki})^\beta, 1-(1-n_{ki})^\beta\right)$ and $(a_{kj})^\eta = \left((1-n_{kj}-m_{kj})^\eta - (1-n_{kj}-m_{kj}-p_{kj})^\eta, (1-n_{kj})^\eta - (1-n_{kj}-m_{kj})^\eta, 1-(1-n_{kj})^\eta\right)$.

Let $r_{ki} = 1-n_{ki}$, $r_{kj} = 1-n_{kj}$, $u_{ki} = 1-n_{ki}-m_{ki}$, $u_{kj} = 1-n_{kj}-m_{kj}$, $v_{ki} = 1-n_{ki}-m_{ki}-p_{ki}$ and $v_{kj} = 1-n_{kj}-m_{kj}-p_{kj}$, then $(a_{ki})^\beta = \left((u_{ki})^\beta - (v_{ki})^\beta, (r_{ki})^\beta - (u_{ki})^\beta, 1-(r_{ki})^\beta\right)$, $(a_{kj})^\eta = \left((u_{kj})^\eta - (v_{kj})^\eta, (r_{kj})^\eta - (u_{kj})^\eta, 1-(r_{kj})^\eta\right)$ and

$$(a_{ki})^\beta \otimes (a_{kj})^\eta = \left((u_{ki})^\beta (u_{kj})^\eta - (v_{ki})^\beta (v_{kj})^\eta, (r_{ki})^\beta (r_{kj})^\eta - (u_{ki})^\beta (u_{kj})^\eta, 1-(r_{ki})^\beta (r_{kj})^\eta\right),$$

$$\Rightarrow \sum_{i=1}^{|S_k|} \sum_{j=i}^{|S_k|} (a_{ki})^\beta \otimes (a_{kj})^\eta =$$

$$\left(1 - \prod_{i=1,j=i}^{|S_k|} \left(1-(u_{ki})^\beta (u_{kj})^\eta + (v_{ki})^\beta (v_{kj})^\eta\right), \prod_{i=1,j=i}^{|S_k|} \left(1-(u_{ki})^\beta (u_{kj})^\eta + (v_{ki})^\beta (v_{kj})^\eta\right) - \prod_{i=1,j=i}^{|S_k|} \left(1+(v_{ki})^\beta (v_{kj})^\eta - (r_{ki})^\beta (r_{kj})^\eta\right),\right.$$

$$\left. \prod_{i=1,j=i}^{|S_k|} \left(1+(v_{ki})^\beta (v_{kj})^\eta - (r_{ki})^\beta (r_{kj})^\eta\right) - \prod_{i=1,j=i}^{|S_k|} \left((v_{ki})^\beta (v_{kj})^\eta\right)\right).$$

Let $t_k = \frac{2}{|S_k|(|S_k|+1)}$, $A_k^{\beta,\eta} = (v_{ki})^\beta (v_{kj})^\eta$, $B_k^{\beta,\eta} = 1 - (u_{ki})^\beta (u_{kj})^\eta + (v_{ki})^\beta (v_{kj})^\eta$ and $C_k^{\beta,\eta} = 1 + (v_{ki})^\beta (v_{kj})^\eta - (r_{ki})^\beta (r_{kj})^\eta$, then

$$\frac{2}{|S_k|(|S_k|+1)} \sum_{i=1}^{|S_k|} \sum_{j=i}^{|S_k|} (a_{ki})^\beta \otimes (a_{kj})^\eta =$$

$$\left( 1 - \left( \prod_{i=1,j=i}^{|S_k|} B_k^{\beta,\eta} \right)^{t_k}, \left( \prod_{i=1,j=i}^{|S_k|} B_k^{\beta,\eta} \right)^{t_k} - \left( \prod_{i=1,j=i}^{|S_k|} C_k^{\beta,\eta} \right)^{t_k}, \left( \prod_{i=1,j=i}^{|S_k|} C_k^{\beta,\eta} \right)^{t_k} - \left( \prod_{i=1,j=i}^{|S_k|} A_k^{\beta,\eta} \right)^{t_k} \right) \quad \Rightarrow$$

$$\left( \frac{2}{|S_k|(|S_k|+1)} \sum_{i=1}^{|S_k|} \sum_{j=i}^{|S_k|} (a_{ki})^\beta \otimes (a_{kj})^\eta \right)^{\frac{1}{\beta+\eta}} =$$

$$\left( \left( \left( 1 - \left( \prod_{i=1,j=i}^{|S_k|} B_k^{\beta,\eta} \right)^{t_k} + \left( \prod_{i=1,j=i}^{|S_k|} A_k^{\beta,\eta} \right)^{t_k} \right)^{\frac{1}{\beta+\eta}} - \left( \left( \prod_{i=1,j=i}^{|S_k|} A_k^{\beta,\eta} \right)^{t_k} \right)^{\frac{1}{\beta+\eta}}, \left( 1 - \left( \prod_{i=1,j=i}^{|S_k|} C_k^{\beta,\eta} \right)^{t_k} + \left( \prod_{i=1,j=i}^{|S_k|} A_k^{\beta,\eta} \right)^{t_k} \right)^{\frac{1}{\beta+\eta}} - \right.$$

$$\left. \left( 1 + \left( \prod_{i=1,j=i}^{|S_k|} A_k^{\beta,\eta} \right)^{t_k} - \left( \prod_{i=1,j=i}^{|S_k|} B_k^{\beta,\eta} \right)^{t_k} \right)^{\frac{1}{\beta+\eta}}, 1 - \left( 1 - \left( \prod_{i=1,j=i}^{|S_k|} C_k^{\beta,\eta} \right)^{t_k} + \left( \prod_{i=1,j=i}^{|S_k|} A_k^{\beta,\eta} \right)^{t_k} \right)^{\frac{1}{\beta+\eta}} \right) \; .$$

Let $M_k^{\beta,\eta} = \left( \prod_{i=1,j=i}^{|S_k|} A_k^{\beta,\eta} \right)^{t_k}$, $N_k^{\beta,\eta} = \left( \prod_{i=1,j=i}^{|S_k|} B_k^{\beta,\eta} \right)^{t_k}$ and $O_k^{\beta,\eta} = \left( \prod_{i=1,j=i}^{|S_k|} C_k^{\beta,\eta} \right)^{t_k}$, then

$$\sum_{k=1}^{x} \left( \frac{2}{|S_k|(|S_k|+1)} \sum_{i=1}^{|S_k|} \sum_{j=i}^{|S_k|} (a_{ki})^\beta \otimes (a_{kj})^\eta \right)^{\frac{1}{\beta+\eta}} =$$

$$\left( 1 - \prod_{k=1}^{x} \left( 1 - (1 - N_k^{\beta,\eta} + M_k^{\beta,\eta})^{\frac{1}{\beta+\eta}} + (M_k^{\beta,\eta})^{\frac{1}{\beta+\eta}} \right), \prod_{k=1}^{x} \left( 1 - (1 - N_k^{\beta,\eta} + M_k^{\beta,\eta})^{\frac{1}{\beta+\eta}} + (M_k^{\beta,\eta})^{\frac{1}{\beta+\eta}} \right) - \right.$$

$$\left. \prod_{k=1}^{x} \left( 1 + (M_k^{\beta,\eta})^{\frac{1}{\beta+\eta}} - (1 - O_k^{\beta,\eta} + M_k^{\beta,\eta})^{\frac{1}{\beta+\eta}} \right), \prod_{k=1}^{x} \left( 1 + (M_k^{\beta,\eta})^{\frac{1}{\beta+\eta}} - (1 - O_k^{\beta,\eta} + M_k^{\beta,\eta})^{\frac{1}{\beta+\eta}} \right) - \prod_{k=1}^{x} \left( (M_k^{\beta,\eta})^{\frac{1}{\beta+\eta}} \right) \right)'$$

$$\Rightarrow \frac{1}{x} \left( \sum_{k=1}^{x} \left( \frac{2}{|S_k|(|S_k|+1)} \sum_{i=1}^{|S_k|} \sum_{j=i}^{|S_k|} (a_{ki})^\beta \otimes (a_{kj})^\eta \right)^{\frac{1}{\beta+\eta}} \right) =$$

$$\left( 1 - \left( \prod_{k=1}^{x} \left( 1 - (1 - N_k^{\beta,\eta} + M_k^{\beta,\eta})^{\frac{1}{\beta+\eta}} + (M_k^{\beta,\eta})^{\frac{1}{\beta+\eta}} \right) \right)^{\frac{1}{x}}, \left( \prod_{k=1}^{x} \left( 1 - (1 - N_k^{\beta,\eta} + M_k^{\beta,\eta})^{\frac{1}{\beta+\eta}} + (M_k^{\beta,\eta})^{\frac{1}{\beta+\eta}} \right) \right)^{\frac{1}{x}} - \right.$$

$$\left. \left( \prod_{k=1}^{x} \left( 1 + (M_k^{\beta,\eta})^{\frac{1}{\beta+\eta}} - (1 - O_k^{\beta,\eta} + M_k^{\beta,\eta})^{\frac{1}{\beta+\eta}} \right) \right)^{\frac{1}{x}}, \left( \prod_{k=1}^{x} \left( 1 + (M_k^{\beta,\eta})^{\frac{1}{\beta+\eta}} - (1 - O_k^{\beta,\eta} + M_k^{\beta,\eta})^{\frac{1}{\beta+\eta}} \right) \right)^{\frac{1}{x}} - \left( \prod_{k=1}^{x} \left( (M_k^{\beta,\eta})^{\frac{1}{\beta+\eta}} \right) \right)^{\frac{1}{x}} \right) \; .$$

Thus, $PFIPHA^{\beta,\eta}(a_1, a_2, \ldots, a_y) =$

$$\left( 1 - \left( \prod_{k=1}^{x} \left( 1 - (1 - N_k^{\beta,\eta} + M_k^{\beta,\eta})^{\frac{1}{\beta+\eta}} + (M_k^{\beta,\eta})^{\frac{1}{\beta+\eta}} \right) \right)^{\frac{1}{x}}, \left( \prod_{k=1}^{x} \left( 1 - (1 - N_k^{\beta,\eta} + M_k^{\beta,\eta})^{\frac{1}{\beta+\eta}} + (M_k^{\beta,\eta})^{\frac{1}{\beta+\eta}} \right) \right)^{\frac{1}{x}} - \right.$$

$$\left. \left( \prod_{k=1}^{x} \left( 1 + (M_k^{\beta,\eta})^{\frac{1}{\beta+\eta}} - (1 - O_k^{\beta,\eta} + M_k^{\beta,\eta})^{\frac{1}{\beta+\eta}} \right) \right)^{\frac{1}{x}}, \left( \prod_{k=1}^{x} \left( 1 + (M_k^{\beta,\eta})^{\frac{1}{\beta+\eta}} - (1 - O_k^{\beta,\eta} + M_k^{\beta,\eta})^{\frac{1}{\beta+\eta}} \right) \right)^{\frac{1}{x}} - \left( \prod_{k=1}^{x} \left( (M_k^{\beta,\eta})^{\frac{1}{\beta+\eta}} \right) \right)^{\frac{1}{x}} \right) \; . \quad \square$$

## Appendix B. The Proof of Theorem 2

**Proof.** According to the interaction operational laws defined in Definition 7, $w_i a_{ki} = \left( 1 - (1 - p_{ki})^{w_i}, (1 - p_{ki})^{w_i} - (1 - p_{ki} - m_{ki})^{w_i}, (1 - p_{ki} - m_{ki})^{w_i} - (1 - p_{ki} - m_{ki} - n_{ki})^{w_i} \right)$, $w_j a_{kj} = \left( 1 - (1 - p_{kj})^{w_j}, (1 - p_{kj})^{w_j} - (1 - p_{kj} - m_{kj})^{w_j}, (1 - p_{kj} - m_{kj})^{w_j} - (1 - p_{kj} - m_{kj} - n_{kj})^{w_j} \right)$, $(w_{ki} a_{ki})^\beta =$

$$\left( \left( 1 + (1 - p_{ki} - m_{ki} - n_{ki})^{w_i} - (1 - p_{ki})^{w_i} \right)^\beta - \left( (1 - p_{ki} - m_{ki} - n_{ki})^{w_i} \right)^\beta, \left( 1 - (1 - p_{ki} - m_{ki})^{w_i} + (1 - p_{ki} - m_{ki} - n_{ki})^{w_i} \right)^\beta - \right.$$

$$\left. \left( 1 + (1 - p_{ki} - m_{ki} - n_{ki})^{w_i} - (1 - p_{ki})^{w_i} \right)^\beta, 1 - \left( 1 - (1 - p_{ki} - m_{ki})^{w_i} + (1 - p_{ki} - m_{ki} - n_{ki})^{w_i} \right)^\beta \right)$$

and $(w_j a_{kj})^\eta =$

$$\left( \left( 1 + (1 - p_{kj} - m_{kj} - n_{kj})^{w_j} - (1 - p_{kj})^{w_j} \right)^\eta - \left( (1 - p_{kj} - m_{kj} - n_{kj})^{w_j} \right)^\eta, \left( 1 - (1 - p_{kj} - m_{kj})^{w_j} + (1 - p_{kj} - m_{kj} - n_{kj})^{w_j} \right)^\eta - \right.$$

$$\left. \left( 1 + (1 - p_{kj} - m_{kj} - n_{kj})^{w_j} - (1 - p_{kj})^{w_j} \right)^\eta, 1 - \left( 1 - (1 - p_{kj} - m_{kj})^{w_j} + (1 - p_{kj} - m_{kj} - n_{kj})^{w_j} \right)^\eta \right) \; .$$

Let $e_{ki} = 1 - p_{ki}$, $e_{kj} = 1 - p_{kj}$, $f_{ki} = 1 - p_{ki} - m_{ki}$, $f_{kj} = 1 - p_{kj} - m_{kj}$, $v_{ki} = 1 - p_{ki} - m_{ki} - n_{ki}$ and $v_{kj} = 1 - p_{kj} - m_{kj} - n_{kj}$, then

$$(w_{ki}a_{ki})^{\beta} = \left(\left(1+(v_{ki})^{w_i}-(e_{ki})^{w_i}\right)^{\beta}-\left((v_{ki})^{w_i}\right)^{\beta}, \left(1-(f_{ki})^{w_i}+(v_{ki})^{w_i}\right)^{\beta}-\left(1+(v_{ki})^{w_i}-(e_{ki})^{w_i}\right)^{\beta}, 1-\left(1-(f_{ki})^{w_i}+(v_{ki})^{w_i}\right)^{\beta}\right),$$

$$(w_j a_{kj})^{\eta} = \left(\left(1+(v_{kj})^{w_j}-(e_{kj})^{w_j}\right)^{\eta}-\left((v_{kj})^{w_j}\right)^{\eta}, \left(1-(f_{kj})^{w_j}+(v_{kj})^{w_j}\right)^{\eta}-\left(1+(v_{kj})^{w_j}-(e_{kj})^{w_j}\right)^{\eta}, 1-\left(1-(f_{kj})^{w_j}+(v_{kj})^{w_j}\right)^{\eta}\right),$$

and $(w_i a_{ki})^{\beta} \otimes (w_j a_{kj})^{\eta} =$

$$\left(\left(1+(v_{ki})^{w_i}-(e_{ki})^{w_i}\right)^{\beta}\left(1+(v_{kj})^{w_j}-(e_{kj})^{w_j}\right)^{\eta}-\left((v_{ki})^{w_i}\right)^{\beta}\left((v_{kj})^{w_j}\right)^{\eta}, \left(1-(f_{ki})^{w_i}+(v_{ki})^{w_i}\right)^{\beta}\left(1-(f_{kj})^{w_j}+(v_{kj})^{w_j}\right)^{\eta}-\right.$$

$$\left.\left(1+(v_{ki})^{w_i}-(e_{ki})^{w_i}\right)^{\beta}\left(1+(v_{kj})^{w_j}-(e_{kj})^{w_j}\right)^{\eta}, 1-\left(1-(f_{ki})^{w_i}+(v_{ki})^{w_i}\right)^{\beta}\left(1-(f_{kj})^{w_j}+(v_{kj})^{w_j}\right)^{\eta}\right)$$

Let $G_k^{\beta,\eta} = \left((v_{ki})^{w_i}\right)^{\beta}\left((v_{kj})^{w_j}\right)^{\eta}$, $H_k^{\beta,\eta} = \left(1+(v_{ki})^{w_i}-(e_{ki})^{w_i}\right)^{\beta}\left(1+(v_{kj})^{w_j}-(e_{kj})^{w_j}\right)^{\eta}$ and

$I_k^{\beta,\eta} = \left(1-(f_{ki})^{w_i}+(v_{ki})^{w_i}\right)^{\beta}\left(1-(f_{kj})^{w_j}+(v_{kj})^{w_j}\right)^{\eta}$, then $(w_i a_{ki})^{\beta} \otimes (w_j a_{kj})^{\eta} = \left(H_k^{\beta,\eta}-G_k^{\beta,\eta}, I_k^{\beta,\eta}-\right.$

$\left. H_k^{\beta,\eta}, 1-I_k^{\beta,\eta}\right), \Rightarrow \sum_{i=1}^{|S_k|}\sum_{j=i}^{|S_k|}(w_i a_{ki})^{\beta} \otimes (w_j a_{kj})^{\eta} = \left(1-\prod_{i=1,j=i}^{|S_k|}\left(1-H_k^{\beta,\eta}+G_k^{\beta,\eta}\right), \prod_{i=1,j=i}^{|S_k|}\left(1-H_k^{\beta,\eta}+G_k^{\beta,\eta}\right)-\right.$

$$\left.\prod_{i=1,j=i}^{|S_k|}\left(1+G_k^{\beta,\eta}-I_k^{\beta,\eta}\right), \prod_{i=1,j=i}^{|S_k|}\left(1+G_k^{\beta,\eta}-I_k^{\beta,\eta}\right)-\prod_{i=1,j=i}^{|S_k|}\left(G_k^{\beta,\eta}\right)\right).$$

Let $t_k = \dfrac{2}{|S_k|(|S_k|+1)}$, then $\dfrac{2}{|S_k|(|S_k|+1)}\sum_{i=1}^{|S_k|}\sum_{j=i}^{|S_k|}(w_i a_{ki})^{\beta} \otimes (w_j a_{kj})^{\eta} =$

$$\left(1-\left(\prod_{i=1,j=i}^{|S_k|}(1-H_k^{\beta,\eta}+G_k^{\beta,\eta})\right)^{t_k}, \left(\prod_{i=1,j=i}^{|S_k|}(1-H_k^{\beta,\eta}+G_k^{\beta,\eta})\right)^{t_k}-\left(\prod_{i=1,j=i}^{|S_k|}(1+G_k^{\beta,\eta}-I_k^{\beta,\eta})\right)^{t_k}, \left(\prod_{i=1,j=i}^{|S_k|}(1+G_k^{\beta,\eta}-I_k^{\beta,\eta})\right)^{t_k}-\left(\prod_{i=1,j=i}^{|S_k|}G_k^{\beta,\eta}\right)^{t_k}\right) \Rightarrow$$

$$\left(\frac{2}{|S_k|(|S_k|+1)}\sum_{i=1}^{|S_k|}\sum_{j=i}^{|S_k|}(w_i a_{ki})^{\beta} \otimes (w_j a_{kj})^{\eta}\right)^{\frac{1}{\beta+\eta}} =$$

$$\left(\left(1+\left(\prod_{i=1,j=i}^{|S_k|}G_k^{\beta,\eta}\right)^{t_k}-\left(\prod_{i=1,j=i}^{|S_k|}\left(1-H_k^{\beta,\eta}+G_k^{\beta,\eta}\right)\right)^{t_k}\right)^{\frac{1}{\beta+\eta}}-\left(\left(\prod_{i=1,j=i}^{|S_k|}G_k^{\beta,\eta}\right)^{t_k}\right)^{\frac{1}{\beta+\eta}}, \left(1-\left(\prod_{i=1,j=i}^{|S_k|}\left(1+G_k^{\beta,\eta}-I_k^{\beta,\eta}\right)\right)^{t_k}+\left(\prod_{i=1,j=i}^{|S_k|}G_k^{\beta,\eta}\right)^{t_k}\right)^{\frac{1}{\beta+\eta}}-\right.$$

$$\left.\left(1+\left(\prod_{i=1,j=i}^{|S_k|}G_k^{\beta,\eta}\right)^{t_k}-\left(\prod_{i=1,j=i}^{|S_k|}\left(1-H_k^{\beta,\eta}+G_k^{\beta,\eta}\right)\right)^{t_k}\right)^{\frac{1}{\beta+\eta}}, 1-\left(1-\left(\prod_{i=1,j=i}^{|S_k|}\left(1+G_k^{\beta,\eta}-I_k^{\beta,\eta}\right)\right)^{t_k}+\left(\prod_{i=1,j=i}^{|S_k|}G_k^{\beta,\eta}\right)^{t_k}\right)^{\frac{1}{\beta+\eta}}\right)$$

Let $J_k^{\beta,\eta} = \left(\prod_{i=1,j=i}^{|S_k|}G_k^{\beta,\eta}\right)^{t_k}$, $L_k^{\beta,\eta} = \left(\prod_{i=1,j=i}^{|S_k|}\left(1-H_k^{\beta,\eta}+G_k^{\beta,\eta}\right)\right)^{t_k}$ and $Q_k^{\beta,\eta} = \left(\prod_{i=1,j=i}^{|S_k|}\left(1+G_k^{\beta,\eta}-I_k^{\beta,\eta}\right)\right)^{t_k}$,

then $\left(\dfrac{2}{|S_k|(|S_k|+1)}\sum_{i=1}^{|S_k|}\sum_{j=i}^{|S_k|}(w_i a_{ki})^{\beta} \otimes (w_j a_{kj})^{\eta}\right)^{\frac{1}{\beta+\eta}} = \left((1+J_k^{\beta,\eta}-L_k^{\beta,\eta})^{\frac{1}{\beta+\eta}}-(J_k^{\beta,\eta})^{\frac{1}{\beta+\eta}}, (1-Q_k^{\beta,\eta}+J_k^{\beta,\eta})^{\frac{1}{\beta+\eta}}-\right.$

$\left. (1+J_k^{\beta,\eta}-L_k^{\beta,\eta})^{\frac{1}{\beta+\eta}}, 1-(1-Q_k^{\beta,\eta}+J_k^{\beta,\eta})^{\frac{1}{\beta+\eta}}\right), \Rightarrow \sum_{k=1}^{x}\left(\dfrac{2}{|S_k|(|S_k|+1)}\sum_{i=1}^{|S_k|}\sum_{j=i}^{|S_k|}(w_i a_{ki})^{\beta} \otimes (w_j a_{kj})^{\eta}\right)^{\frac{1}{\beta+\eta}} =$

$$\left(1-\prod_{k=1}^{x}\left(1-(1+J_k^{\beta,\eta}-L_k^{\beta,\eta})^{\frac{1}{\beta+\eta}}+(J_k^{\beta,\eta})^{\frac{1}{\beta+\eta}}\right), \prod_{k=1}^{x}\left(1-(1+J_k^{\beta,\eta}-L_k^{\beta,\eta})^{\frac{1}{\beta+\eta}}+(J_k^{\beta,\eta})^{\frac{1}{\beta+\eta}}\right)-\right.$$

$$\left.\prod_{k=1}^{x}\left(1+(J_k^{\beta,\eta})^{\frac{1}{\beta+\eta}}-(1-Q_k^{\beta,\eta}+J_k^{\beta,\eta})^{\frac{1}{\beta+\eta}}\right), \prod_{k=1}^{x}\left(1+(J_k^{\beta,\eta})^{\frac{1}{\beta+\eta}}-(1-Q_k^{\beta,\eta}+J_k^{\beta,\eta})^{\frac{1}{\beta+\eta}}\right)-\prod_{k=1}^{x}\left((J_k^{\beta,\eta})^{\frac{1}{\beta+\eta}}\right)\right) \Rightarrow$$

$$\frac{1}{x}\left(\sum_{k=1}^{x}\left(\frac{2}{|S_k|(|S_k|+1)}\sum_{i=1}^{|S_k|}\sum_{j=i}^{|S_k|}(w_i a_{ki})^{\beta} \otimes (w_j a_{kj})^{\eta}\right)^{\frac{1}{\beta+\eta}}\right) =$$

$$\left(1-\left(\prod_{k=1}^{x}\left(1-(1+J_k^{\beta,\eta}-L_k^{\beta,\eta})^{\frac{1}{\beta+\eta}}+(J_k^{\beta,\eta})^{\frac{1}{\beta+\eta}}\right)\right)^{\frac{1}{x}}, \left(\prod_{k=1}^{x}\left(1-(1+J_k^{\beta,\eta}-L_k^{\beta,\eta})^{\frac{1}{\beta+\eta}}+(J_k^{\beta,\eta})^{\frac{1}{\beta+\eta}}\right)\right)^{\frac{1}{x}}-\right.$$

$$\left.\left(\prod_{k=1}^{x}\left(1+(J_k^{\beta,\eta})^{\frac{1}{\beta+\eta}}-(1-Q_k^{\beta,\eta}+J_k^{\beta,\eta})^{\frac{1}{\beta+\eta}}\right)\right)^{\frac{1}{x}}, \left(\prod_{k=1}^{x}\left(1+(J_k^{\beta,\eta})^{\frac{1}{\beta+\eta}}-(1-Q_k^{\beta,\eta}+J_k^{\beta,\eta})^{\frac{1}{\beta+\eta}}\right)\right)^{\frac{1}{x}}-\left(\prod_{k=1}^{x}\left((J_k^{\beta,\eta})^{\frac{1}{\beta+\eta}}\right)\right)^{\frac{1}{x}}\right)$$

Thus, $PFWIPHA^{\beta,\eta}(a_1, a_2, \ldots, a_y) =$

$$\left(1-\left(\prod_{k=1}^{x}\left(1-(1+J_k^{\beta,\eta}-L_k^{\beta,\eta})^{\frac{1}{\beta+\eta}}+(J_k^{\beta,\eta})^{\frac{1}{\beta+\eta}}\right)\right)^{\frac{1}{x}}, \left(\prod_{k=1}^{x}\left(1-(1+J_k^{\beta,\eta}-L_k^{\beta,\eta})^{\frac{1}{\beta+\eta}}+(J_k^{\beta,\eta})^{\frac{1}{\beta+\eta}}\right)\right)^{\frac{1}{x}}-\right.$$

$$\left.\left(\prod_{k=1}^{x}\left(1+(J_k^{\beta,\eta})^{\frac{1}{\beta+\eta}}-(1-Q_k^{\beta,\eta}+J_k^{\beta,\eta})^{\frac{1}{\beta+\eta}}\right)\right)^{\frac{1}{x}}, \left(\prod_{k=1}^{x}\left(1+(J_k^{\beta,\eta})^{\frac{1}{\beta+\eta}}-(1-Q_k^{\beta,\eta}+J_k^{\beta,\eta})^{\frac{1}{\beta+\eta}}\right)\right)^{\frac{1}{x}}-\left(\prod_{k=1}^{x}\left((J_k^{\beta,\eta})^{\frac{1}{\beta+\eta}}\right)\right)^{\frac{1}{x}}\right) \qquad \square$$

## Appendix C. The proof of Theorem 3

**Proof.** According to the interaction operational laws defined in Definition 7, $\beta a_{ki} \oplus \eta a_{kj} = \left(1 - (1-p_{ki})^\beta (1-p_{kj})^\eta, (1-p_{ki})^\beta (1-p_{kj})^\eta - (1-p_{ki}-m_{ki})^\beta (1-p_{kj}-m_{kj})^\eta, (1-p_{ki}-m_{ki})^\beta (1-p_{kj}-m_{kj})^\eta - (1-p_{ki}-m_{ki}-n_{ki})^\beta (1-p_{kj}-m_{kj}-n_{kj})^\eta\right)$

Let $e_{ki} = 1-p_{ki}, e_{kj} = 1-p_{kj}, f_{ki} = 1-p_{ki}-m_{ki}, f_{kj} = 1-p_{kj}-m_{kj}, v_{ki} = 1-p_{ki}-m_{ki}-n_{ki}$ and $v_{kj} = 1-p_{kj}-m_{kj}-n_{kj}$, then $\beta a_{ki} \oplus \eta a_{kj} = \left(1 - (e_{ki})^\beta (e_{kj})^\eta, (e_{ki})^\beta (e_{kj})^\eta - (f_{ki})^\beta (f_{kj})^\eta, (f_{ki})^\beta (f_{kj})^\eta - (v_{ki})^\beta (v_{kj})^\eta\right)$,

$$\Rightarrow \prod_{i=1,j=i}^{|S_k|} (\beta a_{ki} \oplus \eta a_{kj}) =$$

$$\left( \prod_{i=1,j=i}^{|S_k|} \left(1 + (v_{ki})^\beta (v_{kj})^\eta - (e_{ki})^\beta (e_{kj})^\eta\right) - \prod_{i=1,j=i}^{|S_k|} \left((v_{ki})^\beta (v_{kj})^\eta\right), \prod_{i=1,j=i}^{|S_k|} \left(1 - (f_{ki})^\beta (f_{kj})^\eta + (v_{ki})^\beta (v_{kj})^\eta\right) - \prod_{i=1,j=i}^{|S_k|} \left(1 + (v_{ki})^\beta (v_{kj})^\eta - (e_{ki})^\beta (e_{kj})^\eta\right), 1 - \prod_{i=1,j=i}^{|S_k} \left(1 - (f_{ki})^\beta (f_{kj})^\eta + (v_{ki})^\beta (v_{kj})^\eta\right) \right)$$

Let $t_k = \frac{2}{|S_k|(|S_k|+1)}$, then $\left( \prod_{i=1,j=i}^{|S_k|} (\beta a_{ki} \oplus \eta a_{kj}) \right)^{\frac{2}{|S_k|(|S_k|+1)}} =$

$$\left( \left( \prod_{i=1,j=i}^{|S_k|} \left(1 + (v_{ki})^\beta (v_{kj})^\eta - (v_{ki})^\beta (v_{kj})^\eta\right) \right)^{t_k} - \left( \prod_{i=1,j=i}^{|S_k|} \left((v_{ki})^\beta (v_{kj})^\eta\right) \right)^{t_k}, \left( \prod_{i=1,j=i}^{|S_k} \left(1 - (f_{ki})^\beta (f_{kj})^\eta + (v_{ki})^\beta (v_{kj})^\eta\right) \right)^{t_k} - \left( \prod_{i=1,j=i}^{|S_k|} \left(1 + (v_{ki})^\beta (v_{kj})^\eta - (e_{ki})^\beta (e_{kj})^\eta\right) \right)^{t_k}, 1 - \left( \prod_{i=1,j=i}^{|S_k} \left(1 - (f_{ki})^\beta (f_{kj})^\eta + (v_{ki})^\beta (v_{kj})^\eta\right) \right)^{t_k} \right)$$

Let $R_k^{\beta,\eta} = \prod_{i=1,j=i}^{|S_k|} \left((v_{ki})^\beta (v_{kj})^\eta\right)$, $U_k^{\beta,\eta} = \prod_{i=1,j=i}^{|S_k|} \left(1 + (v_{ki})^\beta (v_{kj})^\eta - (e_{ki})^\beta (e_{kj})^\eta\right)$ and $V_k^{\beta,\eta} = \prod_{i=1,j=i}^{|S_k|} \left(1 - (f_{ki})^\beta (f_{kj})^\eta + (v_{ki})^\beta (v_{kj})^\eta\right)$, then $\frac{1}{\beta+\eta} \left( \prod_{i=1,j=i}^{|S_k|} (\beta a_{ki} \oplus \eta a_{kj}) \right)^{\frac{2}{|S_k|(|S_k|+1)}} =$

$$\left( 1 - \left(1 - (U_k^{\beta,\eta})^{t_k} + (R_k^{\beta,\eta})^{t_k}\right)^{\frac{1}{\beta+\eta}}, \left(1 - (U_k^{\beta,\eta})^{t_k} + (R_k^{\beta,\eta})^{t_k}\right)^{\frac{1}{\beta+\eta}} - \left(1 + (R_k^{\beta,\eta})^{t_k} - (V_k^{\beta,\eta})^{t_k}\right)^{\frac{1}{\beta+\eta}}, \left(1 + (R_k^{\beta,\eta})^{t_k} - (V_k^{\beta,\eta})^{t_k}\right)^{\frac{1}{\beta+\eta}} - \left((R_k^{\beta,\eta})^{t_k}\right)^{\frac{1}{\beta+\eta}} \right),$$

$$\Rightarrow \prod_{k=1}^{x} \left( \frac{1}{\beta+\eta} \left( \prod_{i=1,j=i}^{|S_k|} (\beta a_{ki} \oplus \eta a_{kj}) \right)^{\frac{2}{|S_k|(|S_k|+1)}} \right) =$$

$$\left( \prod_{k=1}^{x} \left(1 + \left((R_k^{\beta,\eta})^{t_k}\right)^{\frac{1}{\beta+\eta}} - \left(1 - (U_k^{\beta,\eta})^{t_k} + (R_k^{\beta,\eta})^{t_k}\right)^{\frac{1}{\beta+\eta}}\right) - \prod_{k=1}^{x} \left(\left((R_k^{\beta,\eta})^{t_k}\right)^{\frac{1}{\beta+\eta}}\right), \prod_{k=1}^{x} \left(1 - \left(1 + (R_k^{\beta,\eta})^{t_k} - (V_k^{\beta,\eta})^{t_k}\right)^{\frac{1}{\beta+\eta}} + \left((R_k^{\beta,\eta})^{t_k}\right)^{\frac{1}{\beta+\eta}}\right) - \prod_{k=1}^{x} \left(1 + \left((R_k^{\beta,\eta})^{t_k}\right)^{\frac{1}{\beta+\eta}} - \left(1 - (U_k^{\beta,\eta})^{t_k} + (R_k^{\beta,\eta})^{t_k}\right)^{\frac{1}{\beta+\eta}}\right), 1 - \prod_{k=1}^{x} \left(1 - \left(1 + (R_k^{\beta,\eta})^{t_k} - (V_k^{\beta,\eta})^{t_k}\right)^{\frac{1}{\beta+\eta}} + \left((R_k^{\beta,\eta})^{t_k}\right)^{\frac{1}{\beta+\eta}}\right) \right) \Rightarrow$$

$$\left( \prod_{k=1}^{x} \left( \frac{1}{\beta+\eta} \left( \prod_{i=1,j=i}^{|S_k|} (\beta a_{ki} \oplus \eta a_{kj}) \right)^{\frac{2}{|S_k|(|S_k|+1)}} \right) \right)^{\frac{1}{x}} =$$

$$\left( \left( \prod_{k=1}^{x} \left(1 + \left((R_k^{\beta,\eta})^{t_k}\right)^{\frac{1}{\beta+\eta}} - \left(1 - (U_k^{\beta,\eta})^{t_k} + (R_k^{\beta,\eta})^{t_k}\right)^{\frac{1}{\beta+\eta}}\right) \right)^{\frac{1}{x}} - \left( \prod_{k=1}^{x} \left(\left((R_k^{\beta,\eta})^{t_k}\right)^{\frac{1}{\beta+\eta}}\right) \right)^{\frac{1}{x}}, \left( \prod_{k=1}^{x} \left(1 - \left(1 + (R_k^{\beta,\eta})^{t_k} - (V_k^{\beta,\eta})^{t_k}\right)^{\frac{1}{\beta+\eta}} + \left((R_k^{\beta,\eta})^{t_k}\right)^{\frac{1}{\beta+\eta}}\right) \right)^{\frac{1}{x}} - \left( \prod_{k=1}^{x} \left(1 + \left((R_k^{\beta,\eta})^{t_k}\right)^{\frac{1}{\beta+\eta}} - \left(1 - (U_k^{\beta,\eta})^{t_k} + (R_k^{\beta,\eta})^{t_k}\right)^{\frac{1}{\beta+\eta}}\right) \right)^{\frac{1}{x}}, 1 - \left( \prod_{k=1}^{x} \left(1 - \left(1 + (R_k^{\beta,\eta})^{t_k} - (V_k^{\beta,\eta})^{t_k}\right)^{\frac{1}{\beta+\eta}} + \left((R_k^{\beta,\eta})^{t_k}\right)^{\frac{1}{\beta+\eta}}\right) \right)^{\frac{1}{x}} \right)$$

$\square$

## Appendix D. The Proof of Theorem 4

**Proof.** Based on the interaction operational laws defined in Definition 7, $(a_{ki})^{w_i} = \left((1-n_{ki}-m_{ki})^{w_i} - (1-n_{ki}-m_{ki}-p_{ki})^{w_i}, (1-n_{ki})^{w_i} - (1-n_{ki}-m_{ki})^{w_i}, 1 - (1-n_{ki})^{w_i}\right)$ and $(a_{kj})^{w_j} = \left((1-n_{kj}-m_{kj})^{w_j} - (1-n_{kj}-m_{kj}-p_{kj})^{w_j}, (1-n_{kj})^{w_j} - (1-n_{kj}-m_{kj})^{w_j}, 1 - (1-n_{kj})^{w_j}\right)$.

Let $r_{ki} = 1-n_{ki}, r_{kj} = 1-n_{kj}, u_{ki} = 1-n_{ki}-m_{ki}, u_{kj} = 1-n_{kj}-m_{kj}, v_{ki} = 1-n_{ki}-m_{ki}-p_{ki}$ and $v_{kj} = 1-n_{kj}-m_{kj}-p_{kj}$, then $\beta(a_{ki})^{w_i} =$

$$\left(1-\left(1-(u_{ki})^{w_i}+(v_{ki})^{w_i}\right)^{\beta},\left(1-(u_{ki})^{w_i}+(v_{ki})^{w_i}\right)^{\beta}-\left(1+(v_{ki})^{w_i}-(r_{ki})^{w_i}\right)^{\beta},\left(1+(v_{ki})^{w_i}-(r_{ki})^{w_i}\right)^{\beta}-\left((v_{ki})^{w_i}{}^{\beta}\right)\right) \text{ and } \eta(a_{kj})^{w_j}$$

$$=\left(1-\left(1-(u_{kj})^{w_j}+(v_{kj})^{w_j}\right)^{\eta},\left(1-(u_{kj})^{w_j}+(v_{kj})^{w_j}\right)^{\eta}-\left(1+(v_{kj})^{w_j}-(r_{kj})^{w_j}\right)^{\eta},\left(1+(v_{kj})^{w_j}-(r_{kj})^{w_j}\right)^{\eta}-\left((v_{kj})^{w_j}{}^{\eta}\right)\right), \Rightarrow$$

$$\beta(a_{ki})^{w_i}\oplus\eta(a_{kj})^{w_j}=$$

$$\left(1-\left(1-(u_{ki})^{w_i}+(v_{ki})^{w_i}\right)^{\beta}\left(1-(u_{kj})^{w_j}+(v_{kj})^{w_j}\right)^{\eta},\left(1-(u_{ki})^{w_i}+(v_{ki})^{w_i}\right)^{\beta}\left(1-(u_{kj})^{w_j}+(v_{kj})^{w_j}\right)^{\eta}-\right.$$

$$\left.\left(1+(v_{ki})^{w_i}-(r_{ki})^{w_i}\right)^{\beta}\left(1+(v_{kj})^{w_j}-(r_{kj})^{w_j}\right)^{\eta},\left(1+(v_{ki})^{w_i}-(r_{ki})^{w_i}\right)^{\beta}\left(1+(v_{kj})^{w_j}-(r_{kj})^{w_j}\right)^{\eta}-\left((v_{ki})^{w_i}{}^{\beta}\right)\left((v_{kj})^{w_j}{}^{\eta}\right)\right).$$

Let $G_k^{\beta,\eta}=\left((v_{ki})^{w_i}{}^{\beta}\right)\left((v_{kj})^{w_j}{}^{\eta}\right)$, $T_k^{\beta,\eta}=\left(1-(u_{ki})^{w_i}+(v_{ki})^{w_i}\right)^{\beta}\left(1-(u_{kj})^{w_j}+(v_{kj})^{w_j}\right)^{\eta}$ and $Z_k^{\beta,\eta}=$

$\left(1+(v_{ki})^{w_i}-(r_{ki})^{w_i}\right)^{\beta}\left(1+(v_{kj})^{w_j}-(r_{kj})^{w_j}\right)^{\eta}$, then

$$\prod_{i=1,j=i}^{|S_k|}\beta(a_{ki})^{w_i}\oplus\eta(a_{kj})^{w_j}=\left(\prod_{i=1,j=i}^{|S_k|}\left(1+G_k^{\beta,\eta}-T_k^{\beta,\eta}\right)-\prod_{i=1,j=i}^{|S_k|}G_k^{\beta,\eta},\prod_{i=1,j=i}^{|S_k|}\left(1-Z_k^{\beta,\eta}+G_k^{\beta,\eta}\right)-\right.$$

$$\left.\prod_{i=1,j=i}^{|S_k|}\left(1+G_k^{\beta,\eta}-T_k^{\beta,\eta}\right),1-\prod_{i=1,j=i}^{|S_k|}\left(1-Z_k^{\beta,\eta}+G_k^{\beta,\eta}\right)\right).$$

Let $t_k=\dfrac{2}{|S_k|(|S_k|+1)}$, then $\left(\prod_{i=1,j=i}^{|S_k|}\beta(a_{ki})^{w_i}\oplus\eta(a_{kj})^{w_j}\right)^{\frac{2}{|S_k|(|S_k|+1)}}=\left(\left(\prod_{i=1,j=i}^{|S_k|}\left(1+G_k^{\beta,\eta}-T_k^{\beta,\eta}\right)\right)^{t_k}-\right.$

$\left(\prod_{i=1,j=i}^{|S_k|}G_k^{\beta,\eta}\right)^{t_k},\left(\prod_{i=1,j=i}^{|S_k|}\left(1-Z_k^{\beta,\eta}+G_k^{\beta,\eta}\right)\right)^{t_k}-\left(\prod_{i=1,j=i}^{|S_k|}\left(1+G_k^{\beta,\eta}-T_k^{\beta,\eta}\right)\right)^{t_k},\left.1-\left(\prod_{i=1,j=i}^{|S_k|}\left(1-Z_k^{\beta,\eta}+G_k^{\beta,\eta}\right)\right)^{t_k}\right),\Rightarrow$

$$\frac{1}{\beta+\eta}\left(\prod_{i=1,j=i}^{|S_k|}\beta(a_{ki})^{w_i}\oplus\eta(a_{kj})^{w_j}\right)^{\frac{2}{|S_k|(|S_k|+1)}}=$$

$\left(1-\left(1-\left(\prod_{i=1,j=i}^{|S_k|}\left(1+G_k^{\beta,\eta}-T_k^{\beta,\eta}\right)\right)^{t_k}+\left(\prod_{i=1,j=i}^{|S_k|}G_k^{\beta,\eta}\right)^{t_k}\right)^{\frac{1}{\beta+\eta}},\left(1-\left(\prod_{i=1,j=i}^{|S_k|}\left(1+G_k^{\beta,\eta}-T_k^{\beta,\eta}\right)\right)^{t_k}+\left(\prod_{i=1,j=i}^{|S_k|}G_k^{\beta,\eta}\right)^{t_k}\right)^{\frac{1}{\beta+\eta}}-\right.$

$\left.\left(1+\left(\prod_{i=1,j=i}^{|S_k|}G_k^{\beta,\eta}\right)^{t_k}-\left(\prod_{i=1,j=i}^{|S_k|}\left(1-Z_k^{\beta,\eta}+G_k^{\beta,\eta}\right)\right)^{t_k}\right)^{\frac{1}{\beta+\eta}},\left(1+\left(\prod_{i=1,j=i}^{|S_k|}G_k^{\beta,\eta}\right)^{t_k}-\left(\prod_{i=1,j=i}^{|S_k|}\left(1-Z_k^{\beta,\eta}+G_k^{\beta,\eta}\right)\right)^{t_k}\right)^{\frac{1}{\beta+\eta}}-\left(\left(\prod_{i=1,j=i}^{|S_k|}G_k^{\beta,\eta}\right)^{t_k}\right)^{\frac{1}{\beta+\eta}}\right).$

Let $O_k^{\beta,\eta}=\left(\prod_{i=1,j=i}^{|S_k|}G_k^{\beta,\eta}\right)^{t_k}$, $P_k^{\beta,\eta}=\left(\prod_{i=1,j=i}^{|S_k|}\left(1+G_k^{\beta,\eta}-T_k^{\beta,\eta}\right)\right)^{t_k}$ and $Y_k^{\beta,\eta}=$

$\left(\prod_{i=1,j=i}^{|S_k|}\left(1-Z_k^{\beta,\eta}+G_k^{\beta,\eta}\right)\right)^{t_k}$, then $\prod_{k=1}^{x}\left(\frac{1}{\beta+\eta}\left(\prod_{i=1,j=i}^{|S_k|}\beta(a_{ki})^{w_i}\oplus\eta(a_{kj})^{w_j}\right)^{\frac{2}{|S_k|(|S_k|+1)}}\right)=$

$\left(\prod_{k=1}^{x}\left(1-(O_k^{\beta,\eta})^{\frac{1}{\beta+\eta}}-\left(1-P_k^{\beta,\eta}+O_k^{\beta,\eta}\right)^{\frac{1}{\beta+\eta}}\right)-\prod_{k=1}^{x}\left((O_k^{\beta,\eta})^{\frac{1}{\beta+\eta}}\right),\prod_{k=1}^{x}\left(1-\left(1+O_k^{\beta,\eta}-Y_k^{\beta,\eta}\right)^{\frac{1}{\beta+\eta}}+(O_k^{\beta,\eta})^{\frac{1}{\beta+\eta}}\right)-\right.$

$\left.\prod_{k=1}^{x}\left(1-(O_k^{\beta,\eta})^{\frac{1}{\beta+\eta}}-\left(1-P_k^{\beta,\eta}+O_k^{\beta,\eta}\right)^{\frac{1}{\beta+\eta}}\right),1-\prod_{k=1}^{x}\left(1-\left(1+O_k^{\beta,\eta}-Y_k^{\beta,\eta}\right)^{\frac{1}{\beta+\eta}}+(O_k^{\beta,\eta})^{\frac{1}{\beta+\eta}}\right)\right),\Rightarrow$

$$\left(\prod_{k=1}^{x}\left(\frac{1}{\beta+\eta}\left(\prod_{i=1,j=i}^{|S_k|}\beta(a_{ki})^{w_i}\oplus\eta(a_{kj})^{w_j}\right)^{\frac{2}{|S_k|(|S_k|+1)}}\right)\right)^{\frac{1}{x}}=$$

$\left(\left(\prod_{k=1}^{x}\left(1-(O_k^{\beta,\eta})^{\frac{1}{\beta+\eta}}-\left(1-P_k^{\beta,\eta}+O_k^{\beta,\eta}\right)^{\frac{1}{\beta+\eta}}\right)\right)^{\frac{1}{x}}-\left(\prod_{k=1}^{x}\left((O_k^{\beta,\eta})^{\frac{1}{\beta+\eta}}\right)\right)^{\frac{1}{x}},\left(\prod_{k=1}^{x}\left(1-\left(1+O_k^{\beta,\eta}-Y_k^{\beta,\eta}\right)^{\frac{1}{\beta+\eta}}+(O_k^{\beta,\eta})^{\frac{1}{\beta+\eta}}\right)\right)^{\frac{1}{x}}\right.$

$\left.-\left(\prod_{k=1}^{x}\left(1-(O_k^{\beta,\eta})^{\frac{1}{\beta+\eta}}-\left(1-P_k^{\beta,\eta}+O_k^{\beta,\eta}\right)^{\frac{1}{\beta+\eta}}\right)\right)^{\frac{1}{x}},1-\left(\prod_{k=1}^{x}\left(1-\left(1+O_k^{\beta,\eta}-Y_k^{\beta,\eta}\right)^{\frac{1}{\beta+\eta}}+(O_k^{\beta,\eta})^{\frac{1}{\beta+\eta}}\right)\right)^{\frac{1}{x}}\right).\ \square$

## Appendix E. Questionnaire of Evaluating Hotels

Please make a choice in each line.

**Table A1.** Questionnaire of evaluation hotels.

| Hotels | Criteria | Evaluation | | |
|---|---|---|---|---|
| $\pi_1$ | $\phi_1$ (Price) | ☐ High | ☐ Medium | ☐ Low |
| | $\phi_2$ (Comfortability) | ☐ Good | ☐ Fair | ☐ Poor |
| | $\phi_3$ (Service) | ☐ Good | ☐ Fair | ☐ Poor |
| | $\phi_4$ (Location) | ☐ Good | ☐ Fair | ☐ Poor |
| | $\phi_5$ (Convenience) | ☐ Good | ☐ Fair | ☐ Poor |
| $\pi_2$ | $\phi_1$ (Price) | ☐ High | ☐ Medium | ☐ Low |
| | $\phi_2$ (Comfortability) | ☐ Good | ☐ Fair | ☐ Poor |
| | $\phi_3$ (Service) | ☐ Good | ☐ Fair | ☐ Poor |
| | $\phi_4$ (Location) | ☐ Good | ☐ Fair | ☐ Poor |
| | $\phi_5$ (Convenience) | ☐ Good | ☐ Fair | ☐ Poor |
| $\pi_3$ | $\phi_1$ (Price) | ☐ High | ☐ Medium | ☐ Low |
| | $\phi_2$ (Comfortability) | ☐ Good | ☐ Fair | ☐ Poor |
| | $\phi_3$ (Service) | ☐ Good | ☐ Fair | ☐ Poor |
| | $\phi_4$ (Location) | ☐ Good | ☐ Fair | ☐ Poor |
| | $\phi_5$ (Convenience) | ☐ Good | ☐ Fair | ☐ Poor |
| $\pi_4$ | $\phi_1$ (Price) | ☐ High | ☐ Medium | ☐ Low |
| | $\phi_2$ (Comfortability) | ☐ Good | ☐ Fair | ☐ Poor |
| | $\phi_3$ (Service) | ☐ Good | ☐ Fair | ☐ Poor |
| | $\phi_4$ (Location) | ☐ Good | ☐ Fair | ☐ Poor |
| | $\phi_5$ (Convenience) | ☐ Good | ☐ Fair | ☐ Poor |

## References


1. Tian, Z.P.; Wang, J.Q.; Zhang, H.Y. An integrated approach for failure mode and effects analysis based on fuzzy best-worst, relative entropy, and VIKOR methods. *Appl. Soft Comput.* **2018**, *72*, 636–646. [CrossRef]
2. Liang, W.Z.; Zhao, G.Y.; Wang, X.; Zhao, J.; Ma, C.D. Assessing the rockburst risk for deep shafts via distance-based multi-criteria decision making approaches with hesitant fuzzy information. *Eng. Geol.* **2019**, *260*, 105211. [CrossRef]
3. Chutia, R.; Saikia, S. Ranking intuitionistic fuzzy numbers at levels of decision-making and its application. *Expert Syst.* **2018**, *35*, e12292. [CrossRef]
4. Liang, W.Z.; Zhao, G.Y.; Wu, H. Assessing the performance of green mines via a hesitant fuzzy ORESTE–QUALIFLEX method. *Mathematics* **2019**, *7*, 788. [CrossRef]
5. Luo, S.Z.; Liang, W.Z.; Xing, L.N. Selection of mine development scheme based on similarity measure under fuzzy environment. *Neural Comput. Appl.* **2019**, 1–12. [CrossRef]
6. Zadeh, L.A. Fuzzy sets. *Inf. Control* **1965**, *8*, 338–356. [CrossRef]
7. Atanassov, K.T. Intuitionistic fuzzy sets. *Fuzzy Set Syst.* **1986**, *20*, 87–96. [CrossRef]
8. Cuong, B.C. Picture fuzzy sets. *J. Comput. Sci. Cybern.* **2014**, *30*, 409–420.
9. Luo, S.Z.; Liang, W.Z.; Zhao, G.Y. Likelihood-based hybrid ORESTE method for evaluating the thermal comfort in underground mines. *Appl. Soft Comput.* **2019**, *87*, 105983. [CrossRef]
10. Son, L.H. Generalized picture distance measure and applications to picture fuzzy clustering. *Appl. Soft Comput.* **2016**, *46*, 284–295. [CrossRef]
11. Singh, P.; Mishra, N.K.; Kumar, M.; Saxena, S.; Singh, V. Risk analysis of flood disaster based on similarity measures in picture fuzzy environment. *Afr. Mat.* **2018**, *29*, 1019–1038. [CrossRef]
12. Wei, G. Picture fuzzy cross-entropy for multiple attribute decision making problems. *J. Bus. Econ. Manag.* **2016**, *17*, 491–502. [CrossRef]
13. Wang, L.; Zhang, H.Y.; Wang, J.Q.; Li, L. Picture fuzzy normalized projection-based VIKOR method for the risk evaluation of construction project. *Appl. Soft Comput.* **2018**, *64*, 216–226. [CrossRef]
14. Wei, G.; Alsaadi, F.E.; Hayat, T.; Alsaedi, A. Projection models for multiple attribute decision making with picture fuzzy information. *Int. J. Mach. Learn. Cybern.* **2018**, *9*, 713–719. [CrossRef]
15. Ashraf, S.; Mahmood, T.; Abdullah, S.; Khan, Q. Different approaches to multi-criteria group decision making problems for picture fuzzy environment. *Bull. Braz. Math. Soc.* **2019**, *50*, 373–397. [CrossRef]

16. Liang, W.Z.; Zhao, G.Y.; Luo, S.Z. An integrated EDAS-ELECTRE method with picture fuzzy information for cleaner production evaluation in gold mines. *IEEE Access* **2018**, *6*, 65747–65759. [CrossRef]
17. Wang, L.; Peng, J.J.; Wang, J.Q. A multi-criteria decision-making framework for risk ranking of energy performance contracting project under picture fuzzy environment. *J. Clean. Prod.* **2018**, *191*, 105–118. [CrossRef]
18. He, Y.; Chen, H.; Zhou, L.; Liu, J.; Tao, Z. Intuitionistic fuzzy geometric interaction averaging operators and their application to multi-criteria decision making. *Inf. Sci.* **2014**, *25*, 142–159. [CrossRef]
19. Rani, D.; Garg, H. Complex intuitionistic fuzzy power aggregation operators and their applications in multicriteria decision-making. *Expert Syst.* **2018**, *35*, e12325. [CrossRef]
20. Luo, S.Z.; Liang, W.Z.; Zhao, G.Y. Linguistic neutrosophic power Muirhead mean operators for safety evaluation of mines. *PLoS ONE* **2019**, *14*, e0224090. [CrossRef]
21. Garg, H. Some picture fuzzy aggregation operators and their applications to multicriteria decision-making. *Arab. J. Sci. Eng.* **2017**, *42*, 5275–5290. [CrossRef]
22. Wang, C.; Zhou, X.; Tu, H.; Tao, S. Some geometric aggregation operators based on picture fuzzy sets and their application in multiple attribute decision making. *Ital. J. Pure Appl. Math.* **2017**, *37*, 477–492.
23. Wei, G.W. Picture fuzzy Hamacher aggregation operators and their application to multiple attribute decision making. *Fundam. Inform.* **2018**, *157*, 271–320. [CrossRef]
24. Xu, Y.; Shang, X.; Wang, J.; Zhang, R.; Li, W.; Xing, Y. A method to multi-attribute decision making with picture fuzzy information based on Muirhead mean. *J. Intell. Fuzzy Syst.* **2019**, *36*, 3833–3849. [CrossRef]
25. Bonferroni, C. Sulle medie multiple di potenze. *Boll. Mat. Ital.* **1950**, *5*, 267–270.
26. Blanco-Mesa, F.; León-Castro, E.; Merigó, J.M.; Xu, Z. Bonferroni means with induced ordered weighted average operators. *Int. J. Intell. Syst.* **2019**, *34*, 3–23. [CrossRef]
27. Sykora, S. *Mathematical Means and Averages: Generalized Heronian Means*; Stan's Library: Castano Primo, Italy, 2009.
28. Liu, Z.; Wang, S.; Liu, P. Multiple attribute group decision making based on q-rung orthopair fuzzy Heronian mean operators. *Int. J. Intell. Syst.* **2018**, *33*, 2341–2363. [CrossRef]
29. Liu, P.D.; Chen, S.M. Group decision making based on Heronian aggregation operators of intuitionistic fuzzy numbers. *IEEE Trans. Cybern.* **2017**, *47*, 2514–2530. [CrossRef]
30. Liang, W.Z.; Zhao, G.Y.; Hong, C. Selecting the optimal mining method with extended multi-objective optimization by ratio analysis plus the full multiplicative form (MULTIMOORA) approach. *Neural Comput. Appl.* **2019**, *31*, 5871–5886. [CrossRef]
31. Liu, P.D.; Liu, J.; Merigó, J.M. Partitioned Heronian means based on linguistic intuitionistic fuzzy numbers for dealing with multi-attribute group decision making. *Appl. Soft Comput.* **2018**, *6*, 395–422. [CrossRef]
32. Liu, P.D.; Liu, Z.; Zhang, X. Some intuitionistic uncertain linguistic Heronian mean operators and their application to group decision making. *Appl. Math. Comput.* **2014**, *230*, 570–586. [CrossRef]
33. Wei, G.W. Picture fuzzy aggregation operators and their application to multiple attribute decision making. *J. Intell. Fuzzy Syst.* **2017**, *33*, 713–724. [CrossRef]
34. Jahan, A.; Ismail, M.Y.; Shuib, S.; Norfazidah, D.; Edwards, K.L. An aggregation technique for optimal decision-making in materials selection. *Mater. Des.* **2011**, *32*, 4918–4924. [CrossRef]

 *mathematics*

*Article*

# A New Divergence Measure of Pythagorean Fuzzy Sets Based on Belief Function and Its Application in Medical Diagnosis

**Qianli Zhou [1], Hongming Mo [2] and Yong Deng [1],***

[1]   Institute of Fundamental and Frontier Science, University of Electronic Science and Technology of China, Chengdu 610054, China; 201921210216@std.uestc.edu.cn
[2]   Library, Sichuan Minzu College, Kangding 626001, China; mhm@scun.edu.cn
*   Correspondence: dengentropy@uestc.edu.cn or prof.deng@hotmail.com

Received: 21 December 2019; Accepted: 16 January 2020; Published: 20 January 2020

**Abstract:** As the extension of the fuzzy sets (FSs) theory, the intuitionistic fuzzy sets (IFSs) play an important role in handling the uncertainty under the uncertain environments. The Pythagoreanfuzzy sets (PFSs) proposed by Yager in 2013 can deal with more uncertain situations than intuitionistic fuzzy sets because of its larger range of describing the membership grades. How to measure the distance of Pythagorean fuzzy sets is still an open issue. Jensen–Shannon divergence is a useful distance measure in the probability distribution space. In order to efficiently deal with uncertainty in practical applications, this paper proposes a new divergence measure of Pythagorean fuzzy sets, which is based on the belief function in Dempster–Shafer evidence theory, and is called PFSDM distance. It describes the Pythagorean fuzzy sets in the form of basic probability assignments (BPAs) and calculates the divergence of BPAs to get the divergence of PFSs, which is the step in establishing a link between the PFSs and BPAs. Since the proposed method combines the characters of belief function and divergence, it has a more powerful resolution than other existing methods. Additionally, an improved algorithm using PFSDM distance is proposed in medical diagnosis, which can avoid producing counter-intuitive results especially when a data conflict exists. The proposed method and the magnified algorithm are both demonstrated to be rational and practical in applications.

**Keywords:** Pythagorean fuzzy set; Dempster–Shafer evidence theory; basic probability assignment; medical diagnosis

---

## 1. Introduction

With the development of fuzzy mathematics, medical diagnosis has addressed more and more attention from the research society of applied computer mathematics. In the diagnostic process of medical profession, improving the processing capacity of various uncertain and inconsistent information and achieving more accurate decision-making have become the major challenges in the development of medical diagnosis [1–3]. Up to now, the process of medical diagnosis is driven by various theoretical studies, such as fuzzy sets theory [4–6], intuitionistic fuzzy sets [7–10], interval-valued intuitionistic fuzzy sets [11–13], and quantum decision [14–16].

It is well known that medical diagnosis is an effective and reasonable way to handle the problem of uncertainty. Decision theory [17–19] has been widely applied in this field, containing a variety of theories and methods. For instance, evidence theory and its extension [20–22], evidential reasoning [23,24], D numbers theory [25], R numbers theory [26,27], Z numbers theory [28,29], and other hybrid methods have been used to research it from lots of aspects. The rationality and practicality of these methods have also been proven by their employment in applications, such as strategy selection [30–34], decision-making [35–39], prediction [40–42], and fault diagnosis [43,44]. In addition, Zadeh presented a useful theory in 1965, the fuzzy set (FS) theory [45], which drives a big step forward in decision theory. Atanassov's intuitionistic fuzzy sets (IFSs) [46] proposed and Yager's Pythagorean fuzzy sets (PFSs) [47] are two extensions of a fuzzy set which make use of the membership degree, non-membership degree, and the hesitancy to preciously express the uncertainty. Among them, Pythagorean fuzzy set has a larger range of expressing the uncertainy than intuitionistic fuzzy set [48,49]. Hence, the Pythagorean fuzzy set was chosen to apply for medical diagnosis in this paper.

Distance measure plays a vital role in pattern recognition, information fusion, decision-making, and other fields. The fuzzy set theory and intuitionistic fuzzy sets have been proposed for many years and their distance measurements [50–52] have matured compared with Pythagorean fuzzy sets. There are some useful distances for IFS after suffering the practice and the application. For instance, the Euclidean distance [53], the Hamming distance [53], and the Hausdorff metric [54] are the most widely applied distances of IFS. The distance measurement of PFSs is still an open issue, which attracts many researchers to explore the distance measurement of PFSs and its related applications. A general distance measurement of PFSs was proposed by Chen [55], is an extension of Euclidean distance and Hamming distance, and generates a reasonable result in multiple-criteria decision analysis. Wei and Wei [56] proposed a PFS measurement method based on cosine function and applied it to medical diagnosis to achieve an ideal result. Later on, Xiao [57] presented a distance measurement of PFSs based on divergence, called PFSJS distance. Among these methods of measure distance of PFSs, membership, non-membership and hesitancy are calculated based on the same weights. It is well known that the hesitancy expresses the uncertainty of membership and non-membership, so finding a proper method to distribute hesitancy to membership and non-membership can more reasonably handle the distance of PFSs. The basic probability assignment (BPA) in evidence theory presented by Dempster–Shafer [58,59] unifies uncertainty in a new set, and can handle various uncertainty reasonably. Song [60] presented a divergence measure of belief function based on Kullback–Leibler (KL) divergence [61] and Deng entropy [62], which has better results when dealing with the distance between BPAs with greater uncertainty. This paper proposes a new method to describe the PFSs in the form of BPAs, and uses an improverd divergence measurement of BPAs based on Jensen–Shannon divergence to measure the distance of PFSs. This method applied to medical diagnosis can not only get more intuitive results when the high conflicts appear in several symptoms between patients and diseases but also make higher resolution in different results.

According to the above, the structure of this article is as follows:

- The related concepts and properties of Pythagorean fuzzy set (PFS), basic probability assignment (BPA), the Jensen–Shannon divergence, and some widely applied methods of measuring distance are introduced.
- A new divergence measure of PFSs is proposed, which is called PFSDM distance. Its measurement is divided into three stages. The first is establishing a link between PFSs and BPAs. In addition, then, an improved method is proposed to measure the divergence of PFSs represented as BPAs. The third is proving the properties of PFSDM distance and expressing the feasibility and merits of the proposed method using numerical examples.
- The medical diagnosis algorithm based on Xiao's method [57] is magnified and compares the new algorithm with other existing methods to prove its practicability.
- The merits and uses of the PFSDM distance are summarized and the future direction of algorithm improvement is expected.

## 2. Preliminaries

In this section, the first subsection introduced the definition of Pythagorean fuzzy sets [47,48]. The main idea of basic probability assignment (BPA), Dempster–Shafer evidence theory [58,59] is in the second subsection. Some concepts of divergence are in the third subsection. In the last subsection, some existing methods for measuring the distance of PFSs are displayed.

### 2.1. Pythagorean Fuzzy Sets

**Definition 1.** *Let X be a limited universe of discourse. An intuitionistic fuzzy set (IFS) [46] M in X is defined by*

$$M = \{\langle x, \mu_M(x), v_M(x)\rangle \mid x \in X\}, \tag{1}$$

*where $\mu_M(x) : X \to [0, 1]$ represents the degree of support for membership of the $x \in X$ of IFS, and $v_M(x) : X \to [0, 1]$ represents the degree of support for non-membership of the $x \in X$ of IFS, with the condition that $0 \leq \mu_M(x) + v_M(x) \leq 1$. and the hesitancy function $\pi_M(x)$ of IFS reflecting the uncertainty of membership and non-membership is defined by*

$$\pi_M(x) = 1 - \mu_M(x) - v_M(x). \tag{2}$$

**Definition 2.** *Let x be a limited universe of discourse. A Pythagorean fuzzy set (PFS) [47,48] M in X is defined by*

$$M = \{\langle x, M_Y(x),\ M_N(x)\rangle \mid x \in X\}, \tag{3}$$

*where $M_Y(x) : X \to [0, 1]$ represents the degree of support for membership of the $x \in X$ of PFS, and $M_N(x) : X \to [0, 1]$ represents the degree of support for non-membership of the $x \in X$ of PFS, with the condition that $0 \leq M_Y^2(x) + M_N^2(x) \leq 1$. and the hesitancy function $M_H(x)$ of PFS reflecting the uncertainty of membership and non-membership is defined by*

$$M_H(x) = \sqrt{1 - M_Y^2(x) + M_N^2(x)}. \tag{4}$$

The membership and the non-membership of PFS also can be expressed by another way. A pair of values $r(x)$ and $d(x)$ for each $x \in X$ are used to represent the membership and non-membership as follows:

$$M_Y(x) = r_M(x)Cos(\theta_M(x))\quad M_N(x) = r_M(x)Sin(\theta_M(x)), \tag{5}$$

where $\theta(x) = (1 - d(x))\frac{\pi}{2} = Arctan(M_N(x)/M_Y(x))$. Here, we see that $\theta(x)$ is expressed as radians and $\theta(x) \in [0, \frac{\pi}{2}]$. In addition, $M_H(x) = \sqrt{1 - r^2(x)}$.

**Property 1.** *([49]) Let B and C be two PFSs in X, then*

1. $B \subseteq C$ *if* $\forall x \in X$, $B_Y(x) \leq C_Y(x)$ *and* $B_N(x) \geq C_N(x)$,
2. $B = C$ *if* $\forall x \in X$, $B_Y(x) = C_Y(x)$ *and* $B_N(x) = C_N(x)$,
3. $B \cap C = \{\langle x,\ \min[B_Y(x),\ C_Y(x)],\ \max[B_N(x), C_N(x)]\rangle \mid x \in X\}$,
4. $B \cup C = \{(x,\ \max[B_Y(x),\ C_Y(x)],\ \min[B_N(x), C_N(x)]) \mid x \in X]$,
5. $B \cdot C = \left\{\left\langle x, B_Y(x)C_Y(x),\ (B_N^2(x) + C_N^2(x) - B_N^2(x)C_N^2(x))^{\frac{1}{2}}\right\rangle \mid x \in X\right\}$,
6. $B^n = \left\{\left(x,\ B_Y^n(x),\ (1 - (1 - B_N^2(x)^n))^{\frac{1}{2}}\right) \mid x \in X\right\}$.

### 2.2. Dempster–Shafer Evidence Theory

The Dempster–Shafer evidence theory [58,59] is proposed to deal with conditions that are weaker than Bayes probability [63,64], which can directly express uncertainty and unknown information [65,66], so that it has been widely used in various applications, FMEA [67–69], evidential reasoning [70–72], evaluation [73], target recognition [74], industrial alarm system [75,76], and others [77,78].

**Definition 3.** *$\Theta$ is the set of N elements which represent mutually exclusive and exhaustive hypotheses. $\Theta$ is the frame of discernment [58,59]:*

$$\Theta = \{H_1, H_2, \cdots, H_i, \cdots, H_N\}. \tag{6}$$

*The power set of $\Theta$ is denoted by $2^\Theta$, and*

$$2^\Theta = \{\varnothing, \ \{H_1\}, \cdots, \{H_n\}, \ \{H_1, H_2\}, \cdots, \{H_1, \cdots, H_N\}\}, \tag{7}$$

*where $\varnothing$ is an empty set.*

**Definition 4.** *A mass function [58,59] m, also called as BPA, is a mapping of $2^\Theta$, defined as follows:*

$$m : 2^\Theta \to [0, 1], \tag{8}$$

*which satisfies the following conditions:*

$$m(\varnothing) = 0 \quad \sum_{A \in 2^\Theta} m(A) = 1 \quad 0 \leq m(A) \leq 1 \quad A \in 2^\Theta. \tag{9}$$

*The mass $m(A)$ represents how strongly the evidence supports A [79–81].*

*2.3. Divergence Measure*

**Definition 5.** *Given two probabilities distribution $A = \{A(x_1), A(x_2), \ldots, A(x_n)\}$ and $B = \{B(x_1), B(x_2), \ldots, B(x_n)\}$.*
*Kullback–Leibler divergence [61] between A and B is defined as:*

$$Div_{KL}(A, B) = \sum_{i=1}^{n} A(x_i) \log_2 \left( \frac{A(x_i)}{B(x_i)} \right) \tag{10}$$

*with $\sum_{i=1}^{n} A(x_i) = \sum_{i=1}^{n} B(x_i) = 1$.*

The Kullback–Leibler also has some disadvantages of its properties, and one of them is that it doesn't satisfy the commutative property:

$$Div_{KL}(A, B) \neq Div_{KL}(B, A). \tag{11}$$

In order to realize the commutation in the distance measure, the Jensen–Shannon divergence is an adaptive choice.

**Definition 6.** *Given two probabilities distribution $A = \{A(x_1), A(x_2), \ldots, A(x_n)\}$ and $B = \{B(x_1), B(x_2), \ldots, B(x_n)\}$.*
*Jensen–Shannon divergence [82] between A and B is defined as:*

$$JS_{AB} = \frac{Div_{KL}(A, \frac{A+B}{2}) + Div_{KL}(B, \frac{A+B}{2})}{2} = H(\frac{A+B}{2}) - \frac{H(A)}{2} - \frac{H(B)}{2}, \tag{12}$$

*where $H(A) = -\sum_{i=1}^{n} A(x_i) \log A(x_i)$ and $\sum_{i=1}^{n} A(x_i) = \sum_{i=1}^{n} B(x_i) = 1$.*

Song's divergence [60] is used to measure the belief function, which is capable of processing uncertainty efficiently in a highly fuzzy environment by applying the thinking of Deng entropy [62].

**Definition 7.** *([60]) Given two basic probability assignments (BPAs) $m_1$ and $m_2$, the divergence between $m_1$ and $m_2$ is defined as follows:*

$$D_{SD}(m_1, m_2) = \sum_i \frac{1}{2^{|F_i|-1}} m_1(F_i) \log\left(\frac{m_1(F_i)}{m_2(F_i)}\right),\tag{13}$$

*where the $F_i$ holds on $\sum_{i=1}^{2^\Theta - 1} m(F_i) = 1$, which is the power subset of frame of discernment $\Theta$ and $|F_i|$ is the cardinal number of $F_i$.*

It is obvious that $D_{SD}(m_1, m_2) \neq D_{SD}(m_2, m_1)$. In order to realize the commutative property, a divergence measurement based on Song's divergence is defined as follows:

$$D_{SDM}(m_1, m_2) = \widetilde{D}_{SDM}(m_2, m_1) = \frac{D_{SD}(m_1, m_2) + D_{SD}(m_2, m_1)}{2}.\tag{14}$$

Because of thinking of the number of subsets of the mass function and averagely distributing the BPAs to these subsets, Song's divergence is more reasonable than others when the basic probability assignments of non-singleton powers sets $F_i$ are larger.

*2.4. Distance Measure of Pythagorean Fuzzy Sets*

The Euclidean distance [53] and the Hamming distance [53] are the most widely applied distances, and Chen proposed a generalized distance measure of PFS [55], which is the extension of Hamming distance and Euclidean distance.

**Definition 8.** *Let X be a limited universe of discourse, and M and N are two PFSs. Chen's distance [55] measure between PFSs M and N denoted as $D_C(M, N)$ is defined as:*

$$D_C(M, N) = \left[\frac{1}{2}(\mid M_Y^2(x) - N_Y^2(x)\mid^\beta + \mid M_N^2(x) - N_N^2(x)\mid^\beta + \mid M_H^2(x) - N_H^2(x)\mid^\beta)\right]^{\frac{1}{\beta}},\tag{15}$$

*where $\beta$ holds on $\beta \geq 1$ is called the distance parameter. As the extension of the Hamming distance and Euclidean distance, if $\beta = 1$ and $\beta = 2$, the Chen's distance is equal to Hamming distance and Euclidean distance, respectively:*

- *if $\beta = 1$,*

  $$D_C(M, N) = D_{Hm}(M, N) = \left[\frac{1}{2}(\mid M_Y^2(x) - N_Y^2(x)\mid + \mid M_N^2(x) - N_N^2(x)\mid + \mid M_H^2(x) - N_H^2(x)\mid)\right],$$

- *if $\beta = 2$,*

  $$D_C(M, N) = D_E(M, N) = \left[\frac{1}{2}((M_Y^2(x) - N_Y^2(x))^2 + (M_N^2(x) - N_N^2(x))^2 + (M_H^2(x) - N_H^2(x))^2)\right]^{\frac{1}{2}}.$$

In the application of distance measure, the universe of discourse always has many properties. Xiao extended them as the normalized distance and proposed a divergence measure of PFSs called PFSJS based on the Jensen–Shannon divergence, which is the first work to calculate the distance of PFSs using divergence.

Let $X = \{x_1, x_2, \cdots, x_n\}$ be a limited universe of discourse, two PFSs
$M = \{\langle x_i, M_Y(x_i), M_N(x_i)\rangle \mid x_i \in X\}$ and $N = \{\langle x_i, N_Y(x_i), N_N(x_i)\rangle \mid x_i \in X\}$ are in X.

**Definition 9.** *The normalized Hamming distance [57] denoted as $\widetilde{D}_{Hm}(M, N)$ is defined as:*

$$\widetilde{D}_{Hm}(M, N) = \frac{1}{2n} \sum_{i=1}^{n} (\mid M_Y^2(x) - N_Y^2(x)\mid + \mid M_N^2(x) - N_N^2(x)\mid + \mid M_H^2(x) - N_H^2(x)\mid).$$

The normalized Euclidean distance [57] denoted as $\widetilde{D}_E(M, N)$ is defined as:

$$\widetilde{D}_E(M,N) = \frac{1}{n}\sum_{i=1}^{n}[\frac{1}{2}((M_Y^2(x) - N_Y^2(x))^2 + (M_N^2(x) - N_N^2(x))^2 + (M_H^2(x) - N_H^2(x))^2)]^{\frac{1}{2}}.$$

The normalized Chen's distance [57] denoted as $\widetilde{D}_C(M, N)$ is defined as:

$$\widetilde{D}_C(M,N) = \frac{1}{n}\sum_{i=1}^{n}[\frac{1}{2}(\mid M_Y^2(x) - N_Y^2(x) \mid^{\beta} + \mid M_N^2(x) - N_N^2(x) \mid^{\beta} + \mid M_H^2(x) - N_H^2(x) \mid^{\beta})]^{\frac{1}{\beta}},$$

where $\beta \geq 1$

**Definition 10.** *The normalized divergence measurements of Pythagorean fuzzy sets denoted as PFSJS [57] of M and N are defined as follows:*

$$\begin{aligned}
\widetilde{D}_x(M, N) &= \frac{1}{n}\cdot\sum_{i=1}^{n} D_x(M, N) \\
&= \frac{1}{n}\sum_{i=1}^{n}\sqrt{\frac{1}{2}\left[\sum_{\kappa} M_{\kappa}^2(x_i)\log\left(\frac{2M_{\kappa}^2(x_i)}{M_{\kappa}^2(x_i)+N_{\kappa}^2(x_i)}\right) + \sum_{\kappa} N_{\kappa}^2(x_i)\log\left(\frac{2N_{\kappa}^2(x_i)}{M_{\kappa}^2(x_i)+N_{\kappa}^2(x_i)}\right)\right]},
\end{aligned} \tag{16}$$

*where $\kappa \in \kappa \in \{Y, N, H\}$.*

According to the existing methods for measuring the PFSs' distance, what they have in common is that the weights of membership $A_Y(x)$ non-membership $A_N(x)$ and hesitancy $A_H(x)$ are considered to be the same when calculating distances. As is well known, the hesitancy represents the uncertainty of membership degree and non-membership degree, and the belief function in evidence theory can handle the uncertainty in a more proper way. Hence, if the ability of evidence theory to handle uncertainty is combined with the high resolution of divergence in distance measurement, the PFSs' distance measurement will be further optimized. In the next section, a new divergence measure of PFSs is proposed based on belief function, which describes the PFSs in the form of BPAs and measures the distance of PFSs by calculating the divergence of BPAs.

### 3. A New Divergence Measure of PFSs

In this section, a new divergence measure of PFSs, called PFSDM distance, is proposed. The first subsection shows how PFS reasonably expressed in the form of BPA. A new improved method of BPAs' divergence measure is introduced in the second subsection, and then the PFSDM distance and its properties is proposed. In the last subsection, some examples are used to prove its properties and demonstrate its feasibility by comparing with existing other methods.

*3.1. PFS Is Expressed in the Form of BPA*

In the evidence theory [58,59], the basic probability assignment (BPA) $m(A)$ represents the degree of evidence supporting $A$, and, according to Equations (6) and (7), the elements of power set of frame of discernment ($\Theta$) should satisfy $\sum_{A \in 2^{\theta}} m(A) = 1$. Thus, the method of representing PFS in the form of BPA is shown as follows:

**Definition 11.** *Let X be a limited universe of discourse, according to the second form of PFS shown in Equation (5), a Pythagorean fuzzy set M in X is $M = \{\langle x, r_M(x)Cos(\theta_M(x)), r_M(x)Sin(\theta_M(x))\rangle \mid x \in X\}$, the frame of discernment $\Theta_M$ of M and their basic probability assignments $m_p$ are defined as:*

$$\Theta_M = \{Y_M, N_M\}, \tag{17}$$

- $m_p(Y_M) = r_M(x)Cos^2\theta_M(x),$

- $m_p(N_M) = r_M(x)Sin^2\theta_M(x),$
- $m_p(Y_M N_M) = 1 - r_M(x),$
- $m_p(\phi) = 0,$

*where $m_p(Y_M)$ represents the degree of evidence supporting membership of M. The $m_p(N_M)$ represents the degree of evidence supporting non-membership of M. The $m_p(Y_M N_M)$ represents the degree of evidence supporting membership and non-membership. Because the basic focal elements $Y_M$ and $N_M$ are totally exclusive, and the sum of them is equal to 1, they conform to the Dempster–Shafer evidence theory.*

According to the meaning of $m_p(Y_M N_M)$ and the method of thinking of Deng entropy [62], it contains the degree of evidence supporting the elements of $\{Y_M\}$, $\{Y_N\}$ and $\{Y_M Y_N\}$, and Song [60] redistributes the degree of supporting $\{Y_M N_M\}$ averagely among the three elements when calculating the distance of BPAs in Equation (13).

*3.2. A New Divergence Measure of PFSs*

Jensen–Shannon divergence is widely used in distance measure of probability distributions, and in this subsection, we propose an improved divergence measure of BPA based on Song's divergence and Jensen–Shannon divergence. In addition, a new divergence measure of PFSs and its properties are proposed, which is capable of distinguishing PFSs better.

**Definition 12.** *Let $\Theta$ be a frame of discernment $\Theta = \{A_1, A_2, \cdots, A_n\}$, and the power set of $\Theta$ is $2^\Theta = \{\varnothing, \{A_1\}, \cdots, \{A_n\}, \{A_1, A2\}, \cdots, \{A_1, A_n\}, \cdots, \{A_1, \cdots, A_n\}\} = \{\varnothing, F_1, F_2, \cdots, F_{2^n-1}\}$. The Jensen–Shannon divergence measure $D_{JS}(m_1, m_2)$ of two BPAs $m_1, m_2$ is defined as:*

$$
\begin{aligned}
D_{JS}(m_1, m_2) &= \frac{1}{2}[D_{SD}(m_1, \frac{m_1 + m_2}{2}) + D_{SD}(m_2, \frac{m_2 + m_1}{2})] \\
&= \frac{1}{2}[\sum_{F_i \in 2^\theta} \frac{1}{2^{|Fi|-1}} m_1(F_i)log(\frac{2m_1(F_i)}{m_1(F_i) + m_2(F_i)}) + \sum_{F_i \in 2^\theta} \frac{1}{2^{|Fi|-1}} m_2(F_i)log(\frac{2m_2(F_i)}{m_1(F_i) + m_2(F_i)})],
\end{aligned}
\tag{18}
$$

*where $| F_i |$ is the cardinal number of $F_i$. In addition, just in case there's a zero in the denominator, $10^{-8}$ is used to replace zero in the calculation.*

The improved method satisfies the symmetry and considers the number of elements in the power set. In addition, then, substituting the PFSs in the form of BPAs into Equation (18) produces the new divergence measure of Pythagorean fuzzy sets.

**Definition 13.** *Let $X$ be a limited universe of discourse, according to the second form of PFS shown in Equation (5), two Pythagorean fuzzy sets M and N in X are $M = \{\langle x, r_M(x)Cos(\theta_M(x)), r_M(x)Sin(\theta_M(x))\rangle \mid x \in X\}$, $N = \{\langle x, r_N(x)Cos(\theta_N(x)), r_N(x)Sin(\theta_N(x))\rangle \mid x \in X\}$. The divergence measure denoted as $D_{PFS}(M, N)$ is defined as follows:*

$$
\begin{aligned}
D_{PFS}(M, N) = \frac{1}{2}\Big[ & r_M(x)Cos^2\theta_M(x)log\big(\frac{2r_M(x)Cos^2\theta_M(x)}{r_M(x)Cos^2\theta_M(x) + r_N(x)Cos^2\theta_N(x)}\big)+ \\
& r_N(x)Cos^2\theta_N(x)log\big(\frac{2r_N(x)Cos^2\theta_N(x)}{r_M(x)Cos^2\theta_M(x) + r_N(x)Cos^2\theta_N(x)}\big)+ \\
& r_M(x)Sin^2\theta_M(x)log\big(\frac{2r_M(x)Sin^2\theta_M(x)}{r_M(x)Sin^2\theta_M(x) + r_N(x)Sin^2\theta_N(x)}\big)+ \\
& r_N(x)Sin^2\theta_N(x)log\big(\frac{2r_N(x)Sin^2\theta_N(x)}{r_M(x)Sin^2\theta_M(x) + r_N(x)Sin^2\theta_N(x)}\big)+ \\
& \frac{1}{3}(1 - r_M(x))log\big(\frac{2(1 - r_M(x))}{(1 - r_M(x)) + (1 - r_N(x))}\big)+ \\
& \frac{1}{3}(1 - r_N(x))log\big(\frac{2(1 - r_N(x))}{(1 - r_M(x)) + (1 - r_N(x))}\big)\Big].
\end{aligned}
\tag{19}
$$

*In order to obtain higher resolution when making distance measurement, the divergence measure of PFSs, PFSDM distance, denoted as $D_{mp}(M, N)$, is defined by*

$$
D_{mp}(M, N) = \sqrt{D_{PFS}(M, N)}. \tag{20}
$$

According to the properties of Jensen–Shannon divergence [82], the larger PFSDM distance, the more different PFSs, and the smaller PFSDM, the more similar PFSs. The properties of the PFSDM distance are displayed as follows:

**Definition 14.** *Let $M$ and $N$ be two PFSs in a limited universe of discourse $X = \{x_1, x_2, \cdots, x_n\}$, where $M = \{\langle x_i, r_M(x_i)Cos(\theta_M(x_i)), r_M(x_i)Sin(\theta_M(x_i))\rangle\}$, $N = \{\langle x_i, r_N(x_i)Cos(\theta_N(x_i)), r_N(x_i)Sin(\theta_N(x_i))\rangle\}$. The normalized PFSDM distance, $\widetilde{D}_{mp}(M, N)$, is defined as follows:*

$$
\begin{aligned}
\widetilde{D}_{mp}(M, N) = \frac{1}{n}\sum_{i=1}^{n}\Big\{\frac{1}{2}\Big[ & r_M(x_i)Cos^2\theta_M(x_i)log\big(\frac{2r_M(x_i)Cos^2\theta_M(x_i)}{r_M(x_i)Cos^2\theta_M(x_i) + r_N(x_i)Cos^2\theta_N(x_i)}\big)+ \\
& r_N(x_i)Cos^2\theta_N(x_i)log\big(\frac{2r_N(x_i)Cos^2\theta_N(x_i)}{r_M(x_i)Cos^2\theta_M(x_i) + r_N(x_i)Cos^2\theta_N(x_i)}\big)+ \\
& r_M(x_i)Sin^2\theta_M(x_i)log\big(\frac{2r_M(x_i)Sin^2\theta_M(x_i)}{r_M(x_i)Sin^2\theta_M(x_i) + r_N(x_i)Sin^2\theta_N(x_i)}\big)+ \\
& r_N(x_i)Sin^2\theta_N(x_i)log\big(\frac{2r_N(x_i)Sin^2\theta_N(x_i)}{r_M(x_i)Sin^2\theta_M(x_i) + r_N(x_i)Sin^2\theta_N(x_i)}\big)+ \\
& \frac{1}{3}(1 - r_M(x_i))log\big(\frac{2(1 - r_M(x_i))}{(1 - r_M(x_i)) + (1 - r_N(x_i))}\big)+ \\
& \frac{1}{3}(1 - r_N(x_i))log\big(\frac{2(1 - r_N(x_i))}{(1 - r_M(x_i)) + (1 - r_N(x_i))}\big)\Big]\Big\}^{\frac{1}{2}}.
\end{aligned}
\tag{21}
$$

**Property 2.** *Let $M, N,$ and $O$ be three arbitrary PFSs in the limited universe of discourse $X$, then*

- *(P1)$D_{mp}(M, N) = 0$ if $M = N$, for $M, N \in X$.*
- *(P2)$D_{mp}(M, N) + D_{mp}(N, O) \geq D_{mp}(M, O)$, for $M, N, O \in X$.*
- *(P3)$D_{mp}(M, N) \in [0, 1]$, for $M, N \in X$.*
- *(P4)$D_{mp}(M, N) = D_{mp}(N, M)$, for $M, N \in X$.*

**Proof. (P1)**

Suppose two Pythagorean fuzzy sets $M$ and $N$ in the limited universe of discourse $X$. In addition, the PFSs of them are given as follows:
$M = \{\langle x, r_M(x)Cos(\theta_M(x)), r_M(x)Sin(\theta_M(x))\rangle \mid x \in X\}$,
$N = \{\langle x, r_N(x)Cos(\theta_N(x)), r_N(x)Sin(\theta_N(x))\rangle \mid x \in X\}$.

- If $M = N$, we can get $r_M(x) = r_N(x)$ and $\theta_N(x) = \theta_M(x)$. According to Equations (19) and (20), it can get a result that $D_{mp}(M, N) = 0$.
- If $D_{mp}(M, N) = 0$, we can get $r_M(x) = r_N(x)$ and $\theta_N(x) = \theta_M(x)$ in terms of Equations (19) and (20). Hence, it can be found that $M = N$.

Thus, the property $(P1)$ in the Property 2 is proven. $\square$

**Proof. (P2)**

Suppose two Pythagorean fuzzy sets M and N in the limited universe of discourse X. In addition, the PFSs of them are given as follows:

$M = \{\langle x, r_M(x)Cos(\theta_M(x)), r_M(x)Sin(\theta_M(x))\rangle \mid x \in X\}$,
$N = \{\langle x, r_N(x)Cos(\theta_N(x)), r_N(x)Sin(\theta_N(x))\rangle \mid x \in X\}$,
$O = \{\langle x, r_O(x)Cos(\theta_O(x)), r_O(x)Sin(\theta_O(x))\rangle \mid x \in X\}$.

Given four assumptions:

1. (A1) $r_M(x)Cos^2(\theta_M(x)) \le r_N(x)Cos^2(\theta_N(x)) \le r_O(x)Cos^2(\theta_O(x))$.
2. (A2) $r_O(x)Cos^2(\theta_O(x)) \le r_N(x)Cos^2(\theta_N(x)) \le r_M(x)Cos^2(\theta_M(x))$.
3. (A3) $r_N(x)Cos^2(\theta_N(x)) \le min\{r_M(x)Cos^2(\theta_M(x)), r_O(x)Cos^2(\theta_O(x))\}$.
4. (A4) $r_N(x)Cos^2(\theta_N(x)) \ge max\{r_M(x)Cos^2(\theta_M(x)), r_O(x)Cos^2(\theta_O(x))\}$.

Let $A = r_M(x)Cos^2(\theta_M(x))$, $B = r_N(x)Cos^2(\theta_N(x))$ and $C = r_O(x)Cos^2(\theta_O(x))$. According to the above, it is obvious that $\mid A - C \mid \le \mid A - B \mid + \mid B - C \mid$ is satisfied under the $(A1)$ and $(A2)$. We can easily find $A - B \ge 0$ and $C - B \ge 0$ in terms of $A3$ and $A4$. Therefore, we have:

$$\mid A - B \mid + \mid B - C \mid - \mid A - C \mid$$

$$= A - B + C - B - A + C, \ if \ A \ge C,$$

$$= A - B + C - B + A - C, \ if \ A \le C,$$

$$= 2(B - max\{A, C\}) \ge 0.$$

Hence, the inequality $\mid A - C \mid \le \mid A - B \mid + \mid B - C \mid$ is valid under $A3$ and $A4$. In the same way, the $r(x)Sin(\theta(x))$ and $1 - r$ also satisfy the $\mid A - C \mid \le \mid A - B \mid + \mid B - C \mid$. Therefore, the inequality in the Property 2 (P2), $D_{mp}(M, N) + D_{mp}(N, O) \ge D_{mp}(M, O)$, has been proven. $\square$

If we let $M = \langle x, 0.3, 0.40\rangle$, $N = \langle x, 038, 0.48\rangle$, $O = \langle x, 0.58, 0.68\rangle$, the $D_{mp}(M, N)$, $D_{mp}(N, O)$, $D_{mp}(M, O)$ can be calculated in terms of Equations (11), (19), and (20):

$$D_{mp}(M, \ N) = 0.0779,$$

$$D_{mp}(N, \ O) = 0.2007,$$

$$D_{mp}(M, \ O) = 0.2754.$$

Obviously, they satisfy $D_{mp}(M, N) + D_{mp}(N, O) \ge D_{mp}(M, O)$.

**Proof. (P3 & P4)**

Given two PFSs $M = \langle x, \alpha, \beta\rangle$ and $N = \langle x, \beta, \alpha\rangle$ in the limited universe of discourse X. The values $\alpha$ and $\beta$ represent membership degree and non-membership degree in two PFSs. According to Equation (5), $M$ and $N$ can be formed as

$$M = \langle r_M(x)Cos(\theta_M(x)), r_M(x)Sin(\theta_M(x))\rangle,$$

$$N = \langle r_N(x)Cos(\theta_N(x)), r_N(x)Sin(\theta_N(x))\rangle,$$

where $r_M(x) = r_N(x) = \sqrt{\alpha + \beta}$; $\theta_M(x) = Arctan(\beta/\alpha)$; $\theta_N(x) = Arctan(\alpha/\beta)$. The $\alpha$ and $\beta$ satisfy the $X \to [0, 1]$ and $0 \le \alpha^2 + \beta^2 \le 1$ in terms of Definition 2. Hence, the PFSDM distance between $M$ and $N$ are shown in Figure 1a when $\alpha$ and $\beta$ meet the requirements.

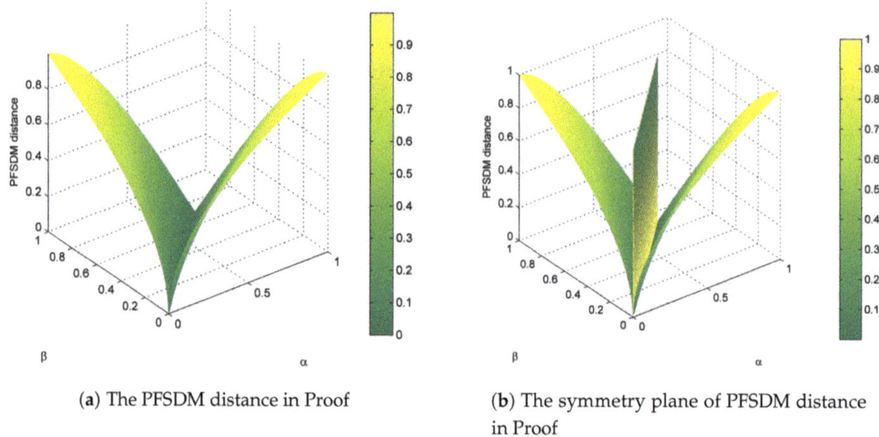

(**a**) The PFSDM distance in Proof

(**b**) The symmetry plane of PFSDM distance in Proof

**Figure 1.** The results in the proof.

- From Figure 1a, we can find that the PFSDM distances $D_{mp}(M, N)$ satisfy that its values are no more than one and no less than zero no matter how the parameters $\alpha$ and $\beta$ change. Thus, Property 2 (P3) has been proven.

- As shown in Figure 1b, let us make a plane when $\alpha = \beta$. According to the distance graph about the plane symmetry, we can demonstrate that the PFSDM meets $D_{mp}(M, N) = D_{mp}(N, M)$ and the Property 2 (P4) has been proven.

□

Up to now, we have demonstrated the four properties of PFSDM distance. In order to fully explore the characteristics and functions of PFSDM, the proposed method is compared with other existing methods such as Hamming distance, Euclidean distance, and PFSJS distance in the next subsection.

### 3.3. Numerical Examples

In this subsection, three numerical examples are used to prove the PFSDM distance's feasibility and merits. The powerful resolution of proposed method is proved in Example 1.

**Example 1.** *Suppose a limited universe of discourse* $X = \{x1, x2\}$*, and the PFSs* $M_i$ *and* $N_i$ *in the discourse under the Case$i$ ($i \in \{1, 2, 3, \cdots, 9\}$), which are given in Table 1.*

**Table 1.** Two PFSs $M_i$ and $N_i$ under different cases.

| PFSs | Case1 | Case2 | Case3 |
|---|---|---|---|
| $M_i$ | $\{\langle x_1, 0.55, 0.45\rangle, \langle x_2, 0.63, 0.55\rangle\}$ | $\{\langle x_1, 0.55, 0.45\rangle, \langle x_2, 0.63, 0.55\rangle\}$ | $\{\langle x_1, 0.55, 0.45\rangle, \langle x_2, 0.63, 0.55\rangle\}$ |
| $N_i$ | $\{\langle x_1, 0.39, 0.50\rangle, \langle x_2, 0.50, 0.59\rangle\}$ | $\{\langle x_1, 0.40, 0.51\rangle, \langle x_2, 0.51, 0.60\rangle\}$ | $\{\langle x_1, 0.67, 0.39\rangle, \langle x_2, 0.74, 0.50\rangle\}$ |

| PFSs | Case4 | Case5 | Case6 |
|---|---|---|---|
| $M_i$ | $\{\langle x_1, 0.71, 0.63\rangle, \langle x_2, 0.63, 0.55\rangle\}$ | $\{\langle x_1, 0.71, 0.63\rangle, \langle x_2, 0.63, 0.55\rangle\}$ | $\{\langle x_1, 0.71, 0.63\rangle, \langle x_2, 0.63, 0.55\rangle\}$ |
| $N_i$ | $\{\langle x_1, 0.63, 0.63\rangle, \langle x_2, 0.71, 0.63\rangle\}$ | $\{\langle x_1, 0.77, 0.55\rangle, \langle x_2, 0.55, 0.45\rangle\}$ | $\{\langle x_1, 0.77, 0.55\rangle, \langle x_2, 0.47, 0.59\rangle\}$ |

| PFSs | Case7 | Case8 | Case9 |
|---|---|---|---|
| $M_i$ | $\{\langle x_1, 0.30, 0.20\rangle, \langle x_2, 0.40, 0.30\rangle\}$ | $\{\langle x_1, 0.30, 0.20\rangle, \langle x_2, 0.40, 0.30\rangle\}$ | $\{\langle x_1, 0.50, 0.40\rangle, \langle x_2, 0.40, 0.30\rangle\}$ |
| $N_i$ | $\{\langle x_1, 0.15, 0.25\rangle, \langle x_2, 0.25, 0.35\rangle\}$ | $\{\langle x_1, 0.12, 0.26\rangle, \langle x_2, 0.22, 0.36\rangle\}$ | $\{\langle x_1, 0.65, 0.35\rangle, \langle x_2, 0.55, 0.25\rangle\}$ |

*The distance measurements are calculated by Hamming distance, Euclidean distance [55], PFSJS distance proposed by Xiao [57], Fei's distance [83], the proposed method, and PFSDM distance are shown in Table 2.*

**Table 2.** The comparison of different methods' results.

| Methods | Case1 | Case2 | Case3 | Case4 | Case5 | Case6 | Case7 | Case8 | Case9 |
|---|---|---|---|---|---|---|---|---|---|
| $\widetilde{D}_{Hm}$ [55] | 0.1487 | 0.1397 | 0.1487 | 0.1444 | 0.1444 | 0.1368 | 0.0825 | 0.1275 | 0.1575 |
| $\widetilde{D}_E$ [55] | 0.1983 | 0.1836 | 0.1951 | 0.1951 | 0.1759 | 0.1724 | 0.1025 | 0.1703 | 0.2226 |
| $\widetilde{D}_{Fei}$ [83] | 0.1074 | 0.1031 | 0.0902 | 0.0683 | 0.0806 | 0.0949 | 0.1118 | 0.1118 | 0.1118 |
| $\widetilde{D}_{Xiao}$ [57] | 0.1449 | 0.1300 | 0.1300 | 0.1771 | 0.1279 | 0.1279 | 0.1352 | 0.1434 | 0.1509 |
| $\widetilde{D}_{mp}$ | 0.1427 | 0.1389 | 0.1159 | 0.0879 | 0.0987 | 0.1193 | 0.1998 | 0.1599 | 0.1482 |

*We can easily find that the PFSDM distance, $D_{mp}$, has a satisfying performance in terms of Table 2, which produces intuitive distances when the PFSJS distance can not distinguish the similar PFSs such as Case2, Case3 and Case5, Case6. In addition, because of its divergence feature, the PFSDM distance can make up for the shortcoming of subtraction to find the distance, which is displayed in Case1, Case3 of Hamming distance and Case3, Case4 of Euclidean distance. The other character of the proposed method for calculating the distance of PFSs is the use of basic probability assignment to express the hesitancy. As the degree of hesitancy decreases, PFSDM distance thinks it has less and less effect on the distance, which is demonstrated in Case7, Case8, and Case9. In general, the PFSDM distance can handle the above cases in a proper way when some existing methods produce counter-intuitive results.*

In order to make the comparison between PFSDM distance and other methods more intuitive, as membership and non-membership change, the distances' figures are displayed in Example 2.

**Example 2.** *Assume two PFSs $M = \langle x, 0.3, 0.4 \rangle$ and $N = \langle x, \alpha, \beta \rangle$.*
*The Hamming distance [55] $D_{Hm}(M, N)$, Euclidean distance [55] $D_E(M, N)$, PFSJS distance [57] $D_{Xiao}(M, N)$, and PFSDM distance $D_{mp}(M, N)$ are displayed in Figure 2a–d.*

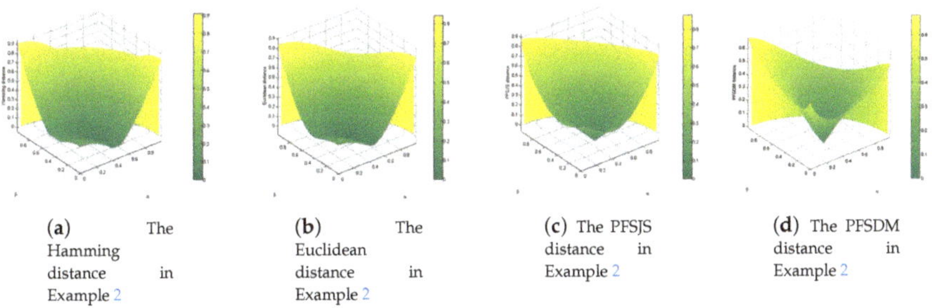

(**a**) The Hamming distance in Example 2

(**b**) The Euclidean distance in Example 2

(**c**) The PFSJS distance in Example 2

(**d**) The PFSDM distance in Example 2

**Figure 2.** The results in Example 2.

*We can know from the observation that Figure 2a–c have similar trends, but the PFSDM distance is different in Figure 2d. It resembles an inverted cone centered on the point $\langle \alpha = 0.3, \beta = 0.4 \rangle$ because it has a more uniform trend as $\alpha$ and $\beta$ change. Suppose four PFSs A1, A2, B1, B2 as follows:*

- $A1 = \langle x, A1_Y(x), A1_N(x) \rangle = \langle x, 0.1200, 0.1600 \rangle$,
- $A2 = \langle x, A2_Y(x), A2_N(x) \rangle = \langle x, 0.1380, 0.1840 \rangle$,
- $B1 = \langle x, B1_Y(x), B1_N(x) \rangle = \langle x, 0.5220, 0.6960 \rangle$,
- $B2 = \langle x, B2_Y(x), B2_N(x) \rangle = \langle x, 0.5400, 0.7200 \rangle$.

*After calculation, we know that*

$$A2_Y(x) - A1_Y(x) = B2_Y(x) - B1_Y(x) = 0.0180 \quad A2_N(x) - A1_N(x) = B2_N(x) - B1_N(x) = 0.0240.$$

*To verify that PFSDM distance changes more evenly, the Hamming distance [55] $D_{Hm}(A1, A2)$ and $D_{Hm}(B1, B2)$, Euclidean distance [55] $D_E(A1, A2)$ and $D_E(B1, B2)$, PFSJS distance [57] $D_{Xiao}(A1, A2)$*

and $D_{Xiao}(B1, B2)$, PFSDM distance $D_{mp}(A1, A2)$, and $D_{mp}(B1, B2)$ are calculated and results are obtained as follows:

- $D_{Hm}(A1, A2) = 0.0129$ $\quad D_{Hm}(B1, B2) = 0.0531$ $\quad \alpha_{Hm} = D_{Hm}(B1, B2)/D_{Hm}(A1, A2) = 4.1163$,
- $D_E(A1, A2) = 0.0113$ $\quad D_E(B1, B2) = 0.0466$ $\quad \alpha_E = D_E(B1, B2)/D_E(A1, A2) = 4.1163$,
- $D_{Xiao}(A1, A2) = 0.0287$ $\quad D_{Xiao}(B1, B2) = 0.0256$ $\quad \alpha_{Xiao} = D_{Xiao}(B1, B2)/D_{Xiao}(A1, A2) = 2.1021$,
- $D_{mp}(A1, A2) = 0.0261$ $\quad D_{mp}(B1, B2) = 0.0548$ $\quad \alpha_{mp} = D_{mp}(B1, B2)/D_{mp}(A1, A2) = 0.8915$.

*In the situation of membership degree and non-membership degree changing equally, we can clearly find that the ratio of distance changes $\alpha$ is different, and the PFSDM distance's $\alpha_{mp}$ is closest to 1 among them, which demonstrate that its variation trend is more even than others when membership and non-membership change the same.*

The previous examples demonstrate the feasibility of PFSDM distance through numerical analysis. It is well known that pattern recognition is one of the important applications of distance measure. Paul [84] modifies Zhang and Xu's distance measure of PFSs and applies them to pattern recognition. In the next example, the proposed method is applied {to pattern recognition to prove its rationality.

**Example 3.** *Suppose a limited universe of discourse $X = \{x_1, x_2, \cdots, x_{10}\}$, and $A_i$ $(i = 1, 2, 3, 4)$ represent four materials of building respectively, and B is an unknown material. Their PFSs are given in Table 3, and we need to recognize the type of B in A by finding the smallest distance $D_(A_i, B)$ from them. The results of Paul's method and proposed method, PFSDM distance, are given in Table 4:*

**Table 3.** The Pythagorean fuzzy sets of building materials in Example 3.

| PFS | $x_1$ | $x_2$ | $x_3$ | $x_4$ | $x_5$ |
|---|---|---|---|---|---|
| $A_1(x)$ | $\langle x, 0.173, 0.524 \rangle$ | $\langle x, 0.102, 0.818 \rangle$ | $\langle x, 0.530, 0.326 \rangle$ | $\langle x, 0.965, 0.008 \rangle$ | $\langle x, 0.420, 0.351 \rangle$ |
| $A_2(x)$ | $\langle x, 0.510, 0.365 \rangle$ | $\langle x, 0.627, 0.125 \rangle$ | $\langle x, 1.000, 0.000 \rangle$ | $\langle x, 0.125, 0.648 \rangle$ | $\langle x, 0.026, 0.823 \rangle$ |
| $A_3(x)$ | $\langle x, 0.495, 0.387 \rangle$ | $\langle x, 0.603, 0.298 \rangle$ | $\langle x, 0.987, 0.006 \rangle$ | $\langle x, 0.073, 0.849 \rangle$ | $\langle x, 0.037, 0.923 \rangle$ |
| $A_4(x)$ | $\langle x, 1.000, 0.000 \rangle$ | $\langle x, 1.000, 0.000 \rangle$ | $\langle x, 0.857, 0.123 \rangle$ | $\langle x, 0.734, 0.158 \rangle$ | $\langle x, 0.021, 0.896 \rangle$ |
| $B_1(x)$ | $\langle x, 0.978, 0.003 \rangle$ | $\langle x, 0.980, 0.012 \rangle$ | $\langle x, 0.798, 0.132 \rangle$ | $\langle x, 0.693, 0.213 \rangle$ | $\langle x, 0.051, 0.876 \rangle$ |

| PFS | $x_6$ | $x_7$ | $x_8$ | $x_9$ | $x_{10}$ |
|---|---|---|---|---|---|
| $A_1(x)$ | $\langle x, 0.008, 0.956 \rangle$ | $\langle x, 0.331, 0.512 \rangle$ | $\langle x, 1.000, 0.000 \rangle$ | $\langle x, 0.215, 0.625 \rangle$ | $\langle x, 0.432, 0.534 \rangle$ |
| $A_2(x)$ | $\langle x, 0.732, 0.153 \rangle$ | $\langle x, 0.556, 0.303 \rangle$ | $\langle x, 0.650, 0.267 \rangle$ | $\langle x, 1.000, 0.000 \rangle$ | $\langle x, 0.145, 0.762 \rangle$ |
| $A_3(x)$ | $\langle x, 0.690, 0.268 \rangle$ | $\langle x, 0.147, 0.812 \rangle$ | $\langle x, 0.213, 0.653 \rangle$ | $\langle x, 0.501, 0.284 \rangle$ | $\langle x, 1.000, 0.000 \rangle$ |
| $A_4(x)$ | $\langle x, 0.076, 0.912 \rangle$ | $\langle x, 0.152, 0.712 \rangle$ | $\langle x, 0.113, 0.756 \rangle$ | $\langle x, 0.489, 0.389 \rangle$ | $\langle x, 1.000, 0.000 \rangle$ |
| $B_1(x)$ | $\langle x, 0.123, 0.756 \rangle$ | $\langle x, 0.152, 0.721 \rangle$ | $\langle x, 0.113, 0.732 \rangle$ | $\langle x, 0.494, 0.368 \rangle$ | $\langle x, 0.987, 0.000 \rangle$ |

**Table 4.** The results generated by two methods in Example 3.

| *Distance* | $D(A_1, B)$ | $D(A_2, B)$ | $D(A_3, B)$ | $D(A_4, B)$ |
|---|---|---|---|---|
| *Paul's method* | 0.5970 | 0.5340 | 0.3240 | **0.0740** |
| *Proposed method* | 0.5296 | 0.5169 | 0.2927 | **0.0486** |

*In the process of calculation, we use a negligible number such as $10^{-8}$ to replace the zero, which is a good method to avoid having a zero denominator. The smallest PFSDM distance of them is $D_{mp}(A_4, B)$, so the material B belongs to $A_4$, which is identical to the result of Paul's modified method. Hence, PFSDM distance is capable of achieving the desired effect when applied to real-life problems.*

In this section, the PFSDM distance for measuring the divergence of PFSs is proposed. Some numerical examples are used to prove its properties and compared with other existing methods. The PFSDM distance has been proven that it not only is rational and practical but has more powerful resolution and a more uniform trend as well, which is because the proposed method combines the properties of BPA handling uncertainty and divergence's character distinguishing similar data.

## 4. Application in Medical Diagnosis

In this section, a modified algorithm for medical diagnosis is designed, which is improved based on Xiao's algorithm [57]. The new algorithm utilizes the PFSDM distance and obtains excellent results in application.

**Problem statement:** Suppose a limited universe of discourse $X = \{x_1, x_2, x_3, \cdots, x_n\}$, and the $x_i$ represent the symptoms of diseases and patients. Assume existing $m$ number of patients listed as $P = \{P_1, P_2, \cdots, P_m\}$ and $o$ number of diseases listed as $D = \{D_1, D_2, \cdots, D_o\}$. The symptoms of diseases and patients represented by Pythagorean fuzzy sets are as follows:

$$D_i = \{\langle x_1, D_Y^i(x_1), D_N^i(x_1)\rangle, \langle x_2, D_Y^i(x_2), D_N^i(x_2)\rangle, \cdots, \langle x_n, D_Y^i(x_n), D_N^i(x_n)\rangle\} \quad (1 \le i \le o),$$

$$P_j = \{\langle x_1, P_Y^j(x_1), P_N^j(x_1)\rangle, \langle x_2, P_Y^j(x_2), P_N^j(x_2)\rangle, \cdots, \langle x_n, P_Y^j(x_n), P_N^j(x_n)\rangle\} \quad (1 \le j \le m).$$

We need to utilize the distance measure to recognize which diseases the patients contract.

Step1   For every symptom, calculate the PFSDM distance $D_{mpt}^i(D_t^i, P_t^j)$ in terms of Equation (20) between $D_t^i$ and $P_t^j$, where $(1 \le t \ge n)$.

Step2   Calculate the weight of every symptom in medical diagnosis:

$$\omega_t = \frac{D_{mpt}^i(D_t^i, P_t^j)}{\sum_1^n D_{mpt}^i(D_t^i, P_t^j)}.$$

Step3   Diagnose the symptoms of the patient and the symptoms of the disease.

Calculate the weighted average PFSDM distance $\bar{D}_{mp}(D_i, P_j) = \sum_{t=1}^n \omega_t D_{mpt}^i(D_t^i, P_t^j)$.

Step4   For each weighted average PFSDM distance, the patients $P_j$ can be classified to diseases $D_i$,

in which $\xi = min_{1 \le i \le m}\{\bar{D}_{mp}(D_i, P_j)\}$, $P_j \to D_\xi$.

When the specific symptoms of the patient conflict with the symptoms of the disease, the proposed algorithm considers that the conflicted symptoms are paid more attention in the medical diagnosis. If all symptoms are diagnosed with the same weight, abnormal symptoms will be ignored when the symptom base is large. In the later examples, the applications of the proposed algorithm and its comparisons with other existing algorithm are shown.

**Example 4.** *([85]) Assume there are four patients, Ram, Mari, Sugu, and Somu, denoted as $P = \{P_1, P_2, P_3, P_4\}$. Five symptoms, Temperature, Headache, Stomach pain, Cough, and Chest pain, are observed, denoted as $S = \{s_1, s_2, s_3, s_4, s_5\}$. Additionally, five diagnoses, Viral fever, Malaria, Typhoid, Stomach problems, and Chest problems, are defined, represented as $D = \{D_1, D_2, D_3, D_4, D_5\}$. Then, the Pythagorean fuzzy relations $P \to S$ and $D \to S$ are displayed in Tables 5 and 6.*

**Table 5.** The symptoms of the patients in Example 4.

| Patient | $s_1$ | $s_2$ | $s_3$ | $s_4$ | $s_5$ |
|---------|-------|-------|-------|-------|-------|
| $P_1$ | $\langle s_1, 0.90, 0.10\rangle$ | $\langle s_2, 0.70, 0.20\rangle$ | $\langle s_3, 0.20, 0.80\rangle$ | $\langle s_4, 0.70, 0.20\rangle$ | $\langle s_5, 0.20, 0.70\rangle$ |
| $P_2$ | $\langle s_1, 0.00, 0.70\rangle$ | $\langle s_2, 0.40, 0.50\rangle$ | $\langle s_3, 0.60, 0.20\rangle$ | $\langle s_4, 0.20, 0.70\rangle$ | $\langle s_5, 0.10, 0.20\rangle$ |
| $P_3$ | $\langle s_1, 0.70, 0.10\rangle$ | $\langle s_2, 0.70, 0.10\rangle$ | $\langle s_3, 0.00, 0.50\rangle$ | $\langle s_4, 0.10, 0.70\rangle$ | $\langle s_5, 0.00, 0.60\rangle$ |
| $P_4$ | $\langle s_1, 0.50, 0.10\rangle$ | $\langle s_2, 0.40, 0.30\rangle$ | $\langle s_3, 0.40, 0.50\rangle$ | $\langle s_4, 0.80, 0.20\rangle$ | $\langle s_5, 0.30, 0.40\rangle$ |

**Table 6.** The symptoms of the diagnoses in Example 4.

| Diagnose | $s_1$ | $s_2$ | $s_3$ | $s_4$ | $s_5$ |
|----------|-------|-------|-------|-------|-------|
| $D_1$ | $\langle s_1, 0.30, 0.00 \rangle$ | $\langle s_2, 0.30, 0.50 \rangle$ | $\langle s_3, 0.20, 0.80 \rangle$ | $\langle s_4, 0.70, 0.30 \rangle$ | $\langle s_5, 0.20, 0.60 \rangle$ |
| $D_2$ | $\langle s_1, 0.00, 0.60 \rangle$ | $\langle s_2, 0.20, 0.60 \rangle$ | $\langle s_3, 0.00, 0.80 \rangle$ | $\langle s_4, 0.50, 0.00 \rangle$ | $\langle s_5, 0.10, 0.80 \rangle$ |
| $D_3$ | $\langle s_1, 0.20, 0.20 \rangle$ | $\langle s_2, 0.50, 0.20 \rangle$ | $\langle s_3, 0.10, 0.70 \rangle$ | $\langle s_4, 0.20, 0.60 \rangle$ | $\langle s_5, 0.20, 0.80 \rangle$ |
| $D_4$ | $\langle s_1, 0.20, 0.80 \rangle$ | $\langle s_2, 0.10, 0.50 \rangle$ | $\langle s_3, 0.70, 0.00 \rangle$ | $\langle s_4, 0.10, 0.70 \rangle$ | $\langle s_5, 0.20, 0.70 \rangle$ |
| $D_5$ | $\langle s_1, 0.20, 0.80 \rangle$ | $\langle s_2, 0.00, 0.70 \rangle$ | $\langle s_3, 0.20, 0.80 \rangle$ | $\langle s_4, 0.10, 0.80 \rangle$ | $\langle s_5, 0.80, 0.10 \rangle$ |

*The results diagnosis by the proposed algorithm and Xiao's method [57] are displayed in Tables 7 and 8. By observing the experimental results of proposed method, it is obvious that the $P_1$ has the least $\bar{D}_{mp}$ 0.4064 for $D_1$, $P_2$ has the least $\bar{D}_{mp}$ 0.2703 for $D_4$, $P_3$ has the least $\bar{D}_{mp}$ 0.2890 for $D_3$, and $P_4$ has the least $\bar{D}_{mp}$ 0.2470 for $D_1$; Hence, the conclusions are obvious in Table 8.*

**Table 7.** The results generated by the Xiao method in Example 4.

| Patient | $D_1$ | $D_2$ | $D_3$ | $D_4$ | $D_5$ | Classification |
|---------|-------|-------|-------|-------|-------|----------------|
| $P_1$ | **0.2583** | 0.3799 | 0.3512 | 0.5295 | 0.5528 | $D_1$ : Viral fever |
| $P_2$ | 0.4348 | 0.4217 | 0.4034 | **0.2394** | 3842 | $D_4$ : Stomach problems |
| $P_3$ | 0.3925 | 0.4787 | **0.2681** | 0.4140 | 0.5166 | $D_3$ : Typhoid |
| $P_4$ | **0.2166** | 0.4133 | 0.3454 | 0.4716 | 0.5382 | $D_1$ : Viral fever |

**Table 8.** The results generated by the proposed method in Example 4.

| Patient | $D_1$ | $D_2$ | $D_3$ | $D_4$ | $D_5$ | Classification |
|---------|-------|-------|-------|-------|-------|----------------|
| $P_1$ | **0.4064** | 0.6139 | 0.4984 | 0.7587 | 0.7776 | $D_1$ : Viral fever |
| $P_2$ | 0.6001 | 0.5971 | 0.5031 | **0.2703** | 5281 | $D_4$ : Stomach problems |
| $P_3$ | 0.4807 | 0.6617 | **0.2890** | 0.7109 | 0.7285 | $D_3$ : Typhoid |
| $P_4$ | **0.2407** | 0.4824 | 0.4475 | 0.5996 | 0.6271 | $D_1$ : Viral fever |

*As shown in Table 9, obviously the proposed method generates the same results as Samuel's method and Xiao's method, which can demonstrate that the proposed algorithm is capable of finishing medical diagnosis problems. In addition to doing this, we can find that the proposed method has better resolution than Xiao's method by observing Tables 7 and 8.*

**Table 9.** The results generated by other methods in Example 4.

| Method | $P_1$ | $P_2$ | $P_3$ | $P_4$ |
|--------|-------|-------|-------|-------|
| Xiao. [57] | Viral fever | Stomach problems | Typhoid | Viral fever |
| Samuel and Rajakumar [86] | Viral fever | Stomach problems | Typhoid | Viral fever |
| Proposed method | Viral fever | Stomach problems | Typhoid | Viral fever |

**Example 5.** *([2,11,57,87–90]) Assume four patients exist, namely, David, Tom, Kim, and Bob, denoted as $P = \{P_1, P_2, P_3, P_4\}$. Five symptoms are observed, in which they are Temperature, Headache, Stomach pain, Cough, and Chest pain, denoted as $S = \{s_1, s_2, s_3, s_4, s_5\}$. Additionally, five diagnoses, namely, Viral Fever, Malaria, Typhoid, Stomach problems, and Chest problems, are defined, represented as $D = \{D_1, D_2, D_3, D_4, D_5\}$. The numerical values respectively represented the membership and non-membership grade of the Pythagorean fuzzy set. Then, through the evaluation of the proposed distance measurement, the Pythagorean fuzzy sets' distance of $P \rightarrow S$ and the Pythagorean fuzzy sets' distance of $D \rightarrow S$ are displayed in the following Tables 10 and 11.*

**Table 10.** The symptoms of the patients in Example 5.

| Patient | $s_1$ | $s_2$ | $s_3$ | $s_4$ | $s_5$ |
|---|---|---|---|---|---|
| $P_1$ | $\langle s_1, 0.80, 0.10\rangle$ | $\langle s_2, 0.60, 0.10\rangle$ | $\langle s_3, 0.20, 0.80\rangle$ | $\langle s_4, 0.60, 0.10\rangle$ | $\langle s_5, 0.10, 0.60\rangle$ |
| $P_2$ | $\langle s_1, 0.00, 0.80\rangle$ | $\langle s_2, 0.40, 0.40\rangle$ | $\langle s_3, 0.60, 0.10\rangle$ | $\langle s_4, 0.10, 0.70\rangle$ | $\langle s_5, 0.10, 0.80\rangle$ |
| $P_3$ | $\langle s_1, 0.80, 0.10\rangle$ | $\langle s_2, 0.80, 0.10\rangle$ | $\langle s_3, 0.00, 0.60\rangle$ | $\langle s_4, 0.20, 0.70\rangle$ | $\langle s_5, 0.00, 0.50\rangle$ |
| $P_4$ | $\langle s_1, 0.60, 0.10\rangle$ | $\langle s_2, 0.50, 0.40\rangle$ | $\langle s_3, 0.30, 0.40\rangle$ | $\langle s_4, 0.70, 0.20\rangle$ | $\langle s_5, 0.30, 0.40\rangle$ |

**Table 11.** The symptoms of the diagnoses in Example 5.

| Diagnose | $s_1$ | $s_2$ | $s_3$ | $s_4$ | $s_5$ |
|---|---|---|---|---|---|
| $D_1$ | $\langle s_1, 0.40, 0.00\rangle$ | $\langle s_2, 0.40, 0.40\rangle$ | $\langle s_3, 0.10, 0.70\rangle$ | $\langle s_4, 0.40, 0.00\rangle$ | $\langle s_5, 0.10, 0.70\rangle$ |
| $D_2$ | $\langle s_1, 0.70, 0.00\rangle$ | $\langle s_2, 0.20, 0.60\rangle$ | $\langle s_3, 0.00, 0.90\rangle$ | $\langle s_4, 0.70, 0.00\rangle$ | $\langle s_5, 0.10, 0.80\rangle$ |
| $D_3$ | $\langle s_1, 0.30, 0.30\rangle$ | $\langle s_2, 0.60, 0.10\rangle$ | $\langle s_3, 0.20, 0.70\rangle$ | $\langle s_4, 0.20, 0.60\rangle$ | $\langle s_5, 0.10, 0.90\rangle$ |
| $D_4$ | $\langle s_1, 0.10, 0.70\rangle$ | $\langle s_2, 0.20, 0.40\rangle$ | $\langle s_3, 0.80, 0.00\rangle$ | $\langle s_4, 0.20, 0.70\rangle$ | $\langle s_5, 0.20, 0.70\rangle$ |
| $D_5$ | $\langle s_1, 0.10, 0.80\rangle$ | $\langle s_2, 0.00, 0.80\rangle$ | $\langle s_3, 0.20, 0.80\rangle$ | $\langle s_4, 0.20, 0.80\rangle$ | $\langle s_5, 0.80, 0.10\rangle$ |

*After calculating the distance by the proposed method and Xiao's method to measure the data, the results are generated in Tables 12 and 13. The results of David (P1) are different because of the data marked red in Tables 10 and 11. Though the D2 is more similar to P1 in other data intuitively, it generates a conflicted data in Headache ($s_2$). We use a weighted average to assign weights to symptoms. Hence, when the patient matches most of the symptoms of the disease, but one of the symptoms conflicts, the proposed algorithm will give greater weight to the effect of this symptom on the diagnosis of the disease, which will make the diagnosis of a complex disease with many symptoms more accurate.*

**Table 12.** The results generated by proposed method in Example 5.

| Patient | $D_1$ | $D_2$ | $D_3$ | $D_4$ | $D_5$ | Classification |
|---|---|---|---|---|---|---|
| $P_1$ | **0.2895** | 0.4163 | 0.4924 | 0.7125 | 0.7958 | $D_1$ : *Viral fever* |
| $P_2$ | 0.6809 | 0.7839 | 0.5058 | **0.1523** | 0.6711 | $D_4$ : *Stomach problems* |
| $P_3$ | 0.3604 | 0.5940 | **0.3197** | 0.7264 | 0.7701 | $D_3$ : *Typhoid* |
| $P_4$ | **0.2867** | 0.3699 | 0.4478 | 0.5839 | 0.3370 | $D_1$ : *Viral fever* |

**Table 13.** The results generated by Xiao's method in Exmple 5.

| Patient | $D_1$ | $D_2$ | $D_3$ | $D_4$ | $D_5$ | Classification |
|---|---|---|---|---|---|---|
| $P_1$ | 0.2511 | **0.2457** | 0.3110 | 0.5191 | 0.5606 | $D_2$ : *Malaria* |
| $P_2$ | 0.3712 | 0.4955 | 0.3389 | **0.1589** | 0.4264 | $D_4$ : *Stomach problems* |
| $P_3$ | 0.3314 | 0.4348 | **0.3126** | 0.4610 | 0.5310 | $D_3$ : *Typhoid* |
| $P_4$ | **0.2543** | 0.3195 | 0.3953 | 0.4605 | 0.5564 | $D_1$ : *Viral fever* |

*By comparing other applied widely methods of medical diagnosis and their results in Table 14, we can find that the diagnosis of David ($P_1$) is controversial and its results in Table 13 are similar between Viral fever ($D_1$) and Malaria ($D_2$). On the contrary, Viral fever ($D_1$) and Malaria ($D_2$) have a bigger difference in Table 12, which means that the new method is quite certain of the result.*

**Table 14.** The results generated by other methods.

| Method | $P_1$ | $P_2$ | $P_3$ | $P_4$ |
|---|---|---|---|---|
| *Szmidt et al.* [88] | *Viral fever* | *Stomach problems* | *Typhoid* | *Viral fever* |
| *Wei etl.* [11] | *Malaria* | *Stomach problems* | *Typhoid* | *Viral fever* |
| *Mondal et al.* [9] | *Malaria* | *Stomach problems* | *Typhoid* | *Viral fever* |
| *Ye* [90] | *Viral fever* | *Stomach problems* | *Typhoid* | *Viral fever* |
| *Vlachos et al.* [89] | *Viral fever* | *Stomachproblems* | *Typhoid* | *Viral fever* |
| *De et al.* [2] | *Malaria* | *Stomach problems* | *Typhoid* | *Viral fever* |
| *Xiao et al.* [57] | *Malaria* | *Stomach problems* | *Typhoid* | *Viral fever* |
| *Proposed method* | *Viral fever* | *Stomach problems* | *Typhoid* | *Viral fever* |

From the above two examples, the magnified method's feasibility and merits have been demonstrated. The feasibility of the new algorithm is illustrated by comparing the results of three methods in Example 4, and it can be intuitively found from Tables 7 and 8 that the new algorithm has higher resolution. In the second Example 5, more methods were compared in Table 14. The new algorithm uses weighted summation to make the previously disputed results more certain, which proves that it considers that the higher conflicted symptoms play a more important role in the diagnostic process.

## 5. Conclusions

In this paper, we propose a new divergence measure, called PFSDM distance, based on belief function, and modify the algorithm based on Xiao's method. The proposed method can produce intuitive results and its feasibility is proven by comparing with the existing method. In addition to this, the new method has more even change trend and better performance when the PFSs have larger hesitancy. In addition, we then apply the new algorithm to medical diagnosis and get the desired effect. The new algorithm has better resolution, which is helpful to ruling thresholds in practical applications. Consequently, the main contributions of this article are as follows:

- A method to express the PFS in the form of BPA is proposed, which is the first time to establish a link between them.
- A new distance measure between PFSs, called the PFSDM distance, based on Jensen–Shannon divergence and belief function, is proposed. Combining the characters of divergence and BPA contributes to more powerful resolution and even more of a change trend than existing methods.
- A modified medical diagnosis algorithm is proposed based on Xiao's method, which utilized the weighted summation to magnify the resolution of algorithm and increase the influence of conflicted data.
- The new divergence measure and the modified algorithm both have satisfying performance in the applications of pattern recognition and medical diagnosis.

It is well known that medical diagnosis is not a 100% accurate procedure and uncertainty is always present in cases. Though we get a definite result in Example 5 by the proposed algorithm, more cases are supposed to be examined in further research to get a safer result. In future research, we will try to explore more PFSs' properties by using the belief function in evidence theory further and apply them to more situations such as the multi-criteria decision-making and pattern recognition.

**Author Contributions:** Methodology, Q.Z. and Y.D.; Writing–original draft, Q.Z.; Writing–review & editing, H.M. and Y.D. All authors have read and agreed to the published version of the manuscript.

**Funding:** The work is partially supported by the National Natural Science Foundation of China (Grant No. 61973332) and the General Natural Research Program of Sichuan Minzu College (Grant No. XYZB18013ZB).

**Acknowledgments:** The work is partially supported by the National Natural Science Foundation of China (Grant No. 61973332) and the General Natural Research Program of Sichuan Minzu College (Grant No. XYZB18013ZB). The authors greatly appreciate the reviews' suggestions and editor's encouragement.

**Conflicts of Interest:** The authors declare no conflicts of interest.

## References


1. Cao, Z.; Ding, W.; Wang, Y.K.; Hussain, F.K.; Al-Jumaily, A.; Lin, C.T. Effects of Repetitive SSVEPs on EEG Complexity using Multiscale Inherent Fuzzy Entropy. *Neurocomputing* **2019**. [CrossRef]
2. De, S.K.; Biswas, R.; Roy, A.R. An application of intuitionistic fuzzy sets in medical diagnosis. *Fuzzy Sets Syst.* **2001**, *117*, 209–213. [CrossRef]
3. Cao, Z.; Lin, C.T.; Lai, K.L.; Ko, L.W.; King, J.T.; Liao, K.K.; Fuh, J.L.; Wang, S.J. Extraction of SSVEPs-based Inherent fuzzy entropy using a wearable headband EEG in migraine patients. *IEEE Trans. Fuzzy Syst.* **2019**. [CrossRef]
4. Ding, W.; Lin, C.T.; Cao, Z. Deep neuro-cognitive co-evolution for fuzzy attribute reduction by quantum leaping PSO with nearest-neighbor memeplexes. *IEEE Trans. Cybern.* **2018**, *49*, 2744–2757. [CrossRef]
5. Xiao, F. EFMCDM: Evidential fuzzy multicriteria decision-making based on belief entropy. *IEEE Trans. Fuzzy Syst.* **2019**. [CrossRef]
6. Tripathy, B.; Arun, K. A new approach to soft sets, soft multisets and their properties. *Int. J. Reason.-Based Intell. Syst.* **2015**, *7*, 244–253. [CrossRef]
7. Song, Y.; Fu, Q.; Wang, Y.F.; Wang, X. Divergence-based cross entropy and uncertainty measures of Atanassov's intuitionistic fuzzy sets with their application in decision making. *Appl. Soft Comput.* **2019**, *84*, 105703. [CrossRef]
8. Feng, F.; Liang, M.; Fujita, H.; Yager, R.R.; Liu, X. Lexicographic Orders of Intuitionistic Fuzzy Values and Their Relationships. *Mathematics* **2019**, *7*, 166. [CrossRef]
9. Mondal, K.; Pramanik, S. Intuitionistic fuzzy similarity measure based on tangent function and its application to multi-attribute decision-making. *Glob. J. Adv. Res.* **2015**, *2*, 464–471.
10. Song, Y.; Wang, X.; Zhu, J.; Lei, L. Sensor dynamic reliability evaluation based on evidence theory and intuitionistic fuzzy sets. *Appl. Intell.* **2018**, *48*, 3950–3962. [CrossRef]
11. Wei, C.P.; Wang, P.; Zhang, Y.Z. Entropy, similarity measure of interval-valued intuitionistic fuzzy sets and their applications. *Inf. Sci.* **2011**, *181*, 4273–4286. [CrossRef]
12. Dahooie, J.H.; Zavadskas, E.K.; Abolhasani, M.; Vanaki, A.; Turskis, Z. A Novel Approach for Evaluation of Projects Using an Interval–Valued Fuzzy Additive Ratio Assessment ARAS Method: A Case Study of Oil and Gas Well Drilling Projects. *Symmetry* **2018**, *10*, 45. [CrossRef]
13. Pan, Y.; Zhang, L.; Li, Z.; Ding, L. Improved Fuzzy Bayesian Network-Based Risk Analysis With Interval-Valued Fuzzy Sets and D-S Evidence Theory. *IEEE Trans. Fuzzy Syst.* **2019**. [CrossRef]
14. Ding, W.; Lin, C.T.; Prasad, M.; Cao, Z.; Wang, J. A layered-coevolution-based attribute-boosted reduction using adaptive quantum-behavior PSO and its consistent segmentation for neonates brain tissue. *IEEE Trans. Fuzzy Syst.* **2017**, *26*, 1177–1191. [CrossRef]
15. Gao, X.; Deng, Y. Quantum Model of Mass Function. *Int. J. Intell. Syst.* **2020**, *35*, 267–282. [CrossRef]
16. Chatterjee, K.; Zavadskas, E.K.; Tamoaitien, J.; Adhikary, K.; Kar, S. A Hybrid MCDM Technique for Risk Management in Construction Projects. *Symmetry* **2018**, *10*, 46. [CrossRef]
17. Cao, Z.; Chuang, C.H.; King, J.K.; Lin, C.T. Multi-channel EEG recordings during a sustained-attention driving task. *Sci. Data* **2019**, *6*. [CrossRef]
18. Palash, D.; Hazarika, G.C. Construction of families of probability boxes and corresponding membership functions at different fractiles. *Expert Syst.* **2017**, *34*, e12202.
19. Talhofer, V.; Hošková-Mayerová, Š.; Hofmann, A. Multi-criteria Analysis. In *Quality of Spatial Data in Command and Control System*; Springer: Cham, Switzerland, 2019; pp. 39–48.
20. Su, X.; Li, L.; Qian, H.; Sankaran, M.; Deng, Y. A new rule to combine dependent bodies of evidence. *Soft Comput.* **2019**, *23*, 9793–9799. [CrossRef]
21. Jiang, W. A correlation coefficient for belief functions. *Int. J. Approx. Reason.* **2018**, *103*, 94–106. [CrossRef]
22. Dutta, P. An uncertainty measure and fusion rule for conflict evidences of big data via Dempster–Shafer theory. *Int. J. Image Data Fusion* **2018**, *9*, 152–169. [CrossRef]
23. Fu, C.; Chang, W.; Xue, M.; Yang, S. Multiple criteria group decision-making with belief distributions and distributed preference relations. *Eur. J. Oper. Res.* **2019**, *273*, 623–633. [CrossRef]
24. Tripathy, B.; Mittal, D. Hadoop based uncertain possibilistic kernelized c-means algorithms for image segmentation and a comparative analysis. *Appl. Soft Comput.* **2016**, *46*, 886–923. [CrossRef]

25. Seiti, H.; Hafezalkotob, A.; Najaf, S.E. Developing a novel risk-based MCDM approach based on D numbers and fuzzy information axiom and its applications in preventive maintenance planning. *Appl. Soft Comput.* **2019**. [CrossRef]

26. Seiti, H.; Hafezalkotob, A.; Martinez, L. R-sets, Comprehensive Fuzzy Sets Risk Modeling for Risk-based Information Fusion and Decision-making. *IEEE Trans. Fuzzy Syst.* **2019**. [CrossRef]

27. Seiti, H.; Hafezalkotob, A. Developing the R-TOPSIS methodology for risk-based preventive maintenance planning: A case study in rolling mill company. *Comput. Ind. Eng.* **2019**, *128*, 622–636. [CrossRef]

28. Jiang, W.; Cao, Y.; Deng, X. A novel Z-network model based on Bayesian network and Z-number. *IEEE Trans. Fuzzy Syst.* **2019**. [CrossRef]

29. Kang, B.; Zhang, P.; Gao, Z.; Chhipi-Shrestha, G.; Hewage, K.; Sadiq, R. Environmental assessment under uncertainty using Dempster–Shafer theory and Z-numbers. *J. Ambient Intell. Humaniz. Comput.* **2019**, 1–20. [CrossRef]

30. Seiti, H.; Hafezalkotob, A.; Fattahi, R. Extending a pessimistic–optimistic fuzzy information axiom based approach considering acceptable risk: Application in the selection of maintenance strategy. *Appl. Soft Comput.* **2018**, *67*, 895–909. [CrossRef]

31. Deng, X.; Jiang, W. Evaluating green supply chain management practices under fuzzy environment: A novel method based on D number theory. *Int. J. Fuzzy Syst.* **2019**, *21*, 1389–1402. [CrossRef]

32. Xiao, F.; Zhang, Z.; Abawajy, J. Workflow scheduling in distributed systems under fuzzy environment. *J. Intell. Fuzzy Syst.* **2019**, *37*, 5323–5333. [CrossRef]

33. Chakraborty, S.; Tripathy, B. Privacy preserving anonymization of social networks using eigenvector centrality approach. *Intell. Data Anal.* **2016**, *20*, 543–560. [CrossRef]

34. Fang, R.; Liao, H.; Yang, J.B.; Xu, D.L. Generalised probabilistic linguistic evidential reasoning approach for multi-criteria decision-making under uncertainty. *J. Oper. Res. Soc.* **2019**, in press. [CrossRef]

35. Liao, H.; Wu, X. DNMA: A double normalization-based multiple aggregation method for multi-expert multi-criteria decision-making. *Omega* **2019**. [CrossRef]

36. Feng, F.; Fujita, H.; Ali, M.I.; Yager, R.R.; Liu, X. Another view on generalized intuitionistic fuzzy soft sets and related multiattribute decision-making methods. *IEEE Trans. Fuzzy Syst.* **2018**, *27*, 474–488. [CrossRef]

37. Fei, L. On interval-valued fuzzy decision-making using soft likelihood functions. *Int. J. Intell. Syst.* **2019**, *34*, 1631–1652. [CrossRef]

38. Liao, H.; Mi, X.; Xu, Z. A survey of decision-making methods with probabilistic linguistic information: Bibliometrics, preliminaries, methodologies, applications and future directions. *Fuzzy Optim. Decis. Mak.* **2019**. [CrossRef]

39. Mardani, A.; Nilashi, M.; Zavadskas, E.K.; Awang, S.R.; Zare, H.; Jamal, N.M. Decision Making Methods Based on Fuzzy Aggregation Operators: Three Decades Review from 1986 to 2017. *Int. J. Inf. Technol. Decis. Mak.* **2018**, *17*, 391–466. [CrossRef]

40. Zhou, D.; Al-Durra, A.; Zhang, K.; Ravey, A.; Gao, F. A robust prognostic indicator for renewable energy technologies: A novel error correction grey prediction model. *IEEE Trans. Ind. Electron.* **2019**, *66*, 9312–9325. [CrossRef]

41. Dutta, P. Modeling of variability and uncertainty in human health risk assessment. *MethodsX* **2017**, *4*, 76–85. [CrossRef]

42. Wu, X.; Liao, H. A consensus-based probabilistic linguistic gained and lost dominance score method. *Eur. J. Oper. Res.* **2019**, *272*, 1017–1027. [CrossRef]

43. Seiti, H.; Hafezalkotob, A.; Najafi, S.; Khalaj, M. A risk-based fuzzy evidential framework for FMEA analysis under uncertainty: An interval-valued DS approach. *J. Intell. Fuzzy Syst.* **2018**, *5*, 1–12. [CrossRef]

44. Liu, B.; Deng, Y. Risk Evaluation in Failure Mode and Effects Analysis Based on D Numbers Theory. *Int. J. Comput. Commun. Control* **2019**, *14*, 672–691.

45. Zadeh, L.A. Fuzzy sets. *Inf. Control* **1965**, *8*, 338–353. [CrossRef]

46. Atanassov, K.T. Intuitionistic fuzzy sets. In *Intuitionistic Fuzzy Sets*; Springer: Heidelberg, Germany, 1999; pp. 1–137.

47. Yager, R.R.; Abbasov, A.M. Pythagorean membership grades, complex numbers, and decision-making. *Int. J. Intell. Syst.* **2013**, *28*, 436–452. [CrossRef]

48. Yager, R.R. Pythagorean membership grades in multicriteria decision-making. *IEEE Trans. Fuzzy Syst.* **2013**, *22*, 958–965. [CrossRef]

49. Yager, R.R. Properties and applications of Pythagorean fuzzy sets. In *Imprecision and Uncertainty in Information Representation and Processing*; Springer: Cham, Switzerland, 2016; pp. 119–136.

50. Xiao, F. A distance measure for intuitionistic fuzzy sets and its application to pattern classification problems. *IEEE Trans. Syst. Man Cybern. Syst.* **2019**. [CrossRef]

51. Song, Y.; Wang, X.; Quan, W.; Huang, W. A new approach to construct similarity measure for intuitionistic fuzzy sets. *Soft Comput.* **2019**, *23*, 1985–1998. [CrossRef]

52. Tripathy, B.; Mohanty, R.; Sooraj, T. On intuitionistic fuzzy soft set and its application in group decision-making. In Proceedings of the 2016 International Conference on Emerging Trends in Engineering, Technology and Science (ICETETS), Pudukkottai, India, 24–26 February 2016; pp. 1–5.

53. Szmidt, E.; Kacprzyk, J. Distances between intuitionistic fuzzy sets. *Fuzzy Sets Syst.* **2000**, *114*, 505–518. [CrossRef]

54. Grzegorzewski, P. Distances between intuitionistic fuzzy sets and/or interval-valued fuzzy sets based on the Hausdorff metric. *Fuzzy Sets Syst.* **2004**, *148*, 319–328. [CrossRef]

55. Chen, T.Y. Remoteness index-based Pythagorean fuzzy VIKOR methods with a generalized distance measure for multiple criteria decision analysis. *Inf. Fusion* **2018**, *41*, 129–150. [CrossRef]

56. Wei, G.; Wei, Y. Similarity measures of Pythagorean fuzzy sets based on the cosine function and their applications. *Int. J. Intell. Syst.* **2018**, *33*, 634–652. [CrossRef]

57. Xiao, F.; Ding, W. Divergence measure of Pythagorean fuzzy sets and its application in medical diagnosis. *Appl. Soft Comput.* **2019**, *79*, 254–267. [CrossRef]

58. Shafer, G. *A Mathematical Theory of Evidence*; Princeton University Press: Princeton, NJ, USA, 1976; Volume 42.

59. Dempster, A.P. Upper and lower probability inferences based on a sample from a finite univariate population. *Biometrika* **1967**, *54*, 515–528. [CrossRef]

60. Song, Y.; Deng, Y. Divergence measure of belief function and its application in data fusion. *IEEE Access* **2019**, *7*, 107465–107472. [CrossRef]

61. Kullback, S. *Information Theory and Statistics*; Courier Corporation: Chelmsford, MA, USA, 1997.

62. Deng, Y. Deng entropy. *Chaos Solitons Fractals* **2016**, *91*, 549–553. [CrossRef]

63. Zhou, M.; Liu, X.B.; Chen, Y.W.; Qian, X.F.; Yang, J.B.; Wu, J. Assignment of attribute weights with belief distributions for MADM under uncertainties. *Knowl.-Based Syst.* **2019**. [CrossRef]

64. Deng, X.; Jiang, W. A total uncertainty measure for D numbers based on belief intervals. *Int. J. Intell. Syst.* **2019**, *34*, 3302–3316. [CrossRef]

65. Xiao, F. Generalization of Dempster–Shafer theory: A complex mass function. *Appl. Intell.* **2019**, in press.

66. Gao, X.; Liu, F.; Pan, L.; Deng, Y.; Tsai, S.B. Uncertainty measure based on Tsallis entropy in evidence theory. *Int. J. Intell. Syst.* **2019**, *34*, 3105–3120. [CrossRef]

67. Wang, H.; Deng, X.; Zhang, Z.; Jiang, W. A New Failure Mode and Effects Analysis Method Based on Dempster–Shafer Theory by Integrating Evidential Network. *IEEE Access* **2019**, *7*, 79579–79591. [CrossRef]

68. Jiang, W.; Zhang, Z.; Deng, X. A Novel Failure Mode and Effects Analysis Method Based on Fuzzy Evidential Reasoning Rules. *IEEE Access* **2019**, *7*, 113605–113615. [CrossRef]

69. Zhang, H.; Deng, Y. Weighted belief function of sensor data fusion in engine fault diagnosis. *Soft Comput.* **2019**. [CrossRef]

70. Zhou, M.; Liu, X.B.; Chen, Y.W.; Yang, J.B. Evidential reasoning rule for MADM with both weights and reliabilities in group decision-making. *Knowl.-Based Syst.* **2018**, *143*, 142–161. [CrossRef]

71. Liu, Z.; Liu, Y.; Dezert, J.; Cuzzolin, F. Evidence combination based on credal belief redistribution for pattern classification. *IEEE Trans. Fuzzy Syst.* **2019**. [CrossRef]

72. Zhou, M.; Liu, X.; Yang, J. Evidential reasoning approach for MADM based on incomplete interval value. *J. Intell. Fuzzy Syst.* **2017**, *33*, 3707–3721. [CrossRef]

73. Gao, S.; Deng, Y. An evidential evaluation of nuclear safeguards. *Int. J. Distrib. Sens. Netw.* **2020**, *16*. [CrossRef]

74. Pan, L.; Deng, Y. An association coefficient of belief function and its application in target recognition system. *Int. J. Intell. Syst.* **2020**, *35*, 85–104. [CrossRef]

75. Xu, X.; Xu, H.; Wen, C.; Li, J.; Hou, P.; Zhang, J. A belief rule-based evidence updating method for industrial alarm system design. *Control Eng. Pract.* **2018**, *81*, 73–84. [CrossRef]

76. Xu, X.; Li, S.; Song, X.; Wen, C.; Xu, D. The optimal design of industrial alarm systems based on evidence theory. *Control Eng. Pract.* **2016**, *46*, 142–156. [CrossRef]

77. Li, Y.; Deng, Y. Intuitionistic Evidence Sets. *IEEE Access* **2019**, *7*, 106417–106426. [CrossRef]

78. Luo, Z.; Deng, Y. A matrix method of basic belief assignment's negation in Dempster–Shafer theory. *IEEE Trans. Fuzzy Syst.* **2019**, *27*. [CrossRef]

79. Jiang, W.; Huang, C.; Deng, X. A new probability transformation method based on a correlation coefficient of belief functions. *Int. J. Intell. Syst.* **2019**, *34*, 1337–1347. [CrossRef]

80. Zhang, W.; Deng, Y. Combining conflicting evidence using the DEMATEL method. *Soft Comput.* **2019**, *23*, 8207–8216. [CrossRef]

81. Wang, Y.; Zhang, K.; Deng, Y. Base belief function: An efficient method of conflict management. *J. Ambient Intell. Humaniz. Comput.* **2019**, *10*, 3427–3437. [CrossRef]

82. Lin, J. Divergence measures based on the Shannon entropy. *IEEE Trans. Inf. Theory* **1991**, *37*, 145–151. [CrossRef]

83. Fei, L.; Deng, Y. Multi-criteria decision-making in Pythagorean fuzzy environment. *Appl. Intell.* **2019**. [CrossRef]

84. Ejegwa, P.A. Modified Zhang and Xu's distance measure for Pythagorean fuzzy sets and its application to pattern recognition problems. *Neural Comput. Appl.* **2019**. [CrossRef]

85. Ngan, R.T.; Ali, M. $\delta$-equality of intuitionistic fuzzy sets: A new proximity measure and applications in medical diagnosis. *Appl. Intell.* **2018**, *48*, 499–525.

86. Samuel, A.E.; Rajakumar, S. Intuitionistic Fuzzy Set with Modal Operators in Medical Diagnosis. *Adv. Fuzzy Math.* **2017**, *12*, 167–175.

87. Peng, X.; Yuan, H.; Yang, Y. Pythagorean fuzzy information measures and their applications. *Int. J. Intell. Syst.* **2017**, *32*, 991–1029. [CrossRef]

88. Szmidt, E.; Kacprzyk, J. Intuitionistic fuzzy sets in intelligent data analysis for medical diagnosis. In *International Conference on Computational Science*; Springer: Berlin/Heidelberg, Germnay, 2001; pp. 263–271.

89. Vlachos, I.K.; Sergiadis, G.D. Intuitionistic fuzzy information–applications to pattern recognition. *Pattern Recognit. Lett.* **2007**, *28*, 197–206. [CrossRef]

90. Ye, J. Cosine similarity measures for intuitionistic fuzzy sets and their applications. *Math. Comput. Model.* **2011**, *53*, 91–97, doi10.1016/j.mcm.2010.07.022. [CrossRef]

 *mathematics*

*Article*

# Separation Axioms of Interval-Valued Fuzzy Soft Topology via Quasi-Neighborhood Structure

**Mabruka Ali [1], Adem Kılıçman [1,2,*] and Azadeh Zahedi Khameneh [1,2]**

[1]   Department of Mathematics, University Putra Malaysia, Serdang 43400 UPM, Selangor, Malaysia;
      altwer2016@gmail.com (M.A.); azadeh503@gmail.com (A.Z.K.)
[2]   Institute for Mathematical Research, University Putra Malaysia, Serdang 43400 UPM, Selangor, Malaysia
*   Correspondence: akilic@upm.edu.my

Received: 27 December 2019; Accepted: 26 January 2020; Published: 2 February 2020

**Abstract:** In this study, we present the concept of the interval-valued fuzzy soft point and then introduce the notions of its neighborhood and quasi-neighborhood in interval-valued fuzzy soft topological spaces. Separation axioms in an interval-valued fuzzy soft topology, so-called $q$-$T_i$ for $i = 0, 1, 2, 3, 4$, are introduced, and some of their basic properties are also studied.

**Keywords:** interval-valued fuzzy soft set; interval-valued fuzzy soft topology; interval-valued fuzzy soft point; interval-valued fuzzy soft neighborhood; interval-valued fuzzy soft quasi-neighborhood; interval-valued fuzzy soft separation axioms

---

## 1. Introduction

In 1999, Molodtsov [1] proposed a new mathematical approach known as soft set theory for dealing with uncertainties and vagueness. Traditional tools such as fuzzy sets [2] and rough sets [3] cannot clearly define objects. Soft set theory is different from traditional tools for dealing with uncertainties. A soft set was defined by a collection of approximate descriptions of an object based on parameters by a given set-valued map. Maji et al. [4] initiated the research on both fuzzy set and soft set hybrid structures called fuzzy soft sets and presented a concept that was subsequently discussed by many researchers. Different extensions of the classical fuzzy soft sets were introduced, such as generalized fuzzy soft sets [5], intuitionist fuzzy soft sets [6,7], vague soft sets [8], interval-valued fuzzy soft sets [9], and interval-valued intuitive fuzzy soft sets [10]. In particular, to alleviate some disadvantages of fuzzy soft sets, interval-valued fuzzy soft sets were introduced where no objective procedure was available to select the crisp membership degree of elements in fuzzy soft sets. Tanya and Kandemir [11] started topological studies of fuzzy soft sets. They used the classical concept of topology to construct a topological space over a fuzzy soft set and named it the fuzzy soft topology. They also studied some fundamental topological properties for the fuzzy soft topology, such as interior, closure, and base. Later, Simsekler and Yuksel [12] studied the fuzzy soft topological space in the case of Tanay and Kandemir [11]. However, they established the concept of the fuzzy soft topology over a fuzzy soft set with a set of fixed parameters and considered some topological concepts for fuzzy soft topological spaces such as the base, subbase, neighborhood, and Q-neighborhood. Roy and Samanta [13] noted a new concept of the fuzzy soft topology. They suggested the notion of the fuzzy soft topology over an ordinary set by adding fuzzy soft subsets of it, where everywhere, the parameter set is supposed to be fixed. Then, in [14], they continued to study the fuzzy soft topology and established a fuzzy soft point definition and various neighborhood structures. Atmaca and Zorlutuna [15] considered the concept of soft quasi-coincidence for fuzzy soft sets. By applying this new concept, they also studied the basic topological notions such as interior and closure for fuzzy soft sets. The concept of the product fuzzy soft topology and the boundary fuzzy soft topology was introduced by Zahedi et al. [16,17], and they studied some of their properties. They also suggested a new definition for the fuzzy soft point and then different neighborhood structures. Separation axioms of the fuzzy topological space and fuzzy soft topological space were studied by many authors, see [18–23] and [24–27]. The aim of this work is to develop interval-valued fuzzy soft separation axioms. We start with preliminaries and then give the definition of the interval-valued fuzzy soft point as a generalization of the interval-valued fuzzy point and fuzzy soft point in order to create different neighborhood structures in the interval-valued fuzzy soft topological space in Sections 3 and 4. Finally, in Section 5, the notion of separation axioms $q$-$T_i, i = 0, 1, 2, 3, 4$ in the interval-valued fuzzy soft topology is introduced, and some of their basic properties are also studied.

## 2. Preliminaries

Throughout this paper, $X$ is the set of objects and $E$ is the set of parameters. The set of all subsets of $X$ is denoted by $P(X)$ and $A \subset E$, showing a subset of $E$.

**Definition 1** ([1]). *A pair $(f, A)$ is called a soft set over $X$, if $f$ is a mapping given by $f : A \to P(X)$. For any parameter $e \in A, f(e) \subset X$ may be considered as the set e-approximate elements of the soft set $(f, A)$. In other words, the soft set is not a kind of set, but a parameterized family of subsets of the set $X$.*

Before introducing the notion of the interval-valued fuzzy soft sets, we give the concept of the interval-valued fuzzy set.

**Definition 2** ([28]). *An interval-valued fuzzy $(IVF)$ set over $X$ is defined by the membership function $f : X \to int([0, 1])$, where $int([0, 1])$ denotes the set of all closed subintervals of $[0, 1]$. Suppose that $x \in X$. Then, $f(x) = [f^-(x), f^+(x)]$ is called the degree of membership of the element $x \in X$, where $f^-(x)$ and $f^+$ are the lower and upper degrees of the membership of $x$ and $0 < f^-(x) < f^+(x) < 1$.*

Yang et al. [9] suggested the concept of interval-valued fuzzy soft set by combining the interval-valued fuzzy set and soft set as below.

**Definition 3** ([9]). *An interval-valued fuzzy soft $(IVFS)$ set over $X$ denoted by $f_E$ or $(f, E)$ is defined by the mapping $f : E \to \mathcal{IVF}(X)$, where $\mathcal{IVF}(X)$ is the set of all interval-valued fuzzy sets over $X$. For any $e \in E, f(e)$ can be written as an interval-valued fuzzy set such that $f(e) = \{\langle x, [f_e^-(x), f_e^+(x)]\rangle : x \in X\}$ where $f_e^-(x)$ and $f_e^+(x)$ are the lower and upper degrees of the membership of $x$ with respect to $e$, where $0 \le f_e^-(x) \le f_e^+(x) \le 1$.*

Note that $\mathcal{IVFS}(X, E)$ shows the set of all $IVFS$ sets over $X$.

**Definition 4** ([9]). *Let $f_A$ and $g_B$ be two IVFS sets over $X$. We say that:*

1.  *$f_A$ is an interval-valued fuzzy soft subset of $g_B$, denoted by $f_A \tilde{\le} g_B$, if and only if:*

    (i)  *$A \le B$,*
    (ii) *For all $e \in A, f_e^-(x) \le g_e^-(x)$ and $f_e^+(x) \le g_e^+(x), \forall x \in X.$*

2.  *$f_A = g_B$ if and only if $f_A \tilde{\le} g_B$ and $g_A \tilde{\le} f_B$.*
3.  *The union of two IVFS sets $f_A$ and $g_B$, denoted by $f_A \tilde{\vee} g_B$, is the IVFS set $(f \vee g, C)$, where $C = A \cup B$, and for all $e \in C$, we have:*

$$(f \vee g)_e(x) = \begin{cases} [f_e^-(x), f_e^+(x)], & e \in A - B \\ [g_e^-(x), g_e^+(x)], & e \in B - A \\ [\max (f_e^-(x), g_e^-(x), \max (f_e^+(x), g_e^+(x))] & e \in A \cap B, \end{cases}$$

*for all $x \in X$.*

4.  The intersection of two IVFS sets $f_A$ and $g_B$, denoted by $f_A \tilde{\wedge} g_B$, is the IVFS set $(f \wedge g, C)$, where $C = A \cap B$, and for all $e \in C$, we have $(f \wedge g)_e(x) = [\min f_e^-(x), g_e^-(x), \min f_e^+(x), g_e^+(x)]$ for all $x \in X$.
5.  The complement of the IVFS set $f_A$ is denoted by $f_A^c(x)$ where for all $e \in A$, we have $f_e^c(x) = [1 - f_e^+(x), 1 - f_e^-(x)]$.

**Definition 5** ([9]). *Let $f_E$ be an IVFS set. Then:*

1.  *$f_E$ is called the null interval-valued fuzzy soft set, denoted by $\varnothing_E$, if $f_e^-(x) = f_e^+(x) = 0$, for all $x \in X, e \in E$.*
2.  *$f_E$ is called the absolute interval-valued fuzzy soft set, denoted by $X_E$, if $f_e^-(x) = f_e^+(x) = 1$, for all $x \in X, e \in E$.*

Motivated by the definition of the soft mapping, discussed in [29], we define the concept of the *IVFS* mapping as the following:

**Definition 6.** *Let $f_A$ be an IVFS set over $X_1$ and $g_B$ be an IVFS set over $X_2$, where $A \subseteq E_1$ and $B \subseteq E_2$. Let $\Phi_u : X_1 \to X_2$ and $\Phi_p : E_1 \to E_2$ be two mappings. Then:*

1.  *The map $\Phi : \mathcal{IVFS}(X_1, E_1) \to \mathcal{IVFS}(X_2, E_2)$ is called an IVFS map from $X_1$ to $X_2$, and for any $y \in X_2$ and $\varepsilon \in B \subseteq E_2$, the lower image and the upper image of $f_A$ under $\Phi$ is the IVFS$\Phi(f_A)$ over $X_2$, respectively, defined as below:*

$$[\Phi(f^-)](\varepsilon)(y) = \begin{cases} \sup_{x \in \Phi_u^{-1}(y)} [\sup_{e \in \Phi_p^{-1} \cap A} f^-(e)](x), & \text{if } \Phi_p^{-1}(\varepsilon) \cap A \neq \phi \text{ and } \Phi_u^{-1}(y) \neq \phi \\ 0, & \text{otherwise,} \end{cases}$$

$$[\Phi(f^+)](\varepsilon)(y) = \begin{cases} \sup_{x \in \Phi_u^{-1}(y)} [\sup_{e \in \Phi_p^{-1} \cap A} f^+(e)](x), & \text{if } \Phi_p^{-1}(\varepsilon) \cap A \neq \phi \text{ and } \Phi_u^{-1}(y) \neq \phi \\ 0, & \text{otherwise.} \end{cases}$$

2.  *Let $\Phi : \mathcal{IVFS}(X_1, E_1) \to \mathcal{IVFS}(X_2, E_2)$ be an IVFS map from $X_1$ to $X_2$. The lower inverse image and the upper inverse image of IVFS $g_B$ under $\Phi$ denoted by $\Phi^{-1}(g_B)$ is an IVFS over $X_1$, respectively, such that for all $x \in X_1$ and $e \in E_1$, it is defined as below:*

$$[\Phi^{-1}(g^-)](e)(x) = \begin{cases} g_{\Phi_{p(e)}}^- \Phi_u(x), & \text{if } \Phi_p(e) \in B \\ 0, & \text{otherwise,} \end{cases}$$

$$[\Phi^{-1}(g^+)](e)(x) = \begin{cases} g_{\Phi_{p(e)}}^+ \Phi_u(x), & \text{if } \Phi_p(e) \in B \\ 0 & \text{otherwise.} \end{cases}$$

**Proposition 1.** *Let $\Phi : \mathcal{IVFS}(X, E) \to \mathcal{IVFS}(Y, F)$ be an IVFS mapping between X and X, and let $\{f_{iA}\}_{i \in J} \subset \mathcal{IVFS}(X, E)$ and $\{g_{iB}\}_{i \in J} \subset \mathcal{IVFS}(Y, F)$ be two families of IVFS sets over X and Y, respectively, where $A \subseteq E$ and $B \subseteq F$, then the following properties hold.*

1.  *$[\Phi(f_{jA})]^c \tilde{\leq} \Phi(f_{jA})^c$ for each $j \in J$.*
2.  *$[\Phi^{-1}(g_{jB})]^c = \Phi^{-1}(g_{jB})^c$ for each $j \in J$.*
3.  *If $g_{iB} \tilde{\leq} g_{jB}$, then $\Phi^{-1}(g_{iB}) \tilde{\leq} \Phi^{-1}(g_{jB})$ for each $i, j \in J$.*
4.  *If $f_{iA} \tilde{\leq} f_{jA}$, then $\Phi(f_{iA}) \tilde{\leq} \Phi(f_{jA})$ for each $i, j \in J$.*
5.  *$\Phi[\tilde{\vee}_{j \in J} f_{jA}] = \tilde{\vee}_{j \in J} \Phi(f_{jA})$ and $\Phi^{-1}[\tilde{\vee}_{j \in J} g_{jB}] = \tilde{\vee}_{j \in J} \Phi^{-1}(g_{jB})$.*
6.  *$\Phi[\tilde{\wedge}_{j \in J} f_{jA}] = \tilde{\wedge}_{j \in J} \Phi(f_{jA})$ and $\Phi^{-1}[\tilde{\wedge}_{j \in J} g_{jB}] = \tilde{\wedge}_{j \in J} \Phi^{-1}(g_{jB})$.*

**Proof.** We only prove Part (5). The other parts follow a similar technique. For any $k \in F, y \in Y$, and $a \in A$, then:

$$
\begin{aligned}
\Phi[\tilde{\vee}_{j \in J} f_{jA}](k)(y) &= \sup_{x \in \Phi_u^-(y)} \left( \sup_{z \in \Phi_p^{-1}(k)} (\tilde{\vee}_{j \in J}) f_{jA})(z)(x) \right. \\
&= \sup_{x \in \Phi_u^{-1}(y)} \left( \sup_{z \in \Phi_p^{-1}(k)} (\max_{j \in J}([f_{ja}^-, f_{ja}^+]))) (k)(y) \right. \\
&= \sup_{x \in \Phi_u^{-1}(y)} (\max_{j \in J}( \sup_{z \in \Phi_p^{-1}(k)} [f_{ja}^-(k), f_{ja}^+(k)]))) (y) \\
&= \max_{j \in J}( \sup_{x \in \Phi_u^{-1}(y)} ( \sup_{z \in \Phi_p^{-1}(k)} [f_{ja}^-(k))(y), f_{ja}^+(k)]))) (y)] \\
&= \max_{j \in J}( \sup_{x \in \Phi_u^-(y)} ( \sup_{z \in \Phi_p^{-1}(k)} f_{jA}(k)(y))) \\
&= \max_{j \in J} \Phi(f_{jA})(k)(y) \\
&= \tilde{\vee}_{j \in J} \Phi(f_{jA})(k)(y).
\end{aligned}
$$

Now, we prove that $\Phi^{-1}[\tilde{\vee}_{j \in J} g_{jB}] = \tilde{\vee}_{j \in J} \Phi^{-1}(g_{jB})$. For any $e \in E, x \in X$ and $b \in B$:

$$
\begin{aligned}
\Phi^{-1}[\tilde{\vee}_{j \in J} g_{jB}](e)(x) &= (\tilde{\vee}_{j \in J}) g_{jB}(\Phi_p(e))(\Phi_u(x)) \\
&= [\max_{j \in J} g_{jb}^-, \max_{j \in J} g_{jb}^+](\Phi_p(e))(\Phi_u(x)) \\
&= [[\max_{j \in J} g_{jb}^-(\Phi_p(e))(\Phi_u(x)), \max_{j \in J} g_{jb}^+(\Phi_p(e))(\Phi_u(x))] \\
&= [\max_{j \in J} \Phi_u^{-1}(g_{jb}^-)(e)(x), \max_{j \in J} \Phi_u^{-1}(g_{jb}^+)(e)(x)] \\
&= \max_{j \in J}[\Phi_u^{-1}(g_{jb}^-)(e)(x), \Phi_u^{-1}(g_{jb}^+)(e)(x)] \\
&= \max_{j \in J} \Phi_u^{-1}(g_{jB})(e)(x) \\
&= \tilde{\vee}_{j \in J} \Phi_u^{-1}(g_{jB})(e)(x).
\end{aligned}
$$

$\square$

### 3. Interval-Valued Fuzzy Soft Topological Spaces

The interval-valued fuzzy topology $IVFT$ was discussed by Mondal and Samanta [30]. In this section, we recall their definition and then present different neighborhood structures in the interval-valued fuzzy soft topology ($IVFST$).

**Definition 7.** *Let X be a non-empty set, and let $\tau$ be a collection of interval valued fuzzy soft sets over X with the following properties:*

*(i)* $\varnothing_E, X_E$ *belong to* $\tau$,
*(ii)* *If $f_{1E}, f_{2E}$ are IVFS sets belong to $\tau$, then $f_{1E} \tilde{\wedge} f_{2E}$ belong to $\tau$,*
*(iii)* *If the collection of IVFS sets $\{f_{jE} | j \in J\}$ where J is an index set, belonging to $\tau$, then $\tilde{\vee}_{j \in J} f_{jE}$ belong to $\tau$.*

*Then, $\tau$ is called the interval-valued fuzzy soft topology over X, and the triplet $(X, E, \tau)$ is called the interval-valued fuzzy soft topological space ($IVFST$).*

As the ordinary topologies, the indiscrete $IVFST$ over $X$ contains only $\varnothing_E$ and $X_E$, while the discrete $IVFST$ over $X$ contains all $IVFS$ sets. Every member of $\tau$ is called an interval-valued fuzzy soft open set ($IVFS$-open) in $X$. The complement of an $IVFS$-open set is called an $IVFS$-closed set.

**Remark 1.** *If $f_e^-(x) = f_e^+(x) = a \in [0, 1]$, then we put $[f_e^-(x), f_e^+(x)] = [a, a] = a$.*

**Example 1.** *Let $X = [0, 1]$ and $E$ be any subset of $X$. Consider the IVFS set $f_E$ over $X$ by the mapping:*

$$f : E \to \mathcal{IVF}([0, 1])$$

*such that for any $e \in E, x \in X$:*

$$\tilde{f}_e(x) = \begin{cases} 1 & 0 \leq x \leq e \\ 0 & e < x \leq 1. \end{cases}$$

*Then, the collection $\tau = \{\Phi_E, X_E, f_E\}$ is an IVFST over $X$.*

1. *Clearly $X_E, \varnothing_E \in \tau$.*
2. *Let $\{f_{jE}\}_{j \in J}$ be a sub-family of $\tau$ where for any $j \in J$ if $x \in X$ such that for all $e \in E$:*

$$f_{je}(x) = \begin{cases} 1 & 0 \leq x \leq e \\ 0 & e < x \leq 1. \end{cases}$$

   *Since:*

$$\vee_j f_{je}(x) = \begin{cases} 1 & 0 \leq x \leq e \\ 0 & e < x \leq 1, \end{cases}$$

   *then $\tilde{\vee}_j f_{jE} \in \tau$.*
3. *Let $f_E, g_E \in \tau$, where:*

$$f_e(x) = \begin{cases} 1 & 0 \leq x \leq e \\ 0 & e < x \leq 1, \end{cases}$$

   *and:*

$$g_e(x) = \begin{cases} 1 & 0 \leq x \leq e \\ 0 & e < x \leq 1. \end{cases}$$

   *Since:*

$$f_e(x) \wedge g_e(x) = \begin{cases} 1 & 0 \leq x \leq e \\ 0 & e < x \leq 1. \end{cases}$$

   *Thus, $f_E \wedge g_E \in \tau$.*

**Example 2** ([23])**.** *Let $\mathbb{R}$ be the set of all real numbers with the usual topology $\tau_u$ where $\tau_u = \langle \{(a, b), a, b \in \mathbb{R}\} \rangle$ and $E$ is a parameter set. Let $U = (a, b) \subset \mathbb{R}$ be an open interval in $\mathbb{R}$; we define IVFS $\tilde{U}_E$ over $\mathbb{R}$ by the mapping:*

$$\tilde{U} : E \to (Int[0, 1])^{\mathbb{R}}$$

*such that for all $x \in \mathbb{R}$:*

$$\tilde{U}_e(x) = \begin{cases} 1 & x \in (a, b) \\ 0 & x \notin (a, b). \end{cases}$$

*Then, the family $\{\tilde{U}_E : (a, b) \subset \mathbb{R}, \forall a, b \in \mathbb{R}\}$ generates an IVFS over $\mathbb{R}$, and we denote it by $\tau_u^{(IVFS)}$:*

1. *Clearly, $\mathbb{R}_E, \varnothing_E \in \tau_u^{(IVFS)}$ where for all $e \in E, k \in \mathbb{R}, \mathbb{R}_E(e)(k) = [1, 1],$ and $\varnothing_e(k) = 0$*
2. *Let $\{\tilde{U}_{jE}\}_{j \in J}$ be a sub-family of $\tau_u^{(IVFS)}$ where for any $j \in J$ if $x \in (a_j, b_j)$ and interval $(a_j, b_j)$ in $\mathbb{R}$ such that for all $e \in E$:*

$$\tilde{U}_{je}(x) = \begin{cases} 1 & x \in (a_j, b_j) \\ 0 & x \notin (a_j, b_j). \end{cases}$$

   *Since $\tilde{\vee}_j \tilde{U}_{jE} = (\widetilde{\cup_j U_j}, E)$ where $\cup_j U_{jE} \in \tau_u$, then $\tilde{\vee}_j \tilde{U}_{jE} \in \tau_u^{(IVFS)}$*
3. *Let $\tilde{U}_E, \tilde{V}_E \in \tau_u^{(IVFS)}$, then $\tilde{U}_E \tilde{\wedge} \tilde{V}_E \in \tau_u^{(IVFS)}$ since $\tilde{U}_E \tilde{\wedge} \tilde{V}_E = (\widetilde{U \cap V}, E)$ where $U \cap V \in \tau_u$.*

**Definition 8.** *Let interval* $[\lambda_e^-, \lambda_e^+] \subseteq [0,1]$ *for all* $e \in E$. *Then,* $\tilde{x}_E$ *is called an interval-valued fuzzy soft point (IVFS point) with support* $x \in X$ *and* $e$ *lower value* $\lambda_e^-$ *and* $e$ *upper value* $\lambda_e^+$, *if for each* $y \in X$:

$$\tilde{x}(e)(y) = \begin{cases} [\lambda_e^-, \lambda_e^+] & y = x \\ 0 & \text{otherwise.} \end{cases}$$

**Example 3.** *Let* $X = [0,1]$ *and* $E$ *be any subset of* $X$. *Consider IVFS point* $\tilde{x}_E$ *with support* $x$, *lower value zero, and upper value* 0.3, *we define IVFS point* $\tilde{x}_E$ *by:*

$$\tilde{x}(e)(c) = \begin{cases} [0,0.3] & c = x \\ 0 & \text{otherwise,} \end{cases}$$

*for any* $e \in E$ *and* $c \in X$.

**Definition 9.** *The IVFS point* $\tilde{x}_E$ *belongs to IVFS set* $f_E$, *denoted by* $\tilde{x}_E \tilde{\in} f_E$, *whenever for all* $e \in E$, *we have* $\lambda_e^- \leq f_e^-(x)$ *and* $\lambda_e^+ \leq f_e^+(x)$.

**Theorem 1.** *Let* $f_E$ *be an IVFS set, then* $f_E$ *is the union of all its IVFS points,*
   *i.e.,* $f_E = \tilde{\vee}_{\tilde{x}_E \tilde{\in} f_E} \tilde{x}_E$.

**Proof.** Let $x \in X$ be a fixed point, $y \in X$ and $e \in E$. Take all $\tilde{x}_E \tilde{\in} f_E$ with different $e$ lower and $e$ upper values $\lambda_{je}^-, \lambda_{je}^+$ where $j \in J$. Then, there exists $\lambda_{je}^- = f_e^-, \lambda_{je}^+ = f_e^+$ where:

$$\begin{aligned} \tilde{\vee}_{\tilde{x}_E \in f_E} \tilde{x}_e(y) &= [\sup \tilde{x}_e^-(y), \sup \tilde{x}_e^+(y)] \\ &= [\sup_{\lambda_{je}^- \tilde{\leq} f^-(x)} \lambda_{je}^-, \sup_{\lambda_{je}^+ \tilde{\leq} f^+(x)} \lambda_{je}^+] \\ &= [f_e^-(x), f_e^+(x)]. \end{aligned}$$

$\square$

**Proposition 2.** *Let* $\{f_{jE}\}_{j \in J}$ *be a family of IVFS sets over* $X$, *where* $J$ *is an index set and* $\tilde{x}_E$ *is an IVFS point with support* $x$, *e lower value* $\lambda_e^-$, *and* $e$ *upper value* $\lambda_e^+$. *If* $\tilde{x} \tilde{\in} \tilde{\wedge}_{j \in J} \{f_{jE}\}$, *then* $\tilde{x}_E \tilde{\in} \{f_{jE}\}$ *for each* $j \in J$.

**Proof.** Let $\tilde{x}_E$ be an *IVFS* point with support $x$, $e$ lower value $\lambda_e^-$, and $e$ upper value $\lambda_e^+$, and let $\tilde{x} \tilde{\in} \tilde{\wedge}_{j \in J} \{f_{jE}\}$. Then, $\lambda_e^- \leq \wedge_{j \in J} \{f_{je}^-\}(x) \leq \{f_{je}^-\}(x)$ for each $e \in E$, $x \in X$ and $\lambda_e^+ \leq \wedge_{j \in J} \{f_{je}^+\}(x) \leq \{f_{je}^+\}(x)$ for each $e \in E$, $x \in X$. Thus, $[\lambda_e^-, \lambda_e^+] \leq [\{f_{je}^-\}(x), \{f_{je}^+\}(x)]$, for each $e \in E$, $x \in X$. Hence, $\tilde{x}_E \tilde{\in} \{f_{jE}\}_{j \in J}$. $\square$

**Remark 2.** *If* $\tilde{x}_E \tilde{\in} f_E \tilde{\vee} g_E$ *does not imply* $\tilde{x}_E \tilde{\in} f_E$ *or* $\tilde{x}_E \tilde{\in} g_E$.

This is shown in the following example.

**Example 4.** *Let* $\tau$ *be an IVFST over* $X$, *where* $\tau = \{\varnothing_E, X_E, f_E, g_E, f_E \tilde{\wedge} g_E\}$, *and* $\tilde{x}_E$ *be the absolute IVFS point with support* $x$, *e lower value* $\lambda_e^-$, *and* $e$ *upper value* $\lambda_e^+$. *If* $f_E$*and* $g_E$ *are two IVFS sets in* $X$ *defined as below:*

$$f : E \to \mathcal{IVF}([0,1])$$

*and:*

$$g : E \to \mathcal{IVF}([0,1])$$

such that for any $e \in E, x \in X$:

$$f_e(x) = \begin{cases} [1, 0.5] & 0 \leq x \leq e \\ 0 & e < x \leq 1 \end{cases}$$

and:

$$g_e(x) = \begin{cases} [0.2, 1] & 0 \leq x \leq e \\ 0 & e < x \leq 1. \end{cases}$$

Since:

$$f_e(x) \vee g_e(x) = \begin{cases} 1 & if\ 0 \leq x \leq e \\ 0 & if\ e < x \leq 1, \end{cases}$$

then $\tilde{x}_E \tilde{\in} f_E \tilde{\vee} g_E$, but $\tilde{x}_E \tilde{\notin} f_E$ and $\tilde{x}_E \tilde{\notin} g_E$.

**Theorem 2.** *Let $\tilde{x}_E$ be an IVFS point with support $x$, $e$ lower value $\lambda_e^-$, and $e$ upper value $\lambda_e^+$ and $f_E$ and $g_E$ be IVFS sets. If $\tilde{x}_E \tilde{\in} f_E \tilde{\vee} g_E$, then there exists IVFS point $\tilde{x}_{1E} \tilde{\in} f_E$ and IVFS point $\tilde{x}_{2E} \tilde{\in} g_E$ such that $\tilde{x}_E = \tilde{x}_{1E} \tilde{\vee} \tilde{x}_{2E}$.*

**Proof.** Let $\tilde{x}_E \tilde{\in} f_E \tilde{\vee} g_E$. Then, $\lambda_e^- \leq f_e^-(x) \vee g_e^-(x)$ and $\lambda_e^+ \leq f_e^+(x) \vee g_e^+(x)$, for each $e \in E, x \in X$. Let us choose

$E_1 = \{e \in E | \lambda_e^- \leq f_e^-(x), \lambda_e^+ \leq f_e^+(x) : x \in X\}$,
$E_2 = \{e \in E | \lambda_e^- \leq g_E^-(x), \lambda_e^+ \leq g_E^+(x) : x \in X\}$

and:

$$\tilde{x}_1(e)(y) = \begin{cases} [\lambda_e^-, \lambda_e^+] & if\ y = x_1,\ e \in E_1 \\ 0, & otherwise, \end{cases}$$

$$\tilde{x}_2(e)(y) = \begin{cases} [\lambda_e^-, \lambda_e^+], & if\ y = x_2, e \in E_2 \\ 0, & otherwise. \end{cases}$$

Since $x_{1e}^- \leq f_{1e}^-(x)$ and $x_{1e}^+ \leq f_{1e}^+(x)$ for each $e \in E_1, x \in X$, that implies $\tilde{x}_{1E} \tilde{\in} f_{1E}$ and also $x_{2e}^- \leq f_{2e}^-(x)$, and $x_{2e}^+ \leq f_{2e}^+(x)$ for each $e \in E_2, x \in X$, that implies $\tilde{x}_{2E} \tilde{\in} f_{2E}$. Consequently, $E_1 \tilde{\vee} E_2 = E$ and $\tilde{x}_E = \tilde{x}_{1E} \tilde{\vee} \tilde{x}_{2E}$. $\square$

**Definition 10.** *Let $(X, E, \tau)$ be an IVFST space and $\tilde{x}_E$ be an IVFS point with support $x$, $e$ lower value $\lambda_e^-$, and $e$ upper value $\lambda_e^+$. The IVFS set $g_E$ is called the interval-valued fuzzy soft neighborhood (IVFSN) of IVFS point $\tilde{x}_E$, if there exists the IVFS-open set $f_E$ in $X$ such that $\tilde{x}_E \tilde{\in} f_E \tilde{\leq} g_E$. Therefore, the IVFS-open set $f_E$ is an IVFSN of the IVFS point $\tilde{x}_E$ if $\forall e \in E, x \in X$ such that $\lambda_e^- < f_e^-(x)$ and $\lambda_e^+ < f_e^+(x)$.*

**Definition 11.** *Let $(X, E, \tau)$ be an IVFST space and $\tilde{x}_E$ be an IVFS point with support $x$, $e$ lower value $\lambda_e^-$, and $e$ upper value $\lambda_e^+$ and $\tilde{x}_E^\star$ be an IVFS point with support $x^\star$, $e$ lower value $\varepsilon_e^-$, and $e$ upper value $\varepsilon_e^+$. $\tilde{x}_E^\star$ is said to be compatible with $\lambda_e^-, \lambda_e^+$, if $\tilde{x}_E^\star$ provides that $0 \leq \varepsilon_e^- \leq \lambda_e^-$ and $0 \leq \varepsilon_e^+ \leq \lambda_e^+$ for each $e \in E$.*

**Proposition 3.**

1. *If $f_E$ is an IVFSN of the IVFS point $\tilde{x}_E$ and $f_E \tilde{\leq} h_E$, then $h_E$ is also an IVFSN of $\tilde{x}_E$.*
2. *If $f_E$ and $g_E$ are two IVFSN of the IVFS point $\tilde{x}_E$, then $f_E \tilde{\wedge} g_E$ is also the IVFSN of $\tilde{x}_E$.*
3. *If $f_E$ is an IVFSN of the IVFS point $\tilde{x}_E^\star$ with support $x^\star$, $e$ lower value $\lambda_e^- - \varepsilon_e^-$, and $e$ upper value $\lambda_e^+ - \varepsilon_e^+$, for all $\varepsilon_e^-$ compatible with $\lambda_e^-$ and $\varepsilon_e^+$ compatible with $\lambda_e^+$, then $f_E$ is an IVFSN of the IVFS point $\tilde{x}_E$.*
4. *If $f_E$ is an IVFSN of the IVFS point $\tilde{x}_{1E}$ and $g_E$ is an IVFSN of the IVFS point $\tilde{x}_{2E}$, then $f_E \tilde{\vee} g_E$ is also an IVFSN of $\tilde{x}_{1E}$ and $\tilde{x}_{2E}$.*
5. *If $f_E$ is an IVFSN of the IVFS point $\tilde{x}_E$, then there exists IVFSN $g_E$ of $\tilde{x}_E$ such that $g_E \tilde{\leq} f_E$ and $g_E$ is IVFSN of IVFS point $\tilde{y}$ with support $y$, $e$ lower value $\gamma_e^-$, and $e$ upper value $\gamma_e^+$, for all $\tilde{y}_E \tilde{\in} g_E$.*

**Proof.**

1. Let $f_E$ be an $IVFSN$ of the $IVFS$ point $\tilde{x}$. Then, there exists the $IVFS$-open set $g_E$ in $X$ such that $\tilde{x}_E \tilde{\in} g_E \tilde{\leq} f_E$. Since $f_E \tilde{\leq} h_E$, $\tilde{x}_E \tilde{\in} g_E \tilde{\leq} f_E \tilde{\leq} h_E$. Thus, $h_E$ is an $IVFSN$ of $\tilde{x}_E$.

2. Let $f_E$ and $g_E$ be two $IVFSN$ of the $IVFS$ point $\tilde{x}_E$. Then, there exists two $IVFS$-open sets $h_E, k_E$ in $X$ such that $\tilde{x}_E \tilde{\in} h_E \tilde{\leq} f_E$ and $\tilde{x}_E \tilde{\in} k_E \tilde{\leq} g_E$. Thus, $\tilde{x}_E \tilde{\in} h_E \wedge k_E \tilde{\leq} f_E \wedge g_E$. Since $h_E \wedge k_E$ is an $IVFS$-open set, $g_E \wedge f_E$ is an $IVFSN$ of $\tilde{x}_E$.

3. Let $f_E$ be an $IVFSN$ of the $IVFS$ point $\tilde{x}_E^\star$ with support $x^\star$, $e$ lower value $\lambda_e^- - \varepsilon_e^-$, and $e$ upper value $\lambda_e^+ - \varepsilon_e^+$, for all $\varepsilon_e^-$ compatible with $\lambda_e^-$ and $\varepsilon_e^+$ compatible with $\lambda_e^+$. Then, there exists $IVFS$-open set $g_E^{x^\star}$ such that $\tilde{x}_E^\star \tilde{\in} g_E^{x^\star} \tilde{\leq} f_E$. Let $g_E = \check{\vee}_{x^\star} g_E^{x^\star}$, then $g_E$ is $IVFS$-open in $X$ and $g_E \tilde{\leq} f_E$. By Theorem 1 and since for all $e \in E$, $\check{\vee} \tilde{x}_E^\star = \tilde{x}_E \tilde{\leq} \check{\vee}_{x^\star} g_E^{x^\star} = g_E \tilde{\leq} f_E$. Hence, $\tilde{x}_E \tilde{\in} g_E \tilde{\leq} f_E$, i.e., $f_E$ is an $IVFSN$ of $\tilde{x}_E$.

4. Let $f_E$ be an $IVFSN$ of the $IVFS$ point $\tilde{x}_{1E}$ with support $x_1$, $e$ lower value $\lambda_{1e}^-$, and $e$ upper value $\lambda_{1e}^+$ and $g_E$ be an $IVFSN$ of the $IVFS$ point $\tilde{x}_{2E}$ with support $x_2$, $e$ lower value $\lambda_{2e}^-$, and $e$ upper value $\lambda_{2e}^+$. Then, there exists $IVFS$-open sets $h_{1E}, h_{2E}$ such that $\tilde{x}_{1E} \tilde{\in} h_{1E} \tilde{\leq} f_E$ and $\tilde{x}_{2E} \tilde{\in} h_{2E} \tilde{\leq} f_E$, respectively. Since $\tilde{x}_{1E} \tilde{\in} h_{1E}$, $\lambda_{1e}^- \leq h_{1e}^-(x), \lambda_{1e}^+ \leq h_{1e}^+(x)$ for each $e \in E$ and $x \in X$. Since $\tilde{x}_{2E} \tilde{\in} h_{2E}$, $\lambda_{2e}^- \leq h_{2e}^-(x), \lambda_{2e}^+ \leq h_{2e}^+(x)$ for each $e \in E$ and $x \in X$. Thus, we have:

$$\max\{[\lambda_{1e}^-, \lambda_{1e}^+], [\lambda_{2e}^-, \lambda_{2e}^+]\} \leq \max\{[h_{1e}^-(x), h_{1e}^+(x)], [h_{2e}^-(x), h_{2e}^+(x)]\}$$

for each $e \in E$, $x \in X$. Therefore, $\tilde{x}_{1E} \check{\vee} \tilde{x}_{2E} \tilde{\in} h_{1E} \check{\vee} h_{2E}$, $h_{1E} \check{\vee} h_{2E} \in \tau$, and $h_{1E} \check{\vee} h_{2E} \tilde{\leq} f_E \check{\vee} g_E$. Consequently, $f_E \check{\vee} g_E$ is an $IVFSN$ of $x_{1E} \check{\vee} x_{2E}$.

5. Let $f_E$ be an $IVFSN$ of the $IVFS$ point $\tilde{x}_E$, with support $x$, $e$ lower value $\lambda_e^-$, and $e$ upper value $\lambda_e^+$. Then, there exists $IVFS$-open set $g_E$ such that $\tilde{x}_E \tilde{\in} g_E \tilde{\leq} f_E$. Since $g_E$ is an $IVFS$-open set, $g_E$ is a neighborhood of its points, i.e., $g_E$ is an $IVFSN$ of $IVFS$ point $\tilde{y}_E$ with support $y$, $e$ lower value $\gamma_e^-$, and $e$ upper value $\gamma_e^+$, for all $e \in E$. Furthermore, $g_E$ is an $IVFSN$ of $IVFS$ point $\tilde{x}_E$ since $\tilde{x}_E \tilde{\in} g_E$. Therefore, there exists $g_E$ that is an $IVFSN$ of $\tilde{x}_E$ such that $g_E \tilde{\leq} f_E$ and $g_E$ is an $IVFSN$ of $\tilde{y}_E$; since $f_E$ is an $IVFSN$ of $\tilde{x}_E$.

$\square$

**Definition 12.** *Let $(X, E, \tau)$ be an $IVFST$ space and $f_E$ be an $IVFS$ set. The $IVFS$-closure of $f_E$ denoted by $Cl f_E$ is the intersection of all $IVFS$-closed super sets of $f_E$. Clearly, $Cl f_E$ is the smallest $IVFS$-closed set over $X$ that contains $f_E$.*

**Example 5** ([23]). *Consider $IVFST$ $\tau_u^{IVFS}$ over $\mathbb{R}$ as introduced in Example 2, and if $\tilde{H}_E$ is an $IVFS$ over $\mathbb{R}$ related of the open interval $H = (a, b) \subset \mathbb{R}$ by mapping:*

$$\tilde{H} : E \to (Int[0, 1])^{\mathbb{R}}$$

$$\tilde{H}_e(x) = \begin{cases} 1 & x \in (a, b) \\ 0 & x \notin (a, b), \end{cases}$$

*where $e \in E$ and $x \in \mathbb{R}$, then the closure of $\tilde{H}_E$ is defined as:*

$$Cl\tilde{H} : E \to (Int[0, 1])^{\mathbb{R}}$$

$$\tilde{H}_e(x) = \begin{cases} 1 & x \in [a, b] \\ 0 & x \notin [a, b]. \end{cases}$$

**Remark 3.** *By replacing $\tilde{x}_E$ for $f_E$, the $IVFS$-closure of $\tilde{x}_E$ denoted by $Cl\tilde{x}_E$ is the intersection of all $IVFS$-closed super sets of $\tilde{x}_E$.*

**Proposition 4.** *Let $(X, E, \tau)$ be an IVFST space and $f_E$ and $g_E$ be two IVFSS over X. Then:*

1. $Cl\varnothing_E = \varnothing_E$ and $Cl\tilde{X}_E = \tilde{X}_E$,
2. $f_E \tilde{\leq} Clf_E$, and $Clf_E$ is the smallest IVFS-closed set containing IVFS$f_E$,
3. $Cl(Clf_E) = Clf_E$,
4. if $f_E \tilde{\leq} g_E$, then $(Clf_E) \tilde{\leq} Clg_E$.
5. $f_E$ is an IVFS-closed set if and only if $f_E = Clf_E$,
6. $Cl(f_E \tilde{\vee} g_E) = Clf_E \tilde{\vee} Clg_E$,
7. $Cl(f_E \tilde{\wedge} g_E) \tilde{\leq} Clf_E \tilde{\wedge} Clg_E$.

**Proof.** We only prove Part (6). A similar technique is used to show the other parts.

Since $f_E \tilde{\leq} f_E \tilde{\vee} g_E$ and $g_E \tilde{\leq} f_E \tilde{\vee} g_E$, by Part (4), we have $Clf_E \tilde{\leq} Cl(f_E \tilde{\vee} g_E)$ and $Clg \tilde{\leq} Cl(f_E \tilde{\vee} g_E)$. Then, $Clf_E \tilde{\vee} Clg_E \tilde{\leq} Cl(f_E \tilde{\vee} g_E)$.

Conversely, we have $f_E \tilde{\leq} Clf_E$ and $g_E \tilde{\leq} Clg_E$, by Part (2). Then, $f_E \tilde{\vee} g_E \tilde{\leq} Clf_E \tilde{\vee} Clg_E$ where $Clf_E \tilde{\vee} Clg_E$ is an IVFS-closed set. Thus, $Cl(f_E \tilde{\vee} g_E) \tilde{\leq} Clf_E \tilde{\vee} Clg_E$.

Therefore, $Cl(f_E \tilde{\vee} g_E) = Clf_E \tilde{\vee} Clg_E$. $\square$

**Definition 13.** *Let $(X_1, E_1, \tau_1)$ and $(X_2, E_2, \tau_2)$ be two IVFSTS and:*

$$\Phi : (X_1, E_1, \tau_1) \to (X_2, E_2, \tau_2)$$

*be an IVFS map. Then, $\Phi$ is called an:*

1. *interval-valued fuzzy soft continuous (IVFSC) map if and only if for each $g_{E_2} \in \tau_2$, we have $\Phi^{-1}(g_{E_2}) \in \tau_1$,*
2. *interval-valued fuzzy soft open (IVFSO) map if and only if for each $f_E \in \tau_1$, we have $\Phi(f_{E_1}) \in \tau_2$.*

**Theorem 3.** *Let $(X_1, E_1, \tau_1)$ and $(X_2, E_2, \tau_2)$ be two IVFST and $\Phi$ be an IVFS mapping from $X_1$ to $X_2$, then the following statements are equivalent:*

1. *$\Phi$ is IVFC,*
2. *For each IVFS point $\tilde{x}_E$ on $X_1$, the inverse of every neighborhood of $\Phi(\tilde{x}_E)$ under $\Phi$ is a neighborhood of $\tilde{x}_E$,*
3. *For each IVFS point $\tilde{x}_E$ on $X_1$ and each neighborhood $g_E$ of $\Phi(\tilde{x}_E)$, there exists a neighborhood $f_E$ of $\tilde{x}_E$ such that $\Phi(f_E) \tilde{\leq} g_E$.*

**Proof.**

(1) $\Rightarrow$ (2) Let $g_E$ be an IVFSN of $\Phi(\tilde{x}_E)$ in $\tau_2$. Then, there exists an IVFS-open set $f_E$ in $\tau_2$ such that $\Phi(\tilde{x}_E) \tilde{\in} f_E \tilde{\leq} g_E$. Since $\Phi$ is IVFSC, $\Phi^{-1}(f_E)$ is an IVFS-open in $\tau_1$, and we have $\tilde{x}_E \tilde{\in} \Phi^{-1}(f_E) \tilde{\leq} \Phi^{-1}(g_E)$.

(2) $\Rightarrow$ (3) Let $g_E$ be an IVFSN of $\Phi(\tilde{x}_E)$. By the hypothesis, $\Phi^{-1}(g_E)$ is an IVFSN of $\tilde{x}_E$. Consider $f_E = \Phi^{-1}(g_E)$ to be an IVFSN of $\tilde{x}_E$. Then, we have $\Phi(f_E) = \Phi(\Phi^{-1}(g_E)) \tilde{\leq} g_E$.

(3) $\Rightarrow$ (1) Let $g_E$ be an IVFS-open set in $\tau_2$. We must show that $\Phi^{-1}(g_E)$ is an IVFS-open set in $\tau_1$. Now, let $\tilde{x}_E \tilde{\in} \Phi^{-1}(g_E)$. Then, $\Phi(\tilde{x}_E) \tilde{\in} g_E$. Since $g_E$ is an IVFS-open set in $\tau_2$, we get that $g_E$ is an IVFSN $\Phi(\tilde{x}_E)$ in $\tau_2$. By the hypothesis, there exists IVFS-open set $f_E$ that is an IVFSN of $\tilde{x}_E$ such that $\Phi(f_E) \tilde{\leq} g_E$. Thus, $f_E \tilde{\leq} \Phi^{-1}[\Phi(f_E)] \tilde{\leq} \Phi^{-1}(g_E)$ for $f_E$ is an IVFSN of $\tilde{x}_E$. From here, $f_E \tilde{\leq} \Phi^{-1}(g_E)$, as $f_E$ is an IVFSN of $\tilde{x}_E$. Hence, $\Phi^{-1}(g_E) \tilde{\in} \tau_1$. $\square$

## 4. Quasi-Coincident Neighborhood Structure of Interval-Valued Fuzzy Soft Topological Spaces

In this section, we present the quasi-coincident neighborhood structure in the interval-valued fuzzy soft topology $(IVFST)$ and its properties.

**Definition 14.** *The IVFS point $\tilde{x}_E$ is called soft quasi-coincident with IVFS $f_E$, denoted by $\tilde{x}_E\tilde{q}f_E$, if there exists $e \in E$ such that $\lambda_e^- + f_e^-(x) > 1$ and $\lambda_e^+ + f_e^+(x) > 1$. If $f_E$ is not soft quasi-coincident with $f_E$, we write $f_E\neg\tilde{q}g_E$.*

**Definition 15.** *The IVFS set $f_E$ is called soft quasi-coincident with IVFS $g_E$, denoted by $f_E\tilde{q}g_E$, if there exists $e \in E$ such that $f_e^-(x) + g_e^-(x) > 1$ and $f_e^+(x) + g_e^+(x) > 1$.*

**Proposition 5.** *Let $\tilde{x}_E$ be an IVFS point with support $x$, $e$ lower value $\lambda_e^-$, and $e$ upper value $\lambda_e^+$ and $f_E, g_E$ two IVFS sets. Then:*

(i) $f_E\tilde{\leq}g_E \Leftrightarrow f_E\neg\tilde{q}g_E^c$,
(ii) $\tilde{x}_E\tilde{\in}f_E \Leftrightarrow \tilde{x}_E\neg\tilde{q}f_E^c$.

**Proof.** We just prove Part (i). A similar technique is used to show Part (ii). For two *IVFS* sets $f_E, g_E$, we have:

$$
\begin{aligned}
f_E\tilde{\leq}g_E \quad &\Leftrightarrow \quad \forall e \in E : [f_e^-(x), f_e^+(x)] \leq [g_e^-(x), g_e^+(x)], \forall x \in X \\
&\Leftrightarrow \quad \forall e \in E : f_e^-(x) \leq g_e^-(x) \text{ and } f_e^+(x) \leq g_e^+(x), \forall x \in X \\
&\Leftrightarrow \quad \forall e \in E : f_e^-(x) + 1 - g_e^-(x) \leq 1 \text{ and } f_e^+(x) + 1 - g_e^+(x) \leq 1, \forall x \in X \\
&\Leftrightarrow \quad \forall e \in E : f_e^-(x) + g^{-c}_e(x) \leq 1 \text{ and } f_e^+(x) + g^{+c}_e(x) \leq 1, \forall x \in X \\
&\Leftrightarrow \quad f_E\neg\tilde{q}g_E^c.
\end{aligned}
$$

$\square$

**Proposition 6.** *Let $\{f_{jE} : j \in J\}$ be a family of IVFS sets over $X$ and $\tilde{x}_E$ be an IVFS point with support $x$, $e$ lower value $\lambda_e^-$, and $e$ upper value $\lambda_e^+$. If $\tilde{x}_E\tilde{q}(\wedge f_{jE})$, then $\tilde{x}_E\tilde{q}f_{jE}$ for each $j \in J$.*

**Proof.** Let $\tilde{x}_E\tilde{q}(\wedge f_{jE})$. Then, $\lambda_e^-\tilde{q}(\wedge_j f_{je}^-)(x)$, $\lambda_e^+\tilde{q}(\wedge_j f_{je}^+)(x)$ for $e \in E$, and $x \in X$. This implies that $\lambda_e^- > 1 - \wedge_j(f_{je}^-)(x)$ and $\lambda_e^+ > 1 - \wedge_j(f_{je}^+)(x)$, $x \in X$. Since $\wedge_j f_{je}^-(x) \leq f_{je}^-(x)$ and $\wedge_j f_{je}^+(x) \leq f_{je}^+(x)$, then $\lambda_e^- > 1 - \wedge_j(f_{je}^-)(x) > 1 - f_{je}^-(x)$ for each $e \in E, x \in X$ and $\lambda_e^+ > 1 - \wedge_j(f_{je}^+)(x) > 1 - f_{je}^+(x)$ for each $e \in E, x \in X$. Hence, $\lambda_e^- > 1 - f_{je}^-(x)$ and $\lambda_e^+ > 1 - f_{je}^+(x)$. Therefore, $[\lambda_e^-, \lambda_e^+] > [1, 1] - [f_{je}^-(x), f_{je}^+(x)]$ implies that $\tilde{x}_E > 1 - f_{jE}$ and $\tilde{x}_E\tilde{q}f_{jE}$ for each $j \in J$. $\square$

**Remark 4.** *$\tilde{x}_E\tilde{q}(f_E \vee g_E)$ does not imply $\tilde{x}_E\tilde{q}f_E$ or $\tilde{x}_E\tilde{q}g_E$. This is shown in the following example.*

**Example 6.** *Let us consider Example 4 where $\tilde{x}_E\tilde{q}(f_E\tilde{\vee}g_E)$, but $\tilde{x}_E\neg\tilde{q}f_E$ and $\tilde{x}_E\neg\tilde{q}g_E$.*

**Theorem 4.** *Let $\tilde{x}_E$ be an IVFS point $\tilde{x}_E$ with support $x$, $e$ lower value $\lambda_e^-$, and $e$ upper value $\lambda_e^+$ and $f_E, g_E$ be IVFS sets over $X$. If $\tilde{x}_E\tilde{q}(f_E \vee g_E)$, then there exists $\tilde{x}_{1E}\tilde{q}f_E$ and $\tilde{x}_{2E}\tilde{q}g_E$ such that $\tilde{x}_E = \tilde{x}_{1E}\tilde{\vee}\tilde{x}_{2E}$.*

The proof is very similar to the proof of Theorem 2.

**Definition 16.** *Let $(X, E, \tau)$ be an IVFSTS and $\tilde{x}_E$ be an IVFS point with support $x$, $e$ lower values $\lambda_e^-$, and $e$ upper values $\lambda_e^+$. The IVFS set $g_E$ is called a quasi-soft neighborhood (QIVFSN) of IVFS point $\tilde{x}_E$ if there exists the IVFS-open set $f_E$ in $X$ such that $\tilde{x}_E\tilde{q}f_E\tilde{\leq}g_E$. Thus, the IVFS-open set $f_E$ is a QIVFSN of the IVFS point $\tilde{x}_E$ if and only if $\exists e \in E, x \in X$ such that $\lambda_e^- + f_e^-(x) > 1$ and $\lambda_e^+ + f_e^+(x) > 1$.*

**Remark 5.** *A quasi-coincident soft neighborhood of an IVFS point generally does not contain the point itself. This is shown by the following:*

**Example 7.** *Let $X = [0, 1]$ and $E$ be any subset of $X$. Consider two IVFS sets $f_E, g_E$ over $X$ by the mapping $f : E \to \mathcal{IVF}([0,1])$ and $f : E \to \mathcal{IVF}([0,1])$ such that for any $e \in E, x \in X$:*

$$\tilde{f}_e(x) = \begin{cases} [0.4, 0.5] & 0 \le x \le e \\ 0 & e < x \le 1, \end{cases}$$

*and:*

$$\tilde{g}_e(x) = \begin{cases} [0.6, 0.7] & 0 \le x \le e \\ 0 & e < x \le 1, \end{cases}$$

*and $\tilde{x}_E$ be any IVFS point defined by:*

$$\tilde{x}_e(c) = \begin{cases} [0.4, 0.5] & c = x \\ 0 & c \ne x. \end{cases}$$

*Let $\tau = \{\varnothing_E, X_E, f_E, g_E\}$. Then clearly, $\tau$ is an IVFST over $X$. Since $f_E \tilde{\le} g_E$ and $\tilde{x}\tilde{q}f_E$, thus $g_E$ is a QIVFSN of $\tilde{x}_E$. However, $\tilde{x}_E \notin g_E$.*

**Proposition 7.**

(1) *If $f_E \tilde{\le} g_E$ and $f_E$ is a QINVSN of $\tilde{x}_E$, then $g_E$ is also a QINVSN of $\tilde{x}_E$,*

(2) *If $f_E, g_E$ are QINVSN of $\tilde{x}_E$, then $f_E \tilde{\wedge} g_E$ is also a QINVSN of $\tilde{x}_E$.*

(3) *If $f_E$ is a QINVSN of $\tilde{x}_{1E}$ and $g_E$ is a QINVSN of $\tilde{x}_{2E}$, then $f_E \tilde{\vee} g_E$ is also a QINVSN of $\tilde{x}_{1E} \tilde{\vee} \tilde{x}_{2E}$.*

(4) *If $f_E$ is a QINVSN of $\tilde{x}_E$, then there exists $g_E$ that is a QINVSN of $\tilde{x}_E$, such that $g_E \tilde{\le} f_E$, and $g_E$ is a QINVSN of $y_E$, $\forall y_E \tilde{q} g_E$.*

**Proof.** (1) and (2) are straightforward.

(3) Let $f_E$ be a $QINVSN$ of $\tilde{x}_{1E}$ and $g_E$ be a $QINVSN$ of $\tilde{x}_{2E}$. Then, there exists an $IVFS$-open set $h_{1E}$ in $X$ such that $\tilde{x}_{1E} \tilde{q} h_{1E} \tilde{\le} f_E$ and $g_E$ is a $QINVSN$ of $\tilde{x}_{2E}$. Thus, there exists an $IVFS$-open set $h_{2E}$ in $X$ such that $\tilde{x}_{2E} \tilde{q} h_{2E} \tilde{\le} g_E$. Since $\tilde{x}_{1E} \tilde{q} h_{1E}$ for each $e \in E, x \in X, \lambda_{1e}^- + h_{1e}^- > 1, \lambda_{1e}^+ + h_{1e}^+ > 1$, this implies that $\lambda_{1e}^- > 1 - h_{1e}^-, \lambda_{1e}^+ > 1 - h_{1e}^+$ for each $e \in E$. Since $\tilde{x}_{2E} \tilde{q} h_{2E}$, for each $e \in E, \lambda_{2e}^- + h_{2e}^- > 1$, $\lambda_{2e}^+ + h_{2e}^+ > 1$, this implies that $\lambda_{2e}^- > 1 - h_{2e}^-, \lambda_{2e}^+ > 1 - h_{2e}^+$ for each $e \in E, x \in X$. From here,

$$\max(\lambda_{1e}^-, \lambda_{2e}^-) > \max(1 - h_{1e}^-(x)), (1 - h_{2e}^-(x)), \max(\lambda_{1e}^+, \lambda_{2e}^+) > \max(1 - h_{1e}^+(x)), (1 - h_{2e}^+(x)).$$

Therefore, $\tilde{x}_{1E} \tilde{\vee} \tilde{x}_{2E} \tilde{q}(h_{1E} \tilde{\vee} h_{2E}) \tilde{\le} f_E \tilde{\vee} g_E$. Consequently, $f_E \tilde{\vee} g_E$ is a $QINVSN$ of $\tilde{x}_{1E} \tilde{\vee} \tilde{x}_{2E}$.

(4) Let $f_E$ be a $QINVSN$ of $\tilde{x}_E$. Then, there exists $g_E$ that is a $QINVSN$ of $\tilde{x}_E$ such that $\tilde{x}_E \tilde{q} g_E \tilde{\le} f_E$. Consider the $g_E = h_E$. Indeed, since $\tilde{x}_E \tilde{q} h_E$ and $h_E$ is an $IVFS$-open set, $h_E$ is a $QINVSN$ of $\tilde{x}_E$. Thus, we obtain $h_E$ that is a $QINVSN$ of $\tilde{y}_E$.

$\square$

**Theorem 5.** *In $IVFST(X, E, \tau)$, the IVFS point $\tilde{x}_E$ belongs to $Clf_E$ if and only if each QIVFS of $\tilde{x}_E$ is soft quasi-coincident with $f_E$.*

**Proof.** Let $IVFS$ point $\tilde{x}_E$ with support $x$, e lower value $\lambda_e^-$, and e upper value $\lambda_e^+$ belong to $Clf_E, i.e, \tilde{x}_E \tilde{\in} Clf_E$. For any $IVFS$-closed $g_E$ containing $f_E$, $\tilde{x}_E \tilde{\in} g_E$, which implies that $\lambda_e^- \le g_e^-(x)$ and $\lambda_e^+ \le g_e^+(x)$, for all $x \in X, e \in E$. Consider $h_E$ to be an $QIVFN$ of the $IVFS$ point $\tilde{x}_E$ and $h_E \neg \tilde{q} f_E$. Then, for any $e \in E$ and $x \in X, h_e^-(x) + f_e^-(x) \le 1, h_e^+(x) + f_e^+(x) \le 1$, and so, $f_E \tilde{\le} h_E^c$. Since $h_E$ is a $QIVFSN$ of the $IVFS$ point $\tilde{x}_E$, by $\tilde{x}_E$, it does not belong to $h_E^c$. Therefore, we have that $\tilde{x}_E$ does not belong to $Clf_E$. This is a contradiction.

Conversely, let any $QIVFSN$ of the $IVFS$ point $\tilde{x}_E$ be soft quasi-coincident with $f_E$. Consider that $\tilde{x}_E$ doe not belong to $Clf_E$, *i.e*, $\tilde{x}_E \notin Clf_E$. Then, there exists an $IVFS$-closed set $g_E$, which contains $f_E$ such that $\tilde{x}_E$ does not belong to $g_E$. We have $\tilde{x}_E \tilde{q} g_E^c$. Then, $g_E^c$ is an $QIVFSN$ of the $IVFS$ point $\tilde{x}_E$ and $f_E \neg \tilde{q} g_E^c$. This is a contradiction with the hypothesis. $\square$

## 5. IVFS Quasi-Separation Axioms

In this section, we develop the separation axioms to $IVFST$, so-called $IVFSQ$ separation axioms ($IVFSq$-$T_i$ axioms) for $i = 0, 1, 2, 3, 4$, and consider some of their properties.

**Definition 17.** *Let $(X, E, \tau)$ be an IVFST space. Let $\tilde{x}_E$ and $\tilde{y}_E$ be IVFS points over $X$, where:*

$$\tilde{x}(e)(z) = \begin{cases} [\lambda_e^-, \lambda_e^+] & z = x \\ 0 & \text{otherwise} \end{cases}$$

*and:*

$$\tilde{y}(e)(z) = \begin{cases} [\gamma_e^-, \gamma_e^+] & z = y \\ 0 & \text{otherwise,} \end{cases}$$

*then $\tilde{x}_E$ and $\tilde{y}_E$ are said to be distinct if and only if $\tilde{x}_E \wedge \tilde{y}_E = \varnothing_E$, which means $x \neq y$.*

**Definition 18.** *Let $(X, E, \tau)$ be an IVFST space. The IVFS point $\tilde{x}_E$ is called a crisp IVFS point $x_E^{[1,1]}$, if $\lambda_e^- = \lambda_e^+ = 1$ for all $e \in E$.*

**Definition 19.** *Let $(X, E, \tau)$ be an IVFST space and $\tilde{x}_E$ and $\tilde{y}_E$ be two IVFS points. If there exists IVFS open sets $f_E$ and $g_E$ such that:*

(a) *when $\tilde{x}_E$ and $\tilde{y}_E$ are two distinct IVFS points with different supports $x$ and $y$, $e$ lower values, and $e$ upper values $\lambda_e^-, \lambda_e^+$ and $\gamma_e^-, \gamma_e^+$, respectively, and $f_E$ is an IVFSN of the IVFS point $\tilde{x}_E$ and $\tilde{y}_E \neg \tilde{q} f_E$ or $g_E$ is an IVFSN of the IVFS point $\tilde{y}_E$ and $\tilde{x}_E \neg \tilde{q} g_E$,*

(b) *when $\tilde{x}_E$ and $\tilde{y}_E$ are two IVFS points with the same supports $x = y$, $e$ value $\lambda_e^- < \gamma_e^-$, and $e$ value $\lambda_e^+ < \gamma_e^+$ and $f_E$ is a QIVFSN of the IVFS point $\tilde{y}_E$ such that $\tilde{x}_E \neg \tilde{q} f_E$,*

*then $(X, E, \tau)$ is an interval-valued fuzzy soft quasi-$T_0$ space (IVFSq-$T_0$ space).*

**Example 8.** *Consider the IVFS set defined in Example 3.1 and $\tilde{x}_E, \tilde{y}_E$ to be any two distinct IVFS points in $X$ defined by:*

$$\tilde{x}(e)(z) = \begin{cases} 1 & z = x \\ 0 & z \neq x \end{cases}$$

*and:*

$$\tilde{y}(e)(z) = \begin{cases} 0 & \text{if } z = y \\ 1 & \text{if } z \neq y. \end{cases}$$

*Then, $f_E$ is an IVFSN of $\tilde{x}_E$ and $\tilde{y}_E \neg \tilde{q} f_E$. Thus, $X$ is an IVFSq-$T_0$ space.*

**Theorem 6.** *$(X, E, \tau)$ is an IVFSq-$T_0$ space if and only if for every two IVFS points $\tilde{x}_E$, $\tilde{y}_E$ and $\tilde{x}_E \notin Cl\tilde{y}_E$ or $\tilde{y}_E \notin Cl\tilde{x}_E$.*

**Proof.** Let $(X, E, \tau)$ be an $IVFSq$-$T_0$ space and $\tilde{x}_E$ and $\tilde{y}_E$ be two $IVFS$ points in $X$.

First consider that $\tilde{x}_E$ and $\tilde{y}_E$ are two distinct $IVFS$ points with different supports $x$ and $y$, $e$ lower values, and $e$ upper values $\lambda_e^-, \gamma_e^-$ and $\lambda_e^+, \gamma_e^+$, respectively. Then, a crisp $IVFS$ point $\tilde{x}_E^{[1,1]}$ has an $IVFSN$ $f_E$ such that $\tilde{y}_E \neg \tilde{q} f_E$ or a crisp $IVFS$ point $\tilde{y}_E^{[1,1]}$ has an $IVFSN$ $g_E$ such that $\tilde{x}_E \neg \tilde{q} f_E$. Consider that the crisp $IVFS$ point $\tilde{x}_E^{[1,1]}$ has an $IVFSN$ $f_E$ such that $\tilde{y}_E \neg \tilde{q} f_E$. Moreover, $f_E$ is an $QINFSN$ of $\tilde{x}_E$

and $\tilde{y}_E \neg \tilde{q} f_E$. Hence, $\tilde{x}_E \notin Cl\tilde{y}_E$. Next, we consider the case $\tilde{x}_E$ and $\tilde{y}_E$ to be two *IVFS* points with the same supports $x = y$, $e$ lower value $\lambda_e^- < \gamma_e^-$, and $e$ upper value $\lambda_e^+ < \gamma_e^+$. Then, $\tilde{y}_E$ has a *QIVFSN* that is not quasi-coincident with $\tilde{x}_E$, and so, by Theorem 5, $\tilde{x}_E \notin Cl\tilde{y}_E$.

Conversely, let $\tilde{x}_E$ and $\tilde{y}_E$ be two *IVFS* points in $X$. Consider without loss of generality that $\tilde{x}_E \notin Cl\tilde{y}_E$. First, consider that $\tilde{x}_E$ and $\tilde{y}_E$ are two distinct *IVFS* points with different supports $x$ and $y$, $e$ lower values, and $e$ upper values $\lambda_e^-, \gamma_e^-$ and $\lambda_e^+, \gamma_e^-$, respectively, since $\tilde{x}_E \notin Cl\tilde{y}_E$ for any $e \in E$, $f_e^-(y) = f_e^+(y) = 0$ and $f_e^-(x) = f_e^+(x) = 1$. Then, $Cl(\tilde{y}_E)^c$ is an *IVFSN* of $\tilde{x}_E$ such that $Cl(\tilde{y}_E)^c \neg \tilde{q}\tilde{y}_E$. Next, let $\tilde{x}_E$ and $\tilde{y}_E$ be two *IVFS* points with the same supports $x = y$, and we must have $e$ lower value $\lambda_e^- > \gamma_e^-$ and $e$ upper value $\lambda_e^+ > \gamma_e^+$, then $\tilde{x}_E$ has a *QIVFSN* that is not quasi-coincident with $\tilde{y}_E$. $\square$

**Definition 20.** *Let* $(X, E, \tau)$ *be an IVFST and* $\tilde{x}_E$ *and* $\tilde{y}_E$ *be two IVFS points, if there exists IVFS open sets* $f_E$ *and* $g_E$ *such that:*

(a) *when* $\tilde{x}_E$ *and* $\tilde{y}_E$ *are two distinct IVFS points with different supports* $x$ *and* $y$, *e lower values, and e upper values* $\lambda_e^-, \gamma_e^-$ *and* $\lambda_e^+, \gamma_e^+$, *respectively,* $f_E$ *is an IVFSN of IVFS points* $\tilde{x}_E$ *and* $\tilde{y}_E \neg \tilde{q} f_E$, *and* $g_E$ *is an IVFSN of IVFS points* $\tilde{y}_E$ *and* $\tilde{x}_E \neg \tilde{q} g_E$,

(b) *when* $\tilde{x}_E$ *and* $\tilde{y}_E$ *are two IVFS points with the same supports* $x = y$, *e value* $\lambda_e^- < \gamma_e^-$, *and e value* $\lambda_e^+ < \gamma_e^+$, $f_E$ *is an QIVFSN of the IVFS point* $\tilde{y}_E$ *such that* $\tilde{x}_E \neg \tilde{q} f_E$,

*then* $(X, E, \tau)$ *is an interval-valued fuzzy soft quasi-$T_1$ space (IVFSq-$T_1$ space).*

**Theorem 7.** $(X, E, \tau)$ *is an IVFSq-$T_1$ space if and only if any IVFS point* $\tilde{x}_E$ *in X is an IVFS-closed set.*

**Proof.** Suppose that each *IVFS* point $\tilde{x}_E$ in $X$ is an *IVFS*-closed set, i.e., $g_E = \tilde{x}_E^c$. Then, $g_E$ is an *IVFS*-open set. Let $x_E$ and $y_E$ be two *IVFS* points as follows: First, consider that $\tilde{x}_E$ and $\tilde{y}_E$ are two distinct *IVFS* points with different supports $x$ and $y$, $e$ lower values, and $e$ upper values $\lambda_e^-, \gamma_e^-$ and $\lambda_e^+, \gamma_e^+$, respectively. Then, $g_E$ is an *IVFS*-open set such that $g_E$ is an *IVFSN* of *IVFS* point $\tilde{y}_E$ and $\tilde{x}_E \neg \tilde{q} g_E$. Similarly, $f_E = \tilde{y}_E^c$ is an *IVFS*-open set and $f_E$ is an *IVFSN* of the *IVFS* points $\tilde{x}_E$ and $\tilde{y}_E \neg \tilde{q} f_E$. Next, we consider the case $\tilde{x}_E$ and $\tilde{y}_E$ to be two *IVFS* points with the same supports $x = y$, $e$ value $\lambda_e^- < \gamma_e^-$, and $e$ value $\lambda_e^+ < \gamma_e^+$. Then, $\tilde{y}_E$ has a *QIVFSN* $g_E$, which is not quasi-coincident with $\tilde{x}_E$. Thus, $X$ is an *IVFSq-$T_1$* space.

Conversely, Let $(X, E, \tau)$ be an *IVFSq-$T_1$* space. Suppose that any *IVFS* point $\tilde{x}_E$ is not an *IVFS*-closet set in $X$, i.e., $f_E \doteq \tilde{x}_E^c$. Then, $\tilde{f}_E \neq Cl\tilde{f}_E$, and there exists $\tilde{y}_E \tilde{\in} Cl\tilde{f}_E$ such that $\tilde{x}_E \neq \tilde{y}_E$.

First, consider that $\tilde{x}_E$ and $\tilde{y}_E$ are two distinct *IVFS* points with different supports $x$ and $y$, $e$ lower values, and $e$ upper values $\lambda_e^-, \gamma_e^-$ and $\lambda_e^+, \gamma_e^+$, respectively. Suppose that $e$ lower value $\lambda_e^- \leq 0.5$ and $e$ upper value $\lambda_e^+ \leq 0.5$. Since $\tilde{y}_E \tilde{\in} Cl f_E$, by Theorem 4.1, any $f_E$ is a *QIVFSN* of $\tilde{y}_E$ and $\tilde{x}_E \tilde{q} f_E$. Then, there exists *IVFS*-open set $h_E$ such that $\tilde{y} \tilde{q} h_E \tilde{\leq} f_E$. Hence, $h_e^-(y) + \gamma_e^- > 1$. Next, let $\tilde{x}_E$ and $\tilde{y}_E$ be two *IVFS* points with the same supports $x = y$, $e$ value $\lambda_e^- < \gamma_e^-$, and $e$ value $\lambda_e^+ < \gamma_e^+$. Since $y_E \tilde{\in} Cl x_E$, by Theorem 5, each $f_E$ is a *QIVFSN* of *IVFS* points $\tilde{y}_E$, $\tilde{x}_E \tilde{q} f_E$. This is a contradiction. $\square$

**Definition 21.** *Let* $(X, E, \tau)$ *be an IVFST and* $\tilde{x}_E$ *and* $\tilde{y}_E$ *be two IVFS points, if there exists IVFS open sets* $f_E$ *and* $g_E$ *such that:*

(a) *when* $\tilde{x}_E$ *and* $\tilde{y}_E$ *are two distinct IVFS points with different supports* $x$ *and* $y$, *e lower values, and e upper values* $\lambda_e^-, \gamma_e^-$ *and* $\lambda_e^+, \gamma_e^+$, *respectively,* $f_E$ *is an IVFSN of the IVFS point* $\tilde{x}_E$ *and* $g_E$ *is an IVFSN of the IVFS point* $\tilde{y}_E$, *such that* $f_E \neg \tilde{q} g_E$,

(b) *when* $\tilde{x}_E$ *and* $\tilde{y}_E$ *are two IVFS points with the same supports* $x = y$, *e value* $\lambda_e^- < \gamma_e^-$, *and e value* $\lambda_e^+ < \gamma_e^+$, $f_E$ *is an IVFSN of IVFS point* $\tilde{x}_E$ *and* $g_E$ *is a QIVFSN of IVFS point* $\tilde{y}_E$,

*then* $(X, E, \tau)$ *is an interval-valued fuzzy soft quasi-$T_2$ space (IVFS q-$T_2$ space).*

**Example 9.** *Suppose that $X = [0,1]$ and $E$ are any proper $(E \subset X)$. Consider IVFS sets $f_E$ and $g_E$ over $X$ defined as below: $f : E \to \mathcal{IVF}([0,1])$ and $g : E \to \mathcal{IVF}([0,1])$, such that for any $e \in E, x \in X$:*

$$f(e)(x) = \begin{cases} 1 & 0 \le x \le e \\ 0 & e < x \le 1 \end{cases}$$

*and:*

$$g(e)(x) = \begin{cases} 0 & 0 \le x \le e \\ 1 & e \le x \le 1. \end{cases}$$

*Let $\tau = \{\varnothing_E, X_E, f_E, g_E\}$. Then clearly, $\tau$ is an IVFST over $X$. Therefore, for any two absolute distinct IVFS points $\tilde{x}_E, \tilde{y}_E$ in $X$ defined by:*

$$\tilde{x}(e)(z) = \begin{cases} 1 & z = x \\ 0 & z \ne x \end{cases}$$

*and:*

$$\tilde{y}(e)(z) = \begin{cases} 0 & \text{if } z = y \\ 1 & \text{if } z \ne y. \end{cases}$$

*Then, $f_E$ is an IVFSN of $\tilde{x}_E$, and $g_E$ is an IVFSN of $\tilde{y}_E$, such that $f_E \neg \tilde{q} g_E$. Then, $X$ is an IVFS q-$T_2$ space.*

**Theorem 8.** *IVFST$(X, E, \tau)$ is an IVFSq-$T_2$ space if and only if for any $x \in X$, we have $\tilde{x}_E = \tilde{\bigwedge}\{Cl f_E : f_E \in IVFSN$ of $\tilde{x}_E\}$.*

**Proof.** Let $(X, E, \tau)$ be a crisp IVFSq-$T_2$ space and $\tilde{x}_E$ be an IVFS point with support $x$, e lower value $\lambda_e^-$, and e upper value $\gamma_e^+$. Let $y_E$ be a crisp IVFS point with support $y$, e lower value $\gamma_e^-$, and e upper value $\lambda_e^+$. If $\tilde{x}_E$ and $\tilde{y}_E$ are two IVFS points with different supports $x$ and $y$, e lower values, and e upper values $\lambda_e^-, \gamma_e^-$ and $\lambda_e^+, \gamma_e^+$, respectively, then there exist two IVFS-open sets $f_E$ and $g_E$ containing IVFS points $\tilde{y}_E$ and $\tilde{x}_E$, respectively, such that $f_E \neg \tilde{q} g_E$. Then, $g_E$ is an IVFSN of IVFS point $\tilde{x}_E$ and $f_E$ is a QIVFSN of $\tilde{y}_E$ such that $f_E \neg \tilde{q} g_E$. Hence, $\tilde{y}_E \notin Cl g_E$. If $\tilde{x}_E$ and $\tilde{y}_E$ are two IVFS points with the same supports $x = y$, then $\gamma_e^- > \lambda_e^-$ and $\gamma_e^- > \lambda_e^+$. Thus, there are QIVFSN $f_E$ of IVFS points $\tilde{y}_E$ and IVFSN $g_E$ such that $f_E \neg \tilde{q} g_E$. Hence, $\tilde{y}_E \notin Cl g_E$.

Conversely, let $\tilde{x}_E$ and $\tilde{y}_E$ be two distinct IVFS points with different supports $x$ and $y$, e lower values, and e upper values $\lambda_e^-, \lambda_e^+$ and $\gamma_e^-, \gamma_e^+$, respectively. Since:

$$\tilde{x}_E = \tilde{\bigwedge}\{Cl f_E : f_E \in IVFSN \text{ of } \tilde{x}_E\}, \text{ and } \tilde{\bigwedge}\{Cl([f_e^-, f_e^+])(y) : f_E \in IVFSN \text{ of } \tilde{x}_E\} = 0.$$

Thus, $\tilde{y}_E \neg \tilde{q} \tilde{\bigwedge}\{Cl f_E : f_E \in IVFSN$ of $\tilde{x}_E\}$. Therefore, there exists $f_E$ that is an IVFSN of $\tilde{x}$ and $\tilde{y}_E \neg \tilde{q} Cl f_E$. Take two $\tau$-IVFS-open sets $f_E$ and $(Cl f_E)^c$. Therefore, $f_E$ is an IVFSN of IVFS point $\tilde{x}_E$, $(Cl f_E)^c$ an IVFSN of IVFS point $\tilde{y}_E$, and $f_E \neg \tilde{q} (Cl f_E)^c$. $\square$

**Definition 22.** *Let $(X, E, \tau)$ be an IVFST. If for any IVFS point $\tilde{x}_E$ with support $x$, e lower values $\lambda_e^-$, and e upper values $\lambda_e^+$ and any IVFS-closed set $f_E$ in $X$ such that $\tilde{x}_E \neg \tilde{q} f_E$, there exists two IVFS-open sets $h_E$ and $k_E$ such that $\tilde{x}_E \tilde{\in} h_E$ and $f_E \tilde{\le} k_E, h_E \neg \tilde{q} k_E$, then $(X, E, \tau)$ is called an interval-valued fuzzy soft quasi regular space (IVFS q-regular space).*

$(X, E, \tau)$ is called an interval-valued fuzzy soft quasi-$T_3$ space, if it is an IVFS q-regular space and an IVFS q-$T_1$ space.

**Theorem 9.** *IVFST$(X, E, \tau)$ is an IVFS q-$T_3$ space if and only if for any IVFSN $g_E$ of IVFS point $\tilde{x}_E$ there exists an IVFS-open set $f_E$ in $X$ such that $\tilde{x}_E \tilde{\in} f_E \tilde{\le} cl f_E \tilde{\le} g_E$.*

**Proof.** Let $g_E$ be an *IVFS* set in $X$ and $\tilde{x}_E$ be an *IVFS* point with support $x$, e lower value $\lambda_e^-$, and e upper value $\lambda_e^+$ such that $\tilde{x}_E \tilde{\in} g_E$. Then, clearly, $g_E^c$ is an *IVFS*-closed set. Since $X$ is an *IVFS* q-$T_3$ space, there exist two *IVFS*-open sets $f_E, h_E$ such that $\tilde{x}_E \tilde{\in} f_E, g_E^c \tilde{\leq} h_E, h_E$ and $f_E \neg \tilde{q} h_E$. Thus, $f_E^c \tilde{\leq} h_E^c$. Therefore, $Cl f_E \tilde{\leq} h_E^c$ implies $Cl f_E \tilde{\leq} g_E$. Hence, $\tilde{x}_E \tilde{\in} f_E \tilde{\leq} Cl f_E \tilde{\leq} g_E$.

Conversely, let $\tilde{x}_E$ be an *IVFS* point with different support $x$, e lower value $\lambda_e^-$, and e upper value $\lambda_e^+$, and let $g_E$ be an *IVFS*-closed set such that $\tilde{x}_E \neg \tilde{q} g_E$. Then, $g_E^c$ is an *IVFS*-open set containing the *IVFS* point $\tilde{x}_E$, i.e., $\tilde{x}_E \tilde{\in} g_E^c$. Thus, there exists an *IVFS*-open set $f_E$ containing $\tilde{x}_E$ such that $\tilde{x}_E \tilde{\in} f_E \tilde{\leq} Cl f_E \tilde{\leq} g_E$ $g_E \tilde{\leq} (Cl f_E)^c$. Therefore, clearly, $(Cl f_E)^c$ is an *IVFS*-open set containing $g_E$ and $f_E \neg \tilde{q} (Cl f_E)^c$. Hence, $X$ is an *IVFS* q-$T_3$ space. $\square$

**Definition 23.** *Let $(X, E, \tau)$ be an IVFST. If for any two IVFS-closed sets $f_E$ and $g_E$ such that $f_E \neg \tilde{q} g_E$, there exists two IVFS-open sets $h_E$ and $k_E$ such that $f_E \tilde{\leq} h_E$ and $g_E \tilde{\leq} k_E$, then $(X, E, \tau)$ is called an interval-valued fuzzy soft quasi-normal space (IVFS q-normal space).*

$(X, E, \tau)$ is called an interval-valued fuzzy soft quasi $T_4$ space if it is an *IVFS* q-normal space and an *IVFSq*-$T_1$ space.

**Theorem 10.** *$IVFST(X, E, \tau)$ is an IVFS q-$T_4$ space if and only if for any IVFS-closed set $f_E$ and IVFS-open set containing $f_E$, there exists an IVFS-open set $h_E$ in $X$ such that $f_E \tilde{\leq} h_E \tilde{\leq} clh_E \tilde{\leq} g_E$.*

**Proof.** Let $f_E$ be an *IVFS*-closed set in $X$ and $g_E$ be an *IVFS*-open set in $X$ containing $f_E$, i.e., $f_E \tilde{\leq} g_E$. Then, $g_E^c$ is an *IVFS*-closed set such that $f_E \neg \tilde{q} g_E^c$.

Since $X$ is an *IVFS* q-$T_4$ space, there exist two *IVFS*-open sets $h_E, k_E$ such that $f_E \tilde{\leq} h_E, g_E^c \tilde{\leq} k_E$, and $h_E \neg \tilde{q} k_E$. Thus, $h_E \tilde{\leq} k_E^c$, but $Cl h_E \tilde{\leq} Cl k_E^c = k_E$. Furthermore, $g_E^c \tilde{\leq} k_E$ implies $k^c \tilde{\leq} g_E$. That is an *IVFS*-closed set over $X$. Therefore, $Cl h_E \tilde{\leq} k_E^c$. Hence, we have $f_E \tilde{\leq} h_E \tilde{\leq} Cl h_E \tilde{\leq} g_E$.

Conversely, let $\tilde{f}_E$ and $g_E$ be any *IVFS*-closed sets such that $f_E \neg \tilde{q} g_E$. Then, $f_E \tilde{\leq} g_E^c$. Thus, there exists an *IVFS*-open set $h_E$ such that $f_E \tilde{\leq} h_E \tilde{\leq} Cl h_E \tilde{\leq} g_E$. Therefore, there are two *IVFS*-open sets $h_E$ and $(Cl h_E)^c$ such that $f_E \tilde{\leq} h_E, g_E \tilde{\leq} (Cl h_E)^c$. This shows that $X$ is an *IVFS* q-$T_4$ space. $\square$

**Theorem 11.** *If $\Phi : (X_1, E_1, \tau_1) \rightarrow (X_2, E_2, \tau_2)$ is an IVFSC and IVFSO map where $\Phi_u X_1 \rightarrow X_2$ and $\Phi_p E_1 \rightarrow E_2$ are two ordinary bijections, then $X_1$ is an IVFSq-$T_i$ space if and only if $X_2$ is an IVFSq-$T_i$ space for $i = 0, 1, 2, 3, 4$.*

**Proof.** We just prove when $i = 2$. The other parts are similar.

Suppose that we have two *IVFS* points $\tilde{k}_{E_2}$ and $\tilde{s}_{E_2}$ with different supports $k$ and $s$, e lowers value, and e upper values $\lambda_e^-, \lambda_e^+$ and $\gamma_e^-, \gamma_e^+$, respectively, for any $e \in E_2$. Then, the inverse lower and upper image of *IVFS* point $\tilde{k}_{E_2}$ under the *IVFSO* map $\Phi$ is an *IVFS* point in $X_1$ with different support $\Phi^{-1}(k)$ as below:

$$\Phi^{-1}(\tilde{k}^-)(e)(x) = \tilde{k}^-(\Phi_p(e))(\Phi_u(x)) \text{ and } \Phi^{-1}(\tilde{k}^+)(e)(x) = \tilde{k}^+(\Phi_p(e))(\Phi_u(x)).$$

Furthermore, the inverse lower and upper image of *IVFS* point $\tilde{s}_{E_2}$ under the *IVFSO* map $\Phi$ is an *IVFS* point in $X_1$ with different support $\Phi^{-1}(s)$ as below:

$$\Phi^{-1}(\tilde{s}^-)(e)(x) = \tilde{s}^-(\Phi_p(e))(\Phi_u(x)) \text{ and } \Phi^{-1}(\tilde{s}^+)(e)(x) = \tilde{s}^+(\Phi_p(e))(\Phi_u(x)).$$

Since $(X_1, E_1, \tau_1)$ is an *IVFSq*-$T_2$ space, there exist two *IVFS*-open sets $f_E$ and $g_E$ in $X_1$ such that $\Phi^{-1}(\tilde{k}_{E_2}) \tilde{\in} f_E$, $\Phi^{-1}(\tilde{s}_{E_2}) \tilde{\in} g_E$, and $f_E \neg \tilde{q} g_E$. Thus, $\tilde{k}_{E_2} \tilde{\in} f_E$ and $\tilde{s}_{E_2} \tilde{\in} g_E$, while $\Phi(f_E) \neg \tilde{q} \Phi(g_E)$. Therefore, $(X_2, E_2, \tau_2)$ is an *IVFSq*-$T_2$ space.

Conversely, suppose that we have two *IVFS* points $\tilde{x}_E$ and $\tilde{y}_E$ with different supports $x, y \in X_1$, e lower value, and e upper value $\lambda_e^-, \lambda_e^+$ and $\gamma_e^-, \gamma_e^+$, respectively. Then, the lower and upper image of an $IVFS$ point $\tilde{x}_E$ under the $IVFSC$ map $\Phi$ is an $IVFS$ point in $X_2$ with different support $\Phi_u(x)$ as below:

$$\Phi(\tilde{x}^-)(\varepsilon)(k) = \sup_{z \in \Phi^{-1}(k)} [\sup_{e \in \Phi_p^{-1}(\varepsilon)} (\tilde{x}^-)(e)](z)$$

$$= \begin{cases} \lambda_e^- & \text{if } k = \Phi_u(x) \\ 0 & \text{otherwise,} \end{cases}$$

and:

$$\Phi(\tilde{x}^+)(\varepsilon)(k) = \sup_{z \in \Phi^{-1}(k)} [\sup_{e \in \Phi_p^{-1}(\varepsilon)} (\tilde{x}^+)(e)](z)$$

$$= \begin{cases} \lambda_e^+ & \text{if } k = \Phi_u(x) \\ 0 & \text{otherwise} \end{cases}$$

and the lower and upper image of an $IVFS$ point $\tilde{y}_E$ under the $IVFSC$ map $\Phi$ is an $IVFS$ point in $X_2$ with different support $\Phi_u(y)$ as below:

$$\Phi(\tilde{y}^-)(\varepsilon)(k) = \sup_{z \in \Phi^{-1}(k)} [\sup_{e \in \Phi_p^{-1}(\varepsilon)} (\tilde{y}^-)(e)](z)$$

$$= \begin{cases} \gamma_e^- & \text{if } k = \Phi_u(y) \\ 0 & \text{otherwise} \end{cases}$$

and:

$$\Phi(\tilde{y}^+)(\varepsilon)(k) = \sup_{z \in \Phi^{-1}(k)} [\sup_{e \in \Phi_p^{-1}(\varepsilon)} (\tilde{y}^+)(e)](z)$$

$$= \begin{cases} \gamma_e^+ & \text{if } k = \Phi_u(y) \\ 0 & \text{otherwise.} \end{cases}$$

Since $(X_2, E_2, \tau_2)$ is an $IVFSq\text{-}T_2$ space, there exist two $IVFS$-open sets $f_{E_2}$ and $g_{E_2}$ in $X_2$ such that $\Phi(\tilde{x}) \tilde{\in} f_{E_2}$, $\Phi(\tilde{y}) \tilde{\in} g_{E_2}$, and $f_{E_2} \neg \tilde{q} g_{E_2}$. Clearly, $\tilde{x}_E \tilde{\in} \Phi^{-1}(f_{E_2})$, $\tilde{y}_E \tilde{\in} \Phi^{-1}(g_{E_2})$ and $\Phi^{-1}(f_{E_2}) \neg \tilde{q} \Phi^{-1}(g_{E_2})$. Then, $(X_1, E_1, \tau_1)$ is an $IVFSq\text{-}T_2$ space. $\square$

## 6. Conclusions

The aim of this study was to develop the interval-valued fuzzy soft separation axioms in order to build a framework that will provide a method for object ranking. Thus, in this paper, we introduced a new definition of the interval-valued fuzzy soft point and then considered some of its properties, and different types of neighborhoods of the $IVFS$ point were studied in interval-valued fuzzy soft topological spaces. The separation axioms of interval-valued fuzzy soft topological spaces were presented, and furthermore, the basic properties were also studied.

**Author Contributions:** Conceptualization, M.A.; methodology, M.A. and A.Z.K.; writing—original draft preparation, M.A.; writing—review and editing, M.A., A.Z.K. and A.K.; supervision, A.K. and A.Z.K.; project administration, A.K. All authors have read and agreed to the published version of the manuscript.

**Funding:** This research was supported by the Fundamental Research Grant Schemes having Ref. No.: FRGS/1/2018/STG06/UPM/01/3 and vot number 5540153.

**Acknowledgments:** The authors would like to thank the referees and Editors for the useful comments and remarks, which improved the present manuscript substantially.

**Conflicts of Interest:** The authors declare they have no conflict of interest.

## References


1. Molodtsov, D. Soft set theory first results. *Comput. Math. Appl.* **1999**, *37*, 19–31. [CrossRef]
2. Zadeh, L. A. Fuzzy sets. *Inf. Control* **1965**, *8*, 338–353. [CrossRef]
3. Pawlak, Z. Rough sets. *Int. J. Comput. Inf. Sci.* **1982**, *11*, 341–356. [CrossRef]
4. Maji, P.K.; Biswas, P.; Roy, A.R. A fuzzy soft sets. *J. Fuzzy Math.* **2001**, *9*, 589–602.
5. Majumdar, P.; Samanta, S.K. Generalised fuzzy soft sets. *Comput. Math. Appl.* **2001**, *59*, 1425–1432. [CrossRef]
6. Maji, P.K.; Biswas, R.; Roy, A.R. Intuitionistic fuzzy soft sets. *J. Fuzzy Math.* **2001**, *9*, 677–692.
7. Maji, P.K.; Roy, A.R.; Biswas, R. On intuitionistic fuzzy soft sets. *J. Fuzzy Math.* **2004**, *12*, 669–684.
8. Xu, W.; Ma, J.; Wang, S.; Hao, G. Vague soft sets and their properties. *Comput. Math. Appl.* **2010**, *59*, 787–794. [CrossRef]
9. Yang, X.; Lin, T.Y.; Yang, J.; Li, Y.; Yu, D. Combination of interval-valued fuzzy set and soft set. *Comput. Math. Appl.* **2009**, *58*, 521–527. [CrossRef]
10. Jiang, Y.; Tang, Y.; Chen, Q.; Liu, H.; Tang, J. Interval-valued intuitionistic fuzzy soft sets and their properties. *Comput. Math. Appl.* **2010**, *60*, 906–918. [CrossRef]
11. Tanay, B.; Kandemir, M.B. Topological structure of fuzzy soft sets. *Comput. Math. Appl.* **2011**, *61*, 2952-2957 [CrossRef]
12. Simsekler, T.; Yuksel, S. Fuzzy soft topological spaces. *Ann. Fuzzy Math. Inform.* **2013**, *5*, 87–96.
13. Roy, S.; Samanta, T.K. A note on fuzzy soft topological spaces. *Ann. Fuzzy Math. Inform.* **2012**, *3*, 305–311.
14. Roy, S.; Samanta, T.K. An introduction to open and closed sets on fuzzy soft topological spaces. *Ann. Fuzzy Math. Inform.* **2013**, *6*, 425–431.
15. Atmaca, S.; Zorlutuna, I. On fuzzy soft topological spaces. *Ann. Fuzzy Math. Inform.* **2013**, *5*, 377–386.
16. Khameneh, A.Z.; Kılıçman, A.; Salleh, A.R. Fuzzy soft product topology. *Ann. Fuzzy Math. Inform.* **2014**, *7*, 935–947.
17. Khameneh, A.Z.; Kılıçman, A.; Salleh, A.R. Fuzzy soft boundary. *Ann. Fuzzy Math. Inform.* **2014**, *8*, 687–703.
18. Wuyts, P.; Lowen, R. On separation axioms in fuzzy topological spaces, fuzzy neighborhood spaces, and fuzzy uniform spaces. *J. Math. Anal. Appl.* **1983**, *93*, 27–41. [CrossRef]
19. Sinha, S.P. Separation axioms in fuzzy topological spaces. *Fuzzy Sets Syst.* **1992**, *45*, 261–270. [CrossRef]
20. El-Latif, A.A. Fuzzy soft separation axioms based on fuzzy $\beta$-open soft sets. *Ann. Fuzzy Math. Inform.* **2015**, *11*, 223–239.
21. Varol, B.P.; Aygünoğlu, A.; Aygün, H. Neighborhood structures of fuzzy soft topological spaces. *J. Intell. Fuzzy Syst.* **2014**, *27*, 2127–2135. [CrossRef]
22. Ghanim, M.H.; Kerre, E.E.; Mashhour, A.S. Separation axioms, subspaces and sums in fuzzy topology. *J. Math. Anal. Appl.* **1984**, *102*, 189–202. [CrossRef]
23. Zahedi Khameneh, A. Bitopological Spaces Associated with fuzzy Soft Topology and Their Applications in Multi-Attribute Group Decision-Making Problems. Doctoral Thesis. Universiti Putra Malaysia, Seri Kembangan, Malaysia, 2017. Available online: http://ethesis.upm.edu.my (accessed on 23 December 2019).
24. El-Latif, A.M.A. Characterizations of fuzzy soft pre separation axioms. *J. New Theory* **2015**, *7*, 47–63.
25. Majumdar, P.; Samanta, S.K. On soft mappings. *Comput. Math. Appl.* **2010**, *60*, 2666–2672. [CrossRef]
26. Onasanya, B.; Hoskova-Mayerova, S. Some Topological and Algebraic Properties of alphalevel Subsets Topology of a Fuzzy Subset. *Analele Stiintifice ale Universitatii Ovidius Constanta* **2018**, *26*, 213–227. [CrossRef]
27. Hoskova-Mayerova, S. An Overview of Topological and Fuzzy Topological Hypergroupoids. *Ratio Math.* **2017**, *33*, 21–38. [CrossRef]
28. Gorzałczany, M.B. A method of inference in approximate reasoning based on interval-valued fuzzy sets. *Fuzzy Sets Syst.* **1987**, *21*, 1–17. [CrossRef]
29. Bonanzinga, M.; Cammaroto, F.; Pansera, B.A. Monotone weakly Lindelofness. *Cent. Eur. J. Math.* **2011**, *9*, 583–592. [CrossRef]
30. Mondal, T.K.; Samanta, S.K. Topology of interval-valued fuzzy sets. *Indian J. Pure Appl. Math.* **1999**, *30*, 23–29.

MDPI

St. Alban-Anlage 66

4052 Basel

Switzerland

Tel. +41 61 683 77 34

Fax +41 61 302 89 18

www.mdpi.com

*Mathematics* Editorial Office

E-mail: mathematics@mdpi.com

www.mdpi.com/journal/mathematics

# Biocomposite Inks for 3D Printing